General Philosophy of Science: Focal Issues

Handbook of the Philosophy of Science

General Editors

Dov M. Gabbay

Paul Thagard

John Woods

Picture left: Rudolf Carnap (from Wikipedia, the free encyclopedia). Picture right: Sir Karl Popper, photograph by Michael von Haugwitz (reprinted with permission).

ELSEVIER

AMSTERDAM • BOSTON • HEIDELBERG • LONDON • NEW YORK • OXFORD
PARIS • SAN DIEGO • SAN FRANCISCO • SINGAPORE • SYDNEY • TOKYO

North-Holland is an imprint of Elsevier

General Philosophy of Science: Focal Issues

Edited by

Theo A.F. Kuipers
Department of Theoretical Philosophy
University of Groningen
Groningen, The Netherlands

ELSEVIER

AMSTERDAM • BOSTON • HEIDELBERG • LONDON • NEW YORK • OXFORD
PARIS • SAN DIEGO • SAN FRANCISCO • SINGAPORE • SYDNEY • TOKYO
North-Holland is an imprint of Elsevier

N·H

North-Holland is an imprint of Elsevier
Radarweg 29, PO Box 211, 1000 AE Amsterdam, The Netherlands
Linacre House, Jordan Hill, Oxford OX2 8DP, UK

First edition 2007

Library of Congress Cataloging-in-Publication Data
A catalog record for this book is available from the Library of Congress

British Library Cataloguing in Publication Data
A catalogue record for this book is available from the British Library

ISBN: 978-0-444-51548-3

Transferred to digital printing 2009

GENERAL PREFACE

Dov Gabbay, Paul Thagard, and John Woods

Whenever science operates at the cutting edge of what is known, it invariably runs into philosophical issues about the nature of knowledge and reality. Scientific controversies raise such questions as the relation of theory and experiment, the nature of explanation, and the extent to which science can approximate to the truth. Within particular sciences, special concerns arise about what exists and how it can be known, for example in physics about the nature of space and time, and in psychology about the nature of consciousness. Hence the philosophy of science is an essential part of the scientific investigation of the world.

In recent decades, philosophy of science has become an increasingly central part of philosophy in general. Although there are still philosophers who think that theories of knowledge and reality can be developed by pure reflection, much current philosophical work finds it necessary and valuable to take into account relevant scientific findings. For example, the philosophy of mind is now closely tied to empirical psychology, and political theory often intersects with economics. Thus philosophy of science provides a valuable bridge between philosophical and scientific inquiry.

More and more, the philosophy of science concerns itself not just with general issues about the nature and validity of science, but especially with particular issues that arise in specific sciences. Accordingly, we have organized this Handbook into many volumes reflecting the full range of current research in the philosophy of science. We invited volume editors who are fully involved in the specific sciences, and are delighted that they have solicited contributions by scientifically-informed philosophers and (in a few cases) philosophically-informed scientists. The result is the most comprehensive review ever provided of the philosophy of science.

Here are the volumes in the Handbook:

General Philosophy of Science: Focal Issues, edited by Theo Kuipers.

Philosophy of Physics, edited by John Earman and Jeremy Butterfield.

Philosophy of Biology, edited by Mohan Matthen and Christopher Stephens.

Philosophy of Mathematics, edited by Andrew Irvine.

Philosophy of Logic, edited by Dale Jacquette.

Philosophy of Chemistry and Pharmacology, edited by Andrea Woody and Robin Hendry.

Philosophy of Statistics, edited by Prasanta S. Bandyopadhyay and Malcolm Forster.

Philosophy of Information, edited by Pieter Adriaans and Johan van Benthem.

Philosophy of Technological Sciences, edited by Anthonie Meijers.

Philosophy of Complex Systems, edited by Cliff Hooker and John Collier.

Philosophy of Earth Systems Science, edited by Bryson Brown and Kent Peacock.

Philosophy of Psychology and Cognitive Science, edited by Paul Thagard.

Philosophy of Economics, edited by Uskali Mäki.

Philosophy of Linguistics, edited by Martin Stokhof and Jeroen Groenendijk.

Philosophy of Anthropology and Sociology, edited by Stephen Turner and Mark Risjord.

Philosophy of Medicine, edited by Fred Gifford.

Details about the contents and publishing schedule of the volumes can be found at http://www.johnwoods.ca/HPS/.

As general editors, we are extremely grateful to the volume editors for arranging such a distinguished array of contributors and for managing their contributions. Production of these volumes has been a huge enterprise, and our warmest thanks go to Jane Spurr and Carol Woods for putting them together. Thanks also to Andy Deelen and Arjen Sevenster at Elsevier for their support and direction.

CONTENTS

INTRODUCTION. EXPLICATION IN PHILOSOPHY OF SCIENCE

Theo A. F. Kuipers

1 AN IMPORTANT, THOUGH LARGELY IMPLICIT, METHOD IN PHILOSOPHY OF SCIENCE

Scientists use concepts, principles and intuitions that are partly specific for their subject matter, but they also share part of them with colleagues working in different fields. Compare, for example, the chemical notion of the 'valence' of an atom with the notion of 'confirmation' of a hypothesis by certain evidence. An important task of 'the philosophy of (the special science of) chemistry' is the explication of the concept of 'valence'. Similarly, an important task of 'general philosophy of science' is the explication of the concept of 'confirmation'. In both cases it is evident that this only makes sense if one tries to do justice, as much as possible, to the actual use of these notions by scientists, without however following this use slavishly. That is, occasionally a philosopher may have good reasons for suggesting to scientists that they should deviate from a standard use. Frequently, this amounts to a plea for differentiation in order to stop debates at cross-purposes due to the conflation of different meanings.

What has been said about concepts, also applies to principles and intuitions of scientists, which may or may not be paradoxical. Compare the subject specific 'principle of the conservation of energy' and the general intuition of 'diminishing returns from repeated tests'. Both aren't crystal clear, they need explication; of course, beginning with the explication of the concepts involved.

Although the term 'explication' is not often used by philosophers, it is clear that when they discuss the meaning of concepts and propose or report specific definitions, characterizations, models, theories, accounts, conceptions, (rational) reconstructions or formalizations of them, they are practicing concept explication in a more or less explicit and rigorous way. Similarly, when philosophers propose or report specific analyses, accounts, reconstructions or formalizations of principles and intuitions, or dissolutions of paradoxes, explication is at stake. Both kinds of activity belong to the dominant ones among (systematic, constructive, analytical) philosophers, and not in the least among philosophers of science. However, explicitly calling these activities 'explication' is not very popular, let alone using the explication terminology in presenting results. There seem to be at least three reasons for the reluctance to use the word 'explication'. First, the word itself may

Handbook of the Philosophy of Science: General Philosophy of Science — Focal Issues, vii–xxiii.
Volume editor: Theo Kuipers. General editors: Dov M. Gabbay, Paul Thagard and John Woods.

be found a bit too affected. Second, making the application of the method explicit may not only lead to rather cumbersome texts, but also appear to be a difficult task. Finally, many philosophers do not like to be associated with the logical empiricists that introduced '(concept) explication' around 1950 as a technical term for this philosophical method, viz. Rudolf Carnap and Carl Hempel.

This general volume of the *Handbook of the Philosophy of Science* deals with focal issues of a general nature, whereas the special volumes address topics relative to a specific discipline. Each volume contains several contributions that illustrate (largely implicitly, as suggested) the use within philosophy of science of the *method of explication* in one form or another. In a sense, it is what is left of the undoubtedly more rigorous 'logical analysis' with which philosophy of science started in the first half of the 20th century and which had to give way to the 'historical approach' (see Aliseda and Gillies, this volume).

Below I will first describe and further articulate the method of explication, paying particular attention to evaluation reports of proposed explications, to the principled possibility of progress, and to explication starting with idealization, followed by successive concretization. Then I will give a survey of the ten chapters in this volume, with emphasis on the most important concepts and intuitions of which the explication is discussed.

2 EXPLICATION OF CONCEPTS AND INTUITIONS IN SCIENCE

The method of concept explication

As explained above, most philosophers apply the method of explication only informally and implicitly. Hence, the reconstruction of this method is itself a kind of meta-explication. Notably Rudolf Carnap [1950, 1966] and Carl Hempel [1950/1966; 1952], but also John Kemeny and Paul Oppenheim [1952] have articulated the method. Here I will freely describe and develop the method in their spirit. As suggested, it has been and can be applied in various degrees of rigor and explicitness.

The point of departure of concept explication in general philosophy of science is an informal, intuitive concept, the *explicandum*, which is frequently used by scientists in different fields. The aim is to define a concept, the *explicatum*, satisfying three desiderata for concept formation in general and some specific ones. The three *general* desiderata are *precision*, *fruitfulness* and *simplicity*. The *specific* desiderata pertain to the *similarity* with the informal concept. This similarity is to be evaluated by two kinds of criteria:

1. The explicatum should apply to evident, undisputed *(types of) examples* of the informal concept and it should not apply to evident, undisputed *(types of) non-examples*.

2. The explicatum should fulfil (other) *conditions of adequacy* that have been derived from the informal concept, and occasionally it should violate *con-*

ditions of inadequacy. Evident (non-)examples may be generalizable to a condition of (in)adequacy.

Due to the general and specific desiderata, concept explication is an active enterprise, as opposed to the passive cognate of concept or meaning *analysis* (and its precursor 'logical analysis'). In the latter activity one merely tries to do justice to the uses of an informal concept, hence without any further desiderata. To be sure, meaning analysis may be very useful, if only to search for evident examples and non-examples and for finding further conditions of (in)adequacy for concept explication.

Confirmation

Let me briefly illustrate the specific desiderata with the famous example of the explication of 'confirmation'. It is generally assumed, at least *prima facie*, that "this entity is a black raven" confirms the hypothesis that "all ravens are black", and hence provides an evident example of a confirmation. This is so-called Nicod's criterion. On the other hand, a white raven is a counterexample and hence an evident non-example. A black or non-black non-raven, e.g. a black or white tie, also seem evident non-examples. However, it also seems plausible that the explicatum of confirmation should satisfy the 'equivalence condition', that is, if certain evidence confirms a certain hypothesis then it confirms any of its logically equivalent versions. From this it would follow that a non-black non-raven, e.g. a white tie, also confirms "all ravens are black", for it is, according to Nicod's criterion, an evident confirming instance of its logically equivalent version "all non-black objects are non-ravens". This is Hempel's so-called paradox of confirmation ([Hempel, 1945/1965]; see also both Ladyman and Niiniluoto, this volume). It is clear that one can accept this consequence, assuming perhaps certain test conditions, or one can start to dispute either what is to be counted as an evident example or how plausible the equivalence condition in fact is.

Following Hempel [1945/1965], scholars accepting the equivalence condition have also argued that the explicatum of confirmation should satisfy the '(special) consequence condition', that is, if certain evidence confirms a certain hypothesis then it confirms any of its consequences. However, it may also be argued that the opposite condition should be taken as a condition of adequacy, that is, if certain evidence confirms a certain hypothesis then it confirms any stronger hypothesis, the so-called 'converse consequence condition'. Although both conditions may have some intuitive appeal, as Hempel already pointed out, it is easily seen that imposing them together leads to absurd conclusions: any evidence confirms any hypothesis as soon as it confirms some hypothesis (if E confirms H, it confirms $H\&H^*$, by the converse consequence condition, hence it confirms H^* by the consequence condition). For Hempel this was a reason to (implicitly) classify the converse consequence condition as a condition of *in*adequacy.

General desiderata

As the example of 'confirmation' illustrates, it may not at all be an easy matter to agree upon the desiderata for similarity. This is not merely a problem among philosophers of science. The claim is that scientists use many concepts of which it is far from clear that they are used in a consistent way. At the same time, both scientists and philosophers of science may have the feeling that one or more specifications of the concept will make it clear that the use of the concept within science is in fact not as confused as it might seem. In this way it has not only become clear that concept explication is far from a straightforward task, but the general desiderata for explication proposals have also come into the picture. I will briefly discuss them.

1. *Precision (or exactness).* This desideratum is far from precise itself, not in the least because it depends very much on the type of concept one wants to explicate and the purposes that one has primarily in mind with this task. In other words, precision is required as far as relevant and possible. However, some aspects can be put forward as desirable in general. The explicatum should neither lead to inconsistent nor to trivial conclusions. It is also plausible to require syntactical and semantical determinateness. In particular, it should be clear to which syntactic category the relevant term belongs: is it a property term, a relational term or a function term? Moreover, is it classificatory (E confirms H, or not), comparative (E confirms H more than H*) or quantitative (the degree of confirmation of H by E is c)? Note that these divisions are incomplete and do not coincide: for example, an equivalence relation is not comparative and not every function is quantitative. Finally, to reach a sufficient degree of precision it is often desirable to introduce at least some kind of formalization.

2. *Fruitfulness.* Besides minimally avoiding trivialization, already captured by the precision desideratum, the proposal should be fruitful in one or more of the following senses. It should throw new light on the problems that motivated the explication enterprise in the first place. This may concern the explication of a puzzling intuition or the dissolving of a paradox. As far as reaching this goal is considered as a condition of adequacy this will be taken into account in the specific evaluation. However, fruitfulness may also concern the desire for a coherent account of related concepts, like providing the solution of a jigsaw puzzle. As a matter of fact, the point of departure of explication frequently is a network of informal concepts, one of which is chosen as the primary target. For example, the concepts of 'confirmation' and 'induction' are related, and it may be fruitful to aim at an explication of 'confirmation' that can be seen as a weak kind of 'induction'. Besides such intended successes, an explication is even more fruitful when it also leads to unintended, extra successes.

3. *Simplicity.* The concept of simplicity is itself an example of a concept one

has tried to explicate through the ages. In particular, the question what scientists mean when they prefer one explanation or one theory to another because the former is claimed to be simpler than the latter. Although various attempts have been made, without leading to a generally accepted answer, there has certainly been made progress, in particular as far as quantitative theories are concerned (see, [Sober, 1998]).

Of course, philosophers (and scientists) may disagree about how important these general desiderata are, and I would not like to take a stance in this. However, one thing is clear: when a proposal satisfies one of these desiderata in a striking way, in particular in comparison with a rivaling explication, it is almost generally conceived as a positive point, other things being equal.

Evaluation report

The evaluation report of an explication makes clear to what extent the proposed explicatum satisfies the general and specific desiderata. As to the three general desiderata, an informal scoring will usually be the best that can be obtained. Of course, for the question whether progress has been made, see below, comparative scoring is more important than separate scoring.

As to the specific desiderata, it might be wise to use in the evaluation report the problem solving terminology of successes and problems that has been introduced by Thomas Kuhn [1962] and Larry Laudan [1977] for the evaluation of scientific theories.

Starting with evident examples and non-examples, on the *success side* of the report we have the 'true positives' and the 'true negatives', that is, the evident examples fitting into the proposed explication and the evident non-examples excluded by it, respectively. On the *problem side* we get the 'false positives' and the 'false negatives', that is, the evident non-examples that nevertheless fit and the evident examples that don't fit, respectively. False positives show that the explication is too wide and false negatives that it is too narrow.

Besides clear-cut classifications of some evident (non-)examples as problems, there may well have been identified other 'evident' (non-)examples of which the question has arisen whether they are really evident. Such disputed (non-)examples may even lead to revision of the connotation of the concept. That is, formerly evident examples and non-examples may become non-examples and examples, respectively, though not necessarily evident ones. They may be classified as (non-evident or) questioned examples and non-examples, respectively. Of course, other (non-) examples about which there were and are no strong feelings may also be classified as questioned. The suggested type of connotation revision need not occur when a problem, a false positive or false negative, can easily be seen as resulting from the fact that the proposal is still very idealized or naïve and, hence, that refinement or concretization of the proposal is on the agenda. I will come back to this.

Regarding the other type of specific desiderata, viz., conditions of (in)adequacy, a similar terminology is called for. Fulfilled conditions of inadequacy and violated conditions of inadequacy are successes, and violated conditions of adequacy and fulfilled conditions of inadequacy are problems. Violated conditions of adequacy show that the explication is too wide and fulfilled conditions of inadequacy that it is too narrow. Again, conditions may be classified as questioned. Unfulfilled conditions of adequacy and fulfilled conditions of inadequacy may become disputed and lead to connotation revision, but such problems may again also be seen as due to initial idealization, requiring concretization of the explication in order to become (un)fulfilled.

In sum, restricted to specific desiderata, an evaluation report of a proposed explication minimally specifies on the success side evident examples that are covered and conditions of adequacy that are fulfilled. On the problem side it at least lists evident examples that are not covered and conditions of adequacy that are not fulfilled. Finally, on both sides there may occur old and new (non-)examples and conditions of which the status still is or has become questioned.

Let me specify part of Hempel's implicit evaluation report for the 'prediction criterion' [Hempel, 1945/1965] or 'conditional deductive confirmation' [Kuipers, 2000] as an explication proposal for confirmation: E confirms H assuming condition C if and only if $H\&C$ logically entails E, but C alone does not, where C and H are non-contradictory and E even contingent. I insert question marks to indicate (non-)examples and conditions of (in)adequacy that have been disputed by others as evident or desirable, respectively. I add the 'exclusion of something like probabilistic confirmation' as a plausible condition of *in*adequacy, and hence as a problem, to be solved by subsequent concretization.

Specific desiderata	*Successes*	*Problems*
Evident (non-)examples		
Evident examples, relative to "All ravens are black"	True positive: - a black raven	False negative:
Evident *non*-examples, relative to "All ravens are black"	True negative: - a white raven - a black tie (?)	False positive: - a white non-raven (?)
Conditions of (in)adequacy		
Conditions of adequacy	Fulfilled: - Nicod's criterion - equivalence condition	Unfulfilled: - consequence condition (?)
Conditions of *in*adequacy	Unfulfilled:	Fulfilled: - converse consequence condition (?) - excluding probabilistic confirmation

Although it is intuitively appealing to continue to take evident non-examples and conditions of inadequacy explicitly into account in the evaluation report, it will be clear that they can implicitly be taken care of by appropriate conditions of adequacy, excluding the evident non-examples, and by appropriate evident examples, breaking the conditions of inadequacy, respectively. For instance, counterexamples to a general hypothesis are excluded as evident non-examples of confirmation by the plausible requirement that confirming evidence at least has to be compatible with the hypothesis. A well known example of the second type is the drawback of Popper's explication of 'closer to the truth', which happened to leave only room for *true* theories closer to the truth than other theories (see Niiniluoto, this volume). This feature is an evident condition of inadequacy, that can be broken by requiring to leave room for *the possibility* that the theory of Einstein is closer to the truth than the theory of Newton, even if the former would turn out to be false.

It is important to note that conditions of adequacy for concept explication in general may be of a formal or an empirical nature, resulting from previous meaning analysis or from previous empirical analysis, respectively [Hempel, 1952]. An example that combines both kinds of conditions is Einstein's explication of simultaneity of events at a distance. It was guided by the need to obey the empirical law of the constancy of the velocity of light in all frames of reference and by the convention to average that speed over outgoing and return paths. As this law is a cornerstone of Einstein's theory of relativity, the example illustrates that concept explication may be strongly laden with empirical laws, theories and conventions. Of course, conditions of adequacy (and evident examples) based on empirical laws and theories may become questioned when they come under attack or when merely different, but compatible perspectives are possible. For example, the empirical notion of 'absolute temperature' is strongly based on some (asymptotic) empirical laws, whereas the theoretical notion of absolute temperature is heavily based on the theory of thermodynamics, and for this reason also is called 'thermodynamic temperature'.

Formal conditions of adequacy may be of a justificatory nature, for example in the sense that the explication should provide some required existence and uniqueness conditions. An important historical example is that the explications (more specifically, the resulting definitions) of the notion of limits of mathematical sequences and series have to satisfy such conditions, at least as much as possible. Formal conditions may also pertain to the intended justification of a certain intuition, for example, the explication of the 'severity' of a test should realize the intuition of 'diminishing returns from repeated tests', see below. Finally, they may pertain to dissolving a paradox, for example Hempel's paradox of confirmation as resulting from a first explication attempt, see above.

Progress in concept explication

This brings us to the comparative evaluation of two explications of the same concept and the possibility of progress in concept explication or, briefly, *conceptual progress.* Concept explication frequently leads to the conclusion that there are in fact two or more concepts that have to be distinguished, leading to a branching of the explication. However, when two explications of a concept are considered as rivals, the plausible question arises whether the one is better than the other. It is easy to see that there is a possibility to define 'strictly better', which however assumes agreement about the relevant evaluation reports, both with respect to the relevant items as to their scores. The latter is the easiest to imagine by one person and, if relevant, from one theoretical perspective. Moreover, it explains why another person frequently disputes the claim by one scholar of having made progress on the basis of a disagreement about the separate evaluation reports. But, as suggested, this does not prevent the following definition, neglecting evident non-examples and conditions of inadequacy, for the reasons indicated above.

Definition. $E2$ is a *strictly better* explication of a concept than E1 if and only if:

1. $E2$ satisfies the general desiderata at least as well as $E1$
2. $E1$ and $E2$ share all questioned examples and conditions of adequacy
3. $E2$ covers all evident examples covered by $E1$
4. $E2$ fulfils all conditions of adequacy fulfilled by $E1$
5. $E2$ covers some more evident examples and/or fulfils some more conditions of adequacy

It is easy to imagine stronger, weaker and refined versions of various kinds, for example, by requiring unintended successes, by merely counting numbers of covered examples and fulfilled conditions, and by assigning weights to them, respectively. Moreover, changes of questioned examples and conditions may be taken into account. In the terms of Rawls, further articulated by Thagard [1988], one may summarize the overall conclusion that one explication is better than another by claiming that the first better approaches a *reflective equilibrium* than the second. However, the most important point is that the ideal of conceptual progress can be defined in a strict sense and hence that it can function as a regulative idea.

The notion of an evaluation report and the definition of conceptual progress are modeled along the lines for evaluating theories and defining empirical progress in terms of counterexamples and explained empirical laws [Kuipers, 2000, Chs 5/6; Kuipers, 2001, Chs 7/8]. Hence, the partial, formal analogy between empirical and conceptual progress need not be surprising. Another reason why such an analogy may be expected is that in both cases we deal with progress in problem solving, albeit of different kinds.

An idealized start followed by successive concretization

As suggested before, a first explication may be a highly idealized way of catching cases and conditions, with the explicit intention to set up subsequent more realistic explications by accounting for cases and aspects that have first been neglected. This is a useful strategy, not only for concept explication, but also for concept formation in general. Moreover, it is largely analogous to what has been explicated for the empirical sciences as the strategy of idealization and concretization or factualization [Nowak, 1980; Krajewski, 1977]; see also [Kuipers, 2002/to appear]. Besides being a strictly better explication, the specific criterion for being a successful concretization of an idealized one is that the latter is an extreme special case of the former.

Well known examples in the sphere of confirmation are, leaving out some technical restrictions, the transition from 'deductive confirmation' (hypothesis H logically entails evidence E) to, on the one hand, 'conditional deductive confirmation' (assuming condition C, H entails E), used above for illustrating the notion of an evaluation report, and to 'probabilistic confirmation' ($p(H/E) > p(H)$), on the other. In the first example of concretization, we get the original back in the extreme special case that the condition C is a tautology, and in the second case when $p(E/H) = 1$. Of course, both types of concretization can be combined (see Niiniluoto, this volume, for further details on confirmation).

Another example of concretization is the transition from simple or Bayesian conditionalization to 'Jeffrey conditionalization', taking into account that the posterior probability of a hypothesis may be based on evidence about which one is not certain. I just quote from the *Stanford Encyclopedia of Philosophy* [Joyce, 2003, 13–4]:

> **Simple Conditioning**:
>
> If a person with a 'prior' such that $0 < \mathbf{P}(E) < 1$ has a learning experience whose sole immediate effect is to raise her subjective probability for E to 1, then her post-learning 'posterior' for any proposition H should be $\mathbf{Q}(H) = \mathbf{P}_E(H)$. [Here $\mathbf{P}_E(H)$ is standardly defined as $\mathbf{P}(H\&E)/\mathbf{P}(E)$, TK]
>
> [...]
>
> **Jeffrey Conditioning**
>
> If a person with a prior such that $0 < \mathbf{P}(E) < 1$ has a learning experience whose sole immediate effect is to change her subjective probability for E to q, then her post-learning posterior for any H should be $\mathbf{Q}(H) = q\mathbf{P}_E(H) + (1 - q)\mathbf{P}_{\neg E}(H)$.
>
> Obviously, Jeffrey conditioning reduces to simple conditioning when $q = 1$. [That is, the latter is an extreme special case of the former, TK]

Explication of intuitions and dissolving paradoxes

Turning briefly to the explication of intuitions and principles and the dissolving of paradoxes, the first task is of course the explication of concepts that are crucial for the formulation of the intuition or paradox. In case of intuition explication the subsequent task is to prove a theorem to the effect that the intuition, if reformulated in explicated terms, becomes justified, demystified or undermined, whatever the case may be. In case of dissolving a paradox, it has to be shown that it can no longer be construed in the explicated terms.

One example is the quantitative and qualitative explication of the intuition that empirical progress is functional for truth approximation, [Niiniluoto, 1987; Kuipers, 2000], respectively; see also Niiniluoto, this volume).

Another example of intuition explication is the 'diminishing returns'- intuition. For whatever my explication of it is worth, it has all indicated features and is easy to present in a short paragraph. Popper expressed the intuition as follows: "There is something like a law of diminishing returns from repeated tests" [Popper, 1963, p. 240]. It is the idea that the returns, in the sense of the relevance, impact or severity of repeated tests decreases in one way or another. The core idea of my 'non-inductive' explication [Kuipers, 1982] is the following. Explicate 'the returns of n repeated tests' as their a priori objective severity: that is, the prior probability that they will generate at least one counterexample of the relevant generalization. Assuming random tests, with fixed probability $(1 - q)$ of a counterexample, this severity is $1 - q^n$. Hence, the extra severity of an extra test, the $n + 1$-th test, is $(1 - q^{n+1}) - (1 - q^n) = q^n(1 - q)$. Note that this equals the prior probability that the $n + 1$-th test generates the *first* counterexample. Now it is easy to prove that this probability diminishes to 0, assuming that $q < 1$. The last assumption implies that the generalization under test is false, which is according to this explication an underlying assumption of the intuition.

Concluding remarks

Ever since logical empiricists like Carnap, Hempel and Kemeny and Oppenheim presented their idea of concept explication in theory and practice, it has become an important, if not dominant, method in analytical philosophy of science. However, presumably because it was considered to be self-evident, philosophers have internalized the method and its applications usually remain largely implicit. As said, there seems to be moreover a certain reluctance to make the method explicit, for several reasons. Recall that one important reason was that it is not always easy to discover, disentangle and classify the specific desiderata. To identify these desiderata in a text, one may start with searching for objections that are claimed to have been met or have to be met according to the author. For this purpose the above-presented apparatus may be helpful for close reading philosophical texts and for doing philosophical research, although it should of course not be used as a Procrustean bed. In the present context of a series of handbooks in philosophy of science, hopefully also with many non-philosophers as readers, for whom the method will be less self-evident, it may also be useful to have spelled it out. Moreover, below it will be used as a selective point of view in presenting a survey of the chapters to come in this general volume.

3 SURVEY OF CHAPTERS

Chapter 1: Laws, theories, and research programs (Theo Kuipers)

Observational laws, theories and research programs are three of the main units in empirical science. The informal distinction between observational (or experimental) laws and proper theories is illustrated with relevant examples and shown to be of crucial importance for the short and long term dynamics of science. However, it is difficult to explicate the two concepts as strictly distinct, mainly because there is no theory-free observation, which suggests a merely gradual distinction. Fortunately, as notably Hempel and Sneed have put forward, a theory-relative explication of the two concepts can save the distinction.

As Kuhn and Lakatos have shown, the concept of paradigm or research program can capture the short and long term development of science as sequences of theories that are bound together by core ideas that are not given up. This strategy is possible due to unavoidable, but revisable auxiliary hypotheses. However, explicating the notion of a research program not only requires branching in at least four kinds of programs, viz., descriptive, explanatory, design and explicative ones, it also requires the introduction of some degrees of strength. Together they yield the view on intra- or interdisciplinary interaction in science as a matter of competing or cooperating research programs.

Since the heydays of the logical empiricists, philosophers of science have felt that the notion of an empirical theory can be explicated such that its structure becomes clear. The structuralist explication of theories, due to Suppes and Sneed, is presented by way of an optional intermezzo. It can be used for reconstructing theories, which have at least some formal aspects, in order to further study their reach and developmental potentials.

Chapter 2: Past and contemporary perspectives on explanation (Stathis Psillos)

The term 'explanation' is perhaps one of the most commonly used terms by scientists and philosophers of science. The explicative endeavors in the past and the present have not only shown that some distinctions have to be introduced, e.g., between deductive, statistical, teleological, and historical explanation, but also that some basic conceptions have fundamentally remained the same. For some scholars the changes are a matter of conceptual progress, for others they are merely a matter of changing emphasis and interests. The first part of the chapter describes how some major thinkers, from Aristotle, through to Descartes, Leibniz, Newton, Hume and Kant, to Mill, conceived of explanation. The second part offers a systematic examination of the most significant and controversial contemporary models of explanation.

The first part opens with Aristotle's conception — the thought that explanation consists in finding out *why* something happened and that answering why-questions requires finding causes — which has set the agenda for almost all subsequent thinking about explanation. This part further discusses the search for a coherent account of laws of nature, as opposed to mere generalizations, causation and explanation in the thought of the early modern philosophers. This culminates in John Stuart Mill's first well-worked out model of scientific explanation, which was based on the idea that there is no necessity in nature and that, ultimately, explanation amounts to unification into a comprehensive deductive system, whose axioms capture the fundamental laws of nature.

The second part starts with the logical empiricists' attempt to explicate and legitimize the concept of causation by subsuming it under the concept of a deductive-nomological explanation, that is, under the concept of explanation as explicated by Hempel and his followers. It moves on to discuss the re-appearance of genuinely causal models of explanation as well as the re-appearance and development of the Millian idea that explanation amounts to *unification*. Next to this it pays attention to models of statistical and probabilistic explanation, with or without the claim to capture causal forms of such explanations. It ends with an examination of historical and teleological approaches to explanation.

Chapter 3: Evaluation of theories (Ilkka Niiniluoto)

In assessing the cognitive merits of science, one might take a hypothetical theory as the basic unit of evaluation. The traditional virtues of a good theory include consistency, truth, prior and posterior probability, information content, empirical content, explanatory and predictive power, problem-solving capacity, simplicity, accuracy, approximate truth, and truthlikeness. All these notions have been subjects of explication. In this chapter some of the most important explications of most of them are discussed.

The chapter further discusses qualitative, comparative, and quantitative explications of the confirmation of theories by means of available evidence. The results formulate conditions under which empirical success inductively supports the truth or truthlikeness of a theory. Particular attention is given to abductive confirmation due to successful deductive or inductive explanation and prediction. Conditions for accepting a theory as true or truthlike are also discussed. For example, attention is paid to the explication of so-called 'inference to the best explanation' and the explication of the intuition of convergence to the truth.

Chapter 4: The role of experiments in the natural sciences. Examples from physics and biology (Allan Franklin)

Whereas the concept of experiment is not very controversial, at least not in the natural sciences, the epistemological role of experiments is controversial, partly because experiments play many roles. One of their important roles is to test theories and to provide the basis for scientific knowledge. They can also call for a new theory, either by showing that an accepted theory is incorrect, or by exhibiting a new phenomenon which needs explanation. Experiment can provide hints toward the structure or mathematical form of a theory and it can provide evidence for the existence of the entities involved in our theories. It can also measure quantities that theory tells us are important. Finally, it may also have a life of its own, independent of theory. Scientists may experimentally investigate a phenomenon just because it looks interesting. This will also provide evidence for a future theory to explain.

In all of this activity, however, we must remember that science is fallible. Theoretical calculations, experimental results, or the comparison between experiment and theory may all be wrong. If experiment is to play the indicated important roles in science then we must have good reasons to believe experimental results. The chapter presents an epistemology of experiment, that is, a set of strategies that provides reasonable belief in experimental results. Scientific knowledge can then be reasonably based on these experimental results. The view is defended that nature, as revealed by experiment, plays an important and legitimate role in science. The examples come, primarily, although not exclusively, from physics, but these episodes seem typical of the natural sciences. Several examples from biology are also included. These examples do not only provide evident cases for the branched explication of the nature and role of experiments, but also evident cases for the explication of notions like confirmation and refutation.

Chapter 5: The role of experiments in the social sciences. The case of economics (Wenceslao Gonzalez)

In the social sciences the notion of an experiment is less clear than in the natural sciences. This chapter discusses the traditional conception of experiments in the social sciences, and its more recent, enlarged vision. Like in the natural sciences, the analysis of the role of experiments is among the central topics in the methodology of the social sciences. However, doing experiments in the social sciences, in general, and in economics, in particular, has not always been accepted, and it is still an issue that raises objections. The case of economics, with a branch explicitly called 'experimental economics', receives special attention.

After the recognition of the transition from observation to experiment, the notion of 'experiment' used in the social sciences is discussed. Thereafter, the focus is on the development of experiments in the social sciences, taking in particular Reinhard Selten's contribution into account. His version of experimental economics touches important philosophical issues, both epistemological and methodological. Among them, prediction is a key notion asking for a suitable explication, especially in relation to the topics of accuracy and precision.

Chapter 6: Ontological, epistemological, and methodological positions (James Ladyman)

Ontological issues in the philosophy of science may be specific to a particular special science, such as questions about the ontological status of biological species. They may also be more general, such as whether or not there are objective natural kinds and laws of nature, which requires in the first place explication of such notions. In the history of science ontological issues have often been of supreme importance; for example, whether or not atoms exist was a question that occupied many scientists in the nineteenth century. The particular epistemological problems raised by science mostly concern inductive inference, since it is widely accepted that substantive knowledge of the world cannot be obtained by deduction alone. The most fundamental of such problems is to explicate the relationship between theory and evidence, leading, for example, to the question whether the notion of 'inference to the best explanation' can be made sense of. Finally, methodology here means the theory of the scientific method. Is there a single such method for all the sciences, and if so what is it? How much should we expect the theory of the scientific method to help with the progress of science? Is the scientific method fixed, or does it change over time?

Chapter 7: Reduction, integration, and the unity of science: natural, behavioral, and social sciences and the humanities (William Bechtel and Andrew Hamilton)

Beginning with a brief review of historical notions of the unity of science, the chapter offers a 'field guide' to modern concepts of unity and reduction that starts with the theory reduction model of the logical positivists and then considers alternatives. In particular, the chapter discusses revisionist accounts of theory reduction, the best systems approach to unity, due to Phillip Kitcher, and the reasons why several thinkers — Suppes, Dupré, and Cartwright — find attempts at unity and reduction wrongheaded. With these notions and counter-notions in place, the chapter reviews arguments for integration instead of unification and for reduction in terms of mechanisms. Mechanisms involve lower-level parts and operations organized to yield higher-level effects and accounts of mechanistic explanation provide a novel perspective permitting integration without locating all causation at the lowest levels. Finally, case studies are offered of putative reduction and integration in thermal physics, molecular and developmental biology, archaeology, and linguistics.

Chapter 8: Logical, historical and computational approaches (Atocha Aliseda and Donald Gillies)

The chapter begins with the logical approach introduced by the Vienna Circle in the philosophy of science. The members of this circle confined the subject of philosophy of science to the question of the justification of scientific theories. The dominance of the logical approach was challenged in the 1960s by the emergence (or perhaps better re-emergence) of the historical approach. This approach allowed the question of scientific discovery to enter philosophy of science. From the mid-1970s, the development of computers began to influence philosophy of science. Investigations in artificial intelligence led to the development of new logics, such as non-monotonic logics, which had been unknown to the Vienna Circle, and which enabled the logical approach to philosophy of science to be developed further. The study of machine learning shed new light on processes such as induction and abduction, enabling renewed explications, and it allowed the question of scientific discovery to be raised in a more formal manner.

Chapter 9: Demarcating science from nonscience (Martin Mahner)

The explication of the distinction between science and nonscience has been one of the main tasks in general philosophy of science. Most contemporary philosophers of science contend that there is no set of both necessary and sufficient criteria to demarcate the two. This chapter deals with the questions of (a) why we should nonetheless distinguish science from nonscience, and in particular from pseudoscience; (b) how we actually can distinguish science and nonscience even if there is no set of necessary and sufficient criteria; and (c) what the appropriate units of such demarcation are (e.g., epistemic fields, theories, methods). To this end, a comprehensive list of descriptive and normative science indicators, characterizing scientific epistemic fields, is proposed. This list allows for an analysis of various fields, such as mathematics, technology and humanities, as well as suspected pseudosciences, as to their actual status as a science (or at least a protoscience) or else as a nonscience, in particular a pseudoscience.

Chapter 10: History of philosophy of science (Friedrich Stadler)

(Post-)modern philosophy of science has been strongly influenced by the direct and indirect contributions of Logical Empiricism, that is, the Vienna Circle around Moritz Schlick and the Berlin Group around Hans Reichenbach, including its critics, Ludwig Wittgenstein and Karl Popper. Since the beginning of the 20th century we can reconstruct a long-term transfer, transformation and interaction of Central European philosophy of science to the Anglo-Saxon world: from *Wissenschaftslogik* (Carnap) to philosophy of science, and back to the (analytic) *Wissenschaftstheorie*. This significant development, brought on by the forced emigration of logical empiricists in the Nazi era, manifests the destruction of a creative network of philosophy of science as well as the intense interaction of scientific philosophy and philosophy of science in Central Europe, including the forgotten 'French connection' that existed with Pierre Duhem and Henri Poincaré, with the scientific community in Great Britain and North America from the 1930s to the 1960s. The latter were represented by neo-pragmatism and operationalism, which centered around Percy W. Bridgman, Willard Van Orman Quine and Charles Morris as well as by linguistic and scientific philosophy of, notably, Bertrand Russell, Susan Stebbing, Frank P. Ramsey and Max Black.

A critical reconstruction of today's history of philosophy of science on the basis of exile studies and history of science, highlights this transatlantic movement and theory dynamics culminating in the long neglected re-transfer of analytic philosophy (of science) back to its roots of the 'third Vienna Circle', around Viktor Kraft, with Arthur Pap, Paul Feyerabend, and Wolfgang Stegmüller. Following the 'linguistic turn', the pragmatic and historical turns in recent philosophy of science (Quine and Kuhn) constitute an essential part of these developments in the period from Hot to Cold War.

4 ACKNOWLEDGEMENTS

On several occasions I had the opportunity to discuss the 'meta-explication of 'explication', for example, in Lisbon (2005), Opole (2006), and a couple of times in Groningen. I like to thank all discussants for their critical and constructive remarks. Moreover, I like to thank David Atkinson and Jeanne Peijnenburg for their detailed remarks on the final manuscript.

BIBLIOGRAPHY

[Carnap, 1950] R. Carnap. On Explication, in: R. Carnap, *Logical foundations of probability*, University of Chicago Press, 1950. Second edition, 1963, pp. 2–8.

[Carnap, 1966] R. Carnap. Three kinds of concepts in science, in: R. Carnap, *An introduction to the philosophy of science*, pp. 51–61. Basic Books, New York, 1966.

[Hempel, 1945/1965] C. G. Hempel. Studies in the logic of confirmation, in: C. Hempel, *Aspects of scientific explanation*, pp. 3–46. The Free Press, London, 1965. First published in *Mind*, 1945.

[Hempel, 1950/1966] C. G. Hempel. The empiricist criterion of meaning, in: *Logical Positivism*, edited by A.J. Ayer, pp. 108–126. New York: The Free Press. (Originally published in 1950.)

[Hempel, 1952] C. G. Hempel. *Fundamentals of concept formation*, University of Chicago Press, Chicago, 1952.

[Joyce, 2003] J. Joyce. Bayes' Theorem, *The Stanford Encyclopedia of Philosophy (Winter 2003 Edition)*, Edward N. Zalta (ed.), 2003. `http://plato.stanford.edu/archives/win2003/entries/bayes-theorem/`

[Kememy and Oppenheim, 1952] J. G. Kemeny and P. Oppenheim. Degrees of factual support, *Philosophy of Science*, Vol. 19, 305–330, 1952.

[Krajewski, 1977] W. Krajewski. *Correspondence principle and growth of science*, Reidel, Dordrecht, 1977.

[Kuhn, 1962] T. Kuhn. *The structure of scientific revolutions*, University of Chicago Press, Chicago, 1962.

[Kuipers, 1983] T. Kuipers. Non-inductive explication of two inductive intuitions, *The British Journal for the Philosophy of Science*, **34**.3, 209-23, 1963.

[Kuipers, 2000] T. Kuipers. *From instrumentalism to constructive realism,* Kluwer, Dordrecht, 2000.

[Kuipers, 2001] T. Kuipers. *Structures in science*, Kluwer, Dordrecht, 2001.

[Kuipers, 2002/to appear] T. Kuipers. O dwóch rodzajach idealizcji I konkretyzacki. Przypadek aproksymacji prawdy, in: J. Brzezinksi *et al.* (eds), *Odwaga Filozofowania. Leszkowi*

Nowakowi w darze. Wydawnictwo Fundacji Humaniora, Poznan, pp. 117-139, 2002. English version, Two types of idealization and concretization, to appear.

[Laudan, 1977] L. Laudan. *Progress and its problems*, University of California Press, Berkeley, 1977.

[Niiniluoto, 1987] I. Niiniluoto. *Truthlikeness*, Reidel, Dordrecht, 1987.

[Nowak, 1980] L. Nowak. *The structure of idealization*, Reidel, Dordrecht, 1980.

[Popper, 1963] K. R. Popper. *Conjectures and refutations*, Routledge and Kegan Paul, London, 1963.

[Sober, 1998] E. Sober. Simplicity (in scientific theories), *Routledge Encyclopedia of Philosophy*, 1998.

[Thagard, 1988] P. Thagard. *Computational philosophy of science*, MIT-press, Cambridge, 1988.

CONTRIBUTORS

Atocha Aliseda
Instituto de Investigaciones Filosóficas, Universidad Nacional Autónoma de México (UNAM), Ciudad Universitaria, Coyoacán, 04510, México, D.F.
atocha@minerva.filosoficas.unam.mx
http://www.filosoficasunam.mx/~atocha/home.html

William Bechtel
Department of Philosophy and Programs in Science Studies and Cognitive Science, University of California, San Diego, CA 92093-0119, USA.
bill@mechanism.uscd.edu
http://mechanism.ucsd.edu/~bill

Allan Franklin
Department of Physics, University of Colorado, CB 390, Boulder, CO 80309, USA.
Allan.Franklin@colorado.edu
http://spot.colorado.edu/~franklia/

Donald Gillies
Department of Science and Technology Studies, University College London, Gower Street, London WC1E 6BT, UK.
donald.gillies@ucl.ac.uk
http://www.ucl.ac.uk/sts/gillies

Wenceslao J. Gonzalez
Faculty of Humanities, University of A Coruña, Dr. Vazquez Cabrera street w/n, 15403 Ferrol, Spain.
wencglez@udc.es
http://www.udc.es/humanidades/html/Wenceslao.htm

Andrew Hamilton
School of Life Sciences, Arizona State University, Tempe, AZ 85287-4501, USA.
ahamilton@asu.edu
http://sols.asu.edu/faculty/ahamilton.php

Theo Kuipers
Department of Theoretical Philosophy, University of Groningen, Oude Boteringestraat 52, 9712 GL Groningen, The Netherlands.
T.A.F.Kuipers@philos.rug.nl
http://www.rug.nl/filosofie/Kuipers

James Ladyman
Department of Philosophy, University of Bristol, 9 Woodland Road, Bristol BS8 1TB, UK.
james.ladyman@bristol.ac.uk
http://www.bristol.ac.uk/philosophy/department/staff/jl.html

Martin Mahner
Center for Inquiry — Europe, Gesellschaft zur wissenschaftlichen Untersuchung von Parawissenschaften (GWUP), Arheilger Weg 11, D-64380 Rossdorf, Germany.
mahner@gwup.org

Ilkka Niiniluoto
Department of Philosophy, University of Helsinki, Finland.
ilkka.niiniluoto@helsinki.fi
http://www.helsinki.fi/filosofia/filo/henk/niiniluoto.htm

Stathis Psillos
Department of Philosophy and History of Science, University of Athens, Panepistimioupolis (University Campus), Athens 15771, Greece.
psillos@phs.uoa.gr
http://www.phs.uoa.gr/~psillos

Friedrich Stadler
University of Vienna, Faculty of Historical-Cultural Studies and Institute Vienna Circle, A-1090 Vienna, Spitalgasse 2, Hof 1, Austria.
Friedrich.Stadler@univie.ac.at
http://www.univie.ac.at/ivc

LAWS, THEORIES, AND RESEARCH PROGRAMS

Theo A. F. Kuipers

INTRODUCTION

Ernest Nagel [1961] has stressed as no one else the importance of the distinction between experimental laws and proper theories, where the latter aim to explain the former by introducing theoretical terms. This 'law-distinction' is one of the main dynamic factors in the empirical sciences. It will be dealt with in Section 1. Since there do not seem to be anything like theory-free or theory-neutral observation terms, the law-distinction is explicated on the basis of a theory-relative explication of theoretical and observation terms. It will also be shown how a similar explication of the main points can be obtained by starting from Popper's so-called empirical basis; this possibility makes it even more surprising that Popper did not pay attention to the law-distinction.

The analysis suggests a disentanglement of the so-called theory-ladenness of observations. In particular, an observation may not only be laden by a theory, even if unladen by it, an observation may nevertheless be relevant to a theory, and even guided by it. After indicating some structural features of proper theories, we will close the first section with a brief presentation of epistemological positions involved in observational and theoretical knowledge claims of increasing strength.

Section 2, as a kind of bridging, but optional, intermezzo of a semi-formal nature, deals with one particular way to represent the structure of scientific theories in some detail. There are two main approaches to the structure of empirical theories. The statement approach conceives theories primarily as sets of statements. This approach has long been considered as the only and obvious approach, e.g., by Carnap and Popper. However, it is also possible to conceive theories primarily as sets of models. One version of this so-called semantic approach is the set-theoretic or structuralist approach. It has been introduced by Suppes [1957] and refined by Sneed [1971], Stegmüller, Balzer, and Moulines.

Its basic idea is that a theory amounts to the specification of classes of set-theoretic structures satisfying certain conditions. After briefly discussing the practical advantages of the structuralist approach in general it will be introduced stepwise; first without the distinction between theoretical and non-theoretical terms, then with that distinction in order to avoid circularity or infinite regress in measurement. The basic outline of the resulting representation of three examples will

Handbook of the Philosophy of Science: General Philosophy of Science - Focal Issues, pp. 1–95.
Volume editor: Theo Kuipers. General editors: Dov M. Gabbay, Paul Thagard and John Woods.

be given: classical particle mechanics, the periodic table of chemical elements, and psychoanalytic theory. Then some further refinements will follow, viz., absolute versus relative empirical content, various ways of determination of intended applications, relations between theories, theory-nets, and constraints. Finally, we will briefly consider the usefulness of the structuralist approach for non-empirical theories.

In Section 3 we will present the more or less generally accepted view, since the 1980s, introduced by Kuhn and Lakatos, that the development of scientific research takes place by means of encompassing cognitive units, called research programs. We will distinguish four kinds of programs: descriptive, explanatory, design, and explicative. Explanatory programs will be given the main attention, followed by descriptive programs. Explicative programs in the philosophy of science are illustrated in this chapter, for example, the structuralist program in Section 2, and several of the other chapters in this handbook (notably, those of Psillos, Niiniluoto, and Mahner). Computational approaches in the philosophy of science illustrate a particular kind of design programs, viz. designing computer programs that can fulfil certain functions (see [Aliseda and Gillies, this volume]).

The main structural and developmental features of programs will be described, using Dalton's atomic theory program to illustrate them. In the development of this explanatory program the law-distinction will turn out to play a crucial role.

Finally, we will address the strategic lessons that may be drawn. They involve in the first place the value of programmatic research as such, as well as some specific strategies for the internal development of programs, in particular, idealization and concretization. However, the strategic lessons concerning the interaction between programs, by competition or cooperation, are at least as important.

Apart from some aspects and specific formulations this chapter is essentially based on the works of others, notably Nagel, Popper, Sneed, Stegmüller, Kuhn and Lakatos. In principle, the three sections can be read independently from each other. In particular, as suggested already, Section 2 is an optional intermezzo for somewhat formally interested readers.

1 OBSERVATIONAL LAWS AND PROPER THEORIES

Introduction

In the empirical sciences the informal distinction between observational laws and proper theories plays a crucial role. Observational laws are supposed to describe observationally, usually experimentally, established regularities. Different names for roughly the same concept are: empirical, experimental or phenomenological laws, reproducible effects, inductive generalizations, general facts. Proper theories or systems of theoretical laws (together with definitions and other conventions), on the other hand, are supposed to explain such laws and to predict new ones, by postulating underlying mechanisms. For easy reference, we will call this distinction between proper and 'improper' theories, that is, observational laws, the law-distinction. The law-distinction forms a crucial construction principle for the hierarchy of knowledge and therefore an important heuristic factor in the dynamics of knowledge development. However, it has occasioned philosophers of science much brain racking to explicate the law-distinction in a defensible way.[1] Without doubt, the distinction is strongly related to the distinction between observational (or empirical or experimental) and theoretical terms. Whereas proper theories introduce theoretical terms, observational laws do not. But how can one make sense of this term-distinction?

The point of departure of the classical logical empiricists [Aliseda and Gillies, this volume] was a theory-free, hence theory-neutral, observational vocabulary. Starting from this postulate their explication of the distinction was obvious. Observational laws were by definition all those laws that could be formulated in this observational vocabulary. Proper theories on the other hand introduce new concepts not belonging to this observational vocabulary. Given their preference for the observational vocabulary the important question remained whether the new terms introduced by these theories could be reduced, in some way or other, to the observational vocabulary. However this may be, the existence of a theory-free observational vocabulary and the law-distinction were interwoven for the logical empiricists.

Gradually it became clear, even in empiricist circles, that the postulate of a neutral observational vocabulary was an unfortunate creation of the empiricist mind, a paradigm of wishful thinking not corresponding to anything in the empirical sciences. Looking backwards, the standard examples of observational laws, such as Galilei's law of free fall or the Balmer series ordering the spectral lines of hydrogen, must have been dubious from the very beginning, for they are, at least *prima facie*, not couched in a pure observational vocabulary. Non-empiricists were eager to embrace the doctrine that all observation was theory-laden. The most popular became the other extreme view, called meaning holism, which states that all terms occurring in a theory are laden with that theory, with the immediate consequence that an interesting distinction between a theory and the observational laws explained by it became impossible.

Empiricists like Nagel [1961], Hempel [1966; 1970] and Sneed [1971], started to elaborate the idea that certain terms occurring in a theory may be laden with that theory, whereas other terms may not. These latter terms may, however, nevertheless be laden with other theoretical connotations and with observational laws. In the relevant literature, however, these 'theory-relative ideas' have been presented or at least understood as just a reinterpretation of the *two-level* distinction between an observational and a theoretical level. These two levels may enable accountability for part of the dynamics in science, the short-term dynamics, in particular the interaction between invention, evaluation and correction of

[1]In the literature at least as much attention has been paid to the explication of the notion of 'law of nature'. In particular, the question how to distinguish that notion from an accidental generalization is very difficult. We confine ourselves in this respect to a few references: [Nagel, 1961, Chapter 4; Bird, 1998, Chapter 1; Johansson, 2005]; and [Psillos, this volume, Section 12].

observational laws and proper theories. However, the picture hides the long-term dynamics. When a proper theory is accepted as (approximately) true, it usually enables the establishment of criteria for the determination of its theoretical terms. In this way it becomes an observation theory, and the corresponding theoretical level transforms into a higher observational level, enabling new observations and hence the establishment of new observational laws, requiring new, 'deeper' theories to explain them. Moreover, the acceptance of a theory enables experimental or technological applications of the theory, that is, applications presupposing that it is true.[2] Of course, such applications will only be overall successful if the theory is in fact (approximately) true.

In this section it will be shown that the theory-relative ideas essentially lead to the suggested *multi-level* picture of knowledge and knowledge development. The two-level picture may then either concern just a fragment of the multi-level picture or it must be the result of a pragmatic contraction of essentially different observational levels. From the multi-level picture it becomes clear that the theory-relative move is a way of rejecting the idea of a neutral observational vocabulary that enables a new explication of the intuitive law-distinction that not only accounts for the short-term dynamics, but also for the indicated long-term dynamics.

The explication of the term- and law-distinction to be presented does not claim to do justice to the way in which some philosophers use the distinctions, but to the ways in which scientists use these distinctions. An impressive exposition of the far-reaching theory-laden character of what scientists call observations is given by Shapere [1982] under the revealing title "The concept of observation in science and philosophy" in which he uses several examples taken from astrophysics. Except when stated otherwise, we will also follow the scientific practice of already saying that a theory explains an observational law when it can (approximately) be derived from the theory. That is, we will say so whether or not one has good reasons to assume that the theory is true; it may even be known to be false. Hence, speaking of an explanation does not imply accepting it as a fully satisfactory explanation.

After the presentation, in Subsection 1.1., of some clear examples of observational laws and related proper theories, and a preliminary inventory of the characteristic differences, as potential conditions of adequacy, we will introduce in Subsection 1.2. the theory-relative distinction between theoretical and non-theoretical terms and use this distinction of terms for an explication of the law-distinction. The explication of the law-distinction will then make the postulate of a multi-level hierarchy of knowledge in terms of observational laws and proper theories highly plausible. The law-distinction will function as the construction principle for this hierarchy.

In Subsection 1.3. we will pay attention to the surprising fact that Popper pays so little attention to the law-distinction. He was not only one of the first proponents of the view that all observation is theory-laden, but by only assuming a theory-laden 'empirical basis', he did so without falling victim to the other extreme of

[2]In terms of Section 3, in both cases we have entered the external phase of the program that generated the accepted theory.

meaning holism. It will be shown that from this 'basis-relative' perspective it is also possible to explicate the law-distinction. Given that this is not a difficult task and given Popper's evident interest in the internal mechanisms of the development of knowledge, his neglect of the distinction is indeed surprising. It will appear to be instructive to dwell upon the good and the problematic reasons that may have been responsible for that neglect.

In Subsection 1.4. it will be shown that the perspective of Subsection 1.2. (and 1.3.) sheds also light on the idea of theory-laden observation. Three related notions can be clearly distinguished: theory-laden, theory-relevant and theory-guided observation.

In Subsection 1.5. we will first give an elementary account of the structure of theories, starting with the important distinction between epistemologically and ontologically stratified theories. In Section 2 we will present the sophisticated structuralist representation of (the structure of) theories. We will close Subsection 1.5. by briefly characterizing, mainly in terms of aspects of theories, the leading epistemological positions: epistemological relativism, along with observational, referential, constructive and essentialistic realism.

1.1 Examples and prima facie characteristics

We start the explication of the law-distinction by listing first a number of evident examples of both entities, and a number of *prima facie* characteristic differences that may serve as conditions of adequacy. Here, and later, we will speak of testing of a (complex) claim when we are only interested in its truth-value, and of evaluation of a claim when we are interested in its merits and failures. The first is usually the case with potential observational laws and the second with proper theories.

1.1.1 Examples of proper theories

In this section we will use 'theory', except when otherwise stated, to refer to a proper theory, a concept which is exemplified by the following theories represented here by a brief statement of their core ideas:

(a) Newton's theory of gravitation. This theory states that all physical objects have a definite mass, that the sum of all forces exerted on an object equals the product of its mass and its acceleration, and that two objects exert an attractive force on each other proportional to their masses and inversely proportional to the square of their distance.

(b) The kinetic theory of gases. This theory postulates that gases consist of particles, called molecules, which exert forces on each other and which move in accordance with Newton's laws of motion.

(c) Dalton's theory of the atom (the example to be elaborated in Subsection 3.2.5.). This theory claims that all chemical substances are composed of indivisible atoms. According to the theory these atoms can group together in certain ways to form molecules. The formation of molecules is associated with chemical reactions. Chemically pure substances are supposed to consist of one type of molecule.

(d) Bohr's theory of the internal structure of the atom. According to this theory atoms are particles consisting of a nucleus and one or more electrons which circulate around the nucleus in fixed orbits. However, the electrons can jump from one orbit to another, absorbing or emitting electromagnetic radiation at the same time.

(e) (1) Mendel's theory of genetics. According to Mendel the characteristics of (sexually reproducing) organisms are inherited by means of discrete genetic factors, called genes. For each gene there are different allelic forms in the game, each individual has a combination of two alleles, of the same or of a different form, and each parent transmits one of them to each of its offspring. That this is a 50-50% chance process amounts to the first law of Mendel's theory, while the fact that the transmission of alleles related to different types of characteristics is independent is known as the second law of that theory.

(2) The theory of chromosomes. This theory states that in (eukaryotic) organisms the nucleus of each cell contains a number of pairs of so-called chromosomes, each consisting of two separate threads, called the chromatids. Each parent transmits by chance, in a very complex process, one chromatid to its descendant. The link with Mendel's theory results of course from the fact that genes are materialized in a linear way in the chromosomes and that the alleles of a gene pair are supposed to be located on the corresponding positions of the two chromatids of a chromosome.

(3) The molecular theory of genetics. This theory tells that the material of the hereditary information consists of DNA-molecules. Moreover, the hereditary information is transformed by a special molecular mechanism to the offspring. The link with the previous theory results of course from the fact that molecular theory analyzes the chemical composition and working of the chromosomes.

(f) Festinger's theory of cognitive dissonance. According to this theory the presence of cognitive dissonance, being psychologically uncomfortable, gives rise to pressures to reduce the dissonance and to achieve consonance. The strength of the pressures is a function of the magnitude of the existing dissonance.

(g) Utility theory or rational-choice theory. According to this theory people choose out of a set of alternative actions the action from which they expect the highest utility.

1.1.2 Examples of observational laws

The mentioned theories are said to be able to explain the following observational laws:

(a*) Galilei's law of free fall stating that falling objects near the earth have constant acceleration.

(b*) The law that the velocity of sound is higher in gases with a lower density.

(c*) Proust's law (or the law of definite proportions) according to which chemical compounds always decompose into component substances with constant weight ratios.

(d*) The Balmer series, which states that the wavelengths of light emitted by glowing hydrogen gas fit in a simple algebraic series.

(e*) Mendel's interbreeding law on the fact that inherited characteristics manifest themselves after two generations in a certain statistical pattern.

(f*) The (quasi-)law stating that when people have made a decision there is active seeking out of information which is consistent with the action taken.

(g*) The macro-economic consumption function, which claims that total national consumption increases with increasing (average, and hence) national income.

1.1.3 Some characteristic differences

Let us now mention a couple of the characteristics of observational laws and proper theories, features that can help to consolidate the intuitive distinction of these two types of statements. We begin with an unimportant difference. To call a statement an observational law also means that it is well enough supported that it may be assumed to be (approximately) true. On the other hand, talking about a theory does not imply any veracity. Here we are essentially concerned with potential observational laws and theories apart from their truth-value, i.e., as hypotheses that may be true or false. Let us now turn our attention to relevant differences.

(i) Whereas an observational law is usually represented as one, possibly complex, statement, a theory is usually presented as a system, a coherent set, of statements (or as a variant of such a system). Of course, this does not exclude the possibility of an artificial representation of a theory as one conjunctive statement. With or without some extra definitions, even a reformulation in an elegant compact statement may be possible, in which case it is again tempting to speak of a law. The ideal gas law (see below) and the law of Archimedes (the upward force exerted on a solid body in a fluid is equal to the weight of the displaced fluid) are examples of this.

(ii) An observational law may specify what will happen under certain experimental conditions. Hence, it gives a partial characterization of what is not only conceptually, but also really possible in the context. The claim of a theory may be stronger: it may not only specify some necessary conditions for being really possible, it may claim to give a complete characterization of what is really possible in the context. But such a (relative) completeness claim is certainly not associated with every theory.

The first two differences do not only leave room for proper theories, but also for 'observational theories', i.e., coherent sets of (potential) observational laws for a certain context. Moreover, it may or may not be possible to summarize any theory in an elegant compact statement, and there may or may not be associated with it a completeness claim. Hence, the first two *prima facie* differences are not acceptable as strict conditions of adequacy.

(iii) Proper theories, however, not only use concepts that are used in the observational laws to be explained, but introduce also new concepts, called 'the theoretical terms' of the theory. For instance, Newton's notions of mass and force do not occur in Galilei's law; Dalton's concepts of atom and molecule do not occur in Proust's law; the notions of subjective utility and probability do not occur in the consumption function, etc.. (Of course, it may be that old terms are used, but then their old meaning is replaced by a new meaning provided by the theory.) On the other hand, observational laws do not introduce such new terms; for all non-logico-mathematical terms occurring in them there are independent application criteria in the form of experimental and argumentative procedures.

(iv) If an observational law can be explained by a theory, it can nevertheless be tested independently from that theory. This is of particular importance when some potential (corrections of) observational laws are predicted by the theory (and hence can be explained by it), and have still to be tested.

(v) The same observational laws can in principle be explained by different theories. It is for example conceivable that there would have been developed a new theory explaining the same laws as explained by Dalton's theory, in which the notion of atom did not occur, although one or more rather different notions did occur. Hence, a theory can be rejected, without the consequence that the observational laws explained by the theory are also dragged down in its fall. When Bohr's theory of the structure of the atom was rejected, this did not imply that the Balmer formula lost its descriptive adequacy.

From the classical logical empiricist point of view it was plausible to think that the fundamental difference between observational laws and theories, responsible for the above mentioned *prima facie* differences, is that observational laws are or at least can be expressed in pure observation terms, free from further assumptions. Although this assumption might be able to explain the differences, it should be stressed that nothing that has been said so far implies that observational laws express regularities that can be (inductively) established on the basis of pure observation, i.e., observation not presupposing instruments or assumptions. On the contrary, it is not difficult to see that the testing of the observational laws mentioned presupposes all kinds of auxiliary assumptions.

Let us consider, as a tribute to Nagel, who used the same example, the testing of the innocently looking law (b*) about the velocity of sound in gases. To test this law we have to know how to produce and to register sound, and how to measure its velocity. Further we should know how to distinguish gases from substances in other aggregation phases, such as a liquid and a solid state, and how to measure the density of gases. All these identification and measurement procedures presumably presuppose the truth of certain theories. The measurement of the (mass-) density for instance requires the measurement of volumes and masses: both presuppose at least some general assumptions of stability and the like, and the first presupposes in principle a (naive or sophisticated) theory of space geometry, the second a theory of mechanics. Moreover, replication-measurements seldom lead to exactly the same results: to arrive at unique values on the basis of the test results presupposes general principles of dealing with 'measurement mistakes'.

But if observational laws have no immediate relation to reality, what then is the fundamental difference between observational laws and proper theories? As suggested, the characteristic differences (iii) - (v) will serve as conditions of adequacy in the explication to follow.

1.2 Theory-relative explications

We will start with the theory-relative explication of the distinction between theoretical and non-theoretical terms, after which the explication of the law-distinction will be possible. This will naturally lead to the epistemological hierarchy of knowledge.

1.2.1 Theory-relative theoretical and non-theoretical terms

Let us consider in some detail the theory that is supposed to explain the law that the velocity of sound is higher in gases with a lower density (b*), viz., the kinetic theory of gases (b). In the context of this theory, sound is associated with wave movements jointly performed by the gas particles under certain conditions, and the velocity of sound then is identified with the velocity of these waves. The mass density of the gas is in this theory identified with the product of the number of gas particles per unit volume (the number density) and the mass of one particle. Theory (b) together with the mentioned auxiliary assumptions explains the law (b*) in the sense that the law is derivable from it.

Let us now look at the (non-logico-mathematical) terms of the theory, i.e., 'gas', 'gas particle', 'sound', 'velocity of sound', 'wave movement performed by gas particles', etc.. It is easy to see that some of these terms can be understood independently from the kinetic theory of gases, viz., 'gas', 'mass density of a gas', 'velocity of sound in a gas', etc.. We know their meaning even if we do not yet know the kinetic theory. But also within the context of this theory these terms still have the same meaning: for example, it is still the case that we indicate with 'gas' a substance which is in certain respects different from liquid and solid substances. The same can be said about terms like 'sound' and 'velocity of sound': they have a clear meaning independent of the theory and they retain this meaning within the context of the theory. They are 'antecedently understood', to use Hempel's phrase.

Let us now turn our attention to terms like 'gas particle', 'wave movement performed by gas particles', etc.. These terms are not antecedently understood. On the contrary, what we have to understand by a gas particle is specified, or implicitly defined by, the theory itself, because it is the theory that introduces the term. By consequence, the correct use of the term presupposes the truth of the theory.

We will use the last point as the basic criterion for a general distinction between two kinds of (non-logico-mathematical) terms in relation to a statement S. We say that term *t is S-laden* if the correct use of t presupposes the truth of S, at least to some extent, and we say that *t is S-free* if t is not S-laden. We assume that this criterion can be made precise in such a way that it can always be applied unambiguously; see Subsection 2.4.2 for a structuralist specification. Note that we do not assume that t occurs in S; in this way we leave room for the case that t may be, in some way or other, indirectly laden with S. Be this as it may, it is also plausible to define that statement $S1$ *is (un-)laden with* $S2$ if $S1$ does (not) contain terms that are laden with $S2$.

Applying this definition to a theory X, conceived as a complex conjunction of statements, X-laden terms are also called *theoretical terms with respect to X* or *X-theoretical* terms, and X-unladen terms are called antecedently understood or *non-theoretical terms of X* or *X-non-theoretical* terms.

It is important to notice that in this way we do not make an absolute distinction between two kinds of terms in scientific language in general, but a *theory-relative* distinction: a term like 'mass-density' is non-theoretical with respect to the kinetic theory of gases. But, as noted before, the correct use of this term, defined as mass per volume unit, presupposes the truth of general assumptions and other theories, concerning the (macroscopic) notions of volume and mass. 'Volume' is theoretical with respect to Euclidean geometry; 'mass' is laden with Newtonian mechanics. With respect to these theories the term 'mass-density' is not antecedently understood but theoretical.

The foregoing definition of X-non-theoretical terms may even be liberalized in two respects. First, it may well be that the theory leads to a meaning enrichment in the sense that the theory may provide new criteria of application of the term to the already existing criteria. A new way of determining the term may be the result. Second, it may even lead to a proper meaning change in the sense that the old criteria of application are changed, but then only in such a way that the new criteria of application, though suggested by the theory, do not invoke it. In the following the concept of X-non-theoretical terms is taken in this liberalized, sophisticated sense.

In the context of a particular theory X the terminology can be simplified by just speaking of theoretical and non-theoretical or even observation terms when the theory-relative qualifications are meant. The qualifications 'non-theoretical' or 'observation(-al)' may then of course not be misunderstood as implying 'not laden with theories'.

1.2.2 Observational laws as improper theories

With the distinction between X-theoretical and X-non-theoretical terms we are close to a general explication of the intuitive distinction between observational laws and proper theories. The following formulation seems adequate at first sight: a theory is only a proper theory when it has at least some theoretical terms of its own, i.e., terms laden with the theory itself. An observational law, on the contrary, is an improper theory, a theory that has no theoretical terms of its own, i.e., no terms laden with the law itself. According to this characterization an observational law does not contain terms for which the correct use depends on the truth of the law.

It is easy to check that law (b*) satisfies this condition, and also that the other examples of observational laws satisfy it. However, there are also examples of laws which are, according to the proposed definition, not observational laws because they have theoretical terms of their own, whereas we are intuitively inclined to qualify them as observational law. A nice example is the ideal gas law $PV=RT$ (P: pressure; V: volume; T: empirical absolute temperature; R: the ideal gas constant). Everyone calls it an observational law (to be sure, highly idealized), whereas it is at the same time generally known that T and R are laden with the law itself in one way or another. As a consequence, according to the above definition the law has to be qualified as a proper theory.

Closer inspection [Kuipers, 1982; 2001, Appendix 1 of Chapter 2] shows that the situation is as follows. Some observational laws can be formulated in the strict sense suggested above (hence, without R and T and also without other theoretical terms of their own) which are together sufficient to define R and T explicitly. Surprisingly enough, their conjunction is precisely equivalent to the, indeed very elegant, ideal gas law. Hence, although the terms R and T are, according to the theory-relative distinction between theoretical and non-theoretical terms, theoretical with respect to the ideal gas law, these terms can be explicitly defined on the basis of observational laws in the strict sense, and hence can be eliminated.[3]

[3]The important restriction to asymptotic behavior has here been neglected. Contrary to Bickle's suggestion [1998, 26], the asymptotic nature of the equivalence between temperature and mean molecular kinetic energy is unproblematic for the reduction of the ideal gas law. As shown in [Kuipers, 2001, Appendix 1 of Chapter 2], although temperature is asymptotically defined, the relevant identification is unrestrictedly that of equal thermal states with equal mean kinetic energy. The standard equivalence (u = (3/2) kT, with u: mean kinetic energy, and

Many other scientific terms are introduced by explicit definition on the basis of observational laws in the strict sense. In particular, such laws may provide the existence and uniqueness conditions enabling the explicit definition, which is also the case in the above example. Another example of this kind is the notion of weight when based on the observational laws of the slide-balance, to be treated in Section 2.

Given the frequency of such definitions it is worthwhile to take them into account in the final definition. In this definition we will use 'lawlike statement' as a primitive term indicating general statements that may be considered for the qualification 'observational law' or 'theoretical law' . Moreover, we will make a distinction between two kinds of inductive jumps that are made by accepting lawlike hypotheses as true. In case of *observational induction* or inductive generalization one essentially remains within the available observational vocabulary. In the case of *theoretical induction*, the acceptance of the relevant statements, among other things based on the observational laws they can explain, implies the conclusion that the new terms refer.[4]

> *Definition*:
>
> **Observational laws, and its cognates**
>
> An *observational hypothesis in the strict sense* is a lawlike statement not containing theoretical terms of its own, i.e., terms laden with the statement itself.
>
> It is called an *observational law in the strict sense* when it is accepted as (approximately) true, a condition which requires *observational induction.*
>
> An *observational hypothesis* is a lawlike statement containing at most theoretical terms of its own which can be eliminated with the aid of an explicit definition based on observational laws in the strict sense.
>
> An observational hypothesis is called an *observational law* when it is accepted as (approximately) true, a condition which requires indirectly the observational inductions necessary for the defining observational laws in the strict sense and perhaps also directly some new observational inductions.
>
> An *observational theory* is a coherent set of observational hypotheses. There may or may not be an associated (relative) completeness claim. That is, a claim to the effect that the conjunction of the observational hypotheses is the strongest true observational law for a certain domain that can be formulated with a given set of terms not laden with this

k: Boltzmann's constant) results from this identity and the indicated observational laws and is usually, somewhat misleadingly, used in expositions of the reduction of the gas law.

[4]For further refinements of the notion of induction, and abduction, see [Kuipers, 2004].

particular potential law in a non-eliminable way. If there is such a completeness claim we speak of a *strong* observational theory.

An observational theory is called an *observational observation theory*[5] when it is accepted as (approximately) true, possibly, but not necessarily, as the strongest (approximately) true observational law, and used for observation.

Proper theories, and its cognates

A *proper theory* is a coherent set of lawlike statements, called *theoretical hypotheses*, their conjunction containing at least one non-eliminable theoretical term of its own; there may or may not be associated a completeness claim with it. If there is such a completeness claim we speak of a *strong* proper theory.

If the theory is accepted as (approximately) true, which requires *theoretical induction*, the constitutive statements are called *theoretical laws*, and the theory itself, when used for observation, a *theoretical observation theory.*

Observational laws with respect to a theory

A lawlike statement is an *observational law with respect to a theory* if it is an observational law that is not laden with that theory.

One might question whether the definition of a 'proper theory' cannot better be replaced by a corresponding definition of 'theoretical hypothesis'. This would however be an unfortunate move, because of the fact that the theoretical hypotheses constituting a theory are usually interwoven in such a way that an isolated evaluation of, for example, the eliminability of terms would be unjustified.

It is also important to note the following. It frequently occurs that certain statements are considered as observational laws (to be) explained by a certain theory X, whereas they are in fact clearly formulated in X-theoretical terms. For example, the results of Wilson chamber experiments, designed for the evaluation of theories in elementary particle physics, are usually couched in terms of orbits described by the particles postulated by the theory under evaluation. But the relevant aspects of these evaluation results can be formulated in terms of traces of water-drops. In such a case, if it is right, a reformulation is possible which avoids the use of X-theoretical terms. The resulting X-non-theoretical statements are, in that case, the genuine observational laws (to be) explained.

From their definitions it is directly clear that observational laws and proper theories satisfy the intuitive characteristic difference (iii), which was proposed as condition of adequacy in Section 1.1.3. However, they do not satisfy (i) and (ii), for an observational law may be a compact reformulation of an observational theory,

[5]Though this terminology has a systematic background, it is certainly not very attractive; one might prefer to speak of a non-theoretical or experimental observation theory.

and such observational laws as well as proper theories may or may not be relatively complete, in contrast to what (i) and (ii), respectively suggested. The defined distinction also satisfies the conditions of adequacy (iv) and (v). An observational law can be tested independently from the theory which is supposed to explain the law, and it can remain well supported even when that theory is falsified, in which case the question then is: by what other theory can it be explained? This is easy to check in the example of the law about the velocity of sound in gases and the kinetic theory. If, however, a statement contains X-theoretical terms, i.e., when it is X-theoretical, it cannot be tested independently from X. Take for instance the statement that gases consist of particles. To test this statement we will have to know what to understand by gas particles, but it is precisely the kinetic theory that specifies this. We will have to presuppose this theory, or at least a part of it, in order to test the statement in question.

It may be true that a potential observational law can be tested independently of the theory that is supposed to explain it, but this fact makes testing not just the unproblematic affair suggested by common scientific parlance. For there are always underlying observational laws and (observational or theoretical) observation theories in the game. In most cases, however, the underlying laws and theories, with which the potential observational law in question is laden, are not in dispute; to put it differently, in most cases the underlying laws and theories belong to the background knowledge, i.e., they are assumed to be true. Against this background, assuming it as 'underground', one wants to know whether the potential observational law itself is true and whether the theory proposed to explain it does indeed imply the law. This brings us to the hierarchy of knowledge.

1.2.3 The epistemological hierarchy of knowledge

Given that one statement may or may not be laden with another there are three possible relationships between two statements. They may be *interwoven* in the sense that the one is laden with the other and vice versa. Two statements are *disconnected* when neither of them is laden with the other. Finally, $S1$ is an *underlying* statement of $S2$ when $S2$ is laden with $S1$, but not vice versa. Of course, being an underlying statement of another is an asymmetric relation and it is safe to assume that it is also transitive, although exceptions to such a case are not inconceivable.

Conceiving observational laws as well as proper theories as (complex conjunctions of) statements, the relation of being an underlying statement of another leads to interesting cases of observational laws and proper theories underlying other observational laws and proper theories.

It is instructive also to consider the relation of an observational law being explained by a proper theory. Such a situation is of course an asymmetric relation. It should be noted that an explanation of an observational law by a proper theory need not be deductive. For example, such an explanation may be of a corrective type, in which case only a deductive explanation of some approximation (a cor-

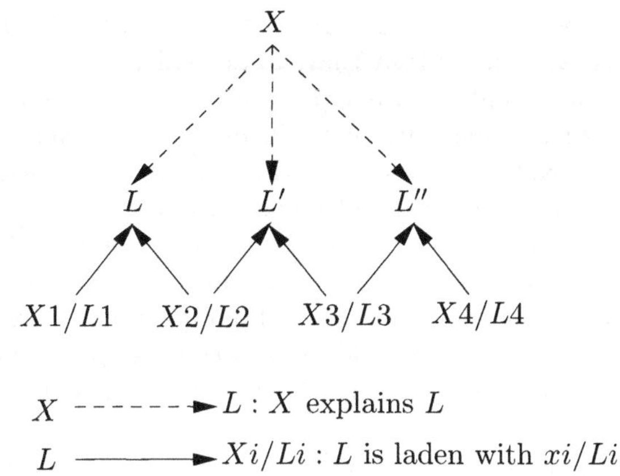

Figure 1. Fragment from the hierarchy of knowledge

rected version) of the law to be explained can be given. In other words, in this case the law itself can only be derived, and hence explained, in this approximate sense. Moreover, explanations of laws by proper theories always need auxiliary hypotheses, but they will not often be mentioned in the following.[6]

If an observational law is explained by a proper theory then it has to be of course an observational law with respect to that theory in the sense defined above and hence the law may not be laden with the theory. The converse, however, is not excluded: an observational law L explained by proper theory X may or may not be an underlying law of that theory, depending on whether or not X uses terms presupposing L. Moreover, if L is explained by X and if, in addition, X^* is an underlying proper theory of L, then X^* will also be an underlying theory of X. Note, moreover, that it is perfectly possible that two observational laws explained by the same proper theory share one or more underlying proper theories.

On the basis of the foregoing asymmetric relations there arises a hierarchy of crucial pieces of knowledge: a proper theory X, the observational laws explained by X, and their underlying proper theories and observational laws.

In Figure 1 we represent the suggested ordering, which we will call the epistemological hierarchy of the context.

The diagram indicates, for instance, that proper theory X explains observational law L and that $X1$ (or $L1$) is an underlying proper theory (or observational law) of L, and hence of X. Of course, Figure 1 presents only a connected (and abstract) fragment of knowledge, which can in principle be extended to all sides. For example, if X has been accepted as (approximately) true, it can then be used as an observation theory with corresponding (new) observation terms. This use of it may lead, in combination with other accepted observation terms, to new observational laws. Furthermore, Figure 1 neither forces us to assume that there

[6]See [Psillos, this volume, Section 13] and [Kuipers 2001, Chapter 3] for further details.

are fundamental observational laws and proper theories without underlying proper theories, neither must we assume that they do not exist.

It is important to note that Figure 1 presents an epistemological order which may not be interpreted ontologically in the sense of the lower the fragment the deeper the concerned level of reality. On the contrary, in particle physics and Mendelian genetics for instance, the ontological whole-part relation will roughly correspond to the upward direction. However, in cosmology for instance, the upward direction will sometimes correspond to the part-whole relation, e.g., from theories about heavenly bodies to theories about galaxies. Hence, there is not supposed to be any standard correspondence between the epistemological hierarchy of knowledge and an intuitive ontological hierarchy of the corresponding objects of knowledge.

The crucial question is of course whether local epistemological hierarchies frequently occur in scientific practice. Inspection teaches us that, for example, the observational law (b*) about the velocity of sound, explained by the kinetic theory (b), does not indeed contain terms that are laden with the kinetic theory. Moreover, we have already noted that the law contains terms, such as 'mass density', that are laden with other theories, such as space geometry and mechanics. It is easy to see that these theories are laden neither with the law nor with the kinetic theory, hence they are indeed underlying theories of the law as well as of the kinetic theory.

Neither is it difficult to verify that the other examples of proper theories (a), (c)-(g) and observational laws (a*), (c*)-(g*) explained by them are in accordance with the hierarchy.

The picture of the hierarchy has been restricted to the most essential elements. The following three features have been left out. First, an explanation of an observational law by a proper theory always needs auxiliary hypotheses, including observational laws and proper theories. Second, domains and subdomains have been omitted. Third and finally, besides observational laws explained by proper theories there are also other important forms of explanation, in particular, an observational theory explaining an observational law or theory, and, last but not least, a proper theory explaining another proper theory.[7]

We confine ourselves to some examples of the last kind: Bohr's theory of the atom (d) explains (a corrected version of) Dalton's theory (c). Mendel's theory (e1) is explained by the theory of chromosomes (e2), which is in its turn explained by the theory of molecular genetics (e3).

It is interesting to dwell further upon Mendel's theory itself. It does not only explain observational laws, it also explains the core theory of population genetics, constituted by Hardy–Weinberg law, which states that, when there are no outside influences and no mutations, the gene ratio in a population remains constant over the generations. It is clear that this law is laden with Mendel's theory, so it certainly is no observational law with respect to Mendel's theory. On the other

[7]In many cases, explanations of laws and theories are even called reduction. On closer inspection this is for various reasons, see [Kuipers, 2001, Chapter 3].

hand it predicts patterns of inheritance of outer characteristics that are, when approximately true, observational laws with respect to Mendel's theory as well as population genetics.

The three elements not included in the above diagram would not alter the hierarchical nature of a more refined picture for the two relations involved: 'to be underlying' and 'to be explained by' remain asymmetric. We are therefore justified in using the simplified figure.

The underlying theories in the diagram essentially represent the proper theories and observational laws with which the terms occurring in the explicitly represented observational laws are laden. But in their turn, these proper theories and observational laws explain (other) observational laws, formulated in terms laden with other proper theories and observational laws, of still lower levels. By way of contraction we can collect together all the terms of all lower levels. Let us call this the contracted observational level, the observational level in short. Hence, determination of the terms of the observational level presupposes the (approximate) truth of all their underlying proper theories and observational laws, in consequence, the truth of all the observational laws explained by them. The combination of these laws and theories is called the background knowledge.

The acceptance of a (general) observational law, however it was generated, logically requires an 'observational inductive jump' from a finite number of singular observation statements, i.e., the data. The acceptance of a proper theory also requires a 'theoretical inductive jump' from the observational laws explained by it. Hence, although the background knowledge may be rather strong, it is not at all a foundation deductively based on data. It is a hierarchically ordered set of assumptions based on observational and theoretical inductions. We may speculate about its lowest level, if that exists at all. We might argue there that all terms are at least laden with some general assumptions, e.g., concerning durability and mutual relations, which guide their application. Apart from that, terms may be indirectly or directly applicable, that is, their application may or may not presuppose (observational or theoretical) inductions.[8]

So far we have neglected a complication that arises due to the fact that one may accept a proper theory as true as far as its observational consequences are concerned, and that one may use it as an observation theory as far as its observation terms are concerned. Such occurrences may be called the *empiricistic* acceptance of a proper theory and the empiricistic use of a proper theory as an observation theory, respectively. Of course, the empiricistic use of a proper theory at least presupposes the empiricistic acceptance of it, and the inductive jump required for the empiricistic acceptance of a proper theory amounts to the observational induction or the theory's observational consequences. In contrast to the empiricistic acceptance and use of a proper theory we may call the above straightforward definitions of the acceptance and use of a proper theory the *realistic* one. That is, the realistic acceptance of a proper theory amounts to the theoretical induction of all consequences of the theory, including its referential claims. And the realistic use

[8]For some more speculations, see [Kuipers 2000, Chapter 13].

of a proper theory as an observation theory amounts to its use for the application of all its terms, which of course presupposes its realistic acceptance.

The foregoing distinction between realist and empiricist attitudes is an initial indication of the different kinds of epistemological positions that are at present taken seriously. In the final subsection we will present a hierarchical survey of such positions.

In this subsection we have discussed the relative distinction between the theoretical and the observational level of a specific theory, and developed a theory-relative distinction of levels. However, these distinctions did not prevent us from giving a theory independent characterization of observational laws. This approach to the level distinction is primarily derived from Sneed [1971], but Nagel [1961] and Hempel [1966] also anticipated it. There is, however, another, completely compatible, approach to the level distinction and to the possible law-distinction, one which was also anticipated by Nagel and Hempel, and which is in addition easy to connect with Popper's work.

1.3 The empirical basis

As mentioned, besides the theory-relative approach, there is another, completely compatible, approach to the level distinction and to the law-distinction, which is easy to connect with Popper's work. We will first present this approach briefly and we will then speculate about Popper's motives for neglecting the law-distinction. The next subsections do not presuppose any details of this subsection.

1.3.1 The basis-relative approach

We start with some of Popper's core concepts. According to Popper [1934/1959] it is possible to reach provisional agreement in every scientific context about what belongs to the level of observation or, to use Popper's favorite term, the (empirical) basis or the basic level. Popper has more than anyone else stressed the theory-laden, swampy character of this observation basis. It is however surprising that he did not make a number of plausible distinctions, let alone exploit them.

Let us call the (non-logico-mathematical) terms occurring at the basic level the *basic terms*. As a matter of fact Popper reserved the term 'basic statement'[9] for a special type of statement that can be formulated in basic terms, viz., so-called singular existential statements, i.e., precisely those statements in basic terms about singular facts which can be in conflict with general statements.

The law-distinction can now be introduced as follows. Calling general statements which can be formulated completely in basic terms (general) observational hypotheses makes it also plausible to call such hypotheses observational laws when they are accepted for the time being after severe testing.

[9] In Subsection 2.2.3. we will present structuralist (non-statement) explications of some of the crucial terms of Popper, notably, 'basic statement', 'counter-example', and 'empirical content'.

We are only concerned with a proper theory and hence with a theoretical level if it concerns a set of statements which, at least partly, breaks through the framework of basic terms. In other words, these statements should postulate new entities or attributes for which new terms have to be introduced that cannot be defined explicitly in terms of the available basic terms.

It is plausible to call the present approach to the level distinction the basis-relative approach. Notice that the corresponding characterization of observational laws is also basis-relative and hence not basis-independent, whereas in the theory-relative approach it was possible to give a theory-independent characterization of observational laws. It is nevertheless clear that the two approaches are essentially the same and that preferences will only depend on one's further purposes.

1.3.2 Why neglect of the law-distinction?

Popper pays little attention to (the possibility of) the law-distinction, let alone to the importance of the distinction for the dynamics of science. We have to guess at Popper's reasons for his lack of interest, because he is not explicit about it. This guessing may, however, be instructive. The only good reason we can think of is the impressive fact[10] that strictly speaking the distinction is not necessary to characterize the logic of theory evaluation in the abstract terms of (singular) basic statements leading to falsification or confirmation (corroboration) of a theory. However, although this shortcut is possible, the distinction has to be introduced in a more realistic and sophisticated characterization of the evaluation of theories, and hence of the structure and development of science.

There is also a reason that has to be respected: Popper does not show any interest worth mentioning in the didactic of scientific textbooks. For someone who does have this interest and thinks that the law-distinction can be made relatively sharply it is clear that the distinction is not yet sufficiently exploited in textbooks.

Poor reasons[11] for Popper's lack of interest are also easy to conceive, however. In the first place, Popper was without doubt a victim of the misunderstanding that the law-distinction itself, or its importance, was related to the assumption of a theory-free observational vocabulary propagated by the classical logical empiricists. It is true that observational laws were considered by them as theory-free universal statements, or at least as reducible to such statements. In the light of the examples that were always mentioned as paradigms of observational laws, such as the laws of Galilei and Kepler, it gradually became clear that two interwoven, but distinguishable claims were involved. We have already shown above that the distinction can be based in a plausible way on a relative distinction of levels, and that Nagel already did this in 1961. The (rightly made) objections to a theory-free observational vocabulary do not provide a good reason for avoiding the distinction.

---

[10]See [Kuipers, 2000, Chapters 5 and 6] or [Kuipers, 2001, Chapters 7 and 8].

[11]From [Hark, ter, 2004] we may conclude that Popper tried to hide his inductive inclinations and psychological inspiration, without causing problematic features of his theory of science. Here we are dealing with a problematic aspect of that theory, the lack of the law-distinction, which partly originates from that same background in view of the three poor reasons.

A second poor reason has perhaps more of a psychological nature and is related to Popper's oversimplified fight against induction. Popper is of course completely right in claiming that induction does not play and cannot play a role in the invention of proper theories. It is however also perfectly clear that observational laws, as far as they are not found by way of prediction by a theory, are frequently found by observational induction; i.e., they are thought of by way of inductive extrapolation. That is also the reason why they are frequently called inductive generalizations. So-called computational philosophy of science [Kuipers, 2001, Chapter 11; Aliseda and Gillies, this volume] exploits and elaborates the various methods of induction. Such methods do not alter the fact that inductively devised potential observational laws still have to be tested in the standard hypothetico-deductive way. A consequence of this is that the recognition of the importance of the law-distinction is almost impossible without simultaneously recognizing that as a matter of fact induction frequently guides the formulation of general observational hypotheses, but again does this not provide a good reason for avoiding the distinction.

The third and last poor reason concerns of course the under-estimation of the importance of the law-distinction for a realistic characterization of the structure and development of science. In general, one can say that Popper placed theories so central, that his attention to the role and nature of experiments became rather one-sided; they were seen from the point of view of theories. 'The neglect of experiment', the telling title of a book by Franklin [1986], was much severer in Popperian circles than in logical empiricist's circles.[12]

Of course, Popper is well aware of the last point and when he is talking about explanation and prediction of facts (e.g., so-called 'novel facts'), he is mostly referring to general facts, hence, observational laws. Many natural scientists also make this elision, but they do not claim, as Popper does, that evaluation of theories is straightforwardly conducted in terms of singular facts. Although such facts may play a crucial role, it is an indirect role, viz., in testing the observational laws predicted by the theory. And even Popper does not always stick to his doctrine. When the evaluation of theories is the main subject, basic statements by definition deal with singular facts, but when talking about basic statements in other contexts, it is often impossible to conclude otherwise than that such statements also include observational laws. This ambiguity, however, has a fortunate aspect, for what is more plausible than to call all statements that can be formulated in basic terms, basic statements?

From the foregoing we conclude that Popper, starting from his own premises, would have improved his analysis considerably by introducing and exploiting the law-distinction.

[12]In this respect the present author is strongly influenced by the logical empiricists, witnessed among others by the topic of the present section and the distinction between descriptive and explanatory research programs in the third section. Finally the fact that [Kuipers 2001, Chapters 3, 7 and 8] reflects that explanation and prediction primarily concern (potential) observational laws, with the explanation and prediction of singular facts as derivatives has an empiricist connotation.

1.4 Theory-ladenness of observation

The foregoing analysis of proper and improper theories can also be used to throw light on the so-called theory-ladenness of observation. Although it may be argued that all observation is in a sense theory-laden, we will see that this position does not imply that all observation is laden with any theory for which that observation is relevant, even if that observation was guided by that theory.

1.4.1 Theory-laden observations

The insight in modern philosophy of science that all statements about the world, however direct and unproblematic they may seem, rest on certain general assumptions, originates to a large extent with Karl Popper. For some other philosophers of science, such as Paul Feyerabend [1962; 1975] and, to a lesser degree, Thomas Kuhn [1962/1969; 1963], this insight has some rather negative implications for the possibility of theory evaluation. So it seems that Feyerabend's ideas about theory-ladenness of terms boil down to the claim that all non-logico-mathematical terms are laden with all theories in which they occur. But this would inevitably lead to the consequence that, for instance, evaluating the theory of Newton with Galilei's law of free fall cannot be more than a circular procedure. For under these circumstances the empirical establishment of the regularity expressed by Galilei's law would already be laden with the theoretical principles of mechanics. Or, as it is also stated, the meaning of the terms occurring in the formulation of the law (distance, time) is determined, among other things, by the principles of mechanics. Some of Kuhn's expositions seem to lead to the same conclusion.

The correct answer to this threatening impasse seems to be the following. We can concede to authors like Feyerabend and Kuhn that such statements as the distance covered by an object near the earth is proportional to the square of the elapsed time required are laden with theories; viz., theories concerning space and time measurement. But such statements therefore do not need to be laden with the specific principles of mechanics. In other words, the meaning of concepts such as position and time, needed for the formulation of such laws, may be fixed in and by theories existing independently from mechanics. In fact, it is in principle possible to reject mechanics, without changing our opinions about space and time measurement. For this reason, the proposed way of evaluating mechanics is not circular, although it is a conditional evaluation: assuming that the theories with which Galilei's law is laden are true, Newtonian mechanics is supported by this law, at least in the sense that it can explain it. As is well known in the history of science, the condition had to be abandoned: Newton's ideas about space and time had to be fundamentally revised, a need that was satisfied by Einstein's theory of relativity. However, this does not exclude the possibility that the described condition could have been fulfilled, and that the suggested positive result of the conditional evaluation is still defensible as a first approximation of actual history.

The above mentioned discussions are frequently put in terms of the theory-ladenness of observations or facts, instead of statements. We shall reformulate the purport of the foregoing in terms of observations. To begin with, it is plausible to make a sharp distinction between the unspecific statement that a certain observation is theory-laden and the specific statement that this observation is laden with theory X or, in short, is X-laden.

We call an observation X-laden when it is, and has to be, formulated in an X-theoretical statement, that is, a statement essentially using an X-theoretical term in the sense of Subsection 1.2.1. Of course, X-laden observations cannot be used to evaluate X without being confronted with the kind of circularity of which Feyerabend and Kuhn were thinking.

Fortunately, however, not all observations that might be relevant to X are laden with X: there are also X-unladen observations, i.e., observations that are or can be phrased in terms of X-non-theoretical statements. Such observations can of course be used to evaluate X without running into circularity problems. Though X-laden observations cannot be used to evaluate X itself, they can perhaps be used to evaluate another theory, in which case it is necessary to presuppose X as an unproblematic background theory, for example as an observation theory.

Let us now also consider the unspecific statement: "All observations are theory-laden". The logical empiricists assumed in the beginning that there are observations that are not laden with general assumptions at all. According to them, the corresponding statements, which would hence be testable without assuming any theory, constituted the class of neutral, theory-free observational statements. The insight that this class is empty might be called Popper's insight: all observations are theory-laden, or at least laden with general assumptions. This, however, does not imply the view that all observations that we can make in order to evaluate a certain theory X are laden with X itself. As mentioned, the latter view was held by Feyerabend and, to a lesser degree, by Kuhn. It has always been severely criticized by Popper, roughly along the suggested lines: for every empirical theory X there are theory-laden but X-unladen observations, which can be used to evaluate X.

1.4.2 Theory-relevant and theory-guided observations

Now we want to discuss two questions that are usually also at stake in discussions about theory-laden observation. The first is that observations may or may not be relevant to or interesting for a certain theory. The second is that (relevant) observations may or may not have been governed or guided by a theory. In both cases we are primarily concerned with observations that are not laden with the theory in question. X-laden observations that are not relevant to X in one way or another are difficult to conceive, and they are by definition guided in a certain sense by X, although it is not necessary that one always realizes this point. X-unladen observations on the contrary may or may not be relevant to X (X-relevant), and if they are relevant to X, they may or may not be guided by X (X-guided). The foregoing is summarized in Figure 2, which gives a classification of observations in relation to theory X.

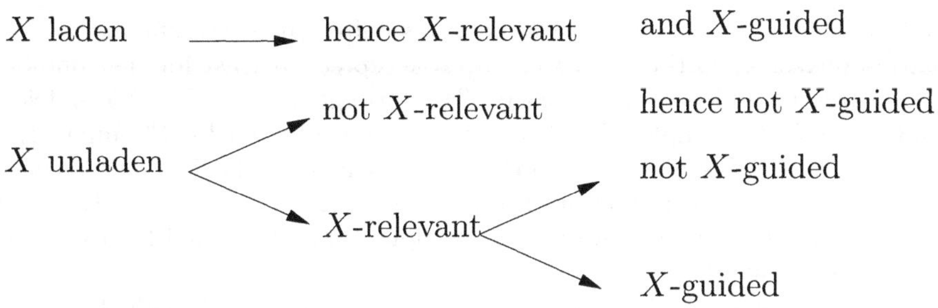

Figure 2. Classification of observations in relation to theory X

As far as relevance is concerned, the foregoing can also be formulated in terms of facts (in the sense of conceptualized facts or data), which may be individual facts or general facts, i.e., observational laws. In this perspective we see that X-laden facts are always relevant to X, but X-unladen facts may or may not be relevant to X.

The notion of relevance that is at stake in the present context is the idea that a certain (X-unladen) fact is relevant to theory X if X is not indifferent with respect to this fact. That is, if X explains this fact, deductively or probabilistically, or if it contradicts it, in both cases with or without some relatively unproblematic auxiliary hypotheses.

There is also, however, a second sense of relevance which may be even more important because there need not be any consensus about it among different scientists. Facts with respect to which a certain theory is indifferent can nevertheless be considered as relevant to that theory in the sense that one may think that the true theory in question should not be indifferent with respect to these facts. However, the proponent of a certain theory can of course also be inclined to consider a fact as irrelevant when his theory is indifferent with respect to that fact.

Newton for instance did not consider it important that his theory neither explained nor contradicted Bode's law, which gives a simple mathematical relation between the radii of the planetary orbits; the indifference of his theory with respect to this law was, according to Newton, no objection to his theory. Kepler, on the contrary, still insisted that the true theory about the solar system should be able to explain Bode's law, and Kepler did have an explanation, which he based on Pythagorean ideas about numerical harmony. Following Newton, present day astronomers also think that Bode's law is irrelevant, an accidental feature of actual orbits with a questionable status as a proper observational law. As a result, not only do we not have any explanation of this law but we also do not feel any need to have one. The phenomenon that a later theory does not give an explanation for a fact that had an explanation before but for which an explanation is no longer considered to be required is called *Kuhn-loss*.

If we want to evaluate a theory X we aim at X-unladen observations that are relevant to X in the first sense: we let ourselves then be guided by X in helping us to decide what to pay attention to. But also after the successful closure of the evaluation phase, when the theory has been accepted, at least for the time being, much research is guided by the theory. The periodic table (PT) of Mendeleev is not only a perfect example of a theory that was evaluated by PT-unladen, but PT-relevant and PT-guided predictions of chemical elements. It later also became an important means for predicting the possibility of artificial production of new elements[13]. All these cases concern observations that are guided by a theory but not laden with that theory.

Observations that are not guided by a theory are frequently called 'accidental observations (or discoveries)'. Accidental observations can of course be perfectly relevant to a theory. A nice example is the Balmer series, which he discovered by trial and error on the basis of data provided by Angström. Hence, the discovery was not guided by a specific theory, but at most by some global Pythagorean ideas. Even so it was recognized as very relevant for the later developed theory of Bohr, for that theory was far from indifferent to the Balmer series: it could explain the series.

1.5 The structure of proper theories and the main epistemological positions

In a first analysis of the structure of theories we will emphasize the distinction between two main types of stratification, viz., epistemological and ontological. We will also pay some attention to non-empirical theories. Finally, we will present the main epistemological positions with respect to proper (empirical) theories and theoretical terms.

1.5.1 Epistemological and ontological stratification

Let us start by summarizing some of the main points made in this chapter so far. A proper theory X has been defined as an epistemologically stratified theory in the sense that it contains terms, and hence statements, that are laden with one or more of its principles: X-theoretical terms. The other terms of X are called X-non-theoretical. In contrast to proper theories, observational hypotheses are defined as improper theories, containing no theoretical terms of their own. A set of connected observational hypotheses is called an observational theory. It should be noted that being X-non-theoretical is a theory-relative, to be precise, an X-relative qualification of a term or a statement: they may well be laden with underlying theories. However, when the theory is clear from the context, a point that we will assume from now on, we will simply speak of theoretical and non-theoretical terms and statements, respectively.

[13] To be sure, the kind of predictions mentioned is of a weak nature. Moreover, one may dispute whether PT is a proper theory or merely a classificatory observational law (see [Mahner and Bunge, 1997, 245-7]). However, we will indicate in Subsection 2.3.3. why PT initially was a proper theory, but transformed into an observational law in the light of quantum mechanics.

The main function of a proper theory is the explanation and prediction of observational laws relative to the theory, i.e., true general hypotheses containing no terms laden with them or the theory. For this function the distinction between observational laws and proper theories is of course crucial.

There is much more than this to say about the structure of proper and observational theories. Here we confine ourselves to some main points. In Section 2 we will present the so-called structuralist way of representing theories in detail.

Besides epistemological stratification there is ontological stratification: they frequently go together, but are essentially independent. A (proper or improper) theory is said to be ontologically stratified when there are two or more kinds of entities involved and when entities of one of these kinds are components of entities of the other kind. It is then plausible to speak of a lower, micro-level and a higher, macro-level. In this case some principles of the theory concern only the micro-entities, and their properties and relations, and are called *micro- or internal principles*, whereas others connect the different kinds of entities, and their properties and relations, and are called *bridge principles*. The example of the atomic theory (dealt with in Subsection 3.2.5.) provides a nice example of an ontologically as well as (along the same lines) epistemologically stratified theory. Of course, auxiliary hypotheses may also have an internal or a bridge character.

Another feature of some theories is that the principles of a theory, whether ontologically and/or epistemologically stratified or not, can frequently be differentiated into core or generic principles, claimed to be true for the whole domain concerned, and special principles, only claimed to be true for a certain subdomain. Of course, a similar distinction can be made for auxiliary hypotheses.

In the case of an epistemologically stratified theory it is plausible to define three types of statements: non-theoretical, purely theoretical and mixed (theoretical) statements. The division of theoretical statements in purely theoretical and mixed ones, however, seems only useful when the epistemological stratification reflects an ontological stratification, in which case the purely theoretical principles, i.e., the internal principles, constitute a clearly separable theory dealing only with the theoretical level. Compare the insightful distinction between the internal and the bridge principles of the atomic theory, in Subsection 3.2.5, with the principles of the (ontologically unstratified) theory of gravitation. In the latter example the distinction between pure principles (e.g., action = minus-reaction, the third law) and mixed principles (e.g., "$f = ma$", the second law; and the special law of gravitation) plays no significant role.

In the case of epistemologically and/or ontologically stratified theories there is a natural distinction of two vocabularies: the complete vocabulary in which the theory is formulated, including theoretical and/or micro-terms, and the sub-vocabulary generated by the non-theoretical and/or macro-terms.

Of course, even if none of both stratifications apply, viz., when we are considering an ontologically unstratified observational theory, it may still be useful to make a distinction between the full theory and the corresponding vocabulary and a sub-theory and the corresponding sub-vocabulary. For an observational theory may be designed to explain an observational sub-theory.

Whatever kind of theory, our discussion more or less implicitly assumes that a theory can be formulated in terms of a finite number of principles. This feature can be conceived of as a very informal type of finite axiomatizability, which is a *conditio sine qua non* to talk about a theory at all. However, this condition should not be confused with the claim of finite axiomatizability in the sense of first or higher order logic. As a matter of fact, in Section 2 we will only illustrate the structuralist claim that it is possible and instructive for many theories to finitely axiomatize them in the set-theoretic sense of set-theoretic structures, defined by a finite number of axiom schemes, using as much mathematical language as necessary.

Explanation and prediction of observational laws have already frequently been mentioned as functions of theories. As additional functions, or at least additional forms of observational success, we should mention: unification, correction and enrichment. A theory may unify, by explanation, a number of *prima facie* rather heterogeneous observational laws. It may predict successfully a corrected version of an observational law, implying that the latter apparently was at most approximately true. Finally, it may predict observational laws concerning new observable phenomena. Of course, theories may also have theoretical success. One example is the conceptual unification of two previous theories into a new theory that is observationally equivalent to their conjunction. Another example is a theory providing a 'deeper' explanation of a proper theory.

1.5.2 Conceptual theories

Theories are up to now understood as empirical theories. Following Popper, we say that a theory is an empirical theory in the strict sense if it is, in combination with certain special or auxiliary hypotheses, falsifiable. And it is an empirical theory if it is intended to become an empirical theory in the strict sense, i.e., if one aims at special or auxiliary hypotheses that make the theory falsifiable. For instance, a generic theory, like Newton's general theory of motion, may well be unfalsifiable as it stands, but become falsifiable together with appropriate special principles, such as the law of gravitation.

However, it also makes sense to leave room for theories that are not intended to be made falsifiable. In Subsection 2.5. we will indicate a number of kinds. Here we will restrict our attention to conceptual theories. A conceptual theory is intended to provide a perspective, a way of looking, at a certain domain without making a general empirical claim. Of course, conceptual theories may or may not be ontologically and/or epistemologically stratified.

The claims that are associated or made with a conceptual theory are either logico-analytic or restricted to individual intended applications. A typical logico-analytic claim is a theorem stating that the instances (models) satisfying the theory can be proven to have a certain explicitly defined property. A typical specific claim states that a certain intended application is (or is not) an instance (model) of the special theory. The very distinction advocated between observational laws and proper theories is an example of a conceptual (meta-)theory for the domain of lawlike statements. This example makes clear at the same time that a conceptual theory may well be the result of concept explication. Of course, the claim that the result of concept explication, a conceptual meta-theory, roughly captures an intuitive concept or distinction is a (quasi-)empirical meta-claim. However, the main point is that, although it may always be possible to formulate a falsifiable general claim with a conceptual theory, the (meta-)claim that all theories are observational theories is usually not intended.

As already suggested, generic theories may well be unfalsifiable as such. They cannot only be made falsifiable, they can also be used as purely conceptual theories.

1.5.3 Epistemological positions[14]

Returning to empirical theories, the core of the ongoing instrumentalism-realism debate concerns the nature of proper theories, or rather the attitude one should have towards them. Here we will briefly sketch the most important epistemological positions in that debate, viz., instrumentalism, constructive empiricism, referential realism and theory realism. In the introductory chapter of [Kuipers, 2000], they are more extensively introduced and ordered according to the ways in which they answer a number of leading questions, where every next question presupposes an affirmative answer to the foregoing one. Moreover, the questions are considered from four perspectives on theories. On the one hand, theories supposedly deal primarily with 'the actual world' or primarily with 'the nomic world', that is, with what is possible in the natural world. On the other hand, one may primarily be interested in whether theories are true or false, or whether they approach 'the truth' regarding the world of interest. It should be stressed that 'the truth' is always to be understood in a domain-and-vocabulary relative way. Hence, no language independent metaphysical or essentialist notion of 'THE TRUTH' is assumed.

The survey of positions and the analysis are restricted to the investigation of the natural world and hence to the natural sciences. Several complications arise if one wants to take the social and cultural world into account. However, the survey of epistemological positions in the natural sciences may well function as a point of departure for discussing epistemological positions in the social sciences and the humanities.[15]

As we have seen, proper theories arise from the two-level distinction between observation and theoretical terms, as opposed to observational laws and observational theories, which only use, by definition, observation terms. Recall that the resulting two-level distinction between observational laws and proper theories

[14]It is plausible to conclude this section with a brief treatment of epistemological positions as they arise from the above treatment of the law- and level-distinction. However, Ladyman deals in his chapter extensively with epistemological positions, including 'structural realism'.

[15]The complications are mainly due to the fact that the social and cultural world is constructed by humans in a sense not applicable to the natural world. It is a topic of increasing interest, e.g., Bhaskar [1979], Giddens [1984], Searle [1995], Tuomela [1995], Balzer and Tuomela [1997], to mention a few.

gives rise to the short-term dynamics in the development of scientific knowledge. Moreover, the long-term dynamics is generated by the transformation of proper theories into observation theories, by accepting them as true. This gives rise to a multi-level distinction according to which proper theories may not only explain or predict a lower level observational law, but also be presupposed by a higher level one. This description of the long-term dynamics typically has a theory realist flavor. However, the other positions have their own way of describing such dynamics. In the following brief survey of questions and answers we restrict ourselves to (the ingredients for) the short-term dynamics as seen from the different positions.

> *Question 0*: Does a natural world that is independent of human beings exist?
>
> *Question 1*: Can we claim to possess true claims to knowledge about the natural world?
>
> *Question 2*: Can we claim to possess true claims to knowledge about the natural world beyond what is observable?
>
> *Question 3*: Can we claim to possess true claims to knowledge about the natural world beyond (what is observable and) reference claims concerning theoretical terms?
>
> *Question 4*: Does there exist a correct or ideal conceptualization of the natural world?

In the following elucidation, we always presuppose an affirmative answer to the foregoing question. *Question 0*, about the existence of an independent natural world, is not an epistemological question, but a preliminary ontological question. The negative answer leads to *ontological idealism*, and the positive one to *ontological realism*. A negative answer to the first epistemological question, *Question 1*, about the possibility of true claims about the natural world, leads to the position of *epistemological relativism* or *skepticism*. It has two forms: *experiential skepticism*, that is, skepticism with respect to claims about sensory and introspective experiences, and *inductive skepticism*, that is, skepticism merely with respect to inductive extrapolations in the sense of inductive predictions and inductive generalizations. The positive answer to *Question 1* leads to epistemological objectivism or *epistemological realism*.

Question 2, about the possibility of more than observational knowledge, brings us to the heart of the distinction between observation and theoretical terms. A negative answer assumes that the notion of observability is relatively fixed. It indicates *observational realism* or just *empiricism*, of which there are two versions. According to *instrumentalism*, advocated for instance by Schlick [1938] and Toulmin [1953], talking about the reference of theoretical terms does not make sense, let alone talking about true or false (proper) theories. The only function of proper theories is to provide good derivation instruments; that is, they need to enable the derivation of as many true observational consequences as possible and as few false observational consequences as possible. Hence, the ultimate aim of the instrumentalist is the best derivation instrument, if any. According to the second type of empiricism, called *(constructive) empiricism* by its inventor and main proponent Van Fraassen [1980; 1989], it may make sense in principle to say that theoretical terms have referential value and that proper theories can be true or false. The problem is that we will never know if such is the case beyond reasonable doubt. Hence, what counts is whether such theories are empirically adequate or inadequate or, to use our favorite terminology, whether they are observationally true or false.

A positive answer to *Question 2* amounts to so-called *scientific realism*, according to which proper theories, or at least theoretical terms, have to be taken seriously. Since the books by Hacking [1983] and Cartwright [1983], there is a weaker version of realism than the traditional one, which amounts to a negative answer to *Question 3* on the possibility of more than (observational and) referential knowledge. Primarily thinking of the referentiality of entity terms, they call their position *entity realism*. However, it seems highly plausible to extrapolate that position to attribute referentiality, in some plausible sense, to many types of terms, and speak of *referential realism*.[16]

The positive answer to *Question 3* brings us to so-called theoretical or *theory realism*, in some version or another advocated by, for instance, Peirce [1934], Popper [1963], and Niiniluoto [1987; 1999]. Theory realism adds to referential realism that theories are claimed to be true and that we have from time to time good reasons to further assume that they are true, that is, to carry out a theoretical induction.

A positive answer to the last *Question 4*, about the existence of a correct or ideal conceptualization, brings us to a position that is not purely epistemologically built on the positive answer to the preliminary, ontological *Question 0* (i.e., ontological realism). It amounts to an extreme kind of metaphysical realism, which we like to call *essentialistic realism*. According to that view, for instance, there must be natural kinds, not only in some pragmatic or nominal sense, but also in the sense of categories in which entities in the natural world *perfectly* fit. Philosophers of science like Boyd [1984] and Harré [1986] seem to come close to this view.

The negative answer to *Question 4* gives rise to what we call *constructive realism*. It combines theory realism with the view that vocabularies are constructed by a human mind guided by previous results. Of course, one set of terms may be more appropriate than another, in the sense that it produces, perhaps in cooperation with other related vocabularies, more and/or more interesting truths about the domain than the other set of terms does. The fruitfulness of alternative vocabularies will usually be comparable, at least in a practical sense, despite the possibility of fundamental incommensurability. There is however no reason to

[16]A kind of antipode of referential realism, and hence of entity realism, arises by denying that referential theoretical claims have truth-values, but lawlike theoretical claims (structural relations) have. This position is known as 'structural realism', see [Niiniluoto, this volume; Ladyman, this volume].

assume that the improvement of vocabularies will ever become impossible.

We summarize the preceding survey in Figure 3

$Q0$:	independent natural world?	$\Rightarrow$ no	ontological idealism
	yes $\Downarrow$ ontological realism		
$Q1$:	true claims about the natural world?	$\Rightarrow$ no	epistemological relativism - experiential skepticism - inductive skepticism
	yes $\Downarrow$ epistemological realism		
$Q2$:	true claims about the natural world beyond the observable?	$\Rightarrow$ no	empiricism (observational realism) - instrumentalism - constructive empiricism
	yes $\Downarrow$ scientific realism		
$Q3$:	beyond reference?	$\Rightarrow$ no	referential realism - entity realism
	yes $\Downarrow$ theory realism		
$Q4$:	ideal conceptualization?	$\Rightarrow$ no	constructive realism
	yes $\Downarrow$ essentialist realism		

Figure 3. The main epistemological positions

The four perspectives, indicated at the beginning of this survey, imply that all (non-relativistic) epistemological positions have an 'actual world version' and a 'nomic world version'. Moreover, they may be restricted to 'true-or-false' claims, or emphasize 'truth approximation claims'. In both cases it is plausible to distinguish between observational, referential, and theoretical claims and corresponding inductions. Instrumentalists, in parallel, speak of theories as 'reliable-or-unreliable' derivation instruments or as 'approaching the best derivation instrument'.

All four perspectives occur, in particular in their realist versions. Standard or traditional realism focuses on 'true/false' claims about the actual world. Giere [1985], who introduced the term 'constructive realism', focuses on the nomic world, but does not take truth approximation into account. Peirce, Popper and Niiniluoto, however, do take truth approximation into account. Moreover, whereas Peirce and Niiniluoto primarily focus on the actual version, Popper and Giere seem to have primarily the nomic version in mind, without excluding the actual version. In our view, the nomic version of constructive realism is best suited to scientific practice.

The epistemological positions of instrumentalism, constructive empiricism, referential realism, and constructive realism have been further characterized in [Kuipers, 2000]. Moreover, with the emphasis on their nomic interpretation, they have been compared in the light of the results of the analysis of confirmation, empirical progress and truth approximation in the rest of that book. The (of course, contestable) conclusions reached in that study are encapsulated in the following summary.

There are good reasons for the instrumentalist to become a constructive empiricist; in his turn, in order to give deeper explanations of success differences, the constructive empiricist is forced to become a referential realist; in his turn, there are good reasons for the referential realist to become a theory realist. The theory realist has good reasons to indulge in constructive realism, since there is no reason to assume that there are essences in the world. As a result, the way leads to constructive realism and amounts to a pragmatic argument for this position, where the good reasons mainly deal with the short-term and the long-term dynamics generated by the nature of, and the relations between, confirmation, empirical progress and truth approximation.

Besides these epistemological conclusions, there are some general methodological lessons to be drawn. There appear to be good reasons for all positions not to use the falsificationist but the instrumentalist or 'evaluation(ist)' methodology. That is, the selection of theories should exclusively be guided by empirical success, even if the better theory has already been falsified. This common methodology, directed at the separate and comparative evaluation of theories, is extensively presented in [Kuipers, 2000, Chapters 5 and 6; Kuipers, 2001, Chapters 7 and 8].

According to the evaluation methodology, the role of falsifications has to be strongly relativized. This does not at all imply that we dispute Popper's claim that falsifiable theories are characteristic for empirical science; on the contrary, only falsifiable theories can obtain empirical success. Moreover, instead of denouncing the hypothetico-deductive method, the evaluation methodology amounts to a sophisticated application of that method. As suggested, the evaluation methodology may also be called the instrumentalist methodology, because the suggested methodology is usually associated with the instrumentalist epistemological position. The reason is, of course, that it is quite natural for instrumentalists not to consider a theory to be seriously disqualified by mere falsification. However, since we will argue that the instrumentalist methodology is also very useful for the other positions, we want to terminologically separate it from the instrumentalist epistemological position, by calling the former the evaluation methodology, and the latter 'instrumentalism'.

We close this subsection with a warning. The suggested hierarchy of the heuristics corresponding to the epistemological positions is, of course, not to be taken in any dogmatic sense. That is, when one is unable to successfully use the constructive realist heuristic, one should not stick to it, but try weaker heuristics: hence first the referential realist, then the empiricist, and finally the instrumentalist heuristic. For, as with other kinds of heuristics, although not everything goes always, *pace* (the suggestion of) Feyerabend's slogan "anything goes", everything goes sometimes. Moreover, after using a weaker heuristic, a stronger heuristic may become applicable at a later stage: "reculer pour mieux sauter".

2 INTERMEZZO. THE STRUCTURALIST APPROACH TO THEORIES[17]

Introduction

This section gives a systematic introduction to the structuralist reconstruction of empirical theories. Although it bridges in a sense the first and the third section, it is an optional intermezzo for readers interested in semi-formal approaches in the philosophy of science (See also [Aliseda and Gillies, this volume]). In Subsection 1.4 we have already presented a first exploration of the nature and structure of empirical theories. Recall that a theory usually has a core of principles and a belt of auxiliary hypotheses. A theory is said to be ontologically stratified when there are two or more kinds of entities involved, where entities of one of these kinds are components of entities of another kind. It is then plausible to speak of a lower micro-level and a higher macro-level. In this case some principles will only concern the micro-entities, and their properties and relations, and are called internal principles, whereas some other principles will connect the different kinds of entities, and their properties and relations, and are called bridge principles. A similar distinction holds for auxiliary hypotheses.

Besides ontological stratification there is epistemological stratification: these frequently go together, but are essentially independent. A proper theory X was defined as an epistemologically stratified theory in the sense that it contains terms, and hence statements, that are laden with one or more of its principles, that are X-laden or X-theoretical, for short. The other terms of X are called X-unladen or X-non-theoretical. In contrast to proper theories, observational hypotheses were defined as improper theories, containing no theoretical terms of their own. A set of connected observational hypotheses was called an observational theory. It should be recalled that being X-non-theoretical is a theory-relative qualification of a term or a statement: such a term or statement may well be laden with underlying theories.

The main function of a proper theory X is the explanation and prediction of X-unladen, observational laws, i.e., true general hypotheses containing neither terms laden with these laws themselves nor terms laden with X. For this purpose the distinction between observational hypotheses and proper theories is of course crucial.

Recall, finally, that the principles of a theory, whether ontologically and/or epistemologically stratified or not, can frequently be distinguished in main or generic principles, claimed to be true for the whole domain concerned, and special principles, only claimed to be true for a certain subdomain.

[17]This section profited a lot from Wolfgang Balzer's detailed criticism.

In this section we will analyze the structure of (improper and proper) theories in more detail. In Subsection 2.1. we will discuss the attractive features of the structuralist approach in general. Starting with the simple example of a slide balance we will first present, in Subsection 2.2., the structuralist representation without making the distinction between theoretical and non-theoretical terms. Reconsidering the slide balance in the light of attempts to measure quantities in a non-circular way, we will introduce, in Subsection 2.3., the structuralist representation with the distinction between theoretical and non-theoretical terms. We will also give the basic outline of this kind of representation for three examples: classical particle mechanics, the periodic table, and psychoanalytic theory. In Subsection 2.4. we will continue with some further refinements of the structuralist approach, viz., the distinction between absolute and relative empirical content, the possibilities of determining the intended applications, relations between theories, theory-nets, and constraints. We will conclude by briefly considering which parts of the structuralist approach are also useful for non-empirical theories, such as metaphysical, mathematical, conceptual and normative theories.

In the previous section, X, Y, and Z were used as variables for theories, whereas in the structuralist presentation M, M', M^* and the like are usual for this purpose, a practice which we also adopt in this section.

2.1 Why the structuralist approach?

There are two main ways of viewing the structure of empirical theories. The *statement* approach conceives theories primarily as sets of statements in a formalized language. In case of an axiomatized theory all these statements are logical consequences of a subset of so-called axioms. This 'logico-linguistic' approach has long been considered as the only and obvious approach, e.g., by Carnap and Popper. Kyburg [1968] illustrates the approach in great detail.

In the *semantic* approach theories are primarily conceived as sets of 'logico-mathematical' models. One version, the state-space version, goes back to Beth and is favored for instance by Suppe and Van Fraassen. Our favorite version, the set-theoretic or *structuralist* approach was introduced by Suppes and refined by Sneed, Stegmüller, Balzer and Moulines. Its basic idea is that theories frequently specify classes of set-theoretic structures satisfying certain conditions. A set-theoretic structure is an ordered set of one or more domain- or base-sets, and one or more properties, relations or functions defined on them, which satisfy certain conditions.

A biological family e.g., can be represented as a structure $\langle A, C\rangle$ with A as the set of members of the family and C as a ternary relation on A. That is, C is a subset of $A\times A\times A$, such that $C(x, y, z)$ states that z is a child of x and y. If there are precisely two members x and y in A such that for all other z in A, $C(x, y, z)$ is true, then this structure can be used to represent a proper two-generation biological family.

According to the structuralist view an axiomatized theory defines such a class of structures and the conditions imposed on the components of the structures are the axioms of the theory. The link with reality is made by the claim, associated with the theory, that the set of set-theoretic representations of the so-called intended applications forms a subset of the class of structures of the theory.

Unfortunately, there has been much debate about what the proper approach to theories is[18], whereas it is easy to see that the two approaches are not at all incompatible. At least for so-called first-order statement theories, i.e., theories formulated as a set of statements of a so-called first-order language this is evident. For the set of models of such a theory, i.e., the structures for which the statements of the theory are true, is precisely a set of structures that might also have been introduced directly in the structuralist way. If we do not restrict ourselves to first-order languages, both approaches are still essentially intertranslatable. Hence, the choice is a pragmatic question.

The main advantage of the structuralist approach is that it is much more a bottom-up approach than the statement approach. It invites us, as it were, to represent and analyze a theory as close to the actual presentations in textbooks as is formally possible. As in scientific practice, all kinds of useful mathematics may be used for that purpose.[19] It does not mean that structuralist reconstruction of theories is an easy task. However, the statement approach is certainly more difficult for specific reconstructions. It is primarily useful in talking about theories in general and in studying logical, in particular model-theoretic, questions about the relation between sentences and their models.[20] These questions and their answers become very complicated as soon as substantial mathematics is involved, e.g., real numbers. Happily enough, not all interesting theoretical questions need logical treatment. For example, as we have demonstrated in Kuipers [2000, Chapter 10], theoretical questions concerning, for instance, idealization and concretization as a truth approximation strategy can be treated relatively easily in structuralist terms.

Another advantage of the structuralist approach is the 'systemic' perspective that takes the world to consist of many systems. Though the statement view could take up this perspective, it has not done so. However this may be, it is plausible that in this respect the structuralist representation is also much simpler than a statement version, if such a version would be made explicit. A third advantage

[18]See, for example, [Mahner and Bunge, 1997, Subsection 9.3.2.]. In my view, they rightly oppose against some naïve aspects of standard structuralist reconstructions of theories. In particular, when structuralists neglect the interpretation of terms and the empirical claims (statements) stating that the so-called intended applications, see below, can be represented as models of the core of the relevant theory. However, they evidently also prefer a 'logico-mathematical' rather than a 'logico-linguistic' axiomatization of theories, witness Bunge's [1967, Chapter 3] axiomatization of classical particle mechanics.

[19]This consideration is rather similar to Giere's [1999] point of view. Enlarging the scope from set-theoretic structures to models of one kind or another, he argues that a 'representational' view of models "is much more adequate to the needs of empirical science" than the standard or 'instantial' view of models. The latter corresponds to the statement view in the sense that models may or may not be instances making a (set of) statement(s) true.

[20]In fact, there is a great distance between (abstract) model theory and models in the empirical sciences, see [Mahner and Bunge, 1997, Section 3.5.].

is that syntactical features, like, for example, the type of a function, can be expressed in a realistic way, talking about functions, and not about building strings of symbols.

Given our preference to be as useful as possible for actual scientific research, we will restrict our attention to the structuralist approach. The best textbooks presenting the structuralist view on theories are Diederich [1981], Balzer [1982], Stegmüller [1973; 1986] and Balzer, Moulines, Sneed [1987]. We will present briefly a number of examples and the main general features, referring to extensive expositions when possible. We will not go into technical details which are not of primary importance for actual practice. Our main goal is to present by way of examples and general exposition the kind of entities that one may be looking for in theory formation and the ways in which standard questions about these entities can be explicated.

2.2 The epistemologically unstratified approach to theories

Starting with the simple example of a slide balance we will first present the structuralist representation of theories and the corresponding terminology without making the epistemological distinction between theoretical and non-theoretical terms.

2.2.1 Example: The slide balance

Consider the slide balance as represented in Figure 4.

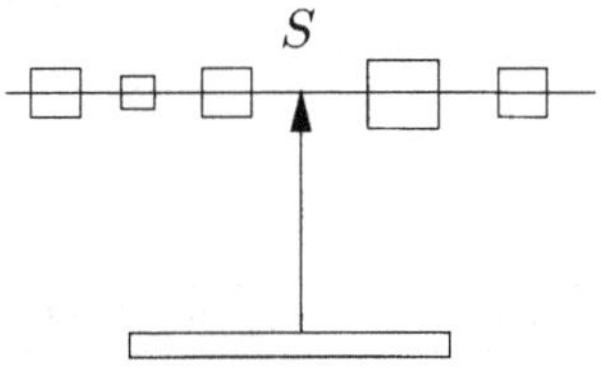

Figure 4. Slide balance

On either side there can be placed a finite number of objects of various weights at all possible distances from the turning-point S. The balance is assumed to be completely symmetric, the equal arms are as long as necessary, and the objects are point masses, i.e., dimensionless particles. Our domain of interest consists of the equilibrium states, i.e., all possible distributions of objects resulting in equilibrium.

A plausible way to represent the equilibrium states, the intended applications, is as follows. We start by characterizing a possible or potential equilibrium state by a structure of the form $\langle P, Pl, d, w \rangle$. Here P is the finite set of particles involved and Pl the subset of P such that Pl and P-Pl represent the particles to the left and to the right of S, respectively. For every particle p in P, $d(p)$ indicates the distance of p from S and $w(p)$ the weight of p. Technically speaking, d and w are positive real-valued functions on P. Let us call the set of structures $\langle P, Pl, d, w \rangle$ satisfying all formal conditions the set of potential equilibrium models of our theory about the equilibrium states of the slide balance, indicated by *SBp*.

Accordingly, our conceptual claim is that the equilibrium states can be represented as members of *SBp*, i.e., there is a subset E of *SBp* representing the nomic, that is, the nomically possible, equilibrium states: the *SBp*-set of intended applications.

The ultimate purpose of theory formation now is to try to characterize E explicitly by one or more additional conditions. As is well known, the adequate condition in the present case is specified by the so-called law of the balance: the sum of distance times weight of the objects on the left should be equal to that sum involving the objects on the right. Let us call the subset of *SBp* of members satisfying this condition the set of equilibrium models of our theory, indicated by *SB*. The proper empirical claim of the law of the balance can now be formulated as "E=*SB*". Assuming some idealizations, e.g., that the objects can be conceived as point-masses, this claim is (generally supposed to be) true. As we will see in other cases the relevant claim need not be as strong as in the present case. The claim might just have been that E is a subset of *SB*.

It will be helpful for later examples to add a more formal presentation of the naive theory of the slide balance.[21]

The naïve theory of the slide balance $\langle SBp, SB, D, E\rangle$		
contains → ↑	$\langle P, Pl, d, w\rangle$ iff	
	1. P is a finite set and Pl is a subset of P	particles the particles left of S
	2. $d: P \rightarrow \mathbb{R}^+$	$d(p)$: the distance of p from S
SBp	3. $w: P \rightarrow \mathbb{R}^+$	$w(p)$: the weight of p
SB	4. $\Sigma_{p \in Pl} d(p).w(p) = \Sigma_{p \in P-Pl} d(p).w(p)$	*the law of the balance*
Concepts and claims		
SBp - *SB*	*empirical content* (to be explained)	
$E \subseteq SBp$	*conceptual claim*: all intended domain of applications D can be represented, by E, as potential models, the intended applications	
$E \subseteq SB$	*(naïve weak) empirical claim*: all intended applications are equilibrium models	
$E = SB$	*(naïve) strong empirical claim*: ... and vice versa	

Later we will see that the empirical claims are still naïve in the sense that it turns out to be impossible to test them in a non-circular way. But first we will present the general structuralistic set-up for unstratified theories.

[21] Set theoretic symbols that are used in this and other schemes: $\in$: element of, $\subseteq$: subset of, $\subset$: proper subset of, $\cap$: intersection, $\cup$: union, $\mathbb{N}^{(+)}$: set of (positive) natural numbers, $\mathbb{R}^{(+)}$: set of (positive) real numbers.

2.2.2 Unstratified theories

Let there be a given domain D of natural phenomena (states, situations, systems) to be investigated. D is supposed to be circumscribed by some informal description and may be called the *intended domain of applications.* Although D is a set, its elements are not yet mutually well distinguished. For this reason we do not yet speak of the domain of intended applications.

In order to characterize the phenomena of D, a set Mp of *conceptual possibilities or potential models* is construed. Technically speaking, Mp is a set of structures of a certain type, a so-called similarity type. In practice Mp will be the conceptual frame of a research program (see Section 3) for D.

The confrontation of D with Mp, i.e., D seen through Mp, is assumed to generate a unique, time-independent subset $Mp(D) =_{def} I$ of all Mp-representations of the members of D, to be called the Mp-set of *intended applications.* Apart from time-independence, this assumption is a conceptual claim. Of course, since nomic impossibilities can, by definition, not be realized, I will be a subset of the (Mp-) set of *nomic possibilities*, but it may be a proper subset, i.e., a more specific set of intended applications satisfying certain additional (more or less precise, but relatively observational) conditions. Assuming that the set of nomic possibilities is a proper subset of Mp, i.e., not everything that is conceivable is nomically possible, I is also a proper subset of Mp. In certain cases I may be a one-element set, in particular when we want to describe 'the actual world' in a certain context, that is, a realized (hence nomic) possibility, e.g., the description of conditions and results of a particular experiment. When dealing with truth approximation [Kuipers, 2000], the attention is focussed on the special case that I is the set of nomic possibilities.

A specific theory about D is concentrated around an explicitly defined subset M of Mp, the *models* of the theory. More specifically, a specific unstratified theory is any combination of the form $UT = \langle Mp, M, D, I\rangle$ with, beside the conceptual claims that M and I are both subsets of Mp, the (weak) empirical claim that I is a subset of M. Sometimes the strong empirical claim is made that I is equal to M, but here we take the weak claim as standard. It is plausible to call UT true when its claim is true, and false otherwise.

The general set-up of the structure of epistemologically unstratified theories will now be presented in a scheme. Such a theory is a meta-structure of the following form:

$\langle Mp, M\rangle$ is sometimes called the theoretical core of the theory, and $\langle D, I\rangle$ may be called the application target of the theory.

The unstratified set-up of theories seems to be rather adequate for observational theories, recall, a combination of one or more observational hypotheses, which contain by definition only terms that are understood independently of the theory concerned.

$\langle Mp, M, D, I\rangle$ is an *epistemologically unstratified theory* iff	
Mp	potential models: a set of structures of a certain type
$M \subseteq Mp$	models: the potential models that satisfy all axioms
Mp - M	empirical content (to be explained)
D	the intended domain of applications
$I \subseteq Mp$	intended applications, resulting from the conceptual claim that D can be represented as a set of members of Mp, i.e. "$I = Mp(D)$"
$I \subseteq M$ $I = M$	(weak) empirical claim strong empirical claim

2.2.3 Basic terminology

Before we go over to stratified theories, we would like to present some useful basic terminology, which can largely be seen as a structuralist explication of Popperian 'statement terminology' [Popper, 1934/1959]. We will neglect all necessary provisos, in particular in regard to the complications arising from underlying theories. To use Lakatos's term [Lakatos, 1978], we explicate naive falsificationism, first unstratified, later stratified.

When the claim of theory $UT = \langle Mp, M, D, I\rangle$ is false $I - M$ is by definition non-empty, in which case it is plausible to call its members instantial mistakes or (empirical) *counter-examples* of *UT*. Note that being a counter-example in this sense does not imply that it has been realized already and registered as such. The set of counter-examples $I - M$ is by definition a subset of Mp - M. Hence, $I - M$ can, whatever I is, only be non-empty when Mp - M is non-empty. In other words, the members of Mp - M may be called the *potential* counter-examples of the theory and, as has already been stated, the set Mp - M itself the *empirical content* of *UT*. From the present point of view, Popper had similar things in mind with his notions of 'potential falsifier' and 'empirical content'.

Other plausible explications of Popperian terminology (which will however not be used in the sequel) are for instance: *UT* is *falsifiable* (or empirical) if and only if Mp - M is non-empty, and UT^* is *better falsifiable* than *UT* when Mp - M is a proper subset of Mp - M^*. The latter condition is equivalent to: $M*$ is a proper subset of M. In its turn, this is equivalent to stating that the claim of UT^* implies that of *UT*, and not conversely, that is, UT^* is stronger than *UT*.

The well-known verification/falsification asymmetry also arises naturally in the present set-up. To *verify* theory *UT* it would be necessary to show that all members of I, that is, all Mp-representations of D, belong to M. In interesting cases, this demonstration will always be an infinite task, even in the case that I is finite, for the task is only finite when D is finite. To *falsify* *UT*, however, it is 'only' necessary to show that there is *at least one* member of I not belonging to M. Hence, if a theory is true, verification will nevertheless not be obtainable if D is infinite. On the other hand, when a theory is false, falsification is attainable in principle, viz., by realizing one counter-example. If an attempt to falsify fails in such a way that the experiment provides an (*empirical*) *example* of UT, i.e., a member of M, this is called *confirmation* (or corroboration) of UT.

In the present set-up Popper's distinction between universal and existential statements gets an adapted interpretation. Here it becomes the distinction between the general claim of the theory ($I \subseteq M$) that all intended applications are models of the theory, and the negation of this claim, the existential claim that at least one intended application is not a model ($I - M$ is non-empty).

A *basic statement* (see Subsection 1.3.1 for Popper's specific idea of a basic statement) becomes a claim to the effect that a certain intended application x in I belongs to a certain subset F of Mp, defined by a certain condition being imposed on potential models, i.e., $x \in I \cap F$. An *accepted* basic statement presupposes of course that the relevant intended application has been realized.

The basic statement $x \in I \cap F$ is in conflict with theory UT if it can be demonstrated on conceptual grounds that $F \cap M$ is empty. Such basic statements may be seen as a more direct explication of Popper's idea of 'potential falsifiers', compared to 'potential counter-examples'. However, it is easy to show that the suggested statement concept of potential falsifier has become essentially redundant. It is readily verified that the discovery of a true potential falsifier of UT, i.e., the discovery of a x in Mp for which '$x \in I \cap F$" is true, implies that $I - M$ is non-empty and hence that x is a counter-example of UT. Conversely, the existence of counter-examples of UT is easily seen to imply that there must be true potential falsifiers. As a consequence, empirically demonstrating the existence of a counter-example, i.e., realizing a potential counter-example, goes hand in hand with demonstrating that there is a true potential falsifier. Hence, the statement concept of potential falsifier is not needed in the face of the concept of a (potential) counter-example.

2.3 The stratified approach to theories

Starting by reconsidering the slide balance in the light of attempts to measure the relevant quantities in a non-circular way, we will introduce the structuralist representation of (*prima facie* proper) theories and the corresponding terminology with the epistemological distinction between theoretical and non-theoretical terms. We will give the basic outline of this kind of representation for three examples: classical particle mechanics, the periodic table, and psychoanalytic theory.

2.3.1 The slide balance reconsidered

The problem with the slide balance is that it might be impossible to test the claim in a non-circular way without leading to an infinite regression. For, to test the claim, it is necessary to measure the distances and the weights. Whereas distance measuring does not require something like a slide balance, weight measuring may not only actually be done by using a slide balance, there might even be no other possibility. If the weight of a particle is measured by a slide balance the law of the balance is obviously presupposed. Hence, assuming that the weight of a particle can only be measured by a slide balance, the concept of weight is SB-theoretical and leads to the so-called *problem of theoretical terms.* A test of the claim of the theory would presuppose that the weights of the particles have been measured before with the same or another slide balance. Hence, we get either circular testing or an infinite regress, if we stick to "$I \subseteq M$" as the empirical claim of the theory. There is, however, a way-out of this dilemma, by restricting the empirical claim to SB-non-theoretical terms. Later we will see that for two reasons the situation in the present example is not as dramatic as suggested, but this did not exclude the fact that the example could, by way of a thought experiment, be transformed into an instructive example of genuine theoretical terms.

In order to formulate a new empirical claim we introduce the set of potential partial equilibrium models *SBpp*, being the structures of *SBp* without the *SB*-theoretical weight component and the corresponding 'status-condition', viz., clause 3). By consequence, there is a restriction or *projection function* π from *SBp* onto *SBpp* projecting every potential model on the potential partial model arising from deleting w and clause 3). Hence, for $x = \langle P, Pl, d, w \rangle \in SBp$, the projection of x, $\pi(x)$, is equal to $\langle P, Pl, d \rangle \in SBpp$. For an arbitrary subset X of *SBp*, πX, the projection of X, is defined as the subset of *SBpp* containing precisely the projections of the members of X.

For stratified theories we assume that the set of intended applications E no longer represents the equilibrium states seen through *SBp*, but seen through *SBpp*. Hence, the corresponding conceptual claim that E is a subset of *SBpp* is not laden with our theory about *SB*.

Now it is also easy to see that the claim that E is a subset of πSB is not laden with the weight term, in the sense that it does not presuppose that the weights of the particles have been empirically determined. Hence, this revised empirical claim can be tested in a non-circular way.

It is plausible to call the members of *E*-πSB, if any, counter-examples of the theory. It is clear that they have to come from *SBpp*-πSB. Hence, it is now plausible to call this set the *empirical content* and its members *potential counter-examples.* Note that the empirical content reduces to the empirical content of the unstratified theory (*SBp*-*SB*) when *SBpp* and *SBp* are identical and π is, by consequence, the identity-function.

Unfortunately, the new claim is not only non-circular, it is also vacuous, for the empirical content is empty. The claim says in fact that all intended applications can be extended to models of the theory. To be precise, the claim is that every $\langle P, Pl, d \rangle$ in E can be supplied with a positive real-valued function w on P such that $\langle P, Pl, d, w \rangle$ is in *SB*. But it is easy to check that this is possible for every member of *SBpp*. In other words the empirical content *SBpp*-πSB is empty.

However, the situation changes when we take so-called constraints into consideration: in the present case we have to require also that the weights assigned to the same particle, occurring in different applications should be the same. In contrast to the distance of the objects from the turning point S, our concept of weight is such that the weight of particles is constant in different applications. The formal treatment of constraints, however, will be postponed to Subsection 2.4.4.

Before we return to the general exposition we will summarize the formal features of the theory of the slide balance. leaving out the plausible specification of the projection function π:

The refined theory of the slide balance $\langle SBp, SBpp, SB, \pi, D, E\rangle$		
contains $\rightarrow$	$\langle P, Pl, d, w\rangle$ iff	
$\uparrow$ contains $\rightarrow$	$\langle P, Pl, d\rangle = \pi\langle P, Pl, d, w\rangle$ iff	
$\uparrow$	1. P is a finite set and Pl is a subset of P	particles the particles on the left of S
$SBpp$	2 $d: P \rightarrow \mathbb{R}^+$	$d(p)$: the distance of p from S
SBp	3 $w: P \rightarrow \mathbb{R}^+$	$w(p)$: the weight of p
SB	4 $\Sigma_{p\in Pl}d(p).w(p) =$ $\Sigma_{p\in P-Pl}d(p).w(p)$	*the law of the balance*
Concepts and claims		
$SBpp$ - πSB	*empirical content* without w-constraint empty, with w-constraint non-empty	
$E \subseteq SBpp$	*conceptual claim*: all intended domain of applications D can be represented, by E, as potential partial models, the intended applications	
$E \subseteq \pi SB$ $E = \pi SB$	*(weak) empirical claim*: all intended applications can be extended to models *strong empirical claim*: ... and vice versa	

By way of digression, it is interesting to note that, assuming the weight-constraint, the SB-theory explains the following observational, i.e., SB-unladen, *factor slide law*: if, starting from an equilibrium, the distances of all objects are multiplied by the same factor, there is again equilibrium. For it follows trivially from the law of the balance.

As a matter of fact, in the present case it is not difficult to formulate an observational law such that the notion of weight can be explicitly defined, apart from a proportionality constant, on its basis. The law referred to states the following: given a unit object at a unit distance at one side of S, every other object p has a 'unique equilibrium distance' $d_u(p)$ at the other side. The weight $w(p)$ is then defined as $1/d_u(p)$, hence, such that in the relevant cases the law of the balance is satisfied by definition. Consequently, for these cases the law cannot be tested in a non-circular way. But there is no regress, let alone infinite regress. For, given the definition, the rest of the law of the balance is a straightforward empirical claim that can be directly tested.

As a consequence, the theory of the slide balance does not, on closer inspection, give rise to the problem of theoretical terms, when certain observational laws are taken into consideration. Of course, this does not affect the instructiveness of the SB-theory as an almost proper theory. Moreover, it illustrates an interesting way in which a seemingly proper theory may on closer inspection be a sophisticatedly formulated observational theory, in the present case: the conjunction of the 'unique equilibrium distance law', the weight-definition on its basis, and the law of the balance.

There is still one other reason why the problem of theoretical terms is not so dramatic in the case of the slide balance: there are other ways of measuring the weight of objects than by using a slide balance. But let us now turn to the general set up of stratified theories, designed for proper theories.

2.3.2 Stratified theories

The general set-up of the structure of epistemologically stratified theories can now directly be presented in a scheme. Such a theory is a meta-structure of the following form:

$\langle Mp, Mpp, M, \pi, D, I\rangle$ is an *epistemologically stratified theory* iff	
Mp	potential models: a set of structures of a certain type
Mpp	potential partial models: the substructures of *Mp* restricted to non-theoretical components
$M \subseteq Mp$	models: the potential models that satisfy all axioms
$\pi : Mp \rightarrow Mpp$	the projection function (from *Mp* onto *Mpp*) $\pi X = \{\pi(x)/x \varepsilon X\}$, for $X \subseteq Mp$, implying $\pi X \subseteq Mpp$
πM	projected models
Mpp - πM	empirical content
D	the intended domain of applications
$I \subseteq Mpp$	intended applications (non-theoretical), resulting from the conceptual claim that D can be represented as a set of members of *Mpp*, i.e. "$I = Mpp(D)$"
$I \subseteq \pi M$ $I = \pi M$	(weak) empirical claim strong empirical claim

Now it is plausible to call $\langle Mp, Mpp, M, \pi\rangle$ the theoretical core of the theory and $\langle D, I\rangle$ remains the application target. Figure 5 illustrates the refined empirical claim: the shaded area, representing I-πM, should be empty. To be precise, I-πM should be empty on conceptual grounds, that is, the conceptual characterization of I and πM should not leave room for conceptual possibilities in I-πM (let alone for actual intended applications).

2.3.3 Examples

In this subsection we will give the theoretical core of the structuralist reconstruction of three well-known theories, viz., Newton's classical (gravitational) particle mechanics, Mendeleev's and the refined theory of the periodic table of chemical elements, and Freud's psycho-analytic theory. The presentation will always start

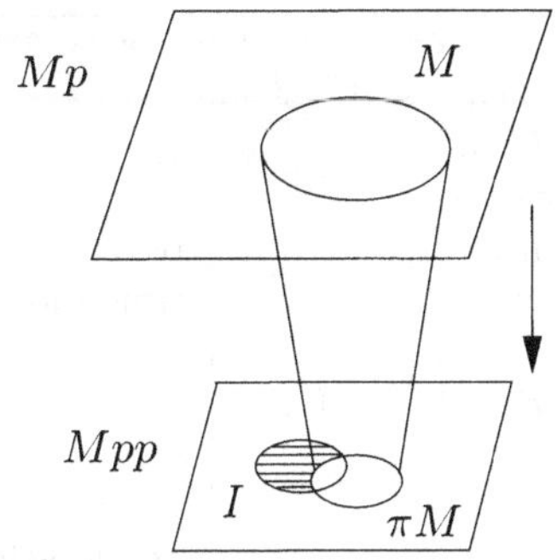

Figure 5. Refined (weak) empirical claim: shaded area empty

with the representation in a table followed by a brief elucidation. For details of the theory and the reconstruction, the reader is referred to the original or other publications of the reconstructions. The theories (more precisely, the theory cores) will be named by their basic class of models.

From the fact that Freud's theory can be reconstructed in the structuralistic way it follows that this way of reconstruction is, like the statement approach, applicable to qualitative, non-mathematical theories. From the other examples, it is evident that the present approach is also well suited for quantitative theories, a kind of theory for which the statement approach leads to all kinds of complications.

In a sense it is a trivial claim that every empirical theory can be reconstructed in structuralist fashion. Hence, there should be additional reasons to do so in particular cases. A general reason frequently is the desire to get a better insight into the theory; besides that, one may be interested in particular questions, such as whether the theory has empirical content, whether it is an observational or a proper theory, what its precise relation is to another theory, etc. The examples to be presented are supplied with some comments to illustrate both reasons of reconstruction. But the main function of getting acquainted with the structuralist approach in general and by way of examples is of course the heuristic role it may play in the construction of new theories.

After the presentation of the three examples we will continue in the next subsection with general matters, such as the distinction between absolute and relative empirical content, the possibilities of determining of the intended applications, relations between theories, theory-nets, and constraints.

Classical particle mechanics

As is well known, Newton's theory of gravitation is based on the generic theory of particle motion, i.e., classical particle mechanics (*CPM*). The core of this theory is formed by three interrelated so-called laws of motion: the first law: the law of inertia, the second law: $F = m.a$, and third law: action is minus reaction. This general or generic theory can be specialized by adding the special law of

Classical particle mechanics for one dimension (with gravitation as specialization) $CPM = \langle CPMp, CPMpp, \pi, CPM, GCPM \rangle$		
contains →	$\langle P, T, s, m, f \rangle$ iff	
↑ contains →	$\langle P, T, s \rangle = \pi \langle P, T, s, m, f \rangle$ iff	
↑ *CPMpp*	1) P is a finite set 2) T is real interval 3) $s : P \times T \to \mathbb{R}$ giving rise to 1st and 2nd time derivatives: $v : P \times T \to \mathbb{R}$ a: $P \times T \to$	particles time-interval position velocity acceleration
  CPMp	4) $m : P \to \mathbb{R}$ 5) $f : P \times T \times P \to \mathbb{R}$ $f(p, t, q)$	mass force force from q on p at t
   CPM	6) *second law* (implying the *first law* in this formulation): for all p in P and t in T $\Sigma_{q \in P}$ $f(p,t,q) = m(p).a(p,t)$ 7) *third law* (action = - reaction): for all p and q in P and all t in T $f(p,t,q) = - f(q,t,p)$	
GCPM	*the law of gravitation:* there is a universal real constant γ such that for all p and q in P and t in T $f(p,t,q)=$ $+/- \gamma$ [m(p).m(q)] / $[s(p,t) - s(q,t)]^2$	

gravitation(al force), but there are other well known specializations, e.g., Hooke's law of spring-force and Coulomb's law of electrostatic force.

The following remarks may elucidate the table to some extent. For detailed expositions the reader is referred to Sneed [1971], Zandvoort [1982] and Balzer, Sneed and Moulines [1987].

It is clear that mass and force are treated as *CPM*-theoretical components; although this treatment may not be strictly necessary, it is always a safe option in case the situation is unclear; but a vacuous claim may be the result.

CPM concerns the generic theory, 6) and 7) are the proper generic laws/principles. *GCPM* is, due to the addition of the law of gravitation, a subset of *CPM*, that is a specialization of *CPM*.

π*CPM* and π*GCPM* provide the projected models of *CPM* and *GCPM*, respectively. *CPMpp*-π*CPM* and *CPMpp*-π*GCPM* constitute the empirical content of *CPM* and *GCPM*, respectively. Note that the former is a subset of the latter, just as it should be, for *GCPM* is stronger than *CPM*.

As long as the identity constraint for mass is not taken into consideration *GCPM* has no empirical content, let alone *CPM*. With the mass constraint *CPM* still lacks empirical content, but *GCPM* gets it.

The intended domain of applications of *CPM* concerns in the first place that of *GCPM*, for instance, planetary orbits, falling stones, paths of projectiles, etc., but also movement of objects by spring or electric forces. Moreover, it contains compound applications, i.e., applications in which two or more force types operate, e.g., three in the case of an electrically charged ball on an isolated vertical spring on a charged table.

Periodic table

Periodic table of chemical elements (naive and refined) $NPT/SPT = \langle PTpp, PTp, \pi, NPT/SPT \rangle$		
contains $\rightarrow$	$\langle E, m, \approx, z\rangle$ iff	
$\uparrow$ contains $\rightarrow$	$\langle E, m, \approx\rangle = \pi\langle E, m, \approx\rangle$ iff	
$\uparrow$	1) E: a finite set	chemical elements
	2) $m : E \rightarrow \mathbb{R}$	atomic mass
$PTpp$	3) $\approx$: equivalence relation on E	chemical similarity
PTp	4) $z : E \rightarrow \mathbb{R}$	atomic number
	5) a. range (z)= 1,2,..., max(z)	z is onto {1,...,max(z)}
	b. $m(e)<m(e')$ iff $z(e)<z(e')$	z increases with m
	c. $z(e)=z(e')$ implies $e = e'$	z is a one-one function
NPT	6N) *naive periodic law* $e \approx e'$ iff $\lvert z(e) - z(e')\rvert$ is a multiple of 8 respectively,	
RPT	6R) *refined periodic law*, elegant, but complicated; core: if $e \approx e'$ and if there is no element with z-number between $z(e)$ and $z(e')$ then $\lvert z(e) - z(e')\rvert$ can be written as $2n^2$, i.e., 2 or 8 or 18 or 32 etc.	

For a detailed exposition the reader is referred to Hettema and Kuipers [1988; 2000][22]. The following remarks highlight some crucial points.

As is well known, Mendeleev developed the periodic table on the basis of the observation that the chemical elements can be classified in groups of elements with chemically similar behavior. Moreover, he noted that the ordering of the elements by increasing atomic mass roughly leads to a matrix in which the groups appear as columns. To explain the system in this matrix he introduced the concept of atomic number and formulated the (naive) periodic law (NPT), which was later refined by others (RPT).

In the present example the intended domain of application concerns the chemical elements taken together, such that the conceptual claim states that this domain can be represented by just one potential projected model, say $\langle E*, m*, \approx *\rangle$.

[22]For a critical discussion of the main historical and reductive claims in [Hettema and Kuipers, 1988] see [Scerri, 1997]. For a continued discussion, see [Hettema and Kuipers, 2000; Scerri, 2005; Kuipers, 2005].

Mendeleev's empirical claim was that this pp-model belongs to πNPT and the modern empirical claim localizes it in πRPT. Or, equivalently, there is $z*$ such that $\langle E*, m*, \approx *, z*\rangle$ belongs to NPT and RPT, respectively.

It is not difficult to verify that both theories have empirical content. In fact both claims are false. To fulfil the claims as much as possible, we must allow counter-examples to the three technical conditions imposed by clause 5). They amount to, using plausible names: 5a) missing elements, which may be discovered later, and some have been, 5b) order disturbers, having greater mass and lower atomic number than others or vice versa, and 5c) isotopes, i.e., different elements with the same atomic number.

Note that the notion of a counter-example is used here on a lower level than in the general set-up. This is possible because there is only one overall intended application, viz., $\langle E*, m*, \approx *\rangle$. If that does not fit into πNPT or πRPT, this failure must be due to lower level counter-examples, i.e., specific elements. There may be systematic or just local counter-examples. In this sense NPT has both types of counter-examples, whereas RPT has only local counter-examples.

The history of PT provides marvelous examples of all four combinations of theory (un)laden and theory (un)guided observation, as described in Subsection 1.4.2. A successful search for missing links, for instance, means theory guided but theory unladen observation.

The quantum mechanical theory of the atom provides a reductive explanation, see [Kuipers, 2001, Chapter 3] for RPT, by means of identification of z with the number of electrons of the atom concerned. In view of the fact that this number can be measured in RPT-independent ways, RPT is in fact an observational theory. Of course, Mendeleev's NPT was a proper theory, with z as a theoretical term. RPT was developed hand in hand with atomic theory, in which process it transformed from a proper theory into an observational theory.

Psychoanalytic theory

Presenting the structure of Freud's theory does of course not mean that we uncritically subscribe to that theory. One may even denounce that theory as totally out of date, and still be interested in its structure. Compare interest in the structure of the phlogiston theory. For a detailed exposition the reader is referred to Balzer [1982], Stegmüller [1986] or, for the most refined one, to Balzer and Marcou [1989]. The general psychoanalytic theory *PA* is intended for all human beings, they are all supposed to repress negative experiences (*PA*-11) and to satisfy the main axiom *PA*-10 that all unconscious impulses are sooner or later realized. Note that *PA*-11 does not use theoretical terms, so it makes sense, as indicated, to introduce (non-theoretical or observational or) *partial models* (*PApart*) as potential partial models satisfying this non-theoretical but substantial axiom. In the next subsection we will generalize this idea and investigate its consequences.

The psychoanalytic theory of neurosis *PAN* is a specialization to people with a neurosis generating experience, as implicitly defined by *PA*-13. It will be illumi-

Psychoanalytic theory $PA = \langle PApp, PAp, \pi, PApart, PA\rangle$		
contains $\rightarrow$	$\langle T, E, L, \leq, ASS, B, N, A, U, REAL\rangle$ iff	
$\uparrow$ contains $\rightarrow$	$\langle T, E, L, \leq, ASS, B, N\rangle = \pi\langle \ldots A, U, REAL\rangle$ iff	
$\uparrow$ *Papp*	1) T is an interval of real numbers; variable t, t^*, etc. 2) E is a non-empty set 3) L is a proper subset of E 4) $\leq$ is a weak linear ordering on T - $<$ by definition $\leq$ and $\neq$ 5) $B(t)$ is a non-empty subset of E 6) $N(t)$ is a subset of $B(t)$ and L 7) ASS is a relation on E - $ASS(e,e)$, i.e., ASS is reflexive	life span of a person experiences painful experiences not later than earlier than consciousness at time t negative experiences at t associated experiences self association
          Pap	8) A is a non-empty set; $A \cap E = \phi$ 9) $U(t)$ is a non-empty subset of A 10) $REAL$ is ternary relation on $E \times A \times T$: $REAL(e,a,t)$ - if $REAL(e,a,t)$ then e in $B(t)$ and a in $U(t)$ - not for all t, e in $B(t)$ and a in $U(t)$: $REAL(e,a,t)$ - if $REAL(e,a,t)$ and $REAL(e',a,t')$ then $ASS(e,e')$	unconscious impulses unconsciousness at t e is a realization of a at t
PApp + *Papart*	11) *repression axiom*: repression of negative experiences, incl. associated ones: if e in $N(t)$ and $ASS(e, e')$ and $t < t^*$ then e' not in $B(t^*)$	
Pap + 11) + *PA*	12) *main axiom*: every unconscious impulse is realized sooner or later: for all t and for all a in $U(t)$ there are e in E and t^* such that $t \leq t^*$ and $REAL(e,a,t^*)$	
PA + *PAN*	13) *neurosis axiom*: having a neurosis generating experience for an impulse: there are t_0, e_0 in E, a_0 in A such that $REAL(e_0, a_0, t_0)$ and e_0 in $N(t_0)$	

nating to define some additional notions:

'e_0 is *repressed* after t_0' iff

e_0 is in $B(t_0)$ and for all $t > t_0$ e_0 is not in $B(t)$

'being *neurotic* with respect to a_0 after t_0' iff

for all $t > t_0$ there is no e in E such that $REAL(a_0, e, t)$

Now it is easy to prove the following

Theorem: if one has had at t_0 a neurosis generating experience e_0 with respect to impulse a_0, e_0 is repressed after t_0 and one is neurotic with respect to a_0 after t_0.

Note that 'being neurotic with respect to a_0 after t_0' is formally almost in conflict with the main axiom, but the point is that the realization of a_0 required by the main axiom has already taken place at t_0, in particular as a result of a negative experience.

A serious problem in the present formulation is what happens when a_0 recurs in the unconsciousness, because the main axiom requires repeated realization. But here we will not deal with the necessary refinements for this and other reasons. We will just mention one other example of a further refinement of the theory of neurosis, which can be obtained by integrating it with another specialization of the general theory, viz., the theory of sublimation.

We conclude this section with references to some further examples: classical and relativistic collision mechanics [Balzer, Sneed and Moulines, 1987]; Lagrangian mechanics [Balzer, Sneed and Moulines, 1987]; special relativity theory [Balzer, 1982]; old quantum theory ([Hettema and Kuipers, 1995], see also [Kuipers, 2000, Chapter 11]); simple equilibrium thermodynamics [Balzer, Sneed and Moulines, 1987]; Daltonian stoichiometry [Balzer, Sneed and Moulines, 1987]; modern genetics [Balzer and Dawe, 1986; 1997]; Jeffrey's theory of decisions [Stegmüller, 1986]; the Arrow-Debreu theory of individual and collective demand [Janssen and Kuipers, 1989], capital structure theory ([Cools, Hamminga, Kuipers, 1994], see also [Kuipers, 2000, Chapter 11]); folk psychology and connectionism [Bickle, 1993; 1998]; Jakobson's theory of literature has also been reconstructed [Stegmüller, 1986]. Several psychological theories are reconstructed in [Westmeyer 1989, 1992]. Finally, [Balzer, Sneed and Moulines, 2000] contains a representative sample of those above, as well as other ones.

Of course, structuralist representation of a theory may not be necessary for the purposes at hand. However, for detailed questions, such as "Does a certain theory have empirical content?" such a representation is almost unavoidable. In the next subsection we will introduce a number refinements of the structuralist approach that enable us to answer such refined questions.

2.4 Refinements

Now we will continue with some further refinements of the structuralist approach, viz., the distinction between absolute and relative empirical content, the possibilities of determination of the intended applications, relations between theories, theory-nets, and constraints.

2.4.1 Absolute and relative empirical content

It is always possible to divide the axioms into, on the one hand analytic (A) and synthetic or substantial (S) axioms and, on the other, non-theoretical (N) and theoretical (T) ones. As a result, there are four types of axioms: NA, TA, NS, and TS. We do not mean to suggest that the two distinctions are unproblematic. We have discussed extensively in Section 1 and the previous subsections how the N/T-distinction can be made. The A/S-distinction is at least as notorious, and it is undoubtedly partly a matter of conventional decision where the boundary is drawn. See Niiniluoto [1999, Chapter 5] and Sober [2000] for lucid accounts, against Quine's well-known challenges, of the tenability of this distinction. However, here, and with respect to the N/T-distinction, it is advisable in case of doubt to choose the cautious classifications, i.e., S and T, respectively.

The following survey of sets and names of their elements will now speak for itself.

Types of models in relation to types of axioms A: analytic, or S: synthetic; N: non-theoretical, or T: theoretical		
Mpp	potential partial models	NA
Mp	potential models	NA + TA
Mpart	partial models	NA + NS
M	models	NA + TA + NS + TS

It is clear that $\langle Mpp, Mpart\rangle$ is the (theoretical) core of a partial theory, i.e., an unstratified, hence observational theory, constituting a substantial part of the full theory. The empirical content of the full theory was defined as $Mpp\text{-}\pi M$, let us call it more specifically the (total or) *absolute empirical content* (AEC). The empirical content of the partial theory, the *partial empirical content* (PEC), is of course *Mpp-Mpart*. Given the trivial fact that πM is a subset of *Mpart*, the partial empirical content is automatically a subset of the absolute empirical content. The interesting question is whether the full theory has something to add to the partial theory, i.e., whether the (extra or) *relative empirical content* (REC), defined as $Mpart\text{-}\pi M$, is non-empty. Figure 6 depicts the three kinds of content.

It is easy to check that the absolute empirical content ($Mpp\text{-}\pi M$) is the union of the partial empirical content (*Mpp-Mpart*) and the relative empirical content ($Mpart\text{-}\pi M$). In consequence, if a theory has relative (and/or partial) empirical content it has absolute empirical content.

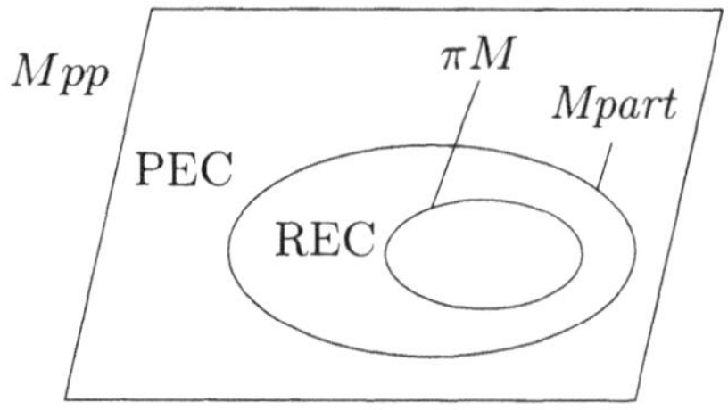

Figure 6. Three contents of a theory: AEC = PEC + REC

Conversely, however, a theory may have absolute empirical content without having relative empirical content, in which case the absolute empirical content coincides with the partial empirical content.

Balzer [1982] claimed that the general psychoanalytic theory has partial, and hence absolute, but no relative empirical content. Stegmüller [1986], however, is able to prove that it also has relative empirical content. Stegmüller then continues with the interesting observations that for this proof it is not necessary to take constraints and/or special laws into consideration and that, as we already noted, classical particle mechanics CPM has only (relative) empirical content when constraints and special laws are taken into consideration. Hence, according to the relative content criterion for empirical impact Freud's theory is in a sense even superior to that of Newton.

But we would like to add that the *CPM* example makes it clear that a generic theory (including constraints) need not have relative empirical content in order to be useful. The important research question is whether a generic theory can be supplemented with special laws (leading to specializations, see below) which have relative empirical content (Cf. [Bunge, 1977]).

It is interesting to note that the status of utility theory (Subsection 1.1) is to some extent comparable to that of classical particle mechanics. The generic versions of both theories have no empirical content. However, although it is quite clear that the latter theory can be specialized so as to have empirical content, this is not beyond dispute for the former.

2.4.2 Intended applications reconsidered

In this subsection we will begin by making a few general remarks about the set of intended applications I, then we will formulate three different ways of determining I, and conclude by elaborating, to a certain extent, the problem of theoretical terms.

I was introduced as 'D seen through Mp' and later revised as 'D seen through Mpp', i.e., I represents the intended domain of applications with the conceptual means of Mpp. We will restrict our formulations to the refined case, when not otherwise stated, for it includes the extreme case that Mp=Mpp.

It is evident that I is Mpp-dependent, and Mpp is manmade. Hence we subscribe to a fundamental form of conceptual relativity. But this need not imply an extreme form of relativism: empirical claims are objectively true or false, for their truth or falsehood depends on nature, assuming that they have empirical content.

In its turn, the objective character of empirical claims does not imply that D, *Mpp* (and hence I) and *Mp* are fixed beforehand, and that the task remains to formulate a subset M of *Mp* leading to a true empirical claim. As a matter of fact, in practice, the determination of D, *Mpp*, *Mp* and M is a complicated dialectical interaction process, guided by the desire to formulate informative and true empirical claims. Unfortunately, it seems difficult to discern general patterns in this interaction process, without making some important idealizations.

However, if we assume, by idealization, that *Mpp* is fixed, the determination of (D and hence) I can be governed by at least three different principles.

If we are interested in all relevant nomic possibilities, I coincides with the set of nomic possibilities at the *Mpp*-level. Let *To* indicate this subset of *Mpp*. Although we may not have an explicit characterization of *To*, there is a clear empirical criterion for membership: x in *Mpp* belongs to I=*To* iff x can be realized. If we are interested in a well-defined subset of the set of *To*, i.e., nomic possibilities satisfying some explicit condition, membership determination is not fundamentally different. In both cases we will speak of the *empirical determination* of I.

In this case, the obvious target of theory development is an explicit characterization of I, i.e., a set of models M is sought for which the strong claim holds: $I = \pi M$, such that M may be called *the true (Mp-)theory about I*, or simply, *the (conceptually relative) truth.* In [Kuipers, 2000, Chapter 7 and 9, respectively], the formal structure of truth approximation by unstratified and stratified theories is studied in detail. Here truth approximation is restricted to revising theories, leaving D, *Mp*, and hence I, fixed. Of course, when I is restricted to a partially well-defined subset of *To*, this also enables keeping the vocabulary *Mp* and the theory Mconstant, and revising (the partial definition of) I, and hence truth approximation by revision of the domain of intended applications (Kuipers, forthcoming b). Instead of speaking of the 'empirical determination' of I, we may then speak of the 'empirical specification' of M.

In [Kuipers, 2000, Chapter 7] the notion of a nomic possibility is presented as an absolute qualification. However, there may well be cases where it makes good sense to distinguish levels of nomic possibility, e.g., as suggested by the following sequence of 'lower' to 'higher' levels: the physical, chemical, biological, psychological, cultural-socio-economical level. Being nomically possible at a higher level then implies being nomically possible at a lower level, but not the converse. Another example concerns the idea of nomically possible states of an artifact, assuming that it remains intact, which means a severe restriction to its physically possible states, including broken ones. Such refinements can also easily be built into the empirical determination of I, as long as the boundaries between the different levels of nomic possibility may be assumed to be sharp.

In many cases, however, the interest is directed to a proper subset of *To*, of which the membership is not sharply defined. One important way in which I can then have been circumscribed is by so-called *paradigmatic determination.*

> *Definition*: *I* is *paradigmatically determined* if there are *PAR* and *SIM* such that

1. *I* is a subset of *To*	the intended applications
2. *PAR* is a finite subset of *I*	the paradigmatic examples
3. *SIM* is a binary relation on *Mpp*	a similarity relation
4. for all x in *I-PAR* there is y in *PAR* such that *SIM(x,y)*	

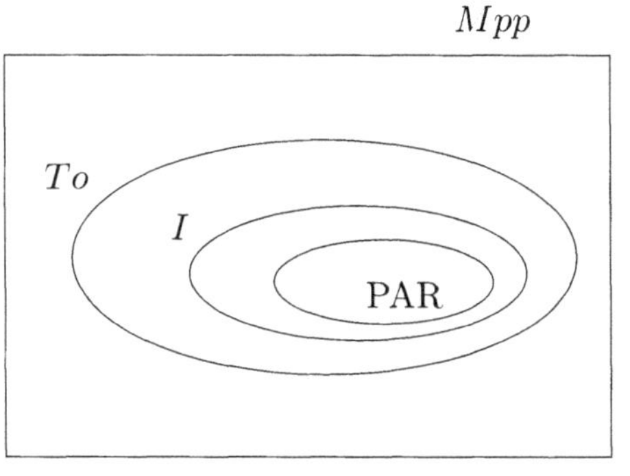

Figure 7. Paradigmatic determination of *I*

Figure 7 depicts the relations between the various sets in the case of paradigmatic determination of *I*. The elements of *PAR* may, for instance, be determined by the founding father of the theory and corresponds to one of the meanings Kuhn [1962] had in mind with the term 'paradigms' and which he later called 'exemplars'. Of course, the main source of vagueness is the notion of similarity, for it will not as a rule be possible to define this notion sharply, at least not at the beginning of the research process. As a matter of fact, the relevant definition of similarity is to be discovered by trying to undertake successful excursions out of *PAR* and a given *M*, roughly in the same way as described in [Kuipers, 2006] for the first way of determination of *I*. Of course, each tentative excursion entails tentative sufficient conditions of similarity, and each definite conclusion not only requires clear-cut sufficient conditions but also a definite *M*, together enabling a definite empirical claim.

In both cases of determination, assuming that the theory has (at least absolute) empirical content, the empirical claim of a stratified theory will not be trivial. As is easy to verify, the empirical claim becomes trivial in the third way of determination of *I*, so-called *auto-determination*: for x in *To*, x belongs to *I* iff x belongs to πM, i.e., is the projection of a model of a stratified theory. In the case of auto-determination, the theory in question is typically not something to be tested, but it will have been designed for other purposes.

In the case of an unstratified theory, the set of intended applications is of course a subset of the set of nomic possibilities on the *Mp*-level, which is then in its turn a subset of *Mp*. For the further determination of I there are again the same three possibilities of empirical, paradigmatic and auto-determination. Where there are proper theoretical terms involved, all three forms of determination result in problems.

Let us briefly restate and elaborate the background of (epistemological) stratification of a theory in T-theoretical and T-non-theoretical terms. Let there be an unstratified theory $UT = \langle Mp, M, D, I\rangle$ and assume that UT has non-empty empirical content *Mp-M*. The term t occurring as a component in *Mp* is said to be T-theoretical iff every known method of measuring t in a specific intended application results in a model of UT. It is otherwise T-non-theoretical. Let UT contain at least one T-theoretical term and let us first assume that I is supposed to be empirically or paradigmatically determined, in which case the empirical claim "I is a subset of M" is non-trivial. However, testing this claim is impossible, for it leads demonstrably either to circularity or to an infinite regress. In the case of auto-determination, the problem is that determination of the membership of M leads to circularity or infinite regress.

The remedy for these problems is the epistemological stratification of the theory in terms of a partial theory containing only and all T-non-theoretical terms. Assuming that the stratified theory has non-empty (absolute) empirical content $Mpp\text{-}\pi M$, the empirical claim or auto-determination is non-trivial, depending on whether I has or has not been fixed in advance.

The indicated definition of T-theoriticity is a pragmatic one, due to the "every known method of measuring"-clause and goes back to Sneed [1971]. Given the fact that the class of known methods can only increase, the definition is perfectly compatible with the advice to classify a term as T-theoretical in case of doubt. However, it is tempting to look for an intrinsic definition of theoriticity. Gähde [1983] has put forward an intrinsic definition. This proposal is not only highly technical and restricted to quantitative terms, it has also been criticized for other reasons (Cf. [Schurz, 1990]). However, Balzer [1996] has rebutted this criticism and has, moreover, proposed an essentially simpler formal criterion than Gähde's is.

2.4.3 Links between theories, and theory-nets

Theories that are roughly about the same domain are frequently related. At each moment they may constitute a network of theories, i.e., an ordered set of theories that are directly or indirectly related. Such a network depicts the synchronic situation, the succession of networks indicates the diachronic development.

Let us first define the main relations between theories, also called (intertheoretical) links. We will presuppose that all theories considered are stratified, hence the theories are of the form $ST = \langle Mp, Mpp, M, \pi, D, I\rangle$. It is easy to derive from the definitions what links result if the stratification is assumed to disappear ($Mp{=}Mpp$ and π becomes the identity function).

We have already indicated some specializations of theories, in the case of the theories of Newton and Freud.

ST^* is *specialization* of ST iff

1. $Mp^*=Mp$ and $Mpp^*=Mpp$ and $\pi^*=\pi$
2. $M*$ is a subset of M and
   $D*$ is a subset of D (and hence $I*$ is a subset of I)
   and at least one of the subsets is proper.

Specialization is one of the main research activities within a research program starting from some basic generic theory.

A new theory may add new (non-)theoretical components. It may be a genuine superposition, such that the old theory remains completely intact. Balzer [1982] describes the example of the classical kinematical theory built on the classical space-time theory. When at least one new theoretical component is introduced this type of link is called (conservative) theoretization and can be defined as follows:

ST^* is a *theoretization* of ST iff

1. all (non-)theoretical components of Mp remain (non-)theoretical components of Mp^*
2. Mp^* adds to Mp one or more new (non-)theoretical components, at least one theoretical one, all of which are stripped off by a function f from Mp^* onto Mp
3. for all $x*$ in $M^* f(x^*)$ belongs to M

Of course, it is possible to define non-conservative links between theories when new components are added.

The third important link between theories is that of reduction.

ST is *reducible* to ST^* iff there is a relation r on $Mp \times Mp^*$ such that

1. for all x in M there is x^* in M^* such that $r(x, x^*)$
2. if $r(x, x^*)$ and x^* in M^* then x in M
3. for all y in I there is y^* in I^* such that $r_\pi(y, y^*)$,
   where r_π indicates the projection of r on $Mpp \times Mpp^*$

Roughly speaking, the definition captures the explanation of one theory or law by another when the models and intended applications can be formally related in a way which is typically possible when one or more of the three basic types of reduction distinguished in [Kuipers, 2001, Chapter 3] apply.

The last type of link between theories to be mentioned is that of (idealization, or conversely) *concretization*, of which a precise definition has been given in [Kuipers, 2000, Subsection 10.4., see also Chapter 11]. There it is shown that concretization plays, for example, a crucial role in the truth approximation analysis of the transition of the theory of ideal gases to that of Van der Waals, and various transitions in the old quantum theory and capital structure theory. An informal explication of 'idealization & concretization' will be given in Subsection 3.3.2.

It is not difficult to check that all defined links generate partial orderings and we assume that all links to be considered are of this type. Let two theories be called directly related when one such link applies and let them be called related when there is a chain of directly related theories between them. Of course, 'being related' is again a partial ordering. A *theory-net* is defined as a set of theories related in this way such that there is a *basic theory* Tb, in the sense that all other theories are directly or indirectly related to this theory. Figure 8 depicts such a theory-net.

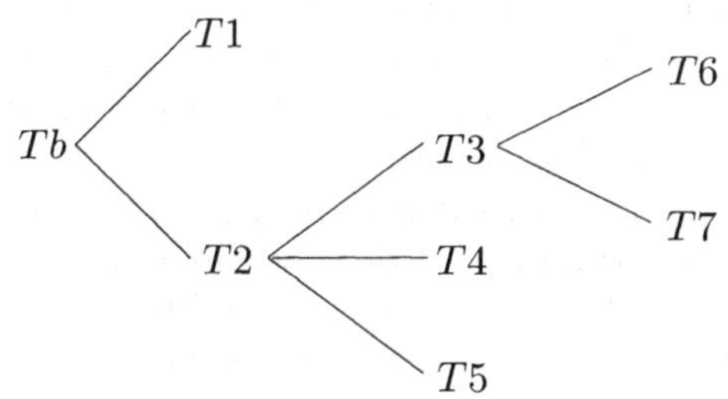

Figure 8. A theory-net

For the succession of theory-nets it is plausible to distinguish two basic types, leaving room for mixed cases. A transition to a new net may be *conservative* in the sense that the new net retains all theories of the old one, but one or more theories to the old net are added in some way. A transition is called *corrective* if one or more theories in the old net are replaced by new theories that are considered to be improvements.

2.4.4 Constraints

We have already referred several times to so-called constraints. Whereas laws and axioms in the normal sense lay down restrictions on individual potential models, constraints impose restrictions on sets of potential models. A particular type of constraint is a so-called identity-constraint, guaranteeing that a function assigns in different potential models, with some common base-sets, the same value to the same individuals. The weight-function in the case of the slide balance as well as the mass-function in the case of classical particle mechanics are cases in point.

A constraint can be formally defined in a very general way.

Definition: C is a *constraint* on the set S iff

1. C is a set of subsets of S
2. the union of the sets in C exhausts S (UC=S)

3. if X is in C and Y is a subset of X then Y is in C (subset-preservation)

It is easy to prove that all singleton sets $\{x\}$, for x in S, belong to C, hence a constraint does not exclude any individual potential model.

Let us now first concentrate on the typical role of a constraint C on *Mp* in a stratified theory $ST = \langle Mp, Mpp, M, C, \pi, D, I \rangle$. The standard empirical claim was "I is subset of πM", which could be paraphrased by saying that all members of I can be extended with theoretical components to genuine models, i.e., there is a subset X of M such that $\pi X{=}I$. Taking the constraint into consideration this claim is strengthened: there is a subset X of M *belonging to* C such that $\pi X{=}I$. Hence, now both M and C restrict the degrees of freedom for the supplementation of theoretical components. In the corresponding versions of the strong claim the clause "X is a subset of M" is simply replaced by "$X = M$". It is clear that a stratified theory may even be a pure *constraint-theory*, in the sense that C is non-trivial and M is trivial, i.e., $M{=}Mp$.

In [Kuipers, 2000] we deal with truth approximation by theories by assuming that theories are sets of structures, with or without the distinction between theoretical and non-theoretical terms. It is not too difficult to check that, when the truth is a constraint-theory, similar (basic and refined) definitions may be given of the claim that one constraint-theory may be closer to the truth than another. Of course, it is then also possible to deal with truth approximation by theories consisting of a 'normal' and a constraint part. However, we will leave the elaboration of the suggested possibilities to the reader.

Taking a constraint into account, the following reformulation of the standard claim is instructive. Let *A(ST)*, the *application space* of *ST*, be defined as the set of projections of all subsets of M satisfying C (formally: $A(ST){=}\ \pi(P(M) \mathrm{x} C)$). The standard claim now comes down to: I is in *A(ST)*.

It is also now plausible to define the *(absolute) empirical content AEC(ST)* as *P(Mpp)-A(ST)*, i.e., the subsets of *Mpp* which are excluded by M and C. Note that *AEC(ST)* reduces to $P(Mpp)\text{-}\pi(P(M))$ when C is trivial, i.e., when $C = P(M)$, and to *P(Mp)-P(M)* when π is trivial, i.e., when π is the identity function. It is easy to verify that *AEC(ST)* is empty in these respective cases iff the originally defined empirical contents $Mpp\text{-}\pi M$ and *Mp-M* are empty. Hence, the suggested new definitions of empirical content reproduce the original ones on the level of sets of sets of potential (partial) models.

Similar relations hold for the plausible definition of the *relative empirical content REC(ST)*: *P(Mpart)-A(ST)*. And again it follows almost trivially that non-empty *REC(ST)* implies non-empty *AEC(ST)*, but not the converse.

Constraints also make sense in other cases, e.g., in partial theories and in unstratified theories. For the first case, let *Cpart* be a constraint on *Mpp*. Like *Mpart*, *Cpart* represents empirical restrictions. We may say that *Mpart* captures the standard observational laws, whereas *Cpart* captures *constraint observational laws*. Of course, the associated claim states that I is a subset of *Mpart* belonging to *Cpart*. It is important to note that many empirical laws are constraint laws, or mixtures of standard and constraint laws.

If the indicated partial theory with constraint is isolated from the full theory, it is clear that we have an unstratified theory with constraint. In general, an unstratified theory with constraint is of course of the form $\langle Mp, M, C, D, I\rangle$, where C is a constraint on Mp.

2.5 Non-empirical Theories

In this section we have dealt with empirical theories, but let us finally consider the question of which structuralist concepts are also useful for non-empirical theories. Following Popper, non-empirical theories are by definition theories which are not (intended to be) falsifiable. One may distinguish at least four types of non-empirical theories:

> *metaphysical theories* are supposed to make claims about reality without assuming any particular conceptualization or, equivalently, they make claims generalizing over conceivable conceptualizations of reality,
>
> *mathematical and logical theories*, some of which deal with defined abstract objects, i.e., mental constructs, e.g., the theory of groups, other ones with 'concrete' mathematical objects, e.g., the theory of natural numbers,
>
> *conceptual theories* concern ways of looking (perspectives) at a certain domain,
>
> *normative theories* deal with what is (supposed to be) ethically, legally, aesthetically (in)admissible.

It is evident that almost all technical ingredients presented for empirical theories are also useful for non-empirical theories. In fact, Suppes [1957] invented the structuralist representation of empirical theories by transferring, as far as possible, the standard way of presenting mathematical theories, such as the theory of groups, to empirical theories. The crucial difference is that non-empirical theories do not make general empirical claims. The claims which are associated with them typically are either conceptual (logical, mathematical etc.) or restricted to individual intended applications. A typical claim around a mathematical theory is a mathematical theorem to the effect that the models of the theory can be proven to have a certain explicitly defined property. A typical claim of a specific conceptual theory is that a certain intended application is (or is not) a model of that special theory. Of course, generic theories, i.e., theories with vacuous empirical claims, are conceptual theories.

The 'structuralist theory of the structures of empirical theories' is a perfect example of a theory that is primarily intended as a conceptual theory (although one may strengthen it to a genuine empirical theory by adding a substantial claim). As a consequence, the foregoing exposition not only provides an elaborate example of a conceptual theory, it can also convince the reader of the usefulness of conceptual theories.

3 RESEARCH PROGRAMS AND RESEARCH STRATEGIES

Introduction

One of the most important insights of the philosophy of science since about 1960 is the awareness that the development of science should not be described in terms of the development of specific hypotheses and theories, but in more encompassing terms. The two main proponents of this insight are Kuhn and Lakatos. Kuhn first preferred the term 'paradigm' and later 'disciplinary matrix' [Kuhn, 1962/1969]. Lakatos [1970; 1978], basically aiming to capture the development of sequences of related theories, introduced the notion of a 'research program'. Half a dozen other terms are used to denote roughly the same concept, although their details may differ.

We prefer Lakatos's term 'research program' for its literal meaning: program of research. Although our favorite conception of research program will be somewhat weaker than that of Lakatos, we will use the same term. However, it should be stressed in advance that nobody means program in the detailed sense of a well-ordered sequence of things to do. At most a program of research in some global sense is meant. From now on, 'program' always means a research program.

Subsection 3.1. distinguishes four types of research programs. Subsection 3.2. presents a necessarily incomplete summary of current insights in the structure and development of research programs and some other global cognitive units. Finally, a number of strategic lessons are suggested, which have been freely derived from both the insights in global cognitive units and the distinction of four types of research programs. Subsection 3.3. deals with program internal strategies and Subsection 3.4. with interaction strategies. In the concluding remarks, we will reconsider the dynamics of descriptive and explanatory programs in the light of the law-distinction treated in Section 1.

3.1 Four types of research programs

Four ideal types of research programs will be described, followed by a survey of the main similarities and differences.

3.1.1 Descriptive, explanatory, design and explicative programs

The four types of programs to be distinguished are the following: descriptive, explanatory, design and explicative programs. They form ideal types. In consequence, mixtures are the rule, rather than the exception. However, it is often possible to describe a mixed program as a cooperative enterprise of two or more programs of an ideal type, with one of them being in some way dominant. To put it still more cautiously, the first characterizations may well be read as descriptions of four types of research, which are, in practice, part of complex undertakings. However, in these four types of research it is then often possible, at least analytically, to identify the underlying programs of an ideal type mentioned above.

Programs of the first three types are usually considered to belong to the empirical sciences. Programs of the fourth type, explicative programs, are not only characteristic for constructive analytic philosophy, but also occur elsewhere, viz., in mathematics and the empirical sciences.

Our claim is that these four types of research and research programs reflect the core of two divisions of labor. One figuratively between the cognitive products of scientific inquiry themselves, viz., descriptions, explanations, products, and concepts. The other literally between their producers, roughly speaking, experimentalists, theoreticians, engineers, and philosophers/mathematicians. Moreover, it is claimed that these related divisions of labor on the level of products and producers, and their interaction, play a crucial role in the dynamics of science, which can be even more fully exploited by understanding their nature in more detail.

Descriptive programs are meant to describe a certain domain of phenomena, primarily in terms of individual facts (individual programs) or primarily in terms of general observable facts (general or inductive programs). Descriptive programs form a certain kind of observation program and may be fundamentally based on experiments, in which case it is plausible to speak of experimental programs. A famous example is Boyle's search for a relation between the pressure and volume of a gas, followed by Charles, Gay-Lussac and others with their quest for the relation with temperature. To mention just one other historical example for the moment, the famous investigation by Durkheim of what he called the social facts about suicide was typically a descriptive program.

Descriptive research takes place by more or less selective (experimentation and successive) observation, and the resulting facts are couched in so-called observation terms. These observation terms are not given by the natural world, but form the specific glasses through which the researcher in that program is looking. At the start of a descriptive program there usually is only some core vocabulary. For the rest it is not altogether clear which further observation terms are to be considered as relevant and precisely how certain observation terms are to be interpreted. Additional terms are only selected and shaped in the course of the development of the program. In line with Section 1, it should also be stressed that, at least as a rule, observation and hence observation terms are, and remain, laden by theoretical presuppositions, which are considered to belong to the so-called unproblematic background knowledge.

Explanatory programs have another aim. Individual and general explanatory programs are directed at the explanation and further prediction of the observable individual and general facts in a certain domain of phenomena. Hence, an explanatory program has a (quasi-) deductive nature and is always built on an underlying descriptive program. For this reason explanatory programs are frequently developed along with underlying descriptive programs, in which case the two types of program can be distinguished only analytically. The kinetic theory of gases on the one hand and the anomy theory of Durkheim on the other provide paradigm cases of explanatory programs built on the previously mentioned descriptive programs. The primary objective of the kinetic program was the explanation and detailed prediction of the precise relation between pressure, volume and temperature by applying Newton's laws to collisions of molecules between each other and with the wall. To illustrate this fact, we confine ourselves to one representative of the many researchers conducting this type of research: Van der Waals. Similarly, Durkheim tried to explain the social facts about suicide with his anomy theory.

Other examples of explanatory programs are Newtonian mechanics, the transformational generative grammar of Chomsky, and the theory of rational choice or general utility theory, the latter providing the foundation of, among other things, neo-classical economics and so-called explanatory sociology.

It is important to be aware of the fact that several explanatory programs may arise on the basis of the same descriptive program. They may be competitive, but need not be.

The most important tools used by explanatory programs are theoretical terms, denoting fundamentally new concepts. The distinction between theoretical and observation terms has already been introduced and studied in the previous sections. In the present context of explanatory programs it is important that theoretical terms have not yet been firmly established as observation terms, neither inside nor outside the program. Of course, the terms as such may have been used before to refer to a related concept. Examples of theoretical concepts are the concept of force in Newtonian mechanics, Chomsky's concept of deep structure, and the concept of utility in utility theory.

The new terms may refer to theoretical properties, relations and functions, as suggested by the examples, but also to newly postulated entities, such as atoms and genes. If an explanatory program introduces theoretical terms, it may also be called a theoretical program. If it does not, which certainly is possible, it belongs to the explanatory subtype of observational programs, to be distinguished from the descriptive subtype.

For most of the empirical sciences the above characterization of descriptive and explanatory programs makes sense and is useful. Although analogous programs occur in the historical sciences, especially programs about individual facts, the characterizations above of descriptive or explanatory programs are not suitable for historical research. In particular, general historical programs are rare, probably due to the fact that general historical facts, i.e., empirical laws and theories, are rare. Unfortunately, it seems that neither type of program in either their individual or general forms have yet been elaborated for the historical sciences.

There remain two further points to make about descriptive and explanatory programs of the general kind. The first may also be called inductive and the second deductive, because induction dominates the first and deduction the second type of program. Moreover, whereas descriptive programs are always observational, explanatory programs may or may not be theoretical.

In the current philosophy of the empirical sciences the main attention is paid to description, explanation and prediction. However, an important part of the empirical sciences is not primarily concerned with any of these three tasks. *Design* or constructive research programs involve the design and actual construction of certain products. Some examples are: programs directed at the production of new medical drugs, the improvement of breeding methods of plants, the design of training programs for certain types of handicaps, the design of so-called expert systems, and the construction of new materials. As the examples illustrate, the products of design programs need not be products in a strict sense but may also be processes, or their improvement. The product targeted by a design program has to satisfy certain previously chosen demands; these demands are of course derived from the intended use of the product being developed.

The examples also illustrate that design programs do not only occur in what are traditionally called the technical or technological sciences but also in other areas of scientific research. This is the reason for not choosing the term 'techn(olog)ical research programs', for that might be too narrowly interpreted.

Since design programs often use knowledge obtained in descriptive and explanatory programs, the design process will only be considered to belong to scientific research if it is not fully based on existing knowledge and techniques. That is, new theories have to be developed or new experiments have to be performed in order for a design program to be scientific in nature.

For philosophy and mathematics the fourth type of program, the *explicative* research program, is the most important type. Such programs are directed at concept explication, i.e., the construction of a simple, precise and useful concept that is, in addition, similar to a given informal concept (cf. [Carnap, 1963^2, 118]). For example, the concepts of 'logical consequence' and 'probability' have given rise to very successful explicative programs in the borderland between philosophy and mathematics. One of the main explicative programs dealt with in [Kuipers, 2000] is intended to explicate the intuitive idea of 'truthlikeness'. In Section 1 we have dealt with the explication of the intuitive conceptual distinction between observational laws and proper theories.

The strategy of concept explication is the following. From the intuitive concept to be explicated and, when relevant, empirical findings one tries to derive conditions of adequacy that the explicated concept will have to satisfy, and evident examples and counter-examples that the explicated concept has to include or exclude.

Explication may go further than the explication of intuitive concepts, it may also aim at the explication of intuitive judgments, i.e., intuitions, including their justification, demystification or even undermining. A main example in [Kuipers, 2000] concerns the intuition about the functionality of choosing empirically more successful theories in order to enhance truth approximation. The strategy of 'intuition explication' is a plausible extension of that involving concept explication.

3.1.2 Similarities and differences

Although the four types of programs distinguished are different in many respects, they also have an important similarity. In all cases we can identify an internal goal, viz., the true description, the true theory, the intended product and the intended concept.

The fully correct observation and registration of the totality of facts observable by the glasses of a descriptive program is called the true description of the domain. All other descriptions of the domain in terms of the program are either incomplete or (partially) false. This true description constitutes the internal goal of the descriptive program. It is important to note that the true description not only depends on reality but also on the program in which the choice of the observation terms delimiting its viewpoint, co-determines what will and what will not be observed. Hence, the true description is a program relative but nonetheless informative characterization of reality.[23] If the program concentrates on individual facts, i.e., object, place and time specific (conceptualized) facts, we will speak of the true individual description, if it concentrates on general facts, i.e., generalizations of (conceptualized) individual facts, we will speak of the true general description. In the latter case, the true description corresponds to the true theory within the (observational) vocabulary of the program.

In the case of a (general) explanatory program there is, usually, supposed to be a unique theory. That is, there is assumed to be a theory in terms of the observation and eventual theoretical terms of the program, which not only explains, and predicts as far as relevant, all observable facts of the domain. It also uses only those theoretical terms in a substantial way that refer to something in reality. This theory, in fact the strongest true hypothesis, will be called the true theory of the domain, constituting the internal goal of the explanatory program. Like the true description, the true theory is determined by the specific combination of program and reality, hence it is program relative. If the vocabulary is observational it corresponds to the true general description.

The intended product, i.e., a product that satisfies the demands put forward, forms of course the internal goal of a design program, the analogue of the sought-after true description or true theory. Finally, in the case of explicative programs the intended concept, i.e., a concept satisfying the conditions of adequacy, constitutes the analogue of the internal goal of the previously considered programs.

Despite the fact that, similar to descriptive and explanatory programs, design programs always have internal goals, such goals differ greatly from those involved in description and explanation. In descriptive and explanatory programs internal goals are only indirectly characterized, and all the efforts are directed at the explicit characterization of the true description or the true theory. In design and explicative programs, the internal goal, the intended product and the intended concept are all explicitly characterized from the beginning, at least to a certain extent.

[23]Of course, not every aspect of this true description may have our interest. It may or may not be possible to restrict the observational vocabulary, and hence the program, to some sub-vocabulary of what we find really interesting.

Another, related difference is the degree of uniqueness of the internal goal. As mentioned above, the true description is in principle uniquely determined jointly by the program and reality. Hence, it cannot change in the course of the program without either changing the program or the domain. The same holds *mutatis mutandis* for the true theory. In the case of design programs, on the contrary, the intended product need not be determined uniquely at all, for there is, as a rule, the possibility of functional equivalents, i.e., different products serving the same purpose, in which case it is also said that the purpose is 'multiply realizable'. Moreover, the desired product has to be determined in more detail in the course of its development, in which the strategic considerations of feasibility, affordability and salability play an important role. As an aside, it should be remarked that the intended product could also be over-determined by the sum of all demands.

Most of the differences mentioned also apply *mutatis mutandis* to explicative programs. This is no accident, for they form a kind of abstract design program, viz., of concepts.[24]

3.2 Structure and development of research programs

We will discuss the structure of programs mainly in terms of five possible components. Then we will discuss the development of programs in terms of an internal and external phase. A brief presentation of the atomic theory as a developing research program will illustrate most of the components and phases. The section concludes with a global survey.

3.2.1 Five components of research programs

So far the descriptions of the four types of programs may well be used to indicate just four types of goal-directed research. However, when we start to discuss the structure of programs it will become clear how programs acquire more identity than defined by their internal goals. We will discern an ordered set of five possible components of a program, viz., domain, problem, idea, heuristic, model. Since each component is supposed to presuppose the foregoing ones, this leads to five qualitative degrees of strength of programs.

A possibly disenchanting reformulation of the similarity between the four types of programs is that they are all directed at the solution of a certain problem, viz., to attain the internal goal of the program. This orientation seems the least one may expect of scientific research, i.e., that it is directed at the solution of a certain problem. Programs satisfying this minimal requirement might be called *programs with a problem.* In the practice of research policy, however, one even speaks of research programs when there is only a more or less well-defined domain of research, without a clear problem, for lack of an internal goal, in which case one might speak of *programs with (only) a domain.*

[24]E.g., the intended concept is usually not uniquely determined, such that there are degrees of freedom in explication, leaving room for diverging explications. For this reason one may even prefer to speak of concept modeling instead of concept explication [Brink, 1989].

From the descriptions of the four types of programs it now follows that a research program is minimally conceived as a program with a problem. However, the prototypical meaning we want to advocate for the term research program is that of a *program based on an idea*, i.e., a program with not only a domain and a problem but also a fundamental idea governing the search for the solution to the problem. It could be called Popper's requirement, because more than anyone else he has stressed the equal importance of problems and ideas in scientific research. Of course, such a fundamental, leading idea is usually a complex idea, i.e., a set of coherent ideas. It will at least include the choice of a core *vocabulary*, and usually it includes one or more *principles* using that vocabulary. The idea should be strong so that it can provide secure footing for a research venture that should be able to withstand some critical blows. In other words, it should be possible to protect the fundamental idea somewhat dogmatically against falsification or similar threats. The standard way to do this is by trying to blame auxiliary hypotheses, but there are several other defense strategies.

According to Lakatos, the leading idea constitutes the hard core of a program. However, the notion of hard core has a double face. Lakatos's primary meaning is that a program is only one and the same program as long as the hard core remains the same. However, it frequently occurs that one feels obliged sooner or later to adjust the fundamental idea of a program, in which case one should strictly speak of a new program. But it seems more adequate to leave room for a semi-hard core of the program, a core that may be adjusted, when no other escape seems possible. We would like to stress another meaning component of the notion of a hard core. An idea, before or after a possible change, may be hard in the sense that it is supposed to be valid for the whole domain. It does not leave room for exceptions. Speaking of a 'core idea' may indicate this quality. Incidentally, one way to retain the goal of no exceptions, in the face of persistent threats is to adjust the core idea, another is to adjust the domain. In sum, we conceive the fundamental idea on which a program can be based as a core idea of a semi-hard character.

To be sure, Lakatos only speaks of a genuine research program when there is, in addition to a hard core, also a so-called positive heuristic providing suggestions for protecting auxiliary hypotheses and their adjustment. Hence, a program, in Lakatos's sense, is a *program with a hard core and a positive heuristic*. That is, a program governed by two ideas, the first one directly bearing on the solution of the problem, and the second one concerning the way in which the first idea can be defended against attacks.[25]

Zandvoort [1984] has convincingly shown that the impressive examples of Lakatosian programs frequently are programs in which the positive heuristic is provided by an analogy or model, where he refers in particular to the notion of analogy as discussed by Nagel [1961]. Such *programs with a hard core and a model as a positive heuristic* are maximally equipped to provide internal guidelines for

[25]Lakatos's concept of negative heuristic coincides with the intention of keeping fixed the hard core as long as possible. Hence, this notion does not add something to the notion of the hard core (in the sense of Lakatos).

research.

Research programs with a core idea and a stable positive heuristic, whether or not in the form of a genuine hard core and a model frequently occur in all forms of empirical science, not only in the natural sciences, but also in the social sciences and the humanities. However, many other programs have a semi-hard core idea about the way in which the goal has to be attained, without having a strong idea about the way in which that fundamental idea has to be protected. In other words, although they have a core idea, they don't have a stable positive heuristic. To put it differently, the historical claim in the beginning of this section can be stated more precisely as follows: the global history of science can best be described in terms of rising, winning and falling programs based on a core idea. For this reason we will henceforth assume the prototypical meaning of the term '(research) program' to be a program based on a semi-hard core idea, but not necessarily equipped with a stable positive heuristic.[26]

3.2.2 *Examples of (the core ideas of) research programs*

We now present a list of examples of ideas forming the core of equally many well-known programs, starting with explanatory programs. These are the idea in kinetic theory that gases consist of molecules which move and collide according to Newton's law; Mendel, starting from the problem whether regularities can be found in the apparent chaos of the manifold variations of observable features in all the hybridizations known to plant breeders at that time, postulated the idea that the male and female hereditary material is not a unit of which one wins out, but that it can be factorized so that part of the male and part of the female hereditary material determines the observable features of the organism; the idea in general utility theory that choices are governed by maximizing expected utility; Chomsky's original idea that the grammatical sentences of languages can be generated by the application of a limited number of transformation rules on an equally limited number of deep structures; and, finally, the central idea in classical computationalism, or symbolism (Newell and Simon), according to which human behavior should be (described and) explained in terms of problem solving.

The last example is a nice borderline case between explanatory and descriptive programs. An example of a purely descriptive program is network analysis, which is based on the idea that schemas using connecting arrows can result in very informative descriptions. Another example is fractal geometry, initiated by Mandelbrot, based on the idea that shapes in nature on different scales may nevertheless be congruent. A coastline of 10 meters looks like a coastline of 10 kilometers. So-called discourse-analysis provides a similar example. Behaviorism can also be viewed as a broad, descriptive program, with the core idea that one should restrict the scientific attention to the description of (patterns in) observable behavior. Finally, the goal of the Human Genome Project, essentially completed in 2000, was the true description of the (almost unique) composition of the 23 human chromo-

[26]Unfortunately, the term 'idea program' is linguistically not very attractive.

somes as pairwise sequences of the four bases C, T, A, and G, that is, the typical vocabulary of DNA. To be sure, although the sequences are almost unique, the individual variations provide an almost perfect means of identification.

The core idea of descriptive programs is frequently formed by a methodological searchlight principle, for instance the principle of causality, functionality or intentionality or by a description or representation principle, as in the cases of network analysis and fractal geometry. Such leading principles usually open, for different domains, the possibility of a specific program directed at that domain. Such a representation principle may be guided, or is at least made available, by accepting a theory as an observation theory, that is, a theory that has become accepted as (approximately) true. This important phenomenon has already been explained in more detail in the first section. A typical example is the functional (descriptive) genomics program, building upon the 'structural' descriptive (Human) Genome Project, identifying the function of fragments of the chromosomes. That is, each fragment may or may not play a crucial role in ('code for') the generation of certain characteristics of organisms. This research is guided by the principle of functionality, rooted in the theory of evolution, and according to which features of organisms have functions. Even more than on the macro-level, there are many exceptions to this principle on the present micro-genetic level. As a matter of fact, most DNA-fragments seem without function.

In the case of a design program the leading idea is frequently called the lead. It is the core idea about the way in which the intended product should be construed and possibly with what material. Some examples are the following, starting with an example of a purely technical, non-scientific, nature. The idea of a bicycle chain was developed at the end of the nineteenth century, and enabled the, still time consuming, design of a riding bike with two wheels of equal size, i.e., the bicycle, which was a very attractive feature of the walkbike of much earlier date (cf. [Bijker, 1995]). The idea of nuclear fission resulted in the development of nuclear power stations, and, to be honest, atomic bombs. The development of power stations based on nuclear fusion is still one of the main challenges of applied physics. One of the main starting points of computer science, the development of the standard Von Neumann architecture of digital computers, began with Turing's idea of a universal computing machine, the so-called Universal Turing Machine. Within the Von Neumann architecture the idea of so-called production systems, containing and generating complex 'if, do then' (production) rules, has turned out to be very successful, in particular for creative computing tasks. According to the central idea of the technological version of connectionism, learning mechanisms can be produced by connections between knots that are strengthened or weakened according to whether the previous response was or was not adequate. One of the leading ideas in cancer research, due to Judah Folkman (see [Boehm, *et al.*, 1997]), is to try to stop the formation of blood vessels leading to the tumor or, alternatively, to try to block their functioning. To mention an example of quite a different nature, the idea of conversation groups made up of people with similar personal problems was only recently introduced in all kinds of therapeutic contexts, with varying but increasing degrees of success. Their development was partly due to a systematic search for the best specific conditions in which to conduct therapy. For example, certain group therapies for breast cancer patients, an idea of Spiegel, Bloom, Kraemer and Gottheil [1989], see also [Spiegel, 1993], seem to improve their immune system. Finally, in drug research [Vos, 1991, 62], the lead can be identified with the set of wished for properties, the wished for profile, and some idea about how to realize it. More specifically, the 'lead compound' comprises a chemical compound with certain operational characteristics, the operational profile, together with the wished for profile. Only if there is already known to be an interesting overlap between the two profiles, it is a serious lead compound, and the challenge is to reduce the differences.

A nice example of an explicative program is the famous idea of Rawls, according to which the determination of the concept of a just society can best be undertaken by way of a thought experiment. In this scenario the future members of the just society to be construed do not know the place they are going to occupy in that society; that information is hidden behind 'the veil of ignorance'. Another example is the core idea of logical model- theory that the intuitive concept of 'logical consequence' should be explicated in terms of the models of the relevant language and according to the principle: the conclusion should be true in at least all models in which the premises are true.

In the last few decades several research programs have been described in detail and, of course, in the terminology preferred by the respective authors. To give one example, Von Eckardt [1995], whose aim is to characterize cognitive science, describes the 'research framework', as a combination of domain-specifying assumptions (domain), basic research questions (problem(s)), and substantive assumptions (core idea(s)). In these terms she has given a lucid description of the framework for cognitive science, to be precise, as far as it is focused on adult, normal, typical cognition. In her approach, symbolism and connectionism appear as two different specifications of the 'computational system (substantive) assumption'.

3.2.3 Additional considerations

A significant problem arises when any attempt is made to identify a respectable or even strong core idea. A plausible procedure to conclude to the existence of a research program based on an idea when there have appeared in the blindly refereed international literature several publications, from one or more authors, in which the idea is exposed, discussed and elaborated. In principle, the existence of international publications coalescing around one idea should be an adequate criterion since science is an international activity in the sense that national borders should not play an important role, in particular when the distribution of strong research ideas is concerned. However, presence in the international literature is certainly not an infallible criterion; it is neither a necessary nor a sufficient one. Referees and journals are necessarily selective, nor are they immune to trends and fashions. Hence, occasionally it may happen that bad ideas are promoted and that good ideas repressed. Moreover, there may well be strong ideas for which there is not very much interest in the discipline itself, but outside that discipline there may be considerable interest from other disciplines or scientific externals, i.e., from society and technology. The last case may particularly apply to ideas in design research, in which case it is not plausible to expect international scientific publications, as they may be prevented by the need for secrecy. Again, strong external interest is not a safe criterion, but the combination of international publications and lasting external interest is the best criterion we can think of for the identification of valuable research programs.

Although our concept of research program resembles Lakatos's concept of research program the most — it is a weakened version — this does not mean that ideas about structure and development of research programs can only be derived from the writings of Lakatos. As already mentioned, other authors have distinguished related cognitive units, and described their structural and dynamic features. Kuhn [1962/1969] speaks first about 'paradigms' and later about 'disciplinary matrices', Fleck [1935/1979] introduced the notion of 'styles of thought', and Laudan [1977] deals with 'research traditions'. We have also seen that Von Eckardt [1995] has more recently dealt with the notion of 'research frameworks'. Additionally, 'theory nets' are distinguished in the structuralist approach to scientific theories that has been presented in the previous section. To conclude this incomplete list, Hamminga [1983] also uses the term 'research program', but gives it a detailed meaning tailored to economic research programs.

In the next subsection we will deal with the main dynamic features of research programs, with emphasis on explanatory programs. We conclude this subsection by mentioning one other structural feature, derived from the structuralist approach. The domain of a research program can frequently be divided into a number of subdomains. In such cases it is possible to make a distinction between the core idea associated with the core vocabulary. That is, to distinguish general or generic principles that are supposed to be valid for the whole domain from special concepts and principles that are only supposed to be at stake for a subdomain. Think of Newton's general laws of motion and the special force laws. In many such cases the division into subdomains is such that it makes a lot of sense to speak of sub-programs, as the crucial idea for a special principle pertaining to a particular subdomain may well constitute a genuine research program in itself.

3.2.4 Phases of developing research programs

Our treatment of the dynamics of programs will mainly concentrate on explanatory programs, and close with a few remarks on the validity of the findings for other types of programs.[27] One might prefer to read first the next subsection, dealing with the development of the atomic theory, and then return to this subsection.

[27] For a detailed analysis of design programs the reader is referred to Chapter 10 of [Kuipers, 2001] and of explicative programs [Kuipers, forthcoming].

We begin by elaborating a previously mentioned relativization of the term 'program'. A program is never fully mapped out in advance. At each moment only a few principal features are established. They enable researchers to look forward no more than a little bit, and depending on the results of their efforts, the program is adjusted and mapped out a bit further. This is the main reason why responsible bureaucratic middle- and long-term planning of research is impossible.

A program can pass through several phases. In cases of successful programs it is frequently possible to make a global distinction between an internal and an external phase.

In the *internal phase* the elaboration and evaluation of the core idea are central. When a program persists for some period of time it is usually possible to divide the internal phase into two subphases, viz., a heuristic and a test or, as we prefer to call it, evaluation phase. In the *heuristic phase* the new idea breaks through and the first auxiliary strategies are invented to protect the idea. This phase may or may not take place against the background of a so-called Kuhnian crisis of another program, for which seemingly unsolvable problems, called anomalies, have accumulated.

Gradually there comes a transition to the *evaluation phase*. The idea is elaborated for a small number of contexts or subdomains into specific theories, and these are evaluated. The core idea now constitutes the so-called core theory or generic theory, common to all specific theories. Evaluating a specific theory implies as a rule, that for the particular subcontext, a sequence of specific theories is developed, each containing auxiliary hypotheses, each resulting ideally in increasing success, and each including a decreasing number of (types of) counter-examples. Usually such a sequence satisfies the pattern of idealization and concretization: the consecutive theories take into account factors neglected by the foregoing ones.[28]

If this way of branched evaluation is not overall successful, the program is not necessarily deemed useless and made to disappear forever into 'the museum of knowledge', exposing, for example, abandoned research programs. A failing program cannot only continue to inspire new research questions, it may also be the case that in a later stage someone succeeds in giving a successful turn to the program.

When the evaluation proceeds successfully, this usually leads to the more or less general acceptance of the core theory of the program and it has become clear for which domain and in what sense and to what extent the core theory can be assumed to be true. It should be stressed that many, if not most, programs in the empirical sciences, not to mention philosophy, do not reach this point. But if this stage is attained, the researchers in that program are left with two options. The first possibility is to look for another program presenting a new challenge. The second possibility is to try to direct the program for the benefit of questions

[28]The branched portrait of evaluation of a research program is mostly a product of the structuralists; the idea of a sequence of improving theories comes from Lakatos, following Popper; the pattern of idealization and concretization originates with Nowak [1974; 1980], and is further developed by others, e.g., Krajewski [1977].

that are *prima facie* independent of the program. The program then enters the *external or application phase.* The so-called Starnberg school [Schäfer, Böhme and Burgess, 1983] calls this finalization, and means by it in particular the application of the core theory to technological or social problems. This is seldom a matter of simple application of the theory. It usually requires highly specialized theory development, and may even lead to the start of a new scientific discipline, e.g., aerodynamics, in the case of the technological goal of airplanes and the like. We will nevertheless simply speak of application, in this case more in particular about the external application of science. Another form of application arises from the fact that an accepted theory may be usable as observation (or measurement) theory.

Zandvoort [1986; 1988; 1995] has convincingly established that research programs in the natural sciences, which have successfully passed the internal phase, are not always directly applied to problems external to science. They are at least as frequently applied to science internal problems. Hence, the terms 'internal' and 'external phase' of a program should be strictly interpreted as program relative: internal and external to the program. As a rule, the science internal application of a successfully established program means that the program is directed at the solution of specific problems generated by other programs, possibly but not necessarily design programs. It may also be used for observations relevant to other programs, requiring the acceptance of the core theory as an observation theory. Zandvoort has shown that Popper, Kuhn, Lakatos and others have unjustly neglected this type of cooperation between programs: it constitutes the main part of successful interdisciplinary research within the natural sciences. Among others, Zandvoort's findings make it clear why it is not only difficult to show the practical, i.e., science external, relevance of natural science research programs that are still in the internal phase, even the practical relevance of programs in the application phase may well be only indirectly so. In the next section we will return to this and other types of cooperation between programs.

So far we have not paid any attention to the question of what is precisely meant when a program is successful or makes progress. To be sure, there are many sorts of success; not all of them need to be sufficient or even relevant for real progress. Success criteria for progress should of course be derived from what scientists themselves count as progress, as for instance expressed by Nobel prizes and other prestigious scientific recognition. According to Popper and Lakatos the factual criterion for progress used for scientific recognition is not just increasing explanatory success[29], predictive success is also required. That is, it is not enough that a program succeeds in explaining new facts, from time to time it should predict, and hence also explain, new facts. To put it differently, there should not only be postdictive but also predictive explanatory success. Although this criterion turns out to be logically too strong as an indicator of truth approximation[30], it

[29]The notion of 'explanatory success' should be taken here in the liberal sense of derivational success, that is, facts that can be derived from the theory.

[30]See [Kuipers, 2000, Chapter 6]. For the purpose of truth approximation, however, it is, in addition to increasing explanatory success, (almost) necessary that there are no new (types of)

should be conceded that explanatory *and* predictive success is in practice the employed criterion, at least as far as the internal phase of (explanatory) programs is concerned.[31]

For the external phase predictive success does not appear to be necessary, although this does not mean that only explanatory success is sufficient for progress. For the external phase another supplementary criterion to explanatory success is obvious: external success. From time to time the program should successfully solve external problems to which it is directed. Indeed, Nobel Prize motivations frequently report, in addition to (new) explanatory success, either predictive or external success.

An important case study undertaken by Zandvoort [1986] concerns the main theory transitions in the nuclear magnetic resonance (NMR-) program, which originates from nuclear physics and is based on quantum mechanics. He had to conclude that in almost all cases theory transition concerned theory accommodation on the basis of newly discovered facts. On closer inspection it also became clear that nobody doubted the possibility that the NMR-program could explain the new facts by some further articulation and hence that such doubts could not be the reason why the program was prolonged. In fact, the program was continued because it simultaneously solved important problems that were very relevant to other programs, in particular in chemistry and biology.

In this short overview of scientific progress we came across the following basic types of success in science: truth approximation, explanatory success, predictive success, external success. At least the following other types of success should also be mentioned: scientific recognition, textbook treatment, financial support, popular-scientific publicity, and institutional power. As is frequently claimed, and regretted, only the first two types are highly correlated with the basic types.

Although the foregoing is generally applicable to all explanatory programs, some distinctions may nevertheless only be made analytically. For instance, the application phase may well have been started when the internal phase has not yet come to an end. However this may be, the proposed distinctions may also be applied in an adapted form to other types of programs. For instance, for descriptive, design and explicative programs similar phases can be distinguished, at least analytically. Moreover, the exposition about progress also applies for example to explicative programs. For such programs the analogue of (intended) increasing explanatory success, is intended increasing explicative success, i.e., succeeding in explicating the informal concept in a more satisfactory way, as determined by the conditions of adequacy, along with the evident examples and counter-examples that have

counter-examples, a condition which Popper and Lakatos just presuppose.

[31]For practical (and ethical) reasons, predictive success may not always be possible. A paradigm case is the string theory program, in which, at least so far, theoretical predictions cannot be tested because "we will never be able to produce energies anywhere near [the required] value" [Atkinson, 2005, 100]. Atkinson considers for this reason the need of introducing a fifth fundamental kind of research program, that "aims at *unification*, that is the bringing together of apparently different explanations into one coherent logical or mathematical framework" (p. 102). See also [Psillos, this volume].

been put forward. However, it is also considered to be very important that the proposed explication turns out to give rise to unintended explications, that is, to satisfactory explications of related concepts and intuitions. This type of success is the analogue of the extra, i.e., predictive or external, success of explanatory programs. Again the question is whether this form of success is formally defensible as a necessary condition for progress, but the fact remains that in practice this type of explicative success plays an important role. For descriptive and design programs it is less clear whether there are similar criteria of 'more than explicitly intended success'. In [Kuipers, 2001, Chapter 10] the plausible claim is developed that a transition in a design program is successful when a modified prototype satisfies more of the desired properties than the foregoing prototype, which evidently is the design analogue of explanatory success. Finally, in [Kuipers, 2000] some refined ideas are introduced in Chapter 5 concerning the successes *and problems* of theories which are particularly relevant for explanatory programs. Moreover, Chapter 7 introduces refined notions of successes and problems of descriptions; they are relevant for explanatory as well as descriptive programs.

3.2.5 The atomic theory as a developing explanatory program

The atomic theory and its development may well serve as an example of a successful research program. The following portrait is a simplification of the reconstructions given by Holton [1973] and Zandvoort [1989].

Dalton [1766-1844] introduced the theory of the atom in order to explain certain laws of chemical reactions and to possibly predict some other ones. Hence, it is an explanatory program. Its *domain* consists of reactions between chemical substances. Along with the development of that program the distinction between pure substances and mixed substances (mixtures) and the division of pure substances into elements and compounds emerged as observational categories. The following exposition will presuppose the idealization that these terms were without problems available to Dalton. The development of the program can be described in three phases.

Phase 1: Dalton's primary *problem* was to explain certain relatively well established (observational) laws of reaction:

> *LR1* (Lavoisier): the total weight of the substances before and after a reaction remains the same.
>
> *LR2* (Proust: the law of definite proportions): compounds always decompose into components with constant weight ratios.

The core ideas of the program initiated by Dalton can be summarized into four principles, with the notions of (types of) atoms and molecules as the program specific theoretical vocabulary. The first two are *internal principles*, only dealing with postulated, hence theoretical micro-entities; the remaining two are *bridge principles*, in fact identity postulates, relating the theoretical terms to the observation terms.

$I1$: *atoms* are indivisible, unchangeable, hence indestructible, small material particles of a certain type.

$I2$: atoms are grouped into *molecules* of a certain type, and they may regroup into other types of molecules.

$B1$: - pure substances consist of one type of molecule;
in the case of elements these molecules consist of one type of atom;
in the case of compounds they consist of more than one type of atom.
- mixed substances consist of more than one type of molecule.

$B2$: chemical reactions amount to systematic regrouping of the molecules of a substance.

Let us indicate the core idea consisting of $I1\&I2\&B1\&B2$ by C, and the specific theory at stage i by Ti, consisting of C + Hi, where Hi indicates the auxiliary hypotheses at stage i. Hence, $T1 = C$, as there are not any substantial auxiliary hypotheses at the start. It is not difficult to check that $T1$ explains the two target observational laws *RL1* and *RL2*.

In agreement with both Popper's and Lakatos's views, Dalton felt obliged to obtain a more impressive result and predicted a third law, viz.,

> *RL3* (the law of multiple proportions): when two different elements unite into two different compounds, the different proportions bear a simple relation to one another.

One successful test tuple of compounds consists of carbonic oxide and carbonic acid, both composed of carbon and oxygen. In the first case the proportion, in terms of weights, of oxygen to carbon is about 4:3, in the second case about 8:3. Hence, the ratio of these proportions is 1:2. For this prediction Dalton needed a strong rule of simplicity concerning the possible composition of molecules of the same type of atoms:

> *A-s* (internal simplicity assumption): if a certain type of molecule exists then all the conceivable more simple types of molecules composed of the same type of atoms exist as well.

In combination with the assumption that the number of existing compounds of two elements is rather limited (an auxiliary hypothesis that will be neglected in the remainder of this section), *RL3* can easily be derived. Although *A-s* is certainly false according to our present knowledge, the derived predictions came true and *RL3* became accepted. Hence according to the progress standards proposed by Popper and Lakatos, the transition from $T1$ to $T2 = C + A\text{-}s$ is progressive: intended (or postdictive) explanatory success is supplemented with predictive explanatory success.

Phase 2: However, in the meantime a severe anomaly was arising: the law of combining volumes, independently established by Gay-Lussac. First an example: two liters of hydrogen gas and one liter of oxygen gas result into two liters of water vapor. In general:

> *RL4*: pure gases combine with simple integer numbers of volume units, into an integer number of volume units of the compound gas, not necessarily equal to the sum total of volume units of the component gases.

It is easy to see that it is not possible to derive *RL4* from *T*2, nor its negation, for the simple reason that *T*2 does not say anything about volumes of atoms and molecules. However, Dalton and some of his followers actually favored an auxiliary (bridge) assumption *A-g* about the nature of gases:

> *A-g* (gas assumption): gases consist of non-moving, contiguous gas particles, in their turn consisting of a molecule and a caloric mantle.

It is quite obvious that the resulting $T3 = C + A\text{-}s + A\text{-}g$, constituting a specific theory for the subdomain of gases, predicts the negation of *RL4*. Indeed, Dalton was confronted with a very big explanatory problem. However, he himself was inclined not to blame his theory, but to put the truth of *RL4* into question. To be sure, the transition from *T*2 to *T*3 cannot be called progressive.

Avogadro (1776-1856) took the explanatory problem more seriously. He not only proposed a modified version of the rule of simplicity $A\text{-}s^*$, but also a totally different and, at first sight, very surprising auxiliary hypothesis about gases:

> $A\text{-}g^*$: (gas assumption proposed by Avogadro; Avogadro's hypothesis): equal volumes of different gases, at equal pressure and temperature, contain equal numbers of molecules.

Note first that $A\text{-}g^*$ typically is an extra bridge principle. Even without specifying $A\text{-}s^*$, it is plausible that Avogadro could show that the resulting specific theory, $T4 = C + A\text{-}s^* + A\text{-}g^*$, is indeed able to explain, in addition to *RL1, 2,* and 3, Gay-Lussac's law of combining volumes, *RL4*. Moreover, it enabled him to produce, using examples of weight and volume ratios in the line of *RL3* and 4, molecular composition formulas and molecular reaction equations. However impressive these results were considered to be, they did not lead to the general acceptance of Avogadro's specific version of the atomic theory. Why? Referring once again to an insight developed by both Popper and Lakatos, the crucial point in this case is that, although *T*4 is explanatorily more successful than *T*2 and *T*3, it has, at this point, not yet achieved (specific) predictive success.

Phase 3: It took about half a century before Cannizaro (1826-1910) was able to derive a new prediction, using an additional auxiliary (bridge) hypothesis concerning dissociation:

A-d (dissociation assumption): large molecules of a certain composition will fall apart in a gaseous state.

From the resulting specific theory $T5 = C + A\text{-}s^* + A\text{-}g^* + A\text{-}d$, using Avogadro's findings on substances that consist of large molecules, the following prediction could be derived:

RL5: substances so and so will dissociate in a gaseous state.

These predictions turned out to be largely correct, hence the transition from *T*2, or *T*3, or *T*4 to *T*5 is progressive in the sense of predictive success.

As a matter of fact, largely due to Cannizaro's results, Avogadro's hypothesis and the whole specific theory became generally accepted, i.e., its internal phase came to an end. It nicely illustrates that a research program can be very successful without a stable overall positive heuristic, which is difficult to discern in the case of the atomic program. At most there is something like a positive heuristic for partial use only, viz., to modify the rule of simplicity when problems arise involving molecular formulas. To be sure, *C* seems to have been really a hard core, in the sense of Lakatos. However, for the sub-program restricted to gases, the transition from *A-g* to *A-g** may well be construed as a fundamental change of the core idea.

As is well-known, the atomic theory has been very successfully applied in other areas of scientific research, and has turned out to be indirectly of great practical use in chemical technology. Moreover, by accepting it, determination criteria for atoms and (the composition of) molecules where also established, and it not only became a background theory, but an observation theory.

This concludes our treatment of the example for the moment, but we will come back to it a number of times.

3.2.6 A pictorial summary

We conclude this section with a pictorial summary of its main content.

The concept of research tradition developed by Laudan [1977] can be interpreted as an even more global conceptual unit than that of a research program. As a matter of fact, a research tradition can be seen as the metaphysical and methodological core of a number of research programs. Behaviorism is a good example; it generated several research programs in psychology and biology. If we include this in a picture we get a double branched description of the state of affairs at a certain moment in a certain tradition, bringing together ideas from Kuhn, Lakatos, Sneed, Laudan, the Starnbergers and Zandvoort. In Figure 9 *RP1/2/3* denote research programs belonging to a certain research tradition, *CT2* the core theory of *RP2*, *T2.i.j* a specific theory of *RP2*, to be precise, the *j*-th attempt for the *i*-th subdomain. Moreover, *RP** denotes a research program which may or, more likely, may not belong to the same research tradition, and *DP* denotes a design program for some science external product. A thin arrow denotes 'gives rise to', a thick arrow denotes a transition with explanatory and predictive-or-external

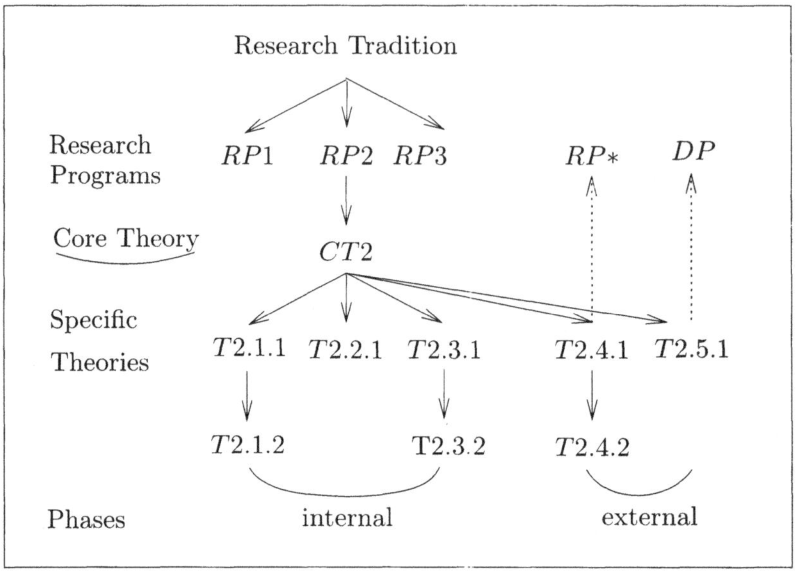

Figure 9. Research programs and their development

success, and, finally, an dashed arrow indicates 'tries to contribute to the solution of a problem of' a certain (research or design) program.

3.3 Program internal research strategies

The above exposition of the structure and development of research programs raises at least two questions:

- how are new specific theories developed within a research program?
- in what ways can research programs interact?

These questions essentially concern research strategies for the (further) development of research programs. On the one hand there are program-bound strategies, that is, strategies within a single program aiming at improving the last specific theory. On the other hand there are strategies directed at interaction between programs for the benefit of at least one of them. Both types of strategies will be discussed in the indicated order. As far as the interactive strategies are concerned no detailed reconstructions of such strategies have been developed, and it is doubtful whether they ever will be. However, knowledge of their global nature is essentially sufficient for application in new contexts. The same is true for the program-bound strategies.

In the present subsection we start with a discussion of the importance of working within research programs, the program strategy itself. Then some more specific strategies will be discussed, research guided by 'idealization and concretization' and by an 'interesting theorem'. We will conclude with some remarks about descriptive and explanatory research programs in the light of truth approximation.

Here, as well as with respect to the interaction of programs in Subsection 3.4., we expose the global lessons to be drawn from the work of Popper, Kuhn, Lakatos, Sneed, Hamminga, the Starnbergers, and Zandvoort. Moreover, we integrate insights taken from Nowak [1980], Krajewski [1977], Darden and Maull [1977], and Bechtel [1988a/b]. See also Bechtel and Hamilton in this volume. It is more or less a strategy oriented synthesis of their insights as far as they are compatible. The reader should understand that the assertive tone concerning strategies should not be taken too seriously. As with all heuristics, you may consider a strategy, you may even try it out, but you cannot blame it.

3.3.1 The program strategy

From the long-term development of the sciences it has become clear that scientific research can be aptly characterized in terms of research programs. This does not alter the fact that the historiography of science can frequently be accused of concentrating too much on the success stories, on the successful research programs. Historiography should also pay much attention to programs that have lost the competition. The results of these programs are of course not added to the body of knowledge, but they rightly deserve a decent place in 'the museum of knowledge'. Unfortunately, the only programs arriving in the museum of knowledge are those that were winning programs until superseded by newer programs. But deposition in the museum of knowledge should for instance also take place when two competing programs started more or less at the same time, with one of them having to give up sooner or later in favor of the other. Hence, the claim is, more precisely, that the main lines of the history of science can well be described in terms of rising, winning and falling research programs. If that history is written with any attention being given to mutual interaction, it will not only become apparent that programs frequently compete[32], but also that they may fruitfully cooperate.

The lessons from these observations seem to be the following. Anyone who wants to undertake frontier research,[33] will in general also aim at getting the results of his research sooner or later incorporated in the international knowledge base, or at least in the museum of knowledge. Taking all things into consideration, and whether one likes it or not, in order to achieve this goal, participation in one or more of the internationally recognized research programs is virtually unavoidable.

Many university researchers consider this point to be obvious. However, there are nevertheless quite a few researchers who think otherwise. One seldom meets

[32]For a brief history of the successive and competing research programs in high-energy physics, see [Cushing 1982]; for the history of the NMR-(nuclear magnetic resonance) program, engaged in asymmetric reductive cooperation with chemical and biological research programs, see [Zandvoort 1986].

[33]For example, my home university, the University of Groningen, advertises with the slogan "to work at the frontiers of knowing".

them in the natural sciences, but regularly in the human sciences and, for sure, in philosophy. As a consequence, the feature of having in a certain domain of inquiry only a few (interacting) research programs is badly developed in the human sciences. In the natural sciences this fruitful characteristic certainly has been partly instigated by the high costs of experimental research.

A frequent objection to program participation is the claim that it inhibits creativity. But the converse seems to be the case. Given the enormous potential of competitors one needs stronger creative talents to deliver a substantial contribution. Moreover, some critics are of the opinion that it should be possible to develop a new program. Hence, in the human sciences many researchers start their own shop, complete with a new publication medium. It is strange that such initiatives are usually taken rather seriously. It is interesting to compare this with a researcher who announces that he is going to make an important new discovery and who will be regarded rather skeptically. But the invention of new ideas that can lay the foundation of new research programs is as rare and difficult as making pioneering empirical discoveries. In fact it concerns pioneering theoretical discoveries. For both types of scientific achievements Nobel prizes are awarded. Moreover, the inventors of ideas for new programs are frequently steered by a fresh look at the severe problems they and others met when conducting research in existing programs.

In the social sciences and the humanities there even seems to be an abundance of research ideas. One therapy against this condition is the foregoing type of plea for program-bound research, the other is a plea, which is to follow in Subsection 3.4., for stimulating interaction between research programs, by cooperation and/or competition.

3.3.2 Program development guided by idealization and concretization

We will now sketch two strategies for the internal development of a research program, particularly for the succession of improving specific theories. To be sure, these strategies can also be used without assuming the boundaries of a research program, the only claim is that they are frequently used within a program. We start with idealization and concretization. Idealization is frequently applied in empirical scientific practice as an unavoidable step in theory formation. This is certainly true in the natural sciences; in the human sciences and also in philosophy the necessity of explicit idealization is not yet generally accepted.

Surprisingly enough, on closer inspection Marx developed his ideas in *Das Kapital* rather systematically according to the method of idealization and successive concretization. Nowak [1974; 1980] has pointed out that this procedure was used by Marx; in particular he shows how Part I and Part III in their succession can be seen as illustrations of what Marx used to call 'rising from the abstract to the concrete'. Another Polish philosopher, Krajewski [1977], freely following Nowak, has also contributed importantly to the growing awareness of the systematic role of what he calls 'idealization and factualization'.

Although idealization-and-concretization (from now on, I&C) also occurs in qualitative theorizing, it is primarily explicated for quantitative theorizing, in particular the succession of specific theories within a research program. The general idea is that it is frequently possible to make an ordering in the degree of importance or relevance of all the factors that influence the value of a certain quantity G, which may even lead to a division of primary and secondary factors. Starting from such an ordering of factors $f0$, $f1$,... *(fm)*, in the n−th stage of concretization factors $f0$ up to *fn* have been accounted for, while the remaining factors are still neglected, leading to the typical I&C-formulation of the n-th specific theory:

if $f0\neq0$, $f1\neq0$,...., $fn\neq0$ and $f(n+1)=0$, $f(n+2)=0$, ...
then $G = Gn(f0,f1,...,fn)$

In the 0-th stage there is maximal idealization and when all factors have been concretized, maximal concretization has been achieved. Note that, although any given functional representation of a factor is allotted the value 0 on a formally arbitrary basis, the neglect of a certain factor is empirically speaking usually not arbitrary, in which case the functional representation can be chosen in accordance with this.

The transition of the ideal gas law to the Law of Van der Waals is a paradigm case. This transition can be represented in a stepwise decomposition, of which the crucial formulas include:

(0) $P = RT/V$

(1) $P = RT/V - a/V^2$ (or, alternatively, $P = RT/(V - b)$)

(2) $P = RT/(V - b) - a/V^2$ (or, standard form: $(P + a/V^2)(V - b) = RT$))

where P, V, T and R indicate pressure, volume, temperature and the ideal gas constant, respectively, and a and b refer respectively to specific gas constants related to mutual attraction between the molecules and the volume of the molecules.

The book series *Poznan Studies for the Philosophy of the Sciences and the Humanities* (since 1990 with a subseries on idealization, e.g. [Nowakowa, 1994]) includes many examples of I&C taken, in particular, from physics, biology, economics and sociology.

I&C can be used to structure theories in their research stage as well as in textbooks. Although it seems very plausible to do so, to say the least, it is very surprising that it seldom is explicitly done. However, in general expositions about what one has been doing or how one should do it, there is frequently reference to I&C. A specific reason for the relative neglect of I&C in the social sciences may be the great social pressure to avoid very strong idealizations: fear of being accused of distorting reality too much seems to be very rampant.

The above mentioned paradigm example raises a very interesting question concerning explanations: is it possible to (re)construct the explanation of a concretized law as a concretization of the explanation of the (more) idealized law? [Kuipers, 2000, Chapter 10] deals with this question in some detail.

Another, at least as important, question is whether and in what sense the I&C-strategy is functional for truth approximation in the empirical sciences. A detailed positive answer is given in [Kuipers, 2000, Subsection 10.4.] and in [Kuipers, forthcoming] that answer is given as a paradigm illustration of concept explication by the I&C-strategy, that is, 'conceptual I&C' as a conceptual twin of 'empirical I&C'. In general, as already mentioned, the I&C-heuristic can also be used in the qualitative theorizing that often occurs in mathematics and philosophy. The ordered textbook presentation of first propositional logic and then predicate logic provides a famous example. In Section 2 the I&C-heuristic has been used to present the structuralist approach to theories.

3.3.3 Program development guided by interesting theorems

Hamminga [1983] made a related strategy of theory development in economics explicit. The time when economists thought that economics could and should in principle be done along naive Popperian lines has passed, but the question remains how economists in fact do their job.

Hamminga studied the development of the theory of international trade in the period 1930-1970 and reached the following diagnosis. Economists direct their attention to theorems that they find interesting and they try to prove their validity for an increasing number of conceivable cases. Probably they have the following motive in the back of their mind: the desire to increase the plausibility that the theorem also holds in the actual world (or the nomic world, see Subsection 1.5.3.). Apart from this motive, the world does not play a clear role: it is all and only mathematics, or so it seems. Nevertheless, or precisely because of this, one can find a large amount of systematics in the details of what theoretical economists do.

To begin with, it is possible to systematize the specific claims of the research program to an even greater extent than Lakatos could have imagined: under such and such conditions it is possible to prove the interesting theorem *(IT)*, or in a formal schematization:

$$V_{lmn}(C_1 \ldots C_i; C_{i+1} \ldots C_j; C_{j+1} \ldots C_k \rightarrow IT)$$

The division of conditions here is as follows. V_{lmn} indicates the *field conditions* that describe the domain of the claim; in the example, l, m, and n indicate, respectively, the number of countries, goods and production factors taken into consideration. C_1,...,C_i indicate the generic or *basic principles*, i.e., the core ideas of the general research program with which the problem area is attacked. In the case of international trade this basic program is that of neo-classical economics, of which the core consists of utility theory. C_{i+1},...C_j indicate the special or *specific principles* for the particular subdomain of international trade, e.g., that the production functions are the same in all countries, while the endowments of production factors may vary greatly. Finally, $C_{j+1},...C_k$ indicate the *technical conditions* of a mathematical nature, e.g., that the production functions are continuous. An example of an interesting theorem is that of factor-price-equalization: the price of a certain factor becomes equal in the dynamic equilibrium with international trade.

Theory development or, more precisely, results that are considered to be important theoretical achievements consist of both new specific claims and their proof. In the latter, field and/or technical conditions have been liberated in one or more of the following ways:

- field extension: increasing the number of countries, goods or factors (2 is for each the point of departure)
- weakening technical conditions
- substituting more plausible conditions for technical conditions
- introducing alternate technical conditions

In [Kuipers, 2000, Subsection 10.4] such developments have been shown to be formally similar to concretization. They are therefore functional for truth approximation.

The picture that Hamminga draws seems representative of neo-classical economics; reconstructions of the theory of the market [Janssen and Kuipers, 1989] and of capital structure theory ([Kuipers, 2000, Subsection 11.2.], based on [Cools, Hamminga, Kuipers, 1994]) confirm this diagnosis. Of course, Hamminga does not provide a complete picture of the whole of the science of economics; in particular applied econometric models are overlooked. His views do however characterize an important part of the discipline, viz., so-called theoretical economics. Moreover, his work suggests the question of what the systematics is, if anything, in theory development in those areas of economics where the picture drawn seems inadequate.

The sketched diagnosis of the mathematical nature of economics may be an important underlying motive for the striking ambivalence of economists about the question of whether economics is an empirical social science or not. Moreover, the diagnosis illustrates that the cognitive aims of the social sciences in general and of economics in particular appear to be less evident than philosophers of science use to assume on the basis of an analogy to the natural sciences.

To be sure, the described strategy is certainly not restricted to economics. For instance, in population biology similar strategies are used with respect to the Law of Hardy-Weinberg [Lastowski, 1977]. In mathematics and philosophy the 'theorem-strategy' is also frequently used. Aiming at soundness and completeness proofs for increasingly complex logical systems is well known. Another example is the 'success theorem' in [Kuipers, 2000, Chapters 7-10]. It amounts to the claim that a theory closer to the truth than another will also be more successful. It can be proved under more and more realistic conditions.

3.3.4 Descriptive and explanatory research programs in the light of truth approximation

As a matter of fact, truth approximation analysis in general ([Kuipers, 2000], see also [Niiniluoto, this volume]) enables a richer perspective on the nature of descriptive and explanatory research programs. Recall that such programs presuppose, by definition, at least a domain, a problem, and a core idea, including a vocabulary, to solve that problem. A *descriptive research program* uses an observational conceptual frame, and may either exclusively aim at one or more true descriptions (as for example in most historiography), or it may also aim at the true (observational) theory. In the latter nomological type of descriptive program, however, the goal of a true theory is supposed to be achieved exclusively by establishing observational laws. Given that this requires (observational) inductive jumps, it is plausible to call such a descriptive program an *inductive research program.*[34] Not surprisingly, such programs 'approach the truth by induction'. For the establishment of observational laws, testing the relevant general observational hypothesis will result in true descriptions that either falsify the hypothesis or are partially derivable from it. According to the basic qualitative definition of 'more truthlike', assuming that accepted observational laws are true, any newly accepted observational law guarantees a step in the direction of the true theory. Hence, inductive research programs are relatively safe strategies of truth approximation: as far as the inductive jumps happen to lead to true accepted laws, the approach not only makes truth approximation plausible, it even guarantees it.

Let us now turn to the explication of the nature of explanatory or theoretical programs, which are by definition of a nomological nature. An *explanatory program* may or may not use a theoretical vocabulary. Even (nomic) empiricists can agree that it is directed at establishing the true observational theory. If there are theoretical terms involved, the referential realists will add that it is also directed at establishing the referential truth. The theory realist will add to this that it is even directed at establishing the theoretical truth. Scientists working within such a program will do so by proposing theories respecting the hard core as long as possible, but hopefully not at any price. They will empirically evaluate these theories separately and comparatively. Theory choice will be governed by being persistently more successful, which is trivially functional for empirical progress. However, although that 'rule of success' is, moreover, demonstrably functional for all distinguished kinds of nomic truth approximation, it cannot guarantee, even assuming correct data, a step in the direction of the relevant truth. Though the basic notions of successfulness and truthlikeness are sufficient to give the above characterization of the typical features of explanatory research programs, they usually presuppose refined means of comparison, which are presented in Part IV of [Kuipers, 2000].

[34]In this terminology, 'inductive' is supposed to exclusively refer to observational induction and not to theoretical induction, as distinguished in Subsection 1.2.2.

3.4 Interaction between programs

Of the two general types of interaction between programs, competition and cooperation, successful interdisciplinary research seems to result as a rule from a special kind of asymmetric cooperation between research programs. Moreover, it may or may not be a matter of cooperation between a holistic and a reductionistic program. Finally, some educational lessons from the program-bound and interactive research strategies will be drawn.

3.4.1 Interaction between programs as a research strategy

It is plausible to distinguish two main types of interaction between research programs, viz., interaction by competition and interaction by cooperation. Of course, it is also possible that after a period of interaction of one kind, the interaction turns into one of the other kind.

We will first concentrate on *competition.* When two programs are directed at the same domain and problem, and both are still in the internal phase, competition will concern the adequacy of the core ideas. When both programs are already in the external phase, competition concerns the question of which program is best suited to solving problems external to science or problems raised by a third program. When one program is still in the internal phase and the other in the external phase, competition usually takes the form of a challenge by the first to the supposed domain of validity or degree of accuracy of the second. A well-known example of the last kind is Einstein's questioning of Newton's theory.

The three indicated types of competition can all be very stimulating. However, when competition occurs, it is seldom seen as an explicit research strategy. Interestingly enough, one is not even always aware of being steered by a competing program, or one is at least not willing to admit that this is the case. These facts explain why the question of whether a further articulation of a competing program may lead to even more stimulating interaction is not always raised.

Population genetics provides a nice example of competition between two programs in the internal phase [Dolman and Gramsbergen, 1981]. Concerning the problem of the origin and dynamics of variations in populations two programs can be distinguished, viz., the classical and the equilibrium program. The development of both programs cannot be described without bringing the stimulating interaction between them into the picture. Moreover, they gradually show a remarkable convergence, with the consequence that the competition increasingly transforms into cooperation in such a way that a fruitful synthesis has emerged. The same development, initiated by, among others, Smolensky [1988], has taken place in the interaction between symbolism and connectionism in cognitive science.

Now let us consider *cooperation.* As in the case of competition, the forms of cooperation can be divided according to the three combinations of phases in which the two programs are situated. We have already seen that a program in the external phase can offer its services to another program, in the internal or external phase, which is confronted with a problem that the program itself is unable to solve. In Zandvoort's appealing terminology [1986; 1988; 1995] the latter program then functions as the *guide program* and the former as *supply program*. The core theory of the supply program may either be specialized (finalized) to the domain of the guide program, or it may be used as observation theory providing relevant observations for the guide program. For the particular problem the cooperation is of a fundamentally asymmetric nature. This character does not exclude the fact that the roles can be interchanged in dealing with another problem, in particular when not only one but both programs are in the external phase.

A typical form of the type of asymmetric interaction is provided by design research programs in the internal phase. They frequently function as guide programs, with descriptive and explanatory programs in the role of supply programs.

Besides the foregoing type of cooperation, in which at least one program is in the application phase, cooperation is possible between two programs in the internal phase, in which case they frequently stimulate each other in rotation, alternating in the role of guide and supply program. In this case, as in the case of two programs in the external phase, the cooperation is (although with respect to specific problems asymmetric) on the whole symmetric: the programs co-evolve [Bechtel, 1986; 1988a; Bechtel and Hamilton, this volume].

For instance, at least on the basis of accounts given in physics textbooks, one easily gets the impression that the interaction between phenomenological thermodynamics and statistical mechanics is a classical example of this type of cooperative co-evolution. However, it is well known that the intentions of the researchers concerned were of a much more competitive nature. Apparently, this does not exclude the fact that the result of competitive interaction can make it plausible that the intention to cooperate could have been at least as productive.

The example shows that researchers themselves may be inclined to perceive the interaction between two explanatory programs as competitive rather than as cooperative. However, when one program evidently is of a descriptive nature and the other of an explanatory nature directed at the first, the interaction between them can easily be conceived by the researchers as cooperative: it is a paradigmatic kind of co-evolution.

In the foregoing the basic aim of cooperation between programs was the solution of a problem encountered in one of the programs by the other. Of course, other goals of cooperation occur. For example, programs may jointly strive to articulate an overarching theory, or a synthesis of theories. Recall that the latter was the case in the example taken from population genetics. Still one other important form of cooperation involves bridging the gap between two theories, requiring a third so-called interfield theory [Darden and Maull, 1977; Bechtel, 1988a; Kuipers, 2001, Chapter 6; Bechtel and Hamilton, this volume].

For all mentioned kinds of interaction the programs may be empirical programs of the same or different type. Moreover, interaction may also involve an empirical program of a certain type and an explicative program of a philosophical or mathematical nature. Current interactive researches within cognitive science and between cognitive and neuroscience are of the latter nature. Some of the current philosophical mind-brain-body theories and theories of representation are not only of an explicative nature, they play at least some interactive role with some empirical programs [Bechtel, 1988a/b; Bechtel and Hamilton, this volume].

3.4.2 Interdisciplinary research

It is worthwhile to consider Zandvoort's model of cooperation in tandem with his model for successful interdisciplinary research [Zandvoort, 1986; 1988; 1995].

> IR-model: interdisciplinary research consists of some research programs, belonging to one or more disciplines, cooperating according to the following rules of the game:
>
> - one program is the guide program which raises problems of a theoretical or experimental nature in the others,
> - the other programs are supply programs, which have successfully passed their evaluation phase and hence can try to solve the problems provided by the guide program.

Compared with the popular ideas about interdisciplinary research, the above model has three fundamental differences. First, interdisciplinary research is not so much a matter of global cooperation between disciplines but, more specifically, cooperation between research programs. Second, it is a matter of asymmetric cooperation: one program poses the problems, the others try to supply solutions, and if successful they have the last word. Third, effective supply programs typically are in the external phase. Note that, if the guide program also has already passed its evaluation phase and is not a design program, then it will usually be directed at a science external problem of a technological or societal nature.

The IR-model suggests that the failure to start successful interdisciplinary research may well be due to the lack of relevant supply programs in the external phase. Moreover, it may be due to the collision of cognitive and social factors; in addition to the necessity of 'cognitive asymmetry' there is an inclination to as much 'social symmetry' as possible: all participants are supposed to deliver contributions of equal importance.

The stress on asymmetric cooperation between programs needs a counterbalance on the level of disciplines. It is conceivable that all interdisciplinary research directed at some science external problem area (e.g., health, environment, education) develops into a state in which one discipline provides all the guide programs, whereas the other participating disciplines provide only supply programs (the hierarchical model). Alternatively, it is also possible that there arises on the level of disciplines a more symmetric situation (the symmetric model). On the level of science and research policy, when setting up long-term strategic interdisciplinary research in some science external problem area, it seems very important to start with the symmetric model. The reason is that it is easy to imagine that starting from the symmetric situation purely scientific reasons may gradually lead to a hierarchical situation, whereas it will be much more difficult to reach a symmetric situation starting from a hierarchical one, let alone to reverse that hierarchy.

For reasons of completeness we conclude by noting that the IR-model does not seem to be appropriate as a point of departure for the investigation of a science external problem area when one wants to obtain short-term practical results. In planning that type of research ad hoc considerations seem to be unavoidable.

3.4.3 Interaction of holistic and reductionistic research programs

One of the most exciting forms of competition and cooperation occurs between reductionistic and holistic research programs involving essentially the same domain. In most cases the interaction can be described in Zandvoort's terms of a guide program and one or more supply programs. Moreover, in one sense or another the guide program is reduced, i.e., there is a reduction of concepts, or laws, or both. In this case, the guide program is called holistic and the supply program reductionistic, terms which are of course relative qualifications. The reduction of laws and concepts may or may not conform to one of the kinds of reduction distinguished in the pluralistic models for the reduction of laws and concepts that has been presented in [Kuipers, 2001, Chapter 3 and 5, respectively].

None of these basic forms of reductive interaction implies the elimination of the guide program. One of them only amounts to one or more corrections in the tentative laws that guided the research. The other two are variants of a non-eliminative reduction of a higher to a lower level. In these cases the laws and concepts of the higher level of the guide program are reduced to laws and concepts of lower levels, without explaining away the higher level.

Below we give a number of illustrative examples of studies of interacting research programs, where reductionistic and holistic perspectives play a major role. All studies mentioned are related to Groningen. Although they represent a type of 'cognitive studies' in philosophy of science which is typical for Groningen (see [Kuipers and Mackor, 1995]), numerous examples from elsewhere could, of course, be given. In a detailed study Janssen [1993] analyses the so-called micro-economic foundation of macro-economic concepts and laws. The current micro-foundation is a non-reductionistic attempt to interaction between the descriptive guide program of macroeconomics and the explanatory program of neo-classical microeconomics, in particular general equilibrium theory. According to Janssen the results are problematic because the supposed individualistic foundation of general equilibrium theory is doubtful. He sketches another way to explain macro- and micro-economic laws and theories in a strictly methodological-individualistic way. In this approach game-theoretic adaptation of utility theory plays a crucial role: it serves as a supply program.

Looijen [1995; 1998/2000] investigates the structure and dynamics of ecological communities. His working hypothesis is that the cooperation between three kinds of research programs could be improved. On the extremes sides one there are holistic guide programs describing the structure and dynamics of communities and radical reductionistic programs that try to explain these patterns in terms of the species composing the communities and their environmental needs could. They could well cooperate with moderately reductionistic (or moderately holistic, for that matter) programs that try to explain these patterns using theories about the interactions between the composing species, such as predation and competition.

Guichard [1995; 1997] starts by documenting that stress researchers, though striving at cooperation between psychological and biological research programs, were unsuccessful in the strategies employed to achieve this end. The reason for this systematic failure seems to have been that these strategies essentially presupposed a dualistic explication of the mind-body problem. Guichard argues that monistic explications, in particular of a materialistic-reductionistic nature, are more appropriate for such cooperation. His intervention uses a philosophical explication program to get new perspectives for cooperation between (relatively holistic) psychological guide programs and (relatively reductionistic) biological supply programs. More specifically, Guichard argues that the proper function theory of Ruth Millikan provides the ideal 'interfield theory' for this purpose.

Mackor [1995; 1997] argues along the same lines, but more generally, that a conceptual unification is possible not only between biology and psychology but that this can be extended to sociology by analyzing meaningful and rule-guided behavior in terms of Millikan's notion of proper functions. The result is a new, naturalistic, mildly reductionistic perspective on the spectrum of disciplines and their possibilities of cooperation. From this perspective, the most important boundary between the sciences runs between physics and chemistry on the one hand and biology on the other, although, successful cooperation between them is never excluded.

Festa [1993; 1995] initiates a confrontation between three research programs that were developing almost independently: inductive logic or inductive probability theory (Carnap), truth approximation (Popper) and Bayesian statistics. In the first place he shows that inductive logic, despite Popper's dismissive attitude about it, can be considered as a part of the truth approximation program, directed at the approximation of the true, objective chances. In the second place Festa elaborates the claim that, using De Finetti's representation theorem, (relatively holistic) systems of inductive probabilities can be reduced to special types of Bayesian statistics, which can hence be used as a supply program for further development of inductive logic. Last but not least, he shows that it is possible to define an optimum inductive system in terms of the available prior information about the domain to be investigated.

The general outline and the examples suggest at least three different research strategies for attacking a domain on the macro-level of some macro-/micro-level distinction. The *radical holistic* strategy is to try not only to describe but also to explain the phenomena at the macro-level in terms of that level, or even higher levels, and to refrain from lower level theories. The *radical reductionistic* strategy is to try not only to explain but also to describe the macro-phenomena in micro-terms as much as possible. The third strategy is a mixed one: according to the *mixed* strategy, one describes the macro-phenomena and their possible relations in macro-terms, and tries to explain them in micro-terms as far as possible, and hence in macro-terms as far as necessary.

These three strategies suggest equally many philosophical (ontological cum epistemological) positions with respect to a certain 'macro-domain'. They are formulated as general statements about the possibility of reduction of the concepts and laws of that domain. *Radical reductionism* is the belief that every macro-concept and macro-law can be reduced. *Radical holism* is the belief that all (interesting) concepts and laws of the domain cannot be reduced. Finally, *restricted reductionism* (and holism!) is the belief that some concepts and laws may be reducible, but others may not be.

It is important to note that the terminologically corresponding strategies and positions are not strictly coupled, except perhaps that radical philosophical holism leaves only room for the radical holistic strategy. The converse is not self-evident. There are excellent examples of research according to the radical holistic strategy, e.g., phenomenological thermodynamics and macroeconomics, where it is seen as a compatible, but separate task to try to reduce the macro-concepts and -laws, i.e., to work according to the radical reductionistic strategy. To be sure, it has to be conceded that reductionistic strategies in general have been very successful in the history of science.

However, the radical reductionistic strategy often leads to impressive minute research, which, however, nobody is waiting for. On the other hand, the radical holistic strategy frequently degenerates into hardly testable and transferable insights.

In line with these roughly formulated impressions, it is plausible to formulate the working hypothesis that the mixed strategy will in many cases be the best strategy. For, to reduce concepts and laws of a certain domain, they have first to be established. In its turn, it is frequently the case that the search for concepts and laws has been considerably stimulated by reductionistic questions.

The mixed strategy provides an important form of interaction of research programs. In the case of reductive interaction the guide program is of course a program on the macro-level, whereas at least one supply program is supposed to deal with the micro-level. When the supply program is, like the guide program, in the internal phase, the reductive interaction is symmetric, when it is in the external phase it is asymmetric.

Zandvoort's paradigm example of a supply program in the natural sciences is the NMR(nuclear magnetic resonance)-program, engaged in asymmetric reductive cooperation with chemical and biological research programs. An important asymmetric example in the social sciences is the utility maximization or rational choice program. It has proved its strength in microeconomics and nowadays it cooperates with guide programs in macroeconomics [Janssen, 1993] and macro-sociology (explanatory sociology). An historical example of the symmetric reductive type is the interaction between phenomenological thermodynamics and statistical mechanics.

3.4.4 Program pluralism as an education and research strategy

Program-bound research has one main disadvantage. One can readily become very indoctrinated with a program. The postgraduate schooling in program-bound research may well degenerate into the delivery of program-bound researchers. As a counterbalance to the importance of program-bound research, a program pluriform education and subsequent research career seems equally important.

Based on the work of Kuhn and Lakatos it may be inferred that mature science frequently consists of dogmatic research, i.e., research sticking to the hard core of ideas, executed by dogmatically inclined researchers. Although such practices can be rational provided one aims at progress within the boundaries of the program, it is also our conviction that science would profit still more if non-dogmatic researchers perform dogmatic research.

The way to learn to do program-bound research, without becoming a prisoner in one program, is to get research experience in at least two programs. They need not be competing programs. According to Kuhn, such program pluralism is almost impossible, due to the Gestalt-switch that it is claimed to require. Although this thesis, together with the so-called incommensurability thesis, has been criticized severely ([Franklin, 1986; Hintikka, 1988], to mention a few), it is important to stress that it may be at least as instructive to undertake research in two programs that might cooperate.

The short-term effects are very positive. To begin with, when one alternatingly does research in two programs, the period engaged in one program may function at the same time as a form of productive breathing space for the work on the other. It may even occur, to say it in popular terms, that the right hemisphere is further stimulated to do its work. For, as suggested by the literature on creativity and serendipity [Van Andel, 1994], successful, or even unsuccessful, attempts at problem solving in one program may be transformed into successful solutions for the problems of the other.

This short-term favorable effect concerns the stimulation of largely unconscious processes, the second to be mentioned results from the conscious stimulation of interaction. If two programs have something to do with each other, knowledge of both leads in a natural way to questions of cooperation or competition. For instance, one may ask whether one program may be of help for the other. Can it solve a problem the other is unable to solve? Additionally, competition questions may arise. If on closer inspection both programs essentially claim to be able to solve the same problem, and if the one has already succeeded in doing so, the other cannot remain behind.

At first sight one may find the second favorable short-term effect of doing research in two programs unimportant. For competition and cooperation questions also may arise in the research team, or in the study of the literature, as specific research questions may be transmitted from one individual researcher to the other. In theory such exchanges should occur, but in practice they are limited, because it is understandable and even productive that one primarily works in program-bound research groups. Moreover, as far as competition questions are concerned, there is the additional fact that it is only attractive to raise such questions when the researcher has acquired affinity with both programs.[35] Hence, it seems plausible that the suggested interaction questions are best stimulated by promoting research training and further practice in more than one program.

The plea for pluralistic research experience should not, however, be misunderstood as a plea for unlimited and diffuse eclecticism. On the contrary, it is a plea for experience within and interaction between a limited number of (analytically) well-distinguished research programs.

The favorable long-term effects of this plural-program-bound research experience also seem advantageous. To join new developments in international research it is of great importance that one has learned about methodological multiplicity and the various perspectives from which one can investigate the world, not only in theory but also in practice. Plural-program-bound research experience stimulates the flexibility of the individual scientist in his further research career. The desirability of this flexibility for scientific research in general and interdisciplinary research in particular is obvious.

CONCLUDING REMARKS

In this last section we have argued that the notion of a research program is very useful in globally describing scientific research and in specifying global research strategies. In Sections 1 and 2 some of the main products of scientific research, usually resulting from isolated or cooperative (descriptive and explanatory) programs, have been studied in some detail, viz., laws and theories. Let us finally come back to the relation between the main subjects of the first and the third section, that is, the distinction between observational laws and proper theories, i.e., the law-distinction, and research programs, respectively.

It is the (relatively) short-term dynamic role of the law-distinction that relates to the development of research programs. The first context in which the law-distinction is used in short-term dynamics is just one explanatory research program, more in particular a (proper) theoretical program. Here a proper theory is revised on the basis of its successes and failures in explaining and predicting observational laws. The second context is the interaction between a theoretical program and a descriptive program. Here the two programs develop hand in hand, each challenging the other with new results. On the one hand, there are potential observational laws that have been predicted by the theoretical program and which have to be tested within the descriptive program. On the other hand, there are observational laws that have been established independently, in particular inductively, within the descriptive program (which may be called to that extent an inductive program) and that have to be explained by the theoretical program.

[35] For example [Kuipers, 2000] grew out of research and teaching in at least two rather different, if not opposing, research programs, the Carnap-Hintikka program of inductive (confirmation) logic and Popper's program of truth approximation.

Finally, the third context is of course the competition between two theoretical programs claiming to be superior with respect to the explanation and prediction of observational laws for a certain domain. Of course, the latter laws may or may not hang together within a descriptive program. In all three contexts the relevant theoretical programs are in the internal phase, whereas the relevant descriptive programs may or may not already be in the external phase.

Although the law-distinction may seem at first sight to be parallel to the different internal goals of descriptive and explanatory research programs, it is actually not so. To be sure, proper theories cannot be the internal goal of descriptive programs. However, the true description, aimed at by a descriptive program, may concern the true general description, i.e., the true observational theory in the sense of the (relative to the program) complete set of observational laws of the domain, which is an improper theory. The complicating fact is that this true observational theory may also be the true theory at which an explanatory program is aiming, in order to explain and further predict certain observational laws, usually those restricted to some sub-vocabulary of the explanatory program. In terms of the distinction between theoretical and (non-)theoretical terms, the failing parallel is even easier to formulate. Whereas a descriptive program by definition can not introduce new theoretical terms, an explanatory program may introduce such terms, and then be called a theoretical program, but it need not introduce such terms.

ACKNOWLEDGMENTS

The Sections 1, 2, and 3 are strongly based on Chapters 2, 12, and 1, respectively, of (Kuipers, 2001). The final version profited a lot from the critical remarks of Martin Mahner.

BIBLIOGRAPHY

[van Andel, 1994] P. van Andel. Anatomy of the unsought finding. *British Journal for the Philosophy of Science*, 45: 631–648, 1994.

[Atkinson, 2005] A. Atkinson. A new metaphysics. Finding a niche for string theory. In R. Festa, A. Aliseda, and J. Peijnenburg (eds.), *Cognitive Structures in Scientific Inquiry. Poznan Studies in the Philosophy of the Sciences and the Humanities*, Vol. 84. Rodopi, Amsterdam/New York, pages 95–102, 2005.

[Balzer, 1982] W. Balzer. *Empirische Theorien: Modelle, 1Strukturen, Beispiele, Vieweg.* Braunschweig/Wiesbaden, 1982.

[Balzer, 1996] W. Balzer. Theoretical terms: recent developments. In W. Balzer and C. U. Moulines eds.), *Structuralist Theory of Science.* De Gruyter Verlag, Berlin, pages 139–166, 1996.

[Balzer and Dawe, 1986] W. Balzer and C. Dawe. Structure and comparison of genetic theories. *The British Journal for the Philosophy of Science*, (1): 37.1: 55–69; (2): 37.2: 177–191, 1986.

[Balzer and Dawe, 1997] W. Balzer and C. Dawe. *Models for Genetics.* Peter Lang, Frankfurt/M, 1997.

[Balzer and Marcou, 1989] W. Balzer and Ph. Marcou. A reconstruction of Sigmund Freud's early theory of the unconscious. In [Westmeyer, 1989], pages 13–31, 1989.

[Balzer *et al.*, 1987] W. Balzer, C. U. Moulines, and J. D. Sneed. *An Architectonic for Science.* Reidel, Dordrecht, 1987.

[Balzer *et al.*, 2000] W. Balzer, J. D. Sneed, and C. U. Moulines. *Structuralist Knowledge Representation. Paradigmatic Examples.* Poznan Studies in the Philosophy of the Sciences and the Humanities, Vol. 75, Rodopi, Amsterdam, 2000.

[Balzer and Tuomela, 1997] W. Balzer and R. Tuomela. A fixed point approach to collective actions. In G. Holmström-Hintikka (eds.), *Contemporary Action Theory.* Kluwer, Dordrecht, pages 115–142, 1997.

[Bechtel, 1986] W. Bechtel (ed.). *Integrating Scientific Disciplines.* Dordrecht, Nijhoff, 1986.

[Bechtel, 1988a] W. Bechtel. *Philosophy of Science. An Overview for Cognitive Science.* Erlbaum, Hillsdale, 1988.

[Bechtel, 1988b] W. Bechtel. *Philosophy of Mind. An Overview for Cognitive Science.* Erlbaum, Hillsdale, 1988.

[Bickle, 1993] J. Bickle. Connectionism, eliminativism, and the semantic view of theories. *Erkenntnis*, 39: 359–382, 1993.

[Bickle, 1998] J. Bickle. *Psychoneural Reduction.* The New Wave, MIT Press, Cambridge, 1998.

[Bird, 1998] A. Bird. *Philosophy of Science.* UCL-Press, London, 1998.

[Bhaskar, 1979] R. Bhaskar. *The Possibility of Naturalism. A Philosophical Critique of the Contemporary Human Sciences.* Harvester Press, Brighton, 1979.

[Bijker, 1995] W. Bijker. *Of Bicycles, Bakelites, and Bulbs.* MIT press, Cambridge Ma., 1995.

[Boehm *et al.*, 1979] T. Boehm, J. Folkman, T. Browder, and M. O'Reilly. Anti-angiogenic therapy of experimental cancer does not induce acquired drug resistance. *Nature*, 390: 404–407, 1979.

[Boyd, 1984] R. Boyd. The current status of scientific realism. In J. Leplin (ed.), *Scientific realism.* University of California Press, Berkeley, pages 41–82, 1984.

[Brink, 1989] C. Brink. Verisimilitude: views and reviews. *History and Philosophy of Logic*, 10: 181–201, 1989.

[Bunge, 1967] M. Bunge. *Foundations of Physics.* Springer Verlag, Heidelberg, 1967.

[Bunge, 1977] M. Bunge. The GST challange to classical philosophies of science. *International Journal of General Systems*, 4: 29–37, 1977.

[Carnap, 1950/1963] R. Carnap. *Logical Foundations of Probability.* University of Chicago Press, Chicago, 1950/1963^2 (1963^2 with a new foreword).

[Cartwright, 1983] N. Cartwright. *How the Laws of Physics Lie.* Clarendon Press, Oxford, 1983.

[Cools *et al.*, 1994] K. Cools, B. Hamminga, and T. Kuipers. Truth approximation by concretization in capital structure theory. In B. Hamminga and N. B. de Marchi (eds.), *Idealization VI: Idealization in Economics.* Poznan Studies in the Philosophy of the Sciences and the Humanities, Vol. 38, Amsterdam, Rodopi, pages 205–228, 1994.

[Cushing, 1982] J. Cushing. Models and methodologies in current theoretical high-energy physics. *Synthese*, 50(1): 5–101, 1982.

[Darden and Maull, 1977] L. Darden and N. Maull. Interfield theories. *Philosophy of Science*, 43: 44–64, 1977.

[Diederich, 1981] W. Diederich. *Strukturalistische Rekonstruktionen.* Vieweg Verlag, Braunschweig, 1981.

[Dolman and Gramsbergen, 1981] H. Dolman and J. B. Gramsbergen. *Kontroverses in de populatiegenetika, een wetenschapsfilosofische analyse.* Masters thesis theoretical biology, University of Groningen, 1981.

[von Eckardt, 1993] B. von Eckardt. *What is Cognitive Science?.* MIT Press, Cambridge, Ma., 1993.

[Festa, 1993] R. Festa. *Optimum Inductive Methods. A Study in Inductive Probability Theory, Bayesian Statistics and Verisimilitude.* Kluwer, Dordrecht, 1993.

[Festa, 1995] R. Festa. Verisimilitude, disorder, and optimum prior probabilities. In [Kuipers and Mackor, 1995], pages 299–320, 1995.

[Feyerabend, 1962] P. Feyerabend. Explanation, reduction, and empiricism. *Minnesota Studies in the Philosophy of Science*, Vol. III, 28–97, 1962.

[Feyerabend, 1975] P. Feyerabend. *Against Method.* NLB, London, 1975.

[Fleck, 1935/1979] L. Fleck. *Entstehung und Entwicklung einer wissenschaftlichen Tatsache.* Franfurt am Main, 1935; translated as *Genesis and Development of a Scientific Fact*, University of Chicago Press, Chicago, 1979.

[van Fraassen, 1980] B. van Fraassen. *The Scientific Image.* Clarendon Press, Oxford, 1980.

[van Fraassen, 1989] B. van Fraassen. *Laws and Symmetry.* Clarendon Press, Oxford, 1989.

[Franklin, 1986] A. Franklin. *The Neglect of Experiment.* Cambridge University Press, Cambridge, 1986.

[Gähde, 1983] U. Gähde. *T-Theoretizität und Holismus.* Peter Lang, Frankfurt, 1983.

[Giddens, 1984] A. Giddens. *The Constitution of Society.* Polity Press, Cambridge, 1984.

[Giere, 1985] R. Giere. Constructive realism. In P. Churchland and C. Clifford (eds.), *Images of science.* The University of Chicago Press, Chicago, pages 75–98, 1985.

[Giere, 1999] R. Giere. Using models to represent reality. In L. Magnani, N. Nersessian, and P. Thagard (eds.), *Model-based reasoning in scientific discovery.* Kluwer, Dordrecht, pages 41–57, 1999.

[Guichard, 1995] L. Guichard. The causal efficacy of propositional attitudes. In [Kuipers and Mackor, 1995], pages 373–392, 1995.

[Guichard, 1997] L. Guichard. *Stress Research and the Mind-Body Problem.* PhD-manuscript, Groningen, 1997.

[Hacking, 1983] I. Hacking. *Representing and Intervening.* Cambridge UP, Cambridge, 1983.

[Hamminga, 1983] B. Hamminga. *Neoclassical Theory Structure and Theory Development.* Springer, Berlin, 1983.

[Hark, 2004] M. ter Hark. *Popper, Otto Selz and the Rise of Evolutionary Epistemology.* Cambridge UP, Cambridge, 2004.

[Harré, 1986] R. Harré. *Varieties of Realism.* Blackwell, Oxford, 1986.

[Hempel, 1966] C. Hempel. *Philosophy of Natural Science.* Prentice-Hall, Englewood Cliffs, 1966.

[Hempel, 1970] C. Hempel. On the 'standard conception' of scientific theories. *Minnesota Studies in the philosophy of science*, Vol.IV, Minneapolis, 1970.

[Hettema and Kuipers, 1988/2000] H. Hettema and T. Kuipers. The periodic table: its formalization, status, and relation to atomic theory. *Erkenntnis*, 28: 387–408. Revised version, entitled "The formalisation of the periodic table", in [Balzer, Sneed and Moulines, 2000], pages 285–305.

[Hettema and Kuipers, 1995] H. Hettema and T. Kuipers. Sommerfeld's *Atombau*: a case study in potential truth approximation. In [Kuipers and Mackor 1995], pages 273–297, 1995.

[Hintikka, 1988] J. Hintikka. On the incommensurability of theories. *Philosophy of Science*, 55(1): 25–38, 1988.

[Holton, 1973] G. Holton. *Introduction to Concepts and Theories in Physical Science.* Second edition, Addison-Wesley, Reading, Ma., 1973.

[Janssen, 1993] M. Janssen. *Micro-foundations.* Routledge, London, 1993.

[Janssen and Kuipers, 1989] M. Janssen and T. Kuipers. Stratification of general equilibrium theory: a synthesis of reconstructions. *Erkenntnis*, 30: 183–205, 1989.

[Johansson, 2005] L-G. Johansson. The nature of natural laws. In J. Faye, P. Needham, U. Scheffler, and M. Urchs (eds.), *Nature's Principles*, Springer, Dordrecht, pages 171–187, 2005

[Krajewski, 1977] W. Krajewski. *Correspondence Principle and Growth of Science.* Reidel, Dordrecht, 1977.

[Kuhn, 1962/1969] T. Kuhn. *The Structure of Scientific Revolutions.* University of Chicago Press, Chicago, 1962/1969.

[Kuhn, 1963] T. Kuhn. The function of dogma in scientific research. In: A. Crombie (ed.), *Scientific Change.* Basic Books, New York, Chapter 11, 1963.

[Kuipers, 1982] T. Kuipers. The reduction of phenomenological to kinetic thermostatics. *Philosophy of Science*, 49(1): 107–119, 1982.

[Kuipers, 2000] T. Kuipers. *From Instrumentalism to Constructive Realism. On some Relations between Confirmation, Empirical Progress, and Truth Approximation.* Synthese Library 287, Kluwer, Dordrecht, 2000.

[Kuipers, 2001] T. Kuipers. *Structures in Science. Heuristic Patterns based on Cognitive Structures. An Advanced Textbook in Neo-Classical Philosophy of Science.* Synthese Library 301, Kluwer, Dorecht, 2001.

[Kuipers, 2004] T. Kuipers. Inference to the best theory, rather than inference to the best explanation. Kinds of abduction and induction. In F. Stadler (ed.), *Induction and Deduction in the Sciences. Proceedings of the ESF-workshop.* Vienna, July, 2002, Dordrecht, Kluwer, pages 25–51, 2004.

[Kuipers, 2005] T. Kuipers. On designing historically adequate formal reconstructions. Reply to Eric Scerri. In R. Festa, A. Aliseda, and J. Peijnenburg (eds.), *Cognitive Structures in Scientific Inquiry*. Poznan Studies in the Philosophy of the Sciences and the Humanities, Vol. 84. Amsterdam/New York, NY: Rodopi, pages 211–216, 2005.

[Kuipers, 2006] T. Kuipers. Theories looking for domains. Fact or fiction? Reversing structuralist truth approximation. In L. Magnani, ed., *Model Based Reasoning in Science and Engineering*, pp. 33–50. Studies in Logic, Volume 2, College Publications, 2006.

[Kuipers, forthcoming] T. Kuipers. Empirical and conceptual idealization and concretization. The case of truth approximation. To appear in a *Liber Amicorum for Leszek Nowak*, forthcoming.

[Kuipers and Mackor, 1995] T. Kuipers and A. R. Mackor (eds.). *Cognitive Patterns in Science and Common Sense*. Poznan Studies in the Philosophy of the Sciences and the Humanities, Vol. 45, Rodopi, Amsterdam, 1995.

[Kyburg, 1968] H. Kyburg. *Philosophy of Science: A Formal Approach*. New York, McMillan, 1968.

[Lakatos, 1970] I. Lakatos. Falsification and the methodology of scientific research programmes. In I. Lakatos and A. Musgrave (eds.), *Criticism and the Growth of Knowledge*. Cambridge UP, pages 91–196; reprinted in [Lakatos, 1978], 8–101, 1970.

[Lakatos, 1978] I. Lakatos. *The Methodology of Scientific Research Programmes*. J. Worrall and G. Currie (eds.). Cambridge University Press, Cambridge, 1978.

[Lastowski, 1977] K. Lastowski. The method of idealization in population genetics. Poznan Studies in the Philosophy of the Sciences and the Humanities, Vol. 3(1-4), Grüner, Amsterdam, pages 199–212, 1977.

[Laudan, 1977] L. Laudan. *Progress and its Problems*. University of California Press, Berkeley, 1977.

[Looijen, 1995] R. Looijen. On the distinction between habitat and niche, and some implications for species' differentiation. In [Kuipers and Mackor, 1995], pages 87–108, 1995.

[Looijen, 1998/2000] R. Looijen. *Holism and Reductionism in Biology and Ecology. The Mutual Dependence of Higher and Lower Level Research Programmes*. PhD-dissertation. Revised version Episteme, Vol. 23, 2000, Kluwer, Dordrecht, 1998/2000.

[Mackor, 1995] A. R. Mackor. Intentional psychology is a biological discipline. In [Kuipers and Mackor, 1995], pages 329–348, 1995.

[Mackor, 1997] A. R. Mackor. *Meaningful and Rule-Guided Behaviour: A Naturalistic Approach. A Teleofunctional Argument against the Alleged Gap between the Natural and the Social Sciences*. PhD-dissertation, University of Groningen, Groningen, 1997.

[Mahner and Bunge, 1997] M. Mahner and M. Bunge. *Foundations of Biophilosophy*. Springer, Berlin, 1997.

[Nagel, 1961] E. Nagel. *The Structure of Science*. Routledge and Kegan Paul, London, 1961.

[Niiniluoto, 1987] I. Niiniluoto. *Truthlikeness*. Reidel, Dordrecht, 1987.

[Niiniluoto, 1999] I. Niiniluoto. *Critical Scientific Realism*. Oxford UP, Oxford, 1999.

[Nowak, 1974] L. Nowak. Galileo of the social sciences. *Revolutionary World*, Vol. 8: 5–11, 1974.

[Nowak, 1980] L. Nowak. *The Structure of Idealization*. Reidel, Dordrecht, 1980.

[Nowakowa, 1994] I. Nowakowa. The dynamics of idealizations. Poznan Studies Vol. 34. Rodopi, Amsterdam, 1994

[Peirce, 1934] C. S. Peirce. *Collected Papers*. Vol. 5, Cambridge University Press, Cambridge Ma., 1934.

[Popper, 1934/1959] K. R. Popper. *Logik der Forschung*. Vienna, 1934; translated as *The Logic of Scientific Discovery*. Hutchinson, London, 1959.

[Popper, 1963] K. R. Popper. *Conjectures and Refutations*. Routledge and Kegan Paul, London, 1963.

[Scerri, 1997] E. Scerri. Has the periodic table been successfully axiomatized? *Erkenntnis*, 47(2): 229–243, 1997.

[Scerri, 2005] E. Scerri. On the formalization of the periodic table. In R. Festa, A. Aliseda and J. Peijnenburg (eds.), *Cognitive Structures in Scientific Inquiry*. Poznan Studies in the Philosophy of the Sciences and the Humanities, Vol. 84. Amsterdam/New York, NY: Rodopi, pages 191–210, 2005.

[Schäfer *et al.*, 1983] W. Schäfer, G. Böhme, and P. Burgess. *Finalization in Science: The Social Orientation of Scientific Progress*. Reidel, Dordrecht, 1983.

[Schlick, 1938] M. Schlick. *Gesammelte Aufsätze*. Gerold, Wien, 1938.

[Schurz, 1990] G. Schurz. Paradoxical consequences of Balzer's and Gähde's criteria of theoriticity. *Erkenntnis*, 32: 161–214, 1990.

[Searle, 1995] J. Searle. *The Construction of Social Reality.* Allan Lane, London, 1995.

[Shapere, 1982] D. Shapere. The concept of observation in science and philosophy. *Philosophy of Science*, 49(4): 485–525, 1982.

[Smolensky, 1988] P. Smolensky. On the proper treatment of connectionism. *Behavioral and Brain Sciences*, 11: 1–74, 1988.

[Sneed, 1971] J. Sneed. *The Logical Structure of Mathematical Physics.* Reidel, Dordrecht, 1971.

[Sober, 2000] E. Sober. Quine's two dogmas. *Aristotelian Society*, Supplementary Volume LXXIV. London, pages 237–280, 2000.

[Spiegel *et al.*, 1989] D. Spiegel, J. Bloom, H. Kraemer, and E. Gottheil. Effects of psychosocial treatment on survival of patients with metastatic breast cancer. *Lancet*, 2: 888–891, 1989.

[Spiegel, 1993] D. Spiegel. *Living beyond Limits.* Time Books, New York, 1993.

[Stegmüller, 1973] W. Stegmüller. *Theorie und Erfahrung.* Band II, Teil D, Springer, Berlin, 1973.

[Stegmüller,, 1986] W. Stegmüller. *Realismus und Strukturalismus, Theorie und Erfahrung.* Band II, Teil H, Springer, Berlin, 1986.

[Suppes, 1957] P. Suppes. *Introduction to Logic.* Van Nostrand, New York, 1957.

[Toulmin, 1953] S. Toulmin. *The Philosophy of Science.* Hutchinson, London, 1953.

[Tuomela, 1995] R. Tuomela. *The Importance of us: A Philosophical Study of basic Social Notions.* Stanford UP, Stanford, 1995.

[Vos, 1991] R. Vos. *Drugs Looking for Diseases.* Kluwer, Dordrecht, 1991.

[Westmeyer, 1989] W. Westmeyer (ed.). *Psychological Theories from a Structuralist Point of View.* Berlin, Springer, 1989.

[Westmeyer, 1992] W. Westmeyer (ed.). *The Structuralist Program in Psychology: Foundations and Applications.* Hogrefe and Huber, Seattle, 1992.

[Zandvoort, 1982] H. Zandvoort. An extension of Sneed's reconstruction of classical particle mechanics to complex applications, and an alternative approach to special force laws. *Erkenntnis*, 18: 39–63, 1982.

[Zandvoort, 1984] H. Zandvoort. Lakatos and Nagel: a fruitful confrontation. *Zeitschrift fur allgemeine Wissenschaftstheorie*, XV-2: 299–307, 1984.

[Zandvoort, 1986] H. Zandvoort., *Models of Scientific Development and the Case of NMR.* Synthese Library 184, Reidel, Dordrecht, 1986.

[Zandvoort, 1988] H. Zandvoort. Nuclear magnetic resonance and the acceptability of guiding assumptions. In A. Donovan, R. Laudan, and L. Laudan (eds.), *Scrutinizing Science: Empirical Studies of Scientific Change.* Reidel, Dordrecht, pages 337–358, 1988.

[Zandvoort, 1989] H. Zandvoort. De struktuur van theorieën en hun rol in verklaringen; De ontwikkeling van theorieën: onderzoeksprogramma's. In M. Korthals (ed.), *Wetenschapsleer.* Open Universiteit, Heerlen, pages 24–47; 48–69, 1989.

[Zandvoort, 1995] H. Zandvoort. Concepts of interdisciplinarity and environmental science. In [Kuipers and Mackor, 1995], pages 45–68, 1995.

PAST AND CONTEMPORARY PERSPECTIVES ON EXPLANATION

Stathis Psillos

> The word *explanation* occurs so continually and holds so important a place in philosophy, that a little time spent in fixing the meaning of it will be profitably employed.
>
> John Stuart Mill

0. INTRODUCTION

In spite of Mill's guarded optimism that fixing the meaning of *explanation* is a task that requires "a little time", the truth is that the time and energy spent on this task in the history of philosophy have been enormous. Though this time and energy have been profitably spent, we do not yet know what exactly explanation is. There is no single and definite meaning attached to the word *explanation*. There is no fully adequate model of explanation that covers everything we think, intuitively, an explanation consists in. We are not even clear on what are the core platitudes that the concept of explanation must satisfy. Yet, we know that the more we think of it, the more the concept of explanation is shown to be indispensable to the ways in which we think and act.

This is just a small sample of the questions and issues that are involved in explaining explanation. Explanation is intimately linked with causation and laws, but how exactly? Does all explanation have to be causal? Does all causal explanation have to be nomological? Do explanations have to be arguments whose conclusion is the fact/event to be explained (the *explanandum*) and whose premises (the *explanans*) have to cite laws of nature? If they do, does explanation consist in the deductive demonstration of the *explanandum* or is an inductive link between the *explanans* and the *explanandum* enough? If explanations are not arguments, what is their logical form? Is it enough to say that they are causal stories linking the purported cause with the effect? Shouldn't these causal stories be backed up with laws? And how are laws themselves explained? Is it enough to say that more fundamental laws explain the less fundamental ones by subsuming the latter under them? What, in the end, does this idea of fundamental laws amount to? Is part of the intuitive meaning of explanation that to explain a number of apparently disparate facts (or laws) is to unify them under a small set of unexplained

Handbook of the Philosophy of Science: General Philosophy of Science - Focal Issues, pp. 97–173.
Volume editor: Theo Kuipers. General editors: Dov M. Gabbay, Paul Thagard and John Woods.

explainers? Are there special patterns of explanation that are suitable for purposeful and intentional behaviour? Or are teleological and intentional patterns of explanations just variants of the causal/nomological patterns? Might it be more profitable or appropriate to focus on the *act* or the *process* of explaining, instead of the *product* of explanation. If an explanation is, ultimately, an answer to a why-question, shouldn't it be the case that the *relevant* answers will depend on the presuppositions or the interests of the questioner, on the space of alternatives, and, in general, on the context of the why-question? If this is so, can there be an objective conception of what an explanation is?

Without aspiring to address all of the above issues and questions, the present essay aims to show how our thinking about explanation has evolved and where it stands now. Its first part presents how some major thinkers, from Aristotle to Mill, conceived of explanation. The second part offers a systematic examination of the most significant and controversial contemporary models of explanation.

Despite the more than two millennia that separate Aristotle's thinking from ours, Aristotle's conception — the thought that explanation consists in finding out *why* something happened and that answering why-questions requires finding causes — set the agenda for almost all subsequent thinking about explanation. For better or worse, causal explanation had been taken to be the model of explanation. The rivalry had been between those who thought that all causal explanation must proceed in terms of efficient causation and those who (following closely on Aristotle's footsteps) thought that there is room (and need for) teleological explanation (that is, for explanation that cites final causes). Most modern philosophers revolted against all but efficient causation. The latter was taken to be the *only* type of causation by (almost) all those who advocated, in one form or another, the mechanical philosophy: in their hands, efficient causation became tantamount to pushings and pullings. Final causes, in particular, were cast to the winds. Where Aristotle saw goals and purposes in nature, mechanical philosophers either excised all purpose from nature (Hobbes, Hume) or placed it firmly in the hands of God (Descartes). It was mainly Gottfried Leibniz who tried to reconcile efficient (mechanical) causation with final causation.

For Aristotle, causal explanation is captured by *demonstrative arguments* of certain sorts: those that respect the asymmetry between cause and effect. In his hands, causal explanation and demonstration became one. What is more, Aristotle embedded his account of explanation into a rich ontological framework that included essences, substantial forms, powers, activities and so on. Most moderns revolted, to varying degrees, against this rich ontological landscape. From René Descartes onwards, the idea gained momentum that causal explanation proceeds by subsuming the events to be explained under general laws. Causation was intertwined with the presence of laws and explanation was taken to consist in a law-based demonstration of the *explanandum*. However, two key Aristotelian ideas, that there is necessity in nature and that this necessity is the same as the logical necessity of a demonstrative argument, remained part of the mainstream philosophical thinking about causation and explanation until David Hume subjected them to severe criticism and undermined them. In a sense, Hume was the first to remove the efficiency from efficient causation: causation just is regular succession — one thing following another. In doing this, he was the first to free *causation* and *explanation* from the metaphysical fetters that his predecessors had used to pin them down. Immanuel Kant reacted to Hume by trying to secure the metaphysical foundations of the fundamental laws of nature, but it was Mill who pushed the Humean project to its extreme by offering a well worked out model of scientific explanation based on the idea that there is no necessity in nature and that, ultimately, explanation amounts to unification into a comprehensive deductive system, whose axioms capture the fundamental laws of nature.

Most of contemporary thinking about explanation has taken place against the backdrop of the views of the Logical Positivists. The key to understanding the Logical Positivist agenda on explanation is this. They thought that by sufficiently explicating the concept of explanation, they could thereby legitimise *causation*. The deep problems they saw in causation stemmed from Hume's critique of it. Ever since Hume's work, philosophers of empiricist persuasions thought that the concept of causation is too mysterious or metaphysical to be taken for granted without any further analysis. They thought that the main culprit was the idea that causation implies the existence of *necessary connections* in nature.

Explanation, the Logical Positivists thought, is different. It can be as transparent as the notions on the basis of which it can analysed, viz., deductive argument and laws of nature. This is an idea they took, almost straightaway, from Mill. Given, as they thought, that both notions (deductive argument and laws of nature) are scientifically respectable, *explanation* becomes a legitimate concept, too. Besides, if whatever is valid in the concept of causation can be captured on the basis of the concept of explanation, a valid residue of causation is preserved and demystified. The project of demystifying causation culminated in the attempts made by Carl Hempel and his followers to articulate the *Deductive-Nomological* model of explanation, the basic kernel of which is that explanation is fully understood as a species of deductive argument, with one of its premises stating a universal law of nature. Later on, Hempel and his followers advanced their project further by enlarging the kind of arguments that can be explanations so as to include *inductive* arguments (and statistical, as opposed to universal, laws), thereby hoping to capture the thrust of probabilistic (or stochastic) causation.

The irony for the Hempelian project is that what came out of the front door seemed to be re-entering from the rear window. For, as Aristotle noted, it seems that one cannot distinguish between good and bad explanations of some phenomena, *unless* one first distinguishes between causal and non-causal explanations, or better between those explanations that reflect the causal connections between the *explanans* and the *explanandum* and those that do not. It appears, then, that we need first to sort out the concept of causation and then talk about causal explanation. If this is right, the empiricist project outlined above gets things the wrong way around. Many alternative models of causal explanation that have seen the light of day since Hempel's rely on this last thought: they start with a preferred account of causation and then try to tailor causal explanation to it. A distinctive trend in this alternative approach is the rejection of the view that explanations are arguments (deductive or inductive). More recent mechanistic or interventionist approaches to causal explanation take the latter to consist in revealing underlying causal mechanisms or relations of invariance among magnitudes.

There have been some plausible ways for modern empiricists to defend the view that explanations are arguments, and in particular, the view that it is the explanatory relations that are primary and not the causal ones. Following Mill, again, it has been argued that explanation amounts to *unification*. This thought ties in well with the best empiricist view on what laws of nature are. As Mill, Frank Ramsey and David Lewis have argued, laws of nature are those regularities that are captured by the axioms and theorem of the best, in terms of simplicity and strength, deductive systematisation of our knowledge of the world. But even this view is not problem-free; and, in any case, the fertile concept of unification resists a fully adequate formulation.

Given that most of this essay is about other people's thinking about explanation, I would like to make a general point about the moral I have drawn from my own thinking about other people's thinking about explanation. Most of the thoughts and arguments in currency during the twentieth century had been put forward, in one form or another, by some past thinker from Aristotle to Mill. Perhaps with very few exceptions, the twentieth century did not add new powerful ideas. But, thanks to the unprecedented level of technical sophistication of the philosophers of the twentieth century, old and powerful ideas got a new lease of life by being more precisely formulated and more carefully worked out. This is by no means a small feat. Most of the time, the devil is in the details. And unless an idea is made precise and sharp, its strengths, limitations and possible flaws do not become visible.

PART I: A (SELECTIVE) HISTORY OF EXPLANATION

1 ARISTOTLE: EXPLANATION AS DEMONSTRATION

Aristotle thought that causal knowledge is a superior type of knowledge, the type that characterises science. He took it that there is a sharp distinction between understanding the fact and understanding the *reason* why. The latter type of understanding, which characterises explanation, is tied to finding the causes (*aitia*) of the phenomena. Though both types of understanding proceed via deductive syllogism, only the latter is characteristic of science because only the latter is tied to the knowledge of causes. He illustrated the difference between these two types of understanding by contrasting the following two instances of deductive syllogism:

> (A): Planets do not twinkle; what does not twinkle is near; therefore, planets are near.

> (B): Planets are near; what is near does not twinkle; therefore, planets do not twinkle.

(A), Aristotle says, demonstrates the fact that planets are near, but does *not* explain it, because it does not state its causes. In contrast, syllogism (B) is explanatory because it gives the *reason why* planets do not twinkle: *because* they are near. Explanatory syllogisms like (B) are formally similar to non-explanatory syllogisms like (A). Both are demonstrative arguments of the form: All Fs are Gs; All Gs are Hs; therefore, all Fs are Hs. The difference between them lies in the "middle term" G. In (B), but not in (A), the middle term states a *cause*. As Aristotle says:

> The middle term is the cause, and in all cases it is the cause that is being sought (90a5-10).

To ask why F is H is to look for a causal link joining F and H. More specifically, the search for causes, which for Aristotle is constitutive of science, is the search for middle terms which will link, like a chain, the major premise of an argument with its conclusion: why is F H? Because F is G and G is H. What Aristotle observed was that, besides being demonstrative, explanatory arguments should also be *asymmetric*: the asymmetric relation between causes and effects should be reflected in the relation between the premises and the conclusion of the explanatory arguments — the premises should explain the conclusion and not the other way around.

How is explanatory (that is, causal) knowledge possible? For Aristotle, scientific knowledge forms a tight deductive-axiomatic system whose axioms are *first principles*, being "true and primary and immediate, and more known than and prior to and causes of the conclusion" (71b19-25). Being an empiricist, Aristotle thought that knowledge of causes has experience as its source. But experience on its own cannot lead, through induction, to the first principles: these are universal and necessary and state the ultimate causes. On pain of either circularity or infinite regress, the first principles themselves cannot be demonstrated either. Something besides experience and demonstration is necessary for the knowledge of first principles. This is a process of abstraction based on intuition, a process that reveals the essences of things, that is the properties by virtue of which the thing is *what it is* (cf. 1140b31-1141a8).

Aristotle calls the first principles "definitions". Yet, they are not verbal: they do not just state what words mean; they also state the essences of things. In the example (B) above, it is of the essence of something's being near that it does not twinkle. In the rich Aristotelian ontology, causes, i.e., middle terms of explanatory arguments, are essential properties of their subjects and necessitate their effects. Accordingly, causal explanation is explanation in terms of essences and essential properties, where "the essence of a thing is what it is said to be in respect of itself" (1029b14). He thought that the logical necessity by which the conclusion follows from the premises of an explanatory argument mirrors the physical necessity by which causes produce their effects.

Though Aristotelian explanations are arguments, that is, ultimately, linguistic constructions, Aristotle favoured an *ontic* conception of explanation. This is because he tied explanation to causation: it is the causes that do the explaining. He distinguishes between four types of causes. The material cause is "the constituent from which something comes to be"; the formal cause is "the formula of its essence"' the efficient cause is "the source of the first principle of change or rest"; and the final cause is "that for the sake of which" something happens (194b23-195a3). For instance, the material cause of a statue is its material (e.g., bronze); its formal cause is its form or shape; its efficient cause is its maker; and its final cause is the purpose for which the statue was made. These different types of a cause correspond to different answers to why-questions. But Aristotle thought that, *ceteris paribus*, a complete causal explanation has to cite all four causes (that is, to answer all four why-questions): the efficient cause is the active agent that puts the form on matter for a purpose. The four causes do not explain the same *feature* of the object (e.g., the material cause of the statue — bronze — explains why it is solid, while its formal cause explains why it is only a bust), yet they all contribute to the explanation of the features of the very same object. All four types of cause can be cast as middle terms in proper causal explanations (cf. 94a20-25).

2 DESCARTES: MECHANICAL EXPLANATION

In *Principles of Philosophy* (1644), Descartes expanded on the Aristotelian idea that explanation consists in demonstrations from first principles. But he gave this idea two important twists. The first is that the basic principles are the fundamental rules or laws of nature. The second was the idea that all explanations of natural phenomena is mechanical. Like Aristotle, Descartes thought that explanation amounts to the search of causes. But unlike Aristotle, he thought that all causation is efficient causation and, in particular mechanical. Though Descartes did not fully abandon the rich Aristotelian philosophical framework, (for instance, he too conceived of the world in terms of substances, natures, essences and necessary connections, the latter being, by and large, a priori demonstrable), he thought that the explanation of natural phenomena proceeds by means of mechanical interactions, and not by reference to violent and natural motions; nor in teleological terms. To be sure, he took God to be "the efficient cause of all things" [1985, 202]. But in line with the scholastic distinction between primary cause (God) and secondary causes (worldly things), he claimed that the secondary and particular causes — "which produce in an individual piece of matter some motion which it previously lacked" [1985, 240] — are the laws of nature and the initial conditions viz., the shapes, sizes and speeds of material corpuscles.

Descartes was not a pure rationalist who thought that *all* science could be done a priori. But he was not, obviously, an empiricist either. He did not think that all knowledge stemmed from experience. As he claimed in *Principia*, the human mind, by the light of reason alone, can arrive at substantive truths about the world, concerning mainly the fundamental laws of nature. These, for instance that the total quantity of motion in the world is conserved, are discovered and justified a priori, as they are supposed to stem directly from the immutability of God. Accordingly, the basic structure of the world is discovered independently of experience, is metaphysically necessary and known with metaphysical certainty; for instance, that the world is a plenum with no vacuum (or atoms) in it, that all bodies are composed of one and the same matter, that the essence of matter is extension etc. It is on the basis of these fundamental laws and principles that all natural phenomena are explained, by being deduced from them. Accordingly, Cartesian causal explanation is nomological explanation. More precisely, causal explanation consists in finding nomologically sufficient causes of the effects. In this sense, causal explanations are demonstrative arguments whose premises include reference to laws of nature.

How is then empirical science possible? Descartes thought that once the basic nomological structure of the world has been discovered by the lights of reason, science must use hypotheses and experiments to fill in the details. This is partly because the basic laws of nature place constraints on whatever else there is and happens in the world, without determining it uniquely. The initial conditions (the shapes, sizes and speeds of corpuscles) can only be determined empirically. That is, among the countless initial conditions that God might have instituted, only experience can tell us which he has actually chosen for the actual world. Besides, though grounded in the fundamental laws, the less fundamental laws of physics are not immediately deducible from them. Further hypotheses are needed to flesh them out. Hence, Descartes thought that the less fundamental laws could be known only with moral certainty.

Indeed, Descartes allowed for the possibility that there are competing systems of hypotheses which, though compatible with the fundamental laws, offer different explanations of the phenomena. He illustrated this possibility by reference to an artisan who produced two clocks that indicate the hours equally well, are externally similar and yet work with different internal mechanisms. In light of this possibility, Descartes wavered between two thoughts, which were to become the two standard responses to the argument from underdetermination of theories by evidence. The first (cf. [1985, §44]) is that it does not really matter which of the two competing systems of hypotheses is true, provided that they are both empirically adequate, that is, they correspond accurately to all the phenomena of nature. The other (cf. [1985 §§44 and 205]) is that theoretical virtues such as simplicity, coherence, unity, naturalness etc. are marks of truth in the sense that it would be very unlikely that a theory possesses them and be false. Interestingly, Descartes put a premium on novel predictions: when postulated causes explain phenomena not previously thought of, there is good reason to think they are their true causes.

Explanatory hypotheses, Descartes claimed, must be mechanical, that is cast in terms of "the shape, size, position and motion of particles of matter" [1985, 279], and that the selfsame mechanical principles should deductively explain the whole of nature, both in the heavens and on the earth. It wouldn't be an exaggeration to claim that Descartes advanced an unificationist account of explanation, where the unifiers are the fundamental laws of nature.

Famously, Descartes distinguished all substances into two sorts: thinking things (*res cogitans*) and extended things (*res extensa*). He took the essence of mind to be thought and of matter extension. Unlike Aristotle, he thought that matter was inert (since its essence is that it occupies space). Yet, there are causal connections between bodies (bits of matter) and between minds and bodies (that is, between different substances). Two big questions, then, emerge within Cartesianism. The first is: how is body-body interaction possible? The second is: how is mind-matter interaction possible? Briefly put, Descartes' answer to the first question is the so-called *transference* model of causation: when x causes y a property of x is communicated to y. He thought that this view is an obvious consequence of the principle "Nothing comes from nothing". As he put it:

> For if we admit that there is something in the effect that was not previously present in the cause, we shall also have to admit that this something was produced by nothing. [1985, Vol. 1, 97]

But Descartes failed to explain how this communication is possible. Indeed, by taking matter to be an inert extended substance, he had to retreat to some external cause of motion and change and ultimately to God himself. This retreat to God cannot save the transference model. Besides, the transference model makes an answer to the second question above (how do mind and matter interact?) metaphysically impossible. Being distinct substances, they have nothing in common which can be communicated between them. In a sense, Descartes was a failed interactionist: there is matter-matter and mind-matter causal interaction but there is no clear idea of how it works.

Descartes' successors were divided into two groups: the occasionalists and those who, following Leibniz reintroduced *activity* into nature. Occasionalism is the view that the only real cause of everything is God and that all causal talk which refers to finite substances is a sham. Nicholas Malebranche drew a distinction between real causes and natural causes (or occasions). As he put it:

> A true cause as I understand it is one such that the mind perceives a necessary connection between it and its effect. Now the mind perceives a necessary connection between the will of an infinite being and its effect. Therefore, it is only God who is the true cause and who truly has the power to move bodies. [1674-5/1997, 450]

Natural causes are then merely the occasions on which God causes something to happen. Malebranche pushed Cartesianism to its extremes: since a body's nature is exhausted by its extension, bodies cannot have the power to move anything, and hence to cause anything to happen. What Malebranche also added was that since causation involves a necessary connection between the cause and the effect (a view that Descartes accepted too), and since no such necessary connection is perceived in cases of alleged worldly causation (where, for instance, it is said that a billiard ball causes another one to move), there is no worldly causation: all there is in the world is regular sequences of events, which strictly speaking are not causal. For Malebranche, all causal explanation must ultimately refer to God and his powers.

3 LEIBNIZ: POWERS AND TELEOLOGY

As noted already, Leibniz tried to reintroduce forces and powers into nature. He thought that the rejection by his contemporaries of the scholastic philosophy had gone too far. Though he too favoured mechanical explanations of natural phenomena and denounced occult qualities and virtues as non-explanatory, he thought wrong the key Cartesian thought that the essence of matter was extension. Extension cannot account for the presence of activity in nature. In *Discourse on Metaphysics* (1686), he argued that the essence of substance is activity. He then found it compelling to appeal to substantial forms as the individuating principles that explain the unity of each substance and the variation among different substances. These substantial forms (which he took them to be vital principles analogous to the souls of human bodies), Leibniz thought, were indispensable in metaphysics (especially in teleological explanations of the phenomena) yet they must not be employed in the explanation of particular events [1686, §X]. The latter should proceed by mechanical demonstrations. But Leibniz was not content with the prevailing mechanistic explanations of phenomena. He thought that the mechanical principles of nature need metaphysical grounding and that they should be supplemented by dynamical explanations in terms of forces and powers. Leibniz insisted that every substance is essentially active; since whatever acts is force, every thing is force or a compound of forces. Indeed, Leibnizian substances (what he called "the monads") are sustained by internal "primitive active forces" which cause their subsequent states.

Like Descartes before him, Leibniz too thought that explanation consists in demonstrations from premises that comprise (descriptions of) the fundamental laws of nature. And like Descartes, he thought that the fundamental laws of nature stemmed directly from God. Yet he drew a distinction between the most fundamental law of nature, viz., that nature is orderly and regular, and "subordinate regularities" such as his three laws of motion. "The universal law of the general order", as he put it, is metaphysically necessary, since in whatever way God might have created the world, "it would always have been regular and in a certain order" [1686, §VI]. It is important to note that the three basic Leibnizian laws of motion are conservation laws.[1] Hence, being invariant, they preserve the fundamental order of nature. The subordinate laws are metaphysically *contingent*, since they might well differ in other possible worlds. Yet, Leibniz thought that among all possible worlds, God has created the most perfect one: the actual world

[1] Translated into modern idiom, these laws state: (i) the conservation of *vis viva* in every impact; (ii) the conservation of the directed quantity of motion in every impact; and (iii) the conservation of relative velocity before and after impact.

is the most perfect of all possible worlds, and it is such that it is the simplest in laws and the richest in phenomena. Hence we may reasonably conclude that Leibniz took laws to be the simplest and strongest set of principles that allow the deduction of all phenomena. Though metaphysically contingent, the subordinate laws are physically necessary (since, as Leibniz put it, denying them would imply an imperfection on the part of God [1973, 139]). Under these laws fall others of an even lower level (cf. [1973, 99])

It was a key Leibnizian thought that all (mechanical) laws of nature need metaphysical grounding. As he put it:

> Nature must always be explained mathematically and mechanically, provided it be kept in mind that the principles or the laws of mechanics and of force do not depend upon mathematical extension alone but have certain metaphysical causes. [Letter to Arnauld, 14 July, 1686]

This point is also made in [1686 §XVIII], where he adds that the general principles of corporeal nature and of mechanics are metaphysical rather than mathematical and "belong rather to certain indivisible forms or natures as the causes of the appearances, than to corporeal mass or to extension". Apart from the universal law of the general order, these principles include "the laws of cause, power and activity" [1973, 140]. They are established a priori and, interestingly, their grounding is, ultimately, teleological: they are grounded in the wisdom of God and in particular in his choice of the best possible plan in creating the actual world. The subordinate laws are then the "most fitted to abstract and metaphysical reasons" [1973, 200]. Ultimately all natural laws are explained by means of two central Leibnizian principles: *the principle of sufficient reason* and *the principle of fitness*. According to the principle of sufficient reason, for everything that happens, there must be a reason, sufficient to bring this about instead of anything else. According to the principle of fitness, the actual world is the fittest or most perfect among all possible worlds that God could have created, a fact that "the wisdom of God permits him to know; his goodness causes him to choose it and his power enables him to produce it" [1698, 55].

Leibniz did admit teleological explanations alongside mechanical ones. Apart from the need of teleological explanations (in terms of God's purposes) in metaphysics, he argued that physical phenomena can be explained by mechanical as well as teleological principles. For instance, he claimed that anatomical phenomena can be best explained in terms of goals and that many other physical phenomena (e.g., Snell's law of reflection) can be explained by teleological principles of least effort or least action [1686, §XXII]. Most interestingly, he thought that, though possible, mechanical explanations in terms of matter in motion are useless in historical explanation, where the aims, desires and intentions of historical actors are most relevant to the explanation of their actions (cf. [1686, §XIX]). Indeed, Leibniz wholeheartedly accepted the Aristotelian final causes alongside efficient causes.

He argued that these two distinct kinds of explanation — efficient and final — are reconcilable (cf. [1686, §XXII]). In the end, all things have efficient and final causes. Things have efficient causes when considered as parts of the material world and final causes when considered as substantial forms. Leibniz's reconciliation is effected by means of a third principle he enunciated, the *principle of pre-established harmony* (cf. [1973, 196]). In all its generality, this principle states that when God created this world as the best among an infinity of possible worlds, he put everything in harmony (the monads and the phenomenal world, the mind and the body, the final and the efficient causes). This principle played an important role in explaining how efficient and final causes, or the body and the mind, are co-ordinated with each other. Each domain (the domain of efficient causes and the domain of final causes; the body and the soul) obeys its own laws and does not interact with the other. Hence, they are independent and yet in accord with each other: it is *as if* they influence each other, though they do not.

It is noteworthy that Leibniz rejected the transference account of causation (what he called the "real influx" model), arguing that no impetus or qualities are transferred from one body to another (or between matter and soul). Instead, each body is moved by an innate force. For Leibniz causes are required to explain why objects exist and change and they do this by providing a reason for this existence and change (cf. [1973, 79]). But the reason (and hence the cause) is to be found in the "primitive active force" of each body, viz., in its power of acting and be acted upon (cf. [1973, 81]). Here again, the *principle of pre-established harmony* plays a key role. It guarantees that there will be a perfect agreement between all bodies (substances), thereby "producing the same effect as would occur if these communicated with one another by means of a transmission of species or qualities (...)" [1973, 123]. Thus, the principle secures that there is something like causal order in the world without the existence of real influences among bodies (or among bodies and souls). In light of this, there is a sense in which Leibniz thought that there is no real causation in nature, since Leibnizian substances do *not* interact. Rather, they are co-ordinated with each other by God's act of pre-established harmony, which confers on them the natural agreement of two very exact clocks. So the only real causation admitted by Leibniz is *within* each finite substance (by means of its primitive active force) and in God who pre-establishes the harmony among substances.[2]

[2] A chief difference between Leibniz and occasionalism concerns the role and nature of miracles. Leibniz claimed that occasionalism was unsatisfactory because, by making God the real cause of every event, it introduced a continuous miracle in nature. This move, Leibniz thought, fails to explain anything since God's will is not sufficient to explain anything: Leibniz's God always has a sufficient reason to act. Leibniz's God would obey the laws of nature even when he intervened in nature. If he did not, what reason would he have to impose these laws? Miracles, for Leibniz, are quite rare. Actually, he thought that, strictly speaking, only one miracle ever happened: the creation of the world by God, and his subsequent imposition of the laws of nature. In his argument with the Newtonians, Leibniz attacked Newton with the claim that the Newtonian gravitational force comes down to a perpetual miracle.

4 NEWTON: DYNAMICAL EXPLANATION

The real break with the Aristotelian philosophical and scientific tradition occurred with the consolidation of empiricism in the seventeenth century. Empiricists attacked the metaphysics of essences and the epistemology of rational intuition, innate ideas and infallible knowledge. Sir Isaac Newton's own influence on empiricism was two-fold.

On the one hand, his own scientific achievements, presented in his monumental *Philosophiae Naturalis Principia Mathematica* (*Mathematical Principles of Natural Philosophy*, 1687), created a new scientific paradigm. The previous paradigm, Cartesianism, was overcome. Down with it went the views that space is a plenum, that there are no atoms, that the planets are carried around by vortices, that the quantity of motion (as distinct from momentum) is conserved etc. Newton extended the mechanical view of nature by systematically using the category of *force* alongside the two traditional mechanical categories, *matter* and *motion*. Force was set in a mechanical framework in which it is measured by the *change* in the quantity of motion it could generate. But Newton insisted that his concept of force was mathematical (cf. *Principia*, Book I, Definition VIII). Mechanical interactions were enriched to include attractive and repulsive forces between particles (where again, these forces were considered not physically but mathematically). The concept *mass* was clearly defined for the first time, by being distinguished from weight. Motion and rest were united: they are relative states of a body. Space became the infinite container in which motion of corpuscles takes place. The mechanics of the earth and the heavens were united: a single, mathematically simple, law of gravity governs all phenomena in the universe.

On the other hand, Newton's methodological reflections became the standard reference point for all subsequent discussion concerning the nature and aim of science and its method. Newton demanded certainty of knowledge but rejected the Cartesian route to it. By placing restrictions on what can be known and on what method should be followed, he thought he secured certainty in knowledge. As he explained, he used the term "hypothesis" "to signify only such a proposition as is not a phenomenon nor deduced from any phenomena, but assumed or supposed — without any experimental proof" (cf. [Letters to Cotes, 1713 in Thayer, 1953, 6]). And he proceeded with his famous dictum *Hypotheses non fingo* (I do not feign hypotheses), which was supposed to act as a constraint on what can be known: it rules out all those metaphysical, speculative and non-mathematical hypotheses that aim to explain, or to provide the ultimate ground of, the phenomena. As he said in the *General Scholium*, [*Principia*, Book III]

> For whatever is not deduced from the phenomena is to be called a hypothesis, and hypotheses, whether metaphysical or physical, whether of occult qualities or mechanical, have no place in experimental philosophy.

Newton took Descartes to be the chief advocate of hypotheses of the sort he was keen to deny. His official suggestion for the method of science was that it is deduction from the phenomena. This was contrasted to the hypothetico-deductive method endorsed by Descartes. Newton's approach was fundamentally mathematical-quantitative. He did not subscribe to the idea that knowledge begins with a painstaking experimental natural history of the sort suggested by Francis Bacon in his *Novum Organum* (1620). The basic laws of motion do, in a sense, stem from experience. They are not a priori true; nor metaphysically necessary. The empirically given phenomena that Newton starts with are laws (e.g. Kelper's laws). Then, by means of mathematical reasoning and the basic laws of motion further conclusions can be drawn, e.g., that the inverse square law of gravity applies to all planets.

Undoubtedly, Newton thought that the explanation of natural phenomena consists in finding the most general principles that account for them, where this relation of 'accounting for' is deductive. These general principles are the fundamental laws of nature. As he stated:

> Natural Philosophy consists in discovering the frame and operations of Nature, and reducing them, as far as may be, to general Rules or Laws — establishing these rules by observations and experiments, and thence deducing the causes and effects of things (...).[3]

But his views led to considerable controversy in connection, in particular, with his account of gravity. Leibniz, for instance, denounced Newtonian gravity as being an occult quality. Indeed, as Newton himself claimed: "But hitherto I have not been able to discover the cause of those properties of gravity from phenomena, and I frame no hypotheses" (ibid.).

Newton's thought was that an explanation cannot be faulted on the grounds that it does not unveil the ultimate causes of the phenomena. On the contrary, since explanations must have empirical content, they must be independently testable. Consequently, the employment of general explanatory hypotheses that transcend the limits of what is observed and inductively generalised in laws is futile and has nothing to do with the mathematical principles of natural philosophy. Newton's defence against Leibniz was that, though he had not explained the cause of gravity, he had established that gravity *is* causal (and hence that it can offer adequate causal explanations of the phenomena). As he stressed (*General Scholium, Principia*, Book III):

> And to us it is enough that gravity does really exist and act according to the laws which we have explained, and abundantly serves to account for all the motions of the celestial bodies and of our sea.

And in Query 31 of *Optics*, he noted:

[3]Quoted by Richard Westfall, *Never at Rest* (Cambridge: Cambridge University Press, 1980, 632).

> To tell us that every species of things is endowed with an occult specific quality by which it acts and produces manifest effects is to tell us nothing, but to derive two or three general principles of motion from the phenomena, and afterward to tell us how the properties and actions of all corporeal things follow from those manifest principles, would be a very great step in philosophy, though the causes of those principles were not yet discovered.

Consequently, it suffices for explanation to subsume the phenomena under universal laws, even if the underlying causal mechanisms are not known. In a recent piece, McMullin [2001] has claimed that Newton offered a dynamical account of explanation placed between an agent-causal account (in terms of the powers of agents to produce effects) and a simple law-based account (in terms of subsumption under a law). Though Newton did emphasise the role of laws in explanation, he also stressed that nomological explanation should be unifying: it should subsume all phenomena under a "single sort of underlying causal agency" [2001, 298] — even if, I should add, this underlying causal agency (e.g., gravity) is not further causally explainable.

5 HUME: AGAINST THE METAPHYSICS OF EXPLANATION

All empiricists of the seventeenth century accepted nominalism and denied the existence of universals.[4] This led them to face squarely the problem of induction. Realists about universals, including Aristotle himself who thought that universals can only exist *in* things, could easily accommodate induction. They thought that after a survey of a relatively limited number of instances, the thought ascended to the universal (what is shared in common by these instances) and thus arrived at truths which are certain and unrevisable. This kind of route was closed for nominalists. They had to rely on experience through and through and inductive generalisations based on experience could not yield certain knowledge. This problem came in sharp focus in Hume's work.

[4]From Plato and Aristotle on, many philosophers thought that a number of philosophical problems (the general applicability of predicates, the unity of the propositions, the existence of similarity among particulars, the generality of knowledge and others) required positing a separate type of entity — the *universal* — along side the particulars. Universals are the features that several distinct particulars share in common (e.g. the colour red or the triangular shape). They are the properties and relations in virtue of which particulars are what they are and resemble other particulars. They are also the referents of predicates. Philosophers who are realists about universals take universals to be really there in the world, as constituents of states-of-affairs. Universals are taken to be the repeatable and recurring features of nature. When we say, for instance, that two apples are both red, we should mean, that the very same property (redness) is instantiated by the two particulars (the apples). Redness is a repeatable constituent of things in the sense that the very same redness — *qua* universal — is instantiated in different particulars. Some realists (like Plato) think that there can be uninstantiated universals (the Platonic forms) while others (like Aristotle) argue that universals can only exist when instantiated *in* particulars. Though there are many varieties of nominalism, they all unite in denying that universals are self-subsistence things. For nominalism, only particulars exist.

The subtitle of Hume's ground-breaking *A Treatise of Human Nature* (1739-40) was: *Being an attempt to introduce the experimental mode of reasoning into moral subjects.* This was a clear allusion to Newton's achievement and method. Hume thought that the moral sciences had yet to undergo their own Newtonian revolution. He took it upon himself to show how the Newtonian rules of philosophising were applicable to the moral sciences. All ideas should come from impressions. Experience must be the arbiter of everything. Hypotheses should be looked at with contempt. His own principles of association by which the mind works, resemblance, contiguity and causation, were the psychological analogue of Newton's laws: "they are really *to us* the cement of the universe" [1740, 662].

Hume focused on causation and aimed to dissolve the issue of its metaphysical nature. Before Hume, here is how Malebranche had characterised the metaphysical state of play:

> There are some philosophers who assert that secondary [i.e., worldly] causes act through their matter, figure, motion (...) others assert that they do so through a substantial form; others through accidents or qualities, and some through matter and form; of these some through form and accidents, others through certain virtues or faculties different from the above. (...) Philosophers do not even agree about the action by which secondary causes produce their effects. Some of them claim that causation must not be produced, for it is what produces. Others would have them truly act through their action; but they find such great difficulty in explaining precisely what this action is, and there are so many different views on the matter that I cannot bring myself to relate to them. [1674-5/1997, 659]

As already noted, Malebranche found in this situation a reason to advocate occasionalism. Hume, on the other hand, presenting the situation in a way strikingly similar to the above, found in it a reason to bury the metaphysical issue altogether, to secularise causation completely and to challenge the distinction between causes and occasions. As he put it, there is "no foundation for that distinction (...) betwixt cause and occasion" [1739, 171]. In effect, Hume made the scientific hunt for causes possible, by freeing the concept of causation from the metaphysical chains that his predecessors had used to pin it down. For Hume, causation, as it is in the world, is regular succession of event-types: one thing invariably following another. His *first* definition of causation runs as follows:

> We may define a CAUSE to be 'An object precedent and contiguous to another, and where all the objects resembling the former are plac'd in like relations of precedency and contiguity to those objects, that resemble the latter'. [1739, 170]

Taking a cue from Malebranche, Hume argued that there was no perception of the supposed necessary connection between the cause and the effect. When a sequence of events that is considered causal is observed (e.g., two billiard balls hitting each other and flying apart), there are impressions of the two balls, of their motions, of their collision and of their flying apart, but there is *no* impression of any alleged necessity by which the cause brings about the effect. Hume went one step further. He found totally worthless his predecessors' appeals to the power of God to cause things to happen, since, as he characteristically said, such appeals give us "no insight into the nature of this power or connection" [1739, 249]. So Hume secularised completely the notion of causation.

But Hume faced a puzzle. According to his empiricist theory of ideas, there were no ideas in the mind unless there were prior impressions (perceptions) (cf. [1739, 4]). Yet, he [1739, 77] did recognise that the ordinary concept of causation involved the idea of *necessary connection*. Where does this idea come from, if there is no perception of necessity in causal sequences? Hume argued that the *source* of this idea is the perception of "a new relation betwixt cause and effect": a "constant conjunction" such that "like objects have always been plac'd in like relations of contiguity and succession" [1739, 88]. The perception of this constant conjunction leads the mind to form a certain habit or custom. As he put it:

> after frequent repetition I find, that upon the appearance of one of the objects, the mind is *determin'd* by custom to consider its usual attendant, and to consider it in a stronger light upon account of its relation to the first object [1739, 156].

And he adds:

> 'Tis this impression, then, or *determination*, which affords me the idea of necessity.

So Hume *does* explain the idea of necessary connection in a way consistent with his empiricism. But instead of ascribing it to a feature of the natural world, he takes it to arise from *within* the human mind, when the latter is conditioned by the observation of a regularity in nature to form an expectation of the effect, when the cause is present. Indeed, Hume went on to offer a *second* definition of causation:

> A CAUSE is an object precedent and contiguous to another, and so united with it, that the idea of the one determines the mind to form the idea of the other, and the impression of the one to form a more lively idea of the other. [1739, 170]

Hume took the two definitions to present "a different view of the same object" [1739, 170]. The idea of necessary connection features in none of them. In fact, he thought that he had unpacked the "essence of necessity": it "is something that exists in the mind, not in the objects" [1739, 165]. He went as far as to claim that the supposed objective necessity in nature is *spread* by the mind onto the world [1739, 167].

Hume placed causation firmly within the realm of experience: all causal knowledge should stem from experience. He revolted against the traditional view that the necessity which links cause and effect is the same as the logical necessity of a demonstrative argument. He argued [1739, 86-7] that there can be *no* a priori demonstration of any causal connection, since the cause can be conceived without its effect and conversely. But his far-reaching observation was that the alleged necessity of causal connection cannot be proved empirically either. As he [1739, 89-90] argued, any attempt to show, based on experience, that a regularity that has held in the past *will* or *must* continue to hold in the future will be circular and question-begging. It will presuppose a *principle of uniformity of nature*, viz., a principle that "instances, of which we have had no experience, must resemble those, of which we have had experience, and that the course of nature continues always uniformly the same" [1739, 89]. But this principle is *not* a priori true. Nor can it be proved empirically without circularity. For any attempt to prove it empirically will have to assume what needs to be proved, viz., that since nature has been uniform in the past it *will* or *must* continue to be uniform in the future. This Humean challenge to any attempt to establish the necessity of causal connections on empirical grounds has become known as his *scepticism* about induction. But it should be noted that Hume never doubted that people think and reason inductively. He just took this to be a fundamental psychological fact about human beings (as well as higher animals) which cannot be accommodated within the confines of the traditional conception of Reason, according to which all beliefs should be justified in order to be rational.

A central target of Hume's criticism is the view that causal action (and interaction) is based on the powers that things have. As we have already seen, this view was resuscitated by Leibniz and was, partly, criticised by Newton. Hume spends quite some time trying to dismiss the view that we can meaningfully talk of powers. His *first* move is that an appeal to "powers" in order to understand the idea of necessary connection would be no good because terms such as "*efficacy, force, energy, necessity, connexion*, and *productive quality*, are all nearly synonimous" [1739, 157]. Hence, an appeal to "powers" would offer no genuine explanation of necessary connection. His *second* move is to look at his opponents' theories: Locke's, Descartes', Malebranche's and others'. The main theme of his reaction is that all these theories have failed to show that there are such things as "powers" or "productive forces". In the end, however, Hume's argument was that we "never have any impression, that contains any power or efficacy. We never therefore have any idea of power" [1739, 161]. He endorsed what might be called the *Manifestation Thesis*: there cannot be unmanifestable "powers", i.e., powers which exist, even though there are no impressions of their manifestations. This thesis should be seen as an instance of *Ockham's Razor*: do not multiply entities beyond necessity. For Hume, positing unmanifestable powers would be a gratuitous multiplication of entities, especially in light of the fact that he can explain the origin of our idea of necessity without any appeal to powers and the like.

Hume articulated the principles on which causal explanation should be based.

These are his well-known "rules by which to judge of causes and effects" [1739, 173]. These principles include:

1. The same cause always produces the same effect, and the same effect never arises but from the same cause.

2. Where several different causes produce the same effect, it must be by means of some quality, which we discover to be common amongst them.

3. The difference in the effects of two resembling causes must proceed from that particular, in which they differ.

4. An object, which exists for any time in its full perfection without any effect, is not the sole cause of that effect, but requires to be assisted by some other principle, which may forward its influence and operation.

These principles are grounded in the first one noted above, viz., *same cause, same effect*. This, Hume thought, is an empirical principle derived from experience. The second and the third principles are early versions of Mill's methods of agreement and difference. Hume's point is that causal explanation (and causal knowledge) does not require the backing of a metaphysical theory of causation. It can proceed by means of principles such as the above. He is adamant that these principles are extremely difficult in their application. But this does not imply that they are inapplicable; nor that they do not yield causal knowledge. After all, Hume denied that knowledge requires certainty.

In Hume then we see the first important philosophical step away from the metaphysics of causal explanation and towards the epistemology or methodology of causal explanation. But Hume made possible what has come to be known as the *Humean* view of causation, viz. the *Regularity View of Causation*. According to this, whether or not a sequence of events is causal depends on things that happen elsewhere and elsewhen in the universe, and in particular on whether or not this particular sequence instantiates a regularity.

6 KANT: THE METAPHYSICAL GROUNDS OF EXPLANATION

It was Hume's critique of necessity in nature that awoke Kant from his "dogmatic slumber", as he famously stated. Kant thought that Hume questioned the very possibility of science and took it upon himself to show how science was possible.

Kant rejected strict empiricism (which denied the active role of the mind in understanding and representing the world of experience) and uncritical rationalism (which did acknowledge the active role of the mind but gave it an almost unlimited power to arrive at substantive knowledge of the world based only on the lights of Reason). He famously claimed that although all knowledge starts with experience it does not arise from it: it is actively shaped by the categories of the understanding and the forms of pure intuition (space and time). The mind, as it were, imposes some conceptual structure onto the world, without which no experience could be possible. There was a notorious drawback, however. Kant thought there could be no knowledge of things as they were in themselves (*noumena*) and only knowledge of things as they appeared to us (*phenomena*). This odd combination, Kant thought, might well be an inevitable price one has to pay in order to defeat empiricist scepticism and to forgo traditional idealism. Be that as it may, his master thought was that some synthetic a priori principles should be in place for experience to be possible. And not just that! These synthetic a priori principles (e.g., that space is Euclidean, that every event has a cause, that nature is law-governed, that substance is conserved, the laws of arithmetic) were necessary for the very possibility of science and of Newtonian mechanics in particular.

Like Hume before him, Kant does not claim that reason alone can discover the connection between any specific cause and any specific effect, nor understand its necessity (cf. [1787, A195; B240-41]). He agrees with Hume that particular causal connections can be discovered only empirically. But unlike Hume, Kant *denies* that the concept of causation arises from experience and in particular that it arises in the same way as does the knowledge of the causes of particular events. In his *Second Analogy of Experience*, Kant tried to demonstrate that the principle of causation, viz., "everything that happens, that is, begins to be, presupposes something upon which it follows by rule", is a precondition for the very possibility of objective experience. He took the principle of causation to be required for the mind to make sense of the temporal irreversibility that there is in certain sequences of impressions. This temporal *order* by which certain impressions appear can be taken to constitute an objective happening *only if* the later event is taken to be necessarily determined by the earlier one (i.e., to follow by rule from its cause). For Kant, objective events are not 'given': they are constituted by the organising activity of the mind and in particular by the imposition of the principle of causation on the phenomena. Consequently, the principle of causation is, for Kant, a synthetic a priori principle.

In *Metaphysical Foundations of Natural Science* (1786), Kant claimed that

> Only that whose certainty is apodeictic can be called science proper; cognition that can contain merely empirical certainty is only improperly called science.

Besides, natural science proper relies on laws that are known a priori and hold with necessity (they are not merely laws of experience). Kant thought all natural science should derive its legitimacy from its pure part, i.e., the part that contains "the a priori principles of all remaining natural explications". He took as his task to show that these a priori principles of pure natural science are certain and necessary for the very possibility of science and experience. This, he thought, was the task of the metaphysics of nature. Unlike Newton, Kant thought that there could not be proper science without metaphysics. Yet, his own understanding of metaphysics was in sharp contrast with that of his predecessors (Leibniz's in particular). Metaphysics, Kant thought, was a science, and in particular *the science of synthetic a priori judgements.* Mathematics was taken to be the key element in the construction of natural science proper: without mathematics no doctrine concerning determinate natural things was possible. The irony, Kant thought, was that though many past thinkers (and Newton in particular) repudiated metaphysics and had relied on mathematics in order to understand nature, they failed to see that this very reliance on mathematics made them unable to dispense with metaphysics. For, in the end, they had to treat matter in abstraction from any particular experiences. They postulated universal laws without inquiring into their a priori sources.

As Kant argued in *Critique of Pure Reason* (1781), the a priori source of the universal laws of nature was the transcendental principles of pure understanding. These constitute the object of knowledge in general. Thought (that is, the understanding) imposes upon objects in general certain characteristics in virtue of which objects become knowable. The phenomenal objects are constituted as objects of experience by the schematised categories of quantity, quality, substance, causation and community. If an object is to be an object of experience, it must have certain necessary characteristics: it must be extended; its qualities must admit of degrees; it must be a substance in causal interaction with other substances. In his three Analogies of Experience, Kant tried to prove that three general principles hold for all objects of experience: that substance is permanent; that all changes take place in conformity with the law of cause and effect; that all substances are in thoroughgoing interaction. These are synthetic a priori principles that make experience possible. They are imposed a priori by the mind on objects.

Yet, these transcendental principles make no reference to any experienceable objects in particular. It was then Kant's aim in *Metaphysical Foundations of Natural Science* to show how these principles could be concretised in the form of laws of matter in motion. These were metaphysical laws in that they determined the possible behaviour of matter in accordance with mathematical rules. Kant thus enunciated the law of conservation of the quantity of matter, the law of inertia and the law of equality of action and reaction and thought that these laws were the mechanical analogues (cases *in concreto*) of his general transcendental principles. They determine the pure and formal structure of motion, where motion is treated purely mathematically *in abstracto*. It is no accident, of course, that the last two of these laws are akin to Newton's law and that the first law was presupposed by Newton too. Kant's metaphysical foundations of (the possibility of) matter in motion were precisely meant to show how Newtonian mechanics was possible. But Kant also thought that there are physical laws that are discovered empirically. Though he held as a priori true that matter and motion arise out of repulsive and attractive forces, he claimed that the particular force-laws, even the law of universal attraction, can only be discovered empirically.

His predecessors, Kant thought, had failed to see this hierarchy of laws that make natural science possible: transcendental laws that determine the object of possible experience in general; metaphysical laws that determine matter in general and physical laws that fill in the actual concrete details of motion. Unlike the third kind, the first two kinds of law require a priori justification and they are necessarily true.

Overall, then, Kant was mostly concerned with the metaphysical foundations of causal explanation, viz., that causal explanation presupposes necessary connections. Given that causation is nomological, Kant's thought amounted to the claim that all causal explanation is nomological explanation. But, especially towards the end of *Critique*, he highlighted another important dimension of explanation, viz., unification. He claimed it to be a "regulative idea" that nature is unified and uniform. He took it that reason aims to systematise its body of knowledge, i.e., "to exhibit the connection of its parts in conformity with a single principle" [A645/B673]. It is this systematic unity of knowledge that shifts it from being "a mere contingent aggregate" to being "a system connected according to necessary laws". This "systematic unity of knowledge" is "*the criterion of the truth* of its rules" [A647/675]. As an example of this, he offered the subsumption of more specific (causal) powers under more fundamental powers. This subsumption, he thought, "claims to have an objective reality, as postulating the systematic unity of the various powers of a substance (...) [A650/B678]. This, to be sure, is a regulative idea (an idea of Reason) and not a principle constitutive of experience. Still, as he put it:

> In all such cases reason presupposes the systematic unity of the various powers, on the ground that special natural laws fall under more general laws, and that parsimony in principles is not only an economical requirement of reason, but is one of nature's own laws. [A650/B678]

By calling "regulative" the idea that nature has an objectively valid and necessary systematic unity (cf. [A651/B679]), he wanted to stress that it is indemonstrable. Yet, without it, Kant thought, there would be no criterion of empirical truth. Besides, it can be confirmed in view of their empirical success in science (cf. [A661/B689]). Then, unification of all phenomena under universal laws of nature emerges as both the ultimate goal of the explanation of natural phenonema and as the criterion for truth. Besides, for Kant, unification confers necessity on certain principles, thereby rendering laws of nature (cf. [Kitcher, 1986]).

Though philosophically impeccable, Kant's architectonic suffered severe blows in the nineteenth and the early twentieth centuries. The blows came, by and large, from science itself. The crisis of the Newtonian mechanics and the emergence of the special and the general theories of relativity, the emergence of non-Euclidean geometries and their application to physics, Gottlob Frege's claim that arithmetic, far from being synthetic a priori, was a body of analytic truths and David Hilbert's arithmetisation of geometry which proved that no intuition was necessary created an explosive mixture that, in the end of a long process, led to the collapse of the Kantian synthetic a priori principles.

7 MILL: EXPLANATION AS UNIFICATION

In his monumental *A System of Logic Ratiocinative and Inductive* (1843), Mill defended the Regularity View of Causation, with the sophisticated addition that in claiming that an effect invariably follows from the cause, the cause should not be taken to be a single factor, but rather the whole conjunction of the conditions that are sufficient and necessary for the effect. For Mill, regular association (or invariable succession) is not sufficient for causation. An invariable succession of events is causal only if it is "unconditional", that is only if its occurrence is *not* contingent on the presence of further factors which are such that, given their presence, the effect would occur even if its putative cause was not present. A clear case in which unconditionality fails is when the events that are invariably conjoined are, in fact, effects of a common cause.

The problem that Mill faced was that there are regularities that are not causal and do not constitute laws. For instance, as Thomas Reid noted, the night always follows the day, but it is not caused by the day. They are both caused by the rotation of the earth around the sun. A similar problem arose in connection with Kant's theory of causation. Arthur Schopenhauer charged Kant with showing the absurd result that all sequence is consequence. As he noted, the tones of a musical composition follow each other in a certain objective order and yet it would be absurd to say that they follow each other according to the law of causation. The problem was that both Hume and Kant seemed to have ended up with a *loose* notion of causation. It was in order to strengthen the concept of causation that Mill introduced the idea of unconditionality. Ultimately, Mill took to be causal those invariable successions that are unconditional. It is these regularities that constitute laws of nature.

Considering how to answer the central problem of "how to ascertain the laws of nature", Mill [1843, 207] noted:

> According to one mode of expression, the question, What are the laws of nature? may be stated thus: What are the fewest and simplest assumptions, which being granted, the whole existing order of nature would result? Another mode of stating it would be thus: What are the fewest general propositions from which all the uniformities which exist in the universe might be deductively inferred?

Mill [1843, 208] was adamant that he was defending a view of laws as regularities:

> for the expression, Laws of Nature, *means* nothing but the uniformities which exist among natural phenomena [...] when reduced to their simpler expression.

Mill's breakthrough (prefigured by Kant, as we have seen) was that the issue of characterising what the laws of nature are cannot be dealt with by looking at individual regularities and by trying to identify when an individual regularity is a law. Rather, it should be dealt with by looking at how the laws form a "web composed of individual threads" (ibid.). "The study of nature", Mill suggested, "is the study of laws, not *a* law; of uniformities in the plural number" (ibid.)

Borrowing Mill's expression, we may call his view 'the web of laws' approach.

What Mill perceived was that there could be no adequate characterisation of the distinction between laws of nature and merely accidentally true generalisations, unless we adopted a holistic view of lawhood. Laws are those *regularities* which are members of a coherent system of regularities, in particular, a system which can be represented as a deductive axiomatic system striking a good balance between *simplicity* and *strength*. As we shall see in section 12, Mill's approach resurfaced in the twentieth century in the writings of Ramsey and Lewis. But we have already seen that it is was a common approach in the age that Mill wrote. Leibniz and Kant held versions of it. Mill's radical twist was that he did not thereby thought that laws are rendered necessary. Nor that, at bottom, laws are anything other than regularities.

Mill was a thoroughgoing inductivist, who took all knowledge to arise from experience through induction. He even held that the law of universal causation, viz., that for every event there is a set of circumstances upon which it is invariably and unconditionally consequent, is an inductively established — and true — principle. Hence, Mill denied that there could be any certain and necessary knowledge.

He should also be credited with the first attempt to articulate the *deductive-nomological model* of explanation, which became prominent in the twentieth century. As he put it:

> An individual fact is said to be explained by pointing out its cause, that is, by stating the law or laws of causation of which its production is an instance. [1843, 305]

Similarly,

> a law of uniformity in nature is said to be explained when another law or laws are pointed out, of which that law is but a case, and from which it could be deduced. (ibid.)

The explanatory pattern that Mill identified is deductive, since the *explananda* (be they individual events or regularities) must be deduced from the *explanans*. And it is nomological, since the *explanans* must include reference to laws of nature. Mill went on to distinguish three patterns within this broad deductive-nomological framework. *First*, an explanation consists in the isolation of the several laws that contribute to the production of a complex effect; more accurately, the law governing a certain effect is explained by being analysed into separate laws that govern its causes. We can call this the *de-compositional pattern* of scientific explanation. Mill frames this pattern of explanation is terms of tendencies. He notes:

> The first mode, then, of explanation of Laws of Causation, is when the law of an effect is resolved into the various tendencies of which it is the result, together with the laws of those tendencies. [1843, 306]

It may then appear that Mill offers a dispositional account of de-compositional explanation. But this is misleading. Mill introduced tendencies to save the universality of laws. He observed that "[a]ll laws of causation are liable to be (...) counteracted, and seemingly frustrated, by coming into conflict with other laws, the separate results of which is opposite to theirs, or more or less inconsistent with it" [1843, 292]. He then restored the universality of laws by claiming that laws describe tendencies: "All laws of causation, in consequence of their liability to be counteracted, require to be stated in words affirmative of tendencies only, and not of actual results" [1843, 293]. So, "all heavy bodies *tend* to fall; and to this there is no exception (...)" [1843, 294]. But Millian tendencies are occurrent qualities.[5] These tendencies are present and manifested even when the laws that govern them are counteracted by other laws. So Mill spoke as if the full effects of two separate causes actually occur and are fused into the resultant.

The *second* explanatory pattern consists in finding the complete causal history of the *explanandum*, as when intermediary causal links between the cause and the effect are found out. Mill thought that this pattern amounts to resolving the law that connects the distal cause and the effect into laws that connect the distal causes with the proximate ones and the proximate causes with the effect. It might be plausibly claimed that this second pattern amounts to *mechanistic* explanation, viz., explanation in terms of the mechanisms through which a cause brings about the effect.

The third pattern of explanation is *unification*:

> The subsumption (...) of one law under another, or (what comes to the same thing) the gathering up of several laws into one more general law which includes them all. [1843, 309]

Unification, according to Mill, is the hallmark of explanation and of laws. Unification is explanatory not because it renders the *explananda* less mysterious than they were before they were subsumed under a law, but because it minimises the number of laws that we take as ultimately mysterious, that is, as inexplicable. This very process of unification, Mill thought, brings us nearer to solving the problem of what the laws of nature are. And, as we have seen, this is no other than the problem of

> What are the fewest general propositions from which all the uniformities existing in nature could be deduced? [1843, 311]

Unification underpins the distinction between fundamental laws of nature (the basic unifiers) and derivative laws. Derivative are those laws that can be resolved into more fundamental ones either by the first or by the second pattern of scientific explanation (cf. [1843, 343]). So de-compositional and mechanistic patterns

[5] Though Mill talks of capacities (dispositions), he takes it that they are not real things existing in objects. He takes dispositional predicates to be 'names' for the claim that objects "will act in a particular manner when certain new circumstances arise" [1843, 220-1]. Mill takes it that dispositions are reducible to the categorical properties of objects.

of explanation are means for the unification pattern, which is the ultimate pattern of explanation. Though, Mill thought, all patterns of explanation are deductive, only the first two are, strictly speaking, patterns of *causal* explanation. Unification is not causal in the sense that the more fundamental laws do *not* cause the less fundamental ones. Rather, they subsume them under them, which amounts to saying that the less fundamental laws are instances, or cases, of the most fundamental ones (cf. [1843, 311]). But, of course, the unified nomological structure of the world captures its causal structure in the sense that the causal structure of the world (viz., the structure of causal laws) is exhausted by its nomological structure.

Like Kant, then, Mill put a premium on unification. It is via unification that regularities are rendered laws of nature. It is via unification that the causal structure of the world is known. But, unlike Kant, he denied that unification confers necessity on laws of nature.

Mill is also famous for his methods by which causes can be discovered. These are known as the *Method of Agreement* and the *Method of Difference.* Briefly put, according to the first, the cause is the common factor in a number of otherwise different cases in which the effect occurs. According to the second, the cause is the factor that is different in two cases, which are similar except that in the one the effect occurs, while in the other doesn't. In effect, Mill's methods encapsulate what is going on in a controlled experiment: we find causes by creating circumstances in which the presence (or the absence) of a factor makes the only difference to the production (or the absence) of an effect. Mill, however, was adamant that his methods (and the scientific method in general) work *only if* certain metaphysical assumptions are already in place. It must be the case that: a) events have causes; b) events have a *limited* number of possible causes; c) same causes have same effects, and conversely; and d) the presence or absence of causes makes a difference to the presence or absence of their effects.

PART II: UNDERSTANDING EXPLANATION

8 THE LOGICAL POSITIVIST LEGACY

The Logical Positivists took Hume to have offered a *reductive* account of causation: one that frees talk about causation from any commitments to a necessary link between cause and effect. Within science, Carnap stressed, "causality means nothing but a functional dependency of a certain sort" [1928, 264]. The functional dependency is between two states of a system, and it can be called a "causal law" if the two states are in temporal proximity and one precedes the other in time. Schlick [1932, 516] expressed this idea succinctly by pointing out that

> the difference between a mere temporal sequence and a causal sequence is the regularity, the uniformity of the latter. If C is *regularly* followed

> by E, then C is the cause of E; if E only 'happens' to follow C now and then, the sequence is called mere chance.

Any further attempt to show that there was a necessary "tie" between two causally connected events, or a "kind of glue" that holds them together, was taken to have been proved futile by Hume, who maintained that "it was impossible to discover any 'impression' of the causal nexus" [Schlick 1932, 522]. The twist that Logical Empiricists gave to this Humean argument was based on their verifiability criterion of meaning: attributing, and looking for, a "linkage" between two events would be tantamount to "committing a kind of nonsense" since all attempts to verify it would be necessarily futile (cf. [Schlick, ibid.]).

For the Logical Positivists, the concept of causation is intimately linked with the concept of law. And the latter is connected with the concept of regular (exceptionless) succession. Given that the reducing concept of regular succession is scientifically legitimate, the reduced concept of causation becomes legitimate, too. Yet, regular succession (or correlation) does not imply causation. How could Schlick and Carnap have missed this point?

The following thought is available on their behalf. The operationalisation of the concept of causation they were after was not merely an attempt to legitimise the concept of causation. It was part and parcel of their view that science aims at *prediction*. If prediction is what *really* matters, then the fact that there can be regularities, which are not causal in the ordinary sense of the word, appears to be irrelevant. A regularity can be used to predict a future occurrence of an event irrespective of whether it is deemed to be causal or not. Correlations can serve prediction, even though they leave untouched some intuitive aspect of causation, according to which not all regularities are causal.

This idea is explicitly present in Schlick [1932]. Carnap too noted that "causal relation means predictability" [1974, 192]. But he was much more careful than Schlick in linking the notion of predictability — and hence, of causation — with the notion of the law of nature. For not all predictions are equally good. Some predictions rely on laws of nature, and hence are more reliable than others which rely on "accidental universals" [1974, 214]. For Carnap, causation is not *just* predictability. It is more akin to subsumption under a universal regularity, i.e., a law of nature. As he [1974, 204] stressed:

> When someone asserts that A caused B, he is really saying that this is a particular instance of a general law that is universal with respect to space and time. It has been observed to hold for similar pairs of events, at other times and places, so it is assumed to hold for any time and place.

It seems reasonable to argue that what Carnap was really after was the connection between causation and *explanation*. When we look for explanations, as opposed to predictions, we look for something more than regularity, and relations of causal dependence might well be what we look for. The thought suggests itself that what distinguishes between a causal regularity and a mere predictive one is their different roles in explanation. It appears, then, that the concept of explanation, and in particular of nomic explanation, is the main tool for an empiricist account of causal dependence.

9 NOMIC EXPECTABILITY

What is it to explain a singular event e, e.g., the explosion of a beer-keg in the pub's basement? The intuitive answer would be to provide the cause of this event: whatever brought about its occurrence. But is it enough to just cite another event c, e.g., the rapid increase of temperature in the basement, in order to offer an adequate explanation of e? Explanation has to do with *understanding*. An adequate explanation of event e (that is, of why e happened) should offer an adequate understanding of this happening. Just citing a cause would not offer an adequate understanding, unless it was accompanied by the citation of a law that connects the two events. According to Hempel [1965] the concept of explanation is primarily *epistemic*: to explain an event is to show how this event would have been *expected* to happen, had one taken into account the laws that govern its occurrence, as well as certain initial conditions. If one expects something to happen, then one is not surprised when it happens. Hence, an explanation amounts to the removal of the initial surprise that accompanied the occurrence of the event e. *Nomic expectability* is the slogan under which Hempel's account of explanation can be placed.

Hempel systematised a long philosophical tradition, going back at least to Mill, by explicating the concept of explanation in terms of his *Deductive-Nomological* model (henceforth, *DN*-model). A singular event e (the *explanandum*) is explained if and only if a description of e is the conclusion of a valid deductive argument, whose premises, the *explanans*, involve essentially a lawlike statement L, and a set C of initial or antecedent conditions. The occurrence of the *explanandum* is thereby subsumed under a natural law. Schematically, to offer an explanation of an event e is to construct a valid deductive argument of the following form:

(*DN*)
Antecedent/Initial Conditions $C_1, \ldots, C_i$
Lawlike Statements $L_1, \ldots, L_j$
____
Therefore, e event/fact to be explained (*explanandum*)

A *DN*-explanation is a special sort of a valid deductive argument — whose logical form is both transparent and objective — and conversely, the species of valid deductive arguments that can be *DN*-explanations can be readily circumscribed, given only their form: the presence of lawlike statements in the premises is the characteristic that marks off an explanation from other deductive arguments. Hempel codified all this by offering 4 conditions of adequacy for an explanation.

Conditions of adequacy:

1. The argument must be deductively valid.

2. The *explanans* must contain essentially a lawlike statement.

3. The *explanans* must have empirical content, i.e., they must be confirmable.

4. The *explanans* must be true.

The first three conditions are called "logical" by Hempel [1965, 247], because they pertain to the *form* of the explanation. The fourth condition is "empirical". Hempel rightly thought that it was an empirical matter whether the premises of an explanation were true or false. He called a *DN*-argument that satisfies the first three conditions a "potential explanation", viz., a valid argument such that, if it were also sound, it would explain the *explanandum*. He contrasted it with an "actual explanation", which is a sound *DN*-argument. The fourth condition is what separates a potential from an actual explanation. The latter is the correct, or the true, explanation of an event. With the fourth condition, Hempel separated what he took it to be the issue of "the logical structure of explanatory arguments" [1965, 249, note 3] from the empirical issue of what is the correct explanation of an event. But the structure of an explanatory argument cannot be purely logical. Indeed, if the issue of whether an argument was a potential explanation of an event was purely logical, it would be an *a priori* matter to decide that it was a potential explanation. But condition three shows that this cannot be a purely *a priori* decision. Without empirical information about the kinds of predicates involved in a lawlike statement, we cannot decide whether the *explanans* have empirical content.

Sometimes, the reference to laws in an explanation is elliptical and should be made explicit. Or, the relevant covering laws are too obvious to be stated. Hempel thought that a proper explanation of an event should use laws, and that unless it uses laws it is, in some sense, defective. It's no accident that the *Deductive-Nomological* model became known as "the covering law model" of explanation. Subsumption under laws is the hallmark of Hempelian explanation. So, a lot turns on what exactly the laws of nature are and this has proved to be a very sticky issue. Without a robust distinction between laws and accidents, the *DN*-model loses most of its putative force as a correct account of explanation.

Hempel took his model to provide the correct account of *causal explanation*. As he put it: "causal explanation is a special type of deductive nomological explanation" [1965, 300]. Let us call this the *Basic Thesis* (*BT*):

(*BT*)

All causal explanations of singular events can be captured by the Deductive-Nomological model.

10 ENTER CAUSATION

It has been a standard criticism of the *Deductive-Nomological* model that, insofar as it aims to offer sufficient and necessary conditions for an argument to count as a *bona fide* explanation, it fails. There are arguments that satisfy the structure of the *DN*-model, and yet fail to be *bona fide* explanations of a certain singular event. Conversely, there are *bona fide* explanations that fail to instantiate the *DN*-model. In what follows, we shall examine the relevant counterexamples and try to see how a Hempelian might escape from them. To get a clear idea of what they try to show, let me state their intended moral in advance: the *DN*-model fails precisely because it leaves out of the explication of the concept of explanation important considerations about the role of *causation* in explanation.

The first class of counter-examples, which aim to show that the *DN*-model is insufficient as an account of explanation, are summarised by the famous flagpole-and-shadow case. Suppose that we construct a *DN*-explanation of why the shadow of a flagpole at noon has a certain length. Using the height of the pole as the initial condition, and employing the relevant nomological statements of geometrical optics (together with elementary trigonometry), we can construct a deductively valid argument with a statement of the length of the shadow as its conclusion. But as Sylvain Bromberger [1966] observed, we can reverse the order of explanation: we can 'explain' the height of the flagpole, using the very same nomological statements, but (this time) the length of the shadow as the initial condition. Surely, this is not a *bona fide* explanation of the height of the pole, although it satisfies the *DN*-model. It is not a *causal* explanation of the height of the pole: although the height of the pole is the *cause* of its shadow at noon, the shadow does not cause the flagpole to have the height it does.

This counter-example can be generalised by exploiting the functional character of some lawlike statements in science: in a functional law, we can calculate the values of each of the magnitudes involved in the equation that expresses the law by means of the others. Given some initial values for the 'known' magnitudes, we can calculate, and hence '*DN*-explain', the value of the 'unknown' magnitude. Suppose, for instance, that we want to explain the period T of a pendulum. This relates to its length l by the functional law: $T = 2\pi\sqrt{l/g}$. We can construct a *DN*-argument whose conclusion is some value of the period T and whose premises are the above law-statement together with some value l of the length as our initial condition. Suppose, instead, that we wanted to explain the length of the pendulum. We could construct a *DN* argument similar to the above, with the length l as its conclusion, using the very same law-statement but, this time, conjoined with a value of the period T as our initial condition. If, in the former case, it is straightforward to say that the length of the pendulum *causes* it to have a period of a certain value, in the latter case, it seems problematic to say that the period causes the pendulum to have the length it does.

Put in more abstract terms, the *Deductive-Nomological* model allows explanation to be a symmetric relation between two statements, viz., the statement that expresses the cause and the statement that expresses the effect. So given the relevant nomological statements, an effect can *DN*-explain the cause and conversely. If we take causation to be an asymmetric relation, the *DN*-model seems unable fully to capture the nature of causal explanation, despite Hempel's contentions to the contrary.

If we wanted to stick to the *DN*-account of explanation and its concomitant claim to cover *all* causal explanation, if, that is, we wanted to defend the *Basic Thesis* (*BT*), what sort of moves would be available?

The counterexamples we have seen so far do not contradict the *Basic Thesis*. They contradict the converse of *BT*, a thesis that might be called (+):

> (+)
>
> All Deductive-Nomological explanations of singular events are causal explanations.

But neither Hempel nor his followers endorse (+). He fully accepted the existence of non-causal *DN*-explanations of singular events (cf. [1965, 353]).[6] The counterexamples do not dispute that causal explanation is a subset of *DN*-explanation. What they claim is that the *DN*-model licenses apparently inappropriate applications of the *DN*-pattern. This claim does *not* contradict *BT*. Still, the above counterexamples do show something important, viz., that unless causal considerations are imported into *DN*-explanatory arguments, they fail to distinguish between legitimate (because causal) and illegitimate (because non-causal) explanations. The task faced by the defender of the *DN*-model is to show what could be added to a *DN*-argument to provide legitimate (causal) explanations. Schematically put, we should look for an extra X such that *DN*-model + X = causal explanation. What could this X be?

One move, made by Hempel [1965, 352] is to take X to be supplied by the law-statements that feature in a *DN*-explanation. To this end, Hempel relied on a distinction, already drawn by Mill, between *laws of coexistence* and *laws of succession*. A *law of co-existence* is the type of law in which an equation links two or more magnitudes by showing how their values are related to one another. Laws of co-existence are *synchronic*: they make no essential reference to time (i.e., to how a system or a state evolves over time); they state how the relevant magnitudes relate to each other at any given time. The law of the pendulum, Ohm's law and the laws of ideal gases are relevant examples. A *law of succession* describes how the state of a physical system changes over time. Galileo's law and Newton's second law would be relevant examples. In general, laws of succession are described by differential equations. Given such an equation, and some initial conditions, one can calculate the values of a magnitude over time. Laws of co-existence display a kind of symmetry in the dependence of the magnitudes involved in them, but

[6]Mathematical explanation is a clear case of non-causal explanation; as is the case in which one explains why an event happened by appealing to conservation laws, or to general non-causal principles (such as Pauli's exclusion principle).

laws of succession do not. Or, at least, they are not symmetric given that *earlier* values of the magnitude determine, via the law, *later* values.

Hempel [1965, 352] argued that only laws of succession could be deemed causal. Laws of co-existence cannot. They do not display the time-asymmetry characteristic of causal laws. Note that the first type of counter-examples to the *DN*-model, where there is explanatory symmetry but causal asymmetry, involves laws of co-existence. In such cases, the explanatory order can be reversed. If these laws are *not* causal, then there is no problem: there is no causally relevant feature of these laws which is not captured by relations of explanatory dependence. So, the extra X that should be added to a *DN*-argument in order to ensure that it is a causal explanation has to do with the asymmetric character of some laws. Only asymmetric laws are causal, and can issue in causal explanations. *DN*-explanation + asymmetric laws (of succession) = causal explanation.

There seems to be something unsatisfactory in Hempel's reply. For, the thought will be, we do make causal ascriptions, even when laws of co-existence are involved. It was, after all, the *compression* of the gas that caused its pressure to rise, even though pressure and volume are two functionally dependent variables related by a law of co-existence. This seems to be a valid objection. However, the following answer is available to someone who wants to remain Hempelian, due basically to von Wright [1973]. Strictly speaking, when laws of co-existence are referred to in a *DN* explanatory argument, the explanation can be symmetric: we can explain the values of magnitude A by reference to the values of magnitude B, and conversely. But, Hempel's defender might go on, in particular *instances* of a *DN* explanatory argument with a law of co-existence, this symmetry can be (and is) broken. How the symmetry is broken — and, hence how the direction of explanation is determined — depends on which of the functionally dependent variables is actually *manipulated.*

When laws of co-existence are involved, the symmetry that *DN* explanations display can be broken in different ways in order to capture what causes what on the particular occasion. An appeal to manipulability can also show how Bromberger-type counter-examples can be avoided. A *DN*-model which cites the length of the shadow as the explanation of the height of the flagpole should not count as a *bona fide* explanation. Although the length of the shadow and the height of the pole are functionally inter-dependent, only the height of the pole is really manipulable. One can create shadows of any desirable length by manipulating the heights of flagpoles, but the converse is absurd. Manipulability can then be seen as the sought after supplement X to the *DN*-model which determines what the causal order is across different symmetric contexts in which a *DN*-argument is employed. *DN*-explanation (with functional laws) + manipulability = causal explanation.

Yet, the concept of manipulation is causal. This means that an advocate of *DN*-explanation who summons von Wright's help can at best have a Pyrrhic victory. For she is forced to employ irreducible causal concepts in her attempt to show how a *DN*-model of explanation can accommodate the intuitive asymmetry that explanatory arguments can possess.

The popular philosophical claim that the *DN*-model leaves important causal considerations out of the picture is supported by a second class of counter-examples. These aim to show that satisfaction of the *DN*-model is not a necessary condition for *bona fide* causal explanations. In fact, these counterexamples aim directly to discredit *BT*. Remember that *BT* says, in effect, that the claim that c causes e will be elliptical, unless it is offered as an abbreviation for a full-blown *DN*-argument. This view has been challenged by Michael Scriven. He made this point by the famous example of the explanation of the ink-stain on the carpet. Citing the fact that the stain on the carpet was *caused* by inadvertently knocking over an ink-bottle from the table, Scriven [1962, 90] argues,

> is the explanation of the state of affairs in question, and there is no nonsense about it being in doubt because you cannot quote the laws that are involved, Newton's and all the others.

His point is that there can be fully legitimate *causal* explanations that are not *DN*. Instead, they are causal stories, i.e., stories that give causally relevant information about how an effect was brought about, without referring to any laws, and without having the form of a deductive argument. Collaterally, it has been a standard criticism of the Hempelian model that it wrongly makes all explanations to be arguments. A main criticism is that citing a causal mechanism can be a legitimate explanation of an event without having the form of a Hempelian deductive-nomological argument (cf. [Salmon 1989, 24].

One can accept Scriven's objection without abandoning the *Deductive-Nomological* model of explanation; nor the *Basic Thesis.* The fact that the relevant nomological connections may not be fully expressible in a way that engenders a proper deductive explanation of the *explanandum* merely shows that, on some occasions, we shall have to make do with what Hempel called "explanation sketches" instead of full explanations. Explanation-sketches can well be ordinary causal stories that, as they stand, constitute incomplete explanations of an event E. But these stories can be completed by taking account of the relevant laws that govern the occurrence of the event E. Scriven's point, however, seems to be more pressing. It is that a causal explanation can be *complete*, without referring to laws (cf. [1962, 94]). So, he directly challenges Hempel's assumption that all causal explanation *has to* be nomological. Scriven insists that explanation is related to understanding and that the latter might, but won't necessarily, involve reference to laws. He proposes [1962, 95]:

> a causal explanation of an event of type [E], in circumstances [R] is exemplified by claims of the following type: there is a comprehensible cause [C] of [E] and it is understood that [C]s can cause [E]s.

But, a Hempelian might argue, it is precisely when we move to the nomological connection between Cs and Es that we understand how Cs can cause Es.

One important implication of the *Deductive-Nomological* model is that *there is no genuine singular causal explanation* (cf. [Hempel 1965, 350 & 361-2]). Scriven's own objection can be taken to imply that a singular causal explanation of an event-token (e.g., the staining of the carpet by ink) is a complete and fully adequate explanation of its occurrence. Since the *DN*-model denies that there can be legitimate singular *causal* explanations of events, what is really at stake is whether causal stories that are not nomological can offer legitimate explanation of singular events. What, then, is at stake is the *Basic Thesis.*

Note that there is an ambiguity in the singularist approach. What does it mean to say that there is *no* nomological connection between two event-tokens c and e that are nonetheless such that c causally explains e? It might mean one of the following two things: (i) there are no relevant event-types under which event-tokens c and e fall such that they are nomologically connected to each other; (ii) even if there is a relevant law, we don't (can't) know it; nor do we have to state it explicitly in order to claim that the occurrence of event-token c causally explains the occurrence of event-token e.

The first option is vulnerable to the following objection. One reason why we are interested in identifying causal facts of the form 'c causes e' (e.g., heating a gas at constant pressure causes its expansion) is that we can then *manipulate* event-type C in order to bring about the event-type E. But the possibility of manipulation requires that *there is* a nomological connection between types C and E. It is this nomological connection that makes possible bringing about the effect e, by manipulating its causes. Hence, if causation is to have any bite, it had better instantiate laws.[7] So the singularist's assertion should be interpreted to mean the second claim above, viz., that even if there is a law connecting event-types C and E, we don't know it; nor do we have to state it explicitly in order to claim that the occurrence of event-token c causally explains the occurrence of event-token e. Given this understanding, it might seem possible to reconcile the singularist approach with a Hempelian one. This is precisely the line taken by Davidson [1967]. On his view, all causation is nomological, *but* stating the law explicitly is not required for causal explanation.

Considering this idea, Hempel noted that when the law is not explicitly offered in a causal explanation, the statement 'c causes e' is incomplete. In making such a statement, one is at least committed to the view that "*there are* certain further unspecified background conditions whose explicit mention in the given statement would yield a truly general law connecting the 'cause' and the 'effect' in question" [1965, 348]. But this purely existential claim does not amount to much. As Hempel carried on to say, the foregoing claim is comparable to having "a note saying that there is a treasure hidden somewhere" [1965, 349]. Such a note would be useless, unless "the location of the treasure is more narrowly circumscribed". So, the alleged reconciliation of the singularist approach with Hempel's will not work, unless there is an attempt to make the covering law explicit. But this will

[7]Even if one were to take the currently popular view that manipulation requires only *invariant relations* among magnitudes or variables, and even if it was admitted that these invariant relations do not hold universally but only for a certain range of interventions/manipulations, one would still be short of a genuinely singularist account of causation.

inevitably take us back to forging a close link between stating causal dependencies and stating laws.

To sum up: if Scriven's counterexample were correct, it would establish the thesis that the *Deductive-Nomological* model is not necessary for causal explanation. Let's call this thesis *UNT*. *UNT* says: if Y is a causal explanation of a singular event, then Y is not necessarily a *DN*-explanation of this event. *UNT*, if true, would contradict the *Basic Thesis*. But we haven't yet found good reasons to accept *UNT*.

11 CAUSAL HISTORIES

In his [1986], Lewis takes causal explanation of a singular event to consist in providing some information about its causal history. In most typical cases, it is hard to say of an effect e that its cause was *the* event c. Lots of things contribute to bringing about a certain effect. All these factors, Lewis says, comprise the *causal history* of the effect. This history is a huge causal net in which the effect is located. To explain why this event happened, we need to offer some information about this causal net. This is "explanatory information" [1986, 185]. A *full* explanation consists in offering the whole causal net. But hardly ever this full explanation is possible. Nor, Lewis thinks, is it necessary. Often, some chunk of the net will be enough to offer an adequate causal explanation of why a certain singular event took place.

Lewis [1986, 221-4] thinks there is no such thing as non-causal explanation of singular events. That is, he endorses the following thesis:

> (*CE*)
>
> All explanation of singular events is causal explanation.

Recall that the *Basic Thesis* (*BT*) says:

> All causal explanation can be captured by the Deductive-Nomological model.

If we added *BT* to *CE*, then it would follow that

> (*CE**)
>
> All explanation of singular events can be captured by the Deductive-Nomological model.

Could a Lewisian accept *BT*, and hence also accept *CE**? That is, does Lewis's account of causal explanation violate the *Basic Thesis*? Or, is his view of causal-explanation-as-information-about-causal-histories compatible with *BT*? Lewis [1986, 235-6] asks:

> is it [...] true that any causal history can be characterised completely by means of the information that can be built into *DN* arguments?

Obviously, if the answer is positive, *BT* is safe. Lewis expresses some scepticism about a fully positive answer to the above question. He thinks that if his theory of causation, based on the notion of counterfactual dependence, is right, there can be genuinely singular causal explanation. Yet, he stresses that in light of the fact that the actual world seems to be governed by a "powerful system of (strict or probabilistic) laws, [...] the whole of a causal history could in principle be mapped by means of *DN*-arguments [...] of the explanatory sort" [1986, 236]. He adds:

> [...] if explanatory information is information about causal histories, as I say it is, then one way to provide it is by means of *DN* arguments. Moreover, under the hypothesis just advanced, [i.e., the hypothesis that the actual world is governed by a powerful system of laws], there is no explanatory information that could not in principle be provided in that way. To that extent the covering-law model is dead right. [ibid.]

So, the *Basic Thesis* is safe for a Lewisian, at least if it is considered as a thesis about causal explanation in the actual world. Then, what is Lewis's disagreement with the *DN*-model? There is a point of principle and a point of detail. The point of principle is this. The *Basic Thesis* has not been discredited. But, if I understand Lewis correctly, he thinks that it has been wounded. It may well be the case that if Y is a causal explanation of a singular event, then Y is also a *DN*-explanation of this event. Lewis does not deny this (cf. [1986, 239-40]). But, in light of the first set of counterexamples above, *BT* might have to be modified to *BT'*:

> (*BT'*)
>
> All causal explanation of singular events can be captured by *suitable instances* of the Deductive-Nomological model.

The modification is important. For it may well be the case that what instances of the *DN*-model are *suitable* to capture causal explanations might well be specifiable only "by means of explicitly causal constraints" [1986, 236]. And if this is so, then the empiricists' aspiration to capture causal concepts by the supposedly unproblematic explanatory concepts seems seriously impaired.

The point of detail is this. Take *BT'* to be unproblematic. It is still the case, Lewis argues, that the *Deductive-Nomological* model has wrongly searched for a "unit of explanation" [1986, 238]. But there is no such unit:

> It's not that explanations are things we may or may not have one of; rather, explanation is something we may have more or less of". [ibid.]

Although Lewis agrees that a full *DN*-explanation of an individual event's causal history is both possible and most complete, he argues that this ideal is chimerical. It is the "ideal serving of explanatory information" [1986, 236]. But, "other shapes and sizes of partial servings may be very much better — and perhaps also better within our reach" [1986, 238]. This is something that the advocate of the *DN*-model need *not* deny.

12 (A BRIEF NOTE ON) LAWS OF NATURE

The *Deductive-Nomological* model of explanation, as well as any attempt to tie causation to laws, faced a rather central conceptual difficulty: how to characterise the laws of nature. Most Humean-empiricists adopted the *Regularity View of Laws*: laws of nature are regularities. Yet, they have had a hurdle to jump: not all regularities are causal. Nor can all regularities be deemed laws of nature. So they were forced to draw a distinction between the good regularities (those that constitute the *laws of nature*) and the bad ones i.e., those that are, as Mill put it, "conjunctions in some sense accidental". Only the former can underpin causation and play a role in explanation. The predicament that Humeans were caught in is this. Something (let's call it the property of lawlikeness) must be added to a regularity to make it a law of nature. But what can this be?

The first systematic attempt to characterise this elusive property of lawlikeness was broadly epistemic. The thought, advanced by A. J. Ayer, Richard Braithwaite and Nelson Goodman among others, was that inquirers have different *epistemic attitudes* towards laws and accidents. Lawlikeness was taken to be the property of those generalisations that play a certain *epistemic* role: they are believed to be true, and they are so believed because they are confirmed by their instances and are used in proper inductive reasoning. In a sense, 'natural law' was taken to be an honorific title that should be given to those regularities that are believed to hold on account of diverse evidence in their favour. But this purely epistemic account of lawlikeness fails to draw a robust line between laws and accidents. Couched in terms of belief, or in terms of a psychological willingness or unwillingness to extend the generalisation to unknown cases, the supposed difference between laws and accidents becomes spurious.

A much more promising attempt to characterise the property of lawlikeness is what we have already called (section 7) the *web of laws* view. According to this view, the regularities that constitute the laws of nature are those that are expressed by the axioms and theorems of an ideal deductive system of our knowledge of the world, and in particular, of a deductive system that strikes the *best* balance between simplicity and strength. Simplicity is required because it disallows extraneous elements from the system of laws. Strength is required because the deductive system should be as informative as possible about the laws that hold in the world. Whatever regularity is not part of this *best system* it is merely accidental: it fails to be a genuine law of nature. The gist of this approach, which, as we have seen, has been advocated by Mill, and in the twentieth century by Ramsey

[1928] and Lewis [1973], is that no regularity, taken in isolation, can be deemed a law of nature. The regularities that constitute laws of nature are determined in a kind of holistic fashion by being parts of a structure.

The Mill-Ramsey-Lewis view has many attractions. It solves the problem of how to distinguish between laws and accidents. It shows, in a non-circular way, how laws can support counterfactuals. For, it identifies laws *independently* of their ability to support counterfactuals. It makes clear the difference between regarding a statement as lawlike and being lawlike. It respects the major empiricist thesis that laws of nature are contingent. For a regularity might be a law in the actual world without being a law in other possible worlds, since in these possible worlds it might not be part of the best system for these worlds. It solves the problem of uninstantiated laws. The latter might be taken to be proper laws insofar as their addition to the best system results in the enhancement of the strength of the best system, without detracting from its simplicity.

Yet, this view faces the charge that it cannot offer a fully *objective* account of laws of nature. For instance, it is commonly argued that how our knowledge of the world is organised into a simple and strong deductive system is, by and large, a subjective matter. Hence, what regularities will be deemed *laws* seems to be based on our subjective attitude towards regularities. But this kind of criticism is overstated. There is nothing in the *web-of-laws* approach that makes laws mind-dependent. The regularities that are laws are fully objective, and govern the world irrespective of our knowledge of them, and of our being able to identify them. In any case, as Ramsey, in effect, pointed out, it is a fact about the world that some regularities form, objectively, a system; that is, that *the world has an objective nomological structure*, in which regularities stand in certain relations to each other; relations that can be captured (or expressed) by relations of deductive entailment in an ideal deductive system of our knowledge of the world. Ramsey's suggestion grounds an objective distinction between laws and accidents in a *worldly* feature: that the world has a certain nomological structure.

In the 1970s, David Armstrong [1983], Fred Dretske [1977] and Michael Tooley [1977] put forward the view that lawhood cannot be reduced to regularity (not even to regularity-plus-something-that-distinguishes-between-laws-and-accidents). Lawhood, they claimed, is a certain contingent necessitating relation among properties (*universals*). Accordingly, it is a law that all Fs are Gs if and only if there is a relation of nomic necessitation $N(F, G)$ between the properties (universals) F-ness and G-ness such that all Fs are Gs. This approach has many attractions. It purports to explain why there are regularities in the world: because there are necessitating relations among properties. It thereby distinguishes between regularities and laws: the regularities that hold in the world do not constitute the laws that hold in the world. Rather, and at best, they are the *symptoms* of the instantiation of laws. It explains the difference between nomic regularities and accidents by claiming that the accidental regularities are not even symptoms of the instantiation of laws. It makes clear how laws can *cause* anything to happen: they do so because they embody causal relations among properties. But the central concept of nomic necessitation is still not sufficiently clear. In particular, it is not clear how the necessitating relation between the property of F-ness and the property of G-ness makes it the case that *All Fs are Gs.* To say that there is a necessitating relation $N(F, G)$ is not yet to explain what this relation is. Nor does it say anything about how the corresponding regularity *All Fs are Gs* obtains. It might seem that $N(F, G)$ *entails* the corresponding regularity *All Fs are Gs*; but it is not clear at all how this entailment goes. If the regularity *All Fs are Gs* is contained in $N(F, G)$ as the sentence 'P' is contained in the sentence '$P\&Q$', the entailment is obvious. But then, there seems to be a mysterious extra 'Q' in $N(F, G)$ over the 'P' (= *All Fs are Gs*). And we are in the dark as to what this might be, and how it ensures that the regularity obtains.

Both the Humeans and the advocates of the Armstrong-Dretske-Tooley view agree that laws of nature are *contingent*. A growing rival thought has been that if laws did not hold with some kind of objective necessity, they could not be robust enough to support either causation or explanation. As a result of this, laws of nature are said to be metaphysically necessary. This amounts to a radical denial of the contingency of laws. Along with it came a resurgence of Aristotelianism in the philosophy of science. The advocates of the view the laws are *contingent* necessitating relations among properties took it to be the case that though an appeal to (natural) properties was indispensable for the explication of lawhood, the properties themselves are passive and freely recombinable. Consequently, there can be a possible world in which some properties are not related in the way they are related in the actual world. The advocates of metaphysical necessity took the stronger line that laws of nature flow from the essences of properties. In so far as properties have essences, and in so far as it is part of their essence to endow their bearers with a certain behaviour, it follows that the bearers of properties *must* obey certain laws, those that are issued by their properties. Essentialism was treated with suspicion in most of the twentieth century, partly because essences were taken to be discredited by the advent of modern science and partly because the admission of essences (and the concomitant distinction between essential and accidental properties) created logical difficulties. Essentialism required the existence of *de re* necessity, that is natural necessity, since if it is of the essence of an entity to be thus-and-so, it is *necessarily* thus-and-so. But before Kripke's [1972] work, the dominant view was that all necessity was *de dicto*, that is, it applies, if at all, to propositions and not to things in the world.

The thought that laws are metaphysically necessary gained support from the (neo-Aristotelian) claim that properties are active powers. The key idea here was introduced by Rom Harré and Edward H. Madden [1975] and strengthened by Sidney Shoemaker [1980]. They argued that properties are best understood as powers since the only way to identify them is via their causal role. Two seemingly distinct properties that have exactly the same powers are, in fact, one and the same property. Similarly, one cannot ascribe different powers to a property without changing this property. It's a short step from these claims that properties are not freely recombinable: there cannot be worlds in which two properties are combined by a different law than the one that unites them in the actual world. On this view, it does not even make sense to say that properties are united by laws. Rather, properties — qua powers — *ground* the laws.

Many philosophers remain unconvinced. A popular claim is that the positing of irreducible powers is, in Mackie's [1977, 366] memorable phrase, the product of metaphysical double-vision. Far from explaining the causal character of certain processes (e.g., the dissolution of a sugar-cube in water), "they just *are* the causal processes which they are supposed to explain seen over again as somehow latent in the things that enter into these processes" [ibid.].[8]

13 UNIFICATION REVISITED

Scientific explanation is centrally concerned with explaining regularities — perhaps more centrally than with explaining particular facts. But when Hempel attempted to extend his *Deductive-Nomological* model to the explanation of laws, he encountered the following difficulty (cf. [1965, 273]). Suppose one wants to explain a low-level law L_1 in a DN-fashion. One can achieve this by simply subsuming L_1 under the more comprehensive regularity $L_1 \& L_2$, where L_2 may be any other law one likes. For instance, one can DN-explain Boyle's law by deriving it from the conjunction of Boyle's law with the law of Adiabatic Change. Although such a construction would meet all the requirements of the DN-model, it wouldn't count as an explanation of Boyle's law. Saying that the conjunction $L_1 \& L_2$ is not more fundamental than L_1 would not help. The issue at stake is precisely what makes a law more *fundamental* than another one. Intuitively, it is clear that the laws of the kinetic theory of gases are more fundamental than the laws of ideal gases. But if what makes them more fundamental *just* is that the latter are derived from the former, the conjunction $L_1 \& L_2$ would also count as more fundamental than its components. Hempel admitted that he did not know how to deal with this difficulty. But this difficulty is very central to his project. The counter-example trivialises the idea that laws can be DN-explained by being deduced from other laws. Hence, the empiricist project should have to deal with 'the problem of conjunction'.

13.1 Reducing the Number of Brute Regularities

An intuitive idea is that a law is more fundamental than others, if it *unifies* them. But how exactly is *unification* to be understood? According to Friedman [1974], explanation is closely linked with understanding. Now, 'understanding' is a slippery notion. It relates, intuitively, to knowing the causes: how the phenomena are brought about. Friedman revived a long-standing empiricist tradition where 'understanding' is linked to conceptual economy.[9] The basic thought is that a

[8]For more on the issue of laws of nature, see my [2002, part II].

[9]This tradition goes back to Mach and Poincaré, but Friedman wants to dissociate the idea of unification from Mach's and Poincaré's phenomenalist or instrumentalist accounts of knowledge.

phenomenon is understood, if it is made to fit within a coherent whole, which is constituted by some basic principles. If a number of seemingly independent regularities are shown to be subsumable under a more comprehensive law, then, the thought is, our understanding of nature is promoted. For, the number of regularities which have to be assumed as 'brute' is minimised. Some regularities, the fundamental ones, should still be accepted as brute. But the smaller the number of regularities that are accepted as brute, and the larger the number of regularities subsumed under them, the more we comprehend the workings of nature: not just what regularities there are, but also why they are and how they are linked to each other. After noting that in important cases of scientific explanation (e.g., the explanation of the laws of ideal gases by the kinetic theory of gases) "we have reduced a multiplicity of unexplained, independent phenomena to one", Friedman [1974, 15] added:

> I claim that this is the crucial property of scientific theories we are looking for; this is the essence of scientific explanation — sciences increases our understanding of the world by reducing the total number of independent phenomena that we have to accept as ultimate or given. A world with fewer independent phenomena is, other things equal, more comprehensible than with more.

Explanation, then, proceeds via unification into a compact theoretical scheme. The basic 'unifiers' are the most fundamental laws of nature. The explanatory relation is still deductive entailment, but the hope is that, suitably supplemented with the idea of 'minimising the number of independently acceptable regularities', it will be able to deal with the conjunction problem.

In outline, Friedman's approach is the following. A lawlike sentence L_1 is acceptable independently of lawlike sentence L_2, if there are sufficient grounds for accepting L_1, which are not sufficient grounds for accepting L_2. This notion of 'sufficient grounds' is not entirely fixed. Friedman [1974, 16] states two conditions that it should satisfy:

i If L_1 implies L_2, then L_1 is not acceptable independently of L_2.

ii If L_1 is acceptable independently of L_2, and L_3 implies L_2, then L_1 is acceptable independently of L_3.

The basic idea is that lawlike sentence L_1 is not acceptable independently of its logical consequences, but it is independently acceptable of other statements logically independent from it. This is not very illuminating, as Friedman admits. But a further step shows how this idea can be put to work in solving 'the problem of conjunction'. Take a lawlike sentence L. Let us call a *partition* of L a set of sentences $L_1, \ldots, L_n$ such that

a their conjunction is logically equivalent to L; and

b each member L_i of the set is acceptable independently of L.

Let us call 'conjunctive' a sentence L which satisfies (a) and (b), and, following Friedman, let us call "atomic" a sentence L which *violates* them. Given this, a lawlike sentence L explains lawlike sentences $L_1, \ldots, L_n$, if L is "atomic". Conversely, a lawlike sentence L *fails* to explain lawlike sentences $L_1, \ldots, L_n$, if L is 'conjunctive'. We can now see how Friedman's account bars the mere conjunction $L_1\&L_2$ of Boyle's law (L_1) with the law of Adiabatic Change (L_2) from explaining Boyle's law: the conjunction of the two laws is not an atomic sentence; it is a 'conjunctive' sentence. It is partitioned into a (logically equivalent) set of independently acceptable sentences, viz. L_1 and L_2. Conversely, we can see why Newton's law of gravity offers a genuine explanation, via unification, of Galileo's law, Kepler's laws, the laws of the tides, etc. On Friedman's account, the difference between Newton's law and the mere conjunction $L_1\&L_2$ is that the content of Newton's law *cannot* be partitioned into a (logically equivalent) set of independently acceptable laws: the sentence which express Newton's law is "atomic".

As Kitcher [1976] has shown, Friedman's account does *not* offer a necessary condition for the explanation-as-unification thesis. His general point is that if, ultimately, explanation of laws amounts to derivation of lawlike statements from other lawlike statements, then in mathematical physics at least, there will be many such derivations that utilise more than one lawlike statement as premises. Hence, ultimately, there are 'conjunctions' that are partitioned into independently acceptable lawlike statements which, nonetheless, explain other lawlike statements.

On the face of it, however, atomicity does offer a *sufficient* condition for genuine unifying, and hence explanatory, power. But can there be atomic lawlike sentences? At a purely syntactic level, there cannot be. *Any* sentence of the form 'All *F*s are *G*s' can be partitioned into a logically equivalent set of sentences such as Àll (*F*s & *H*s) are *G*s' and 'All (*F*s & *not-H*s) are *G*s'}. So the predicate 'is a planet' can be partitioned into a set of logically equivalent predicates: 'is a planet and is between the earth and the sun' (F & H) and 'is a planet and is not between the earth and the sun' (F & *not-H*). Take, then, the statement that expresses Kepler's first law, viz., that all planets move in ellipses. It follows that this can be partitioned into two statements: 'All *inferior planets* move in ellipses'; and 'All *superior planets* move in ellipses'. A perfectly legitimate lawlike statement is partitioned into two other lawlike statements. Is it then "atomic"? Syntactic considerations alone suggest that it is not.

The advocates of "atomicity" might insist that not all syntactic partitions of a lawlike statement will undermine its atomicity, since not all syntactic partitions will correspond to 'natural kind' predicates. The thought might be that whereas F and G in 'All *F*s are *G*s' are 'natural kind' predicates, the predicates, $F\&H$ and *F*&*not-H*, which can be used to form the logically equivalent partition {All (*F*s&*H*s) are *G*s; All (*F*s & *not-H*s) are *G*s}, are not necessarily 'natural kind' predicates. This admission reveals an important weakness of Friedman's approach. In order to be viable, this approach requires a theory of what predicates pick out natural kinds. This cannot be a purely syntactic matter. One standard thought has been that the predicates that pick out natural kinds are the predicates that are constituents of genuine lawlike statements. But on Friedman's approach it seems that this thought would lead to circularity. In order to say what statements are genuinely atomic, and hence what statements express explanatory laws, we first need to show what syntactically possible partitions are *not* acceptable. If we do that by means of a theory of what predicates pick out natural kinds, we cannot, on pain of circularity, say that those predicates pick out natural kinds that are constituents of statements which express explanatory laws. This last objection, however, may not be as fatal as it first sounds. The genuine link that there is between delineating what laws of nature are and what the 'natural-kind' predicates are has led many philosophers to think that the two issues can only be sorted out together. The concept of a law of nature and the concept of a natural-kind predicate form a family: one cannot be delineated without the other.

The basic flaw in Friedman's approach is the following. He defines unification in a syntactic fashion. In this sense, he's very close to the original Hempelian attempt to characterise 'explanation' in a syntactic manner. Hempel run into the problem of how to distinguish between genuine laws and merely accidentally true generalisations. Purely syntactic considerations could not underwrite this distinction. Friedman attempted to solve this problem by appealing to unification. But the old problem re-appears in a new guise. Now it is the problem of how to distinguish between 'good' unifiers (such as Newton's laws) and 'bad' unifiers (such as mere conjunctions). A purely syntactic characterisation is doomed to fail, no less than it failed as a solution to Hempel's original problem.

13.2 Unified Explanatory Store

The failures of Friedman's approach to unification led Kitcher [1981] to advance an alternative view, which changes substantially the characterisation of unification. He calls us to envisage a set K of statements accepted by the scientific community. K is consistent and deductively closed. An "explanatory store $E(K)$" over K is "the best systematisation of K" [1981, 337]. The best systematisation, however, is not what Friedman took it to be. It is not couched in terms of the minimal set of lawlike statements that need to be assumed in order for the rest of the statements in K to follow from them. For Kitcher, the best systematisation is still couched in terms of the derivation of statements of K that best unifies K, but the unification of K is not taken to be a function of the size (cardinality) of its set of axioms. Rather, Kitcher takes unification to be a function of the number of explanatory patterns, or schemata, that are necessary to account for the statements of K. The smaller this number is, the more unified is $E(K)$. Given a small number of explanatory patterns, it may turn out that the number of facts that need to be accepted as brute in the derivations of statements of K might be small too. So, it may be that Kitcher's unification entails (the thrust of) Friedman's unification. But it is important to stress that what bears the burden of unification for Kitcher is the explanatory pattern (schema). As he [1989, 432] put it:

> Science advances our understanding of nature by showing us how to derive descriptions of many phenomena, using the same pattern of derivation again and again, and in demonstrating this, it teaches us how to reduce the number of types of fact that we accept as ultimate.

Before we analyse further Kitcher's central idea, we need to understand his notion of an explanatory schema (or pattern). To fix our ideas, let us use an example (cf. [Kitcher 1989, 445-7]). Take one of the fundamental issues in the post-Daltonian chemistry, viz., the explanation of the fact that the compounds of X and Y always contains X and Y in the weight ratio $m : n$. Kitcher suggests that Dalton's approach can be seen as involving the following explanatory schema.

1. The compound Z between X and Y has an atomic formula of the form: $XpYq$.
2. The atomic weight of X is x and the atomic weight of Y is y.
3. The weight ratio of X to Y is $px : qy (= m : n)$.

This schema can be repeatedly (and successfully) applied to many cases of compounds. Take Z to be water. So, (1), X=H(yrdogen) and Y=O(xygen) and Z is H_2O_1. Then, (2) x=1 and y=16. Then, (3), 2X1:1X16=2:16=1:8 (=$m : n$). The structure of this explanatory schema (general argument-pattern) is an ordered triple: <schematic argument, filling instructions, classification>.

- The *schematic argument* is (1) to (3) above. It is schematic because it consists of schematic sentences. These are sentences in which some nonlogical expressions occurring in them (e.g., names of chemical elements) are replaced by dummy letters, (e.g., Z, X, Y) which can take several values.
- The *filling instructions* are directions for replacing the dummy letters of the schematic sentences with their appropriate values. In the example at hand, the dummy letters X and Y should be replaced by names of elements (e.g., Hydrogen and Oxygen), the dummy letters p and q should take natural numbers as values, and the dummy letters x, y should take real numbers are values.
- The *classification* is a set of statements that describe the inferential structure of the schema. In the case at hand, the classification dictates that (1) and (2) are the premises of the argument while (3) is the conclusion.

Explanatory schemata are the vehicles of explanation. The explanatory store E(K) is "a reserve of explanatory arguments" [1981, 332], whose repeated applications to many phenomena brings order — and hence unifies — K.

A central thought in Kitcher's account is that explanations are arguments, and in particular *deductive* arguments. The best systematisation is still a deductive systematisation, even if what effects the systematisation is the number of deductive patterns that are admissible, and not the number of axioms of the 'best system'. In this sense, Kitcher's approach is a descendant of Hempel's *Deductive-Nomological* model. It shares some of its most important features and consequences. The relation of explanatory dependence is a relation between sentences and it should be such that it instantiates a deductively valid argument with (a description of) the *explanandum* as its conclusion. Yet, we need to be careful here. Kitcher's account, as it now stands, does *not* demand that the premises of explanatory arguments be laws of nature. It does not even demand that they be universally quantified statements. They may be, and yet they may not. So, as it stands, Kitcher's account need not be a way to explicate what the laws of nature are. Nor does it demand that all explanation be *nomological.*

However, it seems that statements that express genuine laws of nature are uniquely apt to do the job that Kitcher demands of explanation. By being genuinely lawlike, these statements can underwrite the power that some schemata have to be repeatedly employed in explanations of singular events. Take the case, discussed also by Hempel, of trying to explain why John Jones is bald. Hempel rightly thought it inadmissible to explain this fact by constructing a *DN*-argument whose premises are the following: "John Jones is a member of the Greenbury School Board for 1964" and "All members of the Greenbury School Board for 1964 are bald". His reason was that the statement "All members of the Greenbury School Board for 1964 are bald" did not express a genuine law. Kitcher agrees with Hempel that this explanation is inadmissible: it rests on an accidentally true generalisation. But how is he to draw the distinction between laws and accidents within his own account? He says that an argument-pattern that aims to explain why certain individuals are bald by employing the sentence "All members of the Greenbury School Board for 1964 are bald" is not "generally applicable" [1981, 341]. On the contrary, an argument-pattern that would aim to explain why certain individuals are bald by reference to some principles of physiology would be generally applicable.

What, however, underwrites the difference in the applicability of argument-patterns such as the above is that the former rests on an accidental generalisation while the latter rests on genuine laws. It's not just that "All members of the Greenbury School Board for 1964 are bald" has a finite number of instances — a fact that would impair its applicability. Kepler's first law has only a finite number of instances, and yet we think that its presence in an argument-pattern would not impair its applicability. So, Kitcher needs to tie the explanatory applicability of an argument-pattern with the presence of genuine lawlike statements in it.

Is Kitcher's account of unification in terms of argument-patterns satisfactory? The notion of an argument-pattern is clear enough and does seem to capture some sense in which a system is unified. But when argument-patterns are applied to several cases, things seem to be more complicated than Kitcher thinks. Take one of his own examples: Newton's second law of motion. Once we are clear on the notion of 'force', Newton's law **F=ma** can be seen as specifying a Kitcher-like argument-pattern. The whole problem, however, is that none of the elements of the triple that specify an argument-pattern, viz., schematic argument, filling instructions, classification, can capture the all-important concept of a force-function. Each specific application of Newton's law requires, as Cartwright has repeatedly stressed, the prior specification of a suitable force-function. So, when we deal with a pendulum, we need to introduce a different force-function (e.g., F=-Kx) than when we are faced with a planet revolving around the sun. It's not part of the schematic argument what force-functions are applicable. Nor can this be added to the filling instructions, simply because the force-functions may be too diverse, or hitherto unspecified.

There is clearly something to the idea that, given a repertoire of force-functions, Newton's second law can be schematised à la Kitcher. But part of explaining a singular event is surely to figure out *what* force-function applies to this particular case. Besides, even when we have chosen the relevant force-function, we need to introduce further assumptions, related to the specific domain of application, which will typically rest on idealisations. All these cannot be part of the explanatory pattern. What really seems to matter in most (if not all) cases is that the phenomena to be explained are traced back to some kind of basic law, such as $\mathbf{F}=\mathbf{ma}$. It's not so much that we can repeatedly apply a certain argument-pattern to derive more specific cases. Instead, more typically, we show how specific cases can be reduced to being instances of some basic principles. That these basic principles will be applicable to many phenomena follows from their universal character. But it seems irrelevant whether or not the repertoire of the arguments from which (descriptions of) several phenomena derive is small or large. Unification consists in minimising the number of *types* of general principles, which are enough to account for the phenomena. Admittedly, this view is closer to Friedman's than to Kitcher's. But so be it.

14 MECHANISTIC EXPLANATION REVISITED

The thought that explanation amounts to identification of causal mechanisms reappeared in the work of Wesley Salmon. He distinguished between three approaches to scientific explanation, which he called the "epistemic conception", the "modal conception" and the "ontic conception".

The *epistemic conception* is the Hempelian approach. It makes the concept of explanation broadly epistemic, since, as we have seen, it takes explanation to be nomic expectability. The *modal conception* differs from the epistemic mostly in its account of necessity. The *explanandum* is said to follow necessarily from the *explanans*, in the sense that it was *not* possible for it not to occur, given the relevant laws. The *ontic conception* takes explanation to be intimately linked to *causation.* As Salmon [1984, 19] put it:

> To give scientific explanations is to show how events [...] fit into the causal structure of the world

Salmon takes the world to have an already built-in causal structure. Explanation is then seen as the process in virtue of which the *explananda* are placed in their right position within this causal structure. On Salmon's ontic conception, causal relations are *prior* to relations of explanatory dependence. What explains what is parasitic on (or determined by) what causes what.

Kitcher [1985, 638] has rightly called Salmon's approach "bottom-up": we first discern causal relations among particular events, and then conceive of the task of explanation as identifying the causal mechanisms that produce the events for which we seek an explanation. To this approach, Kitcher contrasts a "top-down" one: we begin with a unified deductive systematisation of our beliefs; then, we proceed to make ascriptions of causal dependencies (i.e., of relations of cause and effect), which are parasitic on (or determined by) the relations of explanatory dependence that emerge within the best unified system.

A central motivation for Salmon's ontic conception is that causal explanation cannot be captured by the derivationist (top-down) models that have been very popular in the history of thinking about explanation. The commitment to explanation as a species of deductive derivation has been so pervasive that it can hardly be exaggerated. For Salmon, explanation is not a species of *deductive derivation*: explanations are not arguments. It is noteworthy that Salmon is as willing as anyone to adopt unification as the goal of scientific explanation. He [1985, 651] takes it that an advocate of a "mechanistic" view of explanation, (viz., of the view that explanation amounts to the identification of causal mechanisms) is perfectly happy with the idea that there is a small repertoire of causal mechanisms that work in widely different circumstances. He is also perfectly happy with the view that "the basic mechanisms conform to general laws" [ibid.]. Unification, Salmon stresses, promotes our understanding of the phenomena. Nonetheless, he takes it that the causal order of the world is (metaphysically) prior to the explanatory order. He thinks that the 'because' of explanation is always dependent on the 'because' of causation.

What is a causal mechanism? Salmon offers a *generic* account of causal mechanism, based on two key causal concepts: causal process and causal interaction. Though sometimes he talked as if causal processes and causal interactions are two distinct types of causal mechanism, they are really intertwined. According to Salmon, causal processes

> are the mechanisms that propagate structure and transmit causal influence in this dynamic and changing world. (...) [T]hey provide the ties among the various spatiotemporal parts of our universe. [1997, 66]

Examples of causal processes include a light-wave travelling from the sun, or less exotically, the movement of a ball. Using the language of the Special Theory of Relativity, we can say that a process is represented by a world-line in a Minkowski diagram. An important aspect of Salmon's views is that processes are *continuous*. So, a process cannot be represented as a sequence of discrete events. The continuity of the process accounts for the direct link between cause and effect (cf. [1984, 156-7]).

Not all processes are causal. Borrowing an idea of Reichenbach's [1956], Salmon [1984, 142] characterised 'causal' those (and only those) processes that are capable of transmitting a mark. Consequently, non-causal (or "pseudo") processes are those (and only those) that cannot transmit a mark. Intuitively, to mark a process is to interact with it so that a *tag* is put on it. A moving white ball (that is, a process) can be marked by simply painting a red spot on the ball. But it is not enough that the process can be 'markable'. The process should be such that, after the mark has been put on it, by means of a single local interaction, the mark gets *transmitted.* Salmon insists on the *transmission* of the mark because without it, there cannot be an adequate characterisation of causal processes. *Any* process can be marked by means of a single local interaction. In order to avoid the trivialisation of the mark-method, Salmon insists that the mark should be *transmitted* by the process, after the interaction which marked it has taken place (cf. [1984, 142]).

Salmon [1984, 144] goes on to characterise the mark method a bit more formally. A process, be it causal or not, exhibits "a certain structure". A causal process is said to be a process capable of transmitting its own structure. But, Salmon adds, "if a process — a causal process — is transmitting its own structure, then it will be capable of transmitting certain modifications in the structure" [1984, 144]. A mark, then, is a modification of the structure of a process. And a process is causal if it is capable of transmitting the modification of its structure that occurs in a single local interaction. It should be noted, however, that Salmon offers *two* criteria for a process being causal. The *first* is that it is capable of transmitting its own structure, i.e., that it is, in some sense, self-maintaining or self-persisting, or self-determined. This criterion says nothing about marking, unless of course one thinks that the structure that characterises a process is its own mark. But even so, whether a process is causal will depend, on the *first* criterion, on whether the process is capable of *transmitting its own structure.* What pseudo-processes cannot do is *transmit* their structure, unless they are under the influence of some "external agency". The *second* criterion that Salmon offers is that a process is causal if it is capable of transmitting *modifications* of its structure. This modification is, clearly, a marking of the process. Hence, the second criterion is a genuine marking criterion. However, the two criteria are conceptually distinct. They are not even necessarily co-extensive. For instance, a photon might be rightly deemed as causal process according to the first criterion, but it seems that it cannot be a causal process on the second criterion, since it admits of no modification of its structure (assuming that it has one).

Isn't the ability to transmit a mark "a mysterious power"? Salmon's master thought is that there is no mystery in the view that a mark is transmitted from a point A of a process to a subsequent point B, if we take on board Russell's 'at-at' theory of motion. According to this theory — which Russell's developed as a reply to Zeno's paradox of the arrow — "to move from A to B is simply to occupy the intervening points at the intervening instants" [1984, 153]. That is, to move from

A to B is to be *at* the intervening points, *at* the intervening times. Salmon (and Russell) argue that this is a *complete* explanation of the motion since there is no additional question (and hence no extra pressure to explain) why (or how) the object *gets* from point A to point B. Consequently, Salmon [1984, 148] defines mark-transmission (MT) as follows:

> (MT) Let P be a process that, in absence of interactions with other processes, would remain uniform with respect to a characteristic Q, which it would manifest consistently over an interval that includes both of the space-time points A and B ($A \neq B$). Then, a *mark* (consisting of a modification of Q into Q'), which has been introduced into process P by means of a single local interaction at point A, is *transmitted* to point B if P manifests the modification Q' at B and at all stages of the process between A and B without additional interventions.

Note that the first clause of MT strengthens the criteria for a process being causal by introducing a counterfactual characterisation, viz., that "the process P would have continued to manifest the characteristic Q if the specific marking interaction had not occurred" [1984, 148]. This is a considerable strengthening because the two criteria that we have encountered so far (viz., transmission of P's own structure, and transmission of a modification of P's own structure) make no references to counterfactuals. The strengthening, however, is necessary because there can be pseudo-processes which satisfy the second clause of MT.

MT makes extensive reference to the presence and absence of interactions. In his [1984, 171], Salmon defines *Causal Interaction* (CI) as follows:

> (CI): Let P_1 and P_2 be two processes that intersect with one another at the space-time point S, which belongs to the histories of both. Let Q be a characteristic that process P_1 would exhibit throughout an interval if the intersection with P_2 did not occur; let R be a characteristic that process P_2 would exhibit throughout an interval (which includes subintervals on both sides of S in the history of P_2) if the intersection with P_1 did not occur. Then, the intersection of P_1 with P_2 at S constitutes a causal interaction if:
>
> 1. P_1 exhibits the characteristic Q before S, but it exhibits a modified characteristic Q' throughout an interval immediately following S; and
> 2. P_2 exhibits the characteristic R before S, but it exhibits modified characteristic R' throughout an interval immediately following S.

The formulation of CI involves, once more, counterfactuals. This is to secure that intersections between pseudo-processes do not count as causal interactions. Besides, the wording of CI is such that the concept of causal interaction is defined in terms of the geometric (i.e., non-causal) concept of *intersection* of two processes.

It might then appear that Salmon offers an analysis of causal mechanism in non-causal terms. But this is not the case. *CI* makes an essential (if implicit) reference to *marks*, and hence to *causal* processes. Salmon thinks that his appeal to the non-causal concept of *intersection* is enough to ground his theory in non-causal terms. Here is the mature formulation of this theory [1997, 250]:

$S-I$	A process is something that displays consistency of characteristics.
S-II	A mark is an alteration to a characteristic that occurs in a single local intersection.
S-III	A mark is transmitted over an interval when it appears at each spacetime point of that interval, in the absence of interactions.
S-IV	A causal interaction is an intersection in which both processes are marked (altered) and the mark in each process is transmitted beyond the locus of the intersection.
$S-V$	In a causal interaction a mark is introduced into each of the intersecting processes.
S-VI	A causal process is a process that can transmit a mark.

Given this formulation, he is confident that his account is cast in non-causal terms. Yet, even if Salmon is right in this, it's not clear that his account in terms, ultimately, of intersections, is strong enough to characterise causal *interactions*.[10,11]

Suppose we were to leave aside the problems mentioned so far. The question to ask, then, would be the following: is Salmon's mark-method adequate as a theory of causation and hence of causal explanation? The key element of his theory is the idea of mark-transmission. Is, then, mark transmission necessary and sufficient for a process being causal? Kitcher [1985] has argued that it is neither. Take the case of a pseudo-process, e.g., the shadow of a moving car. This can be permanently marked by a single local interaction. The car crashes on a wall and a huge dent appears on its bonnet. The shadow of the car acquires, and transmits, a permanent mark: it is *the shadow of a crashed car*. So, the mark-transmission is not sufficient for a process being causal. Conversely, a process can be causal even if it does *not* transmit a mark. To see how this is possible, consider Salmon's requirement that a process should remain uniform with respect to a characteristic Q for some time. This is necessary in order to distinguish a process (be it causal or not) from what

[10] For more on this see my [2002, 117-8].

[11] An important aspect of Salmon's theory that we shall not discuss, concerns the "production" of causal processes. Salmon's main idea is that the "production of structure and order" in the world is, at least partly, due to the existence of "conjunctive forks", which are exemplified in situations in which a common cause gives rise to two or more effects. The core of this idea goes back to Reichenbach's [1956], though Salmon also adds further cases of causal forks, such as "interactive forks" and "perfect forks", which correspond to different cases of common-cause situations. Salmon uses statistical relations among events to characterise causal forks. He also argues that it is the *de facto* direction of the causal forks from past events to future events that constitutes the direction of causation. For the details of Salmon's views, see [1984, chapter 6; 1997, chapter 18]. For criticisms, see [Dowe, 2000, 79-87].

Kitcher has aptly called "spatiotemporal junk". This requirement however seems to exclude from being causal many genuine processes that are short-lived, e.g., the generation and annihilation of virtual (subatomic) particles (cf. [Dowe 2000, 74]).

A generic problem to which the above counterexamples point is the vagueness of the notion of 'characteristic Q', which gets either transmitted or modified in a causal process. Salmon could block the first of the counterexamples above by denying, for instance, that the modification of the shadow of the car after the crash is a modification of a genuine characteristic of the shadow. In specific cases, we seem to have a pretty clear idea of what this characteristic might be, e.g., the chemical structure of a molecule, or the energy-momentum of a system, or the genetic material of an organism. Once, however, we start thinking about all this in very abstract philosophical terms, it is not obvious that we can say anything other than this characteristic being a *property* of a process. Then again, new problems arise. For at this very abstract level, *any* property of *any* process might well be suitable for offering the markable characteristic of the process. We seem to be in need of an account of which properties are such that their presence or modification marks a causal process.

It has been repeatedly noted that Salmon's theory relies heavily on the truth of certain counterfactual conditionals. This has led some philosophers (e.g., [Kitcher 1989]) to argue that, in the end, Salmon has offered a variant of the counterfactual approach to causation. Such an approach would bring in its tow all the problems that counterfactual analyses face. In particular, it would seem to undermine Salmon's aim to offer an objective analysis of causation. For, it is an open issue whether or not there can be a fully objective theory of the truth-conditions of counterfactuals. In any case, Salmon has always been very sceptical about the objective character of counterfactual assertions. So, as he said, it was "with great philosophical regret", that he took counterfactuals on board in his account of causation (cf. [1997, 18]). The question then is whether his account could be formulated without appeal to counterfactuals.

The short answer to the above question is: *yes, but...* For the mark-method has to be abandoned altogether and be replaced by a variant theory, which seems to avoid the need for counterfactuals. The counterexamples mentioned above, as well as the need to avoid counterfactuals, led Salmon to argue that "the capacity to transmit a mark" is not constitutive of a causal process, but rather a "symptom" of its presence [1997, 253]. So, causal processes, i.e., the "the causal connections that Hume sought, but was unable to find" [1984, 147], should be identified in a different way. The best attempt so far to articulate this different way is Dowe's [2000] theory of *conserved quantities*. According to this theory:

> The central idea is that it is the possession of a conserved quantity, rather than the ability to transmit a mark, that makes a process a causal process. [2000, 89]

We shall not discuss this theory here. It shall only be noted that even if it is granted that it offers a neat account of physical causal mechanisms, it can be generalised as a theory of causal mechanisms *simpliciter* only if it is married to strong reductionistic views that all worldly phenomena (be they social or psychological or biological) are, ultimately, reducible to physical phenomena.

15 EXPLANATION AS MANIPULATION

James Woodward [2003] has put forward a 'manipulationist' account of causal explanation. Briefly put, c causally explains e if e causally depends on c, where the notion of causal dependence is understood in terms of relevant (interventionist) counterfactual, i.e., counterfactuals that describe the outcomes of interventions. A bit more accurately, c causally explains e if, were c to be (actually or counterfactually) manipulated, e would change too. This model ties causal explanation to actual and counterfactual experiments that show how manipulation of factors mentioned in the *explanans* would alter the *explanandum*. It also stresses the role of invariant relationships, as opposed to strict laws, in causal explanation. Explanation in this model consists in answering a network of "what-if-things-had-been-different questions", thereby placing the *explanandum* within a pattern of counterfactual dependencies (cf. [Woodward, 2003, 201]). The law of ideal gases, for instance, is said to be explanatory not because it renders a certain *explanandum* (e.g., that the pressure of a certain gas increased) nomically expected, but because it can tell us how the pressure of the gas would have changed, had the antecedent conditions (e.g., the volume of the gas) been different. The explanation proceeds by locating the *explanandum* "within a space of alternative possibilities" [Woodward 2003, 191]. The key idea, I take it, is that causal explanation shows how the *explanandum* depends on the *explanans* in a stable way.

Let us describe, somewhat sketchily, the two key notions of intervention and invariance. The gist of Woodward's characterisation of an *intervention* is this. A change of the value of X counts as an intervention I if it has the following characteristics:

a) the change of the value of X is entirely due to the intervention I;

b) the intervention changes the value of Y, if at all, only through changing the value of X.

The *first* characteristic makes sure that the change of X does not have causes other than the intervention I, while the *second* makes sure that the change of Y does not have causes other than the change of X (and its possible effects).[12] These characteristics are meant to ensure that Y-changes are exclusively due to X-changes, which, in turn, are exclusively due to the intervention I. As Woodward stresses, there is a close link between intervention and manipulation. Yet, his account makes no special reference to human beings and their (manipulative)

[12]There is a *third* characteristic too, viz., that the intervention I is not correlated with other causes of Y besides X.

activities. In so far as a process has the right characteristics, it counts as an intervention. So interventions can occur 'naturally'.

Woodward links the notion of intervention with the notion of *invariance*. A certain relation (or a generalisation) is invariant, Woodward says, "if it would continue to hold — would remain stable or unchanged — as various other conditions change" [2000, 205]. What really matters for the characterisation of invariance is that the generalisation remains stable under a set of actual and counterfactual *interventions*. So Woodward [2000, 235] notes:

> the notion of invariance is obviously a modal or counterfactual notion [since it has to do] with whether a relationship would remain stable if, perhaps contrary to actual fact, certain changes or interventions were to occur.

Let me highlight three important general elements of Woodward's approach. *First*, causal claims relate variables. Causes should be such that it makes sense to say of them that they could be changed or manipulated. Thinking of them as variables, which can take different values, is then quite natural. But as Woodward goes on to note, it is not difficult to translate talk in terms of changes in the values of variables into talk in terms of events and conversely. For instance, instead of saying that the hitting by the hammer (an event) caused the shattering of the vase (another event), we may say that the change of the value of a certain indicator variable from *not-hit* to *hit* caused the change of the value of another variable from *unshattered* to *shattered*. This strategy, however, will not work in cases in which putative causes cannot be understood as values of variables. But then again, this is fine for Woodward, as he claims that in those cases causal claims will be, to say the least, ambiguous (cf. [2003, 115ff]).

Second, generalisations need not be invariant under *all* possible interventions. Hooke's law, for instance, would 'break down' if one intervened to stretch the spring beyond its breaking point. Still, Hooke's law does remain invariant under some set of interventions. In so far as a generalisation is invariant under a certain range of interventions, it can be explanatorily useful, without being exceptionless (cf. [2000, 227-8]). Woodward [2000, 214] stresses: "[t]here are generalisations that are invariant and that can be used to answer a range of what-if-things-had-been-different questions and that hence are explanatory, even though we may not wish to regard them as laws and even though they lack many of the features traditionally assigned to laws by philosophers". In particular, a generalisation can be *causal* even if it is not universally invariant (cf. [2003, 15]).

Third, Woodward does not aim to offer a reductive account of causation or causal explanation. The notion of intervention is itself causal and, in any case, causal considerations are necessary to specify when a relationship among some variables is causal. For instance, an appropriate intervention I on variable X with respect to variable Y should be such that it is not correlated with other *causes* of Y or does not directly *cause* a change of the value of Y. Woodward [2003, 104-7] insists that his account is not trapped in a vicious circle: an account of causation or causal explanation need not be reductive to be illuminating.

In light of the above, causal explanation proceeds by exploiting the manipulationist element of causation and the invariant element of generalisations. Explanatory information "is information that is potentially relevant to manipulation and control" [Woodward, 2003, 10]. Causal relations are explanatory because they provide information about counterfactual dependencies among causal variables. And invariant generalisations are explanatory because they exhibit stable patterns of counterfactual dependence among causal variables in virtue of which different values of the effect-variable counterfactually depend on different values of the cause-variable.

There have been many significant attempts to offer semantics for counterfactual conditionals. Perhaps the most well-developed, and certainly the most well-known, is Lewis's [1973] account in terms of possible worlds. I will not discuss this theory here.[13] The relevant point is that Woodward offers an account of counterfactuals that tries to avoid the metaphysical excesses of Lewis's theory.

Woodward is very careful in his use of counterfactuals. Not all of them are of the right sort for the evaluation of whether a relation is causal. Only counterfactuals that are related to *interventions* can be of help. An intervention gives rise to an "active counterfactual", that is, to a counterfactual whose antecedent is made true by interventions. In his [2003, 122] he stresses that

> the appropriate counterfactuals for elucidating causal claims are not just any counterfactuals but rather counterfactuals of a very special sort: those that have to do with the outcomes of hypothetical interventions. (...) it does seem plausible that counterfactuals that we do not know how to interpret as (or associate with) claims about the outcomes of well-defined interventions will often lack a clear meaning or truth value.

It follows that the truth-conditions of counterfactual statements (and their truth-values) are not specified by means of an abstract metaphysical theory, e.g., by means of abstract relations of similarity among possible worlds.

The main problem that I see in Woodward's theory relates to the question: what is it that makes a certain counterfactual conditional true? Woodward stresses that causal claims are irreducible:

> According to the manipulationist account, given that C causes E, which counterfactual claims involving C and E are true will always depend on which other *causal* claims involving other variables besides C and E are true in the situation under discussion. For example, it will depend on whether other causes of E besides C are present. [2003, 136]

[13] See my [2002, 92-101].

The idea here, I take it, is that the truth of counterfactuals depend on the truth of certain causal claims, most typically causal claims about the larger causal structure in which the variables that appear in the counterfactuals under examination are embedded. Intuitively, this is a cogent claim. Consider two variables X and Y and examine the counterfactual: if X had changed (that is, if an intervention I had changed the value of X), the value of Y would have changed. Whether this is true or false will depend on whether I causes the value of Y to change by a route independent of X, or on whether some other variable Z causes a direct change of the value of Y. Causal facts such as these are part of the truth-conditions of the foregoing counterfactual. It is clear that they may, or may not, obtain independently of any intervention on X. So whether or not an intervention I on X were to occur, it might be the case that were it to occur, it would not influence the value of Y by a route independent of X. The thought, then, may be that the truth-conditions of a counterfactual are specified by certain causal facts that involve the variables that appear in the counterfactual as well as the variables of the broader causal structure in which the variables of interest are embedded.

It appears, however, that this last thought leads to an unacceptable circle. Causal claims, we are told, should be understood in terms of counterfactual dependence (where the counterfactuals are interventionist). To fix our ideas, let us consider the causal claim

> B_0: X causes Y.

For B_0 to be true, the following counterfactual C_1 should be true.

> C_1: if X had changed (that is, if an intervention I had changed the value of X), the value of Y would have changed.

On the thought we are presently considering, the truth of C_1 will depend, among other things, on the truth of another causal claim:

> B_1: I does not cause a change to the value of Y directly, (that is, by a route independent of X).

How does the truth of B_1 depend on counterfactuals? Let us assume that relations of counterfactual dependence are *part* of the truth-conditions of causal claims. Then, at least *another* (interventionist) counterfactual C_2 would have to be true in order for B_1 to be true.

> C_2: if an(other) intervention I' had changed the value of I, the value of Y would not have changed (by a route independent of X).

But what makes C_2 true? Suppose it is another causal claim B_2.

> B_2: I' does not cause a change to the value of Y directly.

For B_2 to be true, another counterfactual C_3 would have to be true, and so on. Either a regress is in the offing or the truth of some causal claims has to be accepted as a brute fact. In the former case, counterfactuals are part of the truth-conditions of other counterfactuals, with no independent account of what it is for a counterfactual to be true. In the latter case, we are left in the dark as to what causal claims capture brute facts. In particular, why should we not take it as a brute fact that B_0 or B_1 is true?

In any case, it turns out that there are sensible counterfactuals that fail Woodward's criterion of actual and hypothetical interventions. Some of them are discussed by Woodward himself [2003, 127-33]. Consider the true causal claim: Changes in the position of the moon with respect to the earth and corresponding changes in the gravitational attraction exerted by the moon on the earth's surface cause changes in the motion of the tides. As Woodward adamantly admits, this claim cannot be said to be true on the basis of interventionist (experimental) counterfactuals, simply because realising the antecedent of the relevant counterfactual is physically impossible. His response to this is an alternative way for assessing counterfactuals. This is that counterfactuals can be meaningful if there is some "basis for assessing the truth of counterfactual claims concerning what would happen if various interventions were to occur". Then, he adds, "it doesn't matter that it may not be physically possible for those interventions to occur" [2003, 130]. And he sums it up by saying that "an intervention on X with respect to Y will be 'possible' as long it is logically or conceptually possible for a process meeting the conditions for an intervention on X with respect to Y to occur" [2003, 132]. We now have a much more liberal criterion of meaningfulness at play, and it is not clear, to say the least, which counterfactuals end up meaningless by applying it.[14]

Perhaps, the foregoing worries do not affect causal explanation as a practical activity. In many practical cases, we may well have a lot of information about a particular causal structure and this may be enough to answer questions about which (interventionist) counterfactuals are true and what generalisations are invariant under interventions. When we deal with *stable causal or nomological structures* interventionist counterfactuals are meaningful and truth-valuable. The worries raised in this section concern the prospects of the manipulationist account as a philosophical theory of causal explanation. As it stands, Woodward's theory highlights and exploits the *symptoms* of a good causal explanation, without offering a fully-fledged theory of what causal explanation consists in. Invariance-under-interventions is a symptom of causal relations and laws. It is not what causation or lawhood consists in.

[14]For more on Woodward's account of causal explanation and the role of invariant generalisations in it, see my [2002, 182-187].

16 STATISTICAL EXPLANATION

Suppose we want to explain a statistical regularity, viz., the fact that in a large collection of atoms of the radioactive isotope of Carbon-14 ($C14$) approximately three-quarters of them will very probably decay within 11,460 years. This, Hempel [1965, 380-1] observed, can be explained *deductively* in the sense that its description can be the conclusion of a valid deductive argument, whose premises include a statistical nomological statement. The general claim above follows deductively from the statement that every $C14$ atom has a probability of 1/2 of disintegrating within any period of 5,730 year (provided that it is not exposed to external radiation). There is no big mystery here. A valid deductive argument can have a statistical generalisation as its conclusion provided that one of the premises also contains some suitable probabilistic statement. Hempel called this account the *Deductive-Statistical* (DS) model of explanation. Salmon [1989, 53] rightly observes that the DS-model is just a species of the *Deductive-Nomological* model, when the latter is applied to the explanation of statistical regularities.

But, there is more to statistical explanation than the DS-model can cover. We are also interested in explaining *singular events* whose probability of happening is less than unity. Suppose, to exploit one of Hempel's own examples (cf. [1965, 382]), that Jones has suffered from septic sore throat, which is an acute infection caused by bacteria known as *streptococcus hemolyticus*. He takes penicillin and recovers. There is no strict (deterministic) law, which says that whoever is infected by streptococcus and takes penicillin will recover quickly. Hence, we cannot apply the Deductive-Nomological model to account for Jones's recovery. Nor can we apply the DS-model, since what we want to explain is an individual event; not a statistical regularity. How are we to proceed?

Suppose that there is a statistical generalisation of the following form: whoever is infected by streptococcus and takes penicillin has a very high probability of recovery. Let's express this as follows:

$$\text{prob}(R/P\&S) \text{ is very high,}$$

where 'R' stands for quick recovery, 'P' stands for taking penicillin and 'S' stands for being infected by streptococcus germs. We can then say that given this statistical generalisation, and given that Jones was infected by streptococcus and took penicillin, the probability of Jones's quick recovery was high. Hence, we have *inductive* grounds to expect that Jones will recover. We can then construct an inductive argument that constitutes the basis of the explanation of an event whose occurrence is governed by a statistical generalisation. This is the birth of Hempel's *Inductive-Statistical* model (IS). Let a stand for Jones, and let 'R', 'P' and 'S' be as above. Applied to Jones's case, the IS-explanations can be stated thus:

(1)
Sa and Pa
prob($R/P\&S$) is very high
═══════════════ [makes practically certain (very likely)]
Ra

More generally, the logical form of an *Inductive-Statistical* explanation is this:

(IS)
Fa
prob(G/F) $= r$, where r is high (close to 1)
═══════════════ $[r]$
Ga.

The double line before the conclusion indicates that it is an *inductive* argument. The conclusion follows from the premises with high probability. The strength r of the inductive support that the premises lend to the conclusion is indicated in square brackets. As noted by Hempel, the fact an IS-explanation rests on an inductive argument does not imply that its premises cannot explain the conclusion. After all, Ga did occur and we can explain this by saying that, given the premises, we would have expected Ga to occur.

The *Inductive-Statistical* model inherits a number of important features of the *Deductive-Nomological* model. The IS-model makes explanations arguments, albeit inductive. It also understands explanation as nomic expectability. To explain an event is still to show how this event would have been *expected* (with high probability) to happen, had one taken into account the statistical laws that govern its occurrence, as well as certain initial conditions. The IS-model needs an essential occurrence of law-statements in the *explanans*, albeit expressing statistical laws.

Hempel's requirement of high probability is essential to his *Inductive-Statistical* model. It's this requirement that makes the IS-model resemble the DN-model, and it is also this requirement that underwrites the idea that an IS-explanation is a good inductive argument. Yet, this requirement is exactly one of the major problems that the IS-explanation faces. For we also need to explain events whose occurrence is *not* probable, but which, however, do occur. Richard Jeffrey [1969] highlighted this weakness of the IS-model by noting that the requirement of high probability is not a necessary condition for statistical explanation. We must look elsewhere for the hallmark of good statistical explanation. In particular, if the requirement of high probability is relaxed, then statistical explanations are no longer arguments.

Is the requirement of high probability sufficient for a good statistical explanation? The answer is also negative. To see why, we should look at some aspects of the statistical regularities that feature in the *Inductive-Statistical* model. Suppose we explain why Jones recovered from a common cold within a week by saying that he took a large quantity of vitamin C. We can then rely on a statistical law, which says that the probability of recovery from common colds within a week, given taking vitamin C, is very high. The formal conditions for an *IS*-explanation are met and yet the argument offered is not a good explanation of Jones's recovery from common cold. For, the statistical law is no good. It is *irrelevant* to the explanation of recovery since common colds, typically, clear up after a week, irrespective of the administration of vitamin C. This suggests that more stringent requirements should be in place if a statistical generalisation is to be explanatory. High probability is not enough.

It is noteworthy that the specific example brings to light a problem of *IS* that seems to be detrimental. The reason why we think that the foregoing statistical generalisation is not explanatory is that we, rightly, think that it fails to capture a *causal connection* between recovery from common colds and the administering of vitamin C. That two magnitudes (or variables) are connected with a high-probability statistical generalisation does not imply that they are connected causally. Even when the connection is not statistical but deterministic, it still does not follow that they are causally connected. Correlation does not imply causation. To say the least, two magnitudes (or variables) might be connected with a high-probability statistical generalisation (or by a deterministic one) because they are effects of a common cause. So, the causal arrow does not run from one to the other, but instead, from a common cause to both of them.

It might be thought that the *Inductive-Statistical* model is not aimed at causal explanation. Indeed, Hempel refrained from explicitly connecting *IS*-explanation with causal explanation (cf. [1965, sections 3.2 &3.3]). However, in his [1965, 393], he toyed with the idea that the *Inductive-Statistical* model offers

> a statistical-probabilistic concept of 'because' in contradistinction to a strictly deterministic one, which would correspond to deductive-nomological explanation.

But then, it's fair to say that insofar as the *IS*-model aims to capture a sense of statistical (or probabilistic) causation, it fails.

Enough has been said so far to bring to light the grave difficulties of the *Inductive-Statistical* model. But there is another one, which will pave the way for a better understanding of the nature of statistical explanation, and its relation to causation. Hempel [1965, 394] called this problem "the ambiguity of inductive-statistical explanation".

Valid deductive arguments have the property of *monotonicity.* If the conclusion Q follows deductively from a set of premises P, then it will also follow if further premises P^* are added to P. Inductive arguments, no matter how strong they may be, lack this property: they are *non-monotonic.* The addition of extra premises P^* to an inductive argument may even remove the support that the original set of premises P conferred on the conclusion Q. In fact, the addition of extra premises P^* to an inductive argument may be such that the *negation* of the original conclusion becomes probable. Take our stock example of Jones's recovery from streptococcal infection and refer to its *IS*-explanation (1) above. Suppose, now, that Jones was, in fact, infected by a germ of streptococcus that was resistant to penicillin. Then, Jones's taking penicillin cannot explain his recovery. What is now likely to happen is that Jones won't recover from the infection, despite the fact that he took penicillin, and despite the fact that it is a true statistical generalisation that most people who take penicillin recover from streptococcus infection. The addition of the extra premise that Jones was infected by a penicillin-resistant strain (*Ta*) will make it likely that Jones won't recover (*not-Ra*). For now the probability prob(*not-R*/*P*&*S*&*T*) of non-recovery (*not-R*) given penicillin (*P*), strept infection (*S*), and a penicillin-resistant germ (*T*) is very high. So:

(2)
Sa and *Pa*
Ta
prob(*not-R*/*P*&*S*&*T*) is close to 1
———————————— [makes practically certain (very likely)]
not-Ra

The non-monotonic nature of *Inductive-Statistical* explanation makes all this possible. (1) and (2) are two arguments with mutually consistent premises and yet incompatible conclusions. It is this phenomenon that Hempel called the "ambiguity" of *IS*-explanation. What is ambiguous is to what reference class to include the *explanandum*. Given that it may belong to lots of different reference classes, which one shall we choose? The problem is precisely that different specifications of the reference class in which the *explanandum* might be put will lead to different estimations of the probability of its occurrence. Consider the following: what is the probability that an individual lives to be 80 years old? The answer will vary according to which reference class we place him/her.

The problem we are discussing is accentuated if we take into account the fact that, even if there was an objectively correct reference class to which an individual event belongs, in most realistic cases when we need to explain an individual event, we won't be able to know whether the correct identification of the reference class has been made. We will place the individual event in a reference class in order to *IS*-explain its occurrence. But will this be the *right* reference class, and how can we know of it? This is what Hempel [1965, 395] called the *epistemic version* of the ambiguity problem. The result of this is that an *IS*-explanation should always be relativised to a body K of currently accepted (presumed to be true) beliefs.

Note that the problem of ambiguity does not arise in the case of the *Deductive Nomological* explanation. The premises of a *DN*-argument are *maximally specific.* If it is the case that *All Fs are Gs*, then no further specification of *Fs* will change the fact that they are Gs. As the jargon goes, if all Fs are Gs, no further partition of the reference class F can change the probability of an instance of F to be also an instance of G, this probability being already equal to unity. On the contrary, in an *IS*-explanation, further partitions of the reference class F can change the probability that an instance of F is also an instance of G.

This suggests that we may introduce the *Requirement of Maximal Specificity* (*RMS*) to *Inductive-Statistical* explanation. Roughly, to say that the premises of an *IS*-explanation are maximally specific is to say that the reference class to which the *explanandum* is located should be the narrowest one. More formally, suppose that the set P of premises of an *IS*-explanation of an individual event Fa imply that prob(G/F)=r. The set of premises P is maximally specific if, given that background knowledge K tells us that a also belongs to a subclass F_1 of F, and given that prob(G/F_1)=r_1, then r=r_1.[15]

Let's call a reference class *homogeneous* if it cannot be further partitioned into subclasses which violate *RMS*. Clearly, there are two concepts of homogeneity. The *first* is objective: there is no partition of the reference class into subclasses which violate *RMS*.[16] The *second* is epistemic: we don't (currently) know of any partition that violates *RMS*. Hempel's version of *RMS* was the latter. Hence, *IS*-explanation is always relativised to a certain body of background knowledge K, which asserts what partitions of the reference classes are *known* to be relevant to an *IS*-explanation of an individual event. The fact that *IS*-explanations are always epistemically relative has made many philosophers think that the *IS*-model cannot be an adequate model of statistical explanation (cf. [Salmon 1989, 68ff]). What we would need of a statistical explanation is an identification of the relevant features of the world that are nomically connected (even in a statistical sense) with the *explanandum*. The *Inductive-Statistical* model is far from doing that, as the problem with the *Requirement of Maximal Specificity* makes vivid.

The friends of statistical explanation face a dilemma. They might take the view that all genuine explanation is *Deductive-Nomological* (*DN*) and hence treat statistical explanation as *incomplete* explanation. If, indeed, all explanation is *Deductive-Nomological*, the problem of the reference class (and of *RMS*) does not even arise. On this view, an *IS*-explanation is a place-holder for a full *DN*-explanation of an individual event. The statistical generalisations are taken to express our *ignorance* of how to specify the correct reference class in which we should place the *explanandum*. This approach is natural, if one is committed to determinism. According to determinism, every event that occurs has a fully determinate and sufficient set of antecedent causes. Given this set of causes, its probability to happen is unity. If we knew this full set of causes of the *explanandum*, we could use this information to objectively fix its reference class and we would, thereby, establish a true universal generalisation under which the *explanandum* falls. If, for instance, the full set of causes of event-type E was the conjunction of event-types F, G and H, we could simply say that *All* Fs & Gs & Hs *are* Es. So, on the view presently discussed, statistical generalisations simply express our ignorance of the full set of causes of an event. They are by no means useless, but they are not the genuine article. This view is elaborated by Kitcher [1989].

Alternatively, the friends of statistical explanation could take the view that

[15]The exact definition, which is slightly more complicated and accurate than the one offered here, is given by Hempel [1965, 400]. An excellent detailed account of *RMS* is given in Salmon [1989, 55-7].

[16]The concept of objective homogeneity and its implications are discussed in Salmon [1984, chapter 3].

there is *genuine* statistical explanation, which is nonetheless captured by a model different to the *Inductive-Statistical*. In order to avoid the pitfalls of the *IS*-model, they would have to admit that there is a fact of the matter as to the *objectively homogeneous* reference class in which a certain *explanandum* belongs. But this is not enough for genuine statistical explanation, since, as we saw in the previous paragraph, the existence of an objectively homogeneous reference class is compatible with the presence of a universal law. So, the friends of genuine statistical explanation should also accept that even within an objectively homogeneous reference class, the probability of an individual event's occurring is not unity. They have to accept indeterminism: there are no further facts that, were they taken into account, would make this probability equal to unity. An example (cf. [Salmon 1989, 76]) will illustrate what is at issue here. Take a collection of radioactive carbon-14 atoms whose half-life time is 5730 years. This class is as close to being objectively homogeneous as it can be. No further partitions of this class can make a sub-class of carbon-14 atoms have a different half-life time. What is important here is that the law that governs the decay of carbon-14 atoms is *indeterministic*. The explanations that it licenses are genuinely statistical, because the probability that an atom of carbon-14 to decay within 5730 years is irreducibly 1/2. In genuine statistical explanation, there is no room to ask certain why-questions. Why did this *specific* carbon-14 atom decay? If indeterminism is true, there is simply no answer to this question.

17 STATISTICAL RELEVANCE

Take an event-type E whose probability to happen given the presence of a factor C (i.e., prob(E/C)) is r. In judging whether a further factor C_1 is relevant to the explanation of an individual event that falls under type E, we look at how taking C_1 into account affects the probability of E to happen. If prob$(E/C\&C_1)$ is different from prob(E/C), then the factor C_1 is *relevant* to the occurrence of E. Hence, it should be *relevant* to the explanation of the occurrence of an individual event that is E. Let's say that:

- C_1 is positively relevant to E, if prob$(E/C\&C_1) >$ prob(E/C);
- C_1 is negatively relevant to E, if prob$(E/C\&C_1) <$ prob(E/C);
- and C_1 is irrelevant to E, if prob$(E/C\&C_1) =$ prob(E/C).

Judgements such as the above seem to capture the intuitive idea of *causal relevance*. We rightly think, for instance, that the colour of one's eyes is causally irrelevant to one's recovery from streptococcus infection. We would expect that one's probability of recovery (R) given streptococcus infection (S) and penicillin (P), i.e., prob$(R/P\&S)$, will be unaffected, if we take into account the colour of one's eyes (B). So, prob$(R/P\&S)$=prob$(R/P\&S\&B)$. Analogously, we would think that the fact that one is infected by a penicillin-resistant strain of streptococcus (T) is causally relevant to his recovery (in particular, its lack). We would expect that prob($R/P\&S\&T$) $\neq$ prob($R/P\&S$). These thoughts, together with the fact that the requirement of high probability is neither necessary nor sufficient for a good statistical explanation, led Salmon [1971; 1984] to suggest a different conception of statistical explanation. The main idea is that we explain the occurrence of an individual event by citing certain statistical-relevance relations. In particular,

> a factor C explains the occurrence of an event E, if prob(E/C) > prob(E) — which is equivalent to prob(E/C) > prob($E/not\text{-}C$).

This came to be known as the *Statistical-Relevance* model (SR). Where an *Inductive-Statistical* explanation involves just one probability value, the SR-model suggests that explanation compares two probability values. As the jargon goes, we need to compare a *posterior probability* prob(E/C) with a *prior probability* prob(E). Note that the actual values of these probabilities do not matter. Nor is it required that the posterior probability be high. All that is required is that there is a *difference*, no matter how small, between the posterior probability and the prior. Suppose, for example, that the prior probability prob(R) of recovery from streptococcus infection is quite low, say .001. Suppose also that when one takes penicillin, the probability of recovery prob(R/P) is increased by only 10%. So, prob(R/P)=.01. We would not, on the IS-model, be entitled to explain Jones's recovery on the basis of the fact that he took penicillin. Yet, on the SR-model, Jones's taking penicillin is an explanatory factor of his recovery, since prob(R/P) > prob(R). (Equivalently, prob(R/P) > prob($R/not\text{-}P$))

An important feature of the SR-model, which paves the way to the entrance of *causation* in statistical explanation, is this. Suppose that taking penicillin is explanatory relevant to quick recovery from streptococcus infection. That is, prob(R/P) > prob(R). Can we, without further ado, say that taking penicillin causes recovery from streptococcus infection? Not really. For one might be infected by a penicillin-resistant strain (T), thus rendering one's taking penicillin totally ineffective as a cure. So, if we take T into account, it is now the case that prob($R/P\&T$) = prob(R/T). The further fact of infection by a penicillin-resistant germ renders *irrelevant* the fact that penicillin was administered. The probability of recovery given penicillin and infection by a penicillin-resistant germ is equal to the probability of recovery given infection by a penicillin-resistant germ. When a situation like this occurs, we say that factor T *screens off* R from P.

This relation of screening off is very important. Take the standard example in the literature. There is a perfect correlation between well-functioning barometers (B) and upcoming storms (S). The probability prob(S/B) that a storm is coming up given a drop in the barometer is higher than the probability prob(S) that a storm is coming up. So, prob(S/B)>prob(S). It is in virtue of this relationship that barometers can be used to predict storms. Can we then, using the *Statistical-Relevance* model, say that the drop of the barometer explains the storm? Worse, can we say that it causes the storm? No, because the correlation between a drop of the barometer and the storm is *screened off* by the fall of the atmospheric pressure. Let's call this A. It can be easily seen that prob($S/B\&A$)=prob(S/A). The presence of the barometer is rendered irrelevant to the storm, if we take the drop of the atmospheric pressure into account. Instead of establishing a causal relation between B and S, the fact that prob(S/B)>prob(S) points to the further fact that the correlation between B and S exists because of a *common cause.* It is typical of common causes that they screen off the probabilistic relation between their effects. But a factor can screen off a correlation between two others, even if it's not their common cause. Such was the case of infection by penicillin-resistant germ discussed above.

If the probabilistic relations endorsed by the SR-model are to establish genuine explanatory relations among some factors C and E, it's not enough to be the case that prob(E/C) >prob(E). It is also required that his relation not be screened off by further factors. Put more formally:

> C explains E if (i) prob(E/C) >prob(E) [equivalently, prob(E/C) > prob($E/not\text{-}C$)]; *and* (ii) there are no further factors H such that H screens off E from C, i.e., such that prob($E/C\&H$)=prob(E/H).

The moral of all this is that relations of statistical relevance do not imply the presence of causal relations. The converse seems also true, as the literature on the so-called Simpson paradox makes vivid. But we shall not go into this.[17] Correlations that can be screened off are called 'spurious'.

There should be no doubt that the *Statistical-Relevance* model is a definite improvement over the *Inductive-Statistical* model. Of course, if we go for the SR-model, we should abandon the Hempelian idea that explanations are arguments. We should also question the claim that statistical generalisations are really necessary for statistical explanation. For an SR-explanation is not an argument. Nor does it require citing statistical laws. Rather, as Salmon [1984, 45] put it, it is

> an *assembly of facts statistically relevant* to the explanandum, *regardless of the degree of probability* that results.

Besides, the *Statistical-Relevance* model makes clear how statistical explanation can be seen as a species of causal explanation. For if the relevant SR-relations are to be explanatory, they have to capture the right causal dependencies between the *explanandum* and the *explanans*. But it also paves the way for the view *that there is more to causation than relations of statistical dependence.* Salmon himself has moved from the claim that all there is to statistical explanation can be captured by specifying relations of statistical relevance to the claim that, even if we have all of them, we would still need to know something else in order to have genuine

[17]The 'Simpson paradox' suggests that C may cause E, even though C is not statistically co-related with E in the whole population. For more on this see Cartwright [1983, essay 1] and Suppes [1984, 55-7].

explanation, viz., facts about causal relationships. His latest view [1984, 34] is this:

> the statistical relationships specified in the S-R model constitute the *statistical basis* for a bone fide scientific explanation, but [...] this basis must be supplemented by certain *causal factors* in order to constitute a satisfactory scientific explanation.

So, according to Salmon [1984, 22], relations of statistical relevance must be explained by causal relations, and not the other way around. As we have already seen in section 14, his favoured account of causal relations is given in terms of unveiling the causal mechanisms, be they deterministic or stochastic, that connect the cause with its effect.

18 DEDUCTIVE-NOMOLOGICAL-PROBABILISTIC EXPLANATION

Does deductivism and indeterminism mix? Can, that is, one think that although all explanation is, in essence, deductive, there is still space to explain essentially chancy events? Railton's [1981] "Deductive-Nomological-Probabilistic" (*DNP*) model of probabilistic explanation is a very important attempt to show how this can happen.

Being dissatisfied with the epistemic ambiguity of the *Inductive-Statistical* model, and accepting the view that there should be space for the explanation of unlikely events, Railton [1981, 160] suggested that a legitimate explanation of a chancy *explanandum* should consist in

> a) "law-based demonstration that the explanandum had a particular probability of obtaining";
>
> and b) a claim that, "by chance, it did obtain".

Take the case of a very unlikely event such as a Uranium-238 nucleus $u$decaying to produce an alpha-particle. The mean-life of a U-238 nucleus is 6.5 X 10^9 years, which means that the probability p that such a particle will produce an alpha-particle is vanishingly small. Yet, events like this *do* happen, and need to be explained. Railton [1981, 162-3] suggests that we construct the following two-step explanation of its occurrence.

The *first* step is a straightforward *DN*-explanation of the fact that particle u has a probability p to alpha-decay during a certain time-interval Δt.

> (1a) All U-238 nuclei not subjected to external radiation have probability p to emit an alpha-particle during any time-interval Δt.
>
> (1b) u was a U-238 nucleus at time t and was not subjected to any external radiation during time-interval $[t, t + \Delta t]$.
>
> Therefore, (1c) u has a probability p to alpha-decay during time-interval $[t, t + \Delta t]$.

This step does not yet explain why the particular particle u alpha-decayed. It only states the probability of its decay. So, Railton says, the *second* step is to add a "parenthetic addendum" [1981, 163] to the above argument. This addendum, which is put *after* the conclusion (1c), says:

(1d) u did indeed alpha-decay during the time-interval $[t,\ t + \Delta t]$.

If, in addition, the law expressed in premise (1a) is explained (derived) from the underlying theory (quantum mechanics, in this example), then, Railton [1981, 163] says, we have "a full probabilistic explanation of $u's$ alpha-decay". This is an instance of a *DNP*-explanation.

The addendum (1d) is not an extra premise of the argument. If it were, then the explanation of why did particle u alpha-decay would be trivial. So, the addendum has to be placed *after* the conclusion (1c). Still, isn't there a feeling of dissatisfaction? Have we really explained why u did alpha-decay? If we feel dissatisfied, Railton says, it will be because we are committed to determinism. If, on the other hand, we take indeterminism seriously, there is no further fact of the matter as to why particle u did alpha-decay. This is a genuine chancy event. Hence, nothing else could be added to steps (1a)-(1d) above to make them more explanatory than they already are. Note that I have refrained from calling steps (1a)-(1d) an argument because they are not. Better, (1a)-(1c) is a deductively valid argument, but its conclusion (1c) is *not* the *explanandum*. The *explanandum* is the "addendum" (1d). But this does not logically follow from (1a)-(1c). Indeed, Railton defends the view that explanations are not necessarily arguments. Although arguments, (and in particular *DN*-arguments), "play a central role" in explanation, they "do not tell the whole story" [1981, 164]. The *general schema* to which a *DNP*-explanation of a chancy event ($G_{e,t0}$) conforms is this (cf. [1981, 163]):

(2a) For all x and for all t ($F_{x,t} \rightarrow$ Probability ($G_{x,t}$)= r)
(2b) A theoretical derivation of the above probabilistic law
(2c) $F_{e,t0}$

(2d) Probability($G_{e,t0}$)=r
(2e) $G_{e,t0}$.

(2e) is the "parenthetic addendum", which is *not* a logical consequence of (2a)-(2d). As for (2a), Railton stresses that the probabilistic generalisation must be a genuine probabilistic law of nature. The explanation is true if both the premises (2a)-(2c) *and* the addendum (2e) are true.

There are a number of important features of the *Deductive-Nomological-Probabilistic* model that need to be stressed.

- *First*, it shows how the *DN*-model is a limiting case of the *DNP* model. In the case of a *DN*-explanation, (2e) just is the conclusion of the *DN*-argument — so it is no longer a "parenthetic addendum".

- *Second*, it shows that all events, no matter how likely or unlikely they may be, can be explained in essentially the same way. In schema (2) above, the value of probability r is irrelevant. It can be anywhere within the interval (0,1]. That is, it can be anything other than zero.

- *Third*, it shows that single-case probabilities, such as the ones involved in (2a), can be explanatory. No matter what else we might think of probabilities, there are cases, such as the one discussed in Railton's example, in which probabilities can be best understood as fully objective chances.
- *Fourth*, it shows how probabilistic explanation can be fully objective. Since (2a)-(2d) is a valid deductive argument, and since the probability involved in (2a) is "a law-full, physical single-case" probability (cf. 1981, 166), the *DNP*-account does not fall prey to the objections that plagued the *Inductive-Statistical* model. There is no problem of ambiguity, or epistemic relativisation. Single-case probabilities need no reference-classes, and by stating a law, premise (2a) is maximally specific.
- *Fifth*, it shows how probabilistic explanation can be freed of the requirement of nomic expectability as well as of the requirement that the *explanandum had to* occur. So, it accommodates genuinely chancy *explananda*.
- *Sixth*, by inserting premise (2b), it shows in an improved way how explanation can be linked with understanding.

Since this last point is of some special importance, let us cast some more light on it. The Hempelian tradition took explanation to be the prime vehicle for understanding in science. But, as we have already seen, it restricted understanding of why an *explanandum* happened to showing how it should have been expected to happen, given the relevant laws. In particular, it demanded that understanding should proceed via the construction of arguments, be they *Deductive-Nomological*, or *Deductive-Statistical* or *Inductive-Statistical*. Railton's *Deductive-Nomological-Probabilistic* model suggests that understanding of why an *explanandum* happened cannot just consist in producing arguments that show how this event had to be expected. The occurrence of a certain event, be it likely or not, is explained by placing this event within a *web* of

> inter-connected series of law-based accounts of all the nodes and links in the causal network culminating in the explanandum, complete with a fully detailed description of the causal mechanisms involved and theoretical derivations of all covering laws involved. [1981, 174]

In particular, explanation proceeds also with elucidating the *mechanisms* that bring about the *explanandum*, where this elucidation can only be effected if we take into account the relevant theories and models. Railton [1981, 169] rightly protested against the Hempelian view that all this extra stuff, which cannot be captured within a rigorous *DN*-argument, is simply "*marginalia*, incidental to the 'real explanation', the law-based inference to the explanandum". He does not doubt that an appeal to laws and the construction of arguments are important, even indispensable, features of explanation. But he does doubt that they exhaust the nature of explanation.

19 ON HISTORICAL AND TELEOLOGICAL EXPLANATION

Hempel's central idea, we have seen, has been that all explanation is nomological and that all explanations are arguments. This idea was meant to capture all cases of scientific explanation — not only in the natural sciences but also in historical and human sciences as well. Indeed, the very first systematic presentation of the Deductive-Nomological Pattern appeared in a paper titled "The Function of General Laws in History", in the *Journal of Philosophy* in 1942. There, Hempel advanced the DN-model in an attempt to capture *historical* explanation, in particular. This move was a radical break with a whole philosophical tradition that flourished especially on the Continental Europe, a tradition that took historical explanation to be essentially *sui generis.*

The root of this tradition is neo-Kantianism. According to Wilhelm Windelband, there is a fundamental methodological distinction between the natural and the historical sciences. He called 'nomothetic' the method suitable for the natural sciences. They are based on universal and demonstrative judgements. They aim to specify the laws of nature and strive to reveal nomological connections among events. The historical sciences, Windelband thought, are characterised by the 'idiographic' approach. They aim at individual and concrete events. The method of historical sciences, Windelband thought, is based on value-judgements, i.e., on judgements about what events are important and why. Before Windelband, the German historian and philosopher Johann Gustav Droysen introduced a distinction between *explanation* and *understanding* (in German *Erklären* and *Verstehen*) and claimed that while the natural sciences aim to explain the phenomena, the historical sciences aim to understand the phenomena that fall within their purview. Many continental thinkers, most notably Wilhelm Dilthey, took up these ideas and developed them into a whole theory of historical explanation the basis of which is the idea that explanation in history relies on a *sui generis* method of *empathic* understanding: the re-creation in the mind of the historian of the mental milieu, the motivations, the feelings, the reasons for action etc., of the historical subject (that is, the object of the historian's study). In contradistinction to explanation, historical *understanding* was thought to have psychological and intentional elements. It was not supposed to require knowledge of causes, but knowledge of *reasons.* In his influential *The Idea of History* (1946), R. G. Collingwood put forward three basic theses of historical understanding. First, in order for the historian to understand the actions of some historical subject, he must understand the thoughts that these actions express; second, once these thoughts have been grasped, the historian fully understands these actions and hence there is no further requirement for finding the causes that produced them, or the laws, if any, that govern them; and third, understanding these actions in terms of the thoughts they express requires re-thinking of the thoughts of the historical subject by the historian. All history, Collingwood, said, is "the re-enactment of past thought in the historian's own mind".

It was against this tradition that Hempel reacted by claiming that

> Historical explanation, too, aims at showing that the event in question was not 'a matter of chance', but was to be expected in view of certain antecedent or simultaneous conditions. The expectation referred to is not prophecy or divination, but rational scientific anticipation which rests on the assumption of general laws. [1942, 39]

To the obvious charge that many historical explanations fail to state any laws, Hempel replied that, as they stand, they are explanation *sketches*: when the sketches are filled out, the reference to laws (be they historical or psychological) will be made explicit. Occasionally, Hempel thought, the laws will turn out to be probabilistic. Still, the historical explanation, when fully spelt out, will be an inductive-statistical argument. Hempel, then, proposed a unified theory of understanding: there is only one type of understanding and is tantamount to offering a proper nomological explanation of the *explanandum* (be it a natural or a historical event). Understanding, he thought, has nothing to do with empathy and everything to do with nomic expectability.

Hempel, then, was adamant that all explanation consists in the subsumption of the *explanandum* under general laws. In line with the logical empiricist ideal, he thought that he could thereby secure the objectivity of scientific explanation. But is it plausible to think that objectivity requires thinking of explanation as subsumption under laws? Perhaps, Hempel was too quick to classify all explanations that fail this requirement as pseudo-explanations. The problem runs deep. For the issue at stake is precisely whether we can talk of laws of history (or of other special sciences). Hempel was aware of the problem, but he did not face it squarely. He offered only a few examples of historical laws, e.g., that populations tend to migrate to regions that offer better living conditions. But even when these laws are available, the explanation of an individual historical event, e.g., the migration of population X to region Y, will not be a straightforward deduction of the *explanandum* from the *explanans*, the reason being that the law-like statement is too vague or holds only *ceteris paribus*, that is, when other things are equal. His thought was that these laws should be filled out with detailed information that covers the *explanandum*. But it is obvious that the more detailed the law becomes (and hence the more apt to cover the individual case at hand), the less universally applicable it is; hence the less apt it becomes to be deemed a *law*.

This last thought was a key element of William Dray's critique of the Hempelian model of explanation. For Dray the problem is not that historical 'laws' are too complex or vague, but rather that they are not proper laws. On the positive side, he claimed that historical explanation is *rational* (not causal) explanation: to explain (that is, to understand) a historical action is to state its *reasons* and not its causes. In his defence of Collingwood's approach, Dray was sensitive to the charge that since causal explanation requires laws (because the *explanandum* must be necessitated, in some sense, by its cause), if historical explanation is causal explanation, historical explanation must be nomological too. His reply to this was to *deny* that historical explanation is causal explanation. Since, however, explanation must, somehow, necessitate the *explanandum*, he suggested (as the best way to explicate Collingwood's idea) that the necessity involved in historical explanation is *rational* necessity. Hence, we can explain a certain historical action by showing that it was rationally necessary: the action was the rational thing for the agent to do on the occasion under consideration. Dray went further by claiming that rational explanations of the sort he and Collingwood envisaged are complete. Further causal considerations (in terms of natural or historical laws) are irrelevant to the explanation of a historical action.

There are several objections to Dray's idea, but the most telling one comes from Davidson's famous assertion that reasons can be causes. In asserting this, Davidson unravelled the root of the problem faced by the generic pattern of explanation that underwrites Dray's suggestion, viz., the thought that *intentional* explanation is *sui generis* (non causal) explanation. Intentional explanation, a species of which is Dray's rational explanation, refers to the explanation of actions. It has been suggested that it proceeds by citing the intentions and the beliefs of the actors as the *explanans* of certain actions. According to Davidson, intentional explanation is causal explanation, since, as he put it "the primary reason for an action is its cause" [1963, 4]. This might be taken to imply that intentional explanation is *singular* causal explanation. For, it seems, for a certain action A and its cause (or reason) C, there may not be a general law, expressed in the vocabulary of A and C, such that it is the case that whenever C then A. Davidson was ready to grant this last claim, but he denied that it follows from this that causal explanation has to be singular. As we have already seen him arguing (towards the end of section 10), he notes that singular claims of the form c caused e entail that *there is* some causal law that is instantiated in the particular causal sequence of events. This law might not be expressible in the vocabulary of the particular cause (reason) and the particular effect (action). Indeed, as he stressed, the relevant law might be expressed in neurological or physical vocabulary. Davidson's reconciliation of intentional and causal explanation keeps the view (central to the intentional approach) that explanation need not cite laws to be acceptable and good but it also retains the view (central to the Hempelian nomological approach) that causal explanations involve laws. We have already seen that Hempel's reaction to a Davidson-style compromise was that it amounts to claiming that there is a treasure somewhere without giving us any guidance as to how we should find it. The irony is that in the particular case of historical explanation Hempel came quite close to a Davidson-style attitude.

Intentional explanation has been described as a species of teleological explanation. We have already seen Leibniz going for this view. More recently, this view has been defended by von Wright [1971]. Teleological explanations, advanced by Aristotle and defended by Leibniz and Kant among others, are 'future oriented". Whereas in a typical causal explanation the earlier-in-time cause explains the later-in-time effect, in teleological explanations, as traditionally understood, the later-in-time effect (that is, the aim or purpose for which something happened) explains the earlier-in-time cause (that is, why something happened). The typical locution of a teleological explanation is: *this* happened in order that *that* should occur.

A big challenge to our thinking about explanation throughout the centuries has been the issue of whether teleological explanation is *sui generis* or whether it can be subsumed under causal or mechanistic explanation ordinarily understood. To most empiricists the very idea of teleology (that is, the existence of purposes and aims in nature for the sake of which things are done) was an anathema. Vitalism, the view that the explanation of life and living organisms cannot be mechanical but should proceed in terms of vital forces or principles, was taken to be the paradigm-case of a non-scientific theory. Many working biologists and philosophers of science devoted time and energy in trying to show how biological phenomena can be explained mechanically, with no reference to vital forces and the like. But it is fair to say that even if vitalism was neutralised, the idea that biological explanations are, in some sense, teleological survived. The idea was that there is a special type of teleological explanation, viz., *functional* explanation, that is indispensable in biology.

Teleological statements can be classified as goal-ascriptions or as function-ascriptions. Goal-ascriptions state the goal or aim towards which a certain action is directed. Function-ascriptions state the function performed by something (e.g, a biological organ, or an organism, or an artefact). Goal-directed explanation explains actions in terms of their goals, while functional explanation explains the presence of some item in a system in terms of the effects that this item has on the system of which it is a part. I will not discuss these issues in great detail. But it is fair to say that goal-directed explanations can be causal in a very straightforward sense. For instance, as Ernest Nagel has noted, intentional explanation is causal in the sense that the motives, desires and beliefs of an agent explain her actions. So it is not that the goal causes the action. Rather, the agent's desire for a certain goal, together with her belief that certain actions will achieve this goal, bring about the action (as a means to achieve an end). More sophisticated forms of goal-directed behaviour are also explainable causally (see [Nagel, 1977]).

What about, then, functional explanation? In biology, it is typical to explain a feature (a phenotypic characteristic) of a species in terms of its contribution to the enhancement of the chances of survival and reproduction. It is equally commonplace to explain the properties or the behaviour of the parts of an organism in terms of their functions in the whole: they contribute to the adequate functioning, the survival and reproduction of the whole. The explanation of the beating of the heart by appeal to its function to circulate the blood has become a standard example of such a functional explanation. Functional explanations are often characterized by the occurrence of teleological expressions such as 'the function of', 'the role of', 'serves as', 'in order to', 'for the sake of', 'for the purpose of'. It seems, then, that functional explanations explain the presence of an entity by reference to its effects. Hence, they seem to defy a strict causal analysis.

The pervasiveness of functional explanations posed a double problem to all those who denied teleology. Given that they are not present in physics, given, that is, that explanation in physics is nonteleological, the presence of functional explanations in biology suggested that biology was an underdeveloped (or immature) science. But this was absurd given the scientific successes of evolutionary biology. If, on the other hand, it was accepted that there is an indispensable special type of explanation in biology, then the methodological monism that was taken to characterise all science was in danger.

Hempel and Nagel took it upon themselves to show how functional explanation can be understood in a way that has no serious teleological implications. Hempel [1959], to be sure, was sceptical of the possibility of functional explanation. It is no accident that he titled his paper "The Logic of Functional Analysis". One of his main problems was the presence of functional equivalents, i.e., the existence of different ways to perform a certain function (for instance, artificial hearts might circulate the blood). Take then the statement: The heartbeat in vertebrates has the function of circulating blood through the organism. Would it be proper to explain the presence of heartbeat by claiming that it is a necessary condition for the proper working of the organism? If it were, we could construct a proper deductive explanation of the presence of the heartbeat. We could argue thus: The presence of the heartbeat is a necessary condition for the proper working of the organism; the organism works properly; hence, the organism has a heart. But the existence of functional equivalents shows that the intended conclusion does not follow. At best, all we could infer is the presence of one of the several items of a class of items capable of performing a certain function. Hence, Hempel thought that explanation in terms of functions works only in a limited sense and have only heuristic value.

Faced with the problem of functional equivalents, Nagel [1977] suggested that if a sufficiently precise characterisation of the type of organism we deal with is offered, only one kind of mechanism will be apt to fulfil the required function. For instance, given the evolutionary history of *homo sapiens*, the heartbeat is the only mechanism available for the circulation of the blood. If Nagel is right, there can be a proper deductive account of functional explanation. According to him, here is the form that functional explanations have (illustrated by his favourite example):

> *Functional ascription*: During a period when green plants are provided with water, carbon dioxide, and sunlight, the function of chlorophyll is to enable the plants to perform photosynthesis.
>
> *Functional explanation*:
>
> 1. During a stated period, a green plant is provided with water, carbon dioxide, and sunlight.
> 2. During a stated period, and when provided with water, carbon dioxide, and sunlight, the green plant performs photosynthesis.

3. If during a given period a green plant is provided with water, carbon dioxide, and sunlight, then if the plant performs photosynthesis the plant contains chlorophyll.

Conclusion: Chlorophyll is present in the green plant.

More schematically:

(A) This plant performs photosynthesis.

(B) Chlorophyll is a *necessary condition* for plants to perform photosynthesis.

(C) Hence, this plant contains chlorophyll.

This is a deductive-nomological explanation of the presence of chlorophyll in green plants. Premises (1) and (2) state specific conditions, and premise (3) is a law-like statement. Any appearance of teleology in the functional explanation is gone. But, as Nagel [1977, 300] notes, this is *not* a causal explanation of the presence of chlorophyll. The reason is that

> the performance of [photosynthesis] is not an *antecedent* condition for the occurrence of [the chlorophyll], and so the premise [3] is not a causal law.

He concludes:

> Accordingly, if the example is representative of function ascriptions, such explanations are *not* causal — they do not account causally for the presence of the item to which a function is ascribed.

If functional explanations are not causal, what do they achieve? They make evident the role of an item within a system. Nagel was adamant that this is a legitimate role of explanation. To explain is not necessary to cite causes. Explanation is also accomplished by finding the effects or consequences of various items. As he [ibid.] put it:

> inquiries into effects or consequences are as legitimate as inquiries into causes or antecedent conditions; (...) biologists as well as other students of nature have long been concerned with ascertaining effects produced by various systems and subsystems; and (...) a reasonably adequate account of the scientific enterprise must include the examination of both kinds of inquiries [i.e., inquiries into causal antecedents and inquiries into effects or consequences].

Functional explanation is then made to fit within the deductive nomological model, but at the price of ceasing to be causal. Obviously, there are two ways to react to Nagel's suggestion. One is to try to restore the causal character of functional explanation. The other is to deny that explanations have to be arguments. Both ways were put together in Larry Wright's [1973] *etiological* model of functional explanation.

According to Wright [1973, 154], functional ascriptions are explanatory in their own right.

> Merely saying of something, X, that it has a certain function, is to offer an important kind of explanation of X.

For instance, when it is said that the fact that plants have chlorophyll is functionally explained by noting the role that chlorophyll plays in enabling the plants to perform photosynthesis, this is a genuine explanation. It does not have to be an argument of any sort. An explanation is an answer to a why-question and the question 'why do plants have chlorophyll?' is adequately and fully answered by "providing a perfectly respectable etiology; [by] provid[ing] the reason chlorophyll is there" [1976, 100]. For Wright the problem of functional equivalents does not arise, since, as he said, it is based on a false assumption, viz., that an explanation of why a certain item performs a certain function must exclude the possibility that anything else could have performed it. All that is necessary for functional explanation, according to Wright, is that the item (e.g., chlorophyll) was *sufficient* in the circumstances to perform the given function.

'Etiology' means finding the causes. Etiological explanation is *causal* explanation: it concerns the causal background of the phenomenon under investigation. To be sure, Wright's etiological explanation is causal in an extended sense of the term: it explains how "the thing with the function got there" [1973, 156]. The basic pattern of functional explanation is:

(F)

The function of X is Z iff:

- (i) X is there because it does (results in) Z,
- (ii) Z is a consequence (result) of X's being there.

For instance, the function of chlorophyll in plants is to perform photosynthesis iff chlorophyll is there because it performs photosynthesis and photosynthesis is a consequence of the presence of chlorophyll. Clause (ii) is particularly important because it makes explicit the asymmetry of functional explanation: that the function Z is there because of X and not the other way around. An important feature of Wright's account is that it is especially suitable for explanation in biology, where the notion of natural selection looms large. The etiological explanation of natural (biological) functions is in terms of natural selection: they are the results of natural selection because they have endowed their bearers with an evolutionary advantage. Consequently, etiological explanation does not reverse the causal order: a function is performed because it has been causally efficacious *in the past* in achieving a certain goal. Wright, however, insists that etiological explanation is teleological in an important sense: it is future-oriented. As he [1976, 105] put it:

> To deny the propriety of teleological explanation (...) is to deny the obviously right answer to [some] question[s]: namely, that [something] is there because of its (...) consequences.

According to Robert Cummins [1975] approaches like the ones mentioned above fail to capture what is distinctive in functional explanation, viz., that it explains a capacity that a system has in terms of the capacities of its parts. To ascribe a function to an item i which is part of a system S (that is, to say that item i functions as ...) is to ascribe to it some capacity in virtue of which it contributes to the capacities of the whole system S. So, functional explanations explain how a system can perform (that is, has the capacity to perform) a certain complex task by reference to the capacities of the parts of the system to perform a series of subtasks that add up to the system's capacity. For instance, the capacity of an organism to circulate blood is functionally explained by reference to the capacities of certain parts of this organism, viz., the capacity of the blood to carry oxygen, the capacity of the heart to pump the blood, the capacity of the valves to direct the blood from the lungs to the organs etc. We may then say that the heart has the *function*, within the organism, to circulate the blood, (by virtue of its capacity to pump the blood), but we do not thereby explain the presence of the heart in the organism, as the standard conception of functional explanation would have it (cf. [1975, 762]). Cummins sums up his position thus:

> To ascribe a function to something is to ascribe a capacity to it which is singled out by its role in an analysis of some capacity of a containing system. When a capacity of a containing system is appropriately explained by analysing it into a number of other capacities whose programmed exercise yields a manifestation of the analysed capacity, the analysing capacities emerge as functions. [1975, 765]

Cummins's view can be called the *causal role* theory of functional explanation. On this theory functional explanation does not answer why-is-it-there questions at all, but how-does-it-work questions.

This is not, of course, the end of the story for conceptions of functional explanation. The debate as to how exactly functional explanation should be understood, especially in connection with biological explanation, is pretty much alive today.

20 A CONCLUDING THOUGHT

In light of the preceding discussion, which has barely scratched the surface of our thinking about explanation, it should be obvious that there is no consensus on what explanation is. Perhaps, the very task of explaining explanation, if by that we mean the advancement of a single and unified account of what explanation is, is futile and ill-conceived. Perhaps, *explanation* is a loose concept that applies to many things; it is such that it can be partially captured by different models and accounts. Perhaps, the only way to understand explanation is to embed it within a framework of kindred concepts and try to unravel their interconnections. Indeed, the concepts of *causation*, *laws of nature* and *explanation* form a very tight web. As it should be evident by now, hardly any progress can be made in any of those, without relying on, and offering accounts of, some of the others. All we may then hope for is some enlightening accounts of the threads of the web formed by these concepts.

ACKNOWLEDGEMENTS

Many thanks to Theo Kuipers for useful comments. Research for this paper was funded by the framework EPEAEK II in the programme Pythagoras II.

BIBLIOGRAPHY

[Aristotle, 1984] Aristotle. *The Complete Works of Aristotle.* Jonathan Barnes (ed.), 2 vols., Princeton N.J.: Princeton University Press, 1984.

[Armstrong, 1983] D. M. Armstrong. *What Is a Law of Nature?* Cambridge: Cambridge University Press, 1983.

[Bromberger, 1966] S. Bromberger. Why-questions. In R. G. Colodny (ed.), *Mind and Cosmos: Essays in Contemporary Philosophy of Science.* Pittsburgh: Pittsburgh University Press, 1966.

[Carnap, 1928] R. Carnap. *The Logical Structure of the World.* Berkeley: University of California Press, 1928.

[Carnap, 1974] R. Carnap. *An Introduction to the Philosophy of Science.* New York: Basic Books, 1974.

[Cartwright, 1983] N. Cartwright. *How the Laws of Physics Lie.* Oxford: Clarendon Press, 1983.

[Collingwood, 1946] R. G. Collingwood. *The Idea of History.* Oxford: Clarendon Press, 1946.

[Cummins, 1975] R. Cummins. Functional analysis. *The Journal of Philosophy*, 72: 741–765, 1975.

[Davidson, 1967] D. Davidson. Causal relations, *The Journal of Philosophy* 64: 691-703, 1967.

[Davidson, 1963] D. Davidson. Actions, reasons and causes. *The Journal of Philosophy*, 60, 1963. Reprinted in *Essays on Actions and Events*, Oxford: Oxfrod University Press, 1980.

[Descartes, 1644] R. Descartes. *Principles of Philosophy* (1644). In *The Philosophical Writings of Descartes*, Vol. 1. Translated by J. Cottingham, R. Stoothoff, and D. Murdoch, Cambridge and New York: Cambridge University Press, 1985.

[Dowe, 2000] P. Dowe. *Physical Causation.* Cambridge: Cambridge University Press, 2000.

[Dretske, 1977] F. I. Dretske. Laws of nature. *Philosophy of Science*, 44: 248–68, 1977.

[Friedman, 1974] M. Friedman. Explanation and scientific understanding. *Journal of Philosophy*, 71: 5–19, 1974.

[Harré and Madden, 1975] R. Harré and E. H. Madden. *Causal Powers: A Theory of Natural Necessity.* Oxford: Basil Blackwell, 1975.

[Hempel, 1942] C. G. Hempel. The function of general laws in history. *The Journal of Philosophy*, 39: 35–48, 1942.

[Hempel, 1959] C. G. Hempel. The logic of functional analysis. In L. Gross (ed.), *Symposium on Sociological Theory.* New York: Harper and Row Publishers, 1959.

[Hempel, 1965] C. G. Hempel. *Aspects of Scientific Explanation.* New York: The Free Press, 1965.

[Hume, 1739] D. Hume. *A Treatise of Human Nature* (1739). L. A. Selby-Bigge and P. H. Nidditch (eds.), Oxford: Clarendon Press, 1978.

[Hume, 1740] D. Hume. *An Abstract of A Treatise of Human Nature* (1740). L. A. Selby-Bigge and P. H. Nidditch (eds.), Oxford: Clarendon Press, 1978.

[Kant, 1781] I. Kant. *Critique of Pure Reason* (1781). N. Kemp Smith (trans.), New York: St Martin's Press, 1965.

[Kant, 1786] I. Kant. *Metaphysical Foundations of Natural Science* (1786). J. Ellington (trans.), Indianapolis and New York: The Bobbs-Merrill Company, INC, 1970.

[Kitcher, 1976] P. Kitcher. Explanation, conjunction and unification. *The Journal of Philosophy*, 73: 207–12, 1976.

[Kitcher, 1981] P. Kitcher. Explanatory unification. *Philosophy of Science*, 48: 251–81, 1981.

[Kitcher, 1985] P. Kitcher. Two approaches to explanation. *The Journal of Philosophy*, 82: 632–9, 1985.

[Kitcher, 1986] P. Kitcher. Projecting the order of nature. In R. E. Butts (ed.), *Kant's Philosophy of Science*. Dordrecht: D Reidel Publishing Company, pages 201–35, 1986.

[Kitcher, 1989] P. Kitcher. Explanatory unification and causal structure. *Minnesota Studies in the Philosophy of Science*, 13, Minneapolis: University of Minnesota Press, pages 410–505, 1989.

[Kripke, 1972] S. Kripke. *Naming and Necessity*. Oxford: Blackwell, 1972.

[Leibniz, 1686] G. Leibniz. *Discourse on Metaphysics* (1686). In *Discourse on Metaphysics, Correspondence with Arnauld, Monadology*, G. Montgomery (trans.). The Open Court Publishing Company, 1973.

[Leibniz, 1698] G. Leibniz. *Monadology* (1698). In *Discourse on Metaphysics, Correspondence with Arnauld, Monadology*, G. Montgomery (trans.). The Open Court Publishing Company, 1973.

[Leibniz, 1973] G. Leibniz. *Philosophical Writings*. M. Morris and G. H. R. Parkinson (trans.). London: Everyman's Library, 1973.

[Lewis, 1973] D. Lewis. *Counterfactuals*. Cambridge MA: Harvard University Press, 1973.

[Lewis, 1986] D. Lewis. Causal explanation. In his *Philosophical Papers, Vol. II*. Oxford: Oxford University Press, pages 214–40, 1986.

[Mackie, 1977] J. L. Mackie. Dispositions, grounds and causes. *Synthese*, 34: 361–70, 1977.

[Malebranche, 1674–5] N. Malebranche. *The Search After Truth* (1674–5). T. M. Lennon and P. J. Olscamp (trans.). Cambridge and New York: Cambridge University Press, 1997.

[McMullin, 2001] E. McMullin. The impact of Newton's *Principia* on the philosophy of science. *Philosophy of Science*, 68: 279–310, 2001.

[Mill, 1843] J. S. Mill. *A System of Logic: Ratiocinative and Inductive* (1843). London: Longmans, Green and Co., (8th ed.) 1911.

[Nagel, 1977] E. Nagel. Teleology revisited. *The Journal of Philosophy*, 75: 261–301, 1977.

[Psillos, 2002] S. Psillos. *Causation and Explanation*, Chesham and Montreal: Acumen and McGill-Queens University Press, 2002.

[Railton, 1981] P. Railton. Probability, explanation and information. *Synthese*, 48: 233–56, 1981.

[Ramsey, 1928] F. P. Ramsey. Universals of law and of fact (1928). In D. H. Mellor (ed.), *Foundations: Essays in Philosophy, Logic, Mathematics and Economics*. London: Routledge and Kegan Paul, 1978.

[Reichenbach, 1956] H. Reichenbach. *The Direction of Time*. Berkeley and Los Angeles: University of California Press, 1956.

[Salmon *et al.*, 1971] W. Salmon *et al.*. *Statistical Explanation and Statistical Relevance*. Pittsburgh: University of Pittsburgh Press, 1971.

[Salmon, 1984] W. Salmon. *Scientific Explanation and the Causal Structure of the World*. Princeton: Princeton University Press, 1984.

[Salmon, 1985] W. Salmon. Conflicting conceptions of scientific explanation. *Journal of Philosophy*, 82: 651–4, 1985.

[Salmon, 1989] W. Salmon. *Four Decades of Scientific Explanation*. Minneapolis: University of Minnesota Press, 1989.

[Schlick, 1932] M. Schlick. Causation in everyday life and in recent science. In H. L. Mudler and B. F. B. De Velde-Schlick (eds.), *Philosophical Papers*, Vol. 2 (1925–1936). Dordrecht, Netherlands: D. Reidel, 1979.

[Scriven, 1962] M. Scriven. Explanations, predictions and laws. *Minnesota Studies in the Philosophy of Science*, 3, Minneapolis: University of Minnesota Press, 1962.

[Shoemaker, 1980] S. Shoemaker. Causality and properties. In P. van Inwagen (ed.), *Time and Change*. D. Reidel, pages 109–35, 1980.

[Suppes, 1984] P. Suppes. *Probabilistic Metaphysics*. Oxford: Blackwell, 1984.

[Thayer, 1953] H. S. Thayer (ed.). *Newton's Philosophy of Nature: Selections from his Writings.* New York and London: Hafner Publishing Company, 1953.
[Tooley, 1977] M. Tooley. The nature of laws. *Canadian Journal of Philosophy*, 7: 667–98, 1977.
[von Wright, 1971] G. H. von Wright. *Explanation and Understanding.* London: Routledge & Kegan Paul, 1971.
[Woodward, 2000] J. Woodward. Explanation and invariance in the special sciences. *The British Journal for the Philosophy of Science*, 51: 197–254, 2000.
[Woodward, 2003] J. Woodward. *Making Things Happen: A Theory of Causal Explanation.* New York: Oxford University Press, 2003.
[Wright, 1973] L. Wright. Functions. *Philosophical Review*, 82: 139–168, 1973.
[Wright, 1976] L. Wright. *Teleological Explanations: An Etiological Analysis of Goals and Functions.* Berkeley & London: University of California Press, 1976.

EVALUATION OF THEORIES

Ilkka Niiniluoto

1 INTRODUCTION

1.1 Success of science: institution, inquiry, method, knowledge

Science is an institution for promoting the search of knowledge. All those who are engaged in the creative and critical use of the scientific method constitute the scientific community. The research system is composed of researchers, their groups, university departments, research units and networks, stations and laboratories. The funding of research activities is provided by the organs and principles of science policy. The community of investigators has its own normative and ethical standards concerning rewards, quality, and best practices. The ethos of science is to produce systematic and justified new knowledge by inquiry. The task of the universities is to transfer critical thinking and knowledge to new generations by offering education by means of research-based teaching. The innovation system is designed to employ knowledge for the creation of new technologies and commercial applications.

This outline is sufficient to show that science can be viewed from several complementary perspectives. Science is an *institution* supporting a highly organized activity within the research system. Science is also an activity of researchers who are collectively conducting *inquiry*. Within this process, we may distinguish its primary tool, the critical *method* of research, and its primary product, a system of *knowledge*.

All these domains have been in continuous change and development during the history of science. The institutional framework of science has been expanded and reorganized, the method of inquiry has been refined by the introduction of experimental and quantitative techniques, and the body of accepted scientific knowledge has been constantly extended and revised. Thus, the evaluation of the success of science can likewise be discussed and defined on different levels.

The traditional method of assessing scientific quality is by *peer review*, i.e., by judgement of experts in the same scientific specialty. Referees evaluate articles submitted to scientific journals, senior scholars serve as assessors of doctoral dissertations, and selected experts give statements about the applicants of professorships in universities. Peer review methods have been extended also to "quality assurance" in the form of research assessment exercises where external evaluation panels give their judgement about the success of scientific fields or university departments.

Handbook of the Philosophy of Science: General Philosophy of Science - Focal Issues, pp. 175–217.
Volume editor: Theo Kuipers. General editors: Dov M. Gabbay, Paul Thagard and John Woods.

Institutional measures of success have been widely developed in the latter half of the twentieth century by quantitative historians of science and sociologists of science. The new discipline of *scientometrics* employs bibliometric and other statistical techniques in the description of scientific activities [Elkana *et al.*, 1978]. The units of such measures range from individual researchers and research groups to institutes, departments, or whole universities. Various kinds of quantitative *performance indicators* express how well these units have succeeded in the competition of research funding, employment of qualified researchers, acceptance of academic degrees, or publication of articles and monographs. Such indicators may be direct output measures (e.g., the number of articles or patents), or cost-effect measures of efficiency (e.g., the number of articles per professor) (see [Irvine and Martin, 1984]). Some indicators measure *quality* at least indirectly by reflecting recognition, visibility and impact within scientific community (e.g., the number of articles in journals using the referee system, the number of citations in respected journals, international prizes and awards).

As science is a goal-directed activity, ultimately the success of researchers and their institutions should depend on the results that they have achieved. These results should be praised for their *cognitive merits*, as they are expected to give new knowledge. The important task of characterizing knowledge as the goal of inquiry belongs to epistemology and philosophy of science. Plato's classical definition of knowledge (Gr. *episteme*, Lat. *scientia*) as justified true belief has set up the stage for later debates (see [Pappas and Swain, 1978; Niiniluoto *et al.*, 2004]). As Aristotle pointed out, besides *knowing that*, also *knowing why* is an important goal of science. But many philosophers have suggested alternative accounts of the important virtues of science. Whatever analysis of such aims is adopted, to evaluate science on this level one has to consider the contribution of a scientist or a scientific publication to the *content* of the cumulative results of the community of investigators. This presupposes an account of the main kinds of products by means of which the scientists communicate and express their research findings.

1.2 Theories as units of evaluation

Plato's definition suggests three possible bearers of knowledge or knowledge claims: beliefs as mental states, sentences of interpreted languages, and propositions expressed by such sentences. As science strives for public knowledge, analytic philosophers of science have mainly favored a linguistic *statement* view: knowledge claims in science are expressed by more or less hypothetical statements H that can be formulated by interpreted sentences in a scientific language L. Also the evidence E that is hoped to justify such statements, and the relevant background knowledge B employed in such justification, are assumed to be expressible by sentences.

The notion of a *theory* is used as a label of several kinds of scientific statements. Some of them are qualitative or quantitative singular sentences that serve as answers to WH-questions (i.e., what, which, who, where, when -questions).[1] For

[1]Note that the linguists do not include why- and how-questions among WH-questions.

examples, rival scientific "theories" in the fields of natural sciences, humanities, and social sciences may give answers to following types of questions: What is the atomic weight of uranium? When did the dinosaurs disappear from the earth? Where was the original proto-Finnish language spoken? Who wrote the poems of the *Ilias*? What kind of government did the Incas have?

On a higher level, scientific theories may be universal generalizations (All ravens are black, All material particles satisfy Newton's equations), statistical generalizations (10 % of the Finns are left-handed), existential generalizations (There are quarks); in many cases, their quantifier structure is even more complex (Every living creature is born from an egg). Quantitative laws can express relations between quantities (The length of a body is a function of its temperature). Dynamic laws describe the behavior of a system in terms of time (e.g., Galileo's law of free fall).[2]

Empirical and experimental laws, which use only observational predicates and relations, are often distinguished from genuine theories, which use theoretical terms putatively referring to unobservable entities and mechanisms (e.g., atoms, genes, hidden motives). The language of science is then likewise divided into two layers: the observational language L_O is a sublanguage of the language L of the full theory.[3]

Aristotle's ideal was to formulate theories as axiomatic systems. In the statement view, such theories can be formulated by sets of sentences closed under logical deduction. The sentences deducible from the *axioms* are the *theorems* of a theory. This is the standard notion of a theory in mathematical logic and metamathematics [Bell and Slomson, 1969].

Aristotle also emphasized that science should give *explanations* which are answers to why-questions. Theories answering such explanatory questions may include historical narratives (Why was Julius Caesar murdered?) and natural causes (Why did dinosaurs die? Why are ravens black? Why do the planets orbit around the sun?). According to the covering law account, an explanatory question of the form 'Why E?' is typically answered by subsuming the statement E in the framework of a general theory T, i.e., by showing that E is derivable from T and some initial conditions [Hempel, 1965; Salmon, 1984].

In the Aristotelean tradition of logic, scientific theories are claimed to be necessarily true. The distinction between logical and natural necessity was clarified in medieval scholasticism, but after David Hume's criticism of causality scientific laws were generally taken to be generalizations in the indicative mood, i.e., statements about the actual world and its history. Only in the latter part of the twentieth century, when modal logic was reborn, it became again popular to assume that scientific laws and theories involve "nomic necessity": they should be true in all physically or nomically possible worlds, and their "causal powers" [Harré and Madden, 1975] exhibit their counterfactual force (see [Lewis, 1973]).

[2]For classifications of different types of scientific laws, see [Niiniluoto, 1987] and the article of Kuipers in this volume.

[3]The problematic distinction between theoretical and observational language is discussed by Hempel [1965], Tuomela [1973], and Brown [1977].

Logicians and philosophers have proposed many alternative ways of formulating scientific theories and their structure. The *semantic* or *structuralist* view of theories follows the strategy of replacing a proposition by the class of its models (see [Suppe, 1977; Suppes, 1993]). For example, Newton's mechanics applies a set-theoretical predicate K (characterized by Newton's equations) to certain set-theoretical structures W (physical systems). However, even though the proponents of this approach argue against the statement view that theories are not linguistic entities, most of them admit that a theory in their sense makes an "empirical claim". In the simplest case, such a claim asserts that all systems W in a class I can be described by the core predicate K. The main difference to the standard statement view is that a theory is thought to have several intended applications in the set I, instead of one global application to the whole of the actual world. For the purposes of this paper, this subtle difference is not essential.[4] For a simple illustration, let $\mathbf{R}$ be a given class of ravens, and let $\mathbf{B}$ be the predicate applying to classes of black things. Then the structuralist claim

1. $\mathbf{B}$ applies to $\mathbf{R}$

is equivalent to the set-theoretical statement $\mathbf{R} \subseteq \mathbf{B}$, which in turn can be formulated by a generalized conditional about all objects in the universe:

2. $\forall x(Rx \rightarrow Bx)$.

The applications of the formula (2) are widened, if x is allowed to range over systems or structures and the predicates R and B may involve functions and quantitative descriptions (e.g., 'satisfies Newton's equations').

The main novelty of structuralism can be incorporated within the statement view by describing a theory as a pair (T, I), where T is a set of sentences and I a set of structures to which T is claimed to be applicable. By using Tarski's model-theoretical jargon, T is claimed to be *true* in each of the structures W in I, i.e., I is a subset of the set of models of T. If I has only one member W, then the statement view and structuralist view agree that the (core of the) theory T should be applicable to W, i.e., T should be true in structure W (see [Niiniluoto, 1999]).

The next two sections discuss methods and concepts that can be used in the assessment of scientific theories. The basic virtues of a theory are taken to be its consistency, information content, and empirical success in explanation and prediction. We shall also see how the notion of truth can be modified to include approximate truth and truthlikeness. The next two sections about confirmation and acceptance then investigate the conditions under which empirical success is an indicator of the truth or truthlikeness of a theory.

[4]For sophisticated formulations of the structuralist view, see [Stegmüller, 1976] and [Balzer, Moulines, and Sneed, 1987]. Cf. [Niiniluoto, 1984]. See also the article of Kuipers in this volume.

2 DIMENSIONS OF EVALUATION

2.1 Hypotheses: pursuit, confirmation, acceptance

Let C be a cognitive subject, either an individual scientist or the relevant scientific community, and let H be a statement that C considers. Then the cognitive status of H with respect to C may have different stages.

In the weakest sense, C *thinks* or contemplates about H. Such an act of apprehension presupposes that C has been able to formulate H. Perhaps C has *learned* H from some source, or C has *discovered* H.

If C recognizes that she does not believe H and does not believe its negation $\sim H$, then C *suspends judgement* about H. In this situation C may express *interest* about H by raising the whether-question 'H or $\sim H$?'. It may also be the case that H would be promising or relevant to C in providing information that C can use to solve other cognitive or practical problems. In other words, H and $\sim H$ are *potential answers* to some query that C is willing to consider. Then C may start an *inquiry* that is hoped to settle the status of H. We may say that C is *pursuing* investigation about H.

More generally, *cognitive problems* may be represented by sets $B = \{H_1, H_2, \ldots, H_n\}$ of mutually exclusive and jointly exhaustive sets of hypotheses. Then one and only one of these hypothesis is true; this the *target* of the problem B. The hypotheses H_i are potential *complete answers* to problem B, and their disjunctions are *partial answers*. The weakest of these disjunctions $H_1 \vee \ldots \vee H_n$ is a tautology; it represents ignorance about the true answer.

Some statements H may be *provable* by a mathematical proof or *verifiable* by direct empirical observation. In such cases, the cognitive subject C may claim that she is completely justified in believing that H is true, i.e., C *knows* that H. Even though such a guaranteed cognitive status is a regulative ideal of scientific inquiry, the possibility of mistakes should be acknowledged in all areas of factual or empirical research (outside logic and mathematics). This cautious epistemological position was called *fallibilism* by Charles S. Peirce (see [Peirce, 1931-35]). For a fallibilist, all putative knowledge claims in science are temporary and revisable by further evidence.[5]

In particular, this fallibilist attitude applies to all scientific theories. Their truth is always more or less hypothetical. *Inductivists* thought that scientific theories H can be derived from observational data E, but they had to acknowledge that universal empirical laws transcend any finite number of their observed positive instances — this is the root of *the problem of induction*. It is always possible that the next instance to be studied is a counter-example which is sufficient to *falsify* a universal generalization H. Still, a fallibilist may hope that positive instances in evidence E give *support* or *confirmation* to H. This means that E makes H *more*

[5]Fallibilism should be distinguished from the negative doctrine of scepticism — even though the roots of fallibilism go back to the ancient school of Academic scepticism (see [Niiniluoto, 2000]).

credible to C. Eventually, with increasing evidence, C's degree of belief in H may approach *certainty*.

A different but related account of the testing of a hypothesis is given by the *hypothetico-deductive* (HD) model of inquiry. A hypothesis H can be freely discovered or proposed as a potential solution to a cognitive problem. Then H is indirectly *tested* by deducing an empirical consequence E from H (plus some assumed background knowledge B). Assuming B, the observation that E is false leads to the rejection of H by *modus tollens*.[6] The observation that E is true does not prove H, but still it may empirically support H or give confirmation to H.

Some fallibilists acknowledge a further stage of inquiry where H is tentatively *accepted* on the basis of supporting evidence. If the evidence for H is not strong enough, the scientists may suspend judgement about H.[7] But the relevant scientific community may also reach a consensus about the best theories so far proposed and successfully evaluated by critical tests. This body of accepted hypotheses constitutes the current "scientific knowledge", but it is always revisable by new findings.

It may happen that two groups of scientists C and C' pursue different hypothetical theories H and H'. This kind of weak motivational disagreement between C and C' is not yet a real *controversy* between C and C'. A genuine debate starts, if C and C' disagree, on the basis of their investigations, about the confirmation and acceptance of H versus H'.[8]

The background knowledge B expresses the presuppositions of the testing of a hypothesis H. Such presuppositions are paradigms of "normal science" [Kuhn, 1970] or core assumptions of "research programmes" (see [Lakatos and Musgrave, 1970]). Kuhn argued that in normal science only one paradigm dominates the scientific community, but in the period of a crisis there are rival theoretical frameworks competing with each other. Lakatos instead suggested that the existence of competing research programmes is the normal state of science. This means that there may be two or more groups of scientists C and C' who pursue different research programmes B and B'. If the pursuit of interesting theories is not progressive on the basis of B, then the status of assumptions B may be challenged. The revision of B leads to scientific revolutions where scientists are ready to change their research programmes.[9]

[6]Strictly speaking, if B and H together logically entail E, then the falsity of E entails that B or H is false. Thus, H can be rejected only by taking B for granted. This so called Duhem - Quine problem is not systematically discussed in this paper (cf. [Duhem, 1954]).

[7]Thus, there is in general no need to appeal to extra-scientific criteria. Relativist sociologists of science have claimed that scientists in fact tend to choose their favorite theories on non-epistemic social grounds, like personal or class interests and economic profit. This issue is not discussed in this paper (cf. [Longino, 1990; Niiniluoto, 1999]).

[8]An even deeper clash between rival schools appears, if C and C' disagree about the methodological standards and scientific values in assessing hypotheses. For the interplay between theories, methods, and values, see [Laudan, 1984].

[9]See also the article of Kuipers in this volume.

2.2 Explications: qualitative, comparative, quantitative

Philosophers have proposed various kinds of explications for the notions of pursuit, testability, confirmation, support, and acceptance of scientific hypotheses. Methodologically, such explications can be divided into qualitative, comparative, and quantitative.

Qualitative approaches are based on dichotomous yes-or-no concepts: a hypothesis is either promising or not, either confirmed by evidence or not, either acceptable on evidence or not. The same applies to concepts like explanation: theory T either explains data E or not.

Comparative approaches recognize the need to consider the value of rival hypotheses for a cognitive agent C. Let H_1 and H_2 be hypotheses which compete for their status. Here C may prefer H_1 over H_2 as a candidate for a promising testable hypothesis, or C may judge that H_1 is better confirmed by evidence E than H_2, or C may find H_1 to be acceptable instead of H_2. Similarly, C may judge that H_1 is a better explanation of E than H_2. In such cases of preference, we may say in general that

3. H_1 supersedes H_2 for C.

If such a relation of *superseding* is based upon values V accepted by C, then C's preference is *rational* with respect to V. Another agent C' with other values V', even though she might have the same evidence and background assumptions as C, may have different rational preferences. In this sense, comparative evaluations of theories have been *relative* to the variable standards actually adopted in the history of science [Doppelt, 1983]. For example, theories postulating unobservable explanatory entities became viable in the 19th century, when the HD method gained popularity over inductivism which wished to restrict the relevant hypotheses to observational generalizations [Laudan, 1981].

It is an important issue of the axiology of science whether there exist some overarching values or normative standards that all scientists *ought* to accept in their comparative assessments. Some philosophers argue that we are confined to relativism about actually adopted standards, while some claim that philosophers are able to justify atemporal criteria that ideally characterize everything that deserves to be called science (see [Kuhn, 1977; Niiniluoto, 1999]).

Quantitative approaches suggest that testworthiness, testability, confirmation, credibility, and acceptability are matters of degree. Probability was introduced as a quantitative concept already in the 17th century. More recently, various explications have been given to degrees of confirmation, degrees of empirical support, degrees of explanatory power, and degrees of truthlikeness. To generalize this idea, let V be a function which expresses for C the value $V(H, E)$ of hypothesis H with respect to evidence E. Then all cases where H_1 supersedes H_2 for C in the comparative sense of (3) should be covered by V, i.e., we should have

4. $V(H_1, E) > V(H_2, E)$.

On the other hand, it may happen that the quantitative relation (4) has more discriminative power than the comparative one (3): there may be hypotheses that are not comparable by the standards available for (3), so that (4) may lead to richer possibilities of comparison. In compensation, the comparative judgements (3) may be restricted to "safer" cases, and it may be felt that the introduction of quantitative considerations has to rely on more or less "arbitrary" choices.[10]

3 VIRTUES OF THEORIES

In this chapter, we consider various virtues that have been demanded of good theories. Weighted combinations of such virtues define different standards of the evaluation of scientific theories.

3.1 Consistency

A theory should be *consistent*. For an axiomatic theory this requirement seems obvious, since from a contradiction it is possible to deduce any statements. Thus, an inconsistent theory in language L has all the sentences of L as its theorems. Even though consistency is not effectively decidable in first-order logic, as witnessed by axiomatizations of arithmetic, there are many standard logical techniques for checking the consistency of sets of statements.[11] Semantically, consistency of a theory means that it has at least one model [Bell and Slomson, 1969].

The demand of self-consistency should be distinguished from requirements that a theory H ought to be *compatible* with some external body of information. Long-standing controversies between science and religion have been fostered by the demand of the Church that scientific theories should be consistent with the Bible. Today such a view is generally rejected in the name of the autonomy of research.

Often theories are required to be consistent with already established background assumptions and with theories from other neighboring disciplines. In normal situations this requirement is sound: ultimately the results of different scientific specialties should constitute a consistent world view. However, the principle of fallibilism emphasizes that all scientific claims are corrigible and open to criticism. Major scientific revolutions have emerged from the courage of researchers to challenge old and well-established doctrines.

3.2 Truth

The classical definition of knowledge includes the requirement of *truth*. In the dominant tradition, truth is characterized as correspondence between beliefs and

[10]For discussions about this issue in connection with degrees of truthlikeness, see [Niiniluoto, 1987; 1998] and [Kuipers, 2000]. Patrick Maher's paper and Kuipers' reply in [Festa *et al.*, 2005] discuss the relative merits of qualitative and quantitative confirmation.

[11]Paraconsistent logic is an attempt to restrict rules of logical inference so that it becomes possible and interesting to work with inconsistent sets of premises.

facts, or between interpreted statements and reality. According to Alfred Tarski's adequacy condition, the statement 'Snow is white' is true if and only if snow is white. Truth is thus a language — world relation which obtains objectively, i.e., independently of our cognitive states, hopes, and wishes. Tarski's semantical definition, and its further development in model theory, is often understood but also disputed as an explication of the correspondence theory of truth [Tarski, 1956; Kirkham, 1992; Niiniluoto, 1999; David, 2004]. In the possible world semantics, also inspired by Tarski's work, the notions of truth and falsity are extended to modal statements [Lewis, 1973; Hintikka, 1988].

Alternative accounts of truth have been based upon notions of coherence, consensus, ideal acceptability, warranted assertability, and utility. These notions have been proposed by idealists, pragmatists, intuitionists, and verificationist (see [Dummett, 1978; Putnam, 1981]). They are often motivated by the worry that it is not possible to check the correspondence relation between language and reality, and therefore truth should be characterized as an intra-language relation between statements. The correspondence theorist, instead, argues that we should acknowledge that truth might be inaccessible or "recognition transcendent" to us in some circumstances. Hence, the *definition* of truth should be distinguished from the empirical *criteria* or *indicators* of truth. The epistemic and practical criteria of alternative definitions may serve as such indicators, but they should not be confused with the objective truth relation itself.

The issue about truth is an important difference between main schools in philosophy of science. The requirement that truth is an important virtue of a scientific theory has been defended by *scientific realists*. They wish to apply the correspondence notion of truth to all parts of a theory, including its postulates of unobservable entities and its laws about them.[12] *Constructive empiricists* accept that full-blown theories have truth values, but this is deemed to be irrelevant to the goals of science: the demand of truth concerns only what the theory says about the observable phenomena [van Fraassen, 1980; cf. Laudan, 1977]. *Instrumentalists* deny that theoretical statements have truth values — they are simply uninterpreted symbolic devices for the purpose of deriving and systematizing empirical statements [Duhem, 1954]. Besides constructive empiricism, two middle positions between scientific realism and instrumentalism can be noted: according to *structural realism*, laws have but postulates lack truth values [Worrall, 1989]; according to *entity realism*, postulates have but laws lack truth values [Hacking, 1983].

3.3 Probability

The notion of probability has two main interpretations. Physical probabilities or *propensities* are objective features of chance set-ups, like radioactive atoms or

[12] See [Leplin, 1984; Niiniluoto, 1999; Psillos, 1999]. Tuomela's [1984] causal internal realism, which is inspired by Sellars's [1963] inferential semantics, takes explanation to be more basic than truth, and therefore he supports an epistemic notion of truth.

lottery wheels. They express a degree of possibility, i.e., a dispositional tendency of a system to behave in a certain manner in trials of a certain kind [Fetzer, 1981]. In repeated trials, the system exhibits characteristic statistical frequencies. Propensities may be ingredients of probabilistic theories and laws.

In epistemic interpretations, probability $P(H/E)$ is a rational *degree of belief* in the truth of hypothesis H on the basis of evidence E. Here $P(H/E)$ is the *posterior probability* of H given E, while the *prior probability* $P(H)$ can be defined as $P(H/t)$, where t is the empty or tautological evidence. The connection between prior and posterior probabilities is given by Bayes's Theorem:

$$5.\ P(H/E) = \frac{P(H)P(E/H)}{P(E)}$$

$$= \frac{P(H)P(E/H)}{[P(H)P(E/H) + P(\sim H)P(E/\sim H)]},$$

where $P(E/H)$ is the *likelihood* of H relative to E.

Both physical and epistemic probabilities satisfy the basic axioms of mathematical probability. In particular, $0 \leq P(H) \leq 1$ and $P(\sim H) = 1 - P(H)$. Further, when $\vdash$ denotes logical deduction, $P(H/E) = 1$ if $E \vdash H$. The maximum probability 1 corresponds to necessity in the physical interpretation and full certainty in the epistemic interpretation. In this sense, probability is a sort of surrogate or estimate of the truth of a hypothesis.

In the Bayesian tradition, the investigator is free to choose her "subjective" or "personal" prior probabilities. But results proved by Bruno de Finetti show that, under relatively mild conditions, the posterior probabilities of different subjects will converge towards each other with increasing common evidence — independently of their priors, as long as they are not dogmatically chosen to have the extreme values 0 or 1 [Kyburg and Smokler, 1965; Earman, 1992].

More specific recommendations for the choice of epistemic probabilities are given in *inductive logic*, where inductive probabilities are at least partly determined by symmetry assumptions concerning the underlying language [Carnap, 1962; Hintikka and Suppes, 1970; Niiniluoto and Tuomela, 1973; Niiniluoto, 1987].

More precisely, let $Q_1, \ldots, Q_K$ be a K-fold classification system with mutually exclusive predicates, so that every individual in the universe U has to satisfy one and only one Q-predicate. A typical way of creating such a classification system is to assume that our monadic language L contains k basic one-place predicates $M_1, \ldots, M_k$, and each *Q-predicate* is defined by a k-fold conjunction of positive or negative occurrences of the M-predicates: $(\pm)M_1x\& \ldots \&(\pm)M_kx$. Then $K = 2^k$. Each predicate expressible in language L is definable as a finite disjunction of Q-predicates. Rudolf Carnap generalized this approach to the case where the dichotomies $\{M_j, \sim M_j\}$ are replaced by families of mutually exclusive predicates $\mathbf{M}_j$, and each Q-predicate is defined by choosing one element from each family $\mathbf{M}_j$ (see [Jeffrey, 1980]). For example, one family could be defined by color predicates, another by a quantity taking discrete values (e.g., age).

A *state description* relative to individuals $a_1, \ldots, a_m$ tells for each a_i which Q-predicate it satisfies in universe U. A *structure description* tells how many individuals in U satisfy each Q-predicate. Every sentence within the first-order monadic framework L can be expressed as a disjunction of state descriptions. Let e describe a sample of n individuals $a_1, \ldots, a_n$ in terms of the Q-predicates, and let $n_i \geq 0$ be the observed number of individuals in cell Q_i (so that $n_1 + \ldots + n_K = n$). Carnap's λ-continuum takes the posterior probability $P(Q_i(a_{n+1})/e)$ that the next individual a_{n+1} will be of kind Q_i to be

6. $$\frac{(n_i + \lambda/K)}{(n + \lambda)}.$$

The parameter λ indicates the weight given to logical or language-dependent factors over and above purely empirical factors (observed frequencies). The choice $\lambda = K$ gives Carnap's measure c^*, which allocates probability evenly to all structure descriptions; this is a generalization of Laplace's rule of succession. The choice $\lambda = 0$ gives Reichenbach's Straight Rule. The choice $\lambda = \infty$ would give the range measure proposed in Wittgenstein's *Tractatus*, which divides probability evenly to state descriptions, but it makes the inductive probability (6) equal to $1/K$ which is independent of the evidence e and, hence, does not allow for learning from experience.

If the universe U is potentially infinite, so that n may grow without limit, all measures of Carnap's λ-continuum assign the probability zero to universal generalizations on evidence e. Jaakko Hintikka's λ-α-system solves the problem of universal generalization by dividing probability to constituents. A *constituent* C^w tells which Q-predicates are non-empty and which empty in universe U. The number w of non-empty Q-predicates is called the width of C^w. Each generalization H in L (i.e., a quantificational sentence without individual names) can be expressed as a finite disjunction of constituents. Hintikka's parameter $\alpha \geq 0$ regulates the speed in which positive instances increase the probability of a generalization. When α grows without limit, Hintikka's measures approach in the limit the Carnapian values. When α is small, the posterior probability of universal generalizations grows rapidly. In this sense, the choice of a small α is an index of boldness of the investigator, or a regularity assumption about the lawlikeness of the relevant universe U. If background assumptions are expressible by a theory B, possibly in a theoretical language that is richer than L, then posterior probabilities of the form $P(H/e\&B)$ can be calculated relative to observational evidence e and "theoretical evidence" B (see [Niiniluoto and Tuomela, 1973]).

In Hintikka's system, there is one and only one constituent C^c which has asymptotically the probability one when the size n of the sample e grows without limit. This is the constituent C^c which states that the universe U instantiates precisely those c Q-predicates which are exemplified in the sample e:

7. $P(C^c/e) \to 1$, if $n \to \infty$ and c is fixed.

It follows from (7) that a constituent which claims some so far unexemplified Q-predicates to be instantiated in U will asymptotically receive the probability zero. It is natural to assume with Hintikka that the prior probabilities of constituents $C^w$are proportional to their width; as we shall see, this means that the information content of C^w decreases with w. (This holds for $\alpha > 0$; if $\alpha = 0$, all constituents are equally probable.) Then the result (7) means that inductive evidence e asymptotically favors the most informative generalization compatible with e.

The Carnap-Kemeny axiomatization of Carnap's λ-continuum was generalized by Hintikka and Niiniluoto in 1974, who allowed that the inductive probability (6) of the next case being of type Q_i depends on the observed relative frequency of kind Q_i and on the number c of different kinds of individuals in the sample e. The latter factor expresses the *variety* of evidence e, and it also indicates how many universal generalizations e has already falsified. Carnap's system turns out to be biased in the sense that it assigns a priori the probability one to the "atomistic" constituent C^K that claims all Q-predicates to be instantiated in universe U. In this axiomatic way, a system of inductive probability measures is obtained where Carnap's λ-continuum is the only special case with zero probabilities for universal generalizations (see [Jeffrey, 1980; Kuipers, 1978].

Further developments of inductive logic include its modification to problems concerning *analogical reasoning* where the distances between Q-predicates play a significant role in inference. In such cases, the probability (6) depends on the distance of Q_i from the cells exemplified in e.[13]

3.4 Information content

In the traditions of positivism and probabilistic empiricism, high probability in the sense of certainty has taken to be a major virtue of a theory. This view was challenged by Karl Popper in 1934 (see [Popper, 1959]). He pointed out that high probability is a sign of the logical weakness of a hypothesis: logical truths or tautologies have the probability 1. Further, probability decreases with increasing logical strength:

8. If $H \vdash H'$, then $P(H) \leq P(H')$ and $P(H/E) \leq P(H'/E)$ for any E.

Popper's recommendation was that science should seek *improbable* hypotheses. Unlike tautologies, which allow all possible states of affairs, they are informative statements which are in principle *falsifiable* and *testable.* Improbability is thus a measure of the degree of falsifiability and testability of a hypothesis.

Popper's insistence on improbability can be understood as a demand that theories should have a large *information content.* Bar-Hillel, who worked with Carnap, defined in 1952 the semantic content $c(H)$ of H in language L as the set of those states of affairs, describable in L, that H excludes. Then the content of a tautology

[13]See the papers of Kuipers and Niiniluoto in [Helman, 1988].

is empty, and the content of a contradiction is maximal (see [Bar-Hillel, 1964]). A comparative notion of content can be defined by set-theoretic inclusion:

9. H has at least as much content as H' if $c(H') \subseteq c(H)$.

Most statements are incomparable by the criterion (9). If a probability measure P over the statements of L is available, then each state in $c(H)$ can be weighted by its probability. Thereby the *degree of information content* $cont(H)$ of H can be defined as:

10. $cont(H) = P(\sim H) = 1 - P(H)$.[14]

Thus, $cont(H)$ is the complement of H's prior probability $P(H)$. We shall see later how Popper's attack on high probability can to some extent be reconciled with probabilistic empiricism within the theory of semantic information and confirmation.

3.5 Empirical content and empirical success

The Vienna Circle proposed originally very strong principles of Verification (only verifiable statements are meaningful) and Translatability (only statements translatable by explicit definitions to the empirical language are meaningful) [Ayer, 1959]. A more liberal form of empiricism was formulated by Carnap [1936]. He allowed theories including theoretical terms, but required that a theory H should be *empirically testable*: H should logically entail some empirical statements whose truth value can be checked by public observation. This idea is in harmony with the HD method of testing scientific hypotheses.

Popper defined the *empirical content* $EC(H)$ of a theory H as the class of its potential falsifiers, i.e., basic statements forbidden by the theory. A non-empty empirical content guarantees that a theory is empirically testable. For Popper's falsificationism, this condition provides a demarcation criterion between empirical science and metaphysics (see [Popper, 1959]).[15]

A generalization of the notion of empirical content allows that a theory H may have not only deductive but also probabilistic relations to empirical statements. For example, probabilistic and statistical hypotheses H can be tested by observations E, if H has an inductive (rather than a deductive) relation to E. This may be called the *hypothetico-inductive* (HI) method of testing.[16]

The empirical content of a theory includes all of its *empirical potential* about both known and so far unknown empirical consequences. If empirical content is restricted to accepted or verified evidence statements up to the time t, then this notion $EC_t(H)$ expresses the time-dependent notion of the *empirical success* of

[14]The qualitative (9) and quantitative (10) conditions illustrate the difference between (3) and (4).

[15]See also the paper of Mahner in this volume.

[16]See [Niiniluoto and Tuomela, 1973]. Laudan [1996] also argues that a theory T can be confirmed by evidence which is not deducible from T.

H. Comparatively, theory H' can be said to *empirically at least as successful* as theory H (at time t), if H' includes all the empirical successes of H, i.e., $EC_t(H) \subseteq EC_t(H')$ (cf. [Kuipers, 2000]). Again, most theories are incomparable by this criterion, and to gain comparability quantitative considerations about the weights of the successes have to be introduced.

3.6 Explanatory and predictive power

Explanation and prediction are two important tasks of a theory. In his classical account of the method of hypothesis, William Whewell demanded that a theory should explain the known facts, but it should also "fortel new phenomena" [Whewell, 1840; cf. Niiniluoto, 1984]. Carl G. Hempel [1965] combined these requirements by saying that a good theory should have "systematic power". This is a matter largely agreed by scientific realists, constructive empiricists, and instrumentalists (cf., e.g., [van Fraassen, 1980]). But, as we shall see, there is disagreement about the question whether such systematic power indicates anything about the truth of a theory.

It is widely agreed that an explanation needs laws or theories among its premises (*explanans*). The *explanandum* may be a singular or a statistical statement or a law, which is known or assumed to be true. By the logical relation between the explanans and the explanadum, we distinguish *deductive* and *inductive-probabilistic* explanations.[17] More precisely, when a hypothesis H explains evidence E relative to background assumptions B, the following conditions should be distinguished: (a) E is deducible from H and B, and (b) E is inducible from H and B with the probability $P(E/H\&B)$ such that $P(E/B) < P(E/H\&B) < 1$. Case (a) is typical in the HD model of science: theory H is a universal generalization, and $P(E/H\&B) = 1$. Case (b) is typical in situations where alternative causes are connected with their effects only with statistical probabilities. Bayes's Theorem was traditionally applied in such situations to calculate the "probabilities of causes" given an observed effect.

Predictions based on a theory H may have a similar logical structure, but the truth of E is not known in advance but only discovered later. Again one may distinguish deductive and inductive-probabilistic predictions.

A theory T has *actual explanatory power* if there are known statements E that H is able to explain. This allows that T may have also *potential* explanatory power with respect to statements that are not yet known to be true. Similarly, T has actual *predictive power* if there are predictions by T that have turned out to be true. Again, T may have potential predictive power concerning cases that have not yet been verified or even actually derived from T. Theory T has *systematic power* if it has explanatory and predictive power. Theories may reductively explain other theoretical statements. If the explained and predicted statements are empirical, then the notion of potential systematic power comes close to the empirical content

[17] See the paper by Psillos in this volume.

$EC(T)$ of T. When E is taken to include our already accepted or verified evidence statements, systematic power is a measure of the *empirical success* of H so far.

A comparative notion of systematic power can be obtained by requiring that one theory has in the set-theoretical sense more successful explanations and predictions than another (cf. (9)). The alternative is to introduce a quantitative measure of systematic power (cf. (10)). Let $syst(H, E)$ be the *systematic power* of hypothesis H relative to the evidence E. A proposal was derived by Hempel in 1948 (see [Hempel, 1965]). He started by considering how many empirical statements in a given class a theory is able to entail, and generalized this account by applying a probability measure P and the content measure cont (10). If E is the conjunction of all empirical statements to be explained, then the power of theory H with respect to E can be measured by the ratio of the common content of H and E, i.e., $cont(H \vee E)$, to the content of E. This leads to

11. $syst_1(H, E) = (1 - P(H \vee E))/(1 - P(E)) = P(\sim H / \sim E)$.

Hintikka's [1968] proposal

12. $syst_2(H, E) = (P(E/H) - P(E))/(1 - P(E))$

is motivated by the idea that H is expected to transmit information about E. Note that $syst_1(H, E) = syst_2(H, E) = 1$ if $H \vdash E$, i.e., H deductively explains the whole data E. Other known measures are mostly variants of $syst_2$ (see, e.g., [Popper, 1959]).

On the basis of these definitions, two comparative notions of *better$_i$ systematization* can be defined by $syst_i(H', E) > syst_i(H, E)$, for $i = 1, 2$. Then

13. H' is better$_1$ systematization of E than H iff $P(H)(1-P(E/H)) > P(H')(1-P(E/H'))$

14. H' is better$_2$ systematization of E than H iff $P(E/H') > P(E/H)$.

If H and H' have the same prior probability, then these conditions are equivalent in requiring that the better explanation has the greater likelihood. Deductive explanations are not distinguished by (14), but (13) favors a deductive explanation with a higher information content.

3.7 Problem-solving capacity

Some philosophers have argued against scientific realism that science is not a truth-seeking but a problem-solving activity [Kuhn, 1970; Laudan, 1977]. The notion of *problem-solving* is ambiguous, however. Many realists state that science solves cognitive problems by seeking true answers to WH-questions and why-questions. Then the contrast between problem-solving and truth-seeking is misleading. Some philosophers define solutions of problems so that they become identical to successful explanations and predictions. For example, theory T solves the problem 'Why

E?' by deducing E from T; or theory T solves the predictive problem 'E or $\sim E$?' by deriving E or $\sim E$ from T. Then the notion of the problem-solving capacity of theory T becomes identical with the notion of the systematic power of T. Again, just as in the case with empirical content and systematic power, there is an ambiguity between the problem-solving record of T so far, and the overall potential of T for solving empirical problems.

Instead of cognitive problems, it is also possible to consider the relevance of science to problems of action and decision-making (cf. [Rescher, 1977]). Knowledge is power, as Francis Bacon put it. This is not denied by scientific realists: if theories are true or approximately true, they may lead to successful actions and technological applications. In this manner, the results of basic research can be developed via applied research to innovation. But some pragmatist philosophers, working in the field of operations research, have argued that the method of science is essentially and directly a tool of solving *decision problems* (cf. [Niiniluoto, 1984]). The virtues of theories should then be measured by the *practical utilities* (like money and other profits) that result from such successful decisions.

3.8 Simplicity

The instrumentalist tradition in astronomy suggested that a theory should "save the phenomena" (i.e., explain all the apparent observable movements of planets) by making as simple assumptions as possible [Duhem, 1969]. The positivist Ernst Mach argued that science strives for the "economy of thought": theories should be able to handle large domains of empirical data by tools that make our intellectual efforts easier. Also realists may agree that it is useful, especially in applied research, to have simple theories that are manageable and help us in calculations. At least if two theories are logically equivalent, then we should prefer the mathematically simpler formulation of the theory. For strategic purposes, it may be wise to pursue first simpler theories and then to move to the study of more complex cases. Simplicity is thus a traditional virtue of a theory. But, as there is no a priori guarantee that nature itself is simple, there is disagreement whether simplicity is a sign of truth (see [Reichenbach, 1938]).

The notion of simplicity is notoriously complex [Foster and Martin, 1966]. Perhaps the most successful attempts have been in the classification of curves by the complexity of mathematical functions. If the logical complexity $K(H)$ of a theory H could be defined, then Eino Kaila's [1939] notion of *relative simplicity* could be defined by the ratio

15. $RS(H, E) = syst(H, E)/K(H)$.

In other words, a theory has high relative simplicity, if it explains a multitude of empirical data by means of a few independent assumptions (cf. [Niiniluoto, 1999]). This notion is close to the requirement that a good theory should achieve explanatory unification and coherence [Kitcher, 1989]. It is also related to Laudan's [1977] proposal to measure scientific progress by the problem-solving capacity of theory T minus the "conceptual problems" in T.

3.9 Accuracy

Successful prediction is an ideal that is not easily achieved. Especially quantitative laws and theories yield predictions that are only more or less *accurate* relative to our observations. On the other hand, descriptions of data involving numerical measurements always contain observational errors. If a singular prediction x and a numerical measurement y are both expressed as point values (i.e., real numbers), then the *fit* between x and y, or the accuracy of x with respect to y, is simply the geometrical distance $|x - y|$ between x and y. If the measurement value is expressed as an interval (y, y'), then ideally a theory's prediction x should belong to (y, y').

In the more complex case of *curve fitting*, the distance between a curve expressed by the function $y = f(x)$ and the observed points $(x_1, y_1), \ldots, (x_n, y_n)$ can be defined by the Square Difference

(SD) $\sum_{i=1}^{n}(y_i - f(x_i))^2$.

Forster and Sober [1994] measure the *predictive accuracy* of a quantitative curve H relative to observed data E by a function $logP(E/H)$ that, assuming a normal distribution of error probabilities, essentially agrees with SD.

3.10 Approximate truth and truthlikeness

Even though scientific realists accept truth as an important aim of science, they by no means imply that our current theories in fact are true. Most realists are at the same time fallibilists who are aware of the limitations of our cognitive efforts — event the best ones in science. For this reason, critical scientific realists have introduced notions that attempt to explicate the tricky idea that a theory might by false but still "close to the truth".

Qualitatively speaking, a theory is *approximately true* if it is "sufficiently" close to the truth. This idea is easy to understand in the case of quantitative statements. If x^* is the true value of a quantity, then the distance of an estimate x from x^* is measured by the fit $|x - x^*|$ between these real numbers. One estimate is closer to the truth than another if its distance from x^* is smaller, and it is approximately true if this distance is small enough. These simple starting points have been generalized to full first-order theories within the similarity account of truthlikeness [Niiniluoto, 1987].

Popper introduced the notion of *verisimilitude* or *truthlikeness* in 1960 (see [Popper, 1963]). He started from a comparative notion that was hoped to express "the idea of a degree of better (or worse) correspondence to truth". Popper's concept differs from probability, since it attempts to combine truth and information content, while "probability combines truth with lack of content". Verisimilitude is an objective or semantic notion, not an epistemological idea, but sometimes there may be arguments to appraise that "we may have made progress towards the truth".

Popper's qualitative definition is applied to theories treated as deductively closed sets of statements. Let T and F be the sets of the true and false statements, respectively, in some interpreted language L. Then, for a theory A in L, the *truth content* of A is the intersection $A \cap T$, and the *falsity content* of A is the intersection $A \cap F$. According to Popper, theory A is *more truthlike* than theory B if and only if $B \cap T \subseteq A \cap T$ and $A \cap F \subseteq B \cap F$, where one of the set-inclusions is strict. Intuitively, A should have larger truth content than B, but smaller falsity content than B. An equivalent formulation of this set-theoretical criterion states that the symmetric difference $A \Delta T = (A - T) \cup (T - A)$ should be a proper subset of $B \Delta T$.

Popper's definition has some intuitively desirable properties. The complete truth T has the maximal truthlikeness among all theories. A is more truthlike than B, if both A and B are true and A is logically stronger than B (i.e., A logically entails B, but not vice versa). If A is false, then its truth content $A \cap T$ is more truthlike than A itself. However, as David Miller and Pavel Tichý proved in 1974, this definition fails dramatically, since it cannot be used for comparing false theories: if A is more truthlike than B in Popper's sense, then A must be true (see [Miller, 1974]).

As one of the modifications of Popper's approach, Miller and Kuipers have proposed a model-theoretical variant (see [Kuipers, 1987]). Let $Mod(A)$ be the class of models of A, i.e., the L-structures in which all the sentences of A are true. Then define A to be at least as truthlike as B if and only if $Mod(A) \Delta Mod(T) \subseteq Mod(B) \Delta Mod(T)$. This is equivalent to Popper's original definition with a slight but important modification. Popper's requirement for truth content is preserved, but in the requirement for falsity content the class F of false sentences is replaced by the logically weakest false theory Φ in L. If the truth is finitely axiomatizable by a sentence τ in L, then Φ can be defined as the theory axiomatized by $\sim \tau$.

If T is a complete theory, the model-theoretic definition has a very implausible consequence: among false theories, greater logical strength implies higher truhlikeness. It is thus vulnerable for Tichý's "child's play objection": a theory can be improved simply by joining new falsities to it.

Kuipers applies this definition in a context of "nomic truthlikeness" where T is usually not a complete theory: a theory A asserts the physical possibility of the structures in $Mod(A)$. The comparative notion of 'closer to the truth' involves a strong dominance condition: the better theory should have set-theoretically "more" correct models and "less" incorrect models than the worse theory. Kuipers admits that this "naive definition" is simplified in the sense that it treats all mistaken applications of a theory as equally bad. For this reason, Kuipers [2000] has recently developed a "refined definition" which allows that some mistakes are better than other mistakes. The notion of betweenness between structures helps to make sense of the idea that a theory may be improved by replacing worse incorrect models by better incorrect models. He concentrates on a comparative definition with "safe" cases, and avoids quantitative assumptions, with the result that many theories are not comparable by his criteria.

Popper's approach to truthlikeness is a "content definition" rather than a "likeness definition" [Zwart, 2001; Niiniluoto, 2003]: it employs the concepts of truth and logical consequence, but not the notion of similarity. Risto Hilpinen [1976], who represented theories by classes of possible worlds, assumed as a primitive notion the concept of similarity between possible worlds. An alternative is to give an explicit definition of such a similarity: if possible worlds are replaced by maximally informative descriptions of states of affairs in a given first-order language L, and such descriptions are called *constituents* in L, then the basic problem of the "likeness definition" or the similarity approach is to introduce a distance between the constituents of L (see [Niiniluoto, 1987]).[18]

The set of constituents defines a cognitive problem B with mutually exclusive complete answers. The *distance* $\Delta(h_i, h_j) = \Delta_{ij}$ between constituents in B should satisfy $0 \leq \Delta_{ij} \leq 1$ and $\Delta_{ij} = 0$ iff $i = j$. This distance function Δ has to be specified for each type of language separately, but there are canonical ways of doing this for special types of problems. First, as in our discussion of accuracy, Δ may be directly definable by using the metric in the structure of the language. For example, if the set of constituents B is a subclass of the K-dimensional Euclidean space $\mathbf{R}^K$, then the distance between two points is their Euclidean distance. The distance between two real-valued functions f and g on the interval $[a, b]$ can be defined by the Minkowskian distance

16. $[\int_a^b |f(x) - g(x)|^p dx]^{1/p}$.

Secondly, if B is the set of state descriptions, the set of structure descriptions, or the set of constituents of a first-order language L, the distance Δ can be defined by counting the differences in the standard syntactical form of the elements of B. For example, a monadic constituent tells that certain kinds of individuals (given by Q-predicates) exist and others do not exist; the simplest distance between monadic constituents is the relative number of their diverging claims about the Q-predicates. If a monadic constituent C_i is characterized by the class CT_i of Q-predicates that are non-empty by C_i, then the *Clifford distance*[19] between C_i and C_j is the size of the symmetric difference between CT_i and CT_j:

17. $|CT_i \Delta CT_j|/K$.

Given the distance between constituents, the next step is to extend it to arbitrary first-order theories. As Hintikka has shown, each theory H in a first-order

[18] The notion of a monadic constituent C^w, discussed above, is a special case of the broader notion of a constituent applied here.

[19] The "safe" comparative relation, which agrees with the Clifford distance (17), would be the following: letting CT^* be the truly non-empty cells, constituent C_1 is closer to the truth than C_2 if $CT_2 \Delta CT^* \subseteq CT_1 \Delta CT^*$ (cf. [Kuipers, 2000]).

language can be expressed as a disjunction of constituents. Then the distance of a theory H from the truth depends on the distances of the disjuncts of H from the truth. Let C^* be the complete truth τ, i.e., the true constituent of L, and let theory H be a partial answer of B, i.e., a disjunction of the constituents $C_1, C_2, \ldots, C_n$. Let Δ_{i^*} be the distance of C_i from C^*. Then, in the approach of Tichý and Oddie, the distance of H from C^* is defined by the average function $\Sigma\Delta_{i^*}/n$ where $i = 1, \ldots, n$ (see [Oddie, 1986]). This definition does not satisfy Popper's requirement that among true theories truthlikeness should covary with logical strength. In Niiniluoto's approach, truthlikeness is defined by the weighted average of the minimum distance $min\Delta_{i^*}$ and the (normalized) sum $\Sigma\Delta_{i^*}$ of all distances ($i = 1, \ldots, n$) [Niiniluoto, 1987; Kuipers, 1987]:

18. $\Delta_{ms}(C_i, H) = \gamma\Delta_{min}(C_i, H) + \gamma'\Delta_{sum}(C_i, H)$ $(\gamma > 0,\ \gamma' > 0)$,

where the weights γ and γ' indicate our cognitive desire of finding truth and avoiding error, respectively.[20] Here the minimum distance alone serves to define the notion of *approximate truth*: H is approximately true if its minimum distance $\Delta_{min}(C^*, H)$ from C^* is small enough. On the other hand, the additional sum-factor in (18) defines penalties for all the mistakes allowed by the theory. A partial answer g is *truthlike* if its min-sum distance from the target C^* is sufficiently small. One partial answer H' is *more truthlike* than another partial answer H if $\Delta_{ms}(C^*, H') < \Delta_{ms}(C^*, H)$. The *degree of truthlikeness* $Tr(H, C^*)$ of H (relative to the target C^* in B) is now defined by

19. $Tr(H, C^*) = 1 - \Delta_{ms}(C^*, H)$.

This min-sum-definition of degrees of truthlikeness satisfies Popper's basic conditions, but it is also stronger than Popper's approach in the sense that all rival theories (in language L) are comparable with respect to their verisimilitude. The Miller-Tichý refutation and the child's play objection are avoided. Moreover, by this definition it is possible that some false theories are so close to the truth that they are more truthlike than weak true theories (e.g., a tautology).

When the similarity approach is applied to cases where the true constituent is a modal statement with nomic necessity and possibility operators, it gives an explication of what L. J. Cohen [1980] calls *legisimilitude* or "closeness to laws". This modification makes it possible to interpret Kuipers' account of nomic truthlikeness within the framework of the similarity approach.

Miller's [1974] argument about *language-dependence* is the most famous objection against the similarity account of truthlikeness. Miller shows how truthlikeness orderings may be reversed by suitable translations between languages. Comparisons of truthlikeness may thus seem to depend on arbitrary choices of language. This argument turns out to be related to the more general point that metric properties need not be preserved in one-to-one mappings between quantitative spaces.

[20] In typical applications, the parameters should be chosen so that $\gamma \approx 2\gamma'$ (see [Niiniluoto, 2003]).

In this sense, it can be claimed that Miller's demands about the invariance properties of truthlikeness measures are too strong (see [Niiniluoto, 1998; Zwart, 2001]).

The notion of truthlikeness can be applied also to the structuralist concept of a theory.[21] Recall that on this account a theory can be represented as a pair (T, I), where I is the class of the intended applications of T. Another theory (T', I') may have a different class of intended applications, but the theories are rivals about the common applications, i.e., the intersection of I and I'. For simplicity, let us assume that $I = I'$ (cf., however, Laudan's [1977] treatment of rival research traditions). Suppose first that we wish to employ only a comparative notion of truthlikeness. Then theory (T, I) is more truthlike than theory (T', I) if T is more truthlike than T' for each structure W in I, i.e., T is uniformly better than T' for all the intended applications. If a quantitative measure can be used to measure the degree of truthlikeness $Tr(T, W)$ of T relative to the applications W in I, then the *overall degree of truthlikeness* of (T, I) can be defined by the sum of $Tr(T, W)$ over W in I. If there is an agreement in the scientific community about the relative significance of these applications (cf. again [Laudan, 1977]), then these values can be weighted by coefficients indicating their importance.

4 CONFIRMATION

The theory of confirmation attempts to develop formal tools in order to show on what kinds of principles scientific theories can be justified by evidence. We start from qualitative and comparative concepts, and proceed then to quantitative degrees of confirmation. These approaches are usually general in the sense that they apply to all kinds of theories, including ones that use theoretical terms over and above the observational language. However, the specific results of inductive logic have been worked out for more modest situations, but in principle the interesting logical properties of the confirmation relation should be valid in all cases.

4.1 Qualitative and comparative confirmation

Let us start from the situation that has been called the "prediction criterion" [Hempel, 1965], "deductive support" [Stegmüller, 1971] and "deductive confirmation" [Kuipers, 2000]. It is related to typical applications of the HD method: a theoretical hypothesis H is justified by deducing from H (with background assumption B) some empirical statement E, where E is not deducible from B alone. In such cases, H achieves deductive systematization between B and E (see [Hempel, 1965; Niiniluoto and Tuomela, 1973]). If E is now a report of a severe test of H (cf. [Popper, 1959]), then E confirms or supports H relative to B.

The qualitative notion of deductive confirmation satisfies the principles of *Converse Entailment*:

(CE) If hypothesis H logically entails evidence E, then E confirms H,

[21] See Niiniluoto [1984; 1987] and Bonilla [1996]. For a different account, see Kuipers [2000].

and *Converse Consequence*:

(CC) If K logically entails H and E confirms H, then E confirms K.

Thus, if E deductively confirms H, then any theory K that is logically stronger than H also is confirmed by E. This unqualified endorsement of CC is often taken to be an important objection to the notion of deductive confirmation. It might be more plausible to modify CE and CC by replacing the deductive relation with the stronger condition of deductive explanation (see [Niiniluoto and Tuomela, 1973]):

(CE*) If hypothesis H deductively explains evidence E, then E confirms H.

(CC*) If K deductively explains H and E confirms H, then E confirms K.

For example, the kinetic theory of gases K explains the Boyle-Mariotte law, and therefore all the data explained by the latter law confirm K.

Comparative versions of deductive confirmation can be obtained by suggesting that stronger success or stronger logical power implies stronger confirmation [Kuipers, 2000, 25]:

(CE-c) If $H \vdash E \vdash E'$, then E confirms H more than E'

(CC-c) If $H' \vdash H \vdash E$, then E confirms H' more than H.

CE-c is a straightforward way of expressing the principle that a hypothesis is better supported if it has more empirical successes. On the other hand, CC-c is controversial, since the extra strength of H' might be just an irrelevant conjunction added to H without any relation to evidence E.

Another intuition about confirmation starts from the observation that verification should be a special case of support. This is expressed by the principle of *Entailment*:

(5) If evidence E logically entails hypothesis H, then E confirms H.

Further, one way of justifying a scientific theory H is to derive it from a more general or comprehensive theory K. For example, when Newton's mechanics was accepted as a basis of physics, it was possible to derive from it laws concerning various kinds of phenomena, e.g., the movements of celestial bodies, projectiles, and pendulums. In such situations, the independent support of Newton's theory may be though to "flow" to its consequences. This is expressed by the principle of *Special Consequence*:

(SC) If evidence confirms hypothesis H and K logically follows from H, then E confirms K.

Hempel [1965] proposed his own "satisfaction criterion of confirmation" that satisfies E and SC. Again, one might object to SC that the weakening of H to K may be due to the addition of an irrelevant disjunction. It is also difficult to formulate a comparative version of SC.

Hempel's main negative result was the observation that Converse Consequence and Special Consequence are incompatible: if CC and SC are assumed together, then it can be proved that any statement confirms any other statement. It is easy to see that the same holds for CE and SC. Hence, there is no single account where all plausible principles of confirmation could hold together. Smokler [1968] has proposed that CE and CC (or perhaps rather CE* and CC*) are satisfied by *abductive inference* — this is Peirce's term for inferences where a theory is appraised as promising due to its explanatory power. However, CE obviously involves also the idea that predictive power gives support to a theory. Smokler adds that E and SC are typical in *enumerative* and *eliminative* inference — these are names for inductive inferences which typically proceed from the positive instances to a generalization.

4.2 Positive relevance

The qualitative notion of confirmation can be approached from a new perspective, if we assume that an epistemic probability measure P is available for the language including the relevant statements. Then one obvious possibility is to define confirmation by the *Positive Relevance* criterion:

(PR) E confirms H relative to B if and only if $P(H/E\&B) > P(H\&B)$.

Suppressing B, this is equivalent to

(PR′) E confirms H if and only if $P(H/E) > P(H/\sim E)$.

In other words, E *confirms* H by increasing its probability. By the same idea, E *disconfirms* H if $P(H/E) < P(H)$, and E is *irrelevant* to H if $P(H/E) = P(H)$.

The principles of Converse Entailment CE and CE* receive immediately a justification by PR. If H entails E, we have $P(E/H) = 1$. Hence, by Bayes's Theorem (5),

20. If $H \vdash E$, and if $P(H) > 0$ and $P(E) < 1$, then $P(H/E) = P(H)/P(E) > P(H)$.

More generally, as positive relevance is a symmetric relation, it is sufficient for the confirmation of H by E that H is positively relevant to E. This is an instance of the HI model of testing. If inductive explanation is defined by the positive relevance condition, i.e., by requiring that $P(E/H) > P(E)$ (see [Niiniluoto and Tuomela, 1973; Festa, 1999]), then we have the general result:

21. If H deductively or inductively explains E, then E confirms H.

However, PR does not generally satisfy Converse Consequence CC, so it does not perfectly fit Smokler's notion of abductive inference.

Methodologically a more complex situation obtains when the test statement E is not directly verifiable or knowable. Then the probability $P(H/E)$ in (20) cannot be calculated by conditionalization on E, as E itself is uncertain (cf. [Jeffrey, 1965]). Suppose, however, that J is a *reliable witness* or indicator of E. This means that there are constant probabilities $P(J/E) = p$ and $P(J/\sim E) = q$, where $p > q$ (see [Bovens and Hartmann, 2003]). The reliability condition $p > q$ is equivalent to $P(J/E) > P(J)$, so that J is positively relevant to E. As positive relevance does not generally satisfy CC, we cannot conclude from this condition and $H \vdash E$ that J is positively relevant to H as well. But this inference is warranted, if it can be further assumed that H is irrelevant to J relative to E (cf.[*ibid.*, 91]). Hence,

22. If $H \vdash E$, $P(H) > 0$, J is a reliable indicator of E, and $P(J/H\&E) = P(J/E)$, then $P(H/J) > P(H)$.

Another possible probabilistic explication of confirmation is the *High Probability* criterion:

(HP) E confirms H relative to B if and only if $P(H/E\&B) \geq q$,

where the threshold value q satisfies $\frac{1}{2} < q \leq 1$. This definition satisfies principles Entailment E and Special Consequence SC of Smokler's enumerative inference. If H_1 and H_2 are mutually incompatible, then by HP evidence E cannot confirm both H_1 and H_2. Instead, PR allows this: assuming that H_1 is not the negation of H_2, it may happen that E increases their probabilities. Neither of the definitions PR and HP satisfy the conjunction condition: if E confirms H_1 and E confirms H_2, then E confirms $H_1\&H_2$.

HP has some unnatural consequences. For example, we should say by HP that a tautology is confirmed by all statements, since its probability is 1. It appears that high probability should be treated as indicating plausibility and credibility, while confirmation is an incremental notion which means increase of probability in the sense of PR.

4.3 Quantitative confirmation

The "orthodox" school of statistics is the Neyman-Pearson (NP) theory of estimation and testing. Even though this theory speaks about "confidence intervals" and "levels of significance", it attempts to get along without a notion of inductive support — even without inductive inference. Estimation and testing are patterns of "inductive behavior" governed by frequentist error probabilities. In testing a null hypothesis H_0 against its alternative H_1, on the basis of data E, the rejection of H_0 will follow if the *likelihood ratio* $P(E/H_0)/P(E/H_1)$ is small.

The likelihood school of statistics differs from the NP theory by interpreting the likelihood $P(E/H)$ as a measure of the inductive support of H by E [Edwards,

1972]. More generally, support can be measured by the whole likelihood distribution or by average likelihoods (calculated relative to prior probability distributions) (see [Rosenkrantz, 1977]). This is a shift from the physical to the epistemic interpretation of probability. The likelihood school can thereby be regarded as a special case of the *Bayesian* school. They give similar results in cases, where prior information is stable or irrelevant (see [Box and Tiao, 1973]).

The characteristic of Bayesianism is the application of Bayes's Theorem (5) to calculate posterior probabilities on the basis of priors. Many Bayesians take immediately $P(H/E)$ as a measure of the inductive support or confirmation of H by E; the qualitative counterpart of this move is the High Probability criterion HP.

If theories are compared by their posterior probability, then by (5) deductively equally strong hypotheses will be ordered by their prior probabilities:

23. If $H \vdash E$ and $H' \vdash E$, then $P(H/E) > P(H'/E)$ if and only if $P(H) > P(H')$.

More generally, if H and H' are equally good explanations of the data E (i.e., $P(E/H) = P(E/H')$), then the greater posterior credibility of H depends entirely on its greater prior credibility. This result has motivated approaches, where all the "plausibility" considerations (promise as an explanation, simplicity etc.) are placed on prior probabilities (see [Salmon, 1990]). For example, simpler deductive explanations are then also more probable a posteriori. In alternative approaches simplicity will be a feature that is reflected in the confirmability of hypotheses, i.e., in the differences between posterior and prior probabilities (cf. [Rosenkrantz, 1977; Niiniluoto, 1999]).

Following J. M. Keynes and Janina Hosiasson-Lindenbaum, Carnap also initially called the inductive probabilities of his inductive system *degrees of confirmation*. But there was also another idea of confirmation which was important in the debate whether the positive instances of a generalization "confirm" or "support" a universal generalization.

According to Jean Nicod's criterion, only positive instances of the form $\{Fa, Ga\}$ confirm the generalization $\forall x(Fx \rightarrow Gx)$, while negative instances of the form $\{Fa, \sim Ga\}$ disconfirm it, and the cases $\{\sim Fa, Ga\}$ and $\{\sim Fa, \sim Ga\}$ are neutral with respect to it. Hempel argued in 1937 that, as 'All ravens are black' and 'All non-black things are non-ravens' are logically equivalent, both black ravens and white handkerchiefs should be understood to confirm the hypothesis about the color of ravens. Hosiasson-Lindenbaum gave in 1940 the first Bayesian analysis of this "raven paradox" by arguing that an observed black raven increases *more* the inductive probability of the generalization than any non-black non-raven. Confirmation thus concerns incremental changes of probability.

Popper argued in the 1950s against Carnap that inductive logic is inconsistent and impossible (see [Popper, 1959; 1963]). As a reply to Popper's criticism, Carnap [1962] distinguished *degrees of confirmation* in two senses: as the posterior probability

24. $conf_1(H, E) = P(H/E)$,

and as the increase of probability of H due to E

25. $conf_2(H, E) = P(H/E) - P(H)$.

The qualitative concepts of confirmation corresponding to these two alternatives are explicated by the high probability condition HP and the positive relevance condition PR. The corresponding comparative conceptions (cf. [Lakatos, 1968]) 'E confirms H_1 more than H_2' can then be defined either by $P(H_1/E) > P(H_2/E)$ or by $P(H_1/E) - P(H_1) > P(H_2/E) - P(H_2)$. These definitions can also be relativized to background knowledge B.

Other proposals for the degree of confirmation of H given E are usually normalizations of the *difference* measure $conf_2(H, E)$, which is equal to

26. $[P(E/H) - P(E)]P(H)/P(E)$.

They include I. J. Good's 1950 measure for the "weight of evidence":

27. $logP(E/H) - logP(E/\sim H)$,

and Kemeny's and Oppenheim's 1952 measure for "factual support":

28. $[P(E/H) - P(E/\sim H)]/[P(E/H) + P(E/\sim H)]$.

Further variants are given by Popper's 1954 formula for "degrees of corroboration" (cf. [Fetzer, 1981]), and Hintikka's formula for the information transmitted by E on H (cf. (12)) (see [Hintikka, 1968; Niiniluoto and Tuomela, 1973]). All these measures share the property that they are greater than 0 if and only if E is positively relevant to H.

On the other hand, if degrees of confirmation are defined by the *ratio* measure

29. $conf_3(H, E) = P(H/E)/P(H) = P(E/H)/P(E) = P(H\&E)/[P(H)P(E)]$,

or by its logarithm

30. $conf_4(H, E) = logP(H/E) - logP(H)$,

then comparative confirmation satisfies the Likelihood Criterion: E confirms more H_1 than H_2 if and only if $P(E/H_1) > P(E/H_2)$. (See [Milne, 1996; Festa, 1999; Kuipers, 2000].)

All measures $conf_1$, $conf_2$, $conf_3$, and $conf_4$ satisfy the principle that a surprising successful prediction gives more confirmation to a hypothesis than a less surprising one:

31. If $H \vdash E$ and $H \vdash E'$, where $P(E) < P(E')$, then E confirms H more than E' does.

In this sense, severe tests with improbable conclusions "pay" in terms of probabilistic confirmation. It follows from (31) that these measures satisfy (CE-c): a hypothesis receives more confirmation, if it makes more (i.e., stronger) successful predictions.

One of the connecting links between systematic power and confirmation follows directly from (14):

32. If H' is a better$_2$ explanation of E than H, then $conf_3(H'/E) > conf_3(H/E)$ and $conf_4(H'/E) > conf_4(H/E)$.

A similar result for $conf_1$ and $conf_2$ can be proved, if $P(H') > P(H)$ and $P(E/H') > P(E/H) > P(E)$ (see [Niiniluoto, 2004]).

An important difference between the confirmation measures $conf_2$ and $conf_3$ is their behavior with respect to irrelevant additions to a hypothetical explanation (cf. (CC-c)):

33. Assume that H explains E but A is irrelevant to E with respect to H (i.e., $P(E/H\&A) = P(E/H)$ where A is not entailed by H). Then $conf_3(H/E) = conf_3(H\&A/E)$ and $conf_2(H/E) > conf_2(H\&A/E)$.

Measure $conf_2(H/E)$ thus favors a minimal explanation, and allows support only to the part of an explanatory hypothesis that is indispensable for the explanation of the evidence (cf. [Niiniluoto, 1999, 190]). Fitelson [1999] regards this as fatal argument against $conf_3$ (see [Kuipers, 2000], however), but points out that his own favorite measure (27) is not open to this criticism.

Suppose that theory H achieves inductive systematization between mutually independent empirical propositions E and E' (see [Hempel, 1965; Niiniluoto and Tuomela, 1973]). Myrwold [2003] has explicated this idea by measuring the degree of *unification* of E and E' achieved by H by means of the difference $log[P(E'/E\&H)/P(E'/H)] - log[P(E'/E)/P(E')]$ (cf. $conf_4$). Then he argues, by applying the logarithmic ratio measure $conf_4$ again, that the degree of confirmation of H by $E\&E'$ can be divided into three additive parts: confirmation of H by E', confirmation of H by E, and the degree of unification of E and E' achieved by H. This result allows an analysis of Whewell's [1840] idea of the "consilience of inductions".[22]

Glymour [1980], who proposes to replace Bayesianism with his "bootstrap method", has presented the "problem of old evidence" against the positive relevance account of confirmation. Suppose that, besides newly available evidence E, evidence E_0 is known at time t when theory H is introduced. Then at time t we have $P(E_0) = 1$, $P(E_0/H) = 1$, and by (5), $P(H/E_0) = P(H)$. Hence, old evidence cannot confirm a new theory. But this is counterintuitive in the light of many examples from the history of science (cf. [Howson and Urbach, 1989]). There is no agreement of the best way to handle this problem (see [Earman, 1992]). As

[22]See also the discussion on coherence in Bovens and Hartmann [2003] and Dietrich and Moretti [2005].

Glymour himself noted, this problem is related to the idealized assumption that degrees of belief are invariant under logical equivalence. In this sense, they are probabilities for a logically omniscient scientist, and in a more realist treatment they should be replaced by some kind of "surface probabilities" which allow that the discovery of new deductive relations (e.g., that hypothesis H entails the old evidence E) may influence inductive probabilities. On the other hand, the difference measure $conf_2$ and the ratio measure $conf_3$ could be applied so that the confirming power of E_0 (or E_0 together with E) is calculated counterfactually, i.e., without assuming that E_0 is already in our background knowledge.

5 ACCEPTANCE

5.1 Behavioralism and cognitivism

The Bayesians are divided in two schools concerning the acceptance of hypotheses. Partly inspired by the negative results of his own system about inductive generalization, Carnap came to the conclusion that Hume was right about inductive inference: as induction is not necessarily truth-preserving and always involves uncertainty, we are never warranted in accepting the conclusion of an inductive argument (cf. [Jeffrey, 1980]). However, unlike Hume and Popper, Carnap was convinced that inductive logic can give us rational degrees of belief for scientific hypotheses on evidence. When such posterior probabilities are determined, they can be employed in rational decision making by using them in the calculation of expected utilities. In this account, scientists are not truth-seekers but rather decision makers or advisers of decision makers. The values relevant to decisions, related to the good and bad consequences of actions, are practical utilities defined by the decision-maker, the employer or the society. A well-known formulation of Bayesianism, with rules for "probability kinematics" and the "logic of decision", but without acceptance rules, was given by Richard Jeffrey [1965].

Carnap's and Jeffrey's views were influenced by the emergence of Bayesian decision theory within statistics. L. J. Savage's [1954] foundational studies analyzed rationality conditions concerning preferences that are sufficient to prove the existence subjective probability distributions p and utility functions u which satisfy the principle of *Subjective Expected Utility*. Let θ be a real-valued parameter. If $u(a, \theta)$ is the utility of act a when θ is the state of nature, and if $p(\theta/E)$ is a subjective probability distribution over the possible values of θ in $\mathbf{R}$,[23] then the expected utility of act a (relative to p, u, and E) is

$$\text{(EU)} \quad \int_{\mathbf{R}} u(a, \theta) p(\theta/E) d\theta.$$

[23]The posterior distribution $p(\theta/E)$ can be obtained from the prior distribution $p(\theta)$ and the likelihood function $f(E/\theta)$ by means of a formula which is a generalization Bayes's Theorem (5). Non-Bayesian approaches formulate decision principles (e.g., Wald's mini-max rule) without prior and posterior probabilities.

The best act is the one which maximizes this expected utility. Instead of positive gains or utilities, the Bayesian decision rule can be formulated by using losses or negative utilities. For example, a point estimate θ_0 of parameter θ should be chosen so that its posterior loss

34. $\int_{\mathbf{R}} |\theta - \theta_0| p(\theta/E) d\theta$

is minimized.

Savage argued that in statistics hypotheses are not accepted as true. Instead, accepting a hypothesis means only that we are ready to *act* "as if it were true". He called this view, which resembles Neyman's conception of inductive behavior, *behavioralism*. Generalized to all scientific hypotheses, behavioralism would be in harmony with the instrumentalist treatment of theories.

Against Savage and Jeffrey, Isaac Levi's [1967] *cognitivism* argues that scientists tentatively accept hypotheses as parts of the evolving body of scientific knowledge. Therefore, the task of the theory of induction is to analyze the conditions of such a rational acceptance.

Henry Kyburg's lottery paradox shows that high probability cannot be a sufficient condition of acceptance. It is natural to assume that the set of acceptable hypotheses is closed under deduction: if H is acceptable on E, then also H's deductive consequences are acceptable on E (cf. principle SP for qualitative confirmation). Now the conjunction principle is also valid: if H and H' are both acceptable on E, then their conjunction $H \& H'$ is acceptable on E. Then in a fair lottery with many tickets $t_1, \ldots, t_n$ and one lucky winner, it is highly probable that ticket t_i does not win (for all $i = 1, \ldots, n$). By the conjunction principle, it follows that no ticket will win in the lottery.

The notion of acceptance differs structurally from probability and confirmation also in other ways. For example, if hypothesis H is acceptable on evidence E, then $\sim H$ cannot be acceptable on E, even though both H and $\sim H$ may have non-zero probabilities given E. Glenn Shafer's [1976] theory of evidence has been interpreted as being about "degrees of confidence of acceptance". For example, if H has a positive degree on E, then $\sim H$ has the degree 0 on E. Similar ideas have been developed by L. J. Cohen's [1989] non-Bayesian treatment of inductive support, which contrasts "Baconian probabilities" with the ordinary "Pascalian" probabilities.

In Hintikka's treatment of inductive generalization, high posterior probability alone is not sufficient to make a generalization acceptable. But in Hintikka's system one may calculate for the size n of the sample e a threshold value n_0 which guarantees that the informative constituent C^c has a probability exceeding a fixed value $1 - \varepsilon$. Then the following rule avoids the lottery paradox:

35. Let n_0 be the value such that $P(C^c/e) \geq 1 - \varepsilon$ if and only if $n \geq n_0$. Then, given evidence e, accept C^c on e iff $n \geq n_0$.

(See [Hilpinen, 1968].)

5.2 Cognitive decision theory

It was proposed by Hempel [1965] that the virtues of scientific hypotheses serve as *epistemic utilities.* This idea has been worked out in *cognitive decision theory* by Levi and Hintikka. This theory shows that scientific induction can be treated in decision-theoretical terms, without the instrumentalist tone of Savage's behavioralism, but then the expected utilities have to be calculated by means of relevant cognitive goals.

The basic framework is a cognitive problem B, an "ultimate partition" in Levi's terminology, consisting of mutually exclusive and jointly exhaustive alternatives. In the simplest case B has two members $\{H, \sim H\}$, and the relevant formula for the expected utility of accepting hypothesis H is

36. $U(H, E) = P(H/E)u(H, t) + P(\sim H/E)u(H, f)$,

where $u(H, t)$ is the epistemic utility of accepting H, when H is true, and $u(H, f)$ is the utility of accepting H, when H is false. Formula (36) can be immediately generalized to the case, where B has a finite number n of members.[24] The elements of B are complete answers. The class $D(B)$ of partial answers is defined by the disjunctions of members of B. The basic rule of acceptance is then the following:

37. Accept on evidence E that member H of $D(B)$ which maximizes the expected epistemic utility $u(H, E)$.

If the aim of our inquiry is *truth,* and nothing but the truth, the epistemic utility of accepting a hypothesis H on evidence E can be taken to be equal to its truth value (1 for truth, 0 for falsity). Then the expected utility of accepting H is by (36) simply $P(H/E).1 + P(\sim H/E).0 = P(H/E)$. This value is maximized by trivially true tautologies or hypotheses logically entailed by the evidence. The rule (37) leads here to an extremely conservative principle.

Assume then with Popper that our basic aim in science is *truthful information* about the world. Then Levi [1967] points out that we have to "gamble with truth" in order to gain also information as an epistemic utility. We have to risk error, if we wish to be relieved from agnosticism. This is the fallibilist principle that is not realized by those who reject inductive acceptance rules.

Let $|H|$ be the number of elements of B allowed by a partial answer H (i.e., the number of disjuncts in H), and let $|B|$ be the total number of elements in B. Then the information content of H decreases with $|H|/|B|$. Let $0 < q \leq 1$ be an index of boldness, which tells how willing the scientist is to risk error in her attempt to relieve from agnosticism. Levi suggests that $u(H, t) = 1 - q|H|/|B|$ and $u(H, f) = -q|H|/|B|$. This is essentially a weighted average of the truth value of H and the content of H. Levi's choice leads to the expected utility

38. $P(H/E) - q|H|/|B$,

[24]If B is infinite, the sum in (36) will be replaced by integrals (cf. (34)).

and the following rule of acceptance: reject all elements H_i of B with $P(H_i/E) < q/|B|$, and accept the disjunction of all unrejected elements of B as the strongest hypothesis on the basis of E.

If the information content of H is measured by $cont(H) = 1 - P(H)$ (see (10)), the simple rule of directly maximizing $cont(H)$ would lead to the unsatisfactory recommendation of accepting a logical contradiction. If consistency is demanded, the rule would favor improbability independently of evidence. Instead, our gain in accepting H can be taken to be $cont(H)$ when H is true and our loss is $cont(\sim H)$ when H is false. Hence, the expected utility of accepting H is by (36)

$$39. \; u(H, E) = P(H/E)cont(H) - P(\sim H/E)cont(\sim H) = P(H/E) - P(H)$$

(see [Hintikka and Suppes, 1966; Hilpinen, 1968]). This is equal to Carnap's difference measure of confirmation (23). In maximizing (39), it can be written as the sum of $P(H/E)$ and $cont(H)$. In Hintikka's system of inductive logic, the rule of maximizing (39) leads to the acceptance of constituent C^c when the size of the sample is large enough (cf. (35)): it is has the highest content of all constituents compatible with evidence e, and its posterior probability approaches 1 with increasing n.

These results show that it is possible to combine and balance the Popperian demand that science strives for bold (informative, a priori improbable) hypotheses and the traditional Bayesian demand for well-supported (a posteriori probable) hypotheses.

5.3 Inference to the best explanation

Gilbert Harman [1965] has suggested that the basic rule of inductive inference is *inference to the best explanation*:

> (IBE) A hypothesis H may be inferred from evidence E when H is a better explanation of E than any other rival hypothesis.

In particular, if H is the only available explanation of E, then IBE warrants its acceptance on E. This idea is related to Peirce's notion of *abduction*, or inference from an effect to its explanatory causes (cf. [Niiniluoto, 2004; 2005a]). Even if Peirce held that the conclusion of abduction is sometimes "compelling", his usual weaker formulation of its conclusion was that "there is reason to suspect that H is true".

IBE is sometimes questioned by assuming an important difference between "accommodation" and "prediction". This contrast is relevant, but not very significant (cf. [Howson and Urbach, 1989]): in a situation of testing a theory H, there are credits of H due to its earlier ability to explain some initial evidence and to predict some new data. The initial probability H at the time of testing reflects these successes in explanation and prediction, and the later successful predictions will become parts of the total evidence for T. Therefore, it seems appropriate to formulate IBE in terms of systematic power, which combines the ideas of explanatory and predictive success.

IBE can be explicated in cognitive decision theory by analyzing the notion of "the best explanation" in terms of measures of systematic power. We have already noted that there are connections between measures of explanatory power and confirmation, so that the highest degree of confirmation of H on E will be assigned to that H which best explains E. The measure (12) $syst_2(H, E)$ and its variants lead directly to the well-known principle of Maximum Likelihood:

40. Given evidence E, accept the hypothesis H that maximizes the likelihood $P(E/H)$.

(See [Hintikka, 1968].) But the rule of directly maximizing Hempel's measure (11) $syst_1(H, E) = P(\sim H/ \sim E)$ would again recommend the acceptance of a logical contradiction.[25] But if our gain is taken to be $syst_1(H, E)$ when H is true and $-syst_1(\sim H, E)$ when H is false, then the best hypothesis H is the one which maximizes $P(H/E) - P(H)$ (see [Pietarinen, 1970; Niiniluoto, 1999]).

In curve fitting problems, the best explanation of the data is approximate.[26] Given a finite number of observed points E, it is always possible to define a curve which goes through them, but this hypothesis would commit the sin of *overfit*: due to observational errors it would probably be false and give bad predictions. Therefore, in regression analysis one usually fixes first the class of relevant curves (e.g., linear or quadratic functions) and then chooses from this class the most accurate — and hopefully least false — curve by the LSD criterion. More refined methods of balancing simplicity and accuracy in curve fitting are discussed by Forster and Sober [1994].

Hitchcock and Sober [2004] argue that false models often make more accurate predictions than true models. They conclude that for models one should adopt the instrumentalist position which regards predictive accuracy as a more important scientific goal than truth or approximate truth. By "models" in the context of curve fitting they mean polynomials with adjustable parameters. A model is true if it specifies correctly the mathematical function (i.e., the degree of the polynomial). Such models make predictions, when the parameters have been estimated by the maximum likelihood method. The same points could no doubt be repeated for more or less accurate explanations. However, the situation discussed by Hitchcock and Sober is familiar from the theory of truthlikeness: a true hypothesis or explanation of the form $T = C^* \vee C_1 \vee C_2$ may be less truthlike than a false hypothesis C_3, if C_3 is much closer to the truth than the false alternatives C_1 and C_2 allowed by the true hypothesis T. Similarly, models in the sense of Hitchcock

[25]Given the close connection of Laudan's [1977] notion problem-solving capacity and Hempel's systematic power, this is also the reason why Laudan has to substract "conceptual problems" from the number of problems positively solved by a theory. But it is not evident that such a move saves Laudan from a variant of the child's play objection (see [Niiniluoto, 1999, 187]).

[26]Theory T approximately explains E if T explains E' where E and E' are close to each other.

and Sober are disjunctions of hypotheses, and the true non-fitted model contains many false disjuncts. Hence, there may exist a simpler hypothesis which is more accurate than the true model. However, instrumentalism need not be supported here at all, since the merits of the simpler hypothesis may result from its greater truthlikeness.

5.4 Expected verisimilitude

The notions of epistemic probability, confirmation, and acceptance are related to truth as a goal of inquiry. But what can be said about situations where a scientist has to operate with theories that are known to false — e.g., theories involving simplifications, idealizations, and approximations? In such cases, where the theory is known to be false, its degree of credibility and confirmation are zero. A critical scientific realist, who holds that approximate truth and truthlikeness are important goals of science, has to find ways of using evidence to evaluate comparative and quantitative claims about "closeness to the truth".

Kuipers [2000] proposes an alternative to IBE which he calls *inference to the best theory*:

> (IBT) If a theory has so far proven to be the best one among the available theories, then conclude for the time being that it is the closest to the truth of the available theories.

The best theory is allowed to be inconsistent with the evidence. This formulation has the advantage that it covers also cases where the best available theory gives an *approximate explanation* of the data (see [Niiniluoto, 2004]).

For the purpose of IBT, Kuipers explicates the phrase 'closest to the truth' on three levels: closest to observational truth, referential truth, or theoretical truth, and uses his own theory of truth approximation. The phrase 'the best theory' in turn is defined in terms of empirical success. One theory is empirically more successful than another relative to the available data if it has "more" correct consequences and "less" counter-examples than the other theory. With these definitions, Kuipers is able to prove a *Success Theorem*: if theory Y is at least as similar to the truth as theory X, then Y will always be at least as successful as X relative to correct empirical data [Kuipers, 2000, 160]. Thus, higher truthlikeness explains greater empirical success. This means also that in our attempt to approximate the truth it is functional to use a method which is based on the *Rule of Success*: if theory Y has so far proven to be empirically more successful than theory X, accept the "comparative success hypothesis" that Y will remain to be more successful than X relative to all future data, and eliminate X in favor of Y [*ibid.*, 114]. In other words, it is rational to favor a theory which has so far proven to be empirically more successful than its rivals. This gives "a straightforward justification" of IBT in terms of truth approximation.

The results of Kuipers, including his Success Theorem, depend essentially on his way of explicating truthlikeness. By his criteria, if theory Y has been so far more successful than theory X, then X can never become more successful than Y in the future — the best prospect for X is to become incomparable with Y. Further, in many cases there will be no single theory which is better than all the available alternatives, so that a rule like IBT is inapplicable (see [Niiniluoto, 2005a]).

The quantitative treatment of truthlikeness has the methodological advantage that it allows a way of combining Popperian and Bayesian elements in the theory of scientific inference (see [Niiniluoto, 1987]). Following the basic idea of cognitive decision theory, science can be viewed as the project of *maximizing expected verisimilitude.* The mini-sum-measure of truthlikeness (19) can be understood as a generalization of Levi's assignment of epistemic utility (cf. (38)): for a disjunctive hypothesis, truth value is replaced by the minimum distance from the truth, and information content by the normalized sum of the distances of disjuncts from the truth. If all false basic alternatives are equally distant from the truth, this measure (19) reduces to Levi's proposal.

Some typical methods of Bayesian decision theory can be directly reinterpreted in terms of maximization of expected verisimilitude (see [Niiniluoto, 1987; Festa, 1993]). For example, as $|\theta - \theta_0|$ measures the distance of estimate θ_0 from the value of θ, formula (29) is clearly the expected distance of θ_0 from the truth. The Bayes rule then recommends the use of the median of the posterior distribution $p(\theta/e)$ as the best estimate. Alternatively, the posterior loss can be defined by using a quadratic loss function:

$$\int_{\mathbf{R}} \mathrm{p}(\theta/\mathrm{e})(\theta - \theta_o)^2 \mathrm{d}\theta.$$

This value is equal to

$$D^2[\mathrm{p}(\theta/\mathrm{e})] + (\mathrm{E}[\mathrm{p}(\theta/\mathrm{e})] - \theta_o)^2$$

(see [Festa, 1993, 39]). By this criterion, the best point estimate is the mean of the posterior distribution. The same treatment can be generalized to interval hypotheses.

More generally, as the true constituent is typically the unknown target C^* of a cognitive problem B, the value of $Tr(H, C^*)$ cannot be directly calculated by our formulas (18) and (19). However, there is a method of making rational comparative judgments about verisimilitude, if we have — instead of certain knowledge about the truth — rational degrees of belief about the location of truth. Thus, to *estimate* the degree $Tr(H, C^*)$, where C^* is unknown, assume that there is an epistemic probability measure P defined on B, so that $P(C_i/E)$ is the rational degree of belief in the truth of C_i given evidence E. The *expected degree of verisimilitude* of H in $D(B)$ given evidence E is then defined by

41. $ver(H/E) = \sum_{i \in I} P(C_i/E) Tr(H, C_i)$.

(41) gives us a comparative notion of estimated verisimilitude: H' *seems more truthlike* than H on evidence E if and only if $ver(H/E) < ver(H'/E)$. A variant of the concept of confirmation can be defined by the condition that evidence E

ver-confirms hypothesis H relative to B if and only if $ver(H/E\&B) > ver(H/B)$ (cf. [Festa, 1999]).

The comparison of ver-values of constituents (complete answers of B) reduces to the comparison of their posterior probabilities on E. But there are many situations where the evidence favors stronger partial answers to weaker ones. In particular, note that $ver(t/E) = 1 - \gamma'$ for a tautology t. An important difference to probabilistic measures of confirmation and corroboration is the possibility that a hypothesis H that is known to be refuted by evidence E may nevertheless be judged to be highly truthlike by ver, i.e., we may have $P(H/E) = 0$ but $ver(H/E) \approx 1$.

In Hintikka' s system of inductive logic, the result (7) guarantees that asymptotically it is precisely the boldest constituent compatible with the evidence that will have the largest degree of estimated verisimilitude:

42. $ver(H/e) \to Tr(H, C^c)$, when $n \to \infty$ and c is fixed.

43. $ver(H/e) \to 1$ iff $\vdash H \equiv C^c$, when $n \to \infty$ and c is fixed.

Here (43) follows from (42) by the fact that C^c is the only hypothesis H for which $T(H, C^c) = 1$.

Expected verisimilitude ver is not the only way of combining the notions of epistemic probability and closeness to the truth (see [Niiniluoto, 1987; 2004; 2005b]). Let H in $D(B)$ be a partial answer, and $\varepsilon > 0$ a small real number. Define

44. $V_\varepsilon(H) = \{C_i \text{ in } B | \Delta_{min}(C_i, H) \leq \varepsilon\}$.

Denote by H^ε the "blurred" version of g which contains as disjuncts all the members of the neighborhood $V_\varepsilon(H)$. Then $H \vdash H^\varepsilon$, and H is approximately true (within degree ε) if and only if H^ε is true. The probability that the minimum distance of H from the truth C^* is not larger than ε, given evidence E, defines at the same time the posterior probability that the degree of approximate truth $AT(H, C^*)$ of H is at least $1 - \varepsilon$:

45. $PAT_{1-\varepsilon}(H/E) = P(C^* \in V_\varepsilon(H)/E) = \sum_{C_i \in V_\varepsilon(H)} P(C_i/E)$.

PAT defined by (45) is thus a measure of *probable approximate truth*. Clearly we have always $P(H/E) \leq PAT_{1-\varepsilon}(H/E)$. When ε decreases toward zero, in the limit we have $PAT_1(H/E) = P(H/E)$. Further, $PAT_{1-\varepsilon}(H/E) > 0$ if and only if $P(H^\varepsilon) > 0$. Unlike ver, PAT shares with P the property (8) that logically weaker answers will have higher PAT-values than stronger ones.

An important feature of probable approximate truth is that its value can be non-zero even for hypotheses with a zero probability on evidence: it is possible that $PAT_{1-\varepsilon}(H/E) > 0$ even though $P(H/E) = 0$. This suggests that PAT-measure gives an alternative to Abner Shimony's [1970] "tempered personalism" where all seriously entertained hypotheses (even points on a real line) are assigned non-zero probabilities. This also opens the possibility of answering van Fraassen's criticism of abduction by generalizing the important result (20) about confirmation to hypotheses with a zero probability.

Another kind of situation where a hypothesis H has the probability 0 relative to our evidence E is when H is known to contain a counterfactual idealization I. In such cases, H may be tested by E if we first develop its concretization by removing the idealization I (cf. [Niiniluoto, 1999]). Alternatively, we may try to judge what evidence E would have been in the idealized situation I [Suppes, 1993]. This can be done theoretically, or by trying to realize experimentally conditions that resemble I.[27] Then we may study the confirmation $conf(H/E')$ of H or the expected verisimilitude $ver(H/E')$ relative to the counterfactual evidence statement E' [Niiniluoto, 2005a].

5.5 *Convergence to the truth*

Given the starting point of fallibilism, it is always logically possible to doubt the truth of our best hypotheses. But how far can we proceed in the successful confirmation of hypotheses by increasing empirical evidence? Do the convergence results concerning the estimation of verisimilitude mean that there is after all some sort of guarantee in the long run that our best theories are true?

Note first that the gap between a theory H and empirical evidence E may result from the richer terminology employed by H. Such a richer language is needed when theory H postulates unobservable entities. If the language L of H contains theoretical terms, which do not occur in the language L_O of E, then independently of the size of E there may be several relevant hypotheses in L that are compatible with E. Even with indefinitely increasing observational evidence there may be several theories receiving asymptotically non-zero posterior probabilities. In this sense, theories may be *underdetermined* by observational data.

Suppose that the choice between theories H and H' is underdetermined by available evidence E. This gap can be narrowed to some extent (but not fully closed) by two moves. One is to develop the rival theories and their testability by additional assumptions: for example, in the background knowledge B new principles can be introduced as links between the theoretical terms in L and the observational language L_O. The second is improved technology, which may widen the scope of observational evidence E: for example, some of the old theoretical terms can be made liable to observational or experimental criteria.

Whewell [1840] argued that the ability of a hypothesis to account for novel facts that were not "contemplated in its construction" gives us "a criterion of its reality, which has never yet been produced in favour of falsehood". We have already seen in (31) that surprising facts give support to hypotheses that are able to explain or predict them. However, from Bayes's Theorem (5) we know that, given that H entails E, $P(H/E) = 1$ if and only if $P(\sim H)P(E/\sim H) = 0$, i.e., $P(H) = 1$ or

[27] For example, if I states that resistance of air has no effect on the movements of bodies, experiments may be made in conditions approximating a vacuum. For the role experiments in science, see the paper of Franklin and Gonzalez in this volume.

E entails H. These conditions do not hold, if H is a hypothesis in a theoretical language and E is observational evidence. But it is possible that $P(H/E)$ is high, even close to one, if $P(E/\sim H)$ is sufficiently small. This will happen, if H is the only explanation of E, i.e., no alternative to H (included as a disjunct in the negation $\sim H$) is able to explain E. Another possibility is that E may be a novel prediction from H which is also unique for H: no viable rival of H provides a basis for predicting E (see [Leplin, 2004]). Hence,

46. Let H be a theory such that $P(H) > 0$ and $P(E/H) = 1$. Then $P(H/E)$ is close to one, if H is the only explanation of E or E is a unique novel prediction of H.

According to (46), genuine theoretical hypotheses — if they are clearly more successful than any available rivals — can be tested, confirmed, and even accepted on observational evidence. But, to be sure, the risk of error of such theories cannot be eliminated: high posterior probability does not logically guarantee that the hypothesis is true.

Let us then ask what happens in the most favorable situation where H is expressed in the same language as E. This question can be clarified by the systems of inductive logic. Recall that in Hintikka's system the posterior probability $P(C^c/e)$ of constituent C^c approaches one when the number of observed kinds c is fixed and the number of observed individuals n grows without limit. This result (7) states only that our degrees of belief about C^c *converge to certainty* on the basis of inductive evidence (cf. (46)). It does not yet guarantee that C^c is identical with the *true* constituent C^*. Similarly, by (43) we know that the expected verisimilitude $ver(C^c/e)$ converges to one when c is fixed and n grows without limit. Again this does not guarantee that the "real" truthlikeness $Tr(C^c, C^*)$ of C^c is maximal, i.e., that C^c is true. For these stronger results, which link estimated verisimilitude and objective truthlikeness, an additional *evidential success condition* is needed:

(ES) Evidence e is true and fully informative about the variety of the world w.

ES means that e is exhaustive in the sense that it exhibits (relative to the expressive power of the given language L) all the kinds of individuals that exist in the world (see [Niiniluoto, 1987, 276]). With ES as a sufficient condition, convergence to C^c can be replaced by convergence to C^*, so that our results may be reformulated in terms of *convergence to the truth*:

(7′) Given that ES holds, $P(C^*/e) \rightarrow 1$, when c is fixed and $n \rightarrow \infty$.

(7″) Given that ES holds, $P(H/e) \rightarrow 1$ iff H is true, when c is fixed and $n \rightarrow \infty$.

(42′) Given that ES holds, $ver(H/e) \rightarrow Tr(H, C^*)$, when $n \rightarrow \infty$ and c is fixed.

(43′) Given that ES holds, $ver(H/e) \to 1$ iff $\vdash H \equiv C^*$, when $n \to \infty$ and c is fixed.

(See [Niiniluoto, 2005b].)[28] The assumption of complete variety in ES is also necessary for these results, since otherwise some kinds of individuals would remain unobserved and the probability would concentrate on a wrong constituent.

The chance that ES is correct in particular situations can be improved by systematic methods of observation and experimentation. By manipulating nature the researcher can realize otherwise hidden possibilities and thereby increase the variety of the available evidence.[29] ES can be known to be correct in the special case, where all possible kinds of individuals have been observed and thereby the atomistic constituent C^K is verified. However, as the following argument indicates, ES cannot be proved to be generally valid in inductive learning situations.

In Hintikka's system, *epistemic* probabilities are used in the prior distribution $P(C^w)$ and likelihoods $P(e/C^w)$. The convergence result (7) does not depend on any assumption that behind these likelihoods there are some objective conditions concerning the sampling method. But it is possible to combine a system of inductive logic with the assumption that the evidence arises from a *fair* sampling procedure, so that each kind of individual has an objective non-zero *propensity* of appearing in the evidence e [Kuipers, 1978]. Propensities as probabilistic dispositions do not satisfy the notorious Principle of Plenitude, which claims that all genuine possibilities will sometimes be actualized. Therefore, they do not exclude infinite sequences which violate ES, even though such sequences are extremely improbable by the convergence theorems of probability calculus (cf. [Festa, 1993, 76]). If the objective probability of picking out an A is r in a series of independent trials, the Strong Law of Large Numbers states that the observed relative frequency k/n of As converges *with probability one* to the unknown value of r. Such "almost sure" convergence is weaker than convergence in the ordinary mathematical sense. No principle of logic excludes such non-typical sequences of observations that violate ES, even though their measure among all possible sequences is zero.

ES is precisely the reason why inductive inference is always non-demonstrative or fallible even in the ideal limit: there are no logical reasons for excluding the possibility that ES might be incorrect. Hence, the strongest success results for a fallibilist "convergent realist" assert convergence to the truth with probability one.[30]

---

[28]Similar modifications can be made in the convergence results about probable approximate truth. Assumptions similar to ES are made also formal learning theory (cf. [Niiniluoto, 2005b]). Its convergence results about approach to the truth are not stronger than those of inductive logic, since the data streams are assumed to be "complete in that they exhaust the relevant evidence" [Earman, 1992, 210] or "perfect" in that "all true data are presented and no false datum is presented" and all objects are eventually described [Kelly, 1996, 270].

[29]The possibility of realizing counterfactual possibilities is the reason why observation assisted by experimentation is the most powerful method of confirming genuine laws of nature.

[30]It is remarkable that Peirce seemed to be aware of this important fact (see [Niiniluoto, 1984, 82]).

6 CONCLUDING REMARK

Bradie [2005] argues that Niiniluoto's [1999] rejection of Laudan's [1977] problem-solving criterion as a measure of scientific progress has a "presumption that the aim of science is to arrive at the truth". In a sense, this is quite correct: in the axiology of a scientific realist, truth is a central or primary goal of inquiry, even though it has to be balanced by the demand of information content. For a fallibilist realist, truthlikeness is a combination of these goals of truth and information. Some of the results surveyed in this paper indeed have a presumption about the aim of science, i.e., their form is the following: if our goal is high truthlikeness, then these methods are effective tools in promoting or approaching it.

But this is not the whole story about the dialectical situation between scientific realists and anti-realists. It is clear that an instrumentalist, who denies that theories have truth values at all, cannot be persuaded to think that the best theories are likely to be true or truthlike. This does not preclude the possibility that in many situations the preference rankings of theories by realists and instrumentalists agree, even though they disagree about the import of such rankings. But many of our results about the evaluation of theories concern rational scientists who admit that theories have truth values (this much is agreed by van Fraassen and Laudan) and whose belief systems are coherent so that their degrees of belief can be represented by epistemic probabilities. Then we have given arguments to show that, for such a rational scientist, empirical success in explanation and prediction is a fallible indicator of the truth or truthlikeness of a theory. The position which denies such a power of confirmation can be presented within our framework by the dogmatic assumption that the prior probability of a theory is zero. Such an assumption is dogmatic in the sense that it denies for ever the possibility that the epistemic probability of a theory could rise over zero — but we have also seen that even this does not preclude the possibility of estimating that the theory is highly truthlike. So the main point of our results is not that we as realists have presumed that science aims at truth, but rather we have tried to force the anti-realists to a corner where they have to admit (perhaps against their will) that their view is a priori based upon dogmatic scepticism about theories.

BIBLIOGRAPHY

[Ayer, 1959] A. J. Ayer (ed.). *Logical Positivism.* The Free Press, New York, 1959.

[Balzer *et al.*, 1987] W. Balzer, C. U. Moulines, and J. D. Sneed. *An Architectonic for Science: The Structuralist Program.* D. Reidel, Dordrecht, 1987.

[Bar-Hillel, 1964] Y. Bar-Hillel. *Language and Information.* Addison Wesley, Reading, Mass., 1964.

[Bell and Slomson, 1969] J. Bell and A. Slomson. *Models and Ultraproducts.* North-Holland, Amsterdam, 1969.

[Bonilla, 1996] J. P. Z. Bonilla. Verisimilitude, structuralism, and scientific progress. *Erkenntnis*, 44: 25–47, 1996.

[Box and Tiao, 1973] G. Box and G. Tiao. *Bayesian Inference in Statistical Analysis.* Addison-Wesley, Reading, MA., 1973.

[Bovens and Hartmann, 2003] L. Bovens and S. Hartmann. *Bayesian Epistemology*. Oxford University Press, Oxford, 2003.

[Bradie, 2005] M. E. Bradie. Scientific progress. In S. Sarkar and J. Pfeiffer (eds.), *The Philosophy of Science: An Encyclopedia 1-2*, Routledge, New York and London, pages 749–753, 2005.

[Brown, 1977] H. I. Brown. *Perception, Theory, and Commitment: The New Philosophy of Science*. The University of Chicago Press, Chicago, 1977.

[Carnap, 1936-1937] R. Carnap. Testability and meaning. *Philosophy of Science*, 3: 419–471, 1936; 4: 1–40, 1937.

[Carnap, 1949] R. Carnap. Truth and confirmation. In Feigl and Sellars, pages 119–127, 1949.

[Carnap, 1962] R. Carnap. *The Logical Foundations of Probability*. 2nd ed., The University of Chicago Press, Chicago, 1962.

[Cohen, 1980] L. J. Cohen. What has science to do with truth? *Synthese*, 45: 489–510, 1980.

[Cohen, 1989] L. J. Cohen. *An Introduction to the Philosophy of Induction and Probability*. Oxford University Press, Oxford, 1989.

[David, 2004] M. David. Theories of truth. In Niiniluoto *et al.*, pages 331–414, 2004.

[Dietrich and Moretti, 2005] F. Dietrich and L. Moretti. On coherent sets and the transmission of confirmation. *Philosophy of Science*, 72: 403–424, 2005.

[Doppelt, 1983] G. Doppelt. Relativism and recent pragmatic conceptions of scientific rationality. In N. Rescher (ed.), *Scientific Explanation and Understanding*, University Press of America, Lanham, pages 107–142, 1983.

[Duhem, 1954] P. Duhem. *The Aim and Structure of Physical Theory*. Princeton University Press, Princeton, NJ, 1954.

[Duhem, 1969] P. Duhem. *To Save the Phenomena: An Essay on the Idea of Physical Theory from Plato to Galileo*. The University of Chicago Press, Chicago, 1969.

[Dummett, 1978] M. Dummett. *Truth and Other Enigmas*. Duckworth, London, 1978.

[Earman, 1992] J. Earman. *Bayes or Bust?: A Critical Examination of Bayesian Confirmation Theory*. A Bradford Book, The MIT Press, Cambridge, MA, 1992.

[Edwards, 1972] A. Edwards. *Likelihood*. Cambridge University Press, Cambridge, 1972.

[Elkana *et al.*, 1978] Y. Elkana *et al.* (eds.). *Toward a Metric of Science: The Advent of Science Indicators*. Wiley and Sons, New York, 1978.

[Festa, 1993] R. Festa. *Optimum Inductive Methods*. Kluwer, Dordrecht, 1993.

[Festa, 1999] R. Festa. Bayesian confirmation. In M. Galavotti and A. Pagnini (eds.), *Experience, Reality, and Scientific Explanation*. Dordrecht: Kluwer, pages 55–87, 1999.

[Festa *et al.*, 2005] R. Festa, A. Aliceda, and J. Peijnenburg, (eds.). *Confirmation, Empirical Progress, and Truth Approximation*, Rodopi, Amsterdam, 2005.

[Fetzer, 1981] J. Fetzer. *Scientific Knowledge: Causation, Explanation, and Corroboration*. D. Reidel, Dordrecht, 1981.

[Fitelson, 1999] B. Fitelson. The plurality of Bayesian measures of confirmation and the problem of measure sensitivity. *Philosophy of Science* (*Proceedings*): 66: S362–S378, 1999.

[Foster and Martin, 1966] M. H. Foster and M. L. Martin (eds.). *Probability, Confirmation, and Simplicity*. The Odyssey Press, New York, 1966.

[Forster and Sober, 1994] M. Forster and E. Sober. How to tell when simpler, more unified, or less *ad hoc* theories will provide more accurate predictions. *British Journal for the Philosophy of Science*, 45: 1–36, 1994.

[Glymour, 1980] C. Glymour. *Theory and Evidence*. Princeton University Press, Princeton, 1980.

[Hacking, 1983] I. Hacking. *Representing and Intervening*. Cambridge University Press, Cambridge, 1983.

[Harré and Madden, 1975] R. Harré and E. H. Madden. *Causal Powers*. Blackwell, Oxford, 1975.

[Helman, 1988] D. Helman (ed.). *Analogical Reasoning*. Dordrecht, D. Reidel, 1988.

[Hempel, 1965] C. G. Hempel. *Aspects of Scientific Explanation*. The Free Press, New York, 1965.

[Hilpinen, 1968] R. Hilpinen. *Rules of Acceptance and Inductive Logic*. Acta Philosophica Fennica 22, North-Holland, Amsterdam, 1968.

[Hintikka, 1968] J. Hintikka. The varieties of information and scientific explanation. In B. van Rootselar and J. E. Staal (eds.), *Logic, Methodology and Philosophy of Science III*. North-Holland, Amsterdam, 151–171, 1968.

[Hintikka, 1988] J. Hintikka. On the development of the model-theoretic viewpoint in logical theory. *Synthese*, 77: 1–36, 1988.

[Hintikka and Suppes, 1970] J. Hintikka and P. Suppes, (eds.), *Information and Inference*. Reidel, Dordrecht, 1970.

[Hitchcock and Sober, 2004] C. Hitchcock and E. Sober. Prediction versus accommodation and the risk of overfitting. *The British Journal for the Philosophy of Science*, 55: 1–34, 2004.

[Howson and Urbach, 1989] C. Howson and P. Urbach. *Scientific Reasoning: The Bayesian Approach*. Open Court, La Salle, 1989.

[Irvine and Martin, 1984] J. Irvine and B. Martin. *Foresight in Science: Picking the Winners*. Frances Pinter, London and Dover, 1984.

[Jeffrey, 1965] R. Jeffrey. *The Logic of Decision*. McGraw-Hill, New York, 1965.

[Jeffrey, 1980] R. Jeffrey (ed.). *Studies in Inductive Logic and Probability*. Vol. 2, University of California Press, Berkeley, 1980.

[Kaila, 1939] E. Kaila. *Inhimillinen tieto*. Otava, Helsinki, 1939.

[Kaila, 1979] E. Kaila. *Reality and Experience*. D. Reidel, Dordrecht, 1979.

[Kelly, 1996] K. Kelly. *The Logic of Reliable Inquiry*. Oxford University Press, New York, 1996.

[Kirkham, 1992] R. L. Kirkham. *Theories of Truth: A Critical Introduction*. The MIT Press, Cambridge, MA, 1992.

[Kitcher, 1989] P. Kitcher. Explanatory unification and the causal structure of the world. In P. Kitcher and W. Salmon (eds.), *Scientific Explanation*. University of Minnesota Press, Minneapolis, pages 410–505, 1989.

[Kuhn, 1962] T. S. Kuhn. *The Structure of Scientific Revolutions*. University of Chicago Press, Chicago, 1962. 2nd enlarged ed. 1970.

[Kuhn, 1977] T. S. Kuhn. *The Essential Tension*. The University of Chicago Press, Chicago, 1977.

[Kuipers, 1978] T. Kuipers. *Studies in Inductive Probability and Rational Expectation*. D. Reidel, Dordrecht, 1978.

[Kuipers, 1987] T. Kuipers (ed.), *What-Is-Closer-to-the-Truth?* Poznan Studies in the Philosophy of Science, Rodopi, Amsterdam, 1987.

[Kuipers, 2000] T. Kuipers. *From Instrumentalism to Constructive Realism: On Some Relations between Confirmation, Empirical Progress, and Truth Approximation*. Dordrech, Kluwer, 2000.

[Kyburg and Smokler, 1965] H. Kyburg and H. Smokler (eds.), *Studies in Subjective Probability*. Wiley and Sons, New York, 1965.

[Lakatos, 1968] I. Lakatos (ed.). *The Problem of Inductive Logic*, North-Holland, Amstrerdam, 1968.

[Lakatos and Musgrave, 1970] I. Lakatos and A. Musgrave (eds.), *Criticism and the Growth of Knowledge*. Cambridge University Press, Cambridge, 1970.

[Laudan, 1977] L. Laudan. *Progress and Its Problems: Toward a Theory of Scientific Growth*. Routledge and Kegan Paul, London, 1977.

[Laudan, 1981] L. Laudan. *Science and Hypothesis*. D. Reidel, Dordrech, 1981.

[Laudan, 1984] L. Laudan. *Science and Values: The Aims of Science and Their Role in Scientific Debate*. University of California Press, Berkeley, 1984.

[Laudan, 1996] L. Laudan. *Beyond Positivism and Relativism: Theory, Method, and Evidence*. Westview Press, Boulder, 1996.

[Leplin, 1984] J. Leplin (ed.), *Scientific Realism*. University of California Press, Berkeley, 1984.

[Leplin, 2004] J. Leplin. A theory's predictive success can warrant belief in the unobservable entities it postulates. In C. Hitchcock (ed.), *Contemporary Debates in Philosophy of Science*, Blackwell, Oxford, pages 117–132, 2004.

[Levi, 1967] I. Levi. *Gambling With Truth: An Essay on Induction and the Aims of Science*. Harper & Row, New York, 1967; 2nd ed. The MIT Press, Cambridge, Mass., 1973.

[Lewis, 1973] D. Lewis. *Counterfactuals*. Blackwell, Oxford, 1973.

[Longino, 1990] H. Longino. *Science as Social Knowledge*. Princeton University Press, Princeton, 1990.

[Miller, 1974] D. Miller. Popper's qualitative theory of verisimilitude. *The British Journal for the Philosophy of Science*, 25: 166–177, 1974.

[Milne, 1996] P. Milne. $log[p(h/eb)/p(h/b)]$ is the one true measure of confirmation. *Philosophy of Science*, 63: 21–26, 1996.

[Myrwold, 2003] W. C. Myrwold. A Bayesian account of the virtue of unification. *Philosophy of Science*, 70: 399–423, 2003.

[Niiniluoto, 1984] I. Niiniluoto. *Is Science Progressive?* D. Reidel, Dordrecht, 1984.

[Niiniluoto, 1987] I. Niiniluoto. *Truthlikeness*. D. Reidel, Dordrecht, 1987.

[Niiniluoto, 1998] I. Niiniluoto. Verisimilitude: The third period. *The British Journal for the Philosophy of Science*, 49: 1–29, 1998.

[Niiniluoto, 1999] I. Niiniluoto. *Critical Scientific Realism*. Oxford University Press, Oxford, 1999.

[Niiniluoto, 2000] I. Niiniluoto. Scepticism, fallibilism, and verisimilitude. In J. Sihvola (ed.), *Ancient Scepticism and the Sceptical Tradition*. Acta Philosophica Fennica 66, The Philosophical Society of Finland, Helsinki, pages 145–169, 2000.

[Niiniluoto, 2003] I. Niiniluoto. Content and likeness definitions of truthlikeness. In J. Hintikka, T. Czarnecki, K. Kijania-Placek, T. Placek and A. Rojszczak (eds.), *Philosophy and Logic: In Search of the Polish Tradition*, Kluwer, Dodrecht, pages 27–35, 2003.

[Niiniluoto, 2004] I. Niiniluoto. Truth-seeking by abduction. In F. Stadler (ed.), *Induction and Deduction in the Sciences*. Kluwer, Dordrecht, pages 57–82, 2004.

[Niiniluoto, 2005a] I. Niiniluoto. Abduction and truthlikeness. In Festa *et al.*, pages 255–275, 2005

[Niiniluoto, 2005b] I. Niiniluoto. Inductive logic, verisimilitude, and machine learning. In P. Hájek, L. Valdés-Villanueva and D. Westerståhl (eds.), *Logic, Methodology and Philosophy of Science*, King's College Publications, London, pages 295–314, 2005.

[Niiniluoto and Tuomela, 1973] I. Niiniluoto and R. Tuomela. *Theoretical Concepts and Hypothetico-Inductive Inference*. D. Reidel, Dordrecht, 1973.

[Niiniluoto *et al.*, 2004] I. Niiniluoto, M. Sintonen, and J. Wolenski, (eds.), *Handbook of Epistemology*, Kluwer, Dordrecht, 2004.

[Oddie, 1986] G. Oddie. *Likeness to Truth*. D. Reidel, Dordrecht Boston, 1986.

[Pappas and Swain, 1978] G. Pappas and M. Swain, (eds.). *Essays on Knowledge and Justification*. Cornell University Press, Ithaca, 1978.

[Peirce, 1931-35] C. S. Peirce. *Collected Papers*. Ed. by C. Hartshorne and P. Weiss, vols. 1–6, Harvard University Press, Cambridge, MA, 1931-35.

[Pietarinen, 1970] J. Pietarinen. Quantitative tools for evaluating scientific systematizations. In Hintikka and Suppes, pages 123–147, 1970.

[Popper, 1959] K. Popper. *The Logic of Scientific Discovery*. Hutchinson, London, 1959.

[Popper, 1963] K. Popper. *Conjectures and Refutations: The Growth of Scientific Knowledge*. Hutchinson, London, 1963.

[Psillos, 1999] S. Psillos. *Scientific Realism: How Science Tracks Truth*. Routledge, London, 1999.

[Putnam, 1981] H. Putnam. *Reason, Truth and History*. Cambridge University Press, Cambridge, MA, 1981.

[Reichenbach, 1938] H. Reichenbach. *Experience and Prediction*. University of Chicago Press, Chicago, 1938.

[Rescher, 1977] N. Rescher. *Methodological Pragmatism*. Blackwell, Oxford, 1977.

[Rosenkrantz, 1977] R. Rosenkrantz. *Inference, Method and Decision*. D. Reidel, Doirdrecht, 1977.

[Salmon, 1984] W. C. Salmon. *Scientific Explanation and the Causal Structure of the World*. Princeton University Press, Princeton, NJ, 1984.

[Salmon, 1990] W. C. Salmon. Rationality and objectivity in science or Tom Kuhn meets Tom Bayes. In C. W. Savage (ed.), *Scientific Theories*. The University of Minnesota Press, Minneapolis, pages 175–204, 1990.

[Sellars, 1963] W. Sellars. *Science, Perception, and Reality*. Routledge and Kegan Paul, London, 1963.

[Shafer, 1976] G. Shafer. *A Mathematical Theory of Evidence*, Princeton University Press, Princeton, 1976.

[Shimony, 1970] A. Shimony. Scientific inference. In R. C. Colodny (ed.), *The Nature and Function of Scientific Theories*. The University of Pittsburgh Press, Pittsburgh, pages 79–172, 1970.

[Smokler, 1968] H. Smokler. Conflicting conceptions of confirmation. *Journal of Philosophy*, 65: 300–312, 1968.

[Stegmüller, 1971] W. Stegmüller. Das Problem der Induktion: Humes Herausforderung und modernen Antworten. In H. Lenk (ed.), *Neue Aspekte der Wissenschaftstheorie*, Vieweg, Braunschweig, pages 13-74, 1971.

[Stegmüller, 1976] W. Stegmüller. *The Structure and Dynamics of Theories*. Springer-Verlag, New York Heidelberg Berlin, 1976.

[Suppe, 1977] F. Suppe, (ed.). *The Structure of Scientific Theories*. 2nd ed., University of Illinois Press, Urbana, 1977.

[Suppes, 1993] P. Suppes. *Models and Methods in the Philosophy of Science: Selected Essays*. Kluwer, Dordrecht, 1993.

[Tarski, 1956] A. Tarski. *Logic, Semantics, Metamathematics*. Oxford University Press, Oxford, 1956.

[Tuomela, 1973] R. Tuomela. *Theoretical Concepts*. Springer-Verlag, Berlin, 1973.

[Tuomela, 1985] R. Tuomela. *Science, Action and Reality*. Reidel, Dordrecht, 1985.

[van Fraassen, 1980] B. van Fraassen. *The Scientific Image*. Oxford University Press, Oxford, 1980.

[Whewell, 1840] W. Whewell. *The Philosophy of the Inductive Sciences*. Parker and Sons, London, 1840. A new edition, ed. by G. Buchdahl and L. Laudan, 1967.

[Worrall, 1989] J. Worrall. Structural realism: The best of both worlds. *Dialectica*, 43: 99–124, 1989.

[Zwart, 2001] S. Zwart. *Refined Verisimitude*. Kluwer, Dordrecht, 2001.

THE ROLE OF EXPERIMENTS IN THE NATURAL SCIENCES: EXAMPLES FROM PHYSICS AND BIOLOGY

Allan Franklin

Science is a reasonable enterprise based on experimental evidence, criticism, and rational discussion. It provides us with knowledge of the physical world and it is experiment that provides the evidence which grounds that knowledge. As the late Richard Feynman, one of the leading theoretical physicists of the twentieth century, wrote, "The principle of science, the definition, almost, is the following: *The test of all knowledge is experiment.* Experiment is the *sole judge* of scientific 'truth'[Feynman, *et al.*, 1963, p. 1–1]." This is not, of course, to denigrate the role of theory in science. Correct theories of nature are an important goal of science. Theory can also provide guidance to experiment and, as discussed below, help to validate an experimental result. Nevertheless, as Feynman notes, experiment can provide us with good reasons to believe in our theories.

Experiment plays many roles in science. One of its important roles is to test theories and to provide the basis for scientific knowledge. It can also call for a new theory, either by showing that an accepted theory is incorrect, or by exhibiting a new phenomenon which needs explanation. Experiment can provide hints toward the structure or mathematical form of a theory and it can provide evidence for the existence of the entities involved in our theories. It can also measure quantities that theory tells us are important. Finally, it may also have a life of its own, independent of theory. Scientists may investigate a phenomenon just because it looks interesting. This will also provide evidence for a future theory to explain.

In all of this activity, however, we must remember that science is fallible. Theoretical calculations, experimental results, or the comparison between experiment and theory may all be wrong. Science is more complex than "The scientist proposes, Nature disposes." It may not always be clear what the scientist is proposing. Theories often need to be articulated and clarified. It also may not be clear how Nature is disposing. Experiments may not always give clear-cut results, and may even disagree for a time. Sometimes they can be incorrect.

If experiment is to play these important roles in science then we must have good reasons to believe experimental results. I will present below an epistemology of experiment, a set of strategies that provides reasonable belief in experimental results. Scientific knowledge can then be reasonably based on these experimental results.

Handbook of the Philosophy of Science: General Philosophy of Science - Focal Issues, pp. 219–274.
Volume editor: Theo Kuipers. General editors: Dov M. Gabbay, Paul Thagard and John Woods.

In this essay I will argue for the view that nature, as revealed by experiment, plays an important and legitimate role in science. My examples will come, primarily, although not exclusively, from physics, because that is the science I know best, but I believe that these episodes are typical of the natural sciences. Several examples from biology are also included. I will begin by offering my own version of an epistemology of experiment, a set of strategies used by scientists to argue for the correctness of an experimental result.

1 EXPERIMENTAL RESULTS

1.1 The Case for Learning From Experiment

1.1.1 An Epistemology of Experiment

It has been more than two decades since Ian Hacking asked, "Do we see through a microscope?" [Hacking, 1981]. Hacking's question really asked how do we come to believe in an experimental result obtained with a complex experimental apparatus? How do we distinguish between a valid result and an artifact created by that apparatus? If experiment is to play all of the important roles in science mentioned above and to provide the evidential basis for scientific knowledge, then we must have good reasons to believe in those results. Hacking provided an extended answer in the second half of *Representing and Intervening* (1983). He pointed out that even though an experimental apparatus is laden with, at the very least, the theory of the apparatus, observations remain robust despite changes in the theory of the apparatus or in the theory of the phenomenon. His illustration was the continuous belief in microscope images despite the major change in the theory of the microscope when Abbe pointed out the importance of diffraction in its operation. One reason Hacking gave for this is that in making such observations the experimenters intervened. They manipulated the object under observation. Thus, in looking at a cell through a microscope one might inject fluid into the cell or stain the specimen. One expects the cell to change shape or color when this is done. Observing the predicted effect strengthens our belief in both the proper operation of the microscope and in the observation. This is true in general. Observing the predicted effect of an intervention strengthens our belief in both the proper operation of the experimental apparatus and in the observations made with it.

Hacking also discussed the strengthening of one's belief in an observation by independent confirmation. The fact that the same pattern of dots, dense bodies in cells, is seen with "different" microscopes, i.e. ordinary, polarizing, phase-contrast, fluorescence, interference, electron, acoustic etc., argues for the validity of the observation. Hacking correctly argues that it would be a preposterous coincidence if the same pattern of dots were produced in two totally different kinds of physical systems. Different apparatuses have different backgrounds and systematic errors, making the coincidence, if it is an artifact, most unlikely. If it is a correct result, and the instruments are working properly, the agreement of results is understandable.

Hacking's answer is correct as far as it goes. It is, however, incomplete. What happens when one can perform the experiment with only one type of apparatus, such as an electron microscope or a radio telescope, or when intervention is either impossible or extremely difficult? Other strategies are needed to validate the observation. These may include (see Table 1):

1. Experimental checks and calibration, in which the experimental apparatus reproduces known phenomena. For example, if we wished to argue that the spectrum of a substance obtained with a new type of spectrometer is correct, we might check that this new spectrometer could reproduce the known Balmer Series in hydrogen. If we correctly observe the Balmer Series then we strengthen our belief that the spectrometer is working properly. This also strengthens our belief in the results obtained with that spectrometer. If the check fails then we have good reason to question the results obtained with that apparatus.

2. Reproducing artifacts that are known in advance to be present. An example of this comes from experiments to measure the infrared spectra of organic molecules [Randall, *et al.*, 1949]. It was not always possible to prepare a pure sample of such material. Sometimes one had to place the substance in an oil paste or in solution. In such cases, one expects to observe, superimposed on the spectrum of the substance, the spectrum of the oil or the solvent, which one can compare with the known spectrum of the oil or the solvent. Observation of this artifact gives confidence in other measurements made with the spectrometer.

3. Elimination of plausible sources of error and alternative explanations of the result (the Sherlock Holmes strategy).[1] Thus, when scientists claimed to have observed electric discharges in the rings of Saturn, they argued for their result by showing that it could not have been caused by defects in the telemetry, by interaction with the environment of Saturn, by lightning, or by dust. The only remaining explanation of their result was that it was due to electric discharges in the rings. There was no other plausible explanation of the observation. In addition, the same result was observed by both Voyager 1 and Voyager 2. This provided independent confirmation. Often, several epistemological strategies are used in the same experiment.

4. Using the results themselves to argue for their validity. Consider the problem of Galileo's telescopic observations of the moons of Jupiter. Although one might very well believe that his early telescope might have created spots

[1] As Holmes remarked to Watson, "How often have I said to you that when you have eliminated the impossible, whatever remains, *however improbable,* must be the truth Conan Doyle [1967, p. 638].

Table 1. Examples of epistemological strategies used by experimentalists in evolutionary biology, from H.B.D. Kettlewell's [1955; 1956; 1958] investigations of industrial melanism. (See Rudge [1998]).

Epistemological strategies Examples from Kettlewell		
1.	Experimental checks and calibration, in which the apparatus reproduces known phenomena	Use of the scoring experiment to verify that the proposed scoring methods would be feasible and objective
2.	Reproducing artifacts that are known in advance to be present.	Analysis of recapture figures for endemic *betularia* populations.
3.	Elimination of plausible sources of background and alternative result	Use of natural barriers to minimize explanations of the migration.
4.	Using the results themselves to argue for their validity.	Filming the birds preying on the moths.
5.	Using an independently well-corroborated theory of th ephenomenon to explain the results.	Use of Ford's theory of the spread of industrial melanism.
6.	Using an apparatus based on a well-corroborated theory	Use of Fisher, Ford, and Shepard techniques. [The mark-release-capture method had been used in several earlier experiments]
7.	Using statistical arguments.	Use and analysis of large numbers of moths.
8.	Blind analysis	Not used.
9.	Intervention, in which the experimenter manipulates the object under observation.	Not present
10.	Independent confirmation using different experiments.	Use of two different types of traps to recapture the moths.

of light, it would have been extremely implausible that the telescope would create them so that they would appear to be a small planetary system with eclipses and other consistent motions. It would have been even more implausible to believe that the created spots would satisfy Kepler's Third Law (R^3/T^2 = constant).[2] In this case one is arguing that there was no plausible malfunction of the apparatus, or background, that would explain the observations.

5. Using an independently well-corroborated theory of the phenomena to explain the results. This was illustrated in the discovery of the $W^{\pm}$, the charged intermediate vector boson required by the Weinberg-Salam unified theory of electroweak interactions. Although these experiments used very complex apparatuses and used other epistemological strategies (see Franklin [1986, pp. 170–172] for details), I believe that the agreement of the observations with the theoretical predictions of the particle properties helped to validate the experimental results. In this case the particle candidates were observed in exactly the type of events that theory predicted. In addition, the measured particle mass of $81 \pm 5GeV/c^2$ and $80^{+10}_{-6}GeV/c^2$, found in the two experiments (note the independent confirmation), was in good agreement with the theoretical prediction of $82 \pm 2.4GeV/c^2$. It was very improbable that any background effect, which might mimic the presence of the particle, would be in agreement with theory.

6. Using an apparatus based on a well-corroborated theory. In this case the support for the theory passes on to the apparatus based on that theory. This is the case with both the electron microscope and the radio telescope, whose proper operation is based on a well-supported theory, although other strategies are also used to validate the observations.

7. Using statistical arguments. An interesting example of this arose in the 1960s when the search for new particles and resonances occupied a substantial fraction of the time and effort of those physicists working in experimental high-energy physics. The usual technique was to plot the number of events observed as a function of the invariant mass of the final-state particles and to look for bumps above a smooth background. The usual informal criterion for the presence of a new particle was that it resulted in a three standard-deviation effect above the background, a result that had a probability of 0.27% of occurring in a single bin. This criterion was later changed to four standard deviations, which had a probability of 0.0064% when it was pointed out that the number of graphs plotted each year by high-energy physicists made it rather probable, on statistical grounds, that a three standard-deviation effect would be observed.

[2]Kepler's Third Law was not available when Galileo made his obervations, but it was an argument that could have been used later.

8. Using "blind" analysis, a strategy for avoiding possible experimenter bias. This can include excluding the region of interest when setting the selection criteria that will be applied to the data. One can also add a random number to the measured parameter so that the value of the result does not bias the analysis of the data. (For details see [Franklin, 2002, Chapter 6].)

These strategies along with Hacking's intervention and independent confirmation provide an epistemology of experiment.

Although all of the illustrations of the epistemology of experiment come from physics, David Rudge [1998; 2001] has shown that they are also used in biology. His example is Kettlewell's [1955; 1956; 1958] evolutionary biology experiments on the Peppered Moth, *Biston betularia*. The *typical* form of the moth has a pale speckled appearance and there are two darker forms, f. *carbonaria*, which is nearly black, and f. *insularia*, which is intermediate in color. The *typical* form of the moth was most prevalent in the British Isles and Europe until the middle of the nineteenth century. At that time things began to change. Increasing industrial pollution had both darkened the surfaces of trees and rocks and had also killed the lichen cover of the forests downwind of pollution sources. Coincident with these changes, naturalists had found that rare, darker forms of several moth species, in particular the Peppered Moth, had become common in areas downwind of pollution sources.

Kettlewell attempted to test a selectionist explanation of this phenomenon. E. B. Ford [1937; 1940] had suggested a two-part explanation of this effect: 1) darker moths had a superior physiology and 2) the spread of the melanic gene was confined to industrial areas because the darker color made *carbonaria* more conspicuous to avian predators in rural areas and less conspicuous in polluted areas. Kettlewell believed that Ford had established the superior viability of darker moths and he wanted to test the hypothesis that the darker form of the moth was less conspicuous to predators in industrial areas.

Kettlewell's investigations consisted of three parts. In the first part he used human observers to investigate whether his proposed scoring method would be accurate in assessing the relative conspicuousness of different types of moths against different backgrounds. The tests showed that moths on "correct" backgrounds, *typical* on lichen covered backgrounds and dark moths on soot-blackened backgrounds were almost always judged inconspicuous, whereas moths on "incorrect" backgrounds were judged conspicuous.

The second step involved releasing birds into a cage containing all three types of moth and both soot-blackened and lichen covered pieces of bark as resting places. After some difficulties (see [Rudge, 1998] for details), Kettlewell found that birds prey on moths in an order of conspicuousness similar to that gauged by human observers.

The third step was to investigate whether birds preferentially prey on conspicuous moths in the wild. Kettlewell used a mark-release-recapture experiment in both a polluted environment (Birmingham) and later in an unpolluted wood. He released 630 marked male moths of all three types in an area near Birmingham, which contained predators and natural boundaries. He then recaptured the moths using two different types of trap, each containing virgin females of all three types to guard against the possibility of pheromone differences.

Kettlewell found that *carbonaria* was twice as likely to survive in soot-darkened environments as was *typical*. He worried, however, that his results might be an artifact of his experimental procedures. Perhaps the traps used were more attractive to one type of moth, that one form of moth was more likely to migrate, or that one type of moth just lived longer. He eliminated the first alternative by showing that the recapture rates were the same for both types of trap. The use of natural boundaries and traps placed beyond those boundaries eliminated the second, and previous experiments had shown no differences in longevity. Further experiments in polluted environments confirmed that *carbonaria* was twice as likely to survive as *typical*. An experiment in an unpolluted environment showed that *typical* was three times as likely to survive as *carbonaria*. Kettlewell concluded that such selection was the cause of the prevalence of *carbonaria* in polluted environments.

Rudge also demonstrates that the strategies used by Kettlewell are those described above in the epistemology of experiment. His examples are given in Table 1.

The strategies discussed above, along with Hacking's intervention and independent confirmation, constitute an epistemology of experiment. They provide us with good reasons for belief in experimental results. They do not, however, guarantee that the results are correct. There are many experiments in which these strategies are applied, but whose results are later shown to be incorrect. (For an extended discussion of this issue see Franklin [2002, Chapters 7–10].) Experiment is fallible. Neither are these strategies exclusive or exhaustive. No single one of them, or set of them, guarantees the validity of an experimental result. Scientists use as many of the strategies as they conveniently apply in any given experiment.

1.1.2 Galison's Elaboration

In *How Experiments End* (1987), Peter Galison extended the discussion of experiment to more complex situations. In his histories of the measurements of the gyromagnetic ratio of the electron, of the discovery of the muon, and of the discovery of weak neutral currents, he considered a series of experiments measuring a single quantity, a set of different experiments culminating in a discovery, and two high energy physics experiments performed by large groups with complex experimental apparatus.

Galison's view is that experiments end when the experimenters believe that they have a result that will stand up in court. A result that I believe will include, and has included, the use of the epistemological strategies discussed earlier. Galison points out that major changes in theory and in experimental practice and instruments do not necessarily occur at the same time. This persistence of experimental results provides continuity across these conceptual changes. Thus, the experiments on the gyromagnetic ratio spanned classical electromagnetism, Bohr's old quantum theory, and the new quantum mechanics of Heisenberg and Schrödinger. Robert Ackermann has offered a similar view in his discussion of scientific instruments.

> The advantages of a scientific instrument are that it cannot change theories. Instruments embody theories, to be sure, or we wouldn't have any grasp of the significance of their operation....Instruments create an invariant relationship between their operations and the world, at least when we abstract from the expertise involved in their correct use. When our theories change, we may conceive of the significance of the instrument and the world with which it is interacting differently, and the datum of an instrument may change in significance, but the datum can nonetheless stay the same, and will typically be expected to do so. An instrument reads 2 when exposed to some phenomenon. After a change in theory, it will continue to show the same reading, even though we may take the reading to be no longer important, or to tell us something other than what we thought originally [Ackermann, 1985, p. 33]

Galison also discusses other aspects of the interaction between experiment and theory. Theory may influence what is considered to be a real effect, demanding explanation, and what is considered background. In his discussion of the discovery of the muon, he argues that the calculation of Oppenheimer and Carlson, which showed that showers were to be expected in the passage of electrons through matter, left the penetrating particles, later shown to be muons, as the problem. Prior to their work, physicists thought the showering particles were the problem, whereas the penetrating particles seemed to be understood.

The role of theory as an "enabling theory," one that allows calculation or estimation the size of the expected effect and also the size of expected backgrounds is also discussed by Galison (see also [Franklin, 1995]). Such a theory can help to determine whether or not an experiment is feasible. Galison also emphasizes that elimination of background that might mimic or mask an effect is central to the experimental enterprise, and not a peripheral activity. In the case of the weak neutral current experiments the existence of the currents depended crucially on showing that the event candidates could not all be due to neutron background.[3]

Galison also shows that the theoretical presuppositions of the experimenters may enter into the decision to end an experiment and report the result. Einstein and de Haas ended their search for systematic errors when their value for the gyromagnetic ratio of the electron, $g = 1$, agreed with their theoretical model of orbiting electrons. This effect of presuppositions might cause one to be skeptical of both experimental results and their role in theory evaluation. Galison's history shows, however, that, in this case, the importance of the measurement led to many repetitions of the measurement. This resulted in an agreed upon result that disagreed with theoretical expectations. Scientists do not always find what they are looking for.

[3]For another episode in which the elimination of background was crucial see the discussion of the measurement of the K^+_{e2} branching ratio in Franklin [1990, pp. 115–31] and [2002].

1.2 *The Case Against Learning From Experiment*

1.2.1 *Collins and the Experimenters' Regress*

Collins, Pickering, and others, have raised objections to the view that experimental results are accepted on the basis of epistemological arguments. As Donald MacKenzie remarks

> Recent sociology of science, following sympathetic tendencies in the history and philosophy of science, has shown that no experiment, or set of experiments however large, can on its own compel resolution of a point of controversy, or, more generally acceptance of a particular fact. A sufficiently determined critic can always find reason to dispute any alleged "result." [MacKenzie, 1989, p. 489].

MacKenzie is here raising doubts not only about the validity of experimental results, but also on their use in testing theories or hypotheses. I will begin with the former. There are two points at issue here. The first involves the meaning one assigns to "compel." If one reads it, as Mackenzie seems to, as "entail" then I agree that no finite set of confirming instances can entail a universal statement. No matter how many white swans one observes it does not entail that "all swans are white." Neither can any arguments, no matter how persuasive or valid, establish with absolute certainty the correctness of an experimental result. A more reasonable meaning for "compel" is having good reasons for belief. This is the meaning used in science, and those reasons for belief in an experimental result are provided by the epistemology of experiment.

The second point is a logical one, known to philosophers of science as the Duhem-Quine problem. In the usual *modus tollens* if a hypothesis h entails an experimental result e then $\neg e$ (not e) entails $\neg h$. As Duhem and Quine both pointed out it is not just h that entails e but rather h and b, where b includes background knowledge and auxiliary hypotheses. Thus, $\neg e$ entails $\neg h$ or $\neg b$ and we don't know where to place the blame. I am assuming here a weak form of the Duhem-Quine problem in which one assumes the experimental result $\neg e$ is correct. One can, of course, as Mackenzie does, challenge that experimental result. Let us consider here only the question of experimental results. As Quine pointed out any statement can be maintained come what may provided one is willing to make changes elsewhere in one's background knowledge. The question is when does the price that one has to pay become too high.

MacKenzie's skepticism illustrates one of the underlying principles of what has been called the sociology of scientific knowledge. Advocates of that view argue that because experimental evidence or methodological rules cannot resolve points of controversy, other reasons must be invoked to explain the resolution and those reasons are social.

Harry Collins, for example, is well known for his skepticism concerning both experimental results and evidence. He develops an argument that he calls the "experimenters' regress" [Collins, 1985, Chapter 4]: What scientists take to be a correct result is one obtained with a good, that is, properly functioning, experimental apparatus. But a good experimental apparatus is simply one that gives correct results. Collins claims that there are no formal criteria that one can apply to decide whether or not an experimental apparatus is working properly. In particular, he argues that calibrating an experimental apparatus by using a surrogate signal cannot provide an independent reason for considering the apparatus to be reliable.

In Collins's view the regress is eventually broken by negotiation within the appropriate scientific community, a process driven by factors such as the career, social, and cognitive interests of the scientists, and the perceived utility for future work, but one that is not decided by what we might call epistemological criteria, or reasoned judgment. Thus, Collins concludes that his regress raises serious questions concerning both experimental evidence and its use in the evaluation of scientific hypotheses and theories. Indeed, if no way out of the regress can be found then he has a point.

Collins's strongest candidate for an example of the experimenters' regress is presented in his history of the early attempts to detect gravitational radiation, or gravity waves. (For more detailed discussion of this episode see [Collins, 1985; 1994; Franklin, 1994; 1997a]). In this case, the physics community was forced to compare Weber's claims that he had observed gravity waves with the reports from six other experiments that failed to detect them. On the one hand, Collins argues that the decision between these conflicting experimental results could not be made for epistemological or methodological reasons–He claims that the six negative experiments could not legitimately be regarded as replications and hence become less impressive. On the other hand, Weber's apparatus, precisely because the experiments used a new type of apparatus to try to detect a hitherto unobserved phenomenon,[4] could not be subjected to standard calibration techniques.

The results presented by Weber's critics were not only more numerous, but they had also been carefully cross-checked. The groups had exchanged both data and analysis programs and confirmed their results. The critics had also investigated whether or not their analysis procedure, the use of a linear algorithm, could account for their failure to observe Weber's reported results. They had used Weber's preferred procedure, a nonlinear algorithm, to analyze their own data, and still found no sign of an effect. They had also calibrated their experimental apparatuses by inserting acoustic pulses of known energy and finding that they could detect a signal. Weber, on the other hand, as well as his critics using his analysis procedure, could not detect such calibration pulses.

[4]In more detailed discussions of this episode, Franklin [1994; 1997a], I argued that the gravity wave experiment is not at all typical of physics experiments. In most experiments, as illustrated in those essays, the adequacy of the surrogate signal used in the calibration of the experimental apparatus is clear and unproblematical. In cases where it is questionable considerable effort is devoted to establishing the adequacy of that surrogate signal. Although Collins has chosen an atypical example I believe that the questions he raises about calibration in general and about this particular episode of gravity wave experiments should be, and can be, answered.

There were, in addition, several other serious questions raised about Weber's analysis procedures. These included an admitted programming error that generated spurious coincidences between Weber's two detectors, possible selection bias by Weber, Weber's report of coincidences between two detectors when the data had been taken four hours apart, and whether Weber's experimental apparatus could produce the narrow coincidences claimed.

It seems clear that the critics's results were far more credible than Weber's. They had checked their results by independent confirmation, which included the sharing of data and analysis programs. They had also eliminated a plausible source of error, that of the pulses being longer than expected, by analyzing their results using the nonlinear algorithm and by explicitly searching for such long pulses.[5] They had also calibrated their apparatuses by injecting pulses of known energy and observing the output.

Contrary to Collins, I believe that the scientific community made a reasoned judgment and rejected Weber's results and accepted those of his critics. Although no formal rules were applied, e.g. if you make four errors, rather than three, your results lack credibility; or if there are five, but not six, conflicting results, your work is still credible; the procedure was reasonable.

1.2.2 Pickering: Communal Opportunism, and Plastic Resources

Andrew Pickering has argued that the reasons for accepting results are the future utility of such results for both theoretical and experimental practice and the agreement of such results with the existing community commitments. In discussing the discovery of weak neutral currents, Pickering states, "Quite simply, particle physicists accepted the existence of the neutral current because they could see how to ply their trade more profitably in a world in which the neutral current was real [Pickering, 1984, p. 87] "Scientific communities tend to reject data that conflict with group commitments and, obversely, to adjust their experimental techniques to tune in on phenomena consistent with those commitments [Pickering, 1981, p. 236]." The emphasis on future utility and existing commitments is clear. These two criteria do not necessarily agree. For example, there are episodes in the history of science in which more opportunity for future work is provided by the overthrow of existing theory. (See, for example, the histories of the overthrow of parity conservation and of CP symmetry discussed below.)

Pickering has recently offered a different view of experimental results. In his view the material procedure including the experimental apparatus itself along with setting it up, running it, and monitoring its operation; the theoretical model of that apparatus, and the theoretical model of the phenomena under investigation are all plastic resources that the investigator brings into relations of mutual support [Pickering, 1987; 1989]. His example is Morpurgo's search for free quarks, or fractional charges of $1/3e$ or $2/3e$, where e is the charge of the electron. Mor-

[5]Weber had suggested that the actual gravity wave pulses were longer that expected, and that the nonlinear analysis algorithm was more efficient at detecting such pulses.

purgo used a modern Millikan-type apparatus and initially found a continuous distribution of charge values.

When the apparatus had been built, he attempted to use it to measure charges (on samples of graphite, initially). And he found that it did not work. Instead of finding integral or fractional charges, he found that his samples appeared to carry charges distributed over a continuum. There followed a period of tinkering, of pragmatic, trail and error, material interaction with the apparatus. This came to an end when Morpurgo discovered that if he increased the separation of capacitor plates within his apparatus he obtained integral charge measurements.... After some theoretical analysis, Morpurgo concluded that he now had his apparatus working properly, and reported his failure to find any evidence for fractional charges... [Pickering, 1987, p. 199].

Pickering goes on to note that Morpurgo did not, however, tinker with the two competing theories of the phenomena then on offer, those of integral and fractional charge. "And what motivated the search for a new instrumental model was Morpurgo's eventual success in producing findings in accordance with one of the phenomenal models he was willing to accept [Pickering, 1987, p. 199]."

Pickering has made several important and valid points concerning experiment. Most importantly, he has emphasized that an experimental apparatus is rarely initially capable of producing valid experimental results. He has also recognized that both the theory of the apparatus and the theory of the phenomena can enter into the production of a valid experimental result, although I doubt that he would regard these as epistemological strategies. What I wish to question, however, is the emphasis he places on these theoretical components. I have already suggested that these theoretical components can be among the strategies used to argue for the validity of experimental results. I do not believe, as Pickering seems to, that they are necessary parts of such an argument. As Hacking pointed out, experimenters had confidence in microscope images both before and after Abbe's work fundamentally changed the theoretical understanding of the microscope. This was due to intervention, not theory.

Pickering neglects the fact that prior to Morpurgo's experiment it was known, or there were at least excellent reasons to believe, that electric charge was quantized in units of e, the charge on the electron, and that fractional charges, if they existed, were very rare in comparison with integral charges. From Millikan onward, experiments had strongly supported the existence of a fundamental unit of charge, and of charge quantization. The failure of Morpurgo's apparatus to produce measurements of integral charge indicated that it was not operating properly and that his theoretical understanding of it was faulty. It was the failure to produce measurements in agreement with what was already known, to fail an important experimental check, that caused doubts about Morpurgo's measurements. This was true regardless of the theoretical models available, or those that Morpurgo was willing to accept. It was only when Morpurgo's apparatus could reproduce known measurements that it could be trusted and used to search for fractional charge. To be sure, Pickering has allowed a role for the real in the production of the experimental result, but it doesn't seem to be decisive. I have argued that it is.

1.2.3 Critical Responses to Pickering

Robert Ackermann has offered a modification of Pickering's view. He suggests that the experimental apparatus itself is a less plastic resource then either the theoretical model of the apparatus or that of the phenomenon. "To repeat, changes in A [the apparatus] can often be seen (in real time, without waiting for accommodation by B [the theoretical model of the apparatus]) as improvements, whereas "improvements" in B don't begin to count unless A is actually altered and realizes the improvements conjectured. It's conceivable that this small asymmetry can account, ultimately, for large scale directions of scientific progress and for the objectivity and rationality of those directions [Ackermann, 1991, p. 456]."

Hacking [1992] has also offered a more complex version of Pickering's later view. He suggests that the results of mature laboratory science achieve stability and are self-vindicating when the elements of laboratory science are brought into mutual consistency and support. These are (1) ideas: questions, background knowledge, systematic theory, topical hypotheses, and modeling of the apparatus; (2) things: target, source of modification, detectors, tools, and data generators; and (3) marks and the manipulation of marks: data, data assessment, data reduction, data analysis, and interpretation. "Stable laboratory science arises when theories and laboratory equipment evolve in such a way that they match each other and are mutually self-vindicating [1992, p. 56]." "We invent devices that produce data and isolate or create phenomena, and a network of different levels of theory is true to these phenomena. Conversely we may in the end count them as phenomena only when the data can be interpreted by theory (pp. 57-8)." One might ask whether or not such mutual adjustment between theory and experimental results can always be achieved? What happens when an experimental result is produced by an apparatus on which several of the epistemological strategies, discussed earlier, have been successfully applied, and the result is in disagreement with our theory of the phenomenon? Accepted theories can be refuted.

1.2.4 Pickering and the Dance of Agency

Recently Pickering has offered a somewhat revised account of science. "My basic image of science is a performative one, in which the performances–the doings–of human and material agency come to the fore. Scientists are human agents in a field of material agency which they struggle to capture in machines [Pickering, 1995, p. 21]." He then discusses the complex interaction between human and material agency, which I interpret as the interaction between experimenters, their apparatus, and the natural world. "The dance of agency, seen asymmetrically from the human end, thus takes the form of a *dialectic of resistance and accommodations*, where resistance denotes the failure to achieve an intended capture of agency in practice, and accommodation an active human strategy of response to resistance, which can include revisions to goals and intentions as well as to the material form of the machine in question and to the human frame of gestures and social relations that surround it (p. 22)."

Pickering's idea of resistance is illustrated by Morpurgo's observation of continuous, rather than integral or fractional, electrical charge, which did not agree with his expectations. Morpurgo's accommodation consisted of changing his experimental apparatus by using a larger separation between his plates, and also by modifying his theoretical account of the apparatus. That being done, integral charges were observed and the result stabilized by the mutual agreement of the apparatus, the theory of the apparatus, and the theory of the phenomenon. Pickering notes that "the outcomes depend on how the world is (p. 182)." "In this way, then, *how the material world is* leaks into and infects our representations of it in a nontrivial and consequential fashion. My analysis thus displays an intimate and responsive engagement between scientific knowledge and the material world that is integral to scientific practice (p. 183)."

Nevertheless there is something confusing about Pickering's invocation of the natural world. Although Pickering acknowledges the importance of the natural world, his use of the term "infects" seems to indicate that he isn't entirely happy with this. Nor does the natural world seem to have much efficacy. It never seems to be decisive in any of Pickering's case studies. In his account, Morpurgo's observation of continuous charge is important only because it disagrees with his theoretical models of the phenomenon. The fact that it disagreed with numerous previous observations of integral charge doesn't seem to matter. This is further illustrated by Pickering's discussion of the conflict between Morpurgo and Fairbank. As we have seen, Morpurgo reported that he did not observe fractional electrical charges. On the other hand, in the late 1970s and early 1980s, Fairbank and his collaborators published a series of papers in which they claimed to have observed fractional charges (see, for example [LaRue *et al.*, 1981]). Faced with this discord Pickering concludes, "In Chapter 3, I traced out Morpurgo's route to his findings in terms of the particular vectors of cultural extension that he pursued, the particular resistances and accommodations thus precipitated, and the particular interactive stabilizations he achieved. The same could be done, I am sure, in respect of Fairbank. And these tracings are all that needs to be said about their divergence. It just happened that the contingencies of resistance and accommodation worked out differently in the two instances. Differences like these are, I think, continually bubbling up in practice, without any special causes behind them (pp. 211–212)."

The natural world seems to have disappeared from Pickering's account. There is a real question here as to whether fractional charges exist in nature. The conclusions reached by Fairbank and by Morpurgo about their existence cannot both be correct. It seems insufficient to merely state, as Pickering does, that Fairbank and Morpurgo achieved their individual stabilizations and to leave the conflict unresolved. At the very least, I believe, one should consider the actions of the scientific community. Scientific knowledge is not determined individually, but communally.

The fact that Fairbank believed in the existence of fractional electrical charges, or that Weber strongly believed that he had observed gravity waves, does not make them right. These are questions about the natural world that can be resolved. Either fractional charges and gravity waves exist or they don't, or to be more cautious we might say that we have good reasons to support our claims about their existence, or we do not. The difference between our attitudes toward the resolution of discord is one of the important distinctions between my view of science and Pickering's. I do not believe it is sufficient simply to say that the resolution is socially stabilized. I want to know how that resolution was achieved and what were the reasons offered for that resolution. If we are faced with discordant experimental results and both experimenters have offered reasonable arguments for their correctness, then clearly more work is needed. It seems reasonable, in such cases, for the physics community to search for an error in one, or both, of the experiments.

Another issue neglected by Pickering is the question of whether a particular mutual adjustment of theory, of the apparatus or the phenomenon, and the experimental apparatus and evidence is justified. Pickering seems to believe that any such adjustment that provides stabilization, either for an individual or for the community, is acceptable. I do not. There are cases in which experimenters both excluded data and engaged in selective analysis procedures in producing experimental results. (See [Franklin, 2002, Chapters 1–5] for examples.) These practices are, at the very least, questionable as is the use of the results produced by such practices in science. To take a homey example. Suppose one wished to show empirically that all odd numbers were prime. One looks at the odd numbers and notes that 1, 3, 5, and 7 are all primes, one excludes 9 as an experimental error, finds 11 and 13 are prime and then stops looking. Surely no one would, or should, regard this as a legitimate procedure, or base any conclusion on the result.

There is another point of disagreement between Pickering and myself. He claims to be dealing with the practice of science, and yet he excludes certain practices from his discussions. One scientific practice is the application of the epistemological strategies I have outlined above to argue for the correctness of an experimental results (see also (Franklin 1986, Chapter 7 for other cases)). In fact, one of the essential features of an experimental paper is the presentation of such arguments. I note further that writing such papers, a performative act, is also a scientific practice and it would seem reasonable to examine both the structure and content of those papers.[6]

Thus, there is a rather severe disagreement on the reasons for the acceptance of experimental results. For some, like Galison and myself, it is because of epistemological arguments. For others, like Pickering, the reasons are utility for future practice and agreement with existing theoretical commitments. Although the history of science shows that the overthrow of a well-accepted theory leads to an enormous amount of theoretical and experimental work, proponents of this view

[6]For a philosophical discussion of the structure of scientific papers see [Lipton, 1998; Suppe, 1998a; 1998b; Franklin and Howson, 1998].

seem to accept it as unproblematical that it is always agreement with existing theory that has more future utility. Hacking and Pickering further suggest that experimental results are accepted on the basis of the mutual adjustment of elements which include the theory of the phenomenon.

Nevertheless, everyone seems to agree that a consensus does arise on experimental results.

2 THE ROLES OF EXPERIMENT[7]

2.1 A Life of Its Own

Although experiment often takes its importance from its relation to theory, Hacking pointed out that it often has a life of its own, independent of theory. He notes the pristine observations of Carolyn Herschel's discovery of comets, William Herschel's work on "radiant heat," and Davy's observation of the gas emitted by algae and the flaring of a taper in that gas. In none of these cases did the experimenter have any theory of the phenomenon under investigation. One may also note the nineteenth century measurements of atomic spectra and the work on the masses and properties on elementary particles during the 1960s. Both of these sequences of experiments were conducted without any guidance from theory.

In deciding what experimental investigation to pursue, scientists may very well be influenced by the equipment available and their own ability to use that equipment (McKinney 1992). Thus, when the Mann-O'Neill collaboration was doing high energy physics experiments at the Princeton-Pennsylvania Accelerator during the late 1960s, the sequence of experiments was (1) measurement of the K^+ decay rates, (2) measurement of the K^+_{e3} branching ratio and decay spectrum, (3) measurement of the K^+_{e2} branching ratio, and (4) measurement of the form factor in K^+_{e3} decay. These experiments were performed with basically the same experimental apparatus, but with relatively minor modifications for each particular experiment. By the end of the sequence the experimenters had become quite expert in the use of the apparatus and knowledgeable about the backgrounds and experimental problems. This allowed the group to successfully perform the technically more difficult experiments later in the sequence. We might refer to this as "instrumental loyalty" and the "recycling of expertise" [Franklin, 1997c].

Hacking also remarks on the "noteworthy observations" on Iceland Spar by Bartholin, on diffraction by Hooke and Grimaldi, and on the dispersion of light by Newton. "Now of course Bartholin, Grimaldi, Hooke, and Newton were not mindless empiricists without an 'idea' in their heads. They saw what they saw because they were curious, inquisitive, reflective people. They were attempting to form theories. But in all these cases it is clear that the observations preceded any formulation of theory [Hacking, 1983, p. 156]." In all of these cases we may

[7]The accounts of experiments given below are very much simplified to illustrate the philosophical points. Where possible, I give references to more complete historical accounts.

say that these were observations waiting for, or perhaps even calling for, a theory. The discovery of any unexpected phenomenon calls for a theoretical explanation.

Nevertheless several of the important roles of experiment involve its relation to theory. Experiment may confirm a theory, refute a theory, or give hints to the mathematical structure of a theory.

2.2 *Confirmation and Refutation*

2.2.1 *The Discovery of Parity Nonconservation: A Crucial Experiment*[8]

Let us consider first an episode in which the relation between theory and experiment was clear and straightforward. This was a "crucial" experiment, one that decided unequivocally between two competing theories, or classes of theory, deciding against one and supporting the other. The episode was that of the discovery that parity, mirror-reflection symmetry or left-right symmetry, is not conserved in the weak interactions. Parity conservation was a well-established and strongly-believed principle of physics. As students of introductory physics learn, if we wish to determine the magnetic force between two currents we first determine the direction of the magnetic field due to the first current, and then determine the force exerted on the second current by that field. We use two Right-Hand Rules. We get exactly the same answer, however, if we use two Left-Hand Rules, This is left-right symmetry, or parity conservation, in electromagnetism.

In the early 1950s physicists were faced with a problem known as the "$\tau - \theta$" puzzle. Based on one set of criteria, that of mass and lifetime, two elementary particles (the τ and the θ) appeared to be the same, whereas on another set of criteria, that of spin and intrinsic parity, they appeared to be different. T. D. Lee and C. N. Yang [1956] realized that the problem would be solved, and that the two particles would be different decay modes of the same particle if parity were not conserved in the decay of the particles, a weak interaction. They examined the evidence for parity conservation and found, to their surprise, that although there was strong evidence that parity was conserved in the strong (nuclear) and electromagnetic interactions, there was, in fact, no supporting evidence that it was conserved in the weak interaction. It had never been tested.

Lee and Yang suggested several experiments that would test their hypothesis that parity was not conserved in the weak interactions. One was the beta decay of oriented nuclei (Figure 1).

Consider a collection of radioactive nuclei, all of whose spins point in the same direction. Suppose also that the electron given off in the radioactive decay of the nucleus is always emitted in a direction opposite to the spin of the nucleus In the mirror the electron is emitted in the same direction as the spin. The mirror image of the decay is different from the real decay. This would violate parity conservation, or mirror symmetry. Parity would be conserved only if, in the decay of a collection of nuclei, equal numbers of electrons were emitted in both directions.

[8]For details of this episode see Franklin [1986, Chapter 1].

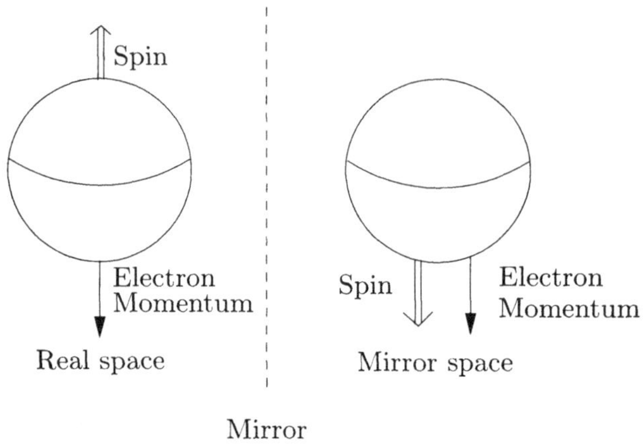

Figure 1. Nuclear spin and momentum of the decay electron in decay in both real space and in mirror space.

This was the experimental test performed by C. S. Wu and her collaborators [1957]. They aligned Cobalt^{60} nuclei and counted the number of decay electrons in the two directions, along the nuclear spin and opposite to the spin. Their results are shown in Figure 2 and indicate clearly that more electrons are emitted opposite to the spin than along the spin and that parity is not conserved . (The arrows give the direction of the magnetic field and the direction of the nuclear spins. The top curve (more electrons emitted) is the one obtained with the electron counter opposite to the spin direction).

Two other experiments, reported at the same time, on the sequential decay $\pi \rightarrow \mu \rightarrow e$ also showed parity nonconservation [Friedman and Telegdi, 1957; Garwin *et al.*, 1957]. These three experiments decided between two classes of theories — that is, between those theories that conserve parity and those that do not. They refuted the theories in which parity was conserved and supported or confirmed those in which it wasn't. These experiments also demonstrated that charge conjugation, or particle-antiparticle, symmetry was violated in the weak interactions and called for a new theory of β decay and the weak interactions. It is fair to say that when physicists learned of the results of these experiments they were convinced that parity was not conserved in the weak interactions.

2.2.2 The Meselson-Stahl Experiment: "The Most Beautiful Experiment in Biology"[9]

In the previous section I discussed a set of crucial experiments that decided between two competing classes of theories. In this section I will discuss an experiment that decided among three competing mechanisms for the replication of DNA, the

[9]For the history of this complex episode and complete references see [Holmes, 2001].

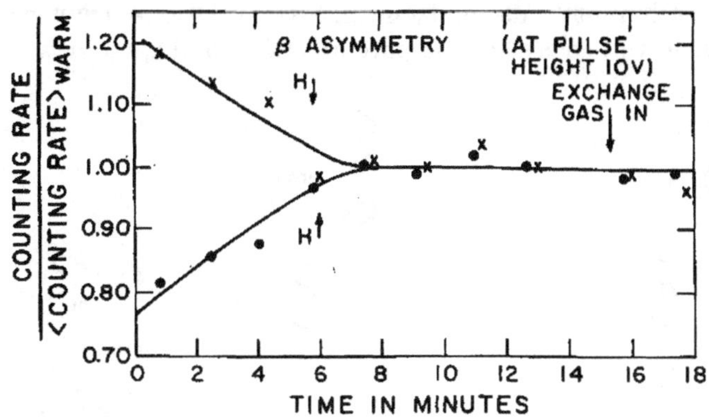

Figure 2. Relative counting rates for particles from the decay of oriented ^{60}Co nuclei for different nuclear orientations (field directions). There is a clear asymmetry with more particles being emitted opposite to the spin direction. From [Wu *et al.*, 1957].

molecule now believed to be responsible for heredity. This is another crucial experiment. It strongly supported one proposed mechanism and argued against the other two.

In 1953 Francis Crick and James Watson proposed a three-dimensional structure for deoxyribonucleic acid (DNA) [Watson and Crick, 1953a]. Their proposed structure consisted of two polynucleotide chains helically wound about a common axis. This was the famous "Double Helix." The chains were bound together by combinations of four nitrogen bases — adenine, thymine, cytosine, and guanine. Because of structural requirements only the base pairs adenine-thymine and cytosine-guanine are allowed. Each chain is thus complementary to the other. If there is an adenine base at a location in one chain there is a thymine base at the same location on the other chain, and vice versa. The same applies to cytosine and guanine. The order of the bases along a chain is not, however, restricted in any way, and it is the precise sequence of bases that carries the genetic information.

The significance of the proposed structure was not lost on Watson and Crick when they made their suggestion. They remarked, "It has not escaped our notice that the specific pairing we have postulated immediately suggests a possible copying mechanism for the genetic material."

If DNA was to play this crucial role in genetics, then there must be a mechanism for the replication of the molecule. Within a short period of time following the Watson-Crick suggestion, three different mechanisms for the replication of the DNA molecule were proposed [Delbruck and Stent, 1957]. These are illustrated in Figure 3. The first, proposed by Gunther Stent and known as conservative replication, suggested that each of the two strands of the parent DNA molecule is replicated in new material. This yields a first generation which consists of the original parent DNA molecule and one newly-synthesized DNA molecule. The second generation will consist of the parental DNA and three new DNAs.

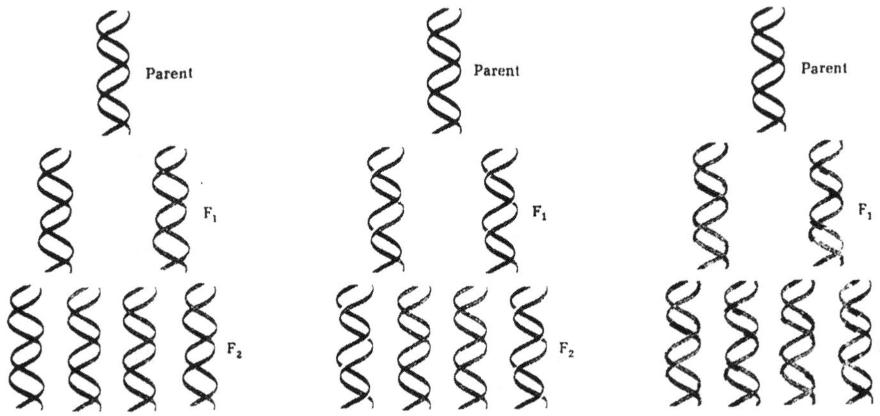

Figure 3. Possible mechanisms for DNA replication. (Left) Conservative replication. Each of the two strands of the parent DNA is replicated to yield the unchanged parent DNA and one newly synthesized DNA. The second generation consists of one parent DNA and three new DNAs. (Center) Semiconservative replication. Each first generation DNA molecule contains one strand of the parent DNA and one newly synthesized strand. The second generation consists of two hybrid DNAs and two new DNAs. (Right) Dispersive replication. The parent chains break at intervals, and the parental segments combine with new segments to form the daughter chains. The darker segments are parental DNA and the lighter segments are newly synthesized DNA. From Lehninger [1975].

The second proposed mechanism, known as semiconservative replication is when each strand of the parental DNA acts as a template for a second newly-synthesized complementary strand, which then combines with the original strand to form a DNA molecule. This was proposed by Watson and Crick [1953b]. The first generation consists of two hybrid molecules, each of which contains one strand of parental DNA and one newly synthesized strand. The second generation consists of two hybrid molecules and two totally new DNAs. The third mechanism, proposed by Max Delbruck, was dispersive replication, in which the parental DNA chains break at intervals and the parental segments combine with new segments to form the daughter strands.

In this section I will discuss the experiment performed by Matthew Meselson and Franklin Stahl, which has been called "the most beautiful experiment in biology," and which was designed to answer the question of the correct DNA replication mechanism [Meselson and Stahl, 1958]. Meselson and Stahl described their proposed method. "We anticipated that a label which imparts to the DNA molecule an increased density might permit an analysis of this distribution by sedimentation techniques. To this end a method was developed for the detection of small density differences among macromolecules. By use of this method, we have observed the distribution of the heavy nitrogen isotope ^{15}N among molecules of DNA following the transfer of a uniformly ^{15}N-labeled, exponentially growing bacterial population to a growth medium containing the ordinary nitrogen isotope ^{14}N. [Meselson and Stahl, 1958, pp. 671–672]"

The experiment is described schematically in Figure 4. Meselson and Stahl placed a sample of DNA in a solution of cesium chloride. As the sample is rotated at high speed the denser material travels further away from the axis of rotation than does the less dense material. This results in a solution of cesium chloride that has increasing density as one goes further away from the axis of rotation. The DNA reaches equilibrium at the position where its density equals that of the solution. Meselson and Stahl grew *E. coli* bacteria in a medium that contained ammonium chloride (NH_4Cl) as the sole source of nitrogen. They did this for media that contained either ^{14}N, ordinary nitrogen, or ^{15}N, a heavier isotope. By destroying the cell membranes they could obtain samples of DNA which contained either ^{14}N or ^{15}N. They first showed that they could indeed separate the two different mass molecules of DNA by centrifugation (Figure 5). The separation of the two types of DNA is clear in both the photograph obtained by absorbing ultraviolet light and in the graph showing the intensity of the signal, obtained with a densitometer. In addition, the separation between the two peaks suggested that they would be able to distinguish an intermediate band composed of hybrid DNA from the heavy and light bands.

Meselson and Stahl then produced a sample of *E coli* bacteria containing only ^{15}N by growing it in a medium containing only ammonium chloride with ^{15}N ($^{15}NH_4Cl$) for fourteen generations. They then abruptly changed the medium to ^{14}N by adding a tenfold excess of $^{14}NH_4CL$. Samples were taken just before the addition of ^{14}N and at intervals afterward for several generations. The cell membranes were broken to release the DNA into the solution and the samples were centrifuged and ultraviolet absorption photographs taken. In addition, the photographs were scanned with a recording densitometer. The results are shown in Figure 6, showing both the photographs and the densitometer traces. The figure shows that one starts only with heavy (fully-labeled) DNA. As time proceeds one sees more and more half-labeled DNA, until at one generation time only half-labeled DNA is present. "Subsequently only half labeled DNA and completely unlabeled DNA are found. When two generation times have elapsed after the addition of ^{14}N half-labeled and unlabeled DNA are present in equal amounts (p. 676)." (This is exactly what the semiconservative replication mechanism predicts). By four generations the sample consists almost entirely of unlabeled DNA. A test of the conclusion that the DNA in the intermediate density band was half labeled was provided by examination of a sample containing equal amounts of generations 0 and 1.9. If the semiconservative mechanism is correct then Generation 1.9 should have approximately equal amounts of unlabeled and half-labeled DNA, whereas

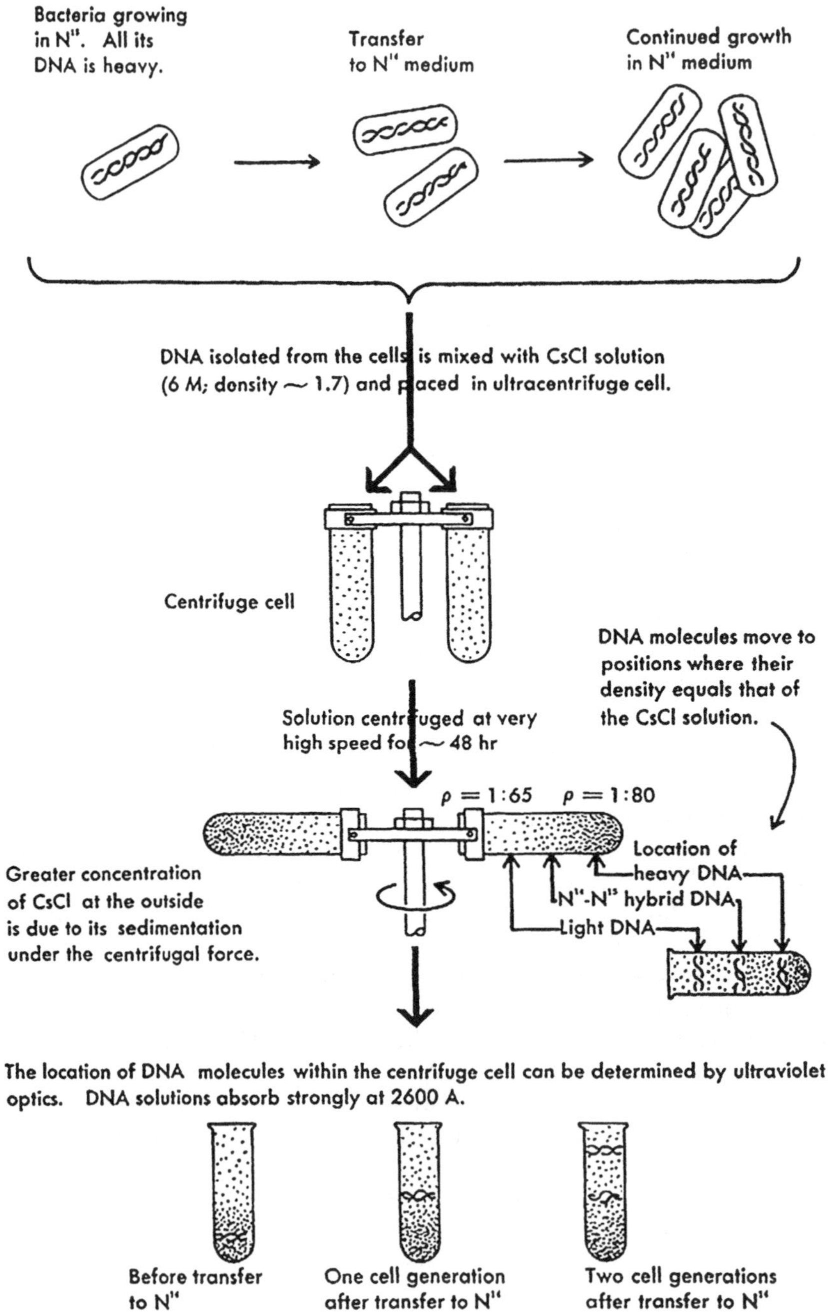

Figure 4. Schematic representation of the Meselson-Stahl experiment. From Watson [1965].

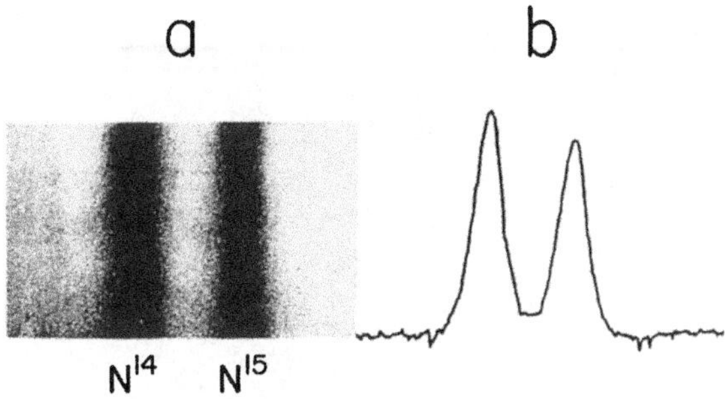

Figure 5. The separation of ^{14}N DNA from ^{15}N DNA by centrifugation. The band on the left is ^{14}N DNA and that on the right is from ^{15}N DNA. From Meselson and Stahl [1958].

Generation 0 contains only fully-labeled DNA. As one can see, there are three clear density bands and Meselson and Stahl found that the intermediate band was centered at (50 $\pm$ 2) percent of the difference between the ^{14}N and ^{15}N bands, shown in the bottom photograph (Generations 0 and 4.1). This is precisely what one would expect if that DNA were half labeled.

Meselson and Stahl stated their results as follows, "The nitrogen of DNA is divided equally between two subunits which remain intact through many generations.... Following replication, each daughter molecule has received one parental subunit (p. 676)."

Meselson and Stahl also noted the implications of their work for deciding among the proposed mechanisms for DNA replication. In a section labeled "The Watson-Crick Model" they noted that, "This [the structure of the DNA molecule] suggested to Watson and Crick a definite and structurally plausible hypothesis for the duplication of the DNA molecule. According to this idea, the two chains separate, exposing the hydrogen-bonding sites of the bases. Then, in accord with base-pairing restrictions, each chain serves as a template for the synthesis of its complement. Accordingly, each daughter molecule contains one of the parental chains paired with a newly synthesized chain..... The results of the present experiment are in exact accord with the expectations of the Watson-Crick model for DNA replication (pp. 677–678)."

It also showed that the dispersive replication mechanism proposed by Delbruck, which had smaller subunits, was incorrect. "Since the apparent molecular weight of the subunits so obtained is found to be close to half that of the intact molecule, it may be further concluded that the subunits of the DNA molecule which are conserved at duplication are single, continuous structures. The scheme for DNA duplication proposed by Delbruck is thereby ruled out (p. 681)." Later work by John Cairns and others showed that the subunits of DNA were the entire single polynucleotide chains of the Watson-Crick model of DNA structure.

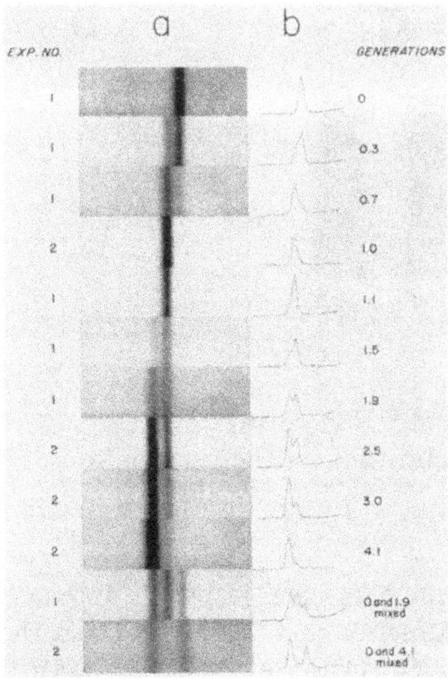

Figure 6. (Left) Ultraviolet absorption photographs showing DNA bands from centrifugation of DNA from *E. Coli* sampled at various times after the addition of an excess of ^{14}N substrates to a growing ^{15}N culture. (Right) Densitometer traces of the photographs. The initial sample is all heavy (^{15}N DNA). As time proceeds a second intermediate band begins to appear until at one generation all of the sample is of intermediate mass (Hybrid DNA). At longer times a band of light DNA appears, until at 4.1 generations the sample is almost all lighter DNA. This is exactly what is predicted by the Watson-Crick semiconservative mechanism. From [Meselson and Stahl, 1958].

The Meselson-Stahl experiment is a crucial experiment in biology. It decided between three proposed mechanisms for the replication of DNA. It supported the Watson-Crick semiconservative mechanism and eliminated the conservative and dispersive mechanisms.

2.2.3 The Discovery of CP Violation: A Persuasive Experiment[10]

After the discovery of nonconservation of parity and of charge conjugation, and following a suggestion by Landau, physicists considered CP (combined parity symmetry and particle-antiparticle symmetry), which was still conserved in the experiments, as the appropriate symmetry. One consequence of CP conservation was

[10]For details of this episode see [Franklin, 1986, Chapter 3]

that the K_1^o meson could decay into two pions, $\pi^+\pi^-$ or $\pi^o\pi^o$, whereas the K_2^o meson could not.[11] Thus, observation of $K_2^o \rightarrow \pi^+\pi^-$ would indicate CP violation.

A group at Princeton University, led by Cronin and Fitch, decided to test CP conservation. The experimental beam contained only K_2^o mesons. (The K_1^o meson has a much shorter lifetime than the K_2^o meson, so that if one starts with a beam containing both types of particles, after a time only the K_2^o mesons will remain). The experimental apparatus detected two charged particles from the decay of the K_2^omeson. The vector momentum of each of the two decay products from the K_2^o beam and the invariant mass m^* were computed assuming that each product had the mass of a pion. If both particles were indeed pions from K_2^o decay, m^* would equal the K_2^o mass. The experimenters also computed the angle θ between the vector sum of the two momenta and the direction of the K_2^o beam. This angle should be zero for two-body decays, but not, in general, for three-body decays.

This was exactly what the Princeton group observed [Christenson *et al.*, 1964]. As seen clearly in Figure 7, there is a peak at the K^o mass, 498 MeV/c^2, for events with cos$\theta \geq 0.9999$ (cos$\theta \approx 1$ means $\theta \approx 0$). No such peak is seen in the mass regions just above or just below the K^o mass. The experimenters reported a total of $45 \pm 9 K_2^o \rightarrow \pi^+\pi^-$ decays out of a total of $22,700 K_2^o$ decays. This was a branching ratio of $(1.95 \pm 0.2) \times 10^{-3}$, or approximately 0.2 percent.

The most obvious interpretation of the Princeton result was that CP symmetry was violated. This was the view taken in three out of four theoretical papers written during the period immediately following the report of that result. The Princeton result had persuaded most of the physics community that CP symmetry was violated. The remaining theoretical papers offered alternative explanations.[12] These alternatives relied on one or more of three arguments: (1) the Princeton results are caused by a CP asymmetry (the local preponderance of matter over antimatter) in the environment of the experiment, (2) $K_2^o \rightarrow \pi^+\pi^-$ does not necessarily imply CP violation, and (3) the Princeton observations did not arise from $K_2^o \rightarrow \pi^+\pi^-$ decay. This last argument can be divided into the assertions that (3a) the decaying particle was not a K_2^o meson, (3b) the decay products were not pions, and (3c) another unobserved particle was emitted in the decay. Included in these alternatives were three suggestions that cast doubt on well-

[11] The K_1^o and K_2^o mesons were elementary particles with the same charge, mass, and intrinsic spin. They did, however, differ with respect to the CP operator. The K_1^o and K_2^o mesons were eigenstates of the CP operator with eigenvalues $CP = +1$ and -1, respectively.

[12] I surveyed eighty such theoretical papers. Sixty accepted the Princeton result as evidence for either CP violation or apparent CP violation. Even those that offered alternative explanations of the result were not necessarily indications that the authors did not accept CP violation. One should distinguish between interesting speculations and serious suggestions. The latter are characterized by a commitment to their truth. I note that T.D. Lee was author, or co-author, of three of these theoretical papers. Two offered alternative explanations of the Princeton result and one proposed a model that *avoided* CP violation. Lee was not seriously committed to the truth of any of them. Bell and Perring, authors of one of the alternatives, remarked, "Before a more mundane explanation is found *it is amusing to speculate* that it might be a local effect due to the dyssymmetry of the environment, namely the local preponderance of matter over antimatter Bell and Perring (1964)." Theirs was an interesting speculation

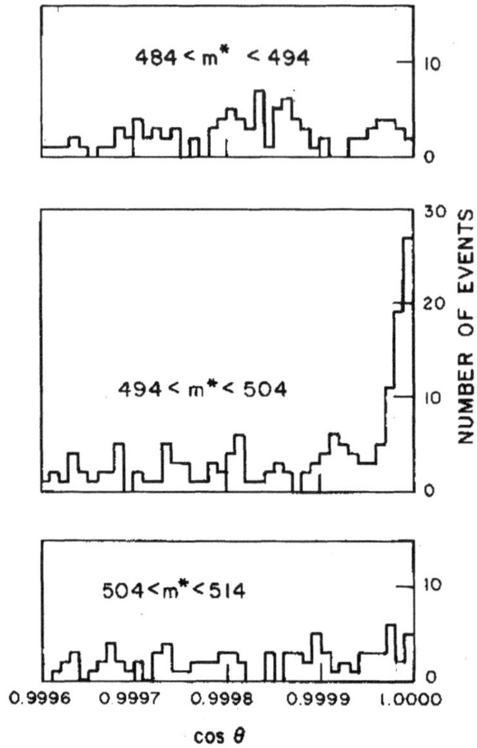

Figure 7. Angular distributions in three mass ranges for events with $\cos\theta > 0.9995$. From [Christenson *et al.*, 1964].

supported fundamental assumptions of modern physics. These were: (1) pions are not bosons, (2) the principle of superposition in quantum mechanics is violated, and (3) the exponential decay law fails. Although by the end of 1967 all of these alternatives had been experimentally tested and found wanting, the majority of the physics community had already accepted CP violation by the end of 1965, even though all the tests had not yet been completed. As Prentki, a theoretical particle physicist, remarked, this was because in some cases "the price one has to pay in order to save CP becomes extremely high," and because other alternatives were "even more unpleasant" [Prentki, 1965].

This is an example of what one might call a pragmatic solution to the Duhem-Quine problem. The alternative explanations and the auxiliary hypotheses were eliminated, leaving CP violation unprotected. One might worry that other plausible alternatives were never suggested or considered. This is not a serious problem in the actual practice of physics. No fewer than ten alternative explanations of the Princeton result were offered, and not all of them were very plausible. Had others been suggested they, too, would have been considered by the physics community. Consider the model of Nishijima and Saffouri [1965]. They explained $K_2^o \to \pi^+\pi^-$ decay by the existence of a "shadow" universe in touch with our "real" universe only through the weak interactions. They attributed the two pion decay observed to the decay of the $K^{o'}$ from the shadow universe. This implausible model was not merely considered, it was also experimentally tested. Everett [1965] noted that if the $K^{o'}$, the shadow K^o postulated by Nishijima and Saffouri existed, then a shadow pion, or $\pi^{o'}$, should also exist, and the decays $K^+ \to \pi^+\pi^o$ and $K^+ \to \pi^+\pi^{o'}$ should occur with equal rates. The presence of the $\pi^{o'}$ could be detected by measuring the $K^+ \to \pi^+\pi^o$ branching ratio in two different experiments, one in which the π^o was detected and one in which it was not. If the $\pi^{o'}$ existed the two measurements would differ. They didn't. There was no $\pi^{o'}$ and thus, no $K^{o'}$.

What was the difference between the episodes of parity nonconservation and CP violation. In the former parity nonconservation was immediately accepted. No alternative explanations were offered. There was a convincing and decisive set of experiments. In the latter at least ten alternatives were proposed, and although CP violation was accepted rather quickly, the alternatives were tested. In both cases there are only two classes of theories, those that conserve parity or CP, and those that do not. The difference lies in the length and complexity of the derivation linking the hypothesis to the experimental result, or to the number of auxiliary hypotheses required for the derivation. In the case of parity nonconservation the experiment could be seen by inspection to violate mirror symmetry (See Figure 1). In the CP episode what was observed was $K_2^o \to \pi^+\pi^-$. In order to connect this observation to CP conservation one had to assume (1) the principle of superposition, (2) that the exponential decay law held to 300 lifetimes, (3) that the decay particles were both "real" pions and that pions were bosons, (4) that no other particle was emitted in the decay, (5) that no other similar particle was produced, and (6) that there were no external conditions present that might regenerate K_1^o mesons. It was these auxiliary assumptions that were tested and eliminated as alternative explanations by subsequent experiments.

The discovery of CP violation called for a theoretical explanation, a call that is still unanswered.

2.2.4 The Discovery of Bose-Einstein Condensation: Confirmation After 70 Years

This section will discuss the discovery of Bose-Einstein condensation (BEC), and will illustrate the confirmation of a specific theoretical prediction 70 years after the theoretical prediction was first made. An interesting aspect of this episode is that the phenomenon in question had never been observed previously. This raises an interesting epistemological problem. How do you know you have observed something that has never been seen before?

Elementary particles can be divided onto two classes: bosons with integral spin (0, 1 ,2, ...), and fermions with half-integral spin (1/2, 3/2, 5/2, ...). Fermions, such as electrons obey the Pauli Exclusion Principle. Two fermions cannot be in the same quantum mechanical state. This explains the shell structure of electrons in atoms and the periodic table. On the other hand, any number of bosons can occupy the same state. At sufficiently low temperatures, when thermal motions are very small, there is a strong tendency for a group of bosons to all go into the same state. Bose [1924] and Einstein [1924; 1925] predicted that a gas of noninteracting bosonic atoms will, below a certain temperature, suddenly develop a macroscopic population in the lowest energy quantum state.[13]

The experiment that first demonstrated the existence of BEC was done by Carl Wieman, Eric Cornell, and their collaborators [Anderson *et al.*, 1995]. In outline the experiment was as follows. A sample of ^{87}Rb atoms was cooled in a magneto-optical trap. It was then loaded into a magnetic trap and further cooled by evaporation. The condensate was formed and the trap removed, allowing the condensate to expand. The expanded condensate was illuminated with laser light and the resulting shadow of the cloud was imaged, digitized, and stored.[14]

The experimental results are shown in Figures 8–10. Figure 8 shows the velocity distribution of the rubidium gas cloud (a) just before the appearance of the condensate, (b) just after, and (c) after further evaporation of the cloud has left a sample of nearly pure condensate. This figure also shows the spatial distribution of the gas. Although the measurement process destroyed the condensate sample, the entire process can be repeated so that one can measure the cloud at different stages. Figure 9 shows the peak density of the gas as a function of v_{evap}(v_{evap} is the frequency of the radiation used to excite the atoms into a non-confined state and to assist the cooling by evaporation). There is a sharp increase in density at v_{evap} = 4.23 MHz. This indicates the appearance of Bose-Einstein condensation. As the sample is further cooled one expects to observe a two-component cloud with a dense central condensate surrounded by a diffuse non-condensate. This is seen clearly in both Figures 9 and 10. Figure 10 shows horizontal sections of the rubidium cloud. At 4.71 MHz, above the transition temperature, one sees only a broad thermal distribution. Beginning at 4.23 MHz one sees the appearance of a sharp central peak, the Bose-Einstein condensate, above the thermal distribution. At 4.11 MHz the cloud is almost a pure condensate.

---

[13]Bose's paper had originally been rejected by the *Philosophical Magazine*. He then sent it, in English, to Einstein with a request that if Einstein thought the paper merited publication that he would arrange for publication in the *Zeitschrift fur Physik*. Einstein personally translated the paper and submitted it to the *Zeitschrift fur Physik*, adding a translator's note, "In my opinion, Bose's derivation of the Planck formula constitutes an important advance. The method used here also yields the quantum theory of the ideal gas, as I shall discuss elsewhere in more detail" Pais [1982]. This discussion appeared in Einstein's own papers of 1924 and 1925. For details see [Pais, 1982, Ch. 23].

[14]One difficulty with using rubidium is that at very low temperatures rubidium should be a solid. (In fact, rubidium is a solid at room temperature). Wieman, Cornell and their collaborators avoided this difficulty by creating a system that does not reach a true equilibrium. The vapor sample created equilibrates to a thermal distribution as a spin polarized gas, but takes a very long time to reach its true equilibrium state as a solid. At the low temperatures and density of the experiment the rubidium remains as a metastable super-saturated vapor for a long time.

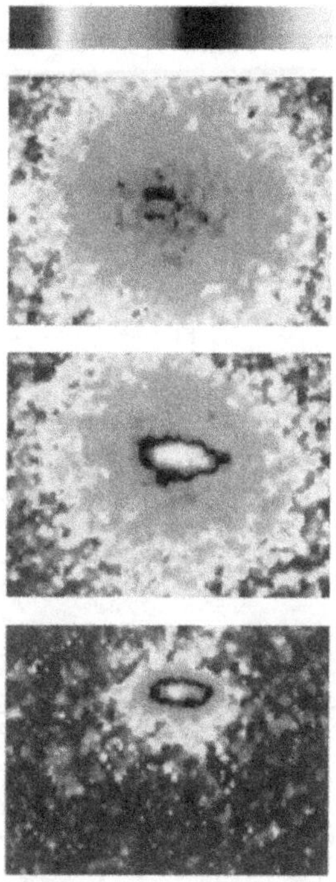

Figure 8. False color images of the velocity distribution of the rubidium BEC cloud: from the left, just before the appearance of the condensate, just after the appearance of the condensate, and after further evaporation has left a sample of nearly pure condensate. From [Anderson *et al.*, 1995].

There are three clear indications of the presence of Bose-Einstein condensation: (1) the velocity distribution of the gas shows two distinct components, (2) the sudden increase in density as the temperature decreases, and (3) the elliptical shape of the velocity distribution (Figure 8). The velocity distribution should be elliptical because for the harmonic trap used, the force in the z direction was eight times larger than in the x and y directions. No phenomenon other than Bose-Einstein condensation could plausibly explain these results.

This result was sufficiently credible that Keith Burnett, an atomic physicist at Oxford University remarked, in the same issue of *Science* in which Wieman and Cornell reported their result, "In short, they have observed the phenomenon called Bose-Einstein condensation (BEC) in a gas of atoms for the first time. The term Holy Grail seems quite appropriate given the singular importance of this discovery [Burnett, 1995, p. 182]."

A theoretical prediction had been confirmed after 70 years

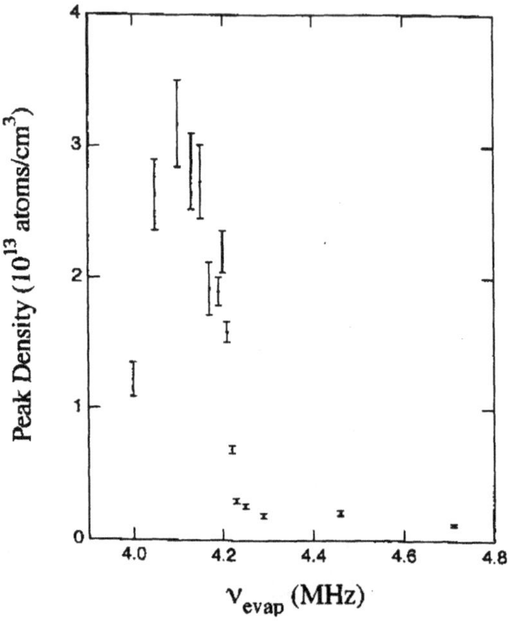

Figure 9. Peak density at the center of the sample as a function of the final depth of the evaporative cut, v_{evap}. As evaporation progresses to smaller values of v_{evap}, the cloud shrinks and cools, causing a modest increase in peak density until v_{evap} reaches 4.23 MHz. The sudden discontinuity at 4.23 MHz indicates the first appearance of the high density condensate as the cloud undergoes a phase transition. From [Anderson *et al.*, 1995].

2.3 Complications

In the four episodes discussed in the previous section, the relation between experiment and theory was clear. The experiments gave unequivocal results and there was no ambiguity about what theory was predicting. None of the conclusions reached has since been questioned. Parity and CP symmetry are violated in the weak interactions, DNA is replicated semiconservatively, and Bose-Einstein condensation is an accepted phenomenon. In the practice of science things are often more complex. Experimental results may be in conflict, or may even be incorrect. Theoretical calculations may also be in error or a correct theory may be incorrectly applied. There are even cases in which both experiment and theory are wrong. As noted earlier, science is fallible. In this section I will briefly discuss several episodes which illustrate these complexities.

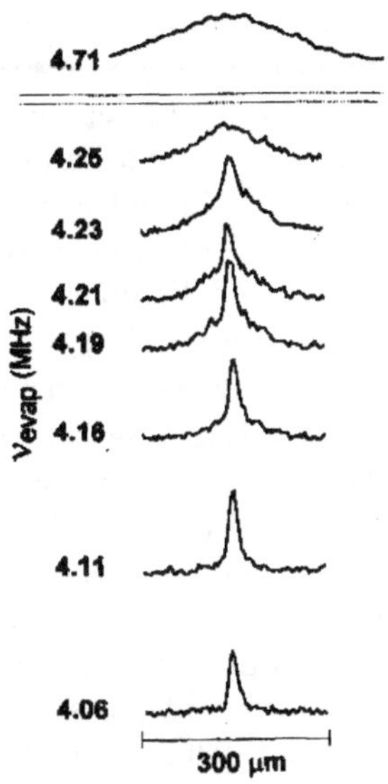

Figure 10. Horizontal sections taken through the velocity distribution at progressively lower values of v_{evap} show the appearance of the condensate fraction. From Anderson *et al.* [1995].

2.3.1 The Fall of the Fifth Force[15]

In this episode I will examine a case of the refutation of a hypothesis, but only after a disagreement between experimental results was resolved. The "Fifth Force" was a proposed modification of Newton's Law of Universal Gravitation. A reanalysis of the original Eötvös experiment[16] by Fischbach and his collaborators [1986] had shown a suggestive deviation from the law of gravity. The Fifth Force, in contrast to the famous Galileo experiment, depended on the composition of the objects. Thus, the Fifth Force between a copper mass and an aluminum mass would differ from that between a copper mass and a lead mass. Fischbach and collaborators also suggested modifying the gravitational potential between two masses from $V = -Gm_1m_2/r$ to $V = -Gm_1m_2/r\ [1 + \alpha e^{-r/\lambda}]$, where the second term gives the Fifth Force with strength α and range λ. The reanalysis also suggested that α was approximately 0.01 and λ was approximately 100m.

In this episode, we have a hitherto unobserved phenomenon along with discordant experimental results. The first two experiments on the Fifth Force gave

[15] For details of this episode see [Franklin, 1993a].

[16] The original Eötvös experiment was designed to measure the ratio of the gravitational mass to the inertial mass of different substances. Eötvös found that these two masses were equal to approximately one part in a million. Fischbach *et al.* reanalyzed Eötvös' data and found a composition-dependent effect, which they interpreted as evidence for the Fifth Force.

contradictory answers. One experiment supported the existence of the Fifth Force, whereas the other found no evidence for it. The first experiment, that of Peter Thieberger [1987a] looked for a composition-dependent force using a new type of experimental apparatus, which measured the differential acceleration between copper and water. The experiment was conducted near the edge of the Palisades cliff in New Jersey to enhance the effect of an intermediate-range force. The experimental apparatus is shown in Figure 11. Thieberger's results are shown in Figure 12. The sphere clearly has a velocity, indicating the presence of a force.

The second experiment, performed by the whimsically named Eöt-Wash group, was also designed to look for a substance-dependent, intermediate range force [Raab, 1987; Stubbs *et al.*, 1987]. The apparatus was located on a hillside on the University of Washington campus, in Seattle (Figure 13). If the hill attracted the copper and beryllium bodies differently, then the torsion pendulum would experience a net torque. This torque could be observed by measuring shifts in the equilibrium angle of the torsion pendulum as the pendulum was moved relative to a fixed geophysical point. Their experimental results are shown in Figure 14. The theoretical curves were calculated with the assumed values of 0.01 and 100m, for the Fifth Force parameters α and λ, respectively, the best values for the parameters at the time. There was no evidence for such a Fifth Force in this experiment.

The problem was, however, that both experiments appeared to be carefully done, with no apparent mistakes in either experiment. Ultimately, the discord between Thieberger's result and that of the Eöt-Wash group was resolved by an overwhelming preponderance of evidence in favor of the Eöt-Wash result (The issue was actually more complex. There were also discordant results on the distance dependence of the Fifth Force. For details see [Franklin, 1993a].

The subsequent history is an illustration of one way in which the scientific community deals with conflicting experimental evidence. Rather than making an immediate decision as to which were the valid results, this seemed extremely difficult to do on methodological or epistemological grounds, the community chose to await further measurements and analysis before coming to any conclusion about the evidence. The torsion balance experiments of EötWash were repeated by others including [Cowsik *et al.*, 1988; Fitch *et al.*, 1988; Adelberger, 1989; Bennett, 1989; Newman *et al.*, 1989; Stubbs *et al.*, 1989; Cowsik *et al.*, 1990; Nelson *et al.*, 1990]. These repetitions, in different locations and using different substances, gave consistently negative results. In addition, Bizzeti and collaborators [1989a; 1989b], using a float apparatus similar to that of Thieberger, also obtained results showing no evidence of a Fifth Force. There is, in fact, no explanation of either Thieberger's original, presumably incorrect, results. The scientific community has chosen, I believe quite reasonably, to regard the preponderance of negative results as conclusive.[17] Experiment had shown that there is no Fifth Force.

[17]It is a fact of experimental life that experiments rarely work when they are initially turned on and that experimental results can be wrong, even if there is no apparent error. It is not necessary to know the exact source of an error in order to discount or to distrust a particular experimental result. Its disagreement with numerous other results can, I believe, be sufficient.

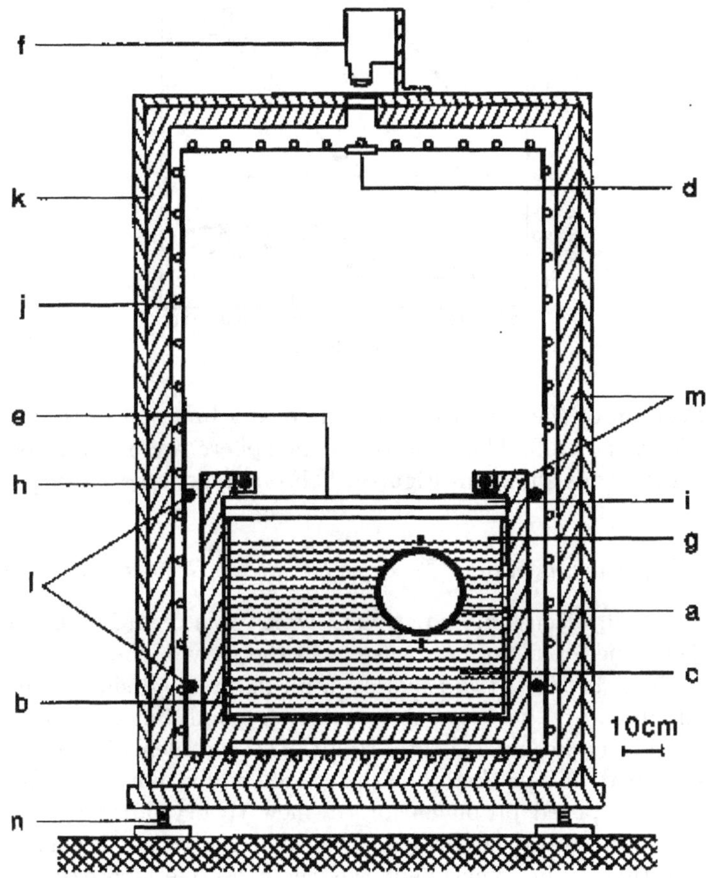

Figure 11. Schematic diagram of the differential accelerometer used in Thieberger's experiment. A precisely balanced hollow copper sphere (a) floats in a copperlined tank (b) filled with distilled water (c). The sphere can be viewed through windows (d) and (e) by means of a television camera (f). The multiplepane window (e) is provided with a transparent xy coordinate grid for position determination on top with a fine copper mesh (g) on the bottom. The sphere is illuminated for one second per hour by four lamps (h) provided with infrared filters (i). Constant temperature is maintained by mea ns of a thermostatically controlled copper shield (j) surrounded by a wooden box lined with Styrofoam insulation (m). The Mumetal shield (k) reduces possible effects due to magnetic field gradients and four circular coils (l) are used for positioning the sphere through forces due to ac produced eddy currents, and for dc tests. From Thieberger [1987].

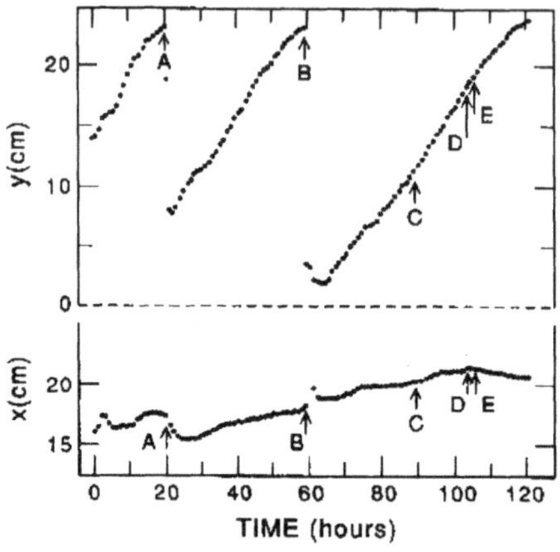

Figure 12. Position of the center of the sphere as a function of time. The y axis points away from the cliff. The position of the sphere was reset at points A and B by engaging the coils shown in Figure 12. From Thieberger [1987].

2.3.2 Right Experiment, Wrong Theory: The Stern-Gerlach Experiment[18]

In this section I will discuss an experiment which was regarded as crucial at the time it was performed, but, in fact, wasn't. In the view of the physics community it decided the issue between two theories, refuting one and supporting the other. In the light of later work, however, the refutation stood, but the confirmation was questionable. In fact, the experimental result posed problems for the theory it had seemingly confirmed. A new theory was proposed and although the Stern-Gerlach result initially also posed problems for the new theory, after a modification of that new theory, the result confirmed it. In a sense, it was crucial after all. It just took some time. The experiment was the Stern-Gerlach experiment, which provides evidence for the existence of electron spin. These experimental results were first published in 1922, although the idea of electron spin wasn't proposed by Uhlenbeck and Goudsmit until 1925 [1925; 1926]. One might say that electron spin was discovered before it was invented.

From the time of Ampere onward, molecular currents were regarded as giving rise to magnetic moments. In the nuclear model of the atom the electron orbits the nucleus. This circular current results in a magnetic moment. The atom behaves as if it were a tiny magnet. In the Stern-Gerlach experiment a beam of silver atoms passed through an inhomogeneous magnetic field (Figure 15). In Larmor's classical theory there was no preferential direction for the direction of the magnetic moment and so one predicted that the beam of silver atoms would show a maximum in

[18]This section is based on the accounts given by Weinert (1995) and Mehra and Rechenberg [1982]. Translations from the German were provided by these authors and are indicated by initials in the text.

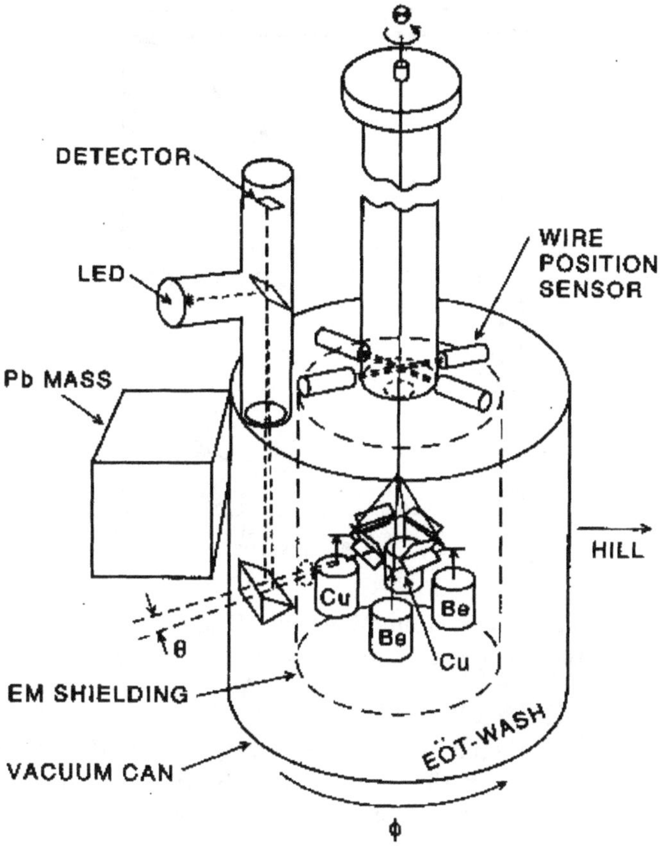

Figure 13. Schematic view of the University of Washington torsion pendulum experiment. The Helmholtz coils are not shown. From Stubbs *et al.* [1987].

the center of the beam. In Sommerfeld's quantum theory an atom in a state with angular momentum equal to one (L = 1) would have a magnetic moment with two components relative to the direction of the magnetic field, $\pm eh/4\pi m_e$. (Bohr had argued that only two spatial components were allowed). In an inhomogeneous magnetic field, H, the force on the magnetic moment μ will be $\mu_z \times \partial\mathrm{H}/\,\partial\mathrm{z}$, where $\mu_z = \pm\ \mathrm{eh}/4\pi\mathrm{m}_e$, e is the charge of the electron, m_e is its mass, h is Planck's constant, and z is the field direction. Thus, depending on the orientation of the magnetic moment relative to the magnetic field there will be either an attractive or repulsive force and the beam will split into two components, exhibiting spatial quantization. There will be a minimum at the center of the beam. "According to quantum theory μ_z can only be $\pm\ (\mathrm{e}/2\mathrm{m}_e)(\mathrm{h}/2\pi)$. In this case the spot on

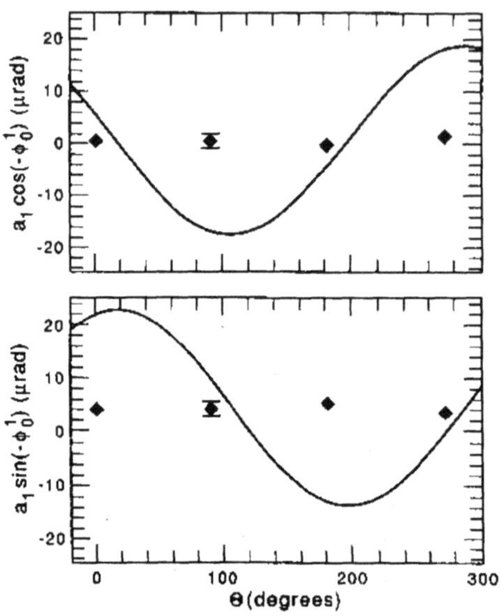

Figure 14. Deflection signal as a function of θ . The theoretical curves correspond to the signal expected for $\alpha = 0.01$ and $\lambda = 100$m. From Raab [1987].

the receiving plate will therefore be split into two, each of them having the same size but half the intensity of the original spot [Stern, 1921, p. 252, JM]." This difference in prediction between the Larmor and Sommerfeld theories was what Stern and Gerlach planned to use to distinguish between the two theories. Stern remarked that "the experiment, if it can be carried out, (will result) in a clear-cut decision between the quantum-theoretical and the classical view [Stern, 1921, FW]."

Sommerfeld's theory also acted as an enabling theory for the experiment. It provided an estimate of the size of the magnetic moment of the atoms so that Stern could begin calculations to see if the experiment was feasible. Stern calculated that a magnetic field gradient of 10^4 Gauss per centimeter would be sufficient to produce deflections that would give detectable separations of the beam components. He asked Gerlach if he could produce such a gradient. Gerlach responded affirmatively, and said he could do even better. The experiment seemed feasible. A sketch of the apparatus is shown in Figure 15. The silver atoms pass through the inhomogeneous magnetic field. If the beam is spatially quantized, as Sommerfeld predicted, two spots should be observed on the screen. (The sketch shows the beam splitting into three components, which would be expected in modern quantum theory for an atom with angular momentum equal to one). I note that Sommerfeld's theory was incorrect, illustrating the point that an enabling theory need not be correct to be useful.

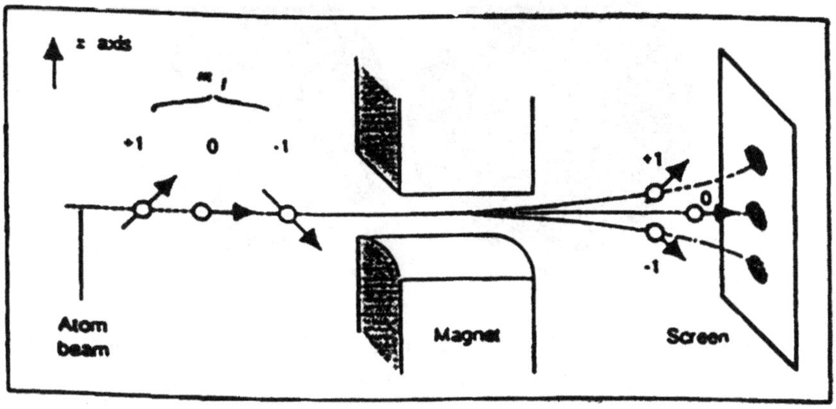

Figure 15. Sketch of the Stern-Gerlach experimental apparatus. The result expected for atoms in an $L = 1$ state (three components) is shown. From Weinert [1995]

A preliminary result reported by Stern and Gerlach did not show splitting of the beam into components. It did, however, show a broadened beam spot. They concluded that although they had not demonstrated spatial quantization, they had provided "evidence that the silver atom possesses a magnetic moment." Stern and Gerlach made improvements in the apparatus, particularly in replacing a round beam slit by a rectangular one that gave a much higher intensity. The results are shown in Figure 16 [Gerlach and Stern, 1922].There is an intensity minimum in the center of the pattern, and the separation of the beam into two components is clearly seen. This result seemed to confirm Sommerfeld's quantum-theoretical prediction of spatial quantization. Pauli, a notoriously skeptical physicist, remarked, "Hopefully now even the incredulous Stern will be convinced about directional quantization" (letter from Pauli to Gerlach 17 February 1922). Pauli's view was shared by the physics community.

Nevertheless the Stern-Gerlach result posed a problem for the Bohr-Sommerfeld theory of the atom. Stern and Gerlach had assumed that the silver atoms were in an angular momentum state with angular momentum equal to one (L = 1). In fact, the atoms are in an L = 0 state, for which no splitting of the beam would be expected in either the classical or the new quantum theory. Stern and Gerlach had not considered this possibility. Had they done so they might not have done the experiment. The later, or new, quantum theory developed by Heisenberg, Schrödinger, and others, predicted that for an L = 1 state the beam should split into three components as shown in Figure 15. The magnetic moment of the atom would be either 0 or $\pm$ eh/4πm. Thus, if the silver atoms were in an L = 1 state

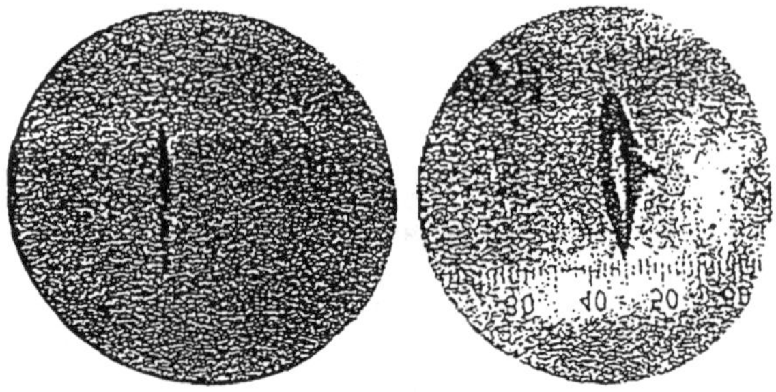

Figure 16. The experimental result of the Stern-Gerlach experiment. The beam has split into two components. From Gerlach and Stern (1922a).

as Stern and Gerlach had assumed, their result, showing two beam components, posed a problem for the new quantum theory. This was solved when Uhlenbeck and Goudsmit [1925; 1926] proposed that the electron had an intrinsic angular momentum or spin equal to $h/4\pi$. This is analogous to the earth having orbital angular momentum about the sun and also an intrinsic angular momentum due to its rotation on its own axis. In an atom the electron will have a total angular momentum $\mathbf{J} = \mathbf{L} + \mathbf{S}$, where $\mathbf{L}$ is the orbital angular momentum and $\mathbf{S}$ is the spin of the electron. For silver atoms in an L = 0 state the electron would have only its spin angular momentum (S = 1/2) and one would expect the beam to split into two components. Although the Stern-Gerlach results were known, and would certainly have provided strong support for the idea of electron spin, Goudsmit and Uhlenbeck made no mention of the result.[19]

The Stern-Gerlach experiment was initially regarded as a crucial test between the classical theory of the atom and the Bohr-Sommerfeld theory. In a sense it was, because it showed clearly that spatial quantization existed, a phenomenon that could be accommodated only within a quantum mechanical theory. It decided between the two classes of theories, the classical and the quantum mechanical. With respect to the particular quantum theory of Bohr and Sommerfeld, however, it wasn't crucial, although it was regarded as such at the time, because that theory predicted no splitting for a beam of silver atoms in the ground state (L = 0). The theory had been wrongly applied. The two-component result was also problematic for the new quantum theory, which also predicts no splitting for an angular momentum zero state and three components for an L = 1 state. Only after the suggestion of electron spin did the Stern-Gerlach result confirm the new theory.

Although the interpretation of the experimental result was incorrect for a time,

[19]The earliest recognition that the Stern-Gerlach experiment shows the existence of electron spin seems to appear in Fraser [1927]. By 1930 physics textbooks were stating the connection. See, for example, Ruark and Urey [1930].

the result itself remained quite robust through the theory change from the old to the new quantum theory. It is important to remember that experimental results do not change when accepted theory changes, although certainly, as we have seen, their interpretation may change. Gerlach and Stern emphasized this point themselves.

Apart from any theory, it can be stated, as a pure result of the experiment, and as far as the exactitude of our experiments allows us to say so, that silver atoms in a magnetic field have only *two discrete* values of the component of the magnetic moment in the direction of the field strength; both have the same absolute value with each half of the atoms having a positive and a negative sign respectively [Gerlach, 1924 #262, pp. 690–691, FW].

Experimental results, as well as experiments, also have a life of their own, independent of theory.

2.3.3 Sometimes Refutation Doesn't Work: The Double-Scattering of Electrons[20]

In the last section we saw some of the difficulty inherent in experiment-theory comparison. One is sometimes faced with the question of whether the experimental apparatus satisfies the conditions required by theory, or conversely, whether the appropriate theory is being compared to the experimental result. In this section I will examine the history of experiments on the double-scattering of electrons by heavy nuclei (Mott scattering) during the 1930s and the relation of these results to Dirac's theory of the electron, an episode in which the question of whether or not the experiment satisfied the conditions of the theoretical calculation was central.

In 1929, Mott [1929, and later; 1931; 1932] calculated, on the basis of Dirac's theory of the electron, that there would be a forward-backward asymmetry of approximately 10% in the double scattering of electrons from heavy nuclei. Mott clearly specified the conditions that would have to be satisfied in order to observe this effect. One had to double scatter relativistic (high velocity) electrons at large angles ($\approx 90^o$) from heavy nuclei (most calculations assumed a nuclear charge Z ≈ 80). The first scatter would polarize the electrons and the second scatter would analyze the produced polarization, giving rise to an asymmetry.

The earliest experiment that discussed Mott's calculation was performed by Chase [1929] He observed a 4% asymmetry in the double scattering of electrons but attributed it to a difference in the path that the electrons followed. His subsequent experiment [Chase, 1930] reported a 1.5% effect, and this time did attribute it to Mott scattering. Most experiments during the early 1930s, showed no polarization effects, although some of them did not satisfy the conditions for Mott scattering. The major positive results were provided by experiments done by Rupp [Rupp, 1929; 1930; 1931; 1932a; 1932b; 1932c; 1934a; 1934b, Rupp and Szilard, 1931]. Although Rupp found positive results, they differed slightly from Mott's predictions. Dymond [1931] also reported a positive result, but one that was five times smaller than the theoretical prediction.

[20]For details of this episode see Franklin [1986, Chapter 2].

Mott and the rest of the electron-scattering community were quite aware of both the confused nature of the experimental results, and of the apparent discrepancy between experiment and theory. Langstroth (1932) reviewed the situation and commented on the difficulty of experiment-theory comparison when one deals with real, as opposed to ideal, experiments. "In view of the fact that practical conditions may be immensely more complicated than those of Mott's theory, it is not surprising that it does not furnish a guide, even in a qualitative way, to all of the above experiments (pp. 566–67)."

The situation became even more confused when Dymond [1932] published a detailed account of his experiment, which restated his positive, but discrepant, result. Adding to the confusion was the fact that Dymond's experiment seemed to satisfy all the conditions for Mott scattering. Rupp [1934a] continued his work and again found a positive result. G.P. Thomson [1933] on the other hand, found no effect. At approximately the same time Sauter [1933] redid Mott's calculations and obtained identical results. If things weren't difficult enough, they got worse when Dymond [1934] published a full repudiation of his earlier results. He had found a considerable and variable experimental asymmetry in his apparatus, and concluded that he had not, in fact, observed any polarization effect. Dymond also considered possible reasons for the theory-experiment discrepancy including inelastic, stray, and plural scattering, and nuclear screening and rejected them all. He concluded, "We are driven to the conclusion that the theoretical results are wrong [Dymond 1932, p. 666]."

G.P. Thomson [1934] also published a comprehensive review of the field. He reported no effects of the type found by Rupp and he found a forward-backward ratio of (0.996 ± 0.01) in comparison to Mott's prediction of 1.15.Thomson also concluded that there was a serious discrepancy between theory and experiment.

Faced with this apparent theory-experiment discrepancy, theorists sought either to modify Dirac's theory or to propose a new theory, and thus accommodate the experimental results. They offered modifications of the Coulomb potential, each of which had the effect that it "annihilates the polarization effect completely." Although each of the theoretical calculations predicted null results from double scattering experiments, they were not regarded as solving the problem. One might speculate that this was because these modifications had no physical or theoretical underpinning. They seemed invented solely for the purpose of explaining the experimental results.

Experimental work also continued. The situation became even more confused when Rupp [1935] withdrew several of his results on electron scattering. This eliminated the most positive results supporting Mott's theory.[21] In 1937 Richter published what he regarded as the definitive experiment on the double scattering of electrons. He claimed to have satisfied the conditions of Mott's calculation exactly and had found no effect. He concluded that "Despite all the favorable

[21]Rupp's work seems to have been fraudulent. His withdrawal included a note from a psychiatrist stating that Rupp had suffered from a mental illness and could not distinguish between fantasy and reality. For details of this episode see French [1999].

conditions of the experiment, however, no sign of the Mott effect could be observed. *With this experimental finding, Mott's theory of the double scattering of electrons from the atomic nucleus can no longer be maintained.* [Richter, 1937, p. 554]." The discrepancy was further confirmed by the theoretical work of Rose and Bethe [1939]. They examined various ways of trying to eliminate the discrepancy and concluded that "the discrepancy between theory and experiment remains — perhaps more glaring than before" (p. 278).

Thus, at the end of 1939 there was a clear discrepancy between Dirac theory, as used by Mott, and the experimental results on the double scattering of electrons. Yet the theory was not regarded as refuted. Why was this? The reason is that, at the time, Dirac theory, and only Dirac theory, predicted the existence of the positron (a positive electron). This particle had been discovered in 1932 and had provided very strong support for Dirac theory. In comparison with this success, the discrepancy in electron scattering, along with another small discrepancy in the spectrum of hydrogen, just did not have sufficient evidential weight. The unique, and confirmed, prediction of the positron outweighed these discrepancies. It isn't easy to refute a strongly confirmed theory. Neither is it impossible to do so, as demonstrated by the histories of both parity nonconservation and CP violation discussed earlier.

Interestingly, it was the experimental results that were wrong. In the early 1940s experimental work showed that the way in which the experiments were performed during the 1930s had precluded the possibility of observing the polarization effects predicted by Mott. In order to avoid problems with multiple scattering the experimenters had scattered the electrons from the front surface of the targets. Unfortunately this made the effects of plural scattering, a few large scatters rather than just one as required by Mott, very large. The symmetric plural scattering swamped the predicted polarization effect. When the experimental apparatuses were changed to eliminate this problem the discrepancy disappeared. Mott's theory was then supported by the experimental evidence.

We have seen here a classic case of the Duhem-Quine problem and how the physics community attempted to solve it. There was a clear discrepancy between the experimental results and the predictions of a well-confirmed theory. The experiments were redone to check the results, with careful attention to the experimental conditions required by the theory. Theorists checked on whether or not other effects might mask the predicted polarization effect. Other theorists offered competing explanations. Ultimately a solution was found

Does the fact that Dirac theory was not regarded as refuted even though experiment clearly disagreed with its predictions mean that physicists disregard negative results whenever it suits their purposes? Do physicists really tune in on existing community commitments, as some social constructivists would have it, and overlook negative evidence? The answer is no. There is no indication in this episode that the negative evidence was disregarded. The physics community examined the theory in the light of all the available experimental evidence, weighed its importance, and then made a decision. I note that even though Dirac theory remained relatively unscathed, both experimental and theoretical work continued until the problem was solved. The discrepancy was not hidden from view, nor was it ignored

2.4 Other Roles

2.4.1 Evidence for a New Entity: J.J. Thomson and the Electron[22]

Experiment can also provide us with evidence for the existence of the entities involved in our theories. In discussing the existence of electrons Ian Hacking has written,"So far as I'm concerned, if you can spray them then they are real" [Hacking, 1983, p. 23]. He went on to elaborate this view. "We are completely convinced of the reality of electrons when we set out to build — and often enough succeed in building — new kinds of device that use various well-understood causal properties of electrons to interfere in other more hypothetical parts of nature" (p. 265).

Hacking worried that the simple manipulation of the first quotation, the changing of the charge on an oil drop or on a superconducting niobium sphere, which involves only the charge of the electron, was insufficient grounds for belief in electrons. His second illustration, which he believed more convincing because it involved several properties of the electron, was that of Peggy II, a source of polarized electrons built at the Stanford Linear Accelerator Center in the late 1970s. Peggy II provided polarized electrons for an experiment that scattered electrons off deuterium to investigate the weak neutral current. Although I agree with Hacking that manipulability can often provide us with grounds for belief in a theoretical entity, his illustration comes far too late. Physicists were manipulating the electron in Hacking's sense in the early twentieth century.[23] They believed in the existence of electrons well before Peggy II, and I will argue that they had good reasons for that belief.[24]

In this section I will discuss the grounds for belief in the existence of the electron by examining J. J. Thomson's experiments on cathode rays. His 1897 experiment on cathode rays is generally regarded as the "discovery" of the electron.

The purpose of J. J. Thomson's experiments was clearly stated in the introduction to his 1897 paper.

> The experiments discussed in this paper were undertaken in the hope of gaining some information as to the nature of Cathode Rays. The most diverse opinions are held as to these rays; according to the almost unanimous opinion of German physicists they are due to some process in the aether to which — inasmuch as in a uniform magnetic field their course is circular and not rectilinear — no phenomenon hitherto

[22]For details see Smith [1997].

[23]Millikan, for example, used the properties of electrons emitted in the photoelectric effect to measure h, Planck's constant. Stern and Gerlach, as discussed below, used the properties of the electron to investigate spatial quantization, and also discovered evidence for electron spin.

[24]For more details of this episode, including a discussion of other early twentieth century experiments, see Franklin [1997b].

> observed is analogous: another view of these rays is that, so far from being wholly aetherial, they are in fact wholly material, and that they mark the paths of particles of matter charged with negative electricity [Thomson, 1897, p. 293].

Thomson's first order of business was to show that the cathode rays carried negative charge. This had presumably been shown previously by Perrin. Perrin placed two coaxial metal cylinders, insulated from one another, in front of a plane cathode. The cylinders each had a small hole through which the cathode rays could pass onto the inner cylinder. The outer cylinder was grounded. When cathode rays passed into the inner cylinder an electroscope attached to it showed the presence of a negative electrical charge. When the cathode rays were magnetically deflected so that they did not pass through the holes, no charge was detected. "Now the supporters of the aetherial theory do not deny that electrified particles are shot off from the cathode; they deny, however, that these charged particles have any more to do with the cathode rays than a rifle-ball has with the flash when a rifle is fired" [Thomson, 1897, p. 294].

Thomson repeated the experiment, but in a form that was not open to that objection. The apparatus is shown in Figure 17. The two coaxial cylinders with holes are shown. The outer cylinder was grounded and the inner one attached to an electrometer to detect any charge. The cathode rays from A pass into the bulb, but would not enter the holes in the cylinders unless deflected by a magnetic field.

> When the cathode rays (whose path was traced by the phosphorescence on the glass) did not fall on the slit, the electrical charge sent to the electrometer when the induction coil producing the rays was set in action was small and irregular; when, however, the rays were bent by a magnet so as to fall on the slit there was a large charge of negative electricity sent to the electrometer.... If the rays were so much bent by the magnet that they overshot the slits in the cylinder, the charge passing into the cylinder fell again to a very small fraction of its value when the aim was true. *Thus this experiment shows that however we twist and deflect the cathode rays by magnetic forces, the negative electrification follows the same path as the rays, and that this negative electrification is indissolubly connected with the cathode rays* (Thomson [1897, p. 294-295], emphasis added).

This experiment also demonstrated that cathode rays were deflected by a magnetic field in exactly the way one would expect if they were negatively charged material particles.[25]

There was, however, a problem for the view that cathode rays were negatively charged particles. Several experiments, in particular those of Hertz, had failed to observe the deflection of cathode rays by an electrostatic field. Thomson proceeded to answer this objection. His apparatus is shown in Figure 18. Cathode

[25]Thomson also demonstrated the magnetic deflection of cathode rays in a separate experiment.

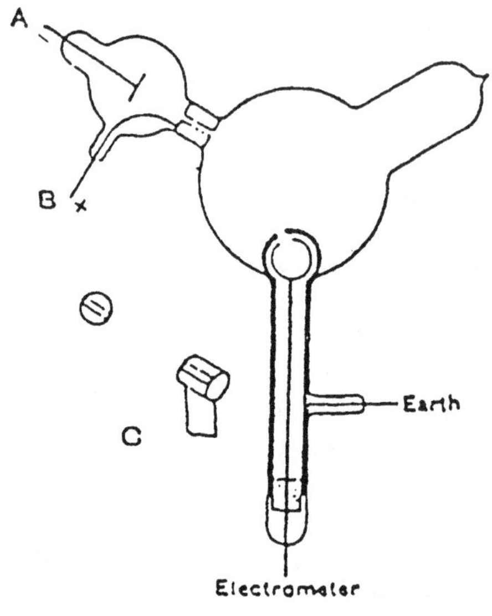

Figure 17. Thomson's apparatus for demonstrating that cathode rays have negative charge. The slits in the cylinders are shown. From Thomson [1897].

rays from C pass through a slit in the anode A, and through another slit at B. They then passed between plates D and E and produced a narrow well-defined phosphorescent patch at the end of the tube, which also had a scale attached to measure any deflection. When Hertz had performed the experiment he had found no deflection when a potential difference was applied across D and E. He concluded that the electrostatic properties of the cathode ray are either *nil* or very feeble. Thomson admitted that when he first performed the experiment he also saw no effect. "on repeating this experiment [that of Hertz] I at first got the same result [no deflection], but subsequent experiments showed that the absence of deflexion is due to the conductivity conferred on the rarefied gas by the cathode rays. On measuring this conductivity it was found that it diminished very rapidly as the exhaustion increased; it seemed that on trying Hertz's experiment at very high exhaustion there might be a chance of detecting the deflexion of the cathode rays by an electrostatic force" [Thomson, 1897, p. 296]. Thomson did perform the experiment at lower pressure [higher exhaustion] and observed the deflection.

Thomson concluded

> *As the cathode rays carry a charge of negative electricity, are deflected by an electrostatic force as if they were negatively electrified, and are acted on by a magnetic force in just the way in which this force would act on a negatively electrified body moving along the path of these rays,*

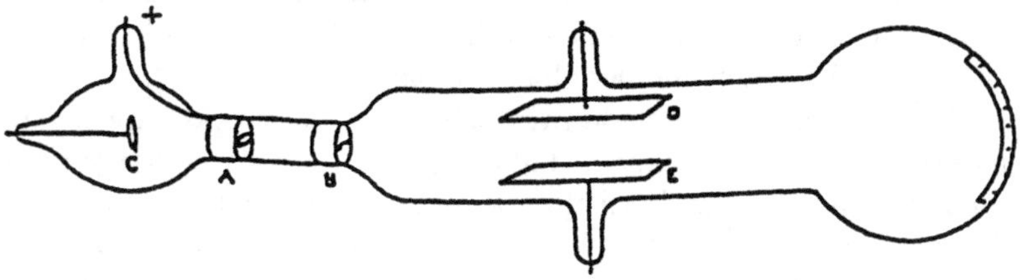

Figure 18. Thomson's apparatus for demonstrating that cathode rays are deflected by an electric field. It was also used to measure m/e. From Thomson [1897].

> *I can see no escape from the conclusion that they are charges of negative electricity carried by particles of matter* Thomson [1897, p. 302], emphasis added).

Thomson's argument is the "duck argument." If it looks like a duck, quacks like a duck, and waddles like a duck, then we have good reason to believe that it is a duck.

Having established that cathode rays were negatively charged material particles, Thomson went on to discuss what the particles were. To investigate this question Thomson made measurements on the charge to mass ratio of cathode rays. Thomson's method used both the electrostatic and magnetic deflection of the cathode rays.[26] The apparatus is shown in Figure 18. It also included a magnetic field that could be created perpendicular to both the electric field and the trajectory of the cathode rays. Thomson adjusted the values of the electric and magnetic fields so that the cathode ray beam was undeflected. This determined the velocity of the beam. He then turned off the magnetic field and measured the beam deflection due to the electric field. From these measurements he couldl calculate the value of m/e for the beam particles.

Thomson found a value of m/e of (1.29 ± 0.17) x 10^{-7}. This value was independent of both the gas in the tube and of the metal used in the cathode, suggesting that the particles were constituents of the atoms of all substances. It was also far smaller, by a factor of 1000, than the smallest value previously obtained, 10^{-4}, that of the hydrogen ion in electrolysis.

Thomson remarked that this might be due to the smallness of m or to the largeness of e. He argued that m was small citing Lenard's work on the range of cathode rays in air. The range, which is related to the mean free path for

[26]Thomson actually used two different methods to determine the charge to mass ratio. The other method used the total charge carried by a beam of cathode rays in a fixed period of time, the total energy carried by the beam in that sane time, and the radius of curvature of the particles in a known magnetic field. Thomson regarded the method discussed in the text as more reliable and this is the method shown in most modern physics textbooks.

collisions, and which depends on the size of the object, was 0.5 cm. The mean free path for molecules in air was approximately 10^{-5} cm. If the cathode ray traveled so much farther than a molecule before colliding with an air molecule, Thomson argued that it must be much smaller than a molecule.

Thomson had shown that cathode rays behave as one would expect negatively charged material particles to behave. They deposited negative charge on an electrometer, and were deflected by both electric and magnetic fields in the appropriate direction for a negative charge. In addition the value for the mass to charge ratio was far smaller than the smallest value previously obtained, that of the hydrogen ion. If the charge were the same as that on the hydrogen ion, the mass would be far less. In addition, the cathode rays traveled farther in air than did molecules, also implying that they were smaller than an atom or molecule. Thomson concluded that these negatively charged particles were constituents of atoms. In other words, Thomson's experiments had given us good reasons to believe in the existence of electrons long before any experiments in which the electron was manipulated..

2.4.2 The Articulation of Theory: Weak Interactions[27]

Radioactivity, the spontaneous decay of a substance, produces alpha particles (positively charged helium nuclei), or beta particles (electrons), or gamma rays (high energy electromagnetic radiation). In this section I will describe the twenty-five-year effort to determine the mathematical form of the theory describing β decay. In 1934 Enrico Fermi proposed a new theory of β decay that incorporated the then newly hypothesized neutrino [Fermi, 1934]. He added a perturbation energy due to the decay interaction to the Hamiltonian describing the nuclear system, which included a mathematical operator O_x.

Pauli [1933] had previously shown that O_x can have only five different forms if the Hamiltonian is to be relativistically invariant. These are S, the scalar interaction; P, pseudoscalar; V, vector; A, axial vector; and T, tensor. Fermi knew this but chose, in analogy with electromagnetic theory, to use only the vector interaction. His theory initially received support from the work of Sargent [1932; 1933] and others. There remained, however, the question of whether or not the other mathematical forms of the interaction also entered into the Hamiltonian.[28] In this episode we shall see how experiment helped to determine the mathematical form of the weak interaction.

Gamow and Teller [1936] soon proposed a modification of Fermi's vector theory. The Gamow-Teller modification required either a tensor or an axial vector form of the interaction. Their theory helped to solve some of the difficulties that arose in assigning nuclear spins using only Fermi's theory. At the end of the 1930s

[27] For details see Franklin [1990].

[28] The actual history is more complex. For a time, an alternative theory of β decay, proposed by Konopinski and Uhlenbeck [1935] was better supported by the experimental evidence than was Fermi's theory. It was subsequently shown that both the experimental results and the theoretical calculations were wrong and that Fermi's theory was , in fact, supported by the evidence. For details see Franklin [1990].

there was support for Fermi's theory with some preference for the Gamow-Teller selections rules and the tensor interaction.

The work of Fierz [1937] helped to restrict the allowable forms of the interaction. He showed that if both S and V interactions were present in the allowed β-decay interaction, or both A and T, then there would be an interference term in the allowed β-decay spectrum. This term vanished if the admixtures were not present. The failure to observe these interference terms showed that the decay interaction did not contain both S and V, or both A and T.

The presence of either the T or A form of the interaction in at least part of the β-decay interaction was shown by Mayer *et al.* [1951]. They found twenty five decays for which $\Delta J = 0, \pm 1$, the change in nuclear spin, with no parity change. These decays could only occur if the A or T forms were present. Their conclusion depended on the correct assignment of nuclear spins which, although reliable, still retained some uncertainty. Further evidence, which did not depend on knowledge of the nuclear spins, came from an examination of the spectra of unique forbidden transitions.[29] These were n-times forbidden transitions in which the change in spin was equal to $n + 1$. These transitions require the presence of either A or T. In addition, only a single form of the interaction makes any appreciable contribution to the decay. This allows the prediction of the shape of the spectrum for such transitions. Konopinski and Uhlenbeck [1941] showed that for an n-times forbidden transition the spectrum would be that of an allowed transition multiplied by an energy dependent term $a_n(W)$. The spectrum for ^{91}Y measured by Langer and Price [1949] showed both the clear presence of either the A or T forms of the interaction as well as the energy-dependent correction.

Evidence in favor of the presence of either the S or V forms of the interaction was provided by Sherr, Muether, and White [1949] and by Sherr and Gerhart [1952]. They observed the decay of ^{14}O to an excited state of ^{14}N, $^{14}N^*$, which was forbidden by the A and T interactions.

Further progress in isolating the particular forms of the interaction was made by examining the spectra of once-forbidden transitions. Here too, interference effects, similar to those predicted by Fierz, were also expected. A. Smith [1951] and Pursey [1951] found that the spectrum for these transitions would contain energy dependent terms of the form $G_V G_T/W$, $G_A G_P/W$, and $G_S G_A/W$, where the G's are the coupling constants for the various interactions, and W is the electron energy. The linear spectrum found for ^{147}Pm demonstrated the absence of these terms [Langer *et al.*, 1950].

Let us summarize the situation. There were five allowable forms of the decay interaction; S, T, A, V, P. The failure to observe Fierz interference showed that the interaction could not contain both S and V or both A and T. Experiments showing the presence of Gamow-Teller selections rules and on unique forbidden transitions had shown that either A or T must be present. The decay of ^{14}O to

[29] Allowed transitions are those for which both the electron and neutrino wavefunctions could be considered constant over nuclear dimensions. Forbidden transitions are those that included higher order terms in the perturbation series expansion of the matrix element.

$^{14}N^*$ had demonstrated that either S or V must also be present. This restricted the forms of the interaction to STP, SAP, VTP, or VAP or doublets taken from these combinations. The absence of interference terms in the once-forbidden spectra eliminated the VT, SA, and AP combinations. VP was eliminated because it did not allow Gamow-Teller transitions. This left only the STP triplet or the VA doublet as the possible interactions.

The spectrum of RaE (^{210}Bi) seemed to provide the decisive evidence. Petschek and Marshak (1952) analyzed the spectrum of RaE and found that the only interaction that would give a good fit to the spectrum was a combination of T and P. This was, in fact, the only evidence favoring the presence of the P interaction. This led Konopinski and Langer (1953) in their 1953 review article on beta decay to conclude that, "As we shall interpret the evidence here, the correct law must be what is known as an STP combination [1953, p. 261]."

Unfortunately, the evidence from the RaE spectrum had led the physics community astray. Further theoretical analysis cast doubt on their assumptions, but all of this became moot when K. Smith[30] measured the spin of RaE and found it to be one, incompatible with the Petschek-Marshak analysis.

The demise of the RaE evidence removed the necessity of including the P interaction in the theory of β decay, and left the decision between the STP and VA combinations unresolved. The dilemma was seemingly resolved by evidence provided by angular-correlation experiments, particularly that from the experiment on ^6He by Rustad and Ruby [1953; 1955].

2.4.2.1 Angular Correlation Experiments Angular correlation experiments are those in which both the decay electron and the recoil nucleus from beta decay are detected in coincidence. The experiments measured the distribution in angle between the electron and the recoil nucleus for a fixed range of electron energy, or measured the energy spectrum of either the electron or the nucleus at a fixed angle between them. These quantities are quite sensitive to the form of the decay interaction and became decisive pieces of evidence in the search for the form of the decay interaction.

The most important of the experiments performed at this time was the measurement of the angular correlation in the decay of ^6He. This decay was a pure Gamow-Teller transition and thus was sensitive to the amounts of A and T present in the decay interaction. The decisive experiment was that of Rustad and Ruby [1953; 1955]. This experiment was regarded as establishing that the Gamow-Teller part of the interaction was predominantly tensor. This was the conclusion reached in several review papers on the nature of β decay. [Ridley, 1954; Kofoed-Hansen, 1955; Wu, 1955]. The experimental apparatus is shown in Figure 19. The definition of the decay volume was extremely important. In order to measure the angular correlation one must know the position of the decay so that one can measure the angle between the electron and the recoil nucleus. The decay volume for

[30] I have been unable to find a published reference to this measurement. It is cited as a private communication in the literature.

the helium gas in this experiment was defined by a 180 microgram/cm^2 aluminum hemisphere and the pumping diaphragm. Rustad and Ruby [1953] presented two experimental results. The first was the coincidence rate as a function of the angle between the electron and the recoil nucleus for electrons in the energy range (2.5–4.0) mc^2. The second was the energy spectrum of the decay electrons with the angle between the electron and the recoil nucleus fixed at 180°. Both results are shown in Figure 20 along with the predicted results for A and T, respectively. The dominance of the tensor interaction is clear. This conclusion was made more emphatic in their 1955 paper which included more details of the experiment and even more data.

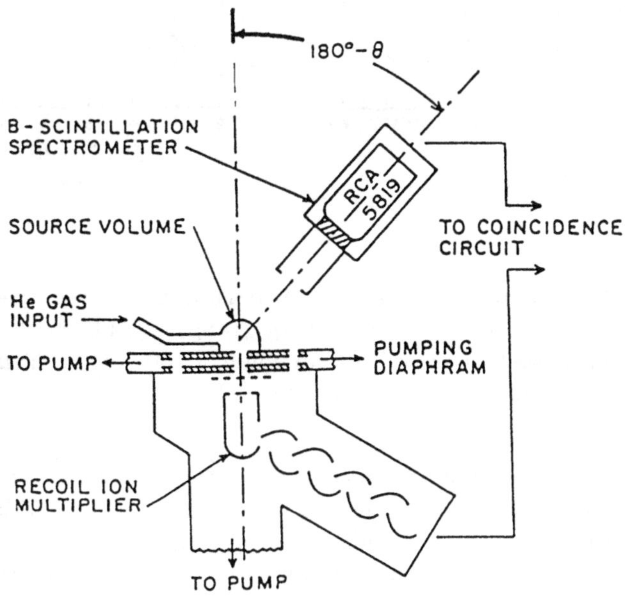

Figure 19. Schematic view of the experimental apparatus for the ^{6}He angular correlation experiment of Rustad and Ruby [1953; 1955]

The Rustad-Ruby result, along with several others, established that the Gamow-Teller part of the decay interaction was tensor and that the decay interaction was STP, or ST, rather than VA. We have seen clearly in this episode the fruitful interaction between experiment and theory. Theoretical predictions became more precise and were tested experimentally until the form of the weak interaction was found. Fermi's theory of β decay had been confirmed. It had also been established that the interaction was a combination of scalar, tensor, and pseudoscalar (STP).

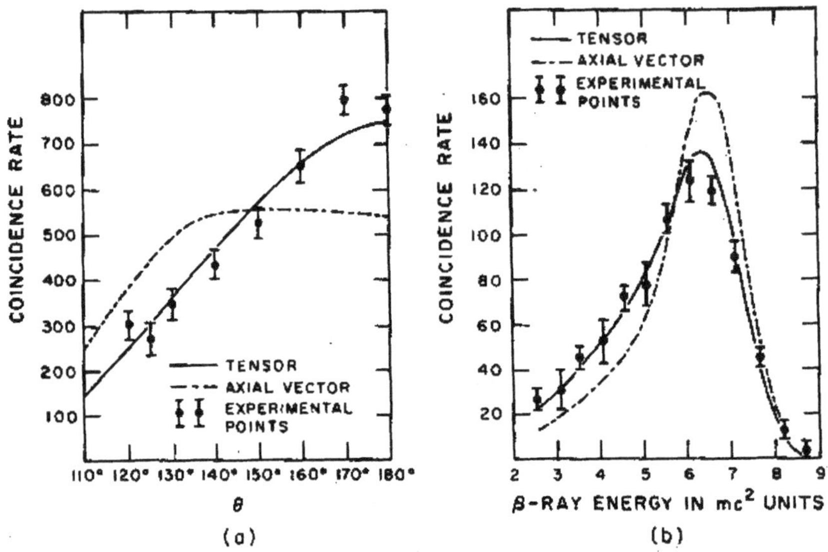

Figure 20. (a) Coincidence counting rate versus angle between the electron and the recoil nucleus, for electrons in the energy range 2.5-4.0 mc^2. (b) Coicidence counting rate versus electron energy for an angle of 180° between the electron and the recoil nucleus. From Rustad and Ruby [1953].

2.4.2.2 Epilogue It would be nice to report that such a simple, satisfying story, with its happy ending was the last word. It wasn't. Work continued on angular correlation experiments and the happy agreement was soon destroyed [Franklin, 1990, Chapter 3]. Things became more complex with the discovery of parity nonconservation in the weak interactions, including β decay. Sudarshan and Marshak [1958] and Feynman and Gell-Mann [1958] showed that only a V-A interaction was compatible with parity nonconservation. If there was to be a single interaction describing all the weak interactions then there was a serious conflict between this work and the Rustad-Ruby result. This led Wu and Schwarzschild [1958] to reexamine and reanalyze the Rustad-Ruby experiment. They found, by calculation and by constructing a physical analogue of the gas system, that a considerable fraction of the helium gas was not in the decay volume. This changed the result for the angular correlation considerably and cast doubt on the Rustad-Ruby result.[31] The ^{6}He angular correlation experiment was redone,

[31]In a post-deadline paper presented at the January 1958 meeting of the American Physical Society, Rustad and Ruby suggested that their earlier result might be wrong. There are no abstracts of post-deadline papers, but the talk was cited in the literature. Ruby remembers the tone of the paper as *mea culpa* (private communication).

correcting the problem with the gas target, and the new result is shown in Figure 21 [Hermannsfeldt *et al.*, 1958]. It clearly favors A, the axial vector interaction. Once again, physics was both fallible and corrigible. This new result on ^{6}He combined with the discovery of parity nonconservation established that the form of the weak interaction was V-A.

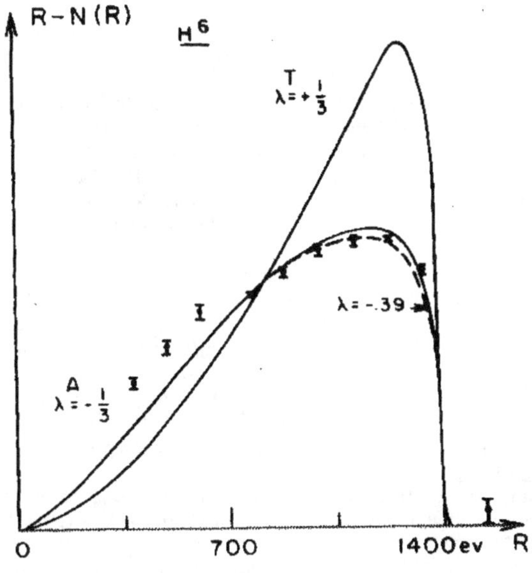

Figure 21. Energy spectrum of recoil ions from ^{35}A decay. From [Hermannsfeldt *et al.*, 1958].

3 CONCLUSION

In this essay varying views on the nature of experimental results have been presented. Some argue that the acceptance of experimental results is based on epistemological arguments, whereas others base acceptance on future utility, social interests, or agreement with existing community commitments. Everyone agrees , however, that for whatever reasons, a consensus is reached on experimental results. These results then play many important roles in physics and we have examined several of these roles, although certainly not all of them. We have seen experiment deciding between two competing theories, calling for a new theory, confirming a theory, refuting a theory, and providing evidence for the existence of an elementary particle involved in an accepted theory. We have also seen that experiment has a life of its own, independent of theory. If, as I believe, epistemological procedures provide grounds for reasonable belief in experimental results, then experiment can legitimately play the roles I have discussed and can provide the basis for scientific knowledge.

BIBLIOGRAPHY

[Ackermann, 1985] R. Ackermann. *Data, Instruments and Theory.* Princeton, N.J.: Princeton University Press, 1985.

[Ackermann, 1991] R. Ackermann. Allan Franklin, Right or Wrong. *PSA 1990, Volume 2.* A. Fine, M. Forbes and L. Wessels. East Lansing, MI: Philosophy of Science Association: 451-457, 1991.

[Adelberger, 1989] E. G. Adelberger. High-Sensitivity Hillside Results from the Eot-Wash Experiment. *Tests of Fundamental Laws in Physics: Ninth Moriond Workshop.* O. Fackler and J. Tran Thanh Van. Les Arcs, France: Editions Frontieres: 485-499, 1989.

[Anderson *et al.*, 1995] M. H. Anderson, J. R. Ensher, M. R. Matthews, *et al.* Observation of Bose-Einstein Condensation in a Dilute Atomic Vapor. *Science* **269**: 198-201, 1995.

[Bell and Perring, 1964] J. S. Bell and Perring. 2π Decay of the K_2o Meson. *Physical Review Letters* **13**: 348-349, 1964.

[Bennett, 1989] W. R. Bennett. Modulated-Source Eotvos Experiment at Little Goose Lock. *Physical Review Letters* **62**: 365-368, 1989.

[Bizzeti *et al.*, 1989a] P. G. Bizzeti, A. M. Bizzeti-Sona, T. Fazzini, *et al.* Search for a Composition Dependent Fifth Force: Results of the Vallambrosa Experiment. *Tran Thanh Van, J.* O. Fackler. Gif Sur Yvette: Editions Frontieres, 511-524, 1989.

[Bizetti *et al.*, 1989b] P. G. Bizzeti, A. M. Bizzeti-Sona, T. Fazzini, *et al.* Search for a Composition-dependent Fifth Force. *Physical Review Letters* **62**: 2901-2904, 1989.

[Bose, 1924] S. Bose. Plancks Gesetz und Lichtquantenhypothese. *Zeitschrift fur Physik* **26**(1924): 178-181, 1924.

[Burnett, 1995] K. Burnett. An Intimate Gathering of Bosons. *Science* **269**: 182-183, 1995.

[Chase, 1929] C. Chase. A Test for Polarization in a Beam of Electrons by Scattering. *Physical Review* **34**: 1069-1074, 1929.

[Chase, 1930] Chase, C. T. (1930). The Scattering of Fast Electrons by Metals. I I. *Physical Review* **36**: 1060-1065.

[Christenson *et al.*, 1964] J. H. Christenson, J. W. Cronin, V. L. Fitch, *et al.* Evidence for the 2π Decay of the K_2^0 Meson. *Physical Review Letters* **13**: 138-140, 1964.

[Collins, 1985] H. Collins. *Changing Order: Replication and Induction in Scientific Practice.* London: Sage Publications, 1985.

[Collins, 1994] H. Collins. A Strong Confirmation of the Experimenters' Regress. *Studies in History and Philosophy of Modern Physics* **25**(3): 493-503, 1994.

[Conan Doyle, 1967] A. Conan Doyle. The Sign of Four. *The Annotated Sherlock Holmes.* W. S. Baring-Gould. New York: Clarkson N. Potter, 1967.

[Cowsik *et al.*, 1988] R. Cowsik, N. Krishnan, S. N. Tandor, *et al.* Limit on the Strength of Intermediate-Range Forces Coupling to Isospin. *Physical Review Letters* **61**: 2179-2181, 1988.

[Cowsik *et al.*, 1990] R. Cowsik, N. Krishnan, S. N. Tandor, *et al.* Strength of Intermediate-Range Forces Coupling to Isospin. *Physical Review Letters* **64**: 336-339, 1990.

[Delbruck and Stent, 1957] M. Delbruck and G. S. Stent. On the Mechanism of DNA Replication. *The Chemical Basis of Heredity.* W. D. McElroy and B. Glass. Baltimore: Johns Hopkins Press: 699-736, 1957.

[Dymond, 1931] E. G. Dymond. Polarisation of a Beam of Electrons by Scattering. *Nature* **128**: 149, 1931.

[Dymond, 1932] E. G. Dymond. On the Polarisation of Electrons by Scattering. *Proceedings of the Royal Society (London)* **A136**: 638-651, 1932.

[Dymond, 1934] E. G. Dymond. On the Polarization of Electrons by Scattering. II. *Proceedings of the Royal Society (London)* **A145**: 657-668, 1934.

[Einstein, 1924] A. Einstein. Quantentheorie des einatomigen idealen gases. *Sitzungsberischte der Preussische Akademie der Wissenschaften, Berlin*: 261-267, 1924.

[Einstein, 1925] A. Einstein. Quantentheorie des einatomigen idealen gases. *Sitzungsberichte der Preussische Akadmie der Wissenschaften, Berlin*: 3-14, 1925.

[Everett, 1965] A. E. Everett. Evidence on the Existence of Shadow Pions in K^+ Decay. *Physical Review Letters* **14**: 615-616, 1965.

[Fermi, 1934] E. Fermi. Attempt at a Theory of β-Rays. *Il Nuovo Cimento* **11**: 1-21, 1934.

[Feynman and Gell-Mann, 1958] R. P. Feynman and M. Gell-Mann. Theory of the Fermi Interaction. *Physical Review* **109**: 193-198, 1958.

[Feynman *et al.*, 1963] R. P. Feynman, R. B. Leighton and M. Sands. *The Feynman Lectures on Physics*. Reading, MA: Addison-Wesley Publishing Company, 196.

[Fierz, 1937] M. Fierz. Zur Fermischen Theorie des β-Zerfalls. *Zeitschrift fur Physik* **104**: 553-565, 1937.

[Fischbach *et al.*, 1986] E. Fischbach, S. Aronson, C. Talmadge, *et al.* Reanalysis of the Eōtvōs Experiment. *Physical Review Letters* **56**: 3-6, 1986.

[Fitch *et al.*, 1988] V. L. Fitch, M. V. Isaila and M. A. Palmer. Limits on the Existence of a Material-dependent Intermediate-Range Force. *Physical Review Letters* **60**: 1801-1804, 1988.

[Ford, 1937] E. B. Ford. Problems of Heredity in the Lepidoptera. *Biological Reviews* **12**: 461-503, 1937.

[Ford, 1940] E. B. Ford. Genetic Research on the Lepidoptera. *Annals of Eugenics* **10**: 227-252, 1940.

[Franklin, 1986] A. Franklin. *The Neglect of Experiment*. Cambridge: Cambridge University Press, 1986.

[Franklin, 1990] A. Franklin. *Experiment, Right or Wrong*. Cambridge: Cambridge University Press, 1990.

[Franklin, 1933a] A. Franklin. *The Rise and Fall of the Fifth Force: Discovery, Pursuit, and Justification in Modern Physics*. New York: American Institute of Physics, 1933.

[Franklin, 1994] A. Franklin. How to Avoid the Experimenters' Regress. *Studies in History and Philosophy of Modern Physics* **25**: 97-121, 1994.

[Franklin, 1995] A. Franklin. Laws and Experiment. *Laws of Nature*. F. Weinert. Berlin: De Gruyter: 191-207, 1995.

[Franklin, 1996] A. Franklin. There Are No Antirealists in the Laboratory. *Realism and Anti-Realism in the Philosophy of Science*. R. S. Cohen, R. Hilpinen and Q. Renzong. Dordrecht: Kluwer Academic Publishers: 131-148, 1996.

[Franklin, 1997a] A. Franklin. Calibration. *Perspectives on Science* **5**: 31-80, 1997.

[Franklin, 1997b] A. Franklin. Are There Really Electrons? Experiment and Reality. *Physics Today* **50**(10): 26-33, 1997.

[Franklin, 1997c] A. Franklin. Recycling Expertise and Instrumental Loyalty. *Philosophy of Science* **64**(4, Supp.): S42-S5), 1997.

[Franklin, 2002] A. Franklin. *Selectivity and Discord: Two Problems of Experiment*. Pittsburgh, PA: University of Pittsburgh Press, 2002.

[Franklin and Howson, 1998] A. Franklin and C. Howson. Comment on 'The Structure of a Scientific Paper' by Frederick Suppe. *Philosophy of Science* **65:** 411-416, 1998.

[Fraser, 1927] R. G. J. Fraser. The Effective Cross Section of the Oriented Hydrogen Atom. *Proceedings of the Royal Society (London)* **114**: 212-221, 1927.

[French, 1999] A. P. French. The Strange Case of Emil Rupp. *Physics in Perspective* **1**: 3-21, 1999.

[Friedman and Telegdi, 1957] J. L. Friedman and V. L. Telegdi. Nuclear Emulsion Evidence for Parity Nonconservation in the Decay Chain pi - mu-e. *Physical Review* **105**: 1681-1682, 1957.

[Galison, 1987] P. Galison. *How Experiments End*. Chicago: University of Chicago Press, 1987.

[Gamow and Teller, 1936] G. Gamow and E. Teller. Selection Rules for the β-Disintegration. *Physical Review* **49**: 895-899, 1936.

[Garwin *et al.*, 1957] R. L. Garwin, L. M. Lederman and M. Weinrich. Observation of the Failure of Conservation of Parity and Charge Conjugation in Meson Decays: The Magnetic Moment of the Free Muon. *Physical Review* **105**: 1415-1417, 1957.

[Gerlach and Stern, 1922] W. Gerlach and O. Stern. Der experimentelle Nachweis der Richtungsquantelung. *Zeitschrift fur Physik* **9**: 349-352, 1922.

[Gerlach and Stern, 1924] W. Gerlach and O. Stern. Uber die Richtungsquantelung im Magnetfeld. *Annalen der Physik* **74**: 673-699, 1924.

[Hacking, 1981] I. Hacking. Do We See Through a Microscope. *Pacific Philosophical Quarterly* **63**: 305-322, 1981.

[Hacking, 1983] I. Hacking. *Representing and Intervening*. Cambridge: Cambridge University Press, 1983.

[Hacking, 1922] I. Hacking. The Self-Vindication of the Laboratory Sciences. *Science as Practice and Culture.* A. Pickering. Chicago: University of Chicago Press: 29-64, 1922.

[Hermannsfeldt *et al.*, 1958] W. B. Hermannsfeldt, R. L. Burman, P. Stahelin, *et al.* Determination of the Gamow-Teller Beta-Decay Interaction from the Decay of Helium-6. *Physical Review Letters* **1**: 61-63, 1958.

[Holmes, 2001] F. L. Holmes. *Meselson, Stahl, and the Replication of DNA, A History of The Most Beautiful Experiment in Biology.* New Haven: Yale University Press, 2001.

[Kettlewell, 1955] H. B. D. Kettlewell. Selection Experiments on Industrial Melanism in the Lepidoptera. *Heredity* **9**: 323-342, 1955.

[Kettlewell, 1956] H. B. D. Kettlewell. Further Selection Experiments on Industrial Melanism in the Lepidoptera. *Heredity* **10**: 287-301, 1956.

[Kettlewell, 1958] H. B. D. Kettlewell. A Survey of the Frequencies of *Biston betularia* (L.) (Lep.) and its Melanic Forms in Great Britain. *Heredity* **12**: 51-72, 1958.

[Kofoed-Hansen, 1955] O. Kofoed-Hansen. Neutrino Recoil Experiments. *Beta- and Gamma-Ray Spectroscopy.* K. Siegbahn. New York: Interscience: 357-372, 1955.

[Konopinski and Uhlenbeck, 1935] E. Konopinski and G. Uhlenbeck. On the Fermi Theory of Radioactivity. *Physical Review* **48**: 7-12, 1935.

[Konopinski and Langer, 1953] E. Konopinski and L. M. Langer. The Experimental Clarification of the Theory of β-Decay. *Annual Reviews of Nuclear Science* **2**: 261-304, 1953.

[Konopinski and Uhlenbeck, 1941] E. Konopinski and G. E. Uhlenbeck. On the Theory of β-Radioactivity. *Physical Review* **60**: 308-320, 1941.

[Langer *et al.*, 1950] L. M. Langer, J. W. Motz and H. C. Price (1950). Low Energy Beta-Ray Spectra: Pm^{147} S^{35}. *Physical Review* **77**: 798-805, 1950.

[Langer and Price, 1949] L. M. Langer and H. C. Price. Shape of the Beta-Spectrum of the Forbidden Transition of Yttrium 91. *Physical Review* **75**: 1109, 1949.

[Langstroth, 1932] G. O. Langstroth. Electron Polarisation. *Proceedings of the Royal Socoety (London)* **A136**: 558-568, 1932.

[LaRue *et al.*, 1981] G. S. LaRue, J. D. Phillips and W. M. Fairbank. Observation of Fractional Charge of $(1/3)e$ on Matter. *Physical Review Letters* **46**: 967-970, 1981.

[Lee and Yang, 1956] T. D. Lee and C. N. Yang. Question of Parity Nonconservation in Weak Interactions. *Physical Review* **104**: 254-258, 1956.

[Lipton, 1998] P. Lipton. The Best Explanation of a Scientific Paper. *Philosophy of Science* **65**: 406-410, 1998.

[MacKenzie, 1989] D. MacKenzie. From Kwajelein to Armageddon? Testing and the Social Construction of Missile Accuracy. *The Uses of Experiment.* D. Gooding, T. Pinch and S. Shaffer. Cambridge: Cambridge University Press: 409-435, 1989.

[Mayer *et al.*, 1951] M. G. Mayer, S. A. Moszkowski and L. W. Nordheim. Nuclear Shell Structure and Beta Decay. I. Odd A Nuclei. *Reviews of Modern Physics* **23**: 315-321, 1951.

[McKinney, 1992] W. McKinney. Plausibility and Experiment: Investigations in the Context of Pursuit. *History and Philosophy of Science.* Bloomington, IN, Indiana (PhD thesis), 1992.

[Mehra and Rechenberg, 1982] J. Mehra and H. Rechenberg. *The Historical Development of Quantum Theory.* New York: Springer-Verlag, 1982.

[Meselson and Stahl, 1958] M. Meselson and F. W. Stahl. The Replication of DNA in Escherichia Coli. *Proceedings of the National Academy of Sciences (USA)* **44**: 671-682, 1958.

[Mott, 1929] N. F. Mott. Scattering of Fast Electrons by Atomic Nuclei. *Proceedings of the Royal Society (London)* **A124**: 425-442, 1929.

[Mott, 1931] N. F. Mott. Polarization of a Beam of Electrons by Scattering. *Nature* **128**: 454. 1931.

[Mott, 1932] N. F. Mott. Tha Polarisation of Electrons by Double Scattering. *Proceedings of the Royal Society (London)* **A135**: 429-458, 1932.

[Nelson *et al.*, 1990] P. G. Nelson, D. M. Graham and R. D. Newman. Search for an Intermediate-Range Composition-dependent Force Coupling to N-Z. *Physical Review D* **42**: 963-976, 1990.

[Newman *et al.*, 1989] R. Newman, D. Graham and P Nelson. A Fifth Force Search for Differential Accleration of Lead and Copper toward Lead. *Tests of Fundamental Laws in Physics: Ninth Moriond Workshop.* O. Fackler and J. Tran Thanh Van. Gif sur Yvette: Editions Frontieres: 459-472, 1989.

[Nishijima and Saffouri, 1965] K. Nishijima and M. J. Saffouri. CP Invariance and the Shadow Universe. *Physical Review Letters* **14**: 205-207, 1965.

[Pais, 1982] A. Pais. *Subtle is the Lord...* Oxford: Oxford University Press, 1982.

[Pauli, 1933] W. Pauli. Die Allgemeinen Prinzipen der Wellenmechanik. *Handbuch der Physik* **24**: 83-272, 1933.

[Petschek and Marshak, 1952] A. G. Petschek and R. E. Marshak. The β-Decay of Radium E and the Pseudoscalar Interaction. *Physical Review* **85**: 698-699, 1952.

[Pickering, 1981] A. Pickering. The Hunting of the Quark. *Isis* **72**: 216-236, 1981.

[Pickering, 1984a] A. Pickering. *Constructing Quarks.* Chicago: University of Chicago Press, 1984.

[Pickering, 1984b] A. Pickering. Against Putting the Phenomena First: The Discovery of the Weak Neutral Current. *Studies in the History and Philosophy of Science* **15**: 85-117, 1984.

[Pickering, 1987] A. Pickering. Against Correspondence: A Constructivist View of Experiment and the Real. *PSA 1986.* A. Fine and P. Machamer. Pittsburgh: Philsophy of Science Association. **2:** 196-206, 1987.

[Pickering, 1989] A. Pickering. Living in the Material World: On Realism and Experimental Practice. *The Uses of Experiment.* D. Gooding, T. Pinch and S. Schaffer. Cambridge: Cambridge University Press**:** 275-297, 1989.

[Pickering, 1995] A. Pickering. *The Mangle of Practice.* Chicago: University of Chicago Press, 1995.

[Prentki, 1965] J. Prentki. *CP Violation.* Oxford International Conference on Elementary Particles, Oxford, England, 1965.

[Pursey, 1951] D. L. Pursey. The Interaction in the Theory of Beta Decay. *Philosophical Magazine* **42**: 1193-1208, 1951.

[Raab, 1987] F. J. Raab. Search for an Intermediate-Range Interaction: Results of the Eot-Wash I Experiment. *New and Exotic Phenomena: Seventh Moriond Workshop.* O. Fackler and T. T. Van. Gif sur Yvette: Editions Frontieres, 567-577, 1987.

[Randall *et al.*, 1949] H. M. Randall, R. G. Fowler, N. Fuson, *et al. Infrared Determination of Organic Structures.* New York: Van Nostrand, 1949.

[Richter, 1937] H. Richter. Zweimalige Streuung schneller Elektronen. *Annalen der Physik* **28**: 553-554, 1937.

[Ridley, 1954] B. W. Ridley. Nuclear Recoil in Beta Decay. *Physics.* Cambridge, Cambridge University (PhD thesis), 1954.

[Rose and Bethe, 1939] M. E. Rose and H. A. Bethe. On the Absence of Polarization in Electron Scattering. *Physical Review* **55**: 277-289, 1939.

[Ruark and Urey, 1930] A. E. Ruark and H. C. Urey. *Atoms, Molecules, and Quanta.* New York: McGraw-Hill, 1930.

[Rudge, 1998] D. W. Rudge. A Bayesian Analysis of Strategies in Evolutionary Biology. *Perspectives on Science* **6**: 341-360, 1998.

[Rudge, 2001] D. W. Rudge. Kettlewell from an Error Statistician's Point of View. *Perspectives on Science* **9**: 59-77, 2001.

[Rupp, 1929] E. Rupp. Versuche zur Frage nach einer Polarisation der Elektronenwelle. *Zeitschrift fur Physik* **53**: 548-552, 1929.

[Rupp, 1930] E. Rupp. Ueber eine unsymmetrische Winkelverteilung zweifach reflektierter Elektronen. *Zeitschrift fur Physik* **61**: 158-169, 1930.

[Rupp, 1931] E. Rupp. Direkte Photographie der Ionisierung in Isolierstoffen. *Naturwissenschaften* **19**: 109, 1931.

[Rupp, 1932a] E. Rupp. Versuche zum Nachweis einer Polarisation der Elektronen. *Physickalsche Zeitschrift* **33**: 158-164, 1932.

[Rupp, 1932b] E. Rupp. Neure Versuche zur Polarisation der Elektronen. *Physikalische Zeitschrift* **33**: 937-940, 1932.

[Rupp, 1932c] E. Rupp. Ueber die Polarisation der Elektronen bei zweimaliger 90° - Streuung. *Zeitschrift fur Physik* **79**: 642-654, 1932.

[Rupp, 1934a] E. Rupp. Polarisation der Elektronen an freien Atomen. *Zeitschrift fur Physik* **88**: 242-246, 1934.

[Rupp, 1934b] E. Rupp. Polarisation der Elektronen in magnetischen Feldern. *Zeitschrift fur Physik* **90**: 166-176, 1934.

[Rupp, 1935] E. Rupp. Mitteilung. *Zeitschrift fur Physik* **95**: 810, 1935.

[Rupp and Szilard, 1931] E. Rupp and L. Szilard. Beeinflussung 'polarisierter' Elektronenstrahlen durch Magnetfelder. *Naturwissenschaften* **19**: 422-423, 1931.

[Rustad and Ruby, 1953] B. M. Rustad and S. L. Ruby. Correlation between Electron and Recoil Nucleus in He^6 Decay. *Physical Review* **89**: 880-881, 1953.

[Rustad and Ruby, 1955] B. M. Rustad and S. L. Ruby. Gamow-Teller Interaction in the Decay of He^6. *Physical Review* **97**: 991-1002, 1955.

[Sargent, 1932] B. W. Sargent. Energy Distribution Curves of the Disintegration Electrons. *Proceedings of the Cambridge Philosophical Society* **24**: 538-553, 1932.

[Sargent, 1933] B. W. Sargent. The Maximum Energy of the b-Rays from Uranium X and other Bodies. *Proceedings of the Royal Society (London)* **A139**: 659-673, 1933.

[Sauter, 1933] F. Sauter. Ueber den Mottschen Polarisationseffekt bei der Streuun von Elektronen an Atomen. *Annalen der Physik* **18**: 61-80, 1933.

[Sherr and Gerhart, 1952] R. Sherr and J. Gerhart. Gamma Radiation of C^{10}. *Physical Review* **86**: 619, 1952.

[Sherr *et al.*, 1949] R. Sherr, H. R. Muether and M. G. White. Radioactivity of C^{10} and O^{14}. *Physical Review* **75**: 282-292, 1949.

[Smith, 1951] A. M. Smith. Forbidden Beta-Ray Spectra. *Physical Review* **82**: 955-956, 1951.

[Smith, 1997] G. E. Smith. J.J. Thomson and the Electron: 1897-1899 An Introduction. *The Chemical Educator* **Vol. 2,** Number 6, 1997.

[Stern, 1921] O. Stern. Ein Weg zur experimentellen Prufung Richtungsquantelung im Magnet feld. *Zeitschrift fur Physik* **7**: 249-253, 1921.

[Stubbs *et al.*, 1989] C. W. Stubbs, E. G. Adelberger, B. R. Heckel, *et al.* Limits on Composition-dependent Interactions using a Laboratory Source: Is There a Fifth Force? *Physical Review Letters* **62**: 609-612, 1989.

[Stubbs *et al.*, 1987] C. W. Stubbs, E. G. Adelberger, F. J. Raab, *et al.* Search for an Intermediate-Range Interaction. *Physical Review Letters* **58**: 1070-1073, 1987.

[Sudarshan and Marshak, 1958] E. C. G. Sudarshan and R. E. Marshak. Chirality Invariance and the Universal Fermi Interaction. *Physical Review* **109**: 1860-1862, 1958.

[Suppe, 1998a] F. Suppe. The Structure of a Scientific Paper. *Philosophy of Science* **65**: 381-405, 1998.

[Suppe, 1998b] F. Suppe. Reply to Commentators. *Philosophy of Science* **65**: 417-424, 1998.

[Thielberger, 1987a] P. Thieberger. Search for a Substance-Dependent Force with a New Differential Accelerometer. *Physical Review Letters* **58**: 1066-1069, 1987.

[Thomson, 1933] G. P. Thomson. Polarisation of Electrons. *Nature* **132**: 1006, 1933.

[Thomson, 1934] G. P. Thomson. Experiment on the Polarization of Electrons. *Philosophical Magazine* **17**: 1058-1071, 1934.

[Thomson, 1897] J. J. Thomson. Cathode Rays. *Philosophical Magazine* **44**: 293-316, 1897.

[Uhlenbeck and Goudsmit, 1925] G. E. Uhlenbeck and S. Goudsmit. Ersetzung der Hypothese von unmechanischen Zwang durch eine Forderung bezuglich des inneren Verhaltens jedes einzelnen Elektrons. *Naturwissenschaften* **13**: 953-954, 1925.

[Uhlenbeck and Goudsmit, 1926] G. E. Uhlenbeck and S. Goudsmit. Spinning Electrons and the Structure of Spectra. *Nature* **117**: 264-265, 1926.

[Watson and Crick, 1953a] J. D. Watson and F. H. C. Crick. A Structure for Deoxyribose Nucleic Acid. *Nature* **171**: 737, 1953.

[Watson and Crick, 1953b] J. D. Watson and F. H. C. Crick. Genetical Implications of the Structure of Deoxyribonucleic Acid. *Nature* **171**: 964-967, 1953.

[Weinert, 1995] F. Weinert. Wrong Theory–Right Experiment: The Significance of the Stern-Gerlach Experiments. *Studies in History and Philosophy of Modern Physics* **26B**(1): 75-86, 1995.

[Winter, 1936] J. Winter. Sur la polarisation des ondes de Dirac. *Academie des Science, Paris, Comptes rendus hebdomadaires des seances* **202**: 1265-1266, 1936.

[Wu, 1955] C. S. Wu. The Interaction in β-Decay. *Beta- and Gamma-Ray Spectroscopy.* K. Siegbahn. New York: Interscience: 314-356, 1955.

[Wu *et al.*, 1957] C. S. Wu, E. Ambler, R. W. Hayward, *et al.* Experimental Test of Parity Nonconservation in Beta Decay. *Physical Review* **105**: 1413-1415, 1957.

[Wu and Schwarzchild, 1958] C. S. Wu and A. Schwarzschild. A Critical Examination of the He^6 Recoil Experiment of Rustad and Ruby. New York, Columbia University, 1958.

THE ROLE OF EXPERIMENTS IN THE SOCIAL SCIENCES: THE CASE OF ECONOMICS

Wenceslao J. Gonzalez

Among the central topics in the methodology of the social sciences is the analysis of the role of experiments. It receives special attention in the case of economics, where there is a branch explicitly called "experimental economics". However, the acceptance of "experiments" in the social sciences, in general, and in economics, in particular, has not been always the case, and it is still an issue that raises objections. Then, after the recognition of the movement from observation to experiment, the notion of "experiment" used in the social sciences will be discussed. Thereafter, the focus will be on the development of experiments in the social sciences, taking into account Reinhard Selten's contribution. In this regard, it will also be a closer look on the role of prediction in experimental economics.

1 FROM OBSERVATIONS TO EXPERIMENTS IN THE SOCIAL SCIENCES

Controlled observations have been commonly accepted as a valid procedure to test and evaluate scientific statements in the social sciences, whereas for a long time a wary attitude has dominated the scene regarding the role of experiments in the social sciences. Moreover, the existence of experiments as a methodological procedure on human affairs has at times been openly questioned or even it has been explicitly neglected.[1] The challenge comes from the methodological starting point: the issue of the possibility itself of "experiments" in this realm of social phenomena. Questioning the *role of experiments* in the social sciences has a long tradition. Originally, it can be connected with the acceptance of a clear methodological gap — a dichotomy — between natural sciences and social sciences. The questioning of the experiments can also be seen in the most developed social sciences, such as economics, and this has even been the case in recent times. Thus, few years ago, in a chapter entitled "Economic Science without Experimentation", Tony Lawson dealt with the problem of "how social scientific research can proceed in the absence of real possibilities of experimental control" [1997, 199]. He maintains a commitment to controlled observations while questioning the possibility of experiments: "despite the lack of opportunities for controlled experimentation

[1]In the case of economics, cf. [Samuelson and Nordhaus, 1983, 8]; and [Morgan, 1990, 9 and 246].

Handbook of the Philosophy of Science: General Philosophy of Science - Focal Issues, pp. 275–301.
Volume editor: Theo Kuipers. General editors: Dov M. Gabbay, Paul Thagard and John Woods.

in the social sciences I remain optimistic about the social scientific prospects" [Lawson, 1997, 199].

The wary attitude — sometimes a sceptic view — on the role of experiments in the social sciences has changed progressively. The dominant perspective has moved steadily towards a clear acceptance of the possibility of experiments in social sciences. Furthermore, it includes a neat attempt to explore new aspects of experimentation in human affairs. The movement has received impulse from two kinds of approaches: a) some *new philosophic-methodological analyses* — on science, in general, and on social sciences, in particular — grounded in the idea of scientific activity,[2] and b) an important amount of contributions in the realm of the *scientific research on human affairs*, such as in the case of economic matters with the experimental branch.

On the one hand, since the mid-1980's the analysis of philosophy and methodology of science has paid the attention to experiments on the basis of a new consideration of the *scientific practice*.[3] The previous emphasis on the contents of science (semantic, logical, epistemological, ...) has given way to a more detailed reflection on how science is made as a *human activity* in a social environment (i.e., the laboratories as institutions where the scientists intervene). This involves a direct reflection on the practice of laboratory experimentation.[4] In addition, the scope of the philosophical and methodological analysis has been enlarged with new light shed on applied science and on applications of science.[5]

Simultaneously, on the other hand, scientific research on social events has widened the original fields in the last decades, mainly in the sciences of psychology and economics. Some *new territories* have been embraced, such as "experimental economics", which in 2002 received public recognition in the form of a Nobel Prize (Vernon Smith[6] and Daniel Kahneman[7]). Experimental economics is a scientific branch which has been the focus of increasing attention since the mid-1980's. Although its first, informal, precedent was set by Alvin Roth as early as Daniel Bernoulli,[8] it is only since the second half of the twentieth century that it has been clearly developed.[9] Several Nobel laureates, who have focused on game the-

[2]This focus sheds more light than the previous emphasis on science as knowledge and opens up more clearly the nexus with the social setting.

[3]A central author in this approach on scientific practice is Ian Hacking, especially since the publication of [Hacking, 1983].

[4]Cf. [Galison, 1987]. On the philosophical and methodological analysis of experiments (characteristics, kinds, ...), cf. [Gooding *et al.*, 1989; Galavotti, 2003; Radder, 2003a].

[5]"It is important to distinguish *applied* science from the *applications* of science. The former is a part of knowledge production, the latter is concerned with the use of scientific knowledge and methods for the solving of practical problems of action (e. g., in engineering of business), where a scientist may play the role of a consult" [Niiniluoto, 1993, 9]. The epistemic and practical aspects can also be seen in the context of the social setting, cf. [Kitcher, 2001, especially 85–91].

[6]Cf. [Knez and Smith, 1987]. Vernon Smith has shown an explicit interest in the philosophy of science, cf. [Smith *et al.*, 1991].

[7]Cf. [Tversky *et al.*, 1990/1993; Kahneman *et al.*, 1990/1993].

[8][Bernoulli, 1738/1954]. Cf. [Roth, 1988, 974/1993, 3].

[9]Even though Volker Häselbarth in 1967 lists 20 publications before 1959, Reinhard Selten stresses that "experimental economics as a field of economic research did not emerge before the

ory (such as the well-known John Nash[10] and Reinhard Selten[11], both in 1994), have also developed economic experiments.

According to this new situation — both in philosophy of science and in the development of social sciences —, there is a methodological framework which is different from the successive versions of the distinction between *Erklären* and *Verstehen*,[12] a discussion which started with a clear dichotomy between natural sciences and social sciences. The present methodological perspectives stress the existence of experiments as a *common ground* between natural sciences and social sciences, even though it is still the case that both kinds of sciences also have differences (in the aims, processes and results). To some extent, the new analyses of philosophy and methodology of the social sciences work with an enlarged vision of experiment, and the sciences themselves use new views on experiments. Thus, the notion of "experiment" is not restricted anymore to a material sort of human intervention based on a previous design.

Moreover, there are conceptions of experimentation which go beyond that point of novelty and accept the possibility of "natural experiments". It is an uncommon view which avoids the feature of *human intervention*, which has been characteristic of the notion of *experiment*. An approach following this route is held, among other authors, by James Woodward: "the important and philosophically neglected category of 'natural experiments' typically involves the occurrence of processes in nature that have the characteristics of an intervention but do not involve human action or at least are not brought about by deliberate human design."[13]

Even though "experiment" and "control experiment" could be considered almost synonymous, sometimes a distinction is used between experiments at large and specifically *controlled* experiments. In this case, the idea is to distinguish a broad sense of "experiment", mainly when the reference is the society as a whole or a large-group, and the strict sense of "experiment", where there is a group which could be controlled in a clearer way. It is a kind of distinction to emphasize the difference between large-scale groups, where a controlled experiment seem hard or even is virtually impossible, and small groups (communities, micro-societies, ...) where the process of control is possible. According to this distinction, large-scale groups rely on comparative analysis ("field research") based on observation, and the level of control is then commonly inferior to the controlled experiments.

1960s" [Selten, 1993, 118].

[10]In 1952 a conference on "The Design of Experiments in Decision Processes" was held in Santa Monica, CA, in order to accommodate the game theorists and experimenters associated with the Rand Corporation. John Nash wrote an important paper with G. K. Kalisch, J. W. Milnor, and E. D. Nering, cf. [Roth, 1995, 10–11].

[11]Cf. [Gonzalez, 2003a].

[12]This methodological controversy has changed several times from the original version to more recent presentations, cf. [Gonzalez, 2003b, especially 34–37].

[13][Woodward, 2003, 94]. A similar idea — "the stream of experiments that Nature is steadily turning our from her own enormous laboratory" — appears in [Haavelmo, 1944, 14–15].

2 THE NOTION OF "EXPERIMENT" USED IN THE SOCIAL SCIENCES: FROM THE TRADITIONAL APPROACH TO THE ENLARGED VISION

When the issue of the characterization of "experiment" arises, the *notion* requires us to consider — in my judgment — several aspects related to central factors of science.[14]. 1) Semantically, experiment originally has a sense and a reference that differs from "observation". 2) Logically, experiment is a structural ingredient of science which is different from "theory" and, in principle, it is also distinct from "model". 3) Epistemologically, experiment is related to a kind of reliable knowledge acquired through a non-immediate process. 4) Methodologically, experiment is connected to a process which should be repeatable and, therefore, it is commonly associated to reproducibility and replicability. 5) Ontologically, experiment is related to the idea of otherness (i.e., something — real or not — which is used to test). 6) Axiologically, the experiments can be oriented through different values according to distinct aims (i.e., experiments in basic science could be diverse from experiments in applied science). 7) Ethically, there is concern on some kinds of experiments, mainly when they are related to certain human affairs (either to the persons as individuals or to the society as a whole).

These aspects of a general characterization of "experiment" can be developed in different ways, according to the *kind of variables* to be controlled and the *type of procedure* used to control them. This variety accounts for the diversity of experiments — in the material and non-material spheres — which may be accepted in the enlarged vision. What many authors have questioned in the past — or even now — is the possibility (in the set of the social sciences and, therefore, in the case of economics) of experiments being understood according to the traditional notion, a view which relies upon "experiment" as a *human intervention* suggested by a theory and on some variables to be controlled in a *repeatable material context.*[15]

2.1 Diversity of Experiments

Basically, the social sciences — and, within these, economics — tend to insist on the epistemological and methodological aspects of experimentation. In this regard, the emphasis is on *reliable knowledge* and *repeatability*: "the idea of a scientific or controlled experiment is to reproduce the conditions required by a theory and then to manipulate the relevant variables in order to make measurements of a particular scientific parameter or to test the theory. When the data are not collected under controlled conditions or are not from repeatable experiments, then the relationship between the data and the theoretical laws is likely to be neither direct nor clear-cut. This problem was not (...) unique to economics; it arose in other social sciences and in natural sciences where controlled experiments were not possible" [Morgan, 1990, 9].

[14]On the central factors of science, cf. [Gonzalez, 2005, especially 10–11].

[15]"In experiments we actively interfere with the material world. In one way or another, experimentation involves the material realization of an experimental process (the object[s] of study, the apparatus, and their interaction)" [Radder, 2003b, 4].

Among the components of the *traditional view* on experiment are also several elements related to a given situation, commonly related to a material setting. i) An experiment is an *intervention* in the world, thus it includes a manipulation of some aspects of reality to identify certain causal mechanisms or to test a theory about those phenomena. ii) Experiments are thought of a way of grasping certain *relatively enduring structures* of the world (some mechanism which acts in a characteristic manner when there are specific circumstances). iii) The experiment is an *active interference* in order to enable or trigger the mechanism which is under investigation, and it could also be used to prevent any countervailing mechanism (cf. [Lawson, 1997, 202–203]).

In this regard, the critics of experiments in the social sciences claim that all of them — and specifically economics — are not in a position to isolate, control and manipulate social (in this case, economic) conditions.[16] Thus, Lawson maintains that "it is certainly reasonable to doubt that controlled experimentation will ever be particularly meaningful in economics due to the impractically of manipulating social structures and mechanisms in order to more clearly identify them." [Lawson, 1997, 203–204]. What he accepts is the existence of social regularities (or "partial regularities") which come from the reproducibility of certain mechanisms of a social world that is open, dynamic and changing.

However, there is a full branch of economics — experimental economics — which focuses on *laboratory experimentation* which assumes the traditional notion of experiment. Moreover, "economic experiments in the laboratory aspire to the standards of laboratory experiments found elsewhere in science. Depending on the experiment in question, economists may focus their design aim on the control of the environment in which the experiment takes place, on controlling the communication between subjects, on setting limits on the range of input behaviour allowed and the variation of output responses and so forth" [Boumans and Morgan, 2001, 17].[17]

Nevertheless, it has also been common in the past to use thought experiment (*Gedanken experiment*) as a heuristic tool, both in natural sciences and in social sciences. Certainly, there is no material intervention in a thought experiment: it could be seen as an imaginative narrative to test a specific theory or a hypothesis. The *thought experiment* considers hypothetical or imaginary test conditions, where some particular instantiations of a real process — it maybe a causal one — that is identified by the theory, and the thought experiment displays a concrete state of affairs that needs an explanation as the end result of the process that is being studied (cf. [Lennox, forthcoming]).

Nowadays we work *de facto* with an enlarged notion of "experiment", even though the experiments with material intervention — laboratory experiments — are still the archetype among them and there is still discussion on the acceptability *as experiment* of some of them. The *criteria* used to distinguish the diversity of experiments are commonly features related to several elements — epistemological,

[16] On Trygve Haavelmo and this issue, cf. [Morgan, 1990, 245–246].

[17] Cf. [Friedman and Sunder, 1994].

methodological and ontological —, mainly i) the range of *controllability* of the variables, ii) the level of materiality of the *processes* employed in the research, and iii) the *sphere* — real, ideal, a hybrid, ... — to be analyzed. Again economics is an interesting case study to see the differences between laboratory experimentation and other experiments, because it is a science that uses a variety of experiments.

i) As for epistemological issue of the *range of controllability* of the variables, there are at least three possibilities: a) direct control; b) indirect control (or statistical control); and c) assumption in model. The last one is the most complex, concerning the kind of *control of variables*, especially if we follow the analysis made by Alan Musgrave on the Milton Friedman approach on the lack of realism of the assumptions. For Friedman, an economic theory should not be criticized for containing "unreal assumptions" because the important point in order to evaluate a theory is successful predictions [Friedman, 1953/1969]. Musgrave considers instead that Friedman's dictum (the so-called "F-twist") is false according to three types of assumption: negligibility assumptions, domain assumptions, and heuristic assumptions (cf. [Musgrave, 1981]).[18]

ii) Concerning the methodological issue of the level of *materiality of the processes* employed in the research, there are differences in the case of economics regarding several possibilities: 1) the empirical domain of *laboratory experimentation* (when a material realm is under direct control), 2) the "passive experimentation" of the *econometric case* (when a material realm receives an indirect control or statistical control), 3) the simulations and, above all, *computer simulations* (when the quasi-material realm or the pseudo-material sphere depends on the assumptions in the model), and 4) the *thought experiments* (when the non-material realm depends upon the assumptions in the model).[19]

iii) Both the range of controllability of the variables and the level of materiality of the processes employed in the research are connected to the *ontological issue* of the sphere — real, ideal, a hybrid, ... — to be analyzed. Certainly the *sphere to be analyzed* by experiments varies from real (a direct tangible object of study or statistical data based on previous evidence) to clearly ideal (a thought experiment or a purely mathematical model). Between these poles — real, although artificially constructed,[20] and ideal — there is the realm of hybrids (e.g., in simulations) which could be quasi-material or pseudo-material.

If these distinctions (epistemological, methodological and ontological) related to the kind of experiments are basically correct, then there are clear differences in the methodological processes to test economic predictions (which is a central

[18] "*Negligibility assumptions* state that some factor has a negligible effect upon the phenomenon under investigation. *Domain assumptions* specify the domain of applicability of the theory. *Heuristic assumptions* are a means of simplifying the logical development of the theory" [Musgrave, 1981, 386].

[19] Cf. [Boumans and Morgan, 2001, 20]. Even though the origin of this distinction is in the analysis of *ceteris paribus* conditions, it seems to me that the main differences which are drawn in this differentiation have a rather general consideration for the methodological process.

[20] "In the laboratory an artificial economic reality is constructed, for example a market or an auction" [Selten, 2003, 63].

issue in the methodology of economics). Experimental economics — mainly in the case of Reinhard Selten — is well aware of this question and tries to emphasize the laboratory experimentation as the most reliable kind of experimentation to deal with the predictive success (and, therefore, with the notions of accuracy and precision). Nevertheless, he also points out limits: "experimental research (...) cannot replace field research. The institutional environment of economic participants must be investigated in the real world. Once we are sure we have modeled such an environment, however, we can and must test or redevelop the behavioral assumptions of theory in the laboratory" [Selten, 2003, 68].

2.2 Laboratory experiments

Although a laboratory experiment is performed in a material setting (a location where the experiment is conducted), the relevant factors are epistemological and methodological: the economic environment should be under the control of the experimenter. "This distinguishes laboratory experiments from 'field' experiments, in which relatively few aspects of the environment can be controlled, and in which only limited access to most of the economic agents may be available. It is precisely this control of the environment, and access to the agents (sufficient to observe and measure attibutes that are not controlled) that give laboratory experiments their power" [Roth, 1988, 974/1993, 3]. The basic level is in the study of some economic processes (bargaining, exchange relations, ...) and the search for stable or structural features of economic behavior associated to them (while bargaining, in auctions, ...).

Much of what is done in laboratory experiments is thought of as addressing problems that are not studied in other branches of economics. Thus, it seems that aims, processes and results of laboratory experiments in economics relate to aspects of economic phenomena that, in principle, are not investigated by other kinds of economic research. The experiment starts with a *design*, which should consider the parameters and the procedures to be used; it follows with the "experiment" itself: the *process of control* of the behavior of trial subjects in laboratory situations which are of interest for economic science; and, eventually, the experiment gets some *results* (e.g., of auction markets or of bargaining) which should be compared with other experiments made.

Laboratory experimentation can be designed with different *aims* in mind. In the case of economics, the design can be originally oriented towards economic theory (thinking of descriptive economics) or to applied economics (mainly towards policy guidance). Thus, on the one hand, an experiment can be designed to *test* some particular formal hypothesis (e.g., on the preference reversal phenomenon), looking for observations that may support a relatively general conclusion; and, on the other hand, the experiment may be designed to resemble some complex reality (such as a market) where the observations which are made seek to be relevant for a particular issue of *policy* (e.g., on market stability), cf. [Roth, 1987, 160]. "Somewhere in between lie experiments designated to collect data on interesting phenomena and important institutions, in the hope of detecting unanticipated regularities" [Roth, 1986, 246].

As regard the *process* itself of experimentation in economics, there are several options. It is clear that the process is often used to test a theory from the point of view of prediction. In this case, there are also several possibilities, according to the aims of the methodological task of testing: "In the case of experiments that test the predictions of existing theories, it will be useful to distinguish between those that test a theory in its specified domain of application, and those that explore its applicability outside the strict confines of this domain. We will also want to know whether the results of these experiments falsify the predictions of the theory or support it" [Roth, 1987, 148] (i.e., evidence that fails to falsify the theory).

Hence, the variety of aims and the diversity of processes can lead us to a large range of *results* which, in principle, are interpreted in the context of the specific aims of each experiment carried out in the laboratory. Moreover, the results obtained by means of laboratory experimentation in economics can show us that well-established conceptions, such as the assumption that economic agents move according to an expected utility function, have serious problems with empirical data. Indeed, a substantial body of empirical work offers a number of systematic ways in which individual preferences of economic agents fail to exhibit the regularities represented by an expected utility function (cf. [Roth, 1986, 248]).[21]

Yet laboratory experiments have been criticized in several ways. a) In laboratory situations trial subjects are commonly university students (and especially economics students), which may offer different results from other kinds of economic agents. b) The process itself of laboratory experiment is clearly artificial, and its resemblance with the real economic world may be questioned (and, consequently, the reliability of the economic knowledge which is obtained). c) The results obtained through experiments might be of limited applicability. Thus, even some important economic experimentalists are cautious when they expect that "patient experimental research will yield new behavioral theories of limited application" [Selten, 2003, 68] (i.e., comprehensive theories can appear only in the long run).

2.3 The Case of Econometrics

Econometrics may be seen as offering a *tertium quid* between the experimentation made in the economic laboratory and the thought experiments. On the one hand, an econometric model shares with laboratory experimentation the *constructed character* of the process and the *artificial nature* of the environment, even though in the laboratory there are real agents (i.e., there is a direct control of variables) whereas in an econometric model there is a package of statistical data related to economic phenomena (i.e., the control of variables is indirect). On the other hand, an econometric model has no clear-cut relation to the circumstances

[21] I recall several conversations with Herbert Simon (in [1993; 1996; 1999]) where he insisted on the idea of gathering empirical data through experiments in order to refute well-established assumptions of the mainstream of economics.

of economic undertakings (mainly when the model relies on a process of generating data) as well as the thought experiments work on a sphere of *possibilities* (and also impossibilities) rather than on an actual environment.

Such a vision of econometrics as *a case of experimentation* between the laboratory experiments made in the context of the material world and the thought experiments related to a non-material world includes a peculiar notion: "passive experimentation", which accompanies the idea of "natural experiments" (and, therefore, a material environment). Thus, "unlike the lab experiment, the 'experiments' of econometrics are not actual ones but statistical ones, conducted on 'passive data': data thrown up by the uncontrolled experiments of Nature (the Economy), incorporating all the multiple variation of the interacting factors with which the econometrician must deal as best he/she can *ex post*" [Boumans and Morgan, 2001, 18].

Compared with the methods of laboratory experimentation, the statistical methods of the econometric experiment — a passive one — might be interpreted as providing a substitute kind of control at the level of measurement process. It is an indirect control of the circumstances, and it requires statistical assumptions which are sometimes unrealistic: econometrics cannot control the circumstances directly and accepts that we are *passive observers* of a stream of experiments that Nature (economic world) is turning out from within, like an enormous laboratory.[22]

Consequently, the case of econometric experimentations relies on the possibility of "natural experiments" and their statistical control: "the econometric model is first built and estimated as a passive experiment and then is used as if it were a mathematical model" [Boumans and Morgan, 2001, 19]. The problem is then the application to the real economic world of those results obtained through this indirect way and, specifically, what are the possible causal inferences regarding economic activity. Not only is causality a debatable question in this context but so are the characteristics of reproducibility and replicability of econometric experiments ("passive experiments") in comparison with laboratory experimentation.

Some well-known econometricians, following a traditional notion of "experiment", have disregarded the possibility of experimentation in econometrics: "econometric theory is the study of the properties of data generation processes, techniques for analysing data, methods of estimating numerical magnitudes of parameters with unkown values and procedures for testing economic hypotheses; it plays an analogous role in primarily non-experimental disciplines to that of statistical theory in inexact experimental sciences (...). As expressed by [Wold, 1969], 'Econometrics is seen as a vehicle for fundamental innovations in scientific method, above all, in the development of operative forecasting procedures in non-experimental situations'. In Wold's view, econometrics needs to overcome both a lack of experimentation (which precludes reproducible knowledge) and the passivity of forecasts based on extrapolative methods" [Hendry, 2000, 13].

---

[22]That is the idea of T. Haavelmo, cf. [Morgan, 1990, 245].

2.4 Simulations and Computer Simulations

Simulations — and, above all, computer simulations[23] — have been characterized as "experiments" or, what is more appropriate, as "virtual experiments."[24] Methodologically they are a hybrid, insofar as simulations mixes mathematical models with experimental ones. "This kind of experimental activity has a comparatively long tradition in economics, predating the computer simulations of the type so familiar nowadays. It consists of statistical or mathematical models that are simulated, or 'run', to generate output series with the aim of mimicking observed economic time-series data. For example, one of the most commonly available, but least understood, sets of economic data is that of stock market prices" [Morgan, 2003, 22–225].[25]

Within this sphere of research — at least in the case of economics — there is an interweaving between the character of social science and the dimension of science of the artificial (in the sense of Herbert A. Simon [1996]). On the one hand, simulations can use empirical information of the world which might be obtained through of observations or even through experiments (including laboratory experiments); and, on the other hand, simulations include a nonmaterial component which seeks to resemble aspects of the world (a virtual representation oriented towards a mimic of the real world). The point of this hybrid of social and artificial is clear: to produce new outcomes in areas where other kinds of research seem infeasible or defective.

Frequently it happens that simulations, in general, and computer simulations, in particular, can play a similar role to thought experiments insofar as they can contribute to establishing some phenomena as possible and to ruling out certain events as impossible. But not all thought experiments are *eo ipso* simulations: they can follow different heuristic routes. Nevertheless both share the consideration of nonmaterial elements (or, at least, non tangible components). Methodologically, simulations are virtual experiments — they work on a hypothetical data stream in the case of economics — and ontologically they rely on constructed items within the artificial world (e.g., a resemblance of the real economic processes).

Alvin Roth has pointed out rightly that there is a "distressing tendency to confuse computer simulations, and the kind of investigations one can do with them, with experiments involving the observation of real people in controlled environments. (...) Computers simulations are useful for creating and exploring theo-

[23]On this issue, cf. [Keller, 2003].

[24]Simulations, in general, and computer simulations, in particular, can be used for quite different purposes, including those related to social life, such as traffic flow and automobile driver behavior under different sorts of conditions. From a descriptive perspective, computer simulation in these cases can be readily compared with real world behavior, and from a prescriptive point of view, road management can improve if the rules take into account that information and are well designed.

[25]She introduces the distinction between "virtual experiments" and "virtually experiments": "Virtual experiments (entirely nonmaterial in object of study and in intervention but which may involve the mimicking of observations) and virtually experiments (almost a material experiment by virtue of the virtually material object of input)" [Morgan, 2003, 233].

retical models, while experiments are useful for observing behaviour" [Roth, 1988, 1000/1993, 29]. In fact, a difference can be pointed out between the conclusions obtained by computer simulation (e.g., in the case of computer "tournaments" reported by R. Axelrod [1980a; 1980b; 1984]) and the experimental results obtained in the laboratory (e.g., in the experiments made by R. Selten and R. Stoecker [1986]).

For Roth, "the difference in results has a great deal to do with the difference between computer simulations and actual experiments. While the computer simulations which produce this result were conducted with an element of experimental flavour that is missing from conventional computer simulations (in that tournament entries were solicited from others), experiments with human subjects introduce a certain amount of open-ended complexity in the form of human behavior, that is absent from a tournament in which individuals are represented by short (or even moderately long) computer programs" [Roth, 1988, 1001].

Obviously, the notion itself of "experiment" is commonly linked to the idea of something *artificial* insofar as there is a human intervention to control a phenomenon or a set of phenomena. But there is — in my judgment — an increasing complexity in the study of the economic events, according to the scale of phenomena. The investigation may start from computer simulations — a clear artificial situation — and, through the analysis of controlled experiments in the laboratory environment — a less artificial case than the previous one —, the research of economic phenomena may reach the following steps of complexity: the real economic activity of human beings.[26] In this search the historical character of economics plays an important role,[27] which should be considered as well.

2.5 Thought Experiments and Mathematical Models

Initially *thought experiments* are a creative procedure related to a non-material domain but their aims, processes and results are oriented towards the real world. Thus, thought experiments are used to show the possibility and the impossibility of natural phenomena and social events as well as their limits in our world. Thought experiments belong to an ideal context when aims are designed, and while processing the information, but their results can be used for other kinds of experiments as well as for theoretical contributions. Moreover, they have been connected to computer experiments and to mathematical models, as can be seen in several sciences, such as economics.

Thought experiments were used a long time before computer experiments were available. However, *computer experiments* and *thought experiments* have some epistemological, methodological and ontological similarities insofar as virtual and mental are different from empirical. In this way computer simulations can be "experimental" in an analogous sense in which a thought experiment is "experi-

[26] On complexity as a typical feature of economic reality, cf. [Gonzalez, 1994, 262].

[27] Alvin Roth sees a parallel between evolutionary biology and economics because they deal "largely with historical data" [Roth, 1986, 270].

mental" [Keller, 2003, 204]. Important differences lie in the process employed: the computer permits the working out of the implications of a hypothesis more rapidly than the speed of thought, and can also consider a large range of variables at a specific moment.

There is also a relevant connection between *thought experiments* and *mathematical models.* Mary Morgan has emphasized the link: "in the post-1950s period, economists have become avid users of mathematical models. (...) I suggested that their usage involved being able to trace through deductively the answers to 'what if' or 'let us assume' type questions about the economic world represented in the model, cf. [Morgan, 2001] (...) We can portray this modern use of mathematical models as extending economists' verbal thought experiments of earlier times that were limited by the capacity of the mind to follow the paths of more than two or three variables in a system. In characterizing such model usage in terms of (...) thought experiments, we can see how asking questions and exploring answers with mathematical models have allowed economists to think through in a consistent and logically deductive way how a large number of variables may interrelate and find the solutions to systems with a large number of units" [Morgan, 2003, 218].

Mathematical model exploration can be related to questions on theories (i.e., economic theory) and issues connected to policy problems from the real world (i.e., applied economics). In the first case, the models can be used to develop the scientific theory, whereas in the second case they can perform a task to resolve concrete problems of the world. But the utilization of mathematical models understood *as experiments* — similar to laboratory experiments — requires us to take into account some important differences.

Fundamentally the differences are three. i) How the experimental *control* is achieved in both cases is different, because the material world has limits on intervention (i.e., control and manipulation) while the mathematical models should always place care on the capacity of their assumptions to represent the world properly; ii) the production of experimental *results* is also different: in laboratory experimentation is material (for the particular situation found in the experimental setup), whereas in the mathematical model experiments are based on the (deductive) reasoning power of mathematics to derive the results; and iii) the range of *potential inference* is different as well for the case of the results of experiments made in the laboratory and those obtained by mathematical model experiments (cf. [Morgan, 2003, 219–221].[28]

[28]On how the results relate to the world in both cases (laboratory and non-material), cf. [Morgan, 2003, 227–232].

3 THE DEVELOPMENT OF EXPERIMENTS IN THE SOCIAL SCIENCES: THE CASE OF ECONOMICS

Economics and psychology are two social sciences where the development of experiments has received more attention.[29] The research has followed frequently quite different lines, both in psychology (psychobiology makes different experiments from social psychology) and in economics (in principle, it is easier to make experiments in microeconomics than in macroeconomics). The fact that the level of development reached by experimental economics as well as the public recognition of the research on experimental economics, through the Nobel Prize, invites a closer analysis of the case of economics.

3.1 Experimental Economics

Experimental economics is a branch of this social science that, in the period between 1975 and 1985, undergoes the transformation from "a seldom encountered curiosity to a well-established part of economic literature" [Roth, 1987, 147]. The process was consolidated around 1985, when the *Journal of Economic Literature* initiated a separate bibliographical category for "Experimental Economic Methods."[30] The list of topics that is under the scrutiny of experimental economics is long: public goods provision, coordination and its failure, bargaining behavior, market organization in the context of competitive equilibrium, auction markets, individual choice behavior, ...[31] Many of the issues belong to microeconomics while others are in the realm of macroeconomics. The research seeks contributions to economic theory on the basis of experimental evidence, but some also look for policy applications of experimental methods (cf. [Plott, 1987]).

Many important economists have developed experimental economics. Among them is Reinhard Selten, who along with John Nash and John Harsanyi, received the Nobel Prize in economics for his work on game theory. Selten's relation with experimental economics appears to have its first expression in 1959, when he published with Heinz Sauermann the paper *Ein Oligopolexperiment* [Sauermann and Selten, 1959/1967]. Since then he has made important contributions to economics. Habitually, his papers include criticisms of mainstream economics, especially of the principle of subjective expected utility maximization. His publications are usually critical of assumptions of mainstream game theory,[32] which is deeply imbued with instrumental rationality. In fact, he links one of his most relevant contributions to

[29]In addition, experiments in psychology are also connected with experiments in economics, e.g., in areas related to decision-making.

[30]The same year — 1985 — the Fifth World Congress of the Econometric Society included a paper on experimental economics, cf. [Roth, 1986, 245].

[31]All these topics can be seen in the *Handbook of Experimental Economics* edited by John Kagel and Alvin Roth. In addition to other topics, they are also analyzed in the papers collected in the volumes on *Recent Developments in Experimental Economics* edited by John D. Hey and Graham Loomes.

[32]John Nash considers that the book *A General Theory of Equilibrium Selection in Games*, written by J. Harsanyi and R. Selten, "is very controversial" [Nash, 1996, 182].

game theory — the chain store paradox — to the need for a bounded rationality supported by experimental evidence. He maintains that the attempts to save the behavioral relevance of full rationality miss the point (cf. [Selten, 1990, 651]).

Furthermore, when Selten develops his economic approach, he presents area theories, such as the theory of equal division payoff bounds, which are based on a limited rationality, cf. [Selten, 1987]. In addition, he offers us a series of phenomena which confirm experimentally the existence of a bounded rationality, cf. [Selten, 1998a]. In this paper, related to the role of experiments in the social sciences, the analysis will follow three steps: firstly, there is — in section 3.2 — a presentation of Selten's approach to experimental economics (built upon his conception of economic rationality as *bounded*); secondly — in section 4 —, his methodological approach to economic predictions is analyzed, taking into account his position on experimentally based bounded rationality; and thirdly — in section 5 —, the role of the experiments is seen in a realm of accuracy and precision, where he provides relevant methodological proposals which are connected with the issue of the success of economic predictions.

3.2 Selten's Epistemological and Methodological Approach to Experimental Economics[33]

Epistemologically, Selten stresses the importance of *empirical knowledge* over theoretical knowledge, a position which is more in tune with an empiricist framework than with a rationalist one. His views, furthermore, differ from the claims of critical rationalism insofar as he is dissatisfied with a negative role of experience and that he highlights the need for experience understood in positive terms. In this regard, he maintains that "we know that Bayesian decision theory is not a realistic description of human economic behavior. There is ample evidence for this, but we cannot be satisfied with negative knowledge — knowledge about what human behavior fails to be. We need more positive knowledge on the structure of human behavior. We need theories of bounded rationality, supported by experimental evidence, which can be used in economic modeling as an alternative to exaggerated rationality assumptions" [Selten, 1991, 21].[34]

Selten's recommendation against the attempts to derive human behavior from a few general principles — either psychological or biological[35] — is the gaining of empirical knowledge. In addition, he is critical of unrealistic principles, thus opposing a view held by influential mainstream economists, and he does not accept criticism

[33]This section is based on [Gonzalez, 2003a, 72–74].

[34]"The application of Bayesian methods makes sense in special contexts. For example, a life insurance company may adopt a utility function for its total assets; subjective probabilities may be based on actuarial tables. However, a general use of Bayesian methods meets serious difficulties. Subjective probabilities and utilities are needed as inputs. Usually these inputs are not readily available" [Selten, 1991a, 19].

[35]"We have to gain empirical knowledge. We cannot derive human economic behavior from biological principles" [Selten, 1991a, 9].

of the use of *ad hoc* assumptions, insofar as they are empirically supported.[36] He maintains that successful explanations of experimental phenomena should be built up along the primacy of empirical knowledge. That knowledge reveals diversity: "experiments show that human behavior is *ad hoc.* Different principles are applied to different decision tasks. Case distinctions determine which principles are used where" [Selten, 1991a, 19]. Moreover, against the dominant position in favor of full rationality, he affirms that the "attempts to save the rationalistic view of economic man by minor modifications have no chance of succeeding" [Selten, 1993, 135].

Methodologically, Selten seems to be sympathetic towards the research on induction[37]. His approach to experimental economics tends to identify some empirical regularities based on experimental data and thereafter to construe a formal theory to explain them, instead of beginning with a formal theory which is submitted to test in the laboratory. This kind of methodological approach is different from other *methodological possibilities* frequent among *experimental economists*, of which there are basically three: 1) experiments designed for testing and modifying formal economic theories; 2) experiments designed to collect data on interesting phenomena and relevant institutions, in the hope of detecting unanticipated regularities; and 3) experiments associated with having a direct impact in the realm of policy-making, cf. [Roth, 1986, 245–246].

The *fourth possibility* on experiments, which can be found in Selten's papers, stems from the dissatisfaction of a present theory in the light of the data and the need for an *alternative theory* based directly on observed behavior. The experimental results are used to identify some empirical regularities (and there is a similitude here with the second possibility). This evidence may suggest theoretical considerations which eventually may lead to the construction of a formal theory (and this is a difference in comparison with the second possibility). This theory is ordinarily of a *limited range*, because experimental results usually support only theories of limited range, whereas an empirical-based general theory of bounded rationality appears as a task for the future, cf. [Selten, 1998a, 414].

This kind of methodological approach, which can be seen in Selten's theory of equal division payoff bounds,[38] is different from the methodological case of theory-oriented experiments insofar as the starting point is different. In one case — the first methodological view — the research starts with a body of formal theory and then proceeds to develop a set of experiments which allow some conclusions to be drawn about the theory, whereas in the other case — the fourth methodological view — the research starts with a body of data from experimental games, which leads to a theory [Roth, 1986, 266–267]. The theory can take "the form of a hypothetical reasoning process which looks at the players in order of their strength" [Selten, 1982/1988, 301].[39]

[36] "It is better to make many empirically supported ad hoc assumptions, than to rely on a few unrealistic principles of great generality and elegance" [Selten, 1991a, 19].

[37] Cf. [Selten, 1990, 656]. He is specially interested in the book Holland *et al.* [1986].

[38] Cf. [Selten, 1982/1988; Selten, 1987, 42–98, especially 64–80].

[39] According to Selten, "typically, game-theoretic solution concepts are based on definitions

Underlying Selten's methodological approach on experimental economics is a rejection of key methodological views of mainstream economics: "the success of the theory of equal division payoff bounds confirms the methodological point of view that the limited rationality of human decision behavior must be taken seriously. It is futile to insist on explanations in terms of subjectively expected utility maximization. The optimization approach fails to do justice to the structure of human decision processes" [Selten, 1987, 95]. This methodological recognition of the need for a bounded rationality approach to human decision making in the economic activity seems to me very relevant.

Nevertheless, we still have certain methodological problems in experimental economics, mainly in the sphere of *methodological limitations*: how much of what is obtained in the economic laboratory can be applied directly to the complex situation of economic activity within the real world? It is not a minor problem, because — as Selten himself recognizes — "also field data are important, but they are more difficult to obtain and harder to interpret" [Selten, 1998a, 414].[40] This aspect can have repercussions in two ways: on the one hand, in the characterization of *economic activity* as such (i.e., in giving the real features of human decision making in ordinary circumstances, instead of in an artificial environment); and, on the other hand, in the analysis of economic activity as *interconnected* with other human activities in a changeable historical setting, because economic activity is *de facto* connected with other human activities, and in a context which is also historical.[41]

3.3 Theory of Equal Division Pay-Off Bounds

Frequently, Selten insists on his theory of equal division pay-off bounds. The original insight was that "equal shares of coalition values have a great significance for the thinking of the players" [Selten, 1993, 120]. He had the idea that players might tend to form a coalition with maximal equal share. In addition, the agreed upon pay offs would be determined by levels of aspiration derived from maximal equal shares of alternative coalitions. He was successful in the prediction of classroom experiments on a specific seven person game, and he tried to generalize his theory to all superadditive characteristic function games.

After studying a great number of plays, Selten saw that "equal share analysis was a better explanation of the data than its alternatives proposed by normative game theory, but it was not really satisfactory. As more data became available,

that describe the proposed solution by inner properties... The theory of equal division payoff bounds has a different character. The payoff bounds are not characterized by inner properties. They are constructively obtained by straightforward commonsense arguments based on easily recognizable features of the strategic situation" [Selten, 1987, 78].

[40]Some experimental economists are highly cautious regarding their work: "we do not go to the basement (laboratory) with the idea of reproducing the world, or a major part of it; that is better done (and it is hoped will be done) through 'field' observation. We go to the laboratory to study, under *relatively* controlled conditions, our *representations* of the world — most particularly our representations of *markets*" [Smith *et al.*, 1991, 197].

[41]On the distinction *economic activity — economics as activity*, cf. [Gonzalez, 1994].

I [Selten] developed a new descriptive theory for zero-normalized supperadditive three-person games in characteristic function form. This theory, called 'equal division pay-off bounds' [1983; 1987],[42] derives lower bounds for the players' aspiration levels based on simple computations involving various equal shares. The improved version of this theory [1987], in particular, has had a remarkable predictive success" [Selten, 1993, 120].

The theory equal division pay-off bounds was thought of as *descriptive* in its character and *procedural* (i.e., it specifies the way in which the solution is determined). Selten considers that it is a theory "which fits the data much better than the bargaining set, at least for three-person games" [1998b, 12]. It was designed as a three-person theory and only for zero-normalized games (i.e., games which have zero pay-offs for one-person coalitions). The theory "describes a boundedly rational reasoning process which arrives at lower bounds $s1$, $s2$, and $s3$ for payoffs of players 1, 2, and 3 respectively, within a two-person coalition. These numbers are called *equal division payoffs bounds*" [Selten, 1998a, 420].

According to Selten's approach, this "theorizing was no longer based on the idea of full rationality, but rather on that of bounded rationality" [Selten, 1998b, 13]. It is also a theory which wants to be empirically based: "the reasoning process starts with the observation that player 1 has better coalition possibilites than player 2, and player 2 has better coalition possibilities than player 3. We express this by saying that the order of strenght is 1>2>3," [Selten, 1998a, 420] where > expresses the sense of "stronger".

Following the rationale of the game, there is a principle: "the stronger member in a two-person coalition should at least get his or her equal share of the coalition value. Player 1 is stronger in 12 and player 2 is stronger in 23. This leads to the lower bounds $s1$ and $s2$ for players 1 and 2, respectively, $s1 = a/2$, $s2 = c/2$. From these lower bounds and upper bounds $h1$ and $h2$ for the payoffs of 1 and 2, respectively, in 12 are derived: $h1 = a - s2$, and $h2 = a - s1$" [Selten, 1998a, 420].[43]

Player 3 is in a difficult situation insofar as "the coalition 12 with the highest equal share is the most attractive one. Moreover, in no two-person coalition player 3 is the stronger number. Therefore, for player 3, a lower bound cannot be derived in the same way as for players 1 and 2. In order to have a chance to be in the final coalition, player 3 may have to be willing to give both players 1 and 2 the upper bounds $h1$ and $h2$ they can obtain in 12. This leaves the minimum of $b - h1$ and $c - h2$ for player 3. However, player 3 also must get at least zero. This leads to the lower bound $s3 = \max\ [0, \min(b - h1, c - h2)]$. We call $s3$ player 3's competitve bound" [Selten, 1998a, 420].

Concerning the issue of *prediction*, Selten's theory of equal division pay-off bounds "predicts that a two-person coalition ij will be formed in which both members receive at least their equal division payoff bounds si and sj, respectively"

[42]Cf. [Selten, 1982/1988; Selten, 1987, especially, 64–80].

[43]The simple case of the fully asymmetric three-person quota game without a grand coalition and with zero payoffs for one-person coalition is illustrated in [Selten, 1998b, 13].

[Selten, 1998a, 420–421]. This kind of descriptive theory is what he considers "casuistic" in the sense "that many case distinctions based on simple criteria are made; simple principles are applied in every single case. Casuistic procedural structures seem to be more adequate for the description of boundedly rational coalition formation than solution concepts based on abstract general principles" [Selten, 1993, 120-121].

4 ROLE OF PREDICTION IN EXPERIMENTAL ECONOMICS: THE INFLUENCE OF GAME THEORY

Prediction is always a key issue in experimental economics as well as in Selten's writings. In this regard, his approach stresses two methodological aspects: a) the idea of prediction as a *significant test* for a theory; and b) the need for a *method for comparing* the predictive success of different theories, cf. [Selten and Krischker, 1982]. The focus is often on designing new experiments to test the predictive value of a theory rather than on merely construing a new theory for describing the data observed. In addition, he proposes statistical tests to compare a new theory with the previously existing ones in order to show the superior predictive power for these experiments.

Reinhard Selten often shows special interest in what he calls "area theory" (such as the theory of equal division payoff bounds), a kind of theory within a specific range of variables which can be checked in order to know whether the prediction is correct or not. This methodological approach improves the revision of the theory to fit the data, taking into account that — for him — there are variations from one case to another: "different theories often aim at different types of predictions" [Selten, 1987, 43].

The *area theory* predicts a range of outcomes, whereas other kinds of economic theories predict only average outcomes, or are even less specific. The methodological advantage of area theory is then clear: for every single playing of the game, one can check the correctness of the prediction. This advantage is useful for improving theories in the light of experimental data. When prediction fails, it is possible to identify what went wrong. Therefore, he accepts a combination of *fallibilism* and *self-correctness* of economics in the methodological use of prediction within experimental economics.

4.1 Prediction as a Significant Test and as a Method for Comparing Theories

Besides the role of prediction as a *significant test* for a theory, which gives the data a crucial role in the revision of the theory, there is, in Selten, a particular emphasis on the need for a *method for comparing* the predictive success of different theories[44]. He proposes a method to solve the problem of different area theories predicting regions of diverse sizes. "A measure of predictive success is defined that is based on the relative frequency of correct predictions and a correction for the size of the predicted region" [Selten, 1987, 44]. It is a distinction reflected in the duality "hit rate"-"area", which seeks to show that the predictive success of a specific area theory (e.g., the theory of equal division payoff bounds) is superior to other theories (such as the theory of united bargaining set). This comparative methodology, based on prediction, seems open to the idea of an *improvement* in the science — in this case economics — which is *objective*, and not merely subjective or intersubjective.

Selten's study of economic prediction is normally guided by game theory. As he has pointed out, game theory is no longer a speciality in economics but rather a common tool of economic theory, cf. [Selten, 1993, 135]. But, even though he has improved game theory throughout his career, he himself recognizes methodological limitations to game theory: "strictly speaking only finite games can be played in the laboratory" [Selten, 1993, 130]. Thus, the models of infinite games cannot be tested as such in the economic laboratory. These kinds of methodological constraints are relevant because Selten has a constant determination to connect game theory with experimental economics — *de facto* two crucial aspects of his career.

Initially, Selten's work on game theory was based on the Nash equilibrium, which belongs to the core of mainstream game theory. Harsanyi has summarized that influence: "one of Reinhard's important contributions was his distinction between *perfect* and *imperfect* Nash equilibria. It was based on his realization that even strategy combinations fully satisfying Nash's definition of Nash equilibria might very well contain some *irrational* strategies. To exclude such imperfect Nash equilibria containing such irrational strategies, at first he [Selten] proposed what now are called *subgame-perfect* equilibria. Later he proposed the even more demanding concept of *trembling-hand* perfect equilibria. Reinhard's work on *evolutionary stable strategies* was likewise based on the concept of Nash equilibria" [Harsanyi, 1996, 160].

These are important contributions to game theory. The question of how to interpret them from a methodological point of view gives rise to some difficulties, because Selten is usually working with a limited rationality instead of a full rationality or strong assumptions of rationality (such as in the case of mainstream game theory). His intellectual attitude is clear when he analyzes his own contributions: "I do not believe in the descriptive relevance of strong rationality assumptions, I prefer to think of game equilibrium in empirically oriented models as the result of adaptive dynamic processes" [Selten, 1993, 127].[45] In addition, he makes critical

[44]This will lead to him to the "accuracy"-"precision" distinction, which is developed in the next section.

[45]For Simon, "the chief contribution of formal game theory to our understanding of rationality has been to demonstrate rather convincingly (if not mathematically) that there is no satisfactory definition of 'optimal' rationality in the presence of opportunities for outguessing and outwitting" [Simon, 2000, 28].

comments on claims of mainstream game theory, for example in the case of quotas, when he is analyzing strategic reasoning: "this kind of circularity is typical for rational game theory. However, subjects in the laboratory usually do not compute quotas. They seem to avoid circular concepts in their strategic reasoning" [Selten, 1998a, 420].

Game theory is interpreted by Selten in terms of *methodological dualism*: normative game theory is very different from descriptive game theory. *Normative* game theory tries to mold a balanced mathematical structure of ideal rationality out of conflicting inherent inclinations of the human mind. Thus, for him, "the problem of normative theory is philosophical, not empirical. Only if you are a naïve rationalist can you think it is empirical" [Selten, 1998b, 23]. *Descriptive* game theory has a different aim: "the explanation of observed *behavior* of men, animals or plants, and has nothing to do with normative game theory. The problem here is empirical and *only empirical* arguments count, nothing else. The need for this distinction arises because there is experimental evidence from human game players which refutes naïve rationalism. Naïve rationalism could have been right, but it is not, it is refuted by experimental evidence" [Selten, 1998b, 23]. Following this sharp distinction, he develops a descriptive game theory on human players and focuses it on cases of microeconomics.

Using the theoretical design of descriptive game theory, Selten conducts economic experiments conceived to *test* the predictive value of the theory which he proposes. This is the case of his theory of equal division payoff bounds, the predictive success of which in the latter version, published in 1987, was understood as a guarantee of the *scientific character* of his theory, cf. [Selten, 1993, 120]. Nevertheless, he considers that the undeniable predictive success of a theory regarding the available data does not mean that the theory has the final answer to the problems which are studied. In other words, a theory should be open to revision, and that is a task which requires experimentation: "the development of successful descriptive theories is a slow process that must be guided by experimental evidence" [Selten, 1987, 96].

4.2 A Difference Between "Prediction" and "Expectation"

What seems to me less clear in Selten's concept of *prediction* in the analysis of experimental games, such as in the case of a solidarity game, is the use of "prediction" and "expectation". Sometimes he uses them as if they were synonymous or interchangeable. Such is the interpretation which can follow from this reflection: "it is possible that someone looks at his or her own behavioral inclinations in order to predict the behavior of others. We refer to this as 'expectations based on own behavior'. On the other hand, it is also possible that someone predicts the behavior of others with the intention of making his or her own behavior dependent on the behavior of the population. We refer to this as 'behavior based on expectations'" [Selten and Ockenfels, 1998, 526].

Here it is possible to introduce some conceptual nuances. This is so not only on account of the possibility of predicting the expectations of others.[46] *Prediction* is a descriptive term: it is a cognitive content which could be related to "novel facts"[47] (in this case, to anticipate a possible behavior of others). It is understood as the expected value of something unknown, given the information available. *Expectation* can have — in my judgment — two senses: a restricted version and a broad one. In the first sense, expectation converges with the idea of prediction[48], because it is the expected value given the information available and has no relation with a subjective process; whereas in the second sense expectation requires the presence of subjective elements (and, in this case, it can include an attitude regarding what is expected).

Accordingly I think that, in the broad version, expectation is more generic than prediction and possesses a more subjective character than prediction. This seems to be recognized by the inspiring author of the "rational expectations" hypothesis: the "expectations of firms (or, more generally, the subjective probability distribution of outcomes) tend to be distributed, for the same information set, about the prediction of the theory (or the 'objective' probability distributions of outcomes)" [Muth, 1961/1981, 4–5].[49]

Therefore, in this economic context, a prediction can be an expectation of some kind, whereas not all expectation is equivalent to a prediction even though it entails a prediction of a future event. At the same time, prediction could be a statement of the future related to a specific value at a particular moment, whereas expectation could be more generic and might include a subjective factor. To sum up, although "prediction" and "expectation" refer to future phenomena, they offer subtle differences of conceptual character which have methodological repercussions.

5 THE MEASURE OF PREDICTIVE SUCCESS: ACCURACY AND PRECISION

When Selten deals with the issue of a measure of predictive success — a method to compare theories —, the focus is placed on an area theory, in which he distinguishes two aspects: i) the hit rate, and ii) the predicted area (or region which covers the dispersion of predictions). The first one is presented in the following context: "a measure of the relative size of the predictive range is subtracted from the relative

[46]This is explicitly accepted by John Muth: "it is often necessary to make sensible predictions about the way expectations would change" [Muth, 1961/1981, 4].

[47]On the different kinds of "novel facts", cf. [Gonzalez, 2001, pp. 505–508].

[48]"I should like to suggest that expectations, since they are informed predictions of future events, are essentially the same as the predictions of the relevant economic theory", [Muth, 1961/1981, 4].

[49]The idea of "rational expectations" has received criticisms from the very beginning on its conception of economic rationality. It is an economic perspective which is completely different from the position based on bounded rationality. In fact, Herbert Simon has criticized it as a new expansion of the rationality principle as optimization. He has also pointed out that the rational expectations position was soon confronted with various conflicting empirical phenomena, cf. [Simon, 2000, 29].

frequency of correct predictions. This yields the measure of predictive success. The term 'hit rate' is used for the relative frequency of correct predictions. If the outcomes are randomly distributed over the whole range of outcomes, the hit rate can be expected to be equal to the relative size of the predictive range. The measure of predictive success can be thought of as the surplus of the observed hit rate over the random hit rate" [Selten, 1987, 80].

In addition to the hit rate, the measurement of the predictive success of economic theories also requires a consideration of the predicted area. The "area theory" has a range of outcomes which takes the form of a non-empty subset of the set of all configurations. Thus, for Selten, "in order to compare the predictive success of two area theories for a body of experimental data, it is not sufficient to examine which theory yields more correct predictions. A theory may produce many correct predictions simply because it predicts a very large range. An extreme example is provided by a theory which we call the *null theory*; the predicted range of the null theory is the set of all configurations" [Selten, 1987, 80]. Therefore, where area theories are to be compared in a meaningful way, the *size of the predicted range* or *area* is another aspect to be taken into account.

Put differently, Selten maintains that "area theories for prediction of experimental results delineate regions of predicted outcomes within the set of all possible outcomes. The difference measure of predictive success for area theories ... is the difference between hit rate and area. The hit rate is the relative frequency of successful predictions and the area is the relative size of the predictive region within the set all possible outcomes" [Selten, 1991b, 153]. He seems to give more weight to getting high hit rates than to obtaining small areas[50]. But, in my judgment, what should be emphasized is that his distinction between *hit rate* and *area* leads to another interesting distinction: "accuracy" and "precision", terms which are usually presented as synonymous and here acquire a new role in clarifying the predictive success of theories.

Accuracy accompanies hit rate, which is the relative frequency of correct predictions. Moreover, "the hit rate is a measure of accuracy" [Selten, 1991b, 153]. But it could be the case that accuracy can reflect the mere fact of getting the predictive results themselves (i.e., achieving the aim of the correct predictions as such instead of the relevant subset of relative frequency). In other words, accuracy may be understood as a poor concept of predictive success in some cases: "no area theory can be more accurate than the trivial one, which simply predicts the set of all possible outcomes. This theory never fails to predict correctly, but it is useless in view of its complete lack of precision" [Selten, 1991b, 153]. Thus, it is not good enough for a theory to achieve the goal of correct predictions: it should be done in a refined manner — high hit rates — in order to be completely suitable.

Precision is related to the area (i.e., the size of the predictive region). It is the achievement regarding the predictive space: it has to be in the appropriate zone.

[50]When Selten analyzes the theory of equal division payoff bounds, he maintains that "in view of the great variance of experimental results, it seems more important to aim for hit rates than for small areas in theory construction" [Selten, 1987, 93].

There is a degree in this kind of exactness: "the precision of an area theory is related to the size of its set of predicted outcomes. The smaller the set is, the more precise is the theory" [Selten, 1991b, 153]. However, for Selten, there could be the case of a theory which is extremely precise but completely inaccurate, cf. [1991b, 161]. This happens when a theory predicts a single point which never occurs as the outcome of an experiment. Thus, the hit rate-area combination for that theory is (0,0), which converts it into a useless theory: it is extremely precise but fully inaccurate. Therefore, for area theories (i.e., specific theories with a subset of predicted outcomes), he stresses the need for accuracy and precision in order to measure the predictive success.

Following this distinction between accuracy and precision, Selten proposes this formula: $m = r - a$, where m= measure of predictive success, r=hit rate (i.e., the relative frequency of correct predictions), and a=the area (i.e., the relative size of the predicted subset compared with the set of all possible outcomes), cf. [Selten, 1991b, 154]. The comparison between two theories from the point of view of predictive success depends on gains and losses in accuracy evaluated by hit rate differences, and gains and losses in precision evaluated by area differences. A new theory T$'$ is more accurate than an older theory T$''$ when the hit rate of T$'$ is greater than the hit rate of T$''$, and the new theory is less precise than the older one when the area of T$'$ is greater than the area of T$''$, cf. [Selten, 1991b, 160].

As a methodological distinction, the difference between "accuracy" and "precision" seems an improvement in clarifying the degree of success of economic predictions, because it could be useful to distinguish between the relative frequency of successful predictions and the relative size of the predicted region within the whole space of possible outcomes. In addition, it contributes to avoiding certain extreme positions: "neither the prediction of a single outcome nor the prediction of almost all outcomes is a reasonable aim in the construction of area theories" [Selten, 1991b, 166].

Even though "accuracy"-"precision" is a useful methodological distinction in comparing theories from the point of view of their predictive success, *two limitations* should be pointed out. On the one hand, this methodological proposal is restricted to measures of predictive success depending on only *two factors* (hit rate and area). On the other hand, due to its *restricted realm*, these types of theories — area theories — are a kind of theory which has some advantages, when compared with other kinds of theories (for example, for every outcome observed it is clear in area theories if the prediction was correct or not). It seems, therefore, that further work should be done on the measure of predictive success of economic theories, in order to take into account the different cases that may arise.

Moreover, it may be the case that there is also a subjective content in Selten's distinction between "accuracy" and "precision": *accuracy* refers to a rate of successes in comparison with the total, whereas *precision* is a quotient between two quantities related to surfaces, the area of an experiment in comparison with the whole area. It may be understood as a uniform behavior for both cases (rates of success and amount of surface), and that is not clear enough.

To sum up, the role of experiments in the social sciences has been questioned or even neglected by those approaches that emphasized the methodological gap between natural sciences and social sciences. Recent contributions in philosophy of science and new developments of the social sciences, such as economics, have led to an increasing presence of experiments. There is a transition from the traditional notion to an enlarged vision. On the one hand, experiments are seen as a human activity oriented towards testing and evaluating scientific theories, both as an intervention in a material setting (i.e., laboratory experimentation) and as a creative procedure (computer simulations, thought experiments, ...); and, on the other hand, experiments have enlarged the capacity of social sciences (mainly economics) to predict. Selten's approach to experimental economics has shown methodological improvements on prediction, such as in the measurement of predictive success in terms of accuracy and precision.

ACKNOWLEDGEMENTS

I am grateful to Nicholas Rescher and Theo Kuipers for their comments on this paper.

BIBLIOGRAPHY

[Axelrod, 1980a] R. Axelrod. Effective choice in the iterated prisoner's dilemma. *Journal of Conflict Resolution*, 24: 3–25, 1980.

[Axelrod, 1980b] R. Axelrod. More effective choice in the prisoner's dilemma. *Journal of Conflict Resolution*, 24: 379–403, 1980.

[Axelrod, 1984] R. Axelrod. *The Evolution of Cooperation.* Basic Books, N. York, 1984.

[Bernoulli, 1738/1954] D. Bernoulli. Specimen theoriae novae de mensura sortis. *Comentarii Academiae Scientiarum Imperialis Petropolitanae*, v. 5, pages 175–192, 1738. English translation by L. Sommer, Exposition of a new theory on the measurement of risk, *Econometrica*, 22: 23–36, 1954.

[Boumans and Morgan, 2001] M. Boumans and M. S. Morgan. *Ceteris paribus* conditions: materiality and the application of economic theories. *Journal of Economic Methodology*, 8(1): 11–26, 2001.

[Davis and Holt, 1993] D. D. Davis and Ch. A. Holt (eds.). *Experimental Economics.* Princeton University Press, Princeton, NJ, 1993.

[Friedman and Sunder, 1994] D. Friedman and S. Sunder (eds.). *Experimental Methods: A Primer for Economists.* Cambridge University Press, Cambridge, 1994.

[Friedman, 1953/1969] M. Friedman. The methodology of positive economics. In M. Friedman, *Essays in Positive Economics.* The University of Chicago Press, Chicago, pages 3–43, 1953 (6th repr., 1969).

[Galison, 1987] P. Galison. *How Experiments End.* The University of Chicago Press, Chicago, 1987.

[Galavotti, 2003] M. C. Galavotti (ed.). *Observation and Experiment in the Natural and the Social Sciences.* Kluwer, Dordrecht, 2003.

[Gooding *et al.*, 1989] D. Gooding, T. Pinch and S. Schaffer (eds.). *The Uses of Experiment.* Cambridge University Press, Cambridge, 1989.

[Gonzalez, 1994] W. J. Gonzalez. Economic prediction and human activity. An analysis of prediction in economics from action theory. *Epistemologia*, 17: 253–294, 1994.

[Gonzalez, 2001] W. J. Gonzalez. Lakatos's approach on prediction and novel facts. *Theoria*, 16(3): 499–518, 2001.

[Gonzalez, 2003a] W. J. Gonzalez. Rationality in experimental economics: an analysis of R. Selten's approach. In M. C. Galavotti (ed.), *Observation and Experiment in the Natural and the Social Sciences*. Kluwer, Dordrecht, pages 71–83, 2003.

[Gonzalez, 2003b] W. J. Gonzalez. From *Erklären–Verstehen* to *Prediction–Understanding*: the methodological framework in economics. In M. Sintonen, P. Ylikoski, and K. Miller (eds.), *Realism in Action: Essays in the Philosophy of Social Sciences*, Kluwer, Dordrecht, pages 33–50, 2003.

[Gonzalez, 2005] W. J. Gonzalez. The philosophical approach to science, technology and society. In W. J. Gonzalez (ed.), *Science, Technology and Society: A Philosophical Perspective.* Netbiblo, A Coruña, pages 3–49, 2005.

[Haavelmo, 1944] T. Haavelmo. The probability approach in econometrics. Supplement to *Econometrica*, 12: 1–118, 1944.

[Hacking, 1983] I. Hacking. *Representing and Intervening.* Cambridge University Press, Cambridge, 1983.

[Hacking, 1988] I. Hacking. Philosophers of experiment. *Proceedings of the Philosophy of Science Association*, 2: 147–156, 1988.

[Harsanyi, 1996] J. C. Harsanyi. The work of John Nash in game theory. *Journal of Economic Theory*, 69: 158–161, 1996.

[Hendry, 2000] D. F. Hendry. *Econometrics: Alchemy or Science? Essays in Econometric Methodology.* New edition, Oxford University Press, Oxford, 2000 (1st edition, 1993).

[Hey and Loomes, 1993] J. D. Hey and G. Loomes (eds.). *Recent Developments in Experimental Economics.* E. Elgar, Aldershot, vol. I and II, 1993.

[Holland *et al.*, 1986] J. H. Holland, K. J. Holyoak, R. E. Nisbett, and P. R. Thagard. *Induction: Processes of Inference, Learning, and Discovery.* The MIT Press, Cambridge, MA, 1986.

[Kahneman *et al.*, 1990] D. Kahneman, J. L. Knetsch, and R. H. Thaler. Experimental tests of the endowment effect and the Coase theorem. *Journal of Political Economy*, 98(6): 1325–1348, 1990. Reprinted in J. D. Hey and G. Loomes (eds.), *Recent Developments in Experimental Economics.* E. Elgar, Aldershot, vol. I, pages 206–229, 1993.

[Keller, 2003] E. F. Keller. Models, simulation, and 'computer experiments'. In H. Radder (ed.), *The Philosophy of Scientific Experimentation.* University of Pittsburgh Press, Pittsburgh, pages 198–215, 2003.

[Kitcher, 2001] Ph. Kitcher. *Science, Truth, and Democracy.* Oxford University Press, Oxford, 2001.

[Knez and Smith, 1987] M. Knez and V. L. Smith. Hypothetical valuations and preference reversals in the context of asset trading. In A. E. Roth (ed.), *Laboratory Experimentation in Economics — Six Points of View.* Cambridge University Press, Cambridge, pages 131–154, 1987.

[Lawson, 1997] T. Lawson. *Economics and Reality.* Routledge, London, 1997.

[Lennox, forthcoming] J. G. Lennox. Thought experiments in evolutionary biology today. In W. J. Gonzalez (ed.), *Evolutionism: Present Approaches.* Netbiblo, A Coruña, forthcoming.

[Morgan, 1990] M. S. Morgan. *The History of Econometric Ideas.* Cambridge University Press, Cambridge, 1990.

[Morgan, 2001] M. S. Morgan. Models, stories and the economic world. *Journal of Economic Methodology*, 8: 361–384, 2001.

[Morgan, 2003] M. S. Morgan. Experiments without material intervention. Models experiments, virtual experiments, and vitually experiments. In H. Radder (ed.), *The Philosophy of Scientific Experimentation.* University of Pittsburgh Press, Pittsburgh, pages 216–235, 2003.

[Morgan and Morrison, 1999] M. S. Morgan and M. Morrison (eds). *Models as Mediators.* Cambridge University Press, Cambridge, 1999.

[Musgrave, 1981] A. Musgrave. 'Unreal assumptions' in economic theory: the F-twist untwisted. *Kyklos*, 34(3): 377–387, 1981.

[Muth, 1961/1981] J. F. Muth. Rational expectations and the theory of price movements. *Econometrica*, 29: 315–335, 1961. Reprinted in R. E. Lucas and Th. J. Sargent (eds.), *Rational Expectations and Econometric Practice.* G. Allen and Unwin, London, pages 3–22, 1981.

[Nash, 1996] J. Nash. The work of John Nash in Game Theory. *Journal of Economic Theory*, 69: 182–183, 1996.

[Niiniluoto, 1993] I. Niiniluoto. The aim and structure of applied research. *Erkenntnis*, 38: 1–21, 1993.

[Plott, 1987] Ch. R. Plott. Some policy applications of experimental methods. In A. E. Roth (ed.), *Laboratory Experimentation in Economics — Six Points of View.* Cambridge University Press, Cambridge, pages 193–219, 1987.

[Radder, 2003a] H. Radder (ed). *The Philosophy of Scientific Experimentation.* University of Pittsburgh Press, Pittsburgh, 2003.

[Radder, 2003b] H. Radder. Towards a More Developed Philosophy of Scientific Experimentation. In H. Radder (ed), *The Philosophy of Scientific Experimentation*, University of Pittsburgh Press, Pittsburgh, pages 1–18, 2003.

[Roth, 1986] A. Roth. Laboratory experimentation in economics. *Economics and Philosophy*, 2:245–273, 1986.

[Roth, 1987] A. Roth. Laboratory experimentation in economics, and its relation to economic theory. In N. Rescher (ed.), *Scientific Inquiry in Philosophical Perspective.* University Press of America, Lanham, pages 147–167, 1987.

[Roth, 1988/1993] A. Roth. Laboratory experimentation in economics: a methodological overview. *The Economic Journal*, 98: 974–1031, 1988; reprinted in J. D. Hey and G. Loomes (eds.), *Recent Developments in Experimental Economics.* Vol. I, E. Elgar, Aldershot, pages 3–60, 1993.

[Roth, 1995] A. Roth. Introduction to experimental economics. In J. H. Kagel and A. Roth (eds.), *Handbook of Experimental Economics.* Princeton University Press, Princeton, NJ, pages 3–109, 1995.

[Samuelson and Nordhaus, 1983] P. Samuelson and W. D. Nordhaus. *Economics.* 12th edition, McGraw-Hill, N. York, 1983.

[Sauermann and Selten, 1959/1967] H. Sauermann and R. Selten. Ein Oligopolexperiment. *Zeitschrift für die gesamte Staatswissenschaft*, 115: 427–471, 1959. Reprinted in H. Sauermann (ed.), *Beiträge zur experimentellen Wirtschaftsforschung.* J. C. B. Mohr (Paul Siebeck), Tübingen, pages 9–59, 1967.

[Schram, 2005] A. Schram. Artificiality: the tension between internal and external validity in economic experiments. *Journal of Economic Methodology*, 12(2): 225–237, 2005.

[Selten, 1982/1988] R. Selten. Equal division payoff bounds for three — person characteristic function experiments. In R. Tietz (ed.), *Aspiration Levels in Bargaining and Economic Decision Making.* Springer, Berlin, pages 255–275, 1982. Reprinted in R. Selten, *Models of Strategic Rationality.* Kluwer, Dordrecht, pages 301–311, 1988.

[Selten, 1987] R. Selten. Equity and coalition bargaining in experimental three — person games. In A. E. Roth (ed.), *Laboratory Experimentation in Economics — Six Points of View.* Cambridge University Press, Cambridge, pages 42–98, 1987.

[Selten, 1990] R. Selten. Bounded rationality. *Journal of Institutional and Theoretical Economics*, 146(4): 649–658, 1990.

[Selten, 1991a] R. Selten. Evolution, learning, and economic behavior. 1989 Nancy Schwartz memorial lecture. *Games and Economic Behavior*, 3(1): 3–24, 1991.

[Selten, 1991b] R. Selten. Properties of a measure of predictive success. *Mathematical Social Sciences*, 21(2): 153–167, 1991.

[Selten, 1993] R. Selten. In search of a better understanding of economic behavior. In A. Heertje (ed.), *Makers of Modern Economics.* Harverstern Wheatsheaf, London, pages 115–139, 1993.

[Selten, 1998a] R. Selten. Features of experimentally observed bounded rationality. *European Economic Review*, 42: 413–436, 1998.

[Selten, 1998b] R. Selten. Game theory, experience, rationality. In W. Leinfellner and E. Köhler (eds.), *Game Theory, Experience, Rationality.* Kluwer, Dordrecht, pages 9–34, 1998.

[Selten, 2001] R. Selten. What is bounded rationality? In G. Gigerenzer and R. Selten (eds.), *Bounded Rationality: The Adaptive Toolbox.* The MIT Press, Cambridge, MA, pages 13–36, 2001.

[Selten, 2003] R. Selten. Emergence and future of experimental economics. In M. C. Galavotti (ed.), *Observation and Experiment in the Natural and the Social Sciences.* Kluwer, Dordrecht, pages 63–70, 2003.

[Selten and Krischker, 1982] R. Selten and W. Krischker. Comparison of two theories for characteristic function experiments. In R. Tietz (ed.), *Aspiration Levels in Bargaining and Economic Decision Making.* Springer, Berlin, pages 259–264, 1982.

[Selten and Ockenfels, 1998] R. Selten and A. Ockenfels. An Experimental Solidarity Game. *Journal of Economic Behavior and Organization*, 34(4):517—539, 1998.

[Selten and Stoecker, 1986] R. Selten and R. Stoecker. End behavior in sequences of finite prisoner's dilemma supergames: a learning theory approach. *Journal of Economic Behavior and Organization*, 7(1): 47–70, 1986.

[Simon, 1996] H. A. Simon. *The Sciences of the Artificial.* 3rd ed., The MIT Press, Cambridge (MA), 1996 (1st ed., 1969; 2nd ed., 1981).

[Simon, 2000] H. A. Simon. Bounded rationality in social science: today and tomorrow. *Mind and Society*, 1(1): 25–39, 2000.

[Smith *et al.*, 1991] V. L. Smith, K. A.McCabe, and S. J. Rassenti. Lakatos and experimental economics. In N. de Marchi and M. Blaug (eds.), *Appraising Economic Theories.* E. Elgar, Aldershot, pages 197–227, 1991.

[Tversky *et al.*, 1990/1993] A. Tversky, P. Slovic, and D. Kahneman. The causes of preference reversal. *American Economic Review*, 80: 204–217, 1990. Reprinted in J. D. Hey and G. Loomes (eds.), *Recent Developments in Experimental Economics.* E. Elgar, Aldershot, vol. I, pages 147–160, 1993.

[Wold, 1969] H. O. A. Wold. Econometrics as pioneering in non-experimental model building. *Econometrica*, 37: 369–381, 1969.

[Woodward, 2003] J. Woodward. Experimentation, causal inference, and instrumental realism. In H. Radder (ed.), *The Philosophy of Scientific Experimentation.* University of Pittsburgh Press, Pittsburgh, pages 87–118, 2003.

[Worrall, 1989] J. Worrall. Fresnel, poisson and the white spot: the role of successful predictions in the acceptance of scientific theories. In D. Gooding, T. Pinch, and S. Schaffer (eds.), *The Uses of Experiment.* Cambridge University Press, Cambridge, pages 135–157, 1989.

ONTOLOGICAL, EPISTEMOLOGICAL, AND METHODOLOGICAL POSITIONS

James Ladyman

INTRODUCTION

This chapter summarises various important ontological, epistemological and methodological issues in the philosophy of science. Ontology is the theory of what exists and is the foremost concern of metaphysics, which is the study of the most fundamental questions about being and the nature of reality. Ontological issues in the philosophy of science may be specific to a particular special science, such as questions about the ontological status of biological species, or they may be more general, such as whether or not there are objective natural kinds. In the history of science ontological issues have often been of supreme importance; for example, whether or not atoms exist was a question that occupied many scientists in the nineteenth century. In what follows, some fundamental questions of ontology are discussed, some of which, such as those concerning laws of nature, are also addressed in analytic metaphysics. A number of these issues also relate to debates in the foundations of physics about the ontological implications of our best physical theories. Readers who wish to know more about the philosophy of space and time and the nature of matter and motion should consult the companion volume to this on the philosophy of physics.

Epistemology is the theory of knowledge and as such is concerned with such matters as the analysis of knowledge and its relationship to belief and truth, the theory of justification, and how to respond to the challenge of local scepticism, say about the past, or global scepticism, which suggests that there is no knowledge. The particular epistemological problems raised by science mostly concern inductive inference, since it is widely accepted that substantive knowledge of the world cannot be obtained by deduction alone. The most fundamental such problem is to explicate the relationship between theory and evidence. There are also epistemological issues that only arise when we reflect on the status of unobservable entities posited by our best scientific theories.

Finally, methodology here means the theory of the scientific method. Is there a single such method for all the sciences, and if so what is it? How much should we expect the theory of the scientific method to help with the progress of science? Is the scientific method fixed, or does it change over time?

Handbook of the Philosophy of Science: General Philosophy of Science - Focal Issues, pp. 303–376.
Volume editor: Theo Kuipers. General editors: Dov M. Gabbay, Paul Thagard and John Woods.

There is obviously a good deal of overlap between the areas discussed in this chapter. For example, accounts of scientific methodology may have implications for the epistemology of science, and vice versa, and the epistemological issues of whether we ought to be scientific realists has a lot of bearing on whether science can help us address ontological issues. There are vast literatures on all these issues and I have offered some advice about further reading that should help the reader find their way into the subject.

1 ONTOLOGICAL POSITIONS

(I) Natural Kinds

We divide the world up into individuals and then we class many individuals together in kinds. Hence, we distinguish between horses and donkeys, gold and silver, apples and pears and so on. One of the main concerns of science is to systematically classify natural phenomena and substances into kinds. A system for dividing things into kinds is called a taxonomy. The progress of chemistry, biology, physics and all the sciences is in part the history of a series of re-classifications, and refinements and reinterpretations of existing classifications. For example, in chemistry the kind acid has evolved from the rough notion of a liquid that would react with a base, such as a metal, and produce a salt and water, to the heavily theoretical idea of a chemical that can donate a hydrogen ion; in biology the living world is no longer divided into plants and animals, rather there is a complex taxonomy of phyla, kingdoms, groups, families, genii, and species; and in physics there have been several taxonomic revolutions from the Newtonian idea of a world of corpuscles and forces, through the late nineteenth century heyday of fields, and into the taxonomy of the four fundamental forces and their associated quantum fields and particles. Nonetheless, there is usually a large degree of retention of taxonomic structure between successive scientific theories. So for example, contemporary chemistry still classifies as acids the most important acids known in the age of alchemy.

A fundamental ontological question is whether taxonomy is a matter or discovery or invention, or, in other words, whether there are any objective natural kinds. It is often taken for granted that there are in the case of the natural sciences, but there has always been controversy about whether kinds picked out by the human sciences are objective, or whether they reflect the values of particular societies to which the science and scientists in question belong. For example, there is a considerable amount of contemporary debate about whether psychiatric classifications are objective. The idea of objective natural kinds in biology was also cast in doubt by the development of the theory of evolution by natural selection. Species came to be seen as historically contingent and defined by ancestry relationships rather than by morphology. Even when it comes to physics, there are those who deny that the natural kind distinctions it makes are objective rather than pragmatic or socially determined (see the section on Truth below).

In the history of philosophy, questions about natural kinds are closely related to the issue of essentialism because it is often thought that all the members of a kind possess some properties necessarily, and that these properties are characteristic of the kind in question. The idea is that some of the properties of some particular object, are properties that it couldn't lack without being a different kind of thing. For example, a piece of copper could be a different shape, but it could not fail to be a good conductor of electricity. The modern debate about natural kinds begins with John Locke's critique of Aristotelian essentialism and the former's distinction between real and nominal essences. Take gold: the nominal essence is the collection of ideas associated with the word gold, such as those of metallic, yellow, malleable and so on. Locke argued that whatever was the real essence, and he thought it would be some kind of characteristic microstructure, could not be known with any degree of certainty. Contemporary science has restored the faith of many philosophers in natural kinds and essences because our understanding of the chemical elements in terms of the periodic table seems to give us knowledge of the microstructure responsible for their sensible properties? Gold is characterised precisely as that atom that has 79 protons in its nucleus (an atomic number of 79). It is not only elements but also compounds that are often thought to have essences revealed by science. For example, it is widely said among philosophers that the essence of water is H_2O. Clearly, if contemporary science gives us knowledge of the essences of natural kinds it does so by empirical rather than a priori means.

W. v. O. Quine argued that appeal to the notion of natural kinds can dissolve two famous paradoxes of confirmation:

(i) Carl Hempel's raven paradox: Intuitively a generalisation like 'all ravens are black' is confirmed by observation of its instances, in other words, by the observation of black ravens. Yet this generalisation is logically equivalent to 'all non-black things are non-ravens' which is confirmed by observation of, say, a green leaf. But how can observing a green leaf confirm 'all ravens are black'?

(ii) Nelson Goodman's grue problem: Suppose that observation of the instances of green emeralds confirms the generalisation 'all emeralds are green'; they are also instances of the generalisation 'all emeralds are grue', where 'grue' means 'green before 2030 and blue afterwards'. Why do we take the one generalisation to be confirmed and not the other?

It seems that there are some predicates, such as 'green', that we are prepared to use to make projections about unobserved objects and others, such as 'grue', that are not projectible. We make judgements of similarity in terms of projectible predicates. Goodman argued that projectible predicates are those that are entrenched in our epistemic community, and denied that similarity judgements among objects are objective.

Quine argued that appeal to natural kinds allows Goodman's problem to be avoided, because not all predicates are projectible and those that are will be those that are true of all and only the things of a kind. Similarly, the Ravens paradox is avoided if predicates like 'is non-black' are ruled out of consideration because they do not refer to natural properties. Clearly the notions of kind and similarity are closely related. The question of whether there are natural kinds is the question of whether there are objective similarities between things, which is the question of whether there are some properties that are natural and in virtue of which similarity and kind membership is definable. But Quine is troubled by this since he thinks that the notion of a kind and the related notion of similarity are of dubious scientific standing. Quine would like to be able to do away with any metaphysical commitments other than to sets of concrete individual things. However, natural properties seem to be intensional while sets are extensional. This is because sets that have all the same members are the same set, but all and only the same things may instantiate two nevertheless distinct properties. There seem to be sets corresponding to the kinds that there are (the set of all the things that belong to the kind), but not all sets correspond to natural kinds.

We use judgements of similarity to learn language not least because we must learn how to decide which similarities in sounds are relevant to meaning. We seem to have innate rankings of similarity. Psychological tests show that people will class a red circle as more similar to a pink ellipse than a blue triangle, even though both red and blue are primary colours but pink is not. The reason for this is the subject matter of evolutionary psychology and allied sciences according to Quine. In order to learn a language we have to map our judgements of similarity onto those of our neighbours, and Quine says, induction itself is essentially only animal expectation or habit formation (he is close to Hume in this and other respects), only now we have to match our judgements of similarity to the world in order to be successful in generalising about the behaviour of things, and we have evolved more and more sophisticated forms of pattern recognition and accuracy as a means of survival.

The notion of kind relates to those of disposition, counterfactual conditionals and causation. We assert counterfactual conditionals like 'if that had been put in water it would have dissolved' when the thing in question is of the same kind as things that did or will behave that way. Quine claims that the notion of kind is what links general and singular causal claims: the singular causal claim can be made because we recognise the events as being of the same kind as those that feature in general causal claims. However, Quine thinks of the notions of causation, disposition, counterfactual conditional and kinds as scientifically disreputable. He is against any form of intensionality, de re modality, natural properties or objective similarity relations, and he thinks that the notions of kinds and property must be accounted for without them. He argues that it is a mark of maturity of a branch of science that the notion of similarity or of a kind is eliminable in favour of the structural properties that give rise to them: there is no need to talk about solubility since we can just talk about the relevant properties of the atomic lattice; there is no need to talk about gold when we can just talk about atoms of atomic number 79. Hence, we progress from our crude 'animal' sense of similarity to a more sophisticated and fine-grained set of similarity relations refined by scientific experimentation. The fact that we can ultimately eliminate talk of kinds and dispositions means that such talk is exonerated. We need only invoke the basic properties of physics and chemistry. Quine then holds an extensional view of these properties according to which they are simply the sets of things that possess them. Whether that view is defensible is another question.

(II) Truth and Mind-independence

Science is widely regarded as our most reliable source of true beliefs about the world. Philosophers quickly disagree about the nature of truth however, and differing views about it inform positions in epistemology and methodology. There is even controversy about the right account of 'truth-bearers', the entities that are true or false. Some philosophers posit propositions as truth-bearers, where propositions are abstract entities that are expressed by sentences, whereas others assert that the only truth-bearers are particular utterances or written sentences. In what follows, the term 'proposition' will be used to mean simply anything that can be true or false.

It is often said that the difference between the truth of a scientific theory, and the 'truth' of a piece of music, is that if the latter is any kind of truth at all, it is truth about the subjective world of emotions, whereas scientific truth concerns matters of objective fact. Objectivity may be taken to be equivalent to mind-independence in the sense that, in general, the objective facts about the world are what they are independently of whatever people happen to believe or desire.

Of course, for lay people looking to science to tell them important truths about the world, there is often no practical difference between the beliefs that are counted as truths by the epistemic authorities at a given time, and the genuine truths. Some sceptically inclined philosophers have therefore suggested that truth is nothing more than a certain kind of legitimacy that is bestowed on beliefs by those with the power to do so. On this view, truth is a social construction in the sense that social processes determine which beliefs are true and which are false. This 'social constructivism' about scientific knowledge suggests that, for example, the Special Theory of Relativity is true because those in the scientific establishment who advocated it overcame the opposition from those who denied it. Hence, social constructivists think that the order of explanation in the history of science goes from social processes to theoretical and experimental facts and not the other way around. They identify what is true with what is believed to be true (by those with the epistemic power), much as Euthythro identified the pious with what is loved by the gods in the famous Platonic dialogue. Socrates disputes the latter identification as follows:

> But if the god-beloved and the pious were the same, my dear Euthythro, and the pious were loved because it was pious, then the god-beloved would be loved because it was god-beloved, and if the god-beloved was god-beloved because it was loved by the gods, the pious

> would also be pious because it was loved by the gods; but now you see that they are in opposite cases as being altogether different from each other; the one is of a nature to be loved because it is loved, the other is loved because it is of a nature to be loved. [1981, 16]

Many philosophers take the contrast drawn by Socrates between the pious and the god-beloved to be analogous to the contrast between the true and what is believed to be true by the scientific establishment. A proposition, such as that the Earth is much more than six thousand years old, is believed and legitimated by the scientific establishment, because it is true.

The view of truth that initially seems appropriate to capture the idea of scientific truth being about facts that are independent of whatever scientists believe is called the 'correspondence theory of truth'. According to it, for a proposition to be true is for it to correspond to the facts. Unfortunately, defenders of the correspondence theory of truth have run into all manner of difficulties, most of which have to do with the nature of the correspondence relation, and with the nature of the ontology that must be posited as the entities to which true propositions correspond. Some philosophers who start out naturally sympathetic to the idea of truth as correspondence become disillusioned with metaphysical theories developed to explain correspondence and which posit facts or states of affairs as existent entities over and above objects and properties.

In practice it seems that whenever we are asked what the evidence for some proposition is, we can do nothing but assert one or more further propositions. Despairing of the project of explaining truth as the relationship between propositions and reality, many philosophers have been inclined to locate truth entirely in the realm of propositions and to treat as a relation, namely coherence, among them. According to coherentism about truth, a proposition is true just in case it coheres with other propositions in a system, and false otherwise. Truth is not a relation of correspondence between beliefs and reality but an internal relation of coherence among a set of beliefs. Many philosophers who have argued for coherentism have been motivated by holist considerations. If for one reason or another it is held that individual beliefs cannot be directly compared to the world, then the only test available for the truth of a belief is whether it coheres with the rest of a system of beliefs. Another source of motivation for coherentism is the fact that whenever we seek to describe the structure of the facts of the world, we always rely upon the structure of our thoughts and sentences. It may be argued that our judgements never confront the world directly but instead further judgement, beliefs and statements. Kant argued that noumenal reality (Ding an sich or what William James later called 'trans-empirical reality') is not accessible by human intuition. If we can never describe the mind-independent world then correspondence between thought and noumenal reality is not possible, and we have to explicate truth in a way which does not make reference to it. The coherence theory of truth is also implicit in some of the writings of rationalists like Leibniz and Spinoza. Other coherentists about truth include those who think the world is purely mental or spiritual in nature (idealists) such as Hegel and Bradley, and some logical positivists such as Neurath and Hempel.

The definition of coherence above seems much too weak since there are surely many sets of beliefs that are internally coherent but are nonetheless not true. There are consistent fictions and fairy-stories. Coherence must mean more than non-contradiction. There are various responses to this. One of the most influential is to argue that consistency must be supplemented by another relation such as explanation. On this view, a set of beliefs is true just in case they are consistent and if they are mutually explanatory. On the other hand, perhaps even consistency is too much to ask of a set of beliefs, since in practice we probably all believe some inconsistent propositions, but it does not follow that all our beliefs are false.

In the face of abstract worries of this kind many philosophers and non-philosophers alike are inclined to try to ground discussion of philosophical questions in our practical lives. One good practical reason to believe what is true is that it is generally a more successful strategy in guiding action than believing what is false. It is often said that knowledge is power and if this is true it is in part because what is known is also true. This motivates the pragmatic theory of truth which crudely put states that what is true is what it is useful to believe, or alternatively, the truth is what works. For example, William James argued that truth is what is expedient in thought, just as the right or the good is what is expedient in behaviour. According to the pragmatists like James, the meaning of a concept is given by the practical or experimental (pragmatic) consequences of its application. Both he and Charles Peirce thought that any difference in meaning must make some possible difference in practice. This is fairly close to the logical positivists verification principle according to which a proposition that cannot be empirically verified is meaningless (unless it is a tautology).

Peirce argued that truth is that set of beliefs which followers of the scientific methods of inquiry will converge upon in the long run (this is often referred to as 'the ideal endpoint of inquiry'). He thought this based on his psychology according to which beliefs are dispositions to behave, and doubts are the negative effect on such dispositions which arise from unruly experiences which subvert our theories. Doubt prompts inquiry because it induces as unpleasant state in us which we try to overcome by seeking stable beliefs. Truth is just the maximally stable state of belief. Note that Peirce thought that truth entailed correspondence with reality but he did not think it consisted in correspondence with reality. James' view was a hybrid of coherentism and pragmatism since he held that our set of beliefs is gradually adjusted to accommodate awkward experiences, and that the limit of this process is truth.

Other philosophers have advocated forms of minimalism about truth, an example of which is the redundancy theory according to which the truth predicate is redundant and so "*p*' is true.' means exactly the same thing as '*p*'. There is also a view of truth known as the identity theory of truth according to which truth-bearers and truth-makers are identical. Finally mention must be made of the famous T-schema of Alfred Tarski according to which, for any proposition '*p*', '*p*' is true if and only if *p*. For example, 'snow is white' is true if and only if snow is white. Some have argued that this exhausts what can correctly be said about truth.

(III) Properties and Universals

The problem of universals is an ancient but perennial philosophical problem about the ontological status of properties. Particulars are entities that only exist in a single instance. They include individual things, but also events. Universals on the other hand are said to be multiply instantiable; they can be multiply realised in space and time. If there are any universals they include properties and relations. Prima facie it seems obvious that the world consists of individual things, and that they have properties and there are relations among them. It is also seems obvious that several things can have the same property. The problem of universals is about whether or not properties themselves are real and hence whether talk of the same property being had by different particulars should be taken literally. Hence, the problem of universals is about the 'One over Many'.

woman, man, woman

How many words are there here? There are either three or two depending on whether we count the types or tokens. Similarly if we had two red apples and two green apples, then we would have either two colours or four instances of colour. The sets of the red and green apples each seem to have a natural unity. The problem of universals has to do with how a many can count as a one. The problem of universals is closely related to the problem of explaining what distinguishes a natural class from an artificial class.

The theory of universals solves the problem of the one over many because universals are capable of being instantiated by more than one particular. This seems to suggest an answer to the problem of similarity judgements; such judgements are correct when the individuals instantiate the same universals (and if those universals include all their essential properties they are members of the same natural kind). As well as objective resemblance, universals have been posited to account for two other important phenomena, namely predication and abstract reference and the meaning of general terms. Predication is exemplified by the following sentence: 'Socrates is a man'. The subject of the sentence is referred to by a singular term, 'Socrates', that denotes a particular thing in the world. It might be supposed that the predicate 'is a man' must also denote something in the world, namely the universal 'Man'. Predication is then analysed by saying that the particular instantiates or participates in the universal. Universals seem to be needed for the predication of relations too, such as Socrates is older than Plato. Abstract reference is when we refer directly to a property or relation, as when we say, for example, red is a colour. Here the meaning of the general terms 'red' and 'colour' is that they refer to the universals Red and Colour.

According to the Platonic theory of forms, universals transcend the reality of concrete particulars in space and time, and indeed there may be universals that are not instantiated in the actual world. On the other hand, the Aristotelian theory of forms is incompatible with the existence of uninstantiated universals, and on that view all universals are found in space and time (they are imminent). Hence the Aristotelian view is compatible with naturalism (the view that everything that exists is revealed to us, if at all, by scientific enquiry), and physicalism (the view that everything that exists is physical). Platonists may suppose that universals can be known a priori whereas Aristotelians insist that universals may only be known a posteriori.

One puzzle about universals is whether or not the universal that corresponds to a property itself has that property. For example, is the universal Man a man? If the answer is yes, and we needed universals to explain what all the men have in common, then it seems we must posit a new universal to explain what the universal Man and all the men have in common (this is called 'the third man argument'). On the other hand, if the answer is no, then it is mysterious how the manliness of men can be explained by their instantiating a universal that is not itself manly. Another puzzle about universals concerns the relation of exemplification that particulars bear to universals which seems to lead to a regress since exemplification must itself be a universal.

There is a good deal of debate among those who do believe in universals about such matters as whether they are abundant — there is a universal for more or less every predicate, or sparse — there are only universals for predicates that name natural properties. Consider, for example, disjunctive predicates, such as 'is red or square': reasons for denying the existence of disjunctive universals include that there is no common feature or common causal power of objects that satisfy such predicates. Similar considerations count against negative universals. Some argue for the special place of physics in saying what universals there are and deny that predicates that refer to non-physical properties name universals.

A state of affairs (or fact) obtains when a particular instantiates a universal. Clearly, there is more to a particular instantiating a universal than the existence of the particular and the universal. So it seems that there is more to a state of affairs than its constituents: as the latter can be 'summed' in different ways they are not merely parts to the state of affairs as whole. Must we then admit that states of affairs exist over and above particulars and universals?

David Lewis influentially defended the idea of universals in the context of his realism about concrete possible worlds, which are distinct since spatio-temporally isolated from each other. According to him, 'actual' is an indexical expression like 'here'; the former picks out the world the speaker is in just as the latter picks out the place where the speaker is. (Even philosophers who do not like Lewis believe in concrete possible worlds find it useful to employ them as models in modal reasoning.) Lewis holds that any mereological sum of any of the objects populating a world is itself an object, and any class of objects is an object. It is then possible to identify properties with classes of possibilia (the property F is the class of all the actual and possible objects that have the property), and propositions with sets of possible worlds (a proposition p is the set of possible worlds at which

p is true). Lewis conceives of universals as repeatable entities, wholly present wherever they are instantiated; hence for Lewis properties and universals differ. Lewis argues that universals may do useful work in various areas of metaphysics, epistemology and the philosophy of mind and science. Unlike properties, universals are present in every world, even those in which they are uninstantiated. There are relatively few universals but very many properties. Take any set of objects: these immediately give rise to as many properties as there are members of the power set of that set (the set of all its subsets). Any two things, actual or possible, share an infinite number of properties, because there are an infinite number of classes of which they are both members. Universals on the other hand are supposed to mark objective resemblances among things, and group things together partly according to whether they share important causal powers. Ockham's razor rules out any further universals as idle and the genuine universals are determined by our best scientific theories, and objective resemblances among particulars are primitive and unanalysable.

Another solution to the problem of universals posits entities that are neither particulars nor universals but property-instances or 'tropes'. Those who deny the existence of universals are called nominalists. They face the problem of accounting for the phenomena described above.

(IV) Identity and Individuality

We are concerned here with numerical identity — being one and the same thing — as opposed to qualitative identity which means being the same with respect to all qualities. There are two fundamental aspects to the problem of elucidating identity, namely identity at a time, or synchronic identity, and identity over time, or diachronic identity. Issues about identity are closely connected to issues about individuation. Problems of individuation concern what it is, if anything, in virtue of which some particular object is the object it is and not any other.

Aristotle defended the theory (hylomorphism) that individuals are the combination of matter and form (properties). Later philosophers argued that individuals are nothing more than a bundle of properties. If the bundle view is to be defensible then it would seem some version of the principle of the identity of indiscernibles (PII) must be true. PII states that there cannot be things with all the same properties, or equivalently, that qualitative identity implies numerical identity. The converse, the indiscernibility of identicals is uncontroversial, since clearly the if a and b have different properties then they are different objects. (Confusingly, both these principles are sometimes called Leibniz' Law.) We can only state these principles by using second-order logic, which quantifies over properties as well as over objects:

PII: $\forall x \forall y[(\forall P(P(x) \leftrightarrow P(y)) \rightarrow (x = y)]$
Converse: $\forall x \forall y[(x = y) \rightarrow (\forall P(P(x) \leftrightarrow P(y))]$

PII has been the subject of much controversy, and even if it is true there is the further question of whether it should be regarded as a necessary or a contingent truth. It is absurd to state PII as the claim that 'two things having all their properties in common are identical', since if they really are identical we don't have two things at all but one thing with two names. However, we can state PII in another logically equivalent form (as above): it is not possible that there be two things that share all their properties

$$(\neg\Diamond(\exists x\exists y[\forall P(P(x) \leftrightarrow P(y))\&\neg(x = y)]).$$

Obviously, if two objects, a and b, are really distinct then a has the property of being identical with a, and the property of being different from b, and b lacks these properties. However, such properties amount to nothing more than that a and b are distinct. (The property of primitive thisness or self-identity is also called 'haecceity'). If identity and difference count as properties PII becomes totally trivial. However, the question of whether PII is true when properties are restricted to be qualitative ones is still interesting and important. Qualitative properties are those properties which can be instantiated by more than one object and do not involve being related to a particular object, for example, being red, being on a brown table, and so on. It is also worth considering whether PII is only true if extrinsic properties are considered. Roughly, an intrinsic property is one which an object may possess even if it is the only thing that exists, for example, mass, charge, height, etc. An extrinsic or relational property is one which is not intrinsic, for example, being in the Northern Hemisphere. Qualitative properties include both intrinsic and extrinsic ones. Recall there is a further distinction between accidental and essential properties; the former are properties that some object can have or not and still be the same object, like being red or being on the Earth, the latter are properties that an object has in virtue of it being what it is and which it must have, like perhaps being charged for an electron or being H_2O for water (see Natural Kinds).

So now we are considering a stronger form of PII namely:

$\forall x\forall y\forall P[(P(x) \leftrightarrow P(y)) \rightarrow (x = y)]$, where P ranges over qualitative properties only. Obviously, the only way to discover that two different things exist is to find out that one has a quality not possessed by the other, or else that one has a relational characteristic that the other lacks. The epistemological question of how it can be known that a is different from b is the question of distinguishability. Verificationists about meaning like the logical positivists argue that if two things possessed all the same qualitative and relational properties, that they were different would be unverifiable and hence meaningless.

It is interesting to note that Leibniz believed that if an individual x is really distinct from an individual y then there is some intrinsic, non-relational property F that x lacks and y has or vice versa. Hence, he held that $\forall x\forall y\forall P[(P(x) \leftrightarrow P(y)) \rightarrow (x = y)]$, where P ranges over non-relational, qualitative properties. This principle does not apply even to classical particles since the ones of a given kind are all supposed to be identical in all their properties except spatio-temporal ones (which are usually taken to be relational). Some have argued that quantum particles can be numerically distinct and nonetheless share all their qualitative properties so that PII is contingently false. Others argue that quantum particles are not individuals. There is also controversy about whether or not spacetime points obey PII.

Consider now the extra problem of identity over time (or genidentity): What is it for a thing, by which we shall understand a concrete particular, to persist, that is, for it to be the same thing at different times? (This question is particularly pressing when we consider the identity of a person over time.) The problem of identity over time arises because things change over time. There are at least two types of change: change in parts and change in properties. The worry is that either of these two types of change construed as change of an individual entity seems to contradict the indiscernibility of identicals. For example, imagine a banana that turns from green to yellow — how can numerically one and the same object possess incompatible properties? Similarly, suppose a table has a leg on it replaced — how can numerically one and the same object have different parts? How many properties or parts can change before one object becomes another?

Theories of persistence divide into two main forms: endurance and perdurance theories. The former have it that one and the same object is wholly present at different times, while the latter have it that what we call the same object at different times are really different temporal parts of the whole object which is extended in time and hence never wholly present at a particular time. So on the perdurantist view identity over time is not really numerical identity at all, and ordinary concrete particulars are in fact aggregates made up of different temporal parts, time-slices or stages. (Perdurantists are divided over the question of whether temporal parts are infinitely divisible or not.)

(V) Matter and Motion

There were ancient philosophers who argued for materialism. This is the view that all that fundamentally exists is matter (there is only one substance), and that there is no immaterial soul beyond the body, the human mind being nothing more than the product of matter in motion. Materialism was advocated by the atomists Leucippus and Democritus. Atomism is the view that matter is ultimately composed of very small objects that are indivisible into further parts, and that all change in the world is attributable to changes in the position (motion) of elementary particles in the void. The only properties that atoms have are their size and shape, and their states of motion. Plato held that matter was essentially illusory and that the real world was the world of universals. On the other hand, for Aristotle the forms (although immaterial) depend on the existence of individual substances, and matter is a central component of Aristotle's theory of the nature of individual substances. Any such thing, for example, a marble statue, consists of matter in some form (see the section on Universals above). Aristotle distinguished between natural and unnatural motion, arguing that unnatural motion always involved some external cause. This gave him a problem in explaining the motion of an arrow in flight some time after it has left the bow. The pre-Socratic philosopher

Parmenides argued that all change was illusory. His follower Zeno tried to support this doctrine by giving a series of arguments to show that motion, being a kind of change, is impossible. Zeno's paradoxes of motion are so-called because they are arguments from seemingly plausible premises to the (unacceptable) conclusion that there is no motion.

Motion may be thought of in two ways, namely as absolute or as relative. Relative motion is only defined with respect to some object or frame or reference, and so the same object has many different states of relative motion at the same time depending in relation to what its relative motion is being considered. For example, when a ball is flying through the air its motion relative to the ground will not be the same as its motion relative to a bird which is flying along next to it. The revolution in physics initiated by Galileo is principally about this difference between absolute and relative motion. In Galileo and Newton's physics only relative (constant) motion is observable, which explains why we don't observe the effects of the Earth's motion around the Sun. In Newtonian mechanics absolute motion has no physical effects, but absolute acceleration, which includes rotation, does have physical effects.

In the seventeenth century, the idea of explaining all natural phenomena in terms of matter in motion became a goal of many of those known as the mechanical philosophers. Locke used the image of a clockwork machine to illustrate the goal of natural philosophy as he saw it: The hands seem to move in a co-ordinated way and the chimes ring out the hours, half hours and so on as appropriate; this corresponds to the appearances of things, the observable properties of, say, a piece of gold. However, the clock has inner workings and this mechanism produces the outer appearance of the clock; similarly the gold has an inner structure that gives rise to its appearance. The goal of natural philosophy is to understand the inner mechanisms responsible for what we observe. The point about a clockwork machine is that the parts all work together in harmony, not because they are co-ordinated by mysterious natural motions or final causes, but because each of them communicates its motion with the part adjacent to it by contact. Mechanists explain the behaviour of things in terms of motions of the particles that compose them, rather than in terms of essences and 'occult forces'. Mechanics, in the hands of Galileo, Descartes and Newton in particular, became a mathematically precise science of matter in motion, and what happens as a result of collisions between bits of matter. (All of them adopted a principle of inertia, which states that a body continues in its state of motion unless a force acts to change it, so only changes in motion require an explanation.)

Newton's theory of gravitation was problematic because Newton offered no explanation for how the force of gravity was transmitted between bodies separated in space. It seemed that gravity was an example of the kind of action at a distance, which mechanist philosophers were trying to avoid. Fields were introduced into physics to solve the problem of action at a distance posed by the force laws of classical mechanics, namely Newton's law of Gravitation and the law of electrostatic attraction and repulsion (Coulomb's law). The prototype for the field was the optical ether, which was thought to be material. Classical fields were replaced by quantum fields, and special relativity introduced the idea that mass and energy were somehow equivalent and that the amount of mass possessed by a body depends on its state of motion relative to the frame of reference in which the mass is being measured. There is much controversy about whether quantum fields and particles can be considered as material. Contemporary physics seems not to describe the world in terms of matter in motion, but rather in terms of spacetime, and fields of potentiality and probability.

(VI) Causation

Causation is apparently fundamental to the scientific understanding of the world. The idea of causation is closely linked with the concepts of laws of nature, dispositions, natural kinds and properties, necessity and possibility, and subjunctive conditionals; all these notions are modal (and, as mentioned above, for some philosophers like Quine therefore dubious).

Aristotle described four types of causation: efficient, material, formal and final. For example, what is the cause of a statue of Socrates? The efficient cause is the sculptor's actions, the material cause is the marble the statue is made of, the formal cause is the idea that the sculptor has of the finished image based on Socrates' appearance, and the final cause is the end for which the statue is made, perhaps to celebrate the intellectual virtues of philosophy. Aristotelian ideas were the subject of much criticism in the Scientific Revolution. The mechanical philosophers argued that science should not search for final causes (teleology), and some went further and argued that such explanations were vacuous; rather science should concentrate on finding the efficient or material causes of phenomena. (Evolutionary biology seems to reintroduce teleology but it is usually claimed that this is legitimate because the teleological talk of function and design is eliminable in favour of efficient cause.) However, there were those philosophers that were sceptical of the notion of causation as it was deployed by materialist and mechanist philosophers. Berkeley famously argued that matter could not be a cause of anything because he identified causation with activity, action and agency, and Malebranche argued that only God could be a true cause.

Hume's empiricist analysis of causation is the starting point for contemporary debates. He questions whether causation has anything to do with necessity. His own theory of causation is sometimes called the regularity theory of causation, according to which instances of the relation A causes B usually have each of the following features:

> Events of type A precede events of type B in time.
>
> Events of type A are constantly conjoined in our experience with events of type B.
>
> Events of type A are spatiotemporally contiguous with events of type B.

Events of type A lead to the expectation that events of type B will follow.

Hume says that we have no 'impression' of a necessary connection.

Hume's analysis maybe okay for generic causation where A and B type events occur lots of times and A is regularly followed by B, but it looks unable to handle single case causation where A and B type events only occur once in the whole history of the universe, so if A causes B this cannot be reduced to a regularity. One option is to deny that there really is any single case causation in the world. Another is to modify Hume's account. An influential account that clarifies the relationship between causation and necessary and sufficient conditions is due to John Mackie who argued that a cause is what he called an 'INUS' condition. Consider, a fire (A) caused by a match (B). A caused B does not imply, either that A is sufficient for B, because the match alone would not have caused a fire if there had been no combustibles, nor necessary for B, because something else could have lit the combustibles. A is an Insufficient but Necessary part of a set of conditions which are together Unnecessary but Sufficient for B (so A is said to be an 'INUS condition' for B).

However, when we consider a particular instance of an event of type A causing an event of type B, it is not enough that A and B happen and A is an INUS condition for B because of two problems:

(i) Epiphenomena

This is where A is a side effect of whatever the causal process is that causes B. A will always occur when B is being caused by the process and so A will be an INUS condition for B but it will not be a cause. For example, the sound of the heart beating is not a cause of the circulation but it is an INUS condition for it.

(ii) Pre-emption

This is where A would have caused B but some other cause of B happens first. For example, a match would have started a fire but another match was lit first and started it.

Notice that when A causes B we are often inclined to say 'if A hadn't happened B wouldn't have happened'(#). There is a close connection between causation and counterfactual or subjunctive conditions. Another way of expressing (#) is by saying 'A is necessary in the circumstances for B'. But how do we pick out the right counterfactuals, in other words, which circumstances do we hold fixed? For example, suppose a match sits next to a matchbox, we are inclined to say that had the match been struck it would have lit, but why don't we say that had the match been struck it would have been damp? One solution is to hold fixed the circumstances around the time when the match was struck. But this will allow us to say had the match been struck it would have been picked up. The solution is to only consider the circumstances prior to the time of the match being on the table or being struck.

(iii) C is an INUS condition for E iff E is an INUS condition for C: problem of the direction of causation (also problem of simultaneous causation)

David Lewis influentially took up an idea from Hume in his proposal of a counterfactual account of causation. The idea is that if (If C hadn't happened then E wouldn't have happened) is true then it is also often true that C caused E. Spelling out precisely how to turn this into a full analysis of causation turns out to be a complex matter that depends crucially on the analysis of counterfactuals, and on finding a way of dealing with the problems of epiphenomena and preemption mentioned above, and the problem of overdetermination, where E would still have happened if C hadn't because some other cause would have taken over.

Finally, there are now many theories of probabilistic causation, since many philosophers believe that there can be genuine causes that do not guarantee the occurrence of their effects. Unfortunately, the simplest account of probabilistic causation, according to which a probabilistic cause must make its effect(s) more likely than not is not true. There are many examples of probabilistic causes whose effects are relatively improbable. It is more plausible to say that probabilistic causes must raise the chances of their effects from what they otherwise would have been, but even this claim turns out to be false.

(VII) Laws of Nature

Discovering the laws of nature and using them for the prediction and explanation of observed phenomena is one, if not the most important job of science. However, it is not always easy to tell what the laws of a particular science are because there seems to be no rule about when to call something a 'law' rather than a 'principle'. Laws sometimes take the form of simple universal generalisations, such as all metals conduct electricity, but more often they have a mathematical form like Kepler's laws of planetary motion. Sometimes laws seem to express deep facts about the unobservable causes of phenomena, like the law that expresses the relationship between the energy and frequency of radiation, whereas other scientific laws seem almost homely by comparison, such as the law that if a gas is kept at a constant volume and its temperature is increased then its pressure will rise. Other ideas associated with laws include those of generalisation, regularity, pattern, stable relationship, symmetry and invariance.

Here are some important different kinds of laws:

(i) laws of motion or state evolution over time such as Newton's second law and the Schrödinger equation

(ii) laws of co-existence that constrain what states of some system are mutually compatible such as the ideal gas laws and Pauli's exclusion principle

(iii) conservation laws, such as the law of conservation of energy

(iv) phenomenological laws that describe the observable phenomena in a particular system, such as the law of the pendulum, versus fundamental laws that purport to explain the underlying unobservable entities and processes, such as the laws of electromagnetism

(v) deterministic laws are those such that given the values of all physical properties at a given time, there is only one possible state of the system at any other time. Probabilistic laws are those that only provide probabilities for the state of the system at other times, like the half-life laws of radioactive substances.

What is a law of nature? There three broad answers to this question. The first is Humeanism which says that a law of nature is a special kind of regularity among properties, events and/or objects in the natural world. The second is necessitarianism which says that a law of nature is a relation of necessity between properties, events and/or objects in the natural world. The third is the sceptical position that there are no laws of nature, or at least that there is no objective distinction between laws and mere regularities.

According to the naïve regularity theory of laws, there is no good reason to think there is a difference between laws and accidents (c.f. Hume on causation and induction). On this view, it is a law that all *A*s are *B*s iff all *A*s are *B*s. If it is correct there are not any regularities that are not laws, nor are there any laws that are not regularities.

A single case occurrence, such as a cat being on a mat at some time, is a trivial kind of regularity. So is it a law of nature that the cat is on the mat at that time? Further problems arise with disjunctions of regularities, and with regularities involving disjunctive predicates and predicates like grue (see (I)–(ii)). Furthermore, vacuous regularities, such as all unicorns love television, are always true (since they are analysed as 'for anything, if it is a unicorn then it loves television' which is true if there are no unicorns). Ought these to be regarded as laws of nature? There also seem to be regularities that are not laws (for example, all the presidents of the USA in the twentieth century were men). On the other hand, there are cases of scientific laws that do not seem to satisfy the regularity account. For example, Newton's first law, which applies to bodies not acted upon by any external forces, is not actually instantiated since there are no such bodies, but it does not seem to be vacuous.

More complex problems besetting the regularity theory of laws include explaining the connection between laws, inference and explanation. Laws are supposed to be explanatory, and to support inductive inferences, but regularities do not seem to be explanatory, nor it is obvious why inductive inferences to the truth of regularities based on the truth of some of their instances are justified. Laws are also closely related to counterfactuals, so for example, it seems that if it is a law of nature that all metals expand when heated, it is true to say of a piece of metal that was not heated, that if it had been heated it would have expanded. But ordinary regularities do not seem to entail counterfactuals in the same way;

for example, that all the coins in my pocket are silver, does not entail that if some copper coin had been in my pocket it would have been silver.

So not all regularities are laws. The sophisticated regularity theorist therefore places restrictions on what regularities are to be counted as laws. These come in two varieties: epistemic restrictions are so-called because it is our cognitive attitudes that determine which regularities are laws. On such views, laws are regularities that play are certain role in our theories, or else they are just regularities to which we attach some significance or importance. The main problem for this account is that it seems plausible that laws can be unknown. Which of the unknown regularities are laws and which are not can only be a matter of what our attitude to them would be if we knew them. This is obviously problematic since such counterfactuals would seem to rely upon laws themselves. What about a world with no minds? There would be no laws either it would seem but surely the laws of nature could have ruled out the possibility of there being minds? Why do we have different attitudes to different regularities? Either this is arbitrary or grounded in some objective difference between them. If the former then this is no good, if the latter then the epistemic view collapses into the systemic view.

The second kind of modified regularity theory places 'systemic restrictions' on which regularities count as laws (this view is associated with Mill, Ramsey, and Lewis). Laws are the propositions we would use as axioms if we knew everything and organised it as simply as possible in a deductive system. On this view, laws are the result of a trade-off between simplicity and strength. Laws are the theorems and axioms of deductive systems that achieve the best combination of simplicity and strength.

Problems with this view include the following:

(i) Arguably, neither simplicity nor strength is an objective notion.

(ii) What achieves the best balance of strength and simplicity may not be agreed upon by all; for example, rationalists might weight simplicity more, whereas empiricists might weight strength more.

(iii) It is possible that the most systematic laws would involve grue or disjunctive predicates.

(iv) It is possible that there be equally systematic but different sets of laws. (Cf. Coherentism about truth.)

(v) The problem of inference and explanation is not obviously explained by the systemic view.

The necessitarian account of laws of nature says that they are relations among universals. On this view, laws of nature differ from mere universal generalisations. Laws of nature express necessary relations among universals (these relations are 2nd order universals). Laws are singular statements about universals not universal generalisations about particulars. Laws imply universal truths but universal truths do not always imply laws.

F-ness $\rightarrow$ G-ness (X-ness is the property of being X, for example, F is being an electron and G is being negatively charged)

'$\rightarrow$' is to be read as 'brings with it' (Dretske), 'nomically necessitates' (Tooley), 'necessitates' (Armstrong).

This approach seems to offer an account of how laws support counterfactual statements, and to deal with the relation between laws, explanation and inference. However, necessitarianism faces a number of further questions:

(i) Are law statements themselves necessary?

(ii) The identification problem: what exactly is the necessitation relation between universals?

(iii) The inference problem: how can we make sense of the inference from 'F-ness $\rightarrow$ G-ness' and 'this is F' to 'this must also be G' if the laws of nature are themselves contingent?

Probabilistic laws raise further problems for all the views discussed above.

Nancy Cartwright argues that phenomenological laws may be true but that fundamental laws are not since their application to the world always involves modelling, idealisation and approximation. She argues that causal powers are more fundamental than laws. On the other hand, Bas van Fraassen argues that there are no laws of nature and that they are features only of the theoretical representation of the world and not the world itself.

Ceteris paribus laws are laws that hold 'all things being equal'. Giving an account of the ceteris paribus clause that does not make the truth of the law trivial, by saying that other things are equal just in case the law is true, turns out to be a difficult task. It is thought by some that the difference between the natural and the social sciences lies in the fact that the former and not the latter are able to find exact laws.

(VIII) Probability, Propensity and Dispositions

The formal theory of probability was invented relatively recently in the history of science and mathematics, but the idea of probabilistic reasoning is commonplace. Probability may be thought to have nothing to do with ontology, but rather to be the science of uncertainty, evidence and estimation, and hence to be part of epistemology and not metaphysics. There are accounts of probability that do indeed claim that probability is an entirely epistemic notion, but to do so is to adopt a position analogous to nominalism in the lively debate about whether there is such a thing as objective chance. Since the advent of quantum mechanics it has been widely thought that it is at least an open question whether the world has fundamentally probabilistic occurrences and causes in it. Probability in the world that does not arise from our ignorance is 'objective chance'.

Objective chance has been identified with:

(i) Finite relative frequencies

(ii) Infinite relative frequencies

(iii) Propensities (these are primitive single case probabilities)

There are problems with all of the above. The finite relative frequency of some occurrence may occasionally depart radically from its probability. For example, if a fair coin is tossed ten times it may well come up heads seven times, yet intuitively the probability of heads is only 50%. Infinite relative frequencies are problematic because the notion of a completed infinity is problematic and transcends the empirical world. It is an interesting question how epistemic and objective probabilities must be related.

Note that determinism is the doctrine that given the state of the world at one time, and the laws of nature, there is only one possible way the world could be at all other times. Indeterminism is the denial of determinism. Determinism is a modal claim about the world rather than a claim about what can be predicted. It is possible for there to be phenomena governed by deterministic laws that we are nonetheless unable to predict. This is the case where very divergent outcomes follow from very small differences in initial conditions, since then the smallest inaccuracy in measurements of the latter will make accurate prediction impossible (this sensitivity to initial conditions characterises chaotic systems).

Dispositions are properties, such as fragility and solubility, that may or not be actualised. Some philosophers hold that dispositions must be reducible to the structural properties of things, while others hold that dispositions may be primitive. Dispositional essentialists argue that the essential properties of physical kinds are dispositional.

(IX) Reductionism, Emergence and Supervenience

There is a great deal of debate in philosophy of science about the relationship between the sciences. How are the domains of physics, chemistry and biology related, and how are the laws, theories and explanations of these sciences related?

Fundamental intuitions of reductionism include:

1. The whole is not greater than the sum of the parts.

2. The behaviour of the whole is caused and explained by the behaviour of the parts.

3. There is a unity to the world and to science.

Reductionism is popular because in general: reduction seems to yield explanatory gain (some theories of explanation assimilate explanatory power to unification); reduction implies ontological unification and so is in keeping with the desire for parsimony in metaphysics and accordance with Occam's razor; and finally, reduction aids conceptual unification.

Here are some examples of different forms of reductionism:

(a) Philosophical/logical behaviourism about the mind that reduces thoughts and other mental states to relations among stimuli and behaviour. This is inspired by verificationism (the idea that all meaningful discourse concerns what can be verified in experience) conjoined with the claim that we can only verify propositions about the mind by observing behaviour.

(b) Logicism about mathematics that regards mathematical theorems as consequences of logical laws.

(c) Set-theoretic reductionism that reduces all mathematical objects to sets. For example, the natural numbers can be identified with a sequences of sets where each successive set contains all the sets that have gone before it.

(d) Semantic reductionism about theoretical terms that reduces sentences involving them into sentences only involving observational terms and logical constants. The Logical Positivists attempted to explicitly define theoretical terms in terms of observational language. For example, 'temperature' would be translated into statements about observable manifestations of it, and statements about mind-independent objects would be translated into statements about observations.

(e) Reductionism about the mind according to which types of mental states are identical to types of brain states.

(f) Reductionism about dispositions according to which the latter are reducible to categorical or structural properties.

(g) Reductionism about colours and sounds according to which they are identical with physical properties.

(h) Reductionism about natural kinds according to which macroscopic kinds, like water, are identical with their microstructural essences (water is identical with H_2O).

Within science there have been reductionist programmes of great significance and some examples are listed below (the first three are intra-science, the rest are inter-science; (iv), (v) and (vi) are broad and programmatic/methodological)

(i) Galileo's law of freefall and Kepler's laws of planetary motion to Newtonian mechanics

(ii) optics to electromagnetism

(iii) thermodynamics to the kinetic theory of gases via statistical mechanics

(iv) laws in the social sciences to laws that only refer to individuals (methodological individualism), for example, laws about the behaviour of economic markets to rational choice theory

(v) social sciences to natural/physical sciences: socio-biology, evolutionary psychology, genetic reductionism

(vi) natural sciences to physics: geology to geophysics and geochemistry, neurophysiology - cell biology - molecular biology - molecular physics - quantum physics (the failure of vitalism/organicism, which posited a special status for living systems encouraged this kind of reductionism)

(vii) genetics to molecular biology

(viii) chemistry to quantum mechanics

There are various kinds of reductionism, notably semantic and having to do with meaning equivalence (a, b, c, d), and ontological (the rest of the above). In the case of the latter, translation is effected by means of 'bridge laws' which correlate terms in reduced theory's vocabulary to those in reducing theory's vocabulary. In the case of the former there must be strict identities between the terms in the reducing theory and the reduced theory.

There are various problems that may arise for reductionist programmes. One is that the bridge laws may turn out to be only partially true. Another is that the reduced theory is usually only approximately true and ends up being corrected rather than recovered exactly by the reducing theory. Reduction also usually relies heavily on idealisation. Finally the most celebrated problem is that of multiple/variable realisation; this is the fact that, for example, 'pain' seems to be realisable in animals with very different kinds of anatomies and physiologies, just as the same word processing programme can be realised by computers with very different internal workings. It is often said that multiple realisability means that mental events are only token identical with physical events, and not type identical with them, where the former means that each mental event is identical with some physical event but that each type of mental event need not be identical with the same type of physical event as the latter requires.

In the light of this, philosophers often think in terms of supervenience rather than reduction. A domain supervenes on another domain, if there can be no changes in the former without changes in the latter, but not necessarily vice versa. For example, arguably the mental state of a person cannot change without their brain state changing, but it is possible for their brain state to change in a way that does not affect their mental state. This is called local supervenience, whereas global supervenience is the claim that all the mental facts about the whole world supervene on the physical facts about the whole world. Dualism, for example, denies even global supervenience of the mental on the physical. On the other hand, emergentism is the doctrine that the whole is not reducible to the sum of the parts and that genuinely new properties and causal powers come into being when parts make up a whole.

The relationship between causation in physics and causation in the special sciences is much discussed in contemporary philosophy of mind. Causal exclusion reasoning proceeds along the following lines: Mental states must either be reducible to physical states, or cannot be the causes of actions, because, for any action A, since A is a physical event and as such, given the causal closure of the physical world, there is some set of physical causes that are sufficient for its occurrence (or at least to fix its objective chance).

Finally, there is the question of whether special science objects, for example, organisms, markets and people, can be identified with the mereological sums of physical objects. Some philosophers conclude that special science objects cannot be so identified and so do not therefore really exist, but realism about the special sciences seems at least if not more plausible than realism about physics. Other suppose that special science objects are individuated by their functional role and are only token identical with collections of physical objects. This is problematic in so far as such objects seem to be actually and not merely potentially multiply realised, for example, a given cat may be identified with numerous subsets of the maximal set of molecules that make it up, since for any set of the molecules that is a candidate for being token identical with the cat, removing a few molecules at random from this set will leave a new set that is also a candidate for being the cat.

(X) Space, Time and Spacetime

The Aristotelian theory of space grants a privileged position to the centre of the Earth, and this induces a privileged direction towards the centre of the Earth. Space is said to be absolute. The Galilean relativity principle entails that absolute position in space and absolute motion through it are physically undetectable. In Newton's theory of space absolute position is nonetheless an objective feature of the world and Newton also posited absolute time. Leibniz rejected the Newtonian ideas of absolute space and time and argued instead for the idea that space and time are nothing more than relations among phenomena. Leibniz appealed to the principle of sufficient reason and the PII to show that there was no such thing as absolute space. The former states that everything that occurs must have a sufficient cause; since position in absolute space and time make no observable difference to anything there could be no cause of why the universe begins in one position in space and time rather than another. PII is in conflict with absolute space and time since different positions for the whole universe in absolute space and time are qualitatively identical. These issues are now discussed in the context of general relativity.

There are several distinct, though often conflated issues in the metaphysics of time:

(i) Are all events, past, present and future, real?

(ii) Is there temporal passage or objective becoming?

(iii) Does tensed language have tenseless truth conditions?

'Eternalism' is the view that all times are real, whereas according to 'presentism' only the present is real (there is also the 'cumulative' view that all past and present events are real). Those who believe in the passage of time or objective becoming often also believe that the process of becoming is that of events coming into existence and going out of existence, but this need not be so; to suppose there is becoming, one need only believe that there is some objective feature of the universe associated with the passage of time. Objective becoming could be like a light shining on events as they are briefly 'present', and is therefore compatible with eternalism. On the other hand, both presentism and cumulative presentism entail a positive answer to question (ii), since if events do come into existence, whether or not they then stay existent or pass out of existence, this is enough to constitute objective becoming. Presentism and becoming have also been associated with the idea that tensed language does not have tenseless truth conditions. However, this is not a necessary connection. So even though the standard opposition is between those who answer 'no' to (i), 'yes' to (ii), and 'no' to (iii) on the one hand (the defenders of McTaggart's '*A*-series'), and those who answer 'yes' to (i), 'no' to (ii), and 'yes' to (iii) (the defenders of McTaggart's '*B*-series'), a variety of more nuanced positions are possible.

There is a further celebrated question about time:

(iv) Does time have a privileged direction?

Clearly if (i) or (ii) are answered positively then that is enough to privilege a particular direction in time. However, eternalism and the denial of objective becoming are also compatible with time having a privileged direction, since there could be some feature of the block universe that has a gradient that always points in some particular temporal direction. For example, the entropy of isolated subsystems of the universe, or the universe itself, might always increase in some direction of time. Another well known possible source of temporal direction was proposed by Reichenbach who argued that temporal asymmetry is grounded in causal asymmetry: in general, the joint effects of a common cause are correlated but the joint causes of a common effect are uncorrelated.

However, it may be that no physical meaning can be attached to the idea of the direction of time in the whole universe, because no global time co-ordinate for the whole universe can be defined. This seems to be implied by special relativity. The status of time in special relativity differs from its status in Newtonian mechanics in that there is no objective global distinction between the dimensions of space and that of time. Spacetime can be split into space and time, but any such foliation is only valid relative to a particular inertial frame, which is associated with the Euclidean space and absolute time of the co-ordinate system of an observer. This seems to imply eternalism, since if there is no privileged foliation of spacetime, then there is no global present, and so the claim that future events are not real does not refer to a unique set of events. Furthermore, many have argued that, since special relativity implies the relativity of simultaneity, whether or not two events are simultaneous is a frame-dependent fact, and therefore there is no such thing as becoming.

On the other hand, special relativity is a partial physical theory that cannot describe the whole universe, even if there is good reason prefer it to its empirically equivalent rivals which some deny. The implications of general relativity for time are not clear. This is because the theory gives us field equations that are compatible with a variety of models having different global topological features, and different topological structures may have very different implications for the metaphysics of time. Clearly we must then turn to cosmological models of the actual universe, of which there are many compatible with the observational data. As yet there is no agreement about which of these is the true one. Highly controversial issues about quantum gravity bear on the question of whether spacetime will turn out to admit of a global foliation, and hence on whether absolute time is physically definable. Even if it does turn out to be definable, there remains the question of whether such a definition ought to be attributed any metaphysical importance.

Non-relativistic many-particle quantum mechanics does not directly bear on the philosophy of time since the status of time in the formalism is not novel in the same way as in relativity. However, it has often been argued that quantum physics is relevant to questions about the openness of the future, becoming, and the direction of time, because of the alleged process of collapse of the wavefunction. Since Heisenberg it has been popular to claim that the modulus squared of the quantum mechanical amplitudes that are attached to different eigenstates in a superposition represent the probabilities of genuinely chancy outcomes, and that when a measurement is made there is an irreversible transition from potentiality to actuality in which the information about the weights of the unactualised possible outcomes is lost forever. Hence, measurement can be seen as constituting irreversible processes of becoming that induce temporal asymmetry. However, quantum measurements need not be so understood. Furthermore, if there is no collapse, as in the Everett interpretation, then again there is no temporal asymmetry in quantum mechanics. The upshot seems to be that the status of the arrow of time in quantum mechanics is open.

There is also a vast literature about whether or not the second law of thermodynamics represents a deep temporal asymmetry in nature. The entropy of an isolated system always increases in time, and so this seems to be an example of the arrow of time being introduced into physics. If the whole universe is regarded as an isolated object, and if it obeys the second law, then it would seem that there is an objective arrow of time in cosmology. However, it is not clear what the status of the second law is with respect to fundamental physics. One possibility is that the second law holds only locally, and that there are other regions of spacetime where entropy is almost always at or very near its maximum. Even if thermodynamics seems to support the arrow of time, it is deeply puzzling how this can be compatible with an underlying physics that is time asymmetric. Conservative solutions to this problem ground the asymmetry of the second law in boundary conditions rather than in any revision of the fundamental dynamics. The most popular response is to claim that the law does indeed hold globally but that its so doing is a consequence of underlying time-reversal invariant laws acting on an initial state of the universe that has very low entropy. It is necessary to posit this because standard arguments in statistical mechanics that show that it is overwhelmingly likely that a typical state of an isolated system will evolve into a higher entropy state in the future, also show that it is overwhelmingly likely that the state in question evolved from a past state that had higher entropy too. A much more radical possibility is that the second law is a consequence of the fact there is a fundamental asymmetry in time built into the dynamical laws of fundamental physics. Given the outstanding measurement problem in quantum mechanics those who propose radical answers to problems in thermodynamics and cosmology often speculate about links between them and the right way of understanding collapse of the wavefunction. Roger Penrose, for example, suggests that gravity plays a role.

(XI) Events and Processes

Philosophers often think about the ontology in terms of what kinds of objects there are. So they ask whether there are only concrete objects or whether abstract objects also exist; they ask whether there are only the fundamental building blocks of the world (mereological atoms), or whether composite objects also exist, and they ask whether there are mental or spiritual objects, as well as physical ones. However, there are other influential accounts in metaphysics that hold that the world consists of entities that are partly temporal in nature, namely events or processes.

Donald Davidson influentially argued that the world consists of events, and that properties like colour and shape are properties of events not of objects (or at least that objects are arrangements or structures of events, rather than the other way round). For Davidson the relata of causal and lawlike relations are events rather than objects or facts. On the other hand, consider what physicist Lee Smolin says: "The universe is made of processes, not things" [2001, 49]. Smolin insists that a lesson of both relativity theory and quantum mechanics is that processes are prior to states. Classical physics seemed to imply the opposite because spacetime could be uniquely broken up into slices of space at a time (states). Relativity theory disrupts this account of spacetime and in quantum mechanics nothing is ever really still it seems, since particles are always subject to a minimum amount of spreading in space and everything is flux in quantum field theory within which even the vacuum is the scene of constant fluctuations.

2 EPISTEMOLOGICAL POSITIONS

(I) Rationalism

Philosophers described as rationalists include Plato, Descartes, Leibniz, and Spinoza. Rationalism is associated with two distinct but often conflated theses. The first is that some of our concepts (ideas) are innate ((vi) below); the second is that some of our knowledge of the world is a priori, that is justified independently of experience or empirical evidence ((v) below). Note that the a priori/a posteriori distinction is an epistemological one. Other related distinctions include the metaphysical distinction between what is necessary (could not have been otherwise), such as that 2+2=4, and contingent (could have been otherwise), such as that the largest mammals are blue whales; and the semantic distinction between the analytic (true or false in virtue of meaning), such as that all bachelors are unmarried, and the synthetic (true or false not merely in virtue of meaning), such as that Paris is the capital of France. Of course, these categories often overlap, for example, that bachelors are unmarried may well be analytic, necessary and a priori, and that Paris is the capital of France may well be synthetic, contingent and a posteriori. However, whether or not this overlap is partial or total is one of the central issues that divides rationalists and empiricists.

Some characteristic doctrines of rationalism (although not held by all rationalists) are as follows:

(i) Sensory knowledge is limited and we should be cautious about it and use reason correctly to overcome these limitations.

(ii) The universe is ordered and accessible to the rational mind.

(iii) Mathematics is general, and Euclidean geometry in particular, provides the model of well-founded and unified system of knowledge. The subject matter is intrinsically clear and knowledge of it is certain and based on reason.

(iv) Basic beliefs (or at least some) are known a priori by the use of pure reason / understanding.

(v) There is a faculty of rational intuition that delivers substantive a priori knowledge.

(vi) Concept innatism: some concepts are not derived from experience, for example those of event, cause, location, time, extension, and substance.

(vii) There are necessary connections in nature. The truth in science and philosophy must refer to what could not be otherwise.

There are various arguments for rationalism. Rationalists claim that certain concepts cannot be derived from experience because nothing that we perceive exemplifies them; for example, identity, equality, perfection, God, power, and cause. Descartes famously argued that even our concept of matter must be a priori. He considers a piece of wax that is heated and so changes its shape, its colour and its other sensible properties. He argues that since we continue to think of it as the same wax, we must be thinking of it as matter or pure extension in space, and that we have no direct experience of it as such and must therefore apply that concept to the world by the use of reason alone. Rationalists also argue that knowledge of the laws of logic (for instance, the law of identity states that everything is identical to itself) that describe which inferential connections among our beliefs are valid, and of mathematics, that apparently describe necessary truths about an abstract realm of mathematical objects, could only be known a priori. Some rationalists argue that metaphysics is knowledge of a priori necessary truths, for example, that every event has a cause or that an object cannot be in two places at the same time. They maintain that such truths, if truths they be, cannot be known by experience. Other domains of possible a priori knowledge include probability theory, decision theory (the theory of action) and mereology (the logic of part/whole relations).

Consider Euclidean geometry. There are primitive and undefined terms such as 'point' and 'line' and then there are a few axioms relating them, such as that any two points define a straight line. The former are alleged by rationalists to be innate (they are examples of Descartes 'clear and distinct ideas'), while the latter are supposed to be self-evident, in the sense that if one entertains the proposition in question one will thereby come to believe it. The rest of the theory is arrived by the use of proof, and the rationalist notions of clear and distinct perception (Descartes) and the 'natural light of reason' (Leibniz) are associated with the state of mind one is in when following a mathematical proof. Thinking about knowledge in terms of the paradigm of the axioms and theorems of Euclid leads naturally to a view about knowledge and justification called foundationalism. This view of knowledge goes back to Aristotle's *Posterior Analytics*, and is attractive because it offers a clear way out of the following famous sceptical problem known as Agrippa's Trilemma, or the Regress Argument: To be justified in believing something is to have a reason for believing it, but then one must have a justification of that reason, and so on ad infinitum. The idea is that this sceptical regress is halted with the intellect as the source of immediate and certain knowledge of foundational truths, upon which the rest of our knowledge is based.

Foundationalism says that there are *basic beliefs* which are justified independently of all other beliefs/non-inferentially justified. There are certain propositions that we seem to be justified in accepting but where that justification does not depend upon our acceptance of any other propositions, for example the aforementioned axiom that two points define a line. According to foundationalism, all justified beliefs are either basic or justified by being supported by basic beliefs, and justification is a 'one-way' relation. On this view, non-basic beliefs are deductively inferred from basic beliefs, and since deduction is truth-preserving, justification is assured. Basic beliefs are supposed by rationalists to be self-evident in the sense that if p is self-evident then if someone entertains it he or she will believe it. They are also required to be indubitable (not capable of being doubted) and incorrigible (not capable of being corrected by further experience). It is important to note however that empiricists may be foundationalists too: the proposition describing the immediate content of one's experience might be thought to be indubitable and incorrigible (although not self-evident).

By the seventeenth century rationalism was discredited in the eyes of many because of the failure of Aristotelian science, since the latter was widely regarded as overly reliant on reason at the expense of experience. Natural philosophers argued that certain knowledge of essences of things, or of substantive necessary truths about the world is not possible. When we consider examples of a priori knowledge, the propositions in question are often either questionable, or they seem to be true just in virtue of the meanings of the terms involved (analytic). Critics of rationalism argue that while there may be some a priori truths, there are no synthetic a priori truths. However, there is still some controversy among contemporary philosophers about whether thought experiments might offer a path to a priori knowledge in science.

(II) Empiricism

Classically empiricism is associated with Locke, Hobbes, Berkeley, and Hume. Empiricists tend to deny the existence of innate concepts and claim instead that the mind is a 'tabula rasa' at birth, and that all ideas are derived from experience. Experience either directly provides us with concepts via sensation, or indirectly via reflection and abstraction. Concepts and ideas are divided into the simple and complex, and the complex ones may not be derived from experience directly but rather composed of simple ideas. Empiricists also argue that there is no innate or a priori knowledge of the world. Rather all knowledge of reality is arrived at directly from particular experiences, or by extrapolating and generalising on the basis of experience.

Empiricists cannot consistently claim to know the truth of empiricism a priori, so they must argue for it on the basis of experience. The emerging natural philosophy led empiricists to make their model of knowledge not mathematics but experimental science. Francis Bacon was an important advocate of a new method of inquiry based on experiment; his vision of *New Atlantis* inspired the creation of *The Royal Society of London for the Improving of Natural Knowledge* (1660-). It is also possible to argue for empiricism from the implausibility and failure of rationalism. In the seventeenth century there were plenty of examples of embarrassing failures of science that was based on pure reason rather than experience and experimentation. The classic examples were Aristotle's theories of motion and cosmology that had been undermined by Galileo and Kepler. The idea of natural philosophers using their reason and intellect to apprehend the forms or essences of substances and processes in nature was discredited. Furthermore, empiricists can point out that there is no guarantee (at least for atheists) that a falsehood will not be self-evident, obvious, indubitable and clearly and distinctly perceived, for example, people might think it is self-evident that the earth is flat and doesn't move.

Hobbes thought he had squared the circle, so even the following of mathematical and logical proofs is subject to errors of reasoning.

Empiricists argue that pure reason cannot produce any useful or substantive knowledge of the world but only of the relations among our concepts. All a priori knowledge is of analytic truths, that is things that are true by definition and that tell us only about how we use words and concepts. This doctrine is often called Hume's Fork: all enquiry is about either, propositions about the 'relations of ideas' that are knowable a priori, for example, mathematics and logic, or, propositions about 'matters of fact and real existence' that are knowable only a posteriori, for example, physics and chemistry. Empiricists often add that all synthetic propositions are contingent and that since only analytic truths are necessary, the only necessity is verbal necessity.

In the nineteenth century an important empiricist movement called positivism came to prominence. The defining characteristic of positivism is that it is extremely in favour of science and opposed to metaphysics and theology. Positivists were also influenced by Hume in their disdain for ideas of necessitation or causation in nature, and their concern with ensuring the meaningfulness of language through an emphasis on verifiability or falsifiability. They also denied the existence of unobservable entities such as atoms.

Logical positivism was a movement of empiricist philosophers (associated with the Vienna Circle), in the twentieth century who used the new methods of mathematical logic to defend many of the traditional tenets of positivism. The logical positivists held that:

(i) Science is the only form of proper knowledge

(ii) All truths are either: (a) analytic, a priori and necessary, or, (b) synthetic, a posteriori and contingent

(iii) Logic is the science of elucidating the relationships among concepts.

(iv) The purpose of philosophy is to explicate the structure or logic of science.

(v) The verifiability criterion of meaning: A statement is held to be literally meaningful if and only if it is either analytic or empirically verifiable.

(vi) The Verification Principle: The meaning of a statement is its method of verification (except tautologies), that is the way in which it is shown to be true.

(vii) Metaphysical propositions are not verifiable and hence not meaningful.

In the light of (iv), the logical positivists held that epistemology just is the philosophy of science. Their projects included:

(a) the analysis of the meanings of theoretical terms in terms of observations or experiences — this is often referred to as operationalism

(b) the explication of the 'logic of confirmation', that is how evidence can confirm a hypothesis or theory

(c) show that a priori knowledge of mathematics and logic is compatible with the verification principle by showing that mathematics is reducible to logic and that logic is analytic.

(b) is discussed in the (III) of 3. Methodological Positions. (c) is beyond the scope of the present work. With respect to (a), an example of such a definition of a theoretical term V_T is:

$$\forall x(V_T x \leftrightarrow [Px \rightarrow Qx])$$

where P is some preparation of an apparatus (known as a test condition) and Q is some observable response of it (so P and Q are describable using only V_O terms). For example, suppose it is the explicit definition of temperature; any object x has a temperature of t iff it is the case that, if x is put in contact with a thermometer then it gives a reading of t. If theoretical terms could be so defined, then this would show that they are convenient devices that are in principle eliminable and need not be regarded as referring to anything in the world (this view is called 'semantic instrumentalism').

It was soon realised that explicit definition of theoretical terms is highly problematic. Perhaps the most serious difficulty is that, according to this definition, if we interpret the conditional in the square brackets as material implication, theoretical terms are trivially applicable when the test conditions do not obtain (because if the antecedent is false the material conditional is always true). In other words, everything that is never put into contact with a thermometer has temperature t.

The natural way to solve this problem is to allow subjunctive assertion into the explicit definitions. That is we define the temperature of object x in terms of what would happen if it were to be put into contact with a thermometer; temperature is understood as a dispositional property. Unfortunately this raises further problems. First, unactualised dispositions, such as the fragility of a glass that is never damaged, seem to be unobservable properties, and they give rise to statements whose truth conditions are problematic for empiricists, namely counterfactual conditionals such as 'if the glass had been dropped it would have broken' where the antecedent is asserted to be false. Dispositions are also modal, that is they pertain to possibility and necessity, and empiricists since Hume have disavowed objective modality. Like laws of nature and causation, dispositions are problematic for empiricists. Secondly, explicit definitions, dispositional or not, for terms like 'spacetime curvature', 'spin' and 'electron' have never been provided and there are no grounds for thinking that they could be.

When it comes to knowledge, many of the logical positivists initially adopted foundationalism about knowledge and justification but they take the foundations of knowledge to be immediate knowledge of our own sensory / perceptual states. The immediate objects of experience are called sense-data or the given, and so it was thought that the foundations of knowledge were to be given in terms of sense-data reports, which are also called protocol statements or basic proposition. These are first person singular, present tense, introspection reports and as such are supposed to be non-inferential, non-necessary, indubitable and incorrigible, and to refer solely to the content of a single experience.

One problem that the logical positivists faced was that of showing how knowledge of other minds and the public world could be built up from knowledge of private sense-data and analytic truths. Phenomenalism is the attempt to solve this problem by reducing all knowledge to knowledge of protocol statements and necessary truths: on this view physical objects are nothing but logical constructions out of actual and possible sense-experiences. Propositions asserting the existence of physical objects are analytically equivalent to ones asserting that subjects would have certain sequences of sensations in certain circumstances. A physical object is a permanent possibility of sensation (Mill).

Other problems concerned the status of the verification principle given that it appears to be neither empirically testable nor analytic, the fact that observation is theory-laden in the sense that all descriptions of observations involve interpretation and classification, and finally the problem of elucidating the logic of confirmation in the face of the problem of induction.

(III) Induction

In the broadest sense induction is any reasoning that is not deductively valid. In a narrower sense it is reasoning to a conclusion about unobserved cases on the basis of observed cases. There is also an even more narrow sense of induction that refers to inferences from finite sets of data to a universal generalisation; this is enumerative induction. The most general problem of induction is to explain when and how ampliative reasoning can be justified. The more specific problem is to explain how reasoning based on knowledge of unobserved cases can be a source of knowledge about unobserved cases.

Hume's problem of induction begins with the observation that all such reasoning is based on our knowledge of cause and effect. Given his analysis of causation, knowledge of cause and effect can only be knowledge that some regularity in has held in the past. Hence, induction is based on the assumption that the behaviour of things in the past is a reliable guide to their behaviour in the future, in other words it is based on the idea that nature is uniform in this respect. Hume then points out that the only reason we have for thinking that nature is uniform in the sense that the past is a good guide to the future is that in the past the past was a good guide to what was then the future. Hence, the justification of induction turns out to depend on circular reasoning and is therefore no justification at all.

There are a number of purported solutions and dissolutions of the problem of induction.

(a) Induction is rational by definition (analytic justification). The idea here is to argue that it is part of the ordinary meaning of the term 'rational' that inference from observed cases to unobserved cases can be rational.

(b) Hume is asking for a deductive defence of induction which is unreasonable. The claim is that just because induction cannot be deductively justified that does not mean induction is not justified.

(c) Induction is justified by the theory of probability. The idea is to construct an inductive logic by analogy with deductive logic. There have been some partial successes in this programme but it is generally agreed that they do not solve the problem of induction. The best that can be said to have been achieved is to show that if any form of non-ampliative rules of reasoning are to be employed then it is best to adopt standard induction. (This is sometimes called the pragmatic defence of induction and is associated with Reichenbach.)

(d) Induction is justified by a principle of induction or of the uniformity of nature. This principle could be claimed to be known a priori, since the claim that we know it a posteriori is denounced by Hume as circular.

(e) Hume's argument is too general. Since it does not appeal to anything specific about our inductive practices, it can only be premised on the fact that induction is not deduction.

(f) Induction is really (a species of) inference to the best explanation (see (IV)), which is justified.

(g) There really are necessary connections and we know that there are such. (It is often claimed that we know this by inference to the best explanation.)

(h) Induction can be inductively justified after all, because even deduction can only be given a circular (in other words, deductive) justifications.

(i) It may be agreed that induction is unjustified and an account of knowledge, in particular scientific knowledge, may be offered which dispenses with the need for inductive inference.

Note that these strategies may be combined.

(IV) Scientific Realism

Realism in the general sense has many faces, and this goes for scientific realism too. Critics differ as to which part of scientific realism it is to which they object, so there is a bewildering complexity of positions. In some contexts the significance of scientific realism is its commitment to the progressive and convergent nature of scientific inquiry and the privileging of the cognitive outcomes of that inquiry. This is a point of contention with some critics of science, but in recent times the debates about scientific realism in analytic philosophy of science have not questioned the success nor indeed the progress of science. Hence, for example, Bas van Fraassen, Larry Laudan, and Arthur Fine will all agree about the rationality of the scientific enterprise and its cumulative production of instrumental knowledge, even though none of them would be happy to be called a scientific realist. Scientific realists are united (and divided from sceptics) in their belief that scientific theories embody real knowledge about the world that goes beyond the observable, and further that the unobservable entities to which scientific theories refer really exist. If scientific theories are sets of statements, including laws, about observable phenomena and unobservable entities, processes and structures, then scientific realism claims that these are approximately true. Hence, according to scientific realism, scientific claims about electrons, quarks, spacetime curvature, and the energy of the vacuum are more or less true, and there really are such things to which these claims refer.

Notice that it has been implicitly supposed above that the language of science is to be taken literally pace the verificationist tradition that attempted to reconstrue theoretical talk as code for complicated sets of conditionals connecting observables. That such a project fails is taken for granted by all the main protagonists in contemporary philosophy of science. However, instead of reconstruing theoretical terms some antirealists reconstrue truth for claims about theoretical entities. So, we have not yet adequately characterised scientific realism. A social constructivist, for example, could assent to all the above, since they need not deny that theoretical terms refer nor that theories are true, but they may insist that truth is internal to our epistemic norms and practices, and that the entities to which we refer are socially constructed. There is no restriction to noncognitivist conceptions of truth in what has been said so far. This raises the question of to what extent a stand on such philosophical issues in defending scientific realism. For some, scientific realism simply amounts to the commitments at the end of the foregoing paragraph. However, usually scientific realists go further and commit themselves to the following claims

(i) the entities or kinds of entities talked about and/or described by theoreticalscientific discourse exist

(ii) their existence is independent of our knowledge and minds

These are the metaphysical requirements.

(iii) the statements of theoretical scientific discourse are irreducible/ineliminable and are genuinely assertoric expressions

(iv) the truth conditions for the statements of theoretical scientific discourse are objective and determine the truth or falsity of those statements depending on how things stand in the world.

These semantic requirements are often cashed out in terms of a correspondence theory of truth, as opposed to a pragmatic or a coherence theory of truth.

(v) truths about theoretical and unobservable entities are knowable and we do in fact know some of them, and hence the terms of theoretical scientific discourse successfully refer to things in the world.

This is the epistemic requirement.

For example, if we are considering electron theory then scientific realism says that:

(i) electrons exist

(ii) mind-independently

(iii) statements about electrons are really about subatomic entities with negative charge, spin 1/2, a certain mass, and so on

(iv) these statements are true or false depending on how the world is

(v) we should believe electron theory and much of it counts as knowledge

So standard scientific realism involves three kinds of philosophical commitment: a metaphysical commitment to the existence of a mind-independent world of observable and unobservable objects; a semantic commitment to the literal interpretation of scientific theories and a correspondence theory of truth; and finally an epistemological commitment to the claim that we can know that our best current theories are approximately true, and that they successfully refer to (most of) the unobservable entities they postulate which do indeed exist. To be an antirealist it is only necessary to reject one of these commitments, and antirealists may have very different motives, so there are a variety of antirealist positions which we ought now to be able to distinguish: Sceptics deny (i), reductive empiricists deny (iii), social constructivists deny (ii), while constructive empiricists like Bas van Fraassen deny only (v), but also don't believe or remain agnostic about (i).

(V) The Duhem-Quine Problem and Underdetermination

It is natural to suppose that scientific theories or hypotheses are tested by predictions being deduced from them. Then the appropriate experiment is performed and if the prediction agrees with what is observed then the theory or hypothesis is confirmed and if not it is falsified. However, in practice it is not possible to deduce statements about what will be observed from a single hypothesis. Rather, hypotheses have to be conjoined with background assumptions about the initial conditions of the system(s) in question, the reliability of the measurement procedures, and other relevant facts. Hence Pierre Duhem argued that experiments cannot confirm or falsify individual laws or hypotheses but only a whole collection of them. Consider the experimental test of Newton's law of gravitation by the observation of the path of a planet. The law of gravitation alone will not issue any prediction without values being given to variables representing the mass of the planet, the mass of the other planets in the solar system and the Sun and their relative positions and velocities, the initial position and velocity of the planet relative to the other planets and the Sun, and the gravitational constant. Newton's other laws of motion will also be needed. Once we have a prediction then we can observe whether it is confirmed or falsified, but suppose that it is the latter that occurs; which of the laws and assumptions we have made should we regard as being falsified? Perhaps none of them have been falsified because a mistake was made in the observation. Hence, Duhem argued that science must be treated as whole when it comes to testing it and considering the evidence for it, because no part of science on its own has determinate empirical content. This is often referred to as 'confirmational holism'.

Quine went further and argued that in principle even mathematics and logic, the laws of which must be used in deriving predictions from scientific theories, must be included in the whole that is confirmed or falsified by the experimental data. Quine argued that it would be reasonable to reject a law of logic, or change the meaning of our terms, if it was more convenient than rejecting a particular theory. Quine therefore rejected the distinction between analytic and synthetic truths that Hume, Kant and the logical positivists believed to be fundamental to epistemology. A trivial example of such a change in the meaning of a term is that of the change in meaning of 'atom' which once meant something indivisible and now refers to a particular type of collection of smaller particles. When physicists discovered that atoms were divisible, they redefined 'atom' rather than abandoning the term altogether.

The Duhem-Quine problem is that no part of science seems to be testable individually, and that therefore it is never possible to say that the a particular hypothesis or law is confirmed or falsified. In practice of course scientists do locate confirmation and falsification at the level of individual hypotheses. Duhem thinks that they use good sense to do so, but that this faculty and the basis on which such judgements are made cannot be fully characterised. Quine is a pragmatist and accepts that scientific knowledge is ultimately conventional. The Duhem-Quine problem is closely related to another problem that particularly undermines scientific realism, namely the underdetermination problem. There are two generic forms of underdetermination, namely weak and strong.

(i) Weak underdetermination:

1. Some theory, T is supposed to be known, and all the evidence is consistent with T.

2. There is another theory $T\#$ which is also consistent with all the available evidence for T. T and $T\#$ are weakly empirically equivalent in the sense that they are both compatible with the evidence we have gathered so far.

3. If all the available evidence for T is consistent with some other hypothesis $T\#$, then there is no reason to believe T to be true and not $T\#$.

Therefore, there is no reason to believe T to be true and not $T\#$.

This kind of underdetermination problem is faced by scientists every day, where T and $T\#$ are rival theories but agree with respect to the classes of phenomena that have so far been observed. What scientists try and do to address it is to find some phenomenon about which the theories give different predictions, so that some new experimental test can be performed to chose between them. For example, T and $T\#$ might be rival versions of the standard model of particle physics which agree about the phenomena that are within the scope of current particle accelerators but disagree in their predictions as to what will happen at even greater energies. The weak underdetermination argument is a form of the problem of induction: T is any empirical law, such as all metals expand when heated, and $T\#$ states that everything observed so far is consistent with T but that the next observation will be different. This form of underdetermination does not undermine scientific realism in particular since it does not entail or rely upon any epistemic differentiation between statements about observables and statements about unobservables.

(ii) Strong underdetermination:

To generate a strong underdetermination problem for scientific theories, we start with a theory H, and generate another theory G, such that H and G have the same empirical consequences, not just for what we have observed so far, but also for any possible observations we could make. If there are always such strongly empirically equivalent alternatives to any given theory, then this might be a serious problem for scientific realism. The relative credibility of two such theories cannot be decided by any observations even in the future and therefore, it is argued, theory choice between them would be underdetermined by all possible evidence. If all the evidence we could possibly gather would not be enough to discriminate between a multiplicity of different theories, then we could not have any rational grounds for believing in the theoretical entities and approximate truth of any particular theory. Hence, scientific realism would be undermined.

The strong form of the undetermination argument for scientific theories is as follows:

1. For every theory there exist an infinite number of strongly empirically equivalent but incompatible rival theories.

2. If two theories are strongly empirically equivalent then they are evidentially equivalent.

3. No evidence can ever support a unique theory more than its strongly empirically equivalent rivals, and theory-choice is therefore radically underdetermined.

Some who accept this argument adopt conventionalism according to which the choice among empirically equivalent rivals is a pragmatic one that involves freely chosen conventions based on simplicity and convenience. However, there are various ways of arguing that strong empirical equivalence is incoherent, or at least ill-defined:

(a) The idea of empirical equivalence requires it to be possible to clearly circumscribe the observable consequences of a theory. However, there is no non-arbitrary distinction between the observable and unobservable.

(b) The observable/unobservable distinction changes over time and so what the empirical consequences of a theory are is relative to a particular point in time.

(c) Theories only have empirical consequences relative to auxiliary assumptions and background conditions. So the idea of the empirical consequences of the theory itself is incoherent.

Furthermore, it may be argued that there is no reason to believe that there will always, or often, exist strongly empirically equivalent rivals to any given theory, either because cases of strong empirical equivalence are too rare, or because the only strongly empirically equivalent rivals available are not genuine theories (against this it is often claimed that quantum physics gives genuine examples of empirical equivalence). Whether or not any of these objections to (1) works, many scientific realists argue that (2) is false. They argue that two theories may predict all the same phenomena, but have different degrees of evidential support. In other words, they think that there are non-empirical features (superempirical virtues) of theories such as simplicity, non-ad hocness, novel predictive power, elegance, and explanatory power, that give us a reason to chose one among the empirically equivalent rivals. Some philosophers agree that superempirical virtues break underdetermination at the level of theory choice, but argue, following van Fraassen, that their value is merely pragmatic, insofar as they encourage us to chose a particular theory with which to work, without giving us any reason to regard it as true. This may motivate the conclusion that science can never give us knowledge of the unobservable world, and that our best scientific theories are empirically adequate rather than true. Strong empirical equivalence shows that theories have extra structure over and above that which describes observable events, so clearly belief in empirical adequacy is logically weaker than belief in truth simpliciter. Note however, that even if the choice among competing ways of embedding empirically equivalent substructures in fundamental theory is a pragmatic one, ultimately different formulations may lead naturally to the discovery of new laws. For example, Newton's force law suggested the mathematical form for Coulomb's law.

The problem that critics of scientific realism, who are not also inductive sceptics, face is how to overcome the weak underdetermination argument. It may be argued that the same superempirical considerations that entitle us to regard a well-tested theory as describing future observations as well as past ones, also entitle us to choose a particular theory among strongly empirically equivalent ones. The particular strong underdetermination problem for scientific realism is that all the facts about observable states of affairs will underdetermine theory-choice between T_0, a full realistically construed theory, and T_1, the claim that T_0 is empirically adequate. However, all the evidence we have available now will underdetermine the choice between T_1 and T_2, the claim that T_0 is empirically adequate before the year 2010 (the problem of induction). Furthermore all the facts about all actually observed states of affairs at all times will underdetermine the choice between T_1 and T_3, the claim that T_0 describes all actually observed events. So, even the judgement that T_0 is empirically adequate is underdetermined by the available evidence, and hence, the advocate of the underdetermination argument against scientific realism must be an inductive sceptic in the absence of a positive solution to the underdetermination problem.

(VI) Inference to the Best Explanation

Inference to the best explanation (IBE) is a (putative) rule of inference according to which, where we have a range of competing hypotheses all of which are empirically adequate to the phenomena in some domain, we should infer the truth of the hypothesis which gives us the best explanation of those phenomena. It is often claimed that IBE gives us justified beliefs and knowledge. It certainly seems that in everyday life when faced with a range of hypotheses that all account for some phenomenon, we usually adopt the one which best explains it. Here is an example from van Fraassen: Suppose you hear scratching in the wall of your house, the patter of little feet at midnight, and cheese keeps disappearing. You would doubtless infer that a mouse has taken up residence [1980, 19]. This inference has the structure, if p then q, q therefore p, in other words, you know that if there is a mouse then there will be droppings, noises and other observable evidence, and you observe the evidence and so infer the existence of a mouse. However, consider the following: if something is a square, then it has four sides, a rectangle has four sides, therefore it is a square; this is deductively invalid because it is possible for the conclusion to be false when the premises are both true, for example, if two of the sides of the rectangle are of different lengths (this is the fallacy called 'affirming the consequent'). Similarly, there is no contradiction in supposing that there is no mouse in the house despite the evidence, so that instance of inference to the best explanation is also deductively invalid.

IBE is usually ampliative and invalid so the problem is to explain what distinguishes justifiable and knowledge-producing instances of IBE, from other invalid inferences that are clearly just bad reasoning. Here are some features that instances of IBE might be required to have:

1. Otherwise surprising phenomena are to be expected if the hypothesis is true.
2. Predictions of empirical consequences must be inferred from the hypothesis and tested and confirmed.
3. Simple and natural hypotheses are to be favoured.

4. Hypotheses which cohere with metaphysical views are to be favoured.

5. Unifying power and wideness of scope of hypotheses are to be favoured.

6. Hypotheses that cohere with other scientific theories are to be favoured.

Inference to the best explanation is used to defend scientific realism in two ways: at the local level, the idea is that if we are to follow the same patterns of inference in philosophy, science and ordinary life then we will be scientific realists since, for example, our best explanation of many phenomena involves the theoretical unobservable entities postulated by science.

IBE is invoked by scientific realists to break the underdetermination of theory by evidence. Recall the second premise of the underdetermination argument: If two theories are empirically equivalent then they are evidentially equivalent. If two theories are empirically equivalent but one of them offers a better explanation of the phenomena, then advocates of IBE will argue that we can infer the truth of the more explanatory theory. Hence advocates of IBE think that the explanatory power of a theory is evidence for its truth and hence that the second premise of the underdetermination argument is false. But van Fraassen argues that explanatory power is a merely pragmatic virtue of theories and does not give us evidence for their truth, and that IBE at the everyday level can always be recast as inference to the empirical adequacy of the best explanation. He also argues that the realist demand for explanation of every regularity leads to infinite regress.

The defence of scientific realism by appeal to IBE at the global level is based on the claim that scientific realism is the best explanation of the overall success of scientific theorising — this is known as 'the no-miracles argument' because the idea is that the success of science would be miraculous on anything but a scientific realist view. In particular, realists (following Richard Boyd) argue that we need to explain the overall instrumental success of scientific methods across the history of science. All parties in the scientific realism debate agree that:

(i) Patterns in data are projectable from the observed to the unobserved using scientific knowledge, which is to say that induction based on scientific theories is reliable.

(ii) The degree of confirmation of a scientific theory is heavily theory-dependent, in the sense that background theories inform judgements about the extent to which different theories are supported by the available evidence.

(iii) Scientific methods are instrumentally reliable, in other words, they are reliable ways of achieving practical goals like prediction and the construction of technological devices.

Scientific realists argue that these features of science would be utterly mysterious if the theories involved were not true or approximately true.

Another feature of scientific practice that realists have long argued cannot be explained by antirealists is the persistent and often successful search for unified theories of diverse phenomena. The well known 'conjunction objection' against antirealism is as follows: Consider two scientific theories, T and T', from different domains of science, say chemistry and physics. That T and T'' are both empirically adequate does not imply that their conjunction T & T' is empirically adequate, however, if T and T' are both true this does imply that T & T' is true. So, the argument goes, only realists are motivated to believe the new empirical consequences obtained by conjoining accepted theories. However, it is claimed that in the course of the history of science the practice of theory conjunction is widespread and a reliable part of scientific methodology. Therefore, if scientists are not irrational, since only realism can explain this feature of scientific practice, then realism must be true.

A fundamental criticism of the use of IBE at the global level was made by Larry Laudan and Arthur Fine, both of whom pointed out that since it is IBE involving unobservables that is in question in the realism debate, it is circular to appeal to the explanatory power of scientific realism at the meta-level to account for the overall success of science because realism is itself a hypothesis involving unobservables. Hence, it is argued that the global defence is question begging. There is a similarity here with the inductive vindication of induction. Richard Braithwaite, and Carnap, defended the view that the inductive defence of induction — induction has worked up to now so it will work in the future — was circular but not viciously so, because it is rule circular not premise circular. In the case of IBE such a view has been defended by David Papineau and Stathis Psillos. The idea is that premise circularity of an argument is vicious because the conclusion is taken as one of the premises; on the other hand rule circularity is when the conclusion of an argument states that a particular rule is reliable, but that conclusion only follows from the premises when that very rule is used. Now notice that the global defence of realism is rule but not premise circular. The conclusion that the use of IBE in science is reliable is not a premise of this defence of realism, but the use of IBE is required to reach this conclusion from the premises that IBE is part of scientific methodology and that scientific methodology is instrumentally reliable. It is conceded that, although it is not viciously circular, this style of argument will not persuade someone who totally rejects IBE. However, what the argument is meant to show is that someone who does make abductive inferences can show the reliability of their own methods. So, it seems that IBE is on a par with inductive reasoning; it cannot be defended by a noncircular argument, but recall that even deduction cannot be defended by a non-circular argument either. Hence, the realist may claim that although they cannot force the non-realist to accept IBE, they can show that its use is consistent and then argue that it forms part of a comprehensive and adequate philosophy of science.

However, an antirealist could agree with the descriptive claim that often our inductive inferences are guided by explanatory considerations, and accept that to be so guided is not prohibited by the canons of rationality. However, it may be argued that nobody is ever rationally compelled to believe something because it is the best explanation of the phenomena. Furthermore, arguably IBE is only pragmatically motivated in general: as it turns out, being guided by explanatory considerations has led us to arrive at empirically adequate theories, and that gives us some reason to search for explanations in the future, but we should not admit explanatory considerations as reasons for belief if we are good empiricists, and we should certainly never regard IBE as rationally compelling. It may be objected that it is capricious to use inference to the best explanation widely, but to always abstain from inferring the truth of the conclusion in the case of unobservable entities. However, there is a salient difference between inferring the existence of an unobserved observable and inferring the existence of an unobservable, namely that the former case is usually the inferring of the existence of an unobserved token of an observed type that is at issue. (In the next section it is shown that the history of science gives us further reasons to be wary of committing ourselves to the existence of the unobservables postulated to explain observable phenomena.)

Van Fraassen offers several arguments against the idea that IBE is a compelling rule of inference:

(i) The Argument from Indifference

The argument from indifference is roughly that, since there are many ontologically incompatible yet empirically equivalent theories, we have no reason to choose among them and indentify one of them as true. This argument appeals to the existence of empirical equivalents to any theory that we have. In the discussion of the underdetermination problem above it was concluded that the antirealist may also be threatened by the existence of empirical equivalents since any finite set of theories that we consider is just as highly unlikely to contain an empirically adequate theory. However, this does not help defend IBE.

(ii) The Argument from the Best of a Bad Lot

This argument is that some 'principle of privilege' is required if we are to think that the collection of hypotheses that we have under consideration will include the true theory. The best explanatory hypothesis we have may just be the best of a bad lot, all of which are false. In other words this argument challenges the proponent of IBE to show how we can know that none of the other possible explanations we have not considered is as good as the best that we have. Unless we know that we have included the best explanation in our set of rival hypotheses, even if it were the case that the best explanation is true, this would not make IBE an acceptable rule of inference.

Realists tend to bite this bullet and argue that scientists do have privilege which issues from background knowledge. Theory choice is informed by background theories which narrow the range of hypotheses under consideration, and then explanatory considerations help select the best hypothesis. Furthermore, they argue that both the realist and the constructive empiricist need privilege, because the constructive empiricist needs to assume that the empirically adequate theory is among the ones considered in order to have warranted belief in the empirical adequacy of the chosen theory. Hence the dispute can only be about the extent of that privilege.

(iii) The argument from Bayesianism

The idea here is that any rule for the updating of belief that goes beyond the rules of Bayesian conditionalisation (see VII of 3) will lead to probabilistic incoherence.

(VII) Arguments from Theory Change

Unlike the underdetermination problem which may seem to be generated a priori, the arguments against scientific realism from theory change are empirically based and their premises are derived from data obtained by examining the practice and history of science. Furthermore, ontological discontinuity across radical changes in theories seems to give us grounds not merely for doubt, but for the positive belief that many central theoretical terms of our best contemporary science will be regarded as non-referring by future science. Hence, the strongest argument from theory change has as its conclusion that scientific realism is not true because it is not even empirically adequate. The argument in question is the 'pessimistic meta-induction', and was anticipated by the ancient Greek sceptics, but in its contemporary form it is due to Larry Laudan. It has the following structure:

(i) There have been many empirically successful theories in the history of science which have subsequently been rejected and whose central theoretical terms do not refer according to our best current theories.

(ii) Our best current theories are no different in kind from those discarded theories and so we have no reason to think they will not ultimately be replaced as well.

So, by induction we have positive reason to expect that our best current theories will be replaced by new theories according to which some of the central theoretical terms of our best current theories do not refer, and hence, we should not believe in the approximate truth or the successful reference of the theoretical terms of our best current theories.

The most common realist response to this argument is to restrict realism to theories with some further properties (usually, maturity, and novel predictive success) so as to cut down the inductive base employed in (i). However, assuming that such an account can be given there are still a couple of cases of mature theories which enjoyed novel predictive success by anyone's standards, namely the ether theory of light and the caloric theory of heat. If their central theoretical terms do not refer, the realist's claim that approximate truth explains empirical success will no longer be enough to establish realism, because we will need some other explanation for success of the caloric and ether theories. If this will do for these theories then it ought to do for others where we happened to have retained the central theoretical terms, and then we do not need the realist's preferred explanation that such theories are true and successfully refer to unobservable entities. To be clear:

(a) Successful reference of its central theoretical terms is a necessary condition for the approximate truth of a theory.

(b) There are examples of theories that were mature and had novel predictive success but which are not approximately true.

(c) Approximate truth and successful reference of central theoretical terms is not a necessary condition for the novel-predictive success of scientific theories

So, the no-miracles argument is undermined since, if approximate truth and successful reference are not available to be part of the explanation of some theories' novel predictive success, there is no reason to think that the novel predictive success of other theories has to be explained by realism.

Hence, we do not need to form an inductive argument based on Laudan's list to undermine the no-miracles argument for realism. Laudan's paper was also intended to show that the successful reference of its theoretical terms is not a necessary condition for the novel predictive success of a theory, and that there are counter-examples to the no-miracles argument.

There are two basic (not necessarily exclusive) responses to this:

(I) Develop an account of reference according to which the abandoned theoretical terms are regarded as referring after all.

(II) Restrict realism to those parts of theories which play an essential role in the derivation of subsequently observed (novel) predictions, and then argue that the terms of past theories which are now regarded as non-referring were non-essential so there is no reason to deny that the essential terms in current theories will be retained.

Realists have used causal theories of reference to account for continuity of reference for terms like 'atom' or 'electron', when the theories about atoms and electrons undergo significant changes. The difference with the terms 'ether' and 'caloric' is that they are no longer used in modern science. In the nineteenth century the ether was usually envisaged as some sort of material solid that permeated all of space. It was thought that light waves had to be waves in some sort of medium and the ether was posited to fulfil this role. Yet if there really is such a medium then we ought to be able to detect the effect of the Earth's motion through it, because light waves emitted perpendicular to the motion of a light source through the ether ought to travel a longer path than light waves emitted in the same direction as the motion of the source through the ether. Of course, various experiments, the most famous being that of Michaelson and Morley, failed to find such an effect. By then Maxwell had developed his theory of the electromagnetic field, which came to be regarded as not a material substance at all. As a result the term 'ether' was eventually abandoned completely.

However, the causal theory of reference may be used to defend the claim that the term 'ether' referred after all, but to the electromagnetic field rather than to a material medium. If the reference of theoretical terms is to whatever causes the phenomena responsible for the terms' introduction, then since optical phenomena are now believed to be caused by the oscillations in the electromagnetic field, than the latter is what is referred to by the term 'ether'. Similarly, since heat is now believed to be caused by molecular motions, then the term 'caloric' can be thought to have referred all along to these rather than to a material substance. The danger with this is that it threatens to make the reference of theoretical terms a trivial matter, since as long as some phenomena prompt the introduction of a term it will automatically successfully refer to whatever is the relevant cause (or causes). Furthermore, this theory radically disconnects what a theorist is talking about from what they think they are talking about. For example, Aristotle or Newton could be said to be referring to geodesic motion in a curved spacetime when, respectively, they talked about the natural motion of material objects, and the fall of a body under the effect of the gravitational force.

The essence of the second strategy is to argue that the parts of theories that have been abandoned were not really involved in the production of novel predictive success. Philip Kitcher says that: "No sensible realist should ever want to assert that the idle parts of an individual practice, past or present, are justified by the success of the whole" [1993, 142]. Kitcher suggests a model of reference according to which some tokens of theoretical terms refer and others do not, but his theory allows that the theoretical descriptions of the theoretical kinds in question may have been almost entirely mistaken, and seems to defend successful reference only for those uses of terms that avoid ontological detail at the expense of reference to something playing a causal role in producing some observable phenomena.

Similarly, Stathis Psillos argues that history does not undermine a cautious scientific realism that differentiates between the evidential support that accrues to different parts of theories, and only advocates belief in those parts that are essentially involved in the production of novel predictive success. This cautious, rather than an all or nothing, realism would not have recommended belief in the parts of the theories to which Laudan draws attention, because if we separate the components of a theory that generated its success from those that did not we find that the theoretical commitments that were subsequently abandoned are the idle ones. On the other hand, argues Psillos: "the theoretical laws and mechanisms that generated the successes of past theories have been retained in our current scientific image" [1999, 108]. Such an argument needs to be accompanied by specific analyses of particular theories which both identify the essential contributors to the success of the theory in question, and show that these were retained in subsequent developments.

Psillos takes up Kitcher's suggestion of (II) and combines it with (I). Laudan claims that if current successful theories are approximately true, then the caloric and ether theories cannot be because their central theoretical terms don't refer (by premise (ii) above). Strategy (I) accepts premise (ii) but Psillos allows that sometimes an overall approximately true theory may fail to refer. He then undercuts Laudan's argument by arguing that abandoned theoretical terms that do not refer, like 'caloric', were involved in parts of theories not supported by the evidence at the time, because the empirical success of caloric theories was independent of any hypotheses about the nature of caloric. Abandoned terms that were used in parts of theories supported by the evidence at the time do refer after all; 'ether' refers to the electromagnetic field. The problem with strategy (II) is that its applications tend to be ad hoc and dependent on hindsight. Furthermore, by disconnecting empirical success from the detailed ontological commitments in terms of which theories were described, it seems to undermine rather than support realism.

As we have seen, in the debate about scientific realism, the no-miracles argument is in tension with the arguments from theory-change. In an attempt to break this impasse, and have "the best of both worlds", John Worrall [1989] introduced structural realism (although he attributes its original formulation to Poincaré). Using the case of the transition in nineteenth century optics from Fresnel's elastic solid ether theory to Maxwell's theory of the electromagnetic field, Worrall argues that:

> There was an important element of continuity in the shift from Fresnel to Maxwell — and this was much more than a simple question of carrying over the successful empirical content into the new theory. At the same time it was rather less than a carrying over of the full theoretical content or full theoretical mechanisms (even in "approximate" form) ... There was continuity or accumulation in the shift, but the continuity is one of form or structure, not of content. [1989, 117]

According to Worrall, we should not accept full blown scientific realism, which asserts that the nature of things is correctly described by the metaphysical and physical content of our best theories. Rather we should adopt the structural realist emphasis on the mathematical or structural content of our theories. Since there is (says Worrall) retention of structure across theory change, structural realism both (a) avoids the force of the pessimistic meta-induction (by not committing us to belief in the theory's description of the furniture of the world) and (b) does not make the success of science (especially the novel predictions of mature physical theories) seem miraculous (by committing us to the claim that the theory's structure, over and above its empirical content, describes the world). A different form of structural realism is also defended by Steven French and James Ladyman in the context of interpreting contemporary physics.

(VIII) Contemporary Empiricism

The constructive empiricism of van Fraassen has provoked renewed debate about scientific realism. Van Fraassen accepts the semantic and metaphysical components of scientific realism, but, he denies the epistemic component. So he thinks that scientific theories about unobservables should be taken literally, and are true or false in the correspondence sense, depending on whether the entities they describe are part of the mind-independent world. However, he argues that acceptance of the best theories in modern science does not require belief in the entities postulated by them, and that the nature and success of modern science relative to its aims can be understood without invoking the existence of such entities.

Van Fraassen defines scientific realism as follows:

> Science aims to give us, in its theories, a literally true story of what the world is like; and acceptance of a scientific theory involves the belief that it is true. [1980, 8]

Constructive empiricism is the view that:

> Science aims to give us theories which are empirically adequate; and acceptance of a theory involves as belief only that it is empirically adequate. [Ibid., 12]

To say that a theory is empirically adequate is to say: "What it says about the observable things and events in this world, is true (ibid.)". In other words:

> the belief involved in accepting a scientific theory is only that it 'saves the phenomena', that is that it correctly describes what is observable. [Ibid., 4]

Note that this means that it saves *all* the *actual* phenomena, past present and future, not just those that have been observed so far, so even to accept a theory as empirically adequate is to believe something more than is logically implied by the data [ibid., 12, 72]. Moreover, for van Fraassen, a phenomenon is simply an *observable* event and not necessarily an observed one. So a tree falling over in a forest is a phenomenon whether or not someone actually witnesses it.

The scientific realist and the constructive empiricist disagree about the purpose of the scientific enterprise: the former thinks that it aims at truth with respect to the unobservable processes and entities that *explain* the observable phenomena; the latter thinks that the aim is merely to tell the truth about what is observable, and rejects the demand for explanation of all regularities in what we observe. Van Fraassen says that explanatory power is not a "rock bottom virtue" of scientific theories whereas consistency with the phenomena is [ibid., 94]. Hence, for the constructive empiricist, empirical adequacy is the internal criterion of success for scientific activity.

Note that

(a) Both doctrines are defined in terms of the aims of science, so constructive empiricism is fundamentally a view about the aims of science and the nature of 'acceptance' of scientific theories, rather than a view about whether electrons and the like exist. Strictly speaking it is possible to be a constructive empiricist and a scientific agnostic, or a scientific realist and scientific agnostic. That said, it is part of van Fraassen's aim to show that abstaining from belief in unobservables is perfectly rational and scientific.

(b) Scientific realism has two components: (i) theories which putatively refer to unobservable entities are to be taken literally as assertoric and truth-apt claims about the world (in particular, as including existence claims about unobservable entities); and (ii) acceptance of these theories (or at least the best of them) commits one to belief in their truth or approximate truth in the correspondence sense (in particular, to belief that tokens of the types postulated by the theories in fact exist). Van Fraassen is happy to accept (i). It is (ii) that he does not endorse. Instead, he argues that acceptance of the best theories in modern science does not require belief in the entities postulated by them, and that the nature and success of modern science relative to its aims can be understood without invoking the existence of such entities.

(c) Empirical adequacy for scientific theories is characterised by van Fraassen in terms of the so-called 'semantic' or 'model-theoretic' conception of scientific theories, the view that theories are fundamentally extra-linguistic entities (models or structures), as opposed to the syntactic account of theories, which treats them as the deductive closure of a set of formulas in first order logic. The semantic view treats the relationship between theories and the world in terms of isomorphism. On this view, loosely speaking, a theory is empirically adequate if it "has at least one model which all the actual phenomena fit inside" [1980, 12].

Initial criticism of van Fraassen's case for constructive empiricism concentrated on three issues:

(i) The line between the observable and the unobservable is vague and the two domains are continuous with one another; moreover the line between the observable and the unobservable changes with time and is an artefact of accidents of human physiology and technology. This is supposed to imply that constructive empiricism grants ontological significance to an arbitrary distinction.

(ii) Van Fraassen eschews the positivist project which attempted to give an a priori demarcation of predicates that refer to observables from those that refer to unobservables, and accepts instead that: (a) all language is theory-laden to some extent; and (b) even the observable world is described using terms that putatively refer to unobservables. Critics argue that this makes van Fraassen's position incoherent.

(iii) The underdetermination of theory by evidence is the only positive argument that van Fraassen has for adopting constructive empiricism instead of scientific realism; but all the data we presently have underdetermine which theory is empirically adequate just as they underdetermine which theory is true (this is the problem of induction), and so constructive empiricism is just as vulnerable to scepticism as scientific realism. This is taken to imply that van Fraassen's advocacy of constructive empiricism is the expression of an arbitrarily selective scepticism.

(i) is rebutted firstly by pointing out that vague predicates abound in natural language but clear extreme cases suffice to render their use acceptable, and secondly by arguing that epistemology ought to be indexical and anthropocentric, and that the distinction between the observable and the unobservable is not to be taken as having direct ontological significance, but rather epistemological significance. Says van Fraassen: "even if observability has nothing to do with existence (is, indeed, too anthropocentric for that), it may still have much to do with the proper epistemic attitude to science" [1980, 19].

For van Fraassen, 'observable' is to be understood as 'observable-to-us": "X is observable if there are circumstances which are such that, if X is present to us under those circumstances, then we observe it" [1980, 16]. What we can and cannot observe is a consequence of the fact that

> the human organism is, from the point of view of physics, a certain kind of measuring apparatus. As such it has certain inherent limitations — which will be described in detail in the final physics and biology. It is these limitations to which the ''able' in 'observable' refers — our limitations, *qua* human beings. [1980, 17]

So we know that, for example, the moons of Jupiter are observable because our current best theories say that were astronauts to get close enough, then they *would* observe them; by contrast the best theories of particle physics certainly do not tell us that we are directly observing the particles in a cloud chamber. Analogous with the latter case is the observation of the vapour trail of a jet in the sky, which does not count as observing the jet itself, but rather as detecting it. Now if subatomic particles exist as our theories say that they do, then we detect them by means of observing their tracks in cloud chambers, but, since we can never experience them directly (as we can jets), there is always the possibility of an empirically equivalent but incompatible rival theory which denies that such particles exist. This fact may give the observable/unobservable distinction epistemic significance. Note, that van Fraassen adopts a direct realism about perception for macroscopic objects: "we can and do see the truth about many things: ourselves, others, trees and animals, clouds and rivers — in the immediacy of experience" [1989, 178].

(ii) is rebutted by showing that there are at least some entities which if they exist are unobservable, for example, quarks, spin states of sub-atomic particles, and light.

(iii) is the most serious problem for van Fraassen. Note first that, contrary to what is often claimed, van Fraassen does not accept that inference to the best explanation is rationally compelling in the case of the observable world while denying it this status for the case of the unobservable world. Furthermore, van Fraassen does not appeal to any global arguments for antirealism such as the underdetermination argument or the pessimistic meta-induction. He rejects realism not because he thinks it irrational but because he rejects the "inflationary metaphysics" which he thinks must accompany it, i.e., an account of laws, causes, kinds and so on, and because he thinks constructive empiricism offers an alternative view that offers a better account of scientific practice without such extravagance [1980, 73]. Empiricists should repudiate beliefs that go beyond what we can (possibly) confront with experience; this restraint allows them to say "good bye to metaphysics" [1989; 1991, 480].

What then is empiricism and why should we believe it? Van Fraassen suggests that to be an empiricist is to believe that "experience is the sole source of information about the world" [1985, 253]. The problem with this doctrine is that it does not itself seem to be justifiable by experience. However, he has argued in recent work that empiricism cannot be reduced to the acceptance of such a slogan, and that empiricism is in fact a stance in Husserl's sense of an orientation or attitude towards the world.

Constructive empiricism is supposed to offer a positive alternative to scientific realism that dispenses with the need for metaphysics. It is a positive view of science which is intended to free us from the need to articulate accounts of laws, causes, and essential properties that take seriously the apparent modal commitments of such notions. This promised liberation from metaphysics is fundamental to van Fraassen's advocacy of a constructive empiricist view of science. Indeed, from his point of view, scepticism about objective modality is partly definitive of an empiricist outlook: "To be an empiricist is to withhold belief in anything that goes beyond the actual, observable phenomena, and to recognise no objective modality in nature" [1980, 202]. However, arguably, to be a constructive empiricist one must, in spite of what van Fraassen says here, recognise objective modality in nature. This is largely because constructive empiricism recommends, on epistemological grounds, belief in the empirical adequacy rather than the truth of theories, and hence requires that there be an objective modal distinction between the observable and the unobservable.

(IX) Pragmatism

Various philosophers have defended forms of pragmatism in philosophy of science, not least because it seems to help scientific realists avoid problems like the underdetermination problem. Brian Ellis says: "scientific realism can be combined with a pragmatic theory of truth: and given such a theory of truth, all of the criteria which we use for the evaluation of theories, including the so-called pragmatic ones, can be seen as being relevant to their truth or falsity" [1985, 41]. Other forms of pragmatism include that of Ian Hacking and Nancy Cartwright who both defend entity realism, which is the idea that the unobservable entities that are posited by science can be known about even if the fundamental theoretical claims of scientific theories are not true, because the entities can be manipulated and play a role in the practical life of experimentalists, engineers and technologists.

Arthur Fine defends what he calls the Natural Ontological Attitude (NOA). NOA is, he claims, a minimalist view that avoids the philosophical conceits of both scientific realism and antirealism, and simply incorporates the 'homely line' that we should regard the pronouncements of science as on a par with everyday talk about objects observed with the senses. Realists have argued in response that NOA is all the scientific realist needs, since it says that the unobservable objects postulated by scientific theories have the same status as tables and chairs, and, in particular, this suggests that we can refer to and know the truth about them. However, Fine argues against all the standard philosophical arguments for scientific realism with some vigour, and his position may therefore seem to be antirealist. For Fine, asking whether electrons really exist is like asking whether tables do, and in both cases he refuses to engage with the question. He seems to have adopted a philosophical quietism that is consistent even with solipsism or Berkelian scepticism. Fine is quite deliberate about this since he claims that precisely what distinguishes NOA from scientific realism is that the latter involves a metaphysics of the external world, and a theory of truth and so on, while the former does not bother with them. Thus, NOA seems to be a recommendation for the abandonment of certain philosophical questions. According to him what marks out the realist or antirealist is what they add on to the everyday talk of scientists. Hence, theories of truth, whether, for example, correspondence (realist) or coherence (antirealist), are equally unnecessary and unhelpful. Rather, it seems Fine proposes that we should simply stop talking about truth per se and accept the homely truths that scientists use just as we accept the truths of everyday life.

(X) Naturalism and Normativity

Naturalism is the view that philosophy is continuous with science. According to naturalists, traditional philosophical questions concerning knowledge ought to be investigated by cognitive science and evolutionary psychology, rather than by a priori reflection. Naturalists also think that metaphysical questions can only be answered by science rather than by thought experiments and other traditional philosophical methods.

Normativity concerns not what is the case but what ought to be the case. The main source of normative claims is ethics, however, logic, rationality and reason also seem to be concerned with what ought to be the case. For example, it seems that we ought to be believe what is true, and that we ought not to be believe what is false. Those opposed to naturalism argue that it will never be possible to explain normativity in scientific terms since science can only describe how the world is and claims about what ought to be the case can never be tested. Hume famously argued that it is not possible to derive an 'ought' from an 'is', and if this is right there would seem to be something to the idea that normativity lies outside the scope of naturalism. However, this leads some philosophers to adopt scepticism about normativity and regard what we ought to do as a mere matter of convention.

3 METHODOLOGICAL POSITIONS

(I) Inductivism

The most general characterisation of inductivism is that it is any position according to which a universal generalisation is positively supported by observation of its instances. Philosophers refer to the idea of evidence positively supporting a law or theory as confirmation. The idea of confirmation is fundamental to most but not all theories of the scientific method.

Naïve inductivism states that the basic means by which scientific knowledge is advanced is generalisation from experience. It is associated with Francis Bacon who criticised the natural philosophy of his day for being insufficiently empirical and experimental. Bacon advocated the influential idea that science in any domain must begin with numerous and wide-ranging observations that are undertaken without prejudice or preconception. Many scientific laws are of the form of universal generalisations (statements that generalise about the properties of all things of a certain kind). For example, 'all metals expand when heated' is a universal generalisation about metals. Induction in the broadest sense is just any form of reasoning which is not deductive, but in the narrower sense which Bacon uses it is the form of reasoning where we generalise from a collection of particular instances to a general conclusion. The simplest form of induction is enumerative induction, which is where we observe that some large number of instances of some phenomenon has some characteristic and then infer that the phenomenon always has that property. Bacon also discussed more involved methods involving the drawing up of tables to compare and contrast different instances of some phenomenon, so that it can be inferred what all such instances have in common.

According to naïve inductivism it is legitimate to infer a universal generalisation from a collection of observation statements when a large number of observations of Xs under a wide variety of conditions have been made, and when all Xs have been found to possess property Y, and when no instance has been found to contradict the universal generalisation 'all Xs possess property Y'. This is known as a Principle of Induction. Once a generalisation has been inductively inferred in accordance with this principle then it assumes the status of a law or theory and we can use deduction to deduce consequences of the law that will be predictions or explanations.

The obvious problem with this is how to make precise the idea of a large number of observations. One common response to this problem is to claim that, given that no amount of evidence of observed cases will ever logically entail a claim about unobserved cases (the problem of induction), then it is never the case that enough observations have been made to establish a hypothesis with certainty, and that therefore we ought instead to think it terms of probabilities, so that that the larger the number of observations that have been made then the higher the probability that the universal generalisation is true. However, it is easy to see that the move to probabilities does not solve the problem since a universal generalisation covers potentially infinitely many cases, so no matter how many instances are observed, if there are a finite number, it would seem that the probability of the universal generalisation will always be small.

In any case this is simplistic, even if it works for some parts of science. For example, it is arguably impossible to engage in observation without any preconceptions or presuppositions, since theory guides the decision as to what to pay attention to when observing, and also theories often suggest experiments that might be performed to test them. Inductivists may appeal to more sophisticated kinds of induction such as Mill's methods for eliminative induction, which attempts to find the right universal generalisation by eliminating the alternatives rather than be enumerative generalisation. Nonetheless, all forms of inductivism face the problem of induction. Naïve inductivism and more sophisticated forms of inductivism face other problems that arise when it is observed that often in the history of science, great advances have been made by people who have not followed inductive methods. In particular, sometimes scientifically valuable hypotheses have been the result of speculation rather than generalisation from experience.

(II) The Context of Discovery and the Context of Justification

There is a fundamental difference between accounts of the scientific method, namely that some are accounts of how to generate scientific theories, and also how to test scientific theories and how to respond to the results of testing them, while others do not attempt to describe how scientific theories should be generated. Clearly, Bacon's inductive logic is an example of the former, since it proscribes how to begin the investigation of some range of phenomena, and the production of generalisations and laws is supposed to be an automatic outcome of the mechanical operation of the method. Examples of the latter include falsificationism which is discussed below, but also some versions of inductivism.

It may be desirable that laws and theories be derived from experimental data (as Newton claimed that he did not speculate but rather deduced the laws of mechanics from the results of observations), but in most interesting cases this is just not possible. The generation of scientific theories is not in general a mechanical procedure, but a creative activity. Scientists have drawn upon many sources of inspiration, such as metaphysical beliefs, dreams, analogies and so on, when trying to formulate a theory. The kind of speculation and imagination which scientists need to employ cannot be formalised or reduced to a set of rules, but once a hypothesis is generated it must be subject to testing by experience, and this must be the final arbiter of any scientific dispute. If this is right then, when we are thinking about scientific methodology, perhaps we ought to make a distinction between the way theories are conceived and the subsequent process of testing them. The scientific method may be silent about where hypotheses should come from, but, faced with two rival hypotheses that equally account for the data, scientists ought to construct an experimental situation (crucial experiment) about which the hypotheses will predict different outcomes.

In general, the evidence in favour of a hypothesis is independent of who believes it, and whether an idea is a good one does not depend on who first thinks of it. So it seems plausible to argue that evaluation of the evidence for a hypothesis ought to take no account of how, why and by whom the hypothesis was conceived. This distinction between the causal origins of scientific theories and their degree of confirmation and scientific status is often thought to be important for the defence of the objectivity of scientific knowledge. Many philosophers of science, who otherwise disagree with each other about fundamental matters, believe that the task of philosophy of science is to logically analyse the testing of scientific theories by observation and experiment. How theories are developed is a matter for psychology not philosophy. Scientists do not need to make presuppositionless observations, nor does it matter if they use background theories to develop new theories.

On this view, there are two contexts in which the history of science may be investigated, namely the context of discovery and the context of justification. The distinction between the context of discovery and the context of justification separates the question of how scientific theories are developed from the question of how to test them against their rivals. The degree of confirmation of a theory is a relationship between it and the evidence and is independent how it was produced.

(III) The Paradoxes of Confirmation

Any theory of confirmation must avoid the following paradoxes (see 1.I Natural Kinds).

(i) The Ravens Paradox

'Nicod's criterion' states that laws are confirmed by observation of their instances. If we assume that logically equivalent propositions are equally confirmed by the same evidence, then the logical equivalence of 'all Ravens are black' and 'all non-black things are non-ravens' implies that observation of a green leaf confirms the law that all ravens are black.

(ii) Goodman's New Riddle of Induction

Consider a law of nature of the form 'all Fs are Gs'. Construct the predicate G^* as follows: a is G^* iff a is G before time t and H after time t, where 'a is H' entails 'it is not the case that a is G'. It seems that all the evidence gathered before time t must equally support the law 'all Fs are G^*s'. Hence, one question Goodman's riddle poses is 'what is the justification for taking generalisations with ordinary predicates to be confirmed by the instances we have so far observed, but not generalisations with predicates like G^*? It seems that appeal to the uniformity of nature alone does not solve the problem of induction because the world may be uniform in different ways; if it is uniform in that all Fs are G^*s then our ordinary inductive inferences will be unreliable.

(iii) The Tacking Paradox

The special consequence condition states that evidence which supports a theory also supports the logical consequences of the theory. This is plausible because it seems to be necessary to explain why evidence that confirms a theory also gives us a reason to believe that the predictions the theory makes are true. The converse consequence condition states that evidence which supports a theory, T, also supports any other theory which entails T. This seems plausible because there are many cases in the history of science where a high level theory entailed a low level law that was already supported by the evidence, and where that evidence was then taken to also support the high-level law. However, it follows from these two conditions that any piece of evidence for an arbitrary theory supports any hypothesis whatsoever. Consider e which supports theory T. Since T is entailed by $T\&G$ for any G, it follows from the converse consequence condition that e supports $T\&G$. But then since $T\&G$ entails G, it follows from the special consequence condition that e supports G.

Each of these paradoxes seems to rely on the assumption that the relation between a theory and the evidence which supports it is a logical one. Some think the paradoxes show that a purely logical theory of confirmation is impossible. Historical theories of confirmation make the history and origin of a theory relevant to its evidential status. For example, Goodman's own response to his problem was to argue that entrenched predicates, ones that have been used in successful inductions in the past, are more confirmed by new evidence than un-entrenched ones like G^*. Historical theories of confirmation collapse the distinction between the context of discovery and the context of justification, according to which the causal history of a theory is quite irrelevant to the extent to which it is supported by the evidence. Many philosophers worry that a historical theory of confirmation is inconsistent with the idea that the evidential basis of scientific knowledge is an objective matter, and hence to invite relativism and subjectivism. On the other hand, many other philosophers have given up on the ideal of an ahistorical theory of confirmation.

(IV) Explanation versus Prediction

Carl Hempel advocates the thesis of structural identity, according to which explanations and predictions have exactly the same structure: they are arguments where the premises state laws of nature and initial conditions. The only difference between them is that, in the case of an explanation we already know that the conclusion of the argument is true, whereas in the case of a prediction the conclusion is unknown. For example, Newtonian physics predicted the return of Halley's comet in December 1758, and the same argument explains its return. However, there are many cases where the observation of one phenomenon allows us to predict the observation of another phenomenon but where the former does not explain the latter. For example, the fall of the needle on a barometer allows us to predict that there will be a storm but doesn't explain it. Similarly, the length of a shadow allows us to predict the height of the building that cast it, and the period of oscillation of a pendulum we can predict its length, but in both these cases the latter explains the former and not the other way round. There also seem to be theories that provide adequate explanations but that cannot make precise predictions. For example, evolutionary theory explains why organisms have the morphology that they do, but it cannot make specific predictions because evolutionary change is subject to random variations in environmental conditions and the genotype of organisms. Furthermore, there are cases of probabilistic explanations where the probability conferred by the explanans on the explanandum is low, so we cannot predict that the explanandum is even likely to happen although we can explain why it did if it does.

According to hypothetico-deductivism, there is also a symmetry between predictions and explanations in respect of confirmation; because an explanation is simply a prediction where the phenomenon predicted has already been observed, the degree of confirmation conferred on a theory is the same for predictions and explanations. Hypothetico-deductivism is a purely logical theory of confirmation, and the origin of a theory, in particular, when it was proposed relative to when the evidence for it was gathered, is irrelevant to its epistemic status. On the other hand, predictivists think that only successful predictions of previously unknown phenomena count as evidence, and explanationists think that only explanations of previously known about phenomena count as evidence. Intermediate positions accord some confirmational power to both predictions and explanations but weight one more highly than the other. Many scientific realists argue that novel predictions of new and unsuspected types of phenomena are of special confirmational status.

The significance of novel predictions was emphasised by Karl Popper. He contrasted the risky predictions of physics with the vague predictions of psychoanalysis, but he also wanted to justify the failure of scientists to abandon Newtonian theory when it was known to be incompatible with certain observations. Often various modifications to background assumptions are made to try and accommodate observed facts that would otherwise refute established theories. Popper, and following him Imre Lakatos and others, argued that this course of action is acceptable only when the new theory produces testable consequences other than the results that motivated it. So for example, the postulation of a new planet to accommodate the observed orbit of a familiar one is legitimate because it ought to be possible to observe the former (or at least its effects on other bodies).

Popper was particularly impressed by the experimental confirmation of Einstein's general theory of relativity in 1917. The latter predicted that light passing close to the Sun ought to have its path bent by the Sun's gravitational field. Another well known example is from optics. In 1818 Fresnel developed a mathematical theory according to which light consists of transverse waves in an optical ether. This theory predicted that in certain circumstances light that was shone on a completely opaque disk would cast a shadow with a bright white spot in its centre. However, Fresnel knew nothing of this phenomenon when he developed his theory, and indeed did not even derive the result himself. This is more striking than the prediction of the existence of an extra planet, because it is a prediction of a completely new and unexpected type of phenomenon.

The most straightforward idea of novelty is that of temporal novelty. A prediction is temporally novel when it is of something that has not yet been observed. The problem with attributing special confirmational status to this kind of novel predictive success is that it seems to introduce an element of arbitrariness into the theory of confirmation. When exactly in time someone first observes some phenomenon entailed by a theory may have nothing to do with how and why the theory was developed. It seems implausible that it should be relevant to the degree of confirmation of a theory provided by some evidence whether or not someone independently observed the evidence before the theory was produced but didn't tell anyone about it. As it turns out, the white spot phenomenon had been observed independently prior to its prediction by Fresnel's theory. A temporal account of novelty would make whether a result was novel for a theory a matter of mere historical accident and that this would undermine the epistemic import novel success is supposed to have for a particular theory.

It is more plausible to argue that what matters in determining whether a result is novel is whether a particular scientist knew about the result before constructing the theory that predicts it. Call this epistemic novelty. The problem with this account of novelty is that, in some cases, that a scientist knew about a result does not seem to undermine the novel status of the result relative to their theory, because they may not have appealed to the former in constructing the latter. For example, many physicists regarded the success of general relativity in accounting for the well-known, previously anomalous orbit of Mercury as highly confirming. Consider again the case of Fresnel. If we say that the fact that the white spot phenomenon was known about is irrelevant, because Fresnel was not constructing his theory to account for it but it still predicted it, then we seem to be saying that the intentions of a theorist in constructing a theory determine in part whether the success of the theory is to be counted as evidence for its truth. Arguably, this undermines the objectivity of theory confirmation.

This motivates the idea of use novelty. A result is use-novel if the scientist did not explicitly build the result into the theory or use it to set the value of some parameter crucial to its derivation. For example, many physicists regard the success of general relativity in accounting for the orbit of Mercury, which was anomalous for Newtonian mechanics, as highly confirming, because the reasoning that led to the theory appealed to general principles and constraints that had nothing to do with the empirical data about the orbits of planets. Even though

Einstein specifically aimed to solve the Mercury problem, the derivation of the correct orbit was not achieved by putting in the right answer by hand.

There is also a modal account of novel prediction, according to which, that a theory could predict some unknown phenomenon is what matters, not whether it actually did so predict. In any case, scientific methodology includes far broader criteria for empirical success, such as providing explanations of previously mysterious phenomena. Indeed, Darwin's theory of evolution and Lyell's theory of uniformitarianism were accepted by the scientific community because of their systematising and explanatory power, and in spite of their lack of novel predictive success.

(V) Falsificationism

Popper argues that it is just too easy to accumulate positive instances which support some theory, especially when the theory is so general in its claims that its seems not to rule anything out. Similarly, some theories that have great explanatory power are scientifically dubious precisely because so much can be explained by them. Popper concludes that the 'confirmation' that a theory is supposed to get from observation of an instance which fits the theory only really counts for anything when it is an instance which was a risky prediction by the theory, that is if it is a potential falsifier of the theory. Even then it doesn't count as positive evidence for the theory, it merely shows that the theory has survived an attempt at refutation. The appropriate response is to try and find another way to try and refute it.

The problem of induction arises because no matter how many positive instances of a generalisation are observed it is still possible that the next instance will falsify it. Popper's solution to the problem of induction is simply to argue that it does not show that scientific knowledge is not justified because science does not depend on induction after all. There is a logical asymmetry between confirmation and falsification of a universal generalisation: a generalisation like all ravens are black would be falsified by a single observation of a raven that is not black. Popper argued that science is fundamentally about falsifying rather than confirming theories, and so he thought that science could proceed without induction because the inference from a falsifying instance to the falsity of a theory is purely deductive. If a theory or hypothesis is in principle unfalsifiable by experience then according to Popper is it unscientific (although it may still be meaningful).

According to falsificationism, science proceeds not by testing a theory and accumulating positive inductive support for it, but by trying to falsify theories. If it is falsified then it is abandoned, but if it is not falsified this just means it ought to be subjected to more attempts to falsify it. Popper says that the scientific method is that of 'conjectures and refutations'. Bold conjectures are those from which novel predictions can be deduced. On this view, science proceeds by natural selection and scientific knowledge is learned only from mistakes. Even the most successful theories could be falsified in the future and so they too should be regarded as conjectures. Popper argued that scientists must state clearly the conditions under which they would give up their theories rather than being committed to them come what may. It is important that on his view, no theories, no matter how well tested, ought to be regarded as even probably true. Popper may have come to this view by thinking about Newtonian mechanics, which must have seemed as well confirmed as a theory could be to a scientist in the early nineteenth century, but by the early twentieth century had been overthrown by special relativity and quantum mechanics, according to which it is quite wrong about the fundamental details of how the world works.

Nonetheless, the falsificationist does not view all scientific theories equally. Some theories are falsifiable but the phenomena they predict are not interesting or surprising. Bold conjectures that make novel predictions are the hypotheses that are scientifically valuable. Popper thought that theories could be ranked according to their degree of falsifiability and that this is the true measure of their empirical content. The more falsifiable a theory is the better it is because if it is highly falsifiable it must make precise predictions about a large range of phenomena. Popper also argued that new theories ought to be more falsifiable than the theories they replace.

Given the Duhem-Quine problem that is discussed in section 2 of this chapter, it is clear that there is no such thing as completely conclusive refutation of a theory by experiment. Popper concedes this point and so claims that as well as a set of observation statements which are potential falsifiers of the theory, there must also be a set of experimental procedures, so that the relevant group of scientists agree on a way in which the truth or falsity of each observation statement can be established. Falsification is only possible in science if there is agreement among scientists about what is being tested on any given occasion. Furthermore, Popper argues that whenever a high-level theoretical hypothesis is in conflict with a basic observation statement, it is the high-level theory that should be abandoned.

There are several problems with Popper's account of falsificationism including the following:

(i) Some legitimate parts of science seem not to be falsifiable

These fall into three categories:

(a) Probabilistic statements

The predictions derived from scientific theories are sometimes statements about the probability of some occurrence. However, such statements cannot be falsified because an improbable experimental outcome is consistent with the original statement. Any statement about the probability of a single event is not falsifiable.

(b) Existential statements

Universal generalisations are part of our scientific knowledge, but so to seem to be statements asserting the existence of things such as black holes, atoms, and viruses. These existential statements cannot be falsified. If a theory asserts the existence of something which is not found this does not deductively entail that the entity does not exist.

(c) Unfalsifiable scientific principles

It is arguable that some unfalsifiable principles may nonetheless be rightly considered part of scientific knowledge. So, for example, the status of the principle of conservation of energy, which states that energy can take different forms but cannot be created or destroyed, is such that it is inconceivable to most scientists that an experiment could falsify it; rather an apparent violation of the principle would be interpreted as revealing that something is wrong with the rest of science and it is likely that a new source, sink or form of energy would be posited.

There are also methodological principles that are arguably central to science but not falsifiable. So, for example, many scientists intuitively regard simple and unifying theories as, all other things being equal, more likely to be true than messy and complex ones. Some people claim that we have inductive grounds for believing in scientific theories that are simple, unified and so on, because in general the search for simple and unifying explanations has been fairly reliable in producing empirically successful theories, but they would add that we should never make simplicity an absolute requirement because sometimes nature is complex and untidy. Another kind of simplicity is that enshrined in the principle known as Ockham's razor, which is roughly the prescription not to invoke more entities in order to explain something than is absolutely necessary (this kind of simplicity is called ontological parsimony). It is not obvious how a falsificationist can justify these methodological rules.

(d) The hypothesis of natural selection

At one time Popper was critical of the theory of evolution because he thought the hypothesis that the fittest species survive was tautological, that is to say true by definition, and therefore not falsifiable, yet evolutionary theory is widely thought to be a prime example of a good scientific theory. Most philosophers of biology would argue that the real content of evolutionary theory lies not in the phrase 'the fittest survive', but in the idea of organisms passing on characteristics, subject to mutation and variation, which either increase or decrease the chances of their offspring surviving long enough to reproduce themselves, and so pass on those characteristics. This is supposed to account for the existence of the great diversity of species and their adaptation to the environment, and also the similarities of form and structure that exist between them. This theory may be indirectly falsifiable but it does not seem to be directly falsifiable.

(ii) Falsificationism is not itself falsifiable

Popper admits this but says that his own theory is not supposed to be because it is a philosophical or logical theory of the scientific method, and not itself a scientific theory so this objection, although often made, misses its target.

(iii) The notion of degree of falsifiability is problematic

The set of potential falsifiers for a universal generalisation is always infinite, so there can be no absolute measure of falsifiability, but only a relative one. The Duhem problem means that judgements about the degree of falsifiability of theories are relative to whole systems of hypotheses, and so our basis for such judgements is past experience and this lets induction in by the back door.

(iv) Falsificationism cannot account for our expectations about the future

Popper says that we are not entitled to believe that our best theories are even probably true. His position is ultimately extremely sceptical, indeed he goes further than Hume, who says induction cannot be justified but that we cannot help but use it, and argues that scientists should avoid induction altogether. But is this really possible, and is it really plausible to say that we never get positive grounds for believing scientific theories?

Our scientific knowledge does not seem to be purely negative and if it were it would be hard to see why we have such confidence in certain scientifically informed beliefs. After all, it is because doctors believe that penicillin fights bacterial infection that they prescribe it for people showing the relevant symptoms. The belief that certain causes do indeed have certain effects is what informs our actions. According to Popper there is no positive inductive support for my belief that if I try to leave the top floor of the building by jumping out the window I will fall hard on the ground and injure myself. If observation of past instances really confers no justification on a generalisation then I am just as rational if I believe that when I jump out of the window I will float gently to the ground. This is an unacceptable consequence of Popper's views for there is nothing more obvious to most of us than that throwing oneself out of high windows when one wishes to reach the ground safely is less rational than taking the stairs. If we adopt Popper's nihilism about induction we have no resources for explaining why people behave the way they do, and furthermore we are obliged to condemn any positive belief in generalisations as unscientific.

Of course, just when and how we can be justified on the basis of experience in believing general laws and their consequences for the future behaviour of the natural world is the problem of induction. But most philosophers think that solving this problem is not a matter of deciding whether it is more rational to take the stairs but why it is more rational to do so. Popper's response to this challenge is to introduce the notion of corroboration; a theory is corroborated if it was a bold conjecture that made novel predictions that were not falsified. Popper says that it is rational to suppose that the most corroborated theory is true because we have tried to prove it false in various ways and failed. The most corroborated theory is not one we have any reason to believe to be true, but it is the one we have least reason to think it is false, so it is rational to use it in making plans for the future, like leaving the building by the stairs and not by jumping. Popper stresses that the fact that a theory is corroborated only means that it invites further challenges.

But the notions of boldness and novelty are historically relative; the former means unlikely in the light of background knowledge and therefore highly falsifiable, and novel means previously unknown, or unexpected given existing corroborated theories, so once again induction based on past experience is smuggled into Popper's account. Furthermore, there is an infinite number of best corroborated theories, because whatever our best corroborated theory is, we can construct an infinite number of theories that agree with what it says about the past, but which say something different about what will happen in the future. The theory that gravity always applies to me when I jump into the air except after today is just as corroborated by all my experience up to now as the alternative that tells me not to jump off tall buildings; again we seem to have no choice but to accept the rationality of at least some inductive inferences despite what Popper says.

(v) Scientists sometimes ignore falsification

In general, contrary to what Popper says, scientists are not prepared to state in advance under what conditions they would abandon their most fundamental assumptions, and indeed many scientists probably would not consider abandoning the idea that species evolve by natural selection, or that ordinary matter is made of atoms. There are also many cases in the history of science where, in the face of falsifying evidence, scientists thought up modifications to save a theory instead of abandoning it. Popper distinguishes between ad hoc and non-ad hoc modifications, where the latter give rise to extra empirical content and the former do not, and argues that only non-ad hoc modifications are acceptable. There are certainly extreme cases where most people will agree that a theory has only been saved from refutation by a gratuitous assumption whose only role or justification is to save the theory.

Unfortunately, it turns out that there are cases in the history of science where a falsifying observation is tolerated for decades despite numerous attempts to account for it. More generally, it often seems to be the case that where scientists have a successful theory, the existence of falsifying observations will not be sufficient to cause the abandonment of the theory in the absence of a better alternative.

For these and other reasons, Popper's falsificationism is probably now more popular among scientists than it is among philosophers.

(VI) Kuhn's Philosophy of Science

The scientific method is supposed to be rational, and to give us objective knowledge of the world. Prior to the work of Thomas Kuhn many philosophers of science agreed with the following statements:

(i) Science is cumulative.

(ii) Science is unified in the sense that there is a single set of fundamental methods for all the sciences, and in the sense that the natural sciences at least are all ultimately reducible to physics.

(iii) There is a epistemologically crucial distinction between the context of discovery and the context of justification.

(iv) There is an underlying logic of confirmation or falsification implicit in all scientific evaluations of the evidence for some hypothesis. Such evaluations are value-free in the sense of being independent of the personal non-scientific views and allegiances of scientists.

(v) There is a sharp distinction (or demarcation) between scientific theories and other kinds of belief systems.

(vi) There is a sharp distinction between observational terms and theoretical terms, and also between theoretical statements and those that describe the results of experiments. Observation and experiment is a neutral foundation for scientific knowledge, or at least for the testing of scientific theories.

(vii) Scientific terms have fixed and precise meanings.

Kuhn argued that many scientists' accounts of the history of their subject considerably simplify and distort the real stories of theory development and change. He argues that the history of science does not consist in the steady accumulation of knowledge, but often involves the wholesale abandonment of past theories. According to Kuhn, the evaluation of theories depends on local historical circumstances, and his analysis of the relationship between theory and observation suggests that theories infect data to such an extent that no way of gathering of observations can ever be theory-neutral and objective. Hence, the degree of confirmation an experiment gives to a hypothesis is not objective, and there is no single logic of theory testing which can be used to determine which theory is most justified by the evidence. He thinks instead that scientists' values help determine, not just how individual scientists develop new theories, but also which theories the scientific community as a whole regards as justified.

There are two closely related ideas of paradigm in Kuhn's work, namely those of paradigm as disciplinary matrix and paradigm as exemplar. Kuhn argues that before scientific inquiry can even begin in some domain, the scientific community in question has to agree upon answers to fundamental questions about, for example: what kinds of things exist in the universe, how they interact with each other and our senses, what kinds of questions may legitimately be asked about these things, what techniques are appropriate for answering those questions, what counts as evidence for a theory, what questions are central to the science, what counts as a solution to a problem, what counts as an explanation of some phenomenon, and so on.

A disciplinary matrix is a set of answers to such questions that are learned by scientists in the course of the education which prepares them for research, and that provide the framework within which the science operates. It is important that different aspects of the disciplinary matrix may be more or less explicit, and some parts of it are constituted by the shared values of scientists, in that they prefer certain types of explanation over others and so on. It is also important that some aspects of it will consist of practical skills and methods that are not necessarily expressible in words.

Exemplars, on the other hand, are those successful parts of science which all beginning scientists learn, and which provide them with a model for the future development of their subject. Anyone familiar with a modern scientific discipline will recognise that teaching by example plays an important role in the training of scientists. Textbooks are full of standard problems and solutions to them, and students are set exercises that require them to adapt the techniques used in the examples to new situations. The idea is that, by repeating this process, eventually, if they have the aptitude for it, students will learn how to apply these techniques to new kinds of problems which nobody has yet managed to solve.

Most science is what Kuhn calls 'normal science', because it is conducted within an established paradigm. It involves elaborating and extending the success of the paradigm, for example, by gathering lots of new observations and accommodating them within the accepted theory, and trying to solve minor problems with the paradigm. Hence, normal science is often said to be a 'puzzle-solving' activity, where the rules for solving puzzles are quite strict and determined by the paradigm. According to Kuhn, most of the everyday practice of science is a fairly conservative activity in so far as, during periods of normal science, scientists do not question the fundamental principles of their discipline. If a paradigm is successful and seems able to account for the bulk of the phenomena in its domain, and if scientists are still able to make progress solving problems and extending its empirical applications, then most scientists will just assume that anomalies that seem intractable will eventually be resolved. They won't give up the paradigm just because it conflicts with some of the evidence.

However, sometimes scientists become aware of anomalies which won't go away no matter how much effort is put into resolving them. These may take the form of conceptual paradoxes or experimental falsifications. Even these will not necessarily cause much serious questioning of the basic assumptions of the paradigm. But when a number of serious anomalies accumulate then, some, often younger or maverick scientists will begin to question some of the core assumptions of the paradigm, and perhaps they will begin speculating about alternatives. This amounts to the search for a new paradigm, which is a new way of thinking about the world. If this happens when successful research within the paradigm is beginning to decline, more and more scientists may begin to focus their attention on the anomalies and the perception that the paradigm is in 'crisis' may begin to take hold of the scientific community. If a crisis happens, and if a new paradigm is adopted by the scientific community, then a 'revolution' or 'paradigm shift' has occurred. On Kuhn's view when a revolution occurs the old paradigm is replaced wholesale. So, for example, the adoption or rejection of each of the examples of paradigms listed above is a scientific revolution.

There are two points about Kuhn's account of this and other scientific revolutions that must be emphasised:

- This is a completely different view of scientific change to the traditional idea of cumulative growth of knowledge, because paradigm shifts or scientific revolutions involve change in scientific theories that is not piecemeal but holistic. In other words, the paradigm does not change by parts of it being changed bit by bit, but rather by a wholesale shift to a new way of thinking about the world, and this will usually mean a new way of practising science as well including new experimental techniques and so on.
- Revolutions only happen when a viable new paradigm is available, and also when there happen to be individual scientists who are able to articulate the new picture to their colleagues.

Kuhn also emphasises the role of psychological and sociological factors in disposing scientists to adopt or a reject a particular paradigm.

Although existing theories guide us in developing new theories, and tell us which observations are significant and so on, the distinction between the context of discovery and the context of justification can be invoked to maintain the idea that scientific theories are tested by observations. Many empiricist philosophers have drawn a sharp distinction between the observational and the theoretical, and both logical positivists, and Popper, at least in his earlier work, assume it. According to the received view, the theory-independence or neutrality of observable facts makes them a suitable foundation for scientific knowledge, or at least for testing theories. The received view incorporated a distinction between observational terms, like 'red', 'heavy', and 'wet', and theoretical terms, like 'electron', 'charge', and 'gravity'. The idea is that the rules for the correct application of observational terms refer only to what a normal human observer perceives in certain circumstances, and that they are entirely independent of theory. So, for example, Ernest Nagel argues that every observational term is associated with at least one overt procedure for applying the term to some observationally identifiable property, when certain specified circumstances are realised. So, for example, the property of being red is applied to an object when it looks red to a normally functioning observer in normal lighting conditions. Many other writers analyse the logic of theory testing relying upon this distinction between observational and theoretical terms.

Incommensurability is a term from mathematics which means 'lack of common measure'. It was adopted by Kuhn, and another philosopher called Paul Feyerabend, both of whom argued that successive scientific theories are often incommensurable with each other in the sense that there is no neutral way of comparing their merits. One of the most radical ideas to emerge from Kuhn's work is that what counts as the evidence in a given domain may depend upon the background paradigm. If this is right then how can it be possible to rationally compare competing paradigms? Kuhn argues that there is no higher standard for comparing theories than the assent of the relevant community, and that, the choice between competing paradigms is a choice between incompatible modes of community life.

In his later work Kuhn sought to distance himself from extreme views which give no role to rationality in the progress of science, and which do not allow for comparison of the merits of theories within different paradigms. He argues that the following five core values are common to all paradigms:

- A theory should be empirically accurate within its domain.
- A theory should be consistent with other accepted theories.
- A theory should be wide in scope and not just accommodate the facts it was designed to explain.
- A theory should be as simple as possible.
- A theory should be fruitful in the sense of providing a framework for ongoing research.

Hence, Kuhn avoids complete irrationalism because these values impose some limits on what theories scientists can rationally accept. On the other hand, these values are not sufficient to determine what decisions they ought to make in most interesting cases, because these values may conflict; a theory may be simple but not accurate, or fruitful but not wide in scope, and so on. Furthermore, a value like simplicity may be understood in different ways depending on background views and so on.

(VII) Bayesianism

Bayesianism is potentially a theory of the relationship between theories and evidence, a theory of rationality, an account of the scientific method, and a theory of probability. It is increasingly being taught and used in place of traditional statistics. Bayesianism employs the mathematical theory of probability. Probabilities of the form $P(A)$ are used to represent a subject's degree of belief that a proposition A is the case. The probabilities must conform to the constraints of the probability calculus ('$\vdash$': entails):

$$0 \leq P(A) \leq 1$$
$$\text{necessarily } A \vdash P(A) = 1$$
$$A \rightarrow \neg B \vdash P(A \text{ or } B) = P(A) + P(B)$$
$$P(\neg A) = 1 - P(A)$$
$$(A \vdash B) \vdash P(B) \geq P(A)$$
$$A \leftrightarrow B \vdash P(A) = P(B)$$
$$A, B \text{ are independent events} \vdash P(A\&B) = P(A) \times P(B)$$

The notation $P(A/B)$ is used to represent the conditional probability of A given B, defined:

$$P(A/B) = P(A\&B)/P(B)$$

which allows Bayes' theorem to be proved:

$$P(h/e) = P(h).P(e/h)/P(e)$$

This equation can be understood as ruling how to update belief in h in the light of new information e. Suppose, h is a scientific hypothesis (or more accurately a combination of hypotheses) and e is the statement that some phenomenon is observed under certain conditions, and that h predicts that e will be true when a test is performed. Let the scientist's prior degree of belief in the hypothesis be $P(h)$, and $P(e)$ the scientist's degree of belief, disregarding h, that the phenomenon will be observed. $P(e/h)$ is the scientist's degree of belief as to how likely e is given that h is true. This is known as 'Bayesian conditionalisation'.

Some intuitive aspects of confirmation are captured by this formalism. Firstly, $P(h/e)$ is proportional to $P(h)$, in other words, the more likely the hypothesis was considered to start with, the more likely it will be considered even in the light of new, possibly disconfirming, evidence. Second, $P(h/e)$ is proportional to $P(e/h)$, in other words, the more closely linked the evidence and the hypothesis are the more observation of the evidence supports the hypothesis. Finally, $P(h/e)$ is inversely proportional to $P(e)$, in other words, the more unlikely the evidence which the hypothesis predicts the more it supports the hypothesis if it is observed. Bayesians claim that evidence confirms a hypothesis if learning the evidence raises the probability of the hypothesis and disconfirms it if learning the evidence lowers that probability.

Bayesianism is alleged to be able to resolve a number of the well-known paradoxes of confirmation, in particular, the Ravens Paradox, the tacking paradox, and Goodman's New Problem of Induction. It is also claimed that Bayesianism deals with the problem of underdetermination.

There are various problems with Bayesianism including the following:

(i) The Problem of Old Evidence.

This is the problem of explaining how it is possible for a theory to be confirmed by evidence that was known about before the theory was formulated. One common response is to argue that the relevant prior probability to be used in Bayes' formula is a counterfactual probability, namely the probability the theory would have been judged to have had the evidence not been known about.

(ii) The Problem of the Priors

According to Bayesianism, how credible a scientific theory is seems to be a function of its prior probability. Must Bayesians therefore provide an account of what the priors ought to be? There are various theorems that show that agents who start with very different prior degrees of belief will nonetheless end up converging in their posterior degrees of belief if they keep updating their degrees of belief on the basis of the same experimental data. So it seems that eventually differences in priors will be irrelevant. On the other hand, these theorems only say what will happen in the (potentially infinitely) long run, and yet we expect scientists to reach agreement about the status of scientific hypotheses in reasonable short amounts of time.

(iii) The Interpretation of Probability

There is some controversy about the nature of the probabilities used in Bayesianism. Some Bayesians believe that they can only represent subjective degrees of belief of agents, while others think that they can be understood as referring to degrees of belief that have to match up to objective probabilities of some kind. Ramsey argued for the former and claimed that degrees of belief can be interpreted as corresponding to the least odds someone would be willing to gamble on.

(iv) The Problem of Psychological Implausibiltity

Consistency with the probability calculus seems to require a lot of agents. Arguably it is psychologically unrealistic to expect people in general and scientists in particular, to have no inconsistencies in their beliefs, and to continually deduce all the logical consequences of their beliefs as they acquire new ones.

(v) The Status of 'Dutch Book' Arguments

There are two kinds of Dutch book arguments namely synchronic and diachronic. The former are intended to show that a rational agents degrees of belief at any given time must satisfy the axioms of the probability calculus, while the latter are intended to show that rational agents ought to update their degrees of belief over time in accordance with Bayesian conditionalisation. The former is relatively uncontroversial and is often regarded as the probabilistic analogue of logical consistency, however the latter has been the subject of intense debate since it appears that the most such arguments can show is that if you reject conditionalisation you would be bound to lose if you bet honestly with someone who knows your strategy for changing your betting quotients.

Bayesianism is actually vastly more complicated than the above discussion suggests since, for example, there are other forms of conditionalisation based on how degrees of belief ought to change in the light of uncertain evidence.

BIBLIOGRAPHY

1. ONTOLOGICAL POSITIONS

(I) Natural Kinds

[Bird, 1998] A. Bird. *Philosophy of Science.* London: UCL Press, chapter 3, 1998.

[Dupre, 1993] J. Dupre. *The Disorder of Things: Metaphysical Foundations of the Disunity of Science.* Cambridge MA: Havard University Press, 1993.

[Dupre, 1981] J. Dupre. Natural kinds and biological taxa. *Philosophical Review*, 90: 66–90, 1981.

[Mellor, 1977] D. H. Mellor. Natural kinds. *The British Journal for the Philosophy of Science*, 28: 299–312, 1977; and in his *Matters of Metaphysics*, Cambridge: Cambridge University Press, 1991.

[Putnam, 1975] H. Putnam. The meaning of meaning. In his *Philosophical Papers, volume ii: Mind, Language and Reality*, Cambridge: Cambridge University Press, 1975.

[Quine, 1969] W. v. O. Quine. Natural kinds. In N. Rescher (ed.), *Essays in Honour of Carl Hempel*, Dordrecht: Reidel, 1969

(II) Truth and Mind-independence

[Blackburn and Simmons, 1999] S. Blackburn and K. Simmons (eds). *Truth*. Oxford: Oxford University Press, 1999.

[Devitt, 1989] M. Devitt. *Realism and Truth*. Oxford: Blackwell, 1989 (second edition 1991).

[Dummett, 1978] M. Dummett. *Truth and other enigmas*. London: Duckworth, 1978.

[Engel, 2002] P. Engel. *Truth*. Chesham: Acumen, 2002.

[Goodman, 1978] N. Goodman. *Ways of Worldmaking*. Indianapolis, IL: Hackett, 1978.

[Horwich, 1990] P. Horwich. *Truth*. Oxford: Oxford University Press, 1990 (second edition 1998).

[Kukla, 2000] A. Kukla. *Social Constructivism and the Philosophy of Science*. London: Routledge, 2000

[James, 1897] W. James. *The Will to Believe and Other Essays*. New York: Dover, 1897.

[James, 1907] W. James. *Pragmatism*. New York: Longmans and Green, 1907.

[Putnam, 1981] H. Putnam. *Reason, Truth and History*. Cambridge: Cambridge University Press, 1981.

[Wright, 1987] C. Wright. *Realism, Meaning and Truth*. Oxford: Oxford University Press, 1987 (second edition 1983).

[Wright, 1992] C. Wright. *Truth and Objectivity*. Harvard: Harvard University Press, 1992.

(III) Properties and Universals

[Armstrong, 1989] D. M. Armstrong. *Universals: An Opinionated Introduction*. Boulder, Colorado: Westview Press, 1989.

[Loux, 2002] M. Loux. *Metaphysics: a contemporary introduction*. London Routledge, chapters 1&2, 2002.

[Mellor and Oliver, 1997] D. H. Mellor and A. Oliver (eds.). *Properties*. Oxford: Oxford University Press, 1997.

[Russell, 1912] B. Russell. *The Problems of Philosophy*. Oxford: Oxford University Press, chapters 9–10, 1912.

[Kim and Sosa, 1999] J. Kim and E. Sosa (eds.). *Metaphysics: An Anthology*. Oxford: Blackwell, part IV, 1999.

(IV) Identity and Individuality

[Black, 1952] M. Black. The identity of indiscernibles. *Mind*, 61: 153–164, 1952.

[Loux, 2002] M. Loux. *Metaphysics: A Contemporary Introduction*. London Routledge, chapter 3, 2002.

[Kim and Sosa, 1999] J. Kim and E. Sosa (eds.). *Metaphysics: An Anthology*. Oxford: Blackwell, part II, 1999.

[Laurence and Macdonald, 1998] S. Laurence and C. Macdonald. *Contemporary Readings in the Foundations of Metaphysics*. Oxford: Blackwell, part IV, 1998.

(V) Matter and Motion

[Salmon, 2001] W. Salmon (ed.). *Zeno's Paradoxes*. Indianapolis, IL: Hackett, 2001.

[Jammer, 1961] M. Jammer. *Concepts of Mass in Classical and Modern Physics*. New York: Dover, 1961.

(VI) Causation

[Hume, 1963] D. Hume. *An Enquiry Concerning Human Understanding.* La Salle: Open Court, section VII, 1963.
[Hume, 1978] D. Hume. *A Treatise of Human Nature.* Oxford: Oxford University Press, I.3, 14/15, 1978.
[Mackie, 1980] J. L. Mackie. *The Cement of the Universe: A Study of Causation.* Oxford: Clarendon Press, 1980.
[Sosa and Tooley, 1993] E. Sosa and M. Tooley (eds.). *Causation.* Oxford; Oxford University Press, 1993.
[Kim and Sosa, 1999] J. Kim and E. Sosa (eds.). *Metaphysics: An Anthology,* Oxford: Blackwell, part VII, 1999.

(VII) Laws of Nature

[Bird, 1998] A. Bird. *Philosophy of Science.* London: UCL Press, chapter 1, 1998.
[Carroll, 2004] J. Carroll (ed.). *Readings on Laws of Nature.* Pittsburgh: University of Pittsburgh Press, 2004.
[Armstrong, 1983] D. Armstrong. *What is a Law of Nature?.* Cambridge: Cambridge University Press, 1983.
[Cartwright, 1983] N. Cartwright. *How the Laws of Physics Lie.* Oxford: Oxford University Press, 1983.

(VIII) Probability, Propensity and Dispositions

[Mellor, 2005] D. H. Mellor. *Probability: a philosophical introduction.* London: Routledge, 2005.
[Ellis, 2001] B. Ellis. *Scientific Essentialism.* Cambridge: Cambridge University Press, 2001.

(IX) Reductionism, Emergence and Supervenience

[Churchland, 1990] P. M. Churchland. *Matter and Consciousness.* Cambridge, MA: MIT Press, chapter 2, 1990.
[Jackson, 1998] F. Jackson. *From Metaphysics to Ethics: A Defence of Conceptual Analysis.* Oxford: Oxford University Press, 1998
[Kim, 1998] J. Kim. *Mind in a Physical World.* Cambridge, MA: MIT Press, 1998.

(X) Space, Time and Spacetime

[Dainton, 2001] B. Dainton. *Time and Space.* Chesham: Acumen, 2001.
[Huggett, 1999] N. Huggett. *Space from Zeno to Einstein: Classic Readings with a Contemporary Commentary.* Cambridge, MA: MIT Press, 1999.
[le Poidevin and McBeath, 1993] R. le Poidevin and M. McBeath (eds.). *The Philosophy of Time.* Oxford: Oxford University Press, 1993.
[Lockwood, 2005] M. Lockwood. *The Labyrinth of Time.* Oxford: Oxford University Press, 2005.
[Mellor, 1998] H. Mellor. *Real Time II.* London: Routledge, 1998.
[Sklar, 1974] L. Sklar. *Space, Time and Spacetime,* Berkeley: University of California Press, 1974.

(XI) Events and Processes

[Casati, and Varzi, 1996] R. Casati, and A. C.Varzi (eds.). *Events.* Dartmouth, Aldershot, 1996.
[Davidson, 1980] D. Davidson. *Essays on Actions and Events.* Oxford: Clarendon Press, 1980.
[Rescher, 1996] N. Rescher. *Process Metaphysics: An Introduction to Process Philosophy.* New York: SUNY Press, 1996.
[Rescher, 2000] N. Rescher. *Process Philosophy: A Survey of Basic issues.* Pittsburgh, Pa.: University of Pittsburgh Press, 2000.

[Whitehead, 1919] A. N. Whitehead. *An Enquiry Concerning the Principles of Natural Knowledge.* Cambridge: Cambridge University Press, 1919; reprinted New York: Kraus Reprints, 1982.

2. EPISTEMOLOGICAL POSITIONS

(I) Rationalism

[Cottingham, 1988] J. Cottingham. *The Rationalists.* Oxford: Oxford University Press, 1988.

[Boghossian and C.Peacock, 2000] P. Boghossian and C. Peacock (eds). *New Essays on the A Priori.* Oxford: Oxford University Press, 2000.

(II) Empiricism

[Ayer, 1952] A. J. Ayer. *Language, Truth and Logic.* New York: Dover, 1952.

[Gower, 1997] B. Gower. *Scientific Method: An Historical and Philosophical Introduction.* London: Routledge, 1997.

[Hanfling, 1981] H. Hanfling, (ed.). *Essential Readings in Logical Positivism.* Oxford: Blackwell, 1981.

[Shapin, 1996] S. Shapin. *The Scientific Revolution.* Chicago: Chicago University Press, 1996.

[Woolhouse, 1988] R. Woolhouse. *The Empiricists.* Oxford: Oxford University Press, 1988.

(III) Induction

[Russell, 1912] B. Russell. *The Problems of Philosophy.* Oxford: Oxford University Press, chapter 6, 1912.

[Goodman, 1973] N. Goodman. *Fact, Fiction and Forecast.* Indianapolis, IL: Bobbs-Merrill, 1973.

[Ladyman, 2002] J. Ladyman. *Understanding Philosophy of Science.* London: Routledge, chapter 2, 2002.

[Hume, 1963] D. Hume. *An Enquiry Concerning Human Understanding.* La Salle: Open Court, section IV, 1963.

[Hume, 1978] D. Hume. *A Treatise of Human Nature.* Oxford: Oxford University Press, part III, 1978.

[Swinburne, 1974] R. Swinburne (ed.). *Justification of Induction.* Oxford: Oxford University Press, 1974.

(IV) Scientific Realism

[Kitcher, 1993] P. Kitcher. *The Advancement of Science: Science without Legend, Objectivity without Illusions.* Oxford: Oxford University Press, 1993.

[Ladyman, 2002] J. Ladyman. *Understanding Philosophy of Science.* London: Routledge, chapter 5, 2002.

[Psillos, 1996] S. Psillos. *Scientific Realism: How Science Tracks Truth.* London: Routledge, 1996.

[Van Fraassen, 1980] B. C. Van Fraassen. *The Scientific Image.* Oxford: Oxford University Press, 1980.

[Churchland and Hooker, 1985] P. Churchland, and C. A. Hooker (eds). *Images of Science.* Chicago: University of Chicago Press, 1985.

(V) The Duhem-Quine Problem and Underdetermination

[Duhem, 1906] P. Duhem. *The Aim and Structure of Physical Theory.* Translated by P. Wiener 1954, Princeton: Princeton University Press, chapter 6, 1906.

[Quine, 1953] W. Quine. Two dogmas of empiricism. In his *From a Logical Point of View*, Cambridge, MA: Harvard University Press, 1953.

[Harding, 1976] S. Harding, (ed.). *Can Theories be Refuted?: Essays on the Duhem-Quine Thesis.* Dordrecht, Netherlands: D.Reidel, 1976.

[Ladyman, 2002] J. Ladyman. *Understanding Philosophy of Science.* London: Routledge, chapter 6, 2002.

[Kukla, 1996] A. Kukla. Does every theory have empirically equivalent rivals? *Erkenntnis*, 44: 137–66, 1996.

[Kukla, 1993] A. Kukla. Laudan, Leplin, empirical equivalence, and underdetermination. *Analysis*, 53: 1–7, 1993.

[Kukla, 1998] A. Kukla. *Studies in Scientific Realism.* Oxford: Oxford University Press, 1998.

[Laudan and Leplin, 1991] L. Laudan, and J. Leplin. Empirical equivalence and underdetermination. *Journal of Philosophy*, 88: 269–85, 1991.

[Laudan and Leplin, 1993] L. Laudan, and J. Leplin. Determination underdeterred. *Analysis*, 53: 8–15, 1993.

[Hoefer and Rosenberg, 1994] C. Hoefer and A. Rosenberg. Empirical equivalence, underdetermination, and systems of the world. *Philosophy of Science*, 61: 592–607, 1994.

(VI) Inference to the Best Explanation

[Harman, 1965] G. Harman. Inference to the best explanation. *Philosophical Review.* 74: 88–95, 1965.

[Ladyman, 2002] J. Ladyman. *Understanding Philosophy of Science.* London: Routledge, chapter 7, 2002.

[Ladyman *et al.*, 1997] J. Ladyman, I. Douven, L. Horsten, and van B. C. Fraassen. In defence of Van Fraassen's critique of abductive reasoning: A reply to Psillos. *Philosophical Quarterly*, 47: 305–321, 1997.

[Lipton, 1991] P. Lipton. *Inference to the Best Explanation.* London: Routledge, 1991.

[Van Fraassen, 1989] B. C. Van Fraassen. *Laws and Symmetry.* Oxford: Oxford University Press, 1989.

[Psillos, 1996] S. Psillos. *Scientific Realism: how science tracks truth.* London: Routledge, chapter 9, 1996.

(VII) Arguments from Theory Change

[Hardin and Rosenberg, 1982] C. L. Hardin and A. Rosenberg. In defence of convergent realism. *Philosophy of Science*, 49: 604–615, 1982.

[Kitcher, 1993] P. Kitcher. *The Advancement of Science: Science without Legend, Objectivity without Illusions.* Oxford: Oxford University Press, 1993.

[Ladyman, 2002] J. Ladyman. *Understanding Philosophy of Science.* London: Routledge, chapter 8, 2002.

[Ladyman, 1998] J. Ladyman. What is structural realism? *Studies in History and Philosophy of Science*, 29: 409–424, 1998.

[Laudan, 1981] L. Laudan A confutation of convergent realism. *Philosophy of Science*, 48: 19–49, 1981; reprinted in D. Papineau (ed.), *Philosophy of Science*, Oxford: Oxford University Press, 1996.

[Laudan, 1984] L. Laudan. Discussion: Realism without the real. *Philosophy of Science*, 51: 156–162, 1984.

[Psillos, 1996] S. Psillos. *Scientific Realism: How Science Tracks Truth.* London: Routledge, 1996.

[Worrall, 1989] J. Worrall. Structural realism: The best of both worlds? *Dialectica*, 3: 99–124; reprinted in D. Papineau (ed.), *Philosophy of Science*, 1996.

(VIII) Contemporary Empiricism

[Ladyman, 2002] J. Ladyman. *Understanding Philosophy of Science.* London: Routledge, chapter 6, 2002.

[Ladyman, 2000] J. Ladyman. What's really wrong with constructive empiricism?: van Fraassen and the metaphysics of modality. *The British Journal for the Philosophy of Science*, 51: 837–856, 2000.

[Monton and van Fraassen, 2003] B. Monton and B. C. van Fraassen. Constructive empiricism and modal nominalism. *The British Journal for the Philosophy of Science*, 54: 405–422, 2003.

[Rosen, 1984] G. Rosen. What is constructive empiricism? *Philosophical Studies*, 74: 143–178, 1984.

[Van Fraassen, 1981] B. C. Van Fraassen. *The Scientific Image.* Oxford: Oxford University Press, 1981.

[Van Fraassen, 1989] B. C. Van Fraassen. *Laws and Symmetry.* Oxford: Oxford University Press, 1989.

[Van Fraassen, 2002] B. C. Van Frassen. *The Empirical Stance.* New Haven: Yale University Press, 2002.

(IX) Pragmatism

[Hacking, 1983] I. Hacking. *Representing and Intervening.* Cambridge: Cambridge University Press, 1983.

(X) Naturalism and Normativity

[Papineau, 1993] D. Papineau. *Philosophical Naturalism.* Oxford: Blackwell, 1993.

3. METHODOLOGICAL POSITIONS

(I) Inductivism

[Achinstein, 1991] P. Achinstein. *Particles and Waves.* Oxford: Oxford University Press, 1991.

[Gower, 1997] B. Gower. *Scientific Method: An Historical and Philosophical Introduction.* London: Routledge, chapter 3, 1997.

[Ladyman, 2002] J. Ladyman. *Understanding Philosophy of Science.* London: Routledge, chapter 1, 2002.

[Urbach, 1987] P. Urbach. *Francis Bacon's Philosophy of Science: An Account and a Reappraisal.* LaSalle, IL: Open Court, 1987.

[Woolhouse, 1988] R. Woolhouse. *The Empiricists.* Oxford: Oxford University Press, chapter 2, 1988.

(II) The Context of Discovery and the Context of Justification

[Popper, 1934-1959] K. Popper. *The Logic of Scientific Discovery.* London: Hutchinson, 1934-1959.

[Newton-Smith, 1987] W. Newton-Smith. *The Rationality of Science.* London: Routledge and Kegan Paul, chapter III, 1987.

(III) The Paradoxes of Confirmation

[Brown, 1977] H. Brown. *Perception, Theory and Commitment.* Chicago: University of Chicago Press, 1977.

[Glymour, 1980] G. Glymour. *Theory and Evidence.* Princeton, N.J.: Princeton University Press, 1980.

(IV) Explanation versus Prediction

[Hempel, 1965] C. Hempel. *Aspects of Scientific Explanation.* New York: Free Press, 1965

[Nagel, 1961] E. Nagel. *The Structure of Science.* New York: Harcourt, Brace, 1961.

[Salmon, 1989] W. C. Salmon Four decades of scientific explanation. In P. Kitcher, and W. C. Salmon (eds), *Scientific Explanation: Minnesota Studies in the Philosophy of Science,* Volume XIII, pages 3–219, 1989.

[Friedman, 1974] M. Friedman. Explanation and scientific understanding. *Journal of Philosophy,* LXXI: 5–19, 1974.

(V) Falsificationism

[Popper, 1934-1959] K. Popper. *The Logic of Scientific Discovery.* London: Hutchinson, 1934-1959.

[Popper, 1969] K. Popper. *Conjectures and Refutations.* London: Routledge and Kegan Paul, 1969.

[Ladyman, 2002] J. Ladyman. *Understanding Philosophy of Science.* London: Routledge, chapter 3, 2002.

[Lakatos, 1968] I. Lakatos. Criticism and the methodology of scientific research programmes. *Proceedings of the Aristotelian Society,* 69: 149–86, 1968.

[Lakatos and Musgrave, 1970] I. Lakatos and A. Musgrave (eds.). *Criticism and the Growth of Knowledge.* Cambridge: Cambridge University Press, 1970.

[Newton-Smith, 1987] W. Newton-Smith. *The Rationality of Science.* London: Routledge and Kegan Paul, chapter III, 1987.

(VI) Kuhn's Philosophy of Science

[Feyerabend, 1977] P. Feyerabend. *Against Method.* London: New Left Books, 1977.

[Hacking, 1981] I. Hacking (ed.). *Scientific Revolutions.* Oxford: Oxford University Press, 1981.

[Hoyningen-Huene, 1993] P. Hoyningen-Huene. *Reconstructing Scientific Revolutions: Thomas Kuhn's Philosophy of Science.* Chicago: University of Chicago Press, 1993.

[Kuhn, 1957] T. S. Kuhn. *The Copernican Revolution: Planetary Astronomy in the Development of Western Thought.* Cambridge, Mass.: Harvard University Press, 1957.

[Kuhn, 1962] T. S. Kuhn. *The Structure of Scientific Revolutions.* Chicago: University of Chicago Press, 1962 (second edition 1970).

[Kuhn, 1977] T. S. Kuhn. *The Essential Tension.* Chicago: University of Chicago Press, 1977.

[Ladyman, 2002] J. Ladyman. *Understanding Philosophy of Science.* London: Routledge, chapter 4, 2002.

[Lakatos and Musgrave, 1970] I. Lakatos and A. Musgrave (eds.). *Criticism and the Growth of Knowledge.* Cambridge: Cambridge University Press, 1970.

[Laudan, 1977] L. Laudan. *Progress and its Problems.* Berkeley: University of California Press, 1977.

(VII) Bayesianism

[Bovens and Hartmann, 2003] L. Bovens and S. Hartmann. *Bayesian Epistemology.* Oxford: Clarendon Press, 2003.

[Horwich, 1982] P. Horwich. *Probability and Evidence.* Cambridge: Cambridge University Press, 1982.

[Howson and Urbach, 1993] C. Howson and P. Urbach. *Scientific Reasoning: The Bayesian Approach.* Open Court, 2nd edition, 1993.

[Glymour, 1980] G. Glymour. *Theory and Evidence.* Princeton, N.J.: Princeton University Press, 1980.

REDUCTION, INTEGRATION, AND THE UNITY OF SCIENCE: NATURAL, BEHAVIORAL, AND SOCIAL SCIENCES AND THE HUMANITIES

William Bechtel and Andrew Hamilton

1 A HISTORICAL LOOK AT UNITY

The notion that science is unified in one way or another dates back at least to Aristotle, though unity claims since then have been diverse and variously motivated. By way of introduction to the modern discussion of unity, disunity, and integration, in this first section we examine five historical attempts to unify knowledge: Aristotle's metaphysical and hierarchical unity; the Enlightenment project of the French Encyclopedists; the systematic unity of *Naturphilosoph* Lorenz Oken; the methodological unity of the Vienna School's *Encyclopedia of Unified Science*; and finally, the organizational unity of cybernetics and general systems theory. We treat these unification projects not only as context, but also because, as we shall see, something of their momentum carries over into the modern discussion.

1.1 Aristotle's Metaphysical and Hierarchical Unity

Aristotle arranged the 'sciences' into three divisions: the theoretical sciences (metaphysics, mathematics, and physics): the practical sciences (e.g., ethics and politics), and the productive sciences (e.g., poetry and rhetoric). That is, he divided sciences according to their purposes. Theoretical sciences are concerned with knowledge alone and for its own sake, practical sciences are for doing, and productive sciences are for making. Despite these divisions, however, Aristotle's image of the sciences was one of a unified hierarchy. In the *Metaphysics*, he made clear that the theoretical sciences — most particularly metaphysics or 'theology' — are at the top of the hierarchy. These are the sciences that investigate first causes, and the people who know them know universally and in the highest degree, as well as "understand... all the underlying subjects" (*Metaphysics* A.2).

Aristotle argued that the theoretical sciences are the most basic. It is by virtue of theoretical knowledge that one has true command of practical and productive matters. Without theory, one merely has experience. With theory, one has art (*techné*). Consider Aristotle's example of the physician who treats Callias.

Handbook of the Philosophy of Science: General Philosophy of Science - Focal Issues, 377–430.
Volume editor: Theo Kuipers. Handbook editors: Dov M. Gabbay, Paul Thagard and John Woods.

Medicine for Aristotle is a practical science, but its practice is enhanced by a grasp of theory. The better physician will not be one who knows only how to treat Callias, or men of a certain age, or those with the specific ailment afflicting Callias. Rather, the better physician will be one who understands disease *qua* disease, according to its principles and causes, and understands people *qua* people.

This consequence of a science's rank in the hierarchy applies even within the theoretical sciences. It is by virtue of doing metaphysics, the highest theoretical science, that one truly grasps the lesser two theoretical sciences. That is, the better physicist or mathematician is one who understands metaphysics. As Aristotle makes clear in the middle books of the *Metaphysics*, he thinks there are causes and substances that are beyond the reach of physics. For him, physics is the science of sensible substances and their causes, but there is a more fundamental substance (*ousia*) as well as a more fundamental source of motion. The study of this substance and of the first motion inform physics rather than the other way around. Mathematics has the same relationship with metaphysics as physics: the study of surfaces and quantity depends upon and is informed by the more universal questions of metaphysics (*Metaphysics M* and *N*): Are there mathematical objects? Do numbers exist? Are numbers causes? Are they substances? Aristotle does not take up these questions by asking what we know about mathematics, but rather by asking what we know universally.

1.2 French Encyclopedists

When we think of comprehensive accounts of knowledge today, we often think of encyclopedias. These modern works have their origin in the period after the scientific revolution, when the integration of knowledge achieved by Aristotle and maintained by the Scholastics was fundamentally undercut. Historically the most famous encyclopedia was *Encyclopédie, ou dictionnaire raisonné des sciences, des arts et des métiers* (Encyclopedia, or Reasoned Dictionary of the Sciences, Arts, and Trades). Its 17 volumes (plus 11 volumes of illustrations) were produced over the period 1751–1772 under the editorship of Denis Diderot along with the mathematician Jean Le Rond d'Alembert. The project had its origins in a French translation undertaken by John Mills in 1743-1745 of Ephraim Chambers's *Cyclopaedia, or Universal Dictionary of Arts and Sciences.* The French publisher wrested control from Mills and, intending speedy publication, engaged two editors in succession who instead expanded the project's contours. The second, Diderot, undertook a monumental effort to outline the present state of knowledge in the sciences, arts, and practical crafts and to make this knowledge widely accessible. Originally each topic was to be covered by a scholar or craftsperson expert in it, and contributors included such prominent Enlightenment figures as Voltaire and Rousseau. In the end, though, Diderot and d'Alembert wrote many of the 71,818 entries themselves.

Although clearly embracing a philosophical perspective, the *Encyclopédie* served more to bring together different domains of knowledge than to unify or even systematize them. To the extent that there was a unifying theme, it lay in the Enlightenment's reliance on reason and empirical observation to provide knowledge. Even religion was presented as an object of human reason, not as a source of knowledge via revelation. The *Encyclopédie* thus stood in opposition to the scholastic tradition, which maintained Aristotle's legacy but subordinated it to Christian theology. The entry on philosophy emphasizes the role of reason:

> Reason is to the *philosopher* what grace is to the Christian. ... Other men are carried away by their passions, without their actions being preceded by reflection: these are men who walk in the shadows; whereas the *philosopher*, even in his passions, acts only after reflection; he walks in the night, but he is preceded by a torch. The *philosopher* forms his principles on the basis of an infinite number of discrete observations. ... He certainly does not confuse it with probability; he takes as true that which is true, as false that which is false, as doubtful that which is doubtful, and as probable that which is only probable. He goes further — & here is a great perfection of the *philosopher* — when he has no proper motive for judging, he remains undecided. (Translation by Dena Goodman from *The Encyclopedia of Diderot and d'Alembert Collaborative Translation Project*, http://www.hti.umich.edu/d/did/)

Not surprisingly, this emphasis on reason and empirical knowledge and criticism of claims for revealed truth ran afoul of the Church, so after the first seven volumes were published in Paris under a royal privilege, the remainder were published under the false imprint of Samuel Faulche, Neuchâtel (in fact they were published in Paris).

Reflecting the great diversity of human pursuits that involve acquisition of knowledge, the *Encyclopédie* represents a compilation of knowledge rather than an integration of it. In many respects, this reflects our contemporary situation. But in the wake of the enlightenment, other theorists resumed the pursuit of systematic unity.

1.3 Oken's Systematic Unity

Lorenz Oken (1779–1851) was an anatomist and a leader of the *Naturphilosophie* movement in Germany. A student and follower of Friedrich Schelling, Oken applied the precepts of *Naturphilosophie* to his thinking about biological systematics. The metaphysics he learned from Schelling — a Pantheistic view by which everything in nature could be deduced from a first principle, namely God — led him not only to treat the biological world as a part of God, but also to articulate a hierarchical classification of everything [Oken, 1809; 1831; Ghiselin and Breidbach, 2002]. Oken treated the organization of the world as a divine code that could be read by understanding the systematic relations between each thing and everything else.

Oken's approach to systematics was essentially that of the *scala natura.* His *Lehrbach der Naturphilosophie* offers an account in which his philosophical, theological, numerological, and biological assumptions were all tied together to produce a single, unified 'anatomy' of the world. There was first an argument that God is nothing, since all comes from nothing. This is just to say, of course, that God is (the source of) everything [Ghiselin, 2004]. After this theological argument was some numerological reasoning relating the four basic elements of the world (fire, air, water, and earth) to processes like electricity and crystallization. The book culminated in an argument that war-making is the highest art.

The thoroughgoing unity of Oken's classification is well illustrated by his theory of color.[1] For numerological, theological, alchemical, and scientific reasons, red corresponds to fire, then to love, and then to God the Father. Blue, as we might expect, corresponds to air, then to faith, and then to God the Holy Spirit. Yellow corresponds to earth, vice, and Satan. The colors of natural entities fit into, and are regarded as explained by, this overarching system. For example, animals are predominantly red because they correspond to fire (and the cosmos). Plants have green leaves because they correspond to water (and the planets). Flowers get a three-way classification: those of lower plants are most often yellow, the intermediate ones blue, and the highest ones red.

1.4 Encyclopedia of Unified Science

Whereas Oken attempted to build unity in terms of conceptual (semantic) ideas, other approaches to systematizing knowledge appealed to logic (syntax) for the bridges between bodies of knowledge. Logical positivism, later known as logical empiricism, developed in the 1920s in Austria (Verein Ernst Mach in Wien, commonly known as *the Vienna Circle*), Germany (Gesellschaft für Wissenschaftliche Philosophie Berlin, commonly known as *the Berlin Circle*), and Poland. The term and basic doctrine of *positivism* originated with August Comte, an early 19th century French philosopher who was skeptical of philosophical systems and metaphysics generally and emphasized positive knowledge — that is, knowledge grounded on observation and experimentation. A more immediate influence was the positivism of Ernst Mach, a professor of physics in Prague and Vienna until his retirement in 1901. He adopted a radical empiricism in which the only source of knowledge was sensory experience, and scientific laws were instrumental, serving to describe and predict phenomena available to the senses. Most of the early logical positivists adopted Mach's emphasis on the experiential grounding of knowledge, although most did not share his extreme instrumentalism. The adjective *logical* identifies the chief resource to which the logical positivists appealed in advancing beyond individual observations to generalized scientific claims. The logic to which they appealed was not traditional Aristotelian logic, but rather the modern mathematical logic developed in the late 19th and early 20th centuries by Frege, Peano, Russell, Whitehead, and others. Many of the logical positivists were themselves scientists who were concerned about clarifying the foundations

[1]This example is due to Michael Ghiselin, and is spelled out in more detail in [Ghiselin, 2004].

of science, especially in light of major developments in physics and other sciences contemporaneous with the rise of mathematical logic.

Although many of the logical positivists focused on physics, their emphasis was on providing a general account of knowledge, which they equated with scientific knowledge. They also, as discussed in more detail below, articulated a vision of how different sciences could be unified into a theoretical whole through theory reduction. One motivation was to counter a view, widespread at the time, that psychology addressed an inner world that was discontinuous with the outer world studied by the other sciences. Initially Rudolf Carnap [1928] proposed to overcome this discontinuity by treating all science as grounded on private experience, from which the world was *constructed.* This project, however, was unsuccessful. An alternative proposal for unification was offered by Moritz Schlick, who distinguished the content of experience (specific sensations) from its structure (relations between experiences). He maintained that the structure of experience was objective and could be investigated empirically. These and other attempts to provide a common account of the methodology of all sciences and link them into a common theoretical edifice gave rise to the *International Encyclopedia of Unified Science*, edited jointly by Otto Neurath, Rudolf Carnap, and Charles Morris.[2]

Neurath envisaged that the encyclopedia would grow to hundreds of volumes, with one entry issued each month in a subscription series. In the end only 20 entries were published in two volumes, the first under the original title and the second under the more modest title *Foundations of the unity of science; toward an international encyclopedia of unified science.* The goal, according to Neurath [1938, 24], was "to integrate the scientific disciplines, so to unify them, so to dovetail them together, that advances in one will bring about advances in the others." The main tool for such dovetailing of different sciences was logical analysis, which would serve to relate the concepts and ultimately the theoretical claims of various sciences. Although the editors envisioned an axiomatized integration of the great body of knowledge provided by the various sciences, they adopted a piecemeal strategy. They fully expected this procedure would uncover inconsistencies whose eventual resolution would improve each science as well as the prospects for their integration.

Since the account of unity advanced by the logical positivists has been the chief focus of philosophical accounts of the unity of science ever since, we will return in greater detail to this account in part 2 of this chapter. First, however, we consider one last proposal for unity which, although receiving less attention in philosophy, had and continues to have considerable influence in the sciences themselves.

[2]The term *unified science* was first invoked in 1938 when *Erkenntnis*, which had been the house organ of the Vienna Circle since 1930, was moved to the Hague and renamed the *Journal of Unified Science.* Just two years later, however, it ceased publication.

1.5 Cybernetics and General Systems Theory

Beginning in the 1940s, cybernetics and general systems theory advanced a very different conception of how to unify science that focused primarily on the organization found in phenomena the sciences seek to explain, especially the biological and social sciences. The term *cybernetics* was coined by mathematician Norbert Wiener from the Greek word for 'helmsperson', and was applied to systems that could steer themselves [Wiener, 1948]. Working during World War II, Wiener initially focused on a practical problem: developing a system for improving the accuracy of anti-aircraft guns. His desired solution invoked feedback control; that is, the accuracy of previous shots would be used to adjust gun controls before taking the next shot. Challenges he faced in getting the idea to work led him to collaborate with an engineer, Julian Bigelow, and a physiologist, Arturo Rosenblueth. In a paper in *Philosophy of Science* [Rosenblueth *et al.*, 1943], the three developed the idea that feedback enabled both biological and artificial systems to be goal-directed. They regarded this as resuscitating a notion that was anathema to the positivists: that of *teleology.* Subsequently Wiener organized a multi-year conference series. He initially called it Conference on Circular Causal and Feedback Mechanisms in Biological and Social Systems but, beginning in 1949, the conference adopted Wiener's term *cybernetics* for its name. As the initial name suggests, the participants regarded the idea of feedback organization as having the potential to unify biological and social systems.

Around the same time, biologist Ludwig von Bertalanffy [1951] advanced General Systems Theory as an antireductionist yet unifying perspective. Rather than focusing on the particular components out of which different things were made, systems theory emphasized the organization of parts into wholes and maintained that the same principles of organization, such as negative feedback, would be found applicable in physics, chemistry, biology, the social sciences, and technology.

Although there is an International Society for Systems Sciences that is still active and runs large international meetings, cybernetics and general systems theory have declined into niche specializations. Today the strongest influence of these approaches is indirect, funneled through successors with new ways to identify general principles of organization and use them towards unifying science. The new work goes under such rubrics as the sciences of complexity, complexity theory, and self-organizing systems and emphasizes systems with non-linear interactions. Tools for describing such systems were first developed in physics by Poincaré and others in the late 19th century, giving rise to Dynamical Systems Theory (DST) in the 20th century. DST was initially applied to physical phenomena such as eddies in a stream [Landau, 1944], but also was used to elucidate phenomena in biology (see [Kaufmann, 1993]) and then psychology. The earliest psychological accounts focused on motor coordination [Kelso, 1995] and its development [Thelen and Smith, 1994], but gradually DST has expanded to other domains. Indeed, some proponents have presented DST as a revolutionary, overarching alternative to other approaches to cognition [Port and van Gelder, 1995; Keijzer, 2001].

One particularly interesting offshoot of complex systems research has been the introduction of a number of important ideas about the structure of networks and how they can be used to characterize phenomena in the world. Most traditional investigations of networks focused either on regular lattices, in which only neighboring units are connected, or on random networks (the focus of pioneering investigations by Erdös and Rényi). Such organization is very different from the "small world" networks first articulated by Stanley Milgram [1967], who discovered empirically that while individual humans are primarily connected to those around them (as in regular lattices; this feature is known as *high clustering*), they are indirectly connected to a vast number of others via relatively short paths of direct connections through people who know each other (as in random networks; this is known as *short path length* and provided the premise for the play and movie *Six Degrees of Separation*). Duncan Watts and Steven Strogatz [1998] showed that minimal changes to a regular lattice can transform it into a small-world network and explored real-world phenomena exhibiting this form of organization — including collaborations between actors in feature films, the electrical power grid of the western U.S., and the neural network in a nematode. Moreover, Albert-Lászlo Barabási and Réka Albert [1999] discovered that many networks in the real world are *scale free*, in that connections exhibit a power-law distribution (the majority of units are connected to only a few others, but a few are connected to a *very* large number of others). (The term *scale free* is used to reflect the fact that power-law distributions lack any intrinsic scale.) Barabási and his collaborators have attempted to account for the occurrence of scale-free networks as a result of historically earlier nodes having a longer time to attract attachments and to new nodes preferentially attaching to already highly connected nodes [Albert and Barabási, 2002]. More recently Cees van Leeuwen and his collaborators have shown how scale-free small-world networks can evolve through coupling of chaotic oscillators [Gong and van Leeuwen, 2003]. These developments potentially provide a powerful set of tools for analyzing organization in a wide variety of natural and social systems.

2 FIELD GUIDE TO MODERN CONCEPTS OF REDUCTION AND UNITY

In the 20th century, claims about unity of science were commonly tied to claims about theory reduction. In particular, the strategy was to reduce the theories of higher-level sciences such as biology to the laws and theories of lower-level sciences such as physics and chemistry. (Spelling out the notion of levels is challenging and we will return to this issue at several junctures below.) Claims about reduction were, in turn, treated as claims about deductive relations between theories. Recently, strong dissent has been raised on both scores, with some philosophers rejecting both reduction (see below, 2.3) and unity of science (see below, Section 4). Other philosophers, more sanguine about unity, have advanced alternative conceptions that emphasize integration more than unity and detach these issues from questions of theory reduction. In addition, accounts of reduction that do not tie it to deductive relations between theories have been advanced. Although the more recent alternative treatments of both integration and reduction offer much promise for providing more adequate accounts of both of these notions, we will start by laying out the traditional accounts of both positions.

2.1 The Theory-Reduction Model

The logical positivists advanced the theory-reduction model as part of their effort to provide an account of science that avoided entanglement with metaphysical issues. To accomplish this they focused on the knowledge claims of science and emphasized the role of logical relations between these. A crucial move was to represent two kinds of knowledge claims in the same format, yielding sets of propositions encompassing both observation statements (reports of empirical observations such as "The marble is rolling down the incline") and theoretical statements like Newton's law of universal gravitation, which says that the attractive force between any two bodies is equal to the product of their masses divided by the square of the distance between them. Nagel identified an intermediate category of *experimental laws*, which provide an empirical summary of the phenomena observed. Galileo's law that the distance a falling object travels is proportional to the square of the time it is in motion is an example. These experimental laws are contrasted with *theoretical laws*, such as Newton's, which go beyond the observed phenomena by positing theoretical entities like forces and masses to account for the experimental laws. The power of laws or theories to explain observations could then be rooted in the ability to derive new observation statements — predictions — from laws. This is the well-known deductive-nomological (D-N) or covering-law model of explanation [Hempel and Oppenheim, 1948; Hempel, 1965]. To account for the relations between the laws or theories of different sciences, the logical empiricists proposed simply generalizing this account, and argued that it should be possible to derive the laws or theories of one discipline or science from those of another [Nagel, 1961; see also Woodger, 1952; Quine, 1964; Kuipers, 2001, chapter 3].[3]

Two fundamental challenges arose in developing this generalization of the D-N model. First, the laws in the different sciences are typically presented in different

[3]Kemeny and Oppenheim [1956], see also [Oppenheim and Putnam, 1958], advanced an alternative account of reduction that did not derive the reduced theory from the reducing theory but only required generating identical observable predictions from the reducing theory as the reduced theory. This account of reduction is far more liberal, since it allows for the reduction of what are regarded as false theories (e.g., phlogiston chemistry) from what are taken to be true theories (e.g., Lavoisier's oxygen-based chemistry) as long as the predictions made by the reducing theory include all those made by the reduced theory. Yet another alternative was put forward by Patrick Suppes, who required an isomorphism between any model (in the model-theoretic sense) of one theory and a model of the reduced theory: "To show in a sharp sense that thermodynamics may be reduced to statistical mechanics, we would need to axiomatize both disciplines by defining appropriate set theoretical predicates, and then show that given any model T of thermodynamics we may find a model of statistical mechanics on the basis of which we may construct a model isomorphic to T" [Suppes, 1957, 271].

vocabularies.[4] Laws in physics, for example, might employ terms such as *mass* and *attractive force*, whereas those in chemistry would invoke names of elements and molecules and types of chemical bonds. But logical inferences are only possible between statements using the same vocabulary, in much the same way as certain algebraic problems can be solved only when the units of time, length, or weight are expressed using the same measure. To address this issue, advocates of the theory reduction model appealed to *bridge principles* (Nagel called them *rules of correspondence*) that equated vocabulary in the two laws. Sklar emphasized that these correspondence claims are really identity claims: "Light waves are not correlated with electromagnetic waves, for they *are* electromagnetic waves" [Sklar, 1967, 120]. Applied to the context of relating psychology to neuroscience, this contention that the terms in the different theories picked out the same entity became the foundation of the celebrated mind-brain identity theory [Place, 1956; Feigl, 1958/1967; Smart, 1959]. Although such bridge principles might seem unproblematic, we will see that they are the target of one of the more powerful objections to unification through reduction.

The second challenge confronting advocates of the theory reduction model is the fact that the regularities captured in higher-level laws arise only under a particular range of conditions. To accommodate this, they proposed that reduction also required statements of boundary conditions. With these components in place, a reduction was then conceived to have the form of the following deduction:

Lower-level laws (in the basic, reducing science)
Bridge principles
Boundary conditions
___
∴ Higher-level laws (in the secondary, reduced science)

An oft-cited example is the derivation of the Boyle-Charles' law from the kinetic theory of gases, as part of an overall reduction of classical thermodynamics to the newer and more basic science of statistical mechanics [Nagel, 1961, 338–366]. This law states that the temperature (T) of an ideal gas in a container is proportional to the pressure (P) of the gas and volume (V) of the container. Because the term *temperature* does not appear in statistical mechanics, to achieve the reduction a linkage to a term in that science is required. This is expressed in a bridge principle (rule of correspondence) stating that the temperature of a gas is proportional to the mean kinetic energy (E) of its molecules. A number of boundary conditions also must be specified, such as those limiting the deduction to monotonic gases in a temperature range far from liquefaction. With the appropriate bridge principles and boundary conditions included as premises, the Boyle-Charles' law can be derived from laws of statistical mechanics. Here is a key part of the full derivation:

[4]Nagel did consider cases in which the same vocabulary was employed in the reducing and reduced theories. He referred to such reductions as *homogeneous*.

Laws of statistical mechanics (including the theorem $PV = 2E/3$)
Bridge principles ($2E/3 = kT$)
Boundary conditions (monotonic gas; T in specified range)

$\therefore$ Boyle-Charles' law ($PV = kT$)

Notice that behind the unity-as-reduction conception is a view of the natural world as comprised of levels, often referred to as 'levels of organization'. Given their reluctance to engage ontological issues, the logical empiricists tended to construe levels in terms of the disciplines that investigate them. On this view there are levels of organization in the world that correspond to such disciplines as physics, chemistry, biology, psychology, and sociology. Unification consists of reducing the theories of each higher discipline to theories of a lower discipline. Some philosophers regard this as thereby achieving a reduction between the disciplines themselves. So, for instance, if biological theory were reduced to physical and chemical theory, the science of biology would also thereby be reduced to the sciences of physics and chemistry, and biology would no longer be an autonomous science.

While embracing many features of the logical empiricists' account of reduction, Robert Causey [1977] advanced a more ontologically committed interpretation of levels wherein higher levels resulted from the structuring of lower-level entities. On this view, theories at the lower level primarily describe the operation of parts of the structured wholes, while those at the higher level focus on the behavior of the structured wholes themselves. For a reduction to be possible, the lower-level theory must itself have the resources to describe the structured wholes and their behavior. (Although this is quite problematic, assume for a moment that it is possible.) We then have two descriptions of the higher-level entity, one as a whole unit in the vocabulary of the higher-level science, and one as an entity structured out of lower-level components. For Causey, reduction then requires bridge principles that relate terms in the higher-level theory referring to the wholes to those terms in the lower-level theory that characterize them as composed, structured wholes. Assuming that the lower-level theory has laws that describe the behavior of the structured wholes, one can try to derive the upper-level theory from the lower-level one.

An important feature of the theory-reduction model is that it requires the lower-level theory (or science) to have all the resources required to derive the upper-level theory (when bridge principles and boundary conditions are supplied). Below we will consider whether this is plausible. But noting this feature of the model allows us to consider what many practitioners of higher-level sciences find problematic in philosophical accounts of reduction: successful reduction apparently obviates any need for any laws or theories specific to the higher-level sciences. At least in a hypothetical final picture of science, higher-level sciences would be expendable or redundant: by supplying the appropriate boundary conditions, any higher level regularity could be derived directly from the lower-level theory. In practice, at a given stage in the development of science, appeals to the higher-level sciences may be required because the reduction base may not yet have been developed. Higher-level sciences may even play a heuristic role in the development of the lower-level sciences; for example, they may reveal regularities (laws) in the behavior of the structured wholes that must be accounted for. In this respect, there may even be a co-evolution of higher- and lower-level sciences [Churchland, 1986]. In the end, however, the theories of the lower-level science will be complete, and the only reason for invoking the vocabulary and laws of the higher-level science will be that they provide a convenient shorthand for referring to what, in the lower-level theory, may be unmanageably complex statements.

2.2 Revisionist Accounts of Theory Reduction

Among the early challenges to the theory-reduction model, one of the most influential focused on the possibility of establishing the appropriate bridge principles. Paul Feyerabend [1962; 1970], as a result of adopting an account that characterized the meaning of scientific vocabulary in terms of the theory in which they were used, argued that words in different theories, even if they have the same form, are *incommensurable* with one another. In classical thermodynamics, for example, *temperature* can be defined in terms of Carnot cycles and its behavior described by the non-statistical version of the second law of thermodynamics. But in statistical thermodynamics, temperature is characterized in statistical terms. Given the important differences in the surrounding theory and hence the different entailments of the meanings of 'temperature', it would seem impossible to construct bridge principles that would adequately relate 'temperature' as used in these two theories. At the same time, Thomas Kuhn [1962/1970] focused on other examples of putative reduction, such as Newtonian to Einsteinian mechanics, and maintained that words like 'mass', were used incommensurably in the two theories. On the basis of such examples, Kuhn challenged the account of progress implicitly assumed by the logical empiricists, in which sciences progress towards better theories through a process of continual extension and refinement. Kuhn argued instead that the history of science is a history of revolutions in which new theories replace, rather than build upon older, incommensurable theories.

One specific context in which Feyerabend maintained that reduction would fail involved attempts to relate psychological theories presented in mentalistic vocabulary to accounts of brain function in neuroscience. Although Feyerabend later came to champion the position that incompatible theories ought both to be maintained [Feyerabend, 1975], in his early writing on mind-brain relations he advanced a position known as *eliminative materialism* [Feyerabend, 1963]. The key claim of Feyerabend and subsequent eliminativists [Rorty, 1970; Churchland, 1981; Churchland, 1986] is that instead of reducing the old (folk) psychological theory to the new neuroscientific theory, the old psychological theory is *replaced* by the new theory and *eliminated* from the corpus of science. The model of such replacement is the replacement of Ptolemy's astronomical theory by Copernicus'. The old, Ptolemaic, theory accounted for the observed motions of the planets by assuming that they moved on epicycles whose centers themselves orbit around the earth (the epicycles explained the apparent retrograde motion of the planets when viewed from earth). Copernican astronomy, as in Figure 1b, explains the same phenomenon by assuming that the earth and other planets orbit the sun. Since the two astronomical accounts are inconsistent and we assume that the Copernican account is fundamentally correct, eliminativists conclude that Ptolemy's account is wrong and should be discarded and replaced. Such replacement befell not only historical theories such as Ptolemaic astronomy, the impetus theory of motion, and phlogiston chemistry but also, on this view, awaits folk psychology and other mentalistic accounts.[5]

Although Feyerabend and Kuhn viewed themselves as opposing reduction, and eliminativists such as the Churchlands held that elimination awaits when reduction fails, other philosophers such as Kenneth Schaffner treated Feyerabend and Kuhn as advancing an alternative account of reduction. On this alternative, even when deduction fails (as it must when the reducing theory is true and the reduced theory is false), one can still relate the old theory to the new one. In the late 16th century, for example, Tyco Brahe developed a way to map the Copernican model of the solar system onto the Ptolemaic one (Figure 1c). This showed that all the empirical observations that had supported Ptolemy's model also fit Copernicus's and offered support to it. Thus, one reason for exploring such relations between an old theory and its replacement is to enable the replacing theory to claim much of the empirical support that had been developed for the old theory.

After construing the discovery of such similarities as a kind of reduction that differs from the traditional model in interesting ways, Schaffner [1967] suggested that these two kinds of reduction need not be regarded as competitors. Instead, he proposed a comprehensive account in which reduction by deduction and reduction by replacement each play a role. In particular, a frequent consequence of a new lower-level theory (T_1) is that an old upper-level theory (T_2) gives way to a revised one (T_2*). T_2* should be deducible from T_1, just as envisaged in the standard theory-reduction model, but Schaffner thought its relation to T_2 should also be recognized. He suggested that the T_2-T_2* relation was one of analogy:

> T_2* corrects T_2 in the sense of providing more accurate experimentally verifiable predictions than T_2 in almost all cases (identical results cannot be ruled out however), and should also indicate why T_2 was incorrect (e.g., crucial variable ignored), and why it worked as well as it did. ... The relations between T_2 and T_2* should be one of strong analogy — that is (in current jargon) they possess a large "positive analogy". [p. 144]

Subsequently Schaffner [1969] amended his model to incorporate revision of an existing lower-level theory (T_1) to obtain a corrected lower-level theory (T_1*) in addition to the revision of old higher-level T_2 into T_2* (see Figure 2).

Schaffner provided little guidance as to what counted as a strong analogy. Thomas Nickles [1973] argued that in many instances these analogies could be understood mathematically as limit relations. At specific limit values for variables in the new theory, he argued, the new theory will yield the older theory, nearly enough. Nickles give the example of Einstein's formula for momentum reducing to the Newtonian formula by taking the limit as velocity approaches zero. Such

[5]Although most commonly the Churchlands have targeted folk psychology for their eliminativist claims, they also on occasion target contemporary cognitive psychology: "There is a tendency to assume that the capacities at the cognitive level are well defined ... As we see in the case of memory and learning, however, the categorial definition is far from optimal, and remembering stands to go the way of impetus" [Churchland, 1986, 373].

(a)

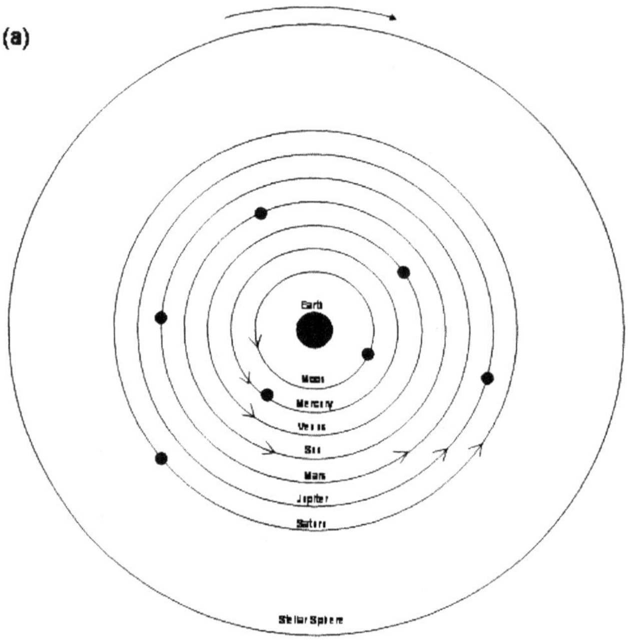

(b)

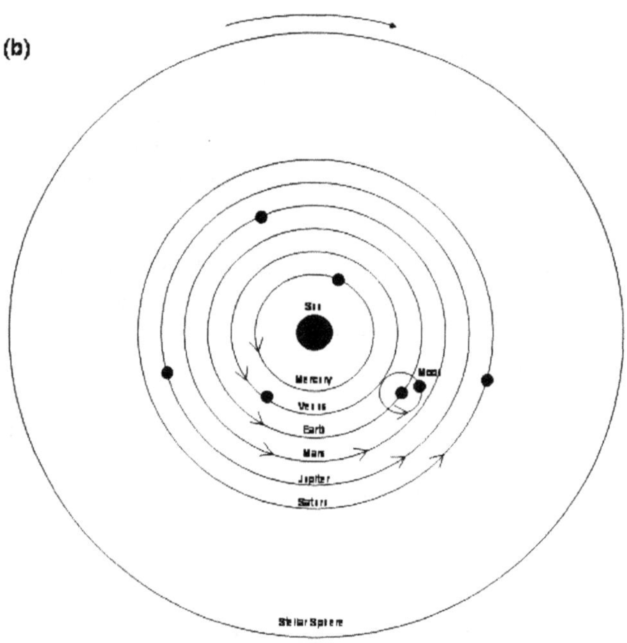

(c)

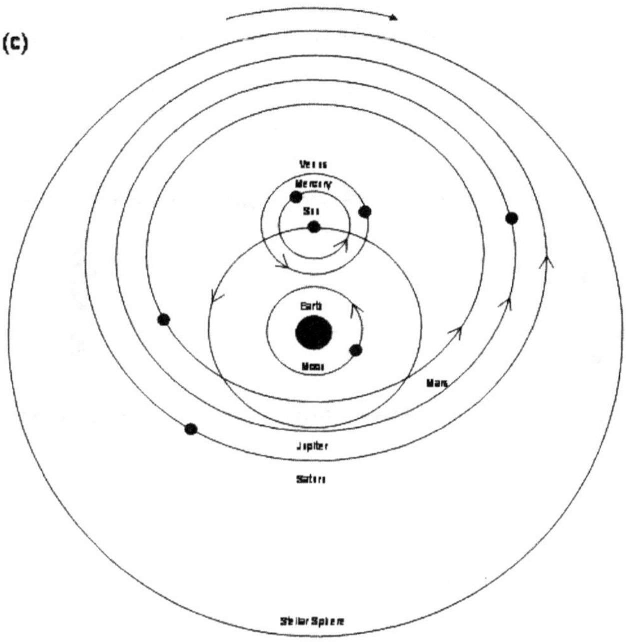

Figure 1. The orbits of the planets according to (a) Ptolemy, (b) Copernicus, and (c) Tycho Brahe. In all cases the planets orbit counter-clockwise while the stellar sphere moves clockwise. In Brahe's account, the earth is at the center of the solar system, as it was for Ptolemy, but the planets other than the moon revolve around the sun, as in Copernicus's account.

limit relations enable researchers to appreciate why the older theory worked as well as it did — most velocities Newtonian scientists considered were sufficiently small that the actual momentum differed only minutely from that predicted by the Newtonian formula.

The strategy of using a limit relation to capture the analogy between a revised theory and its predecessor will not work in all cases, however. William Wimsatt [1976a] argued that Schaffner's conception of strong analogy should be understood in terms of pattern-matching, in which a limit relation is just one of a number of ways to construct a match. Moreover, he extended Nickles' account of the function of such matches by focusing on the differences remaining after the match. These differences not only mark points at which evidence may show the revised theory to be an improvement; they may also, in cases where the predictions from the new theory are not as successful as those from the old theory, point to loci where yet further work is needed to amplify and extend the new theory.

Nickles further argued, convincingly, that advocates of the traditional theory re-

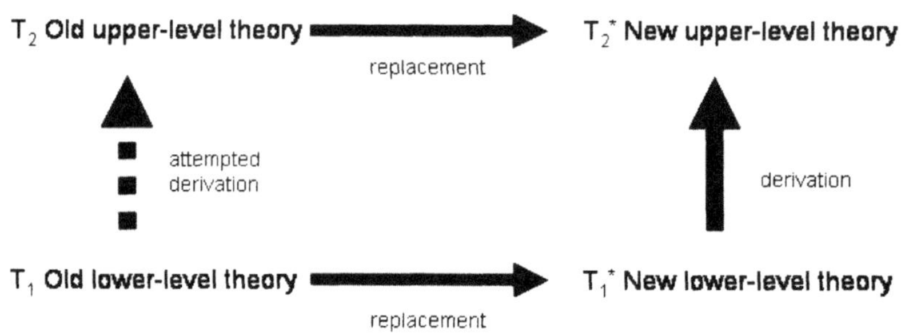

Figure 2. Schaffner's (1969) model of reduction, in which a new upper-level theory (T_2^*) is derived from a new lower-level theory (T_1^*) and each new theory replaces an older theory at the same level.

duction model were talking about a very different relation than were such critics as Kuhn and Feyerabend. He labeled reduction as envisaged by the theory-reduction model *reduction*$_1$ and argued that it is particularly relevant in explaining domain-combining types of reduction (which Wimsatt [1976a] characterized as interlevel reductions). But the relation between predecessor and successor theories is a domain-preserving relation which he labels *reduction*$_2$ (and Wimsatt construed as intralevel reduction). One feature that Nickles identified as distinguishing the two types of reduction is that they tend to be invoked for different reasons: reductions$_2$ serve heuristic and justificatory roles, while reductions$_1$ are unifying and explanatory. He also noted that the two reductions point in opposite directions with respect to theories differing in their generality. In reduction$_1$ the more specific upper-level theory is reduced to the more general lower-level one (e.g., the reduction of gas laws to the more general theory of statistical mechanics). In reduction$_2$ the more general theory is a newer one that reduces to the older theory, now recognized to be incorrect (e.g., the reduction of Einstein's formula for momentum to Newton's). In sum, in reduction$_1$ the move is from specific to general, whereas in reduction$_2$ it is from general to specific. (See Figure 3)

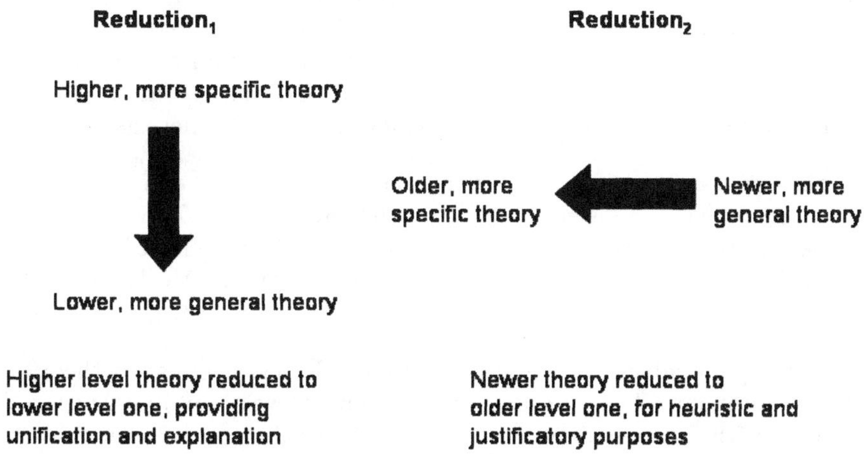

Figure 3. Nickles' two senses of reduction. In reduction_1 a higher-level theory is reduced to a lower-level one, whereas in reduction_2 a new, more general theory is reduced (e.g., in the limit) to an older, more specific theory.

In his development of Nickles' position, Wimsatt offered a novel reading of when new theories eliminate older ones. In cases for which there is a close pattern match between the old theory and the new one, the older theory might well continue to be employed because it is simpler or easier to use. But reductions between successively introduced theories are unlikely to be transitive. Rather, "*intralevel reductions should be intransitive* — ... a number of intralevel *reductions* could 'add up' to an intralevel *replacement*. ... Relativistic [Einsteinian] mechanics may reduce to classical mechanics (etc.) but it clearly replaces (rather than reduces to) Aristotelian physics" [Wimsatt, 1976a, 217–219].

Wimsatt's distinction between interlevel and intralevel reductions reveals interesting consequences for the eliminativist argument as applied to the relation between neuroscience and psychology. Whereas "eliminative materialism seems ... to derive its inspiration from intralevel reduction," Wimsatt contended, "the proper model for the mind-body problem is interlevel reduction" [Wimsatt, 1976a, 215]. This critique was further developed by McCauley [1986; 1996], who showed that historical cases exemplifying the replacement and elimination of an old theory have all involved a revised theory that is at the same level as the old theory.

McCauley suggested that the same would be true in the case of a psychological theory: elimination would be expected only when it was superseded by a replacement theory that lay at the same level — i.e., another psychological theory rather than a neural one). As for interlevel reductions, McCauley distinguished cases in which there is a tight fit between upper- and lower-level theories and cases in which there is not. Loose fit may result from the very nature of theorizing at the upper and lower level. In some cases, the finer grain of an account at the lower level may enable it to explain what appear to be deviations at the higher level. But the advantage is not always with the lower level. In other cases,

> the upper-level theory lays out regularities about a subset of the phenomena that the lower-level theory encompasses but for which it has neither the resources nor the motivation to highlight. That is the price of the lower-level theory's generality and finer grain. [McCauley, 1996, 31]

McCauley thus advocates a pluralistic approach that would allow theorists a fair degree of autonomy. Theories at higher and lower levels could be developed independently, with no immediate need to force the levels to relate in a reductionistic manner.

2.3 Criticism of Theory Reduction

Revisionists presented the difficulty of providing bridge principles as arising principally with cases involving successive theories at the same level (Wimsatt, as we will see below, is an exception), leading them to invoke a different account of intralevel and interlevel relations. Some influential critics, however, see the problem as arising even in the interlevel case and as providing the death-knell for the theory reduction account of interlevel relations. Similar arguments were advanced independently by two such critics, David Hull regarding biology and Jerry Fodor regarding psychology. The strategy in both cases was to maintain that the same term used in the laws of the higher-level theory must be related on different occasions with different terms and fall under different laws at the lower level. As it is sometimes expressed, one type of entity as characterized in the higher-level theory is *realized* by multiple different types of lower level entities on different occasions.

Hull [1972; 1974] focuses on the notion of *gene* as it figures in both Mendelian genetics and molecular genetics. One challenge to providing a reductive account in this case is that genes in Mendelian accounts are characterized in terms of phenotypic traits for which they code (e.g., a pea plant is tall, not short). Genes in molecular genetics are characterized in terms of their molecular constitution. Any one of a number of distinct molecular mechanisms could produce the same phenotypic trait (this is often referred to as *multiple realizability*, to which we will return below). Although the complicated nature of the phenotype-genotype map makes developing the reduction difficult, it does not necessarily block it. To achieve a reduction, what is needed is "to discover one or more molecular mechanisms which correspond to the various predicate terms of Mendelian genetics, such that the resulting classification of traits into types corresponds fairly well with the classification of these traits according to the principles of Mendelian genetics" [Hull, 1972, 497].[6] Hull went on to point out that even this modest goal cannot be reached; instead, scientists have found that "the same molecular mechanism can produce different phenotypic effects". This is just the reverse of multiple realizability, as it involves multiple different effects produced by the same mechanism. The reason is not a mystery: other conditions vary. Which conditions and combinations of conditions produce different phenotypic effects can be determined empirically by researchers, if desired. However, to bring such detailed findings into molecular genetics, so an adequate reduction of Mendelian genetics could be accomplished, would result is a radical expansion in scope: "We are no longer correlating Mendelian predicate terms with molecular mechanisms but with the entire molecular milieu" (p. 498). One possible conclusion is that reduction fails in the case of Mendelian genetics, but Hull pointed the blame instead at the account of reduction offered by philosophers:

> If the logical empiricist analysis of reduction is correct, then Mendelian genetics cannot be reduced to molecular genetics. The long-awaited reduction of a biological theory to physics and chemistry turns out not to be a case of "reduction" after all but an example of replacement. But given our pre-analytic intuitions about reduction, it *is* a case of reduction, a paradigm case. [Hull, 1974, 44]

If a paradigm case of reduction fails to go through on the theory-reduction model, Hull reasoned, the philosophical framework would seem to have failed. However, some philosophers of biology drew a different conclusion from the difficulties identified by Hull: they treat the failure as pointing to fundamental deficiencies in biology. In particular, Alexander Rosenberg [1994] argued that because natural selection selects for function rather than structure, the relations between Mendelian genetics (phenotypic features characterized functionally) and molecular genetics (genotypes characterized structurally) are so complex that any attempt to construct bridge laws between them will yield disjunctions too long to be useful for creatures of our mental capacity. Focusing just on the multiple realizability of traits, not the reverse relation, Rosenberg observes that given an environmental 'problem' to solve, selection can achieve the same phenotypic function by any number of molecular pathways. The phenotypic or functional features 'tallness' and 'roundness' are, in other words, multiply realizable from the point of view of molecular genetics. Offering bridge laws, then, will amount to making a list of all the possible pathways. This process, Rosenberg argues, leads to intractably

[6] As Richardson [1979] noted, Nagel actually allowed for such multiple realizations of the same higher-level property as long as it was possible to explain why the different lower-level properties realized the same higher-level one. Differences in context may determine whether a particular lower-level property realizes a higher-level one.

long lists rather than a better understanding (which is what a true science would provide) of the molecular underpinnings of Mendelian genetics or of the operation of natural selection. Since the theories of functional biology are not reducible to molecular foundations, they provide only problematic access to the biological world.

The other critic of theory reduction, Fodor [1974], focused on psychological predicates and argues that they cannot be linked via bridge principles to neuroscientific ones. Invoking an analogy with finance, he noted that money does not correspond to any natural kind of physical stuff. In the right circumstances, pieces of paper, gold, silver, bronze, or even patterns of electrons can each serve as money; hence, money is multiply realizable The example nicely draws out Fodor's primary point that the factors that determine kinds in behavioral and societal realms, such as finance, are very different from those determining kinds in the physical realm. In particular, Fodor, as well as Hilary Putnam [1978], maintained that psychological kinds should be identified functionally in terms of how they interact in the generation of behavior. For example, hunger will interact with cognitive states, such as beliefs, in generating particular food-seeking behaviors. Given the differences in their nervous systems, a functional state such as hunger will arise as a result of different neural processes in species such as octopi and humans, although in both cases the state will result in food-seeking behavior (this example is due to Putnam). Accordingly, both Fodor and Putnam reject the project of reducing psychology to neuroscience, instead advocating the autonomy of what Fodor refers to as the *special sciences*.[7]

A second response is advocated by Causey and Hooker. They recommend acknowledging multiple realizability and accepting that a different reduction will be needed for different lower-level realizations of a given higher-level law. Far from promoting unity, this response may actually result in greater disunity when phenomena that appear very similar in high-level terms turn out to be reduced to ver y different lower-level theories. Pylyshyn [1984], for example, argued that folk psychology successfully groups diverse behaviors under the same regularities, enabling us to predict behavior effectively, but that virtue would be lost if one tried to reduce it to the diverse behaviors that realized the regularity. For example, in our folk idiom we make generalizations about people's propensity to answer the phone,

[7]Fodor also maintains that in developing their taxonomies and relating states, special sciences will commonly appeal to very different principles than those that are typical in more basic sciences. For example, in seeking a psychological account of human decision making, we will prefer one that renders people and their decisions as rational; whereas this is not an objective in developing neuroscientific accounts. Charles Taylor [1967, 206] made essentially the same argument: "... if human behavior exhibits lawlike regularity, on the physiological level, of the sort which enables prediction and control, and a rougher regularity of a less all-embracing kind on the psychological level, it does not follow that we can discover one-one or even one-many correspondences between the terms which figure in the first regularities and those which figure in the second. For we can talk usefully about a given set of phenomena in concepts of different ranges, belonging to different modes of classification, between which there may be no exact correspondence, without denying that one range yields laws which are far richer in explanatory force than the others."

yet on different occasions that activity can involve different motor systems (e.g., picking up a handpiece and talking; sending a text message). By treating each instance separately, we lose the generality provided in the folk idiom "answering the phone".

Although many philosophers have assumed that multiple realizability is rampant and undermines the prospects of relating higher-level kinds to those of the more basic sciences, drawing such connections has been a key strategy in biological investigation. While recognizing that the mechanisms underlying physiological and psychological processes in different species do differ, investigators nonetheless draw extensively on what they have learned in one taxon to understand others. For example, much of what is now known about mechanisms of visual processing in humans was secured through research on other mammalian species, especially the cat and monkey [Bechtel, 2001]. Although neuroscientists fully realize that there are differences between brains of different organisms, especially of organisms from different taxa, they also expect and have found extensive commonality. This should not be surprising — it has long been known that biological mechanisms at all levels are often highly conserved, attributed in part to the high cost in fitness for large changes. Also — and this point has not received sufficient attention in the philosophical literature — biologists generalize from mechanisms, processes, and features identified in one taxon, to others by means of what might be called *phylogenetic reasoning*. Where such mechanisms, processes, and features can be shown to be carried through lineages, investigators expect fundamental similarities [Hennig, 1966]. Accordingly, the underlying mechanisms are not likely to be as radically different as advocates of multiple realizability assume. Researchers also expect differences between taxa and seek these out, but these will often be variations on a common structure: in the language of cladistics systematics, these similarities (and dissimilarities) will be both *shared* because of membership in a common lineage and *derived* due to the differential influences of evolution.[8] Given the conservative nature of evolution, we should not be surprised that human brains retain much of what is found in cat and monkey brains (and indeed, even the brains of invertebrates).

Those who view multiple realizability as an obstacle for reduction often neglect a further factor — just as there are neural differences between organisms and especially between species, there are psychological differences as well. The behavior of a hungry octopus is very different from that of a hungry human. Putnam ignores these differences when he applies the same psychological predicate to both. But these differences often matter as well in developing psychological theory. In both psychology and neuroscience, researchers can select a coarse-grained analysis, lumping together instances that differ in many respects, or a fine-grained analysis,

[8]There are, of course, examples of convergent evolution in which similar adaptations arise in different lineages (e.g., wings in bats, birds, and pterodactyls). But these are typically readily distinguishable functionally in a variety of ways (e.g., the amount of weight that can be supported or response to turbulence in the case of wings) and so typically do not provide good examples of the *same* function being multiply realized. For further criticisms of the assumption of multiple realizability, see [Bickle, 2003; Polger, 2004; Shapere, 2004].

splitting similar instances into different kinds. For different purposes, they may select one or the other. Putative examples of multiple realizability, however, often trade on invoking coarse-grained analyses of psychological kinds and fine-grained analyses of neural kinds. When the same grain is employed in lumping brains in the same category as is employed in lumping mental states into the same category, the alleged problems induced by multiple realizability for reduction seem to vanish [Bechtel and Mundale, 1999].

Leaving behind these worries about multiple realizability, a critical feature of theory reduction accounts, either in their original or revisionist versions, is the assumption that the lower-level theories have sufficient resources from which to derive all the laws of the higher-level science. This assumption is radically implausible. A first objection is that the lower-level theories to which higher-level ones could be successfully reduced would have to be rather different from those currently under development in the lower-level sciences. We can appreciate this by returning to Causey's version of the theory-reduction model. In his discussion, although not in his formal treatment, Causey suggests that researchers will study the behavior of the components of structured wholes when they are not part of the whole (his *non-bound condition*) and then derive their behavior when part of the structured whole from this information plus specification of the boundary conditions prevailing when they are bound. Yet, in real science, researchers frequently find that what they know about the behavior of entities in their non-bound condition fails to reveal how they will behave in various complex environments. The behavior of atoms as they behave independently reveals little of how they will behave when bound into molecules; likewise, the behavior of amino acid strings reveals little of how they will behave when folded into proteins. Instead, how such entities will behave in bound situations has to be determined empirically. (One indication of this is that when research teams include scientists from both lower-level and higher-level disciplines, the relationship is not one in which the lower-level scientist provide general theories and the higher-level scientist derives the consequences. Rather, all recognize they must discover new information and that what the lower-level scientist often has to offer are techniques that can help reveal how the component parts are behaving in the more complex environment.)

An alternative strategy is simply to incorporate into the lower-level theory everything that is learned about lower-level entities as they are bound into various structured wholes. Clifford Hooker adopts this view:

> First, the mathematical development of statistical mechanics has been heavily influenced precisely by the attempt to construct a basis for the corresponding thermodynamical properties and laws. For example, it was the discrepancies between the Boltzmann entropy and thermodynamical entropy that led to the development of the Gibbs entropies, and the attempt to match mean statistical quantities to thermodynamical equilibrium values which led to the development of ergodic theory. Conversely, however, thermodynamics is itself undergoing a process of enrichment through the injection "back" into it of statistical mechan-

> ical constructs, e.g., the various entropies can be injected "back" into thermodynamics, the differences among them forming a basis for the solution of the Gibbs paradox. [Hooker, 1981, 49]

The idea that lower-level theories need to be enriched to account for what is learned at the higher level leads to a view that reduced and reducing theories co-evolve, a view that Patricia Churchland [1986] espouses for the relation between psychology and neuroscience. The difficulty with this approach is that lower-level accounts of the behavior of entities when they are bound in complex structures may share little with accounts of how they behave in isolation. The resulting lower-level theory may be so complex and its various claims sufficiently unrelated to one another that little unity will have been achieved.

Before leaving criticisms of the theory-reduction account, we should note one feature of the account not often discussed — the role played by boundary conditions. It is only under specific boundary conditions that, on this account, higher-level laws can be derived from lower-level ones. But where do these boundary conditions come from? They are not themselves derived from the lower-level laws. Rather, they must be determined empirically as investigators try to develop the reduction. This has significant consequences for the claims that reduction unifies all higher-level laws in terms of basic-level ones. In fact, the higher-level laws are derived from lower-level theories *plus* bridge principles and boundary conditions. Even if the rest of the theory reduction account proved adequate, it would not promote as much unity between the various sciences as is often suggested.

3 KITCHER'S REVISIONIST ACCOUNT OF UNIFICATION

Pursuing a line of argument first formulated by Michael Friedman [1974], Philip Kitcher has argued for more than two decades that we should be interested in the unity of science because of the tight connection between unification and explanation. Kitcher [1981] defends this view as a means to offering an account of explanation that both builds on the work of some of the logical empiricists (particularly Hempel and Feigl) and overcomes some shortcomings of the covering-law (D-N) model of explanation (and by extension, the theory-reduction model). Three of these inadequacies are of chief importance. First, according to Kitcher [1981, 508], the covering-law model does not make clear just how it is that scientific explanation advances understanding. Second, the covering-law model does not offer a means to weigh the explanatory power of some theory, or of some theory as against another one. Third, the quality of the covering-law model depends on there being a good way to distinguish between laws and accidental generalizations, but this distinction has been famously problematic since Goodman [1955].

Kitcher's emphasis on unification is meant to be a way to retain the logical empiricists' commitment to explanation as derivation. Kitcher is able to avoid the problems discussed above by arguing that successful explanations are part of a "system" or "store" of explanations, such that no putative explanation can be evaluated individually, but rather must be assessed (at least partly) by reference to the rest of the explanations science accepts at a time.

> Science supplies us with explanations whose worth cannot be appreciated by considering them one-by-one but only by seeing how they form part of a systematic picture of the order of nature. [Kitcher, 1989, 430]

The central move here is to accept, with the logical empiricists, that explanations are derivations, but to deny that such derivations can be assessed in a piecemeal fashion. Rather, they must be part of the best systematization of the set of statements accepted by the scientific community at a given time. "Best systematization" here means, roughly, the set of derivations that minimizes the number of argument patterns while maximizing the number of conclusions. The number of argument patterns can be obtained by giving a classification of argument patterns based on inferential characteristics.

The change from individual derivations to a best system of derivations circumvents the three problems noted above by making no use of the law-accidental generalization dichotomy, by providing a means of assessment for the explanatory power of a candidate explanation (a better explanation is one that leads to more conclusions while adding the least number of argument patterns), and finally, by showing how explanations lead to understanding. The unification approach accomplishes the latter by "showing us how to derive descriptions of many phenomena using the same patters of derivation ... and it teaches us to reduce the number of types of facts we have to accept as ultimate (or brute)" [Kitcher, 1989, 432]. On this view, unificatory power is a criterion by which new explanations can be evaluated against old ones, and a means to force explanations to advance our understanding by making them cumulative parts of an over-arching system.

Prompted by critics of unity (see below), Kitcher seems to have softened his view in recent years to one that he calls "modest unificationism" [Kitcher, 1999]. The essential scheme — "finding as much unity as we can by discovering perspectives from which we can fit a large number of apparently disparate empirical results into a small number of schemata" [Kitcher, 1999, 339] — is the same, but Kitcher now acknowledges that the world may indeed be a messy place and that we may have to "employ concepts that cannot be neatly integrated" into a single best system. Still, Kitcher is not willing to abandon unification entirely, as he thinks that explanatory unification functions well as a "regulative ideal".

4 CRITICS OF UNITY

In the late 1970s and on through the early- and mid-1980s, the idea that science is or can be unified even in Kitcher's revisionist sense met with powerful criticisms from a group of philosophers centered around Stanford University. In "The Plurality of Science", Patrick Suppes [1981] offers a short argument to the effect that unity of science theses as conceived by philosophers and scientists down the ages have been poorly supported by theory and practice. The several forms of reductionism upon which these theses rely, Suppes claims, are untenable. What is left is a kind of pluralism of scientific language, practice, and subject matter. These, Suppes argues, are diverging rather than converging, and this is as it should be.

At about the same time as Suppes published his piece on pluralism, Nancy Cartwright was developing her view that the empirical success of our best physical theories argues against, rather than for, the universality of our theories and the unity of science [Cartwright, 1980; 1983; 1999]. John Dupré also [Dupré, 1983; 1993] mounted an attack on the unity of science that was motivated by his understanding of biological science, particularly regarding how natural kinds are identified and differentiated.

Cartwright's opposition to the unity of science works by turning the observations that fund views like the one voiced by Oppenheim and Putnam and Nagel on their heads. Cartwright grants that science can often provide predictions of impressive accuracy and can be used to manipulate certain systems very precisely. She argues that in order to do so, however, the laboratory scientist or mathematical modeler must abstract in crucial ways from the world as we usually encounter it. The charge, at base, is that scientists often describe and model systems that are constituted as much by human engineering as they are by the world. Research systems such as a sealed beaker in a laboratory incubator, or an insulated housing to be sent aloft in a spacecraft, are highly circumscribed and shielded from intrusions. But outside the beaker or box, in the universe at large, the models may very well fail to apply. Cartwright emphasizes that the world is a good deal messier than our theoretical descriptions of carefully and artificially isolated systems in it would lead us to believe.

According to Cartwright, the more restricted relevance of theoretical models suggested by this view should not be cause for concern. We do not usually try to apply models outside their domain of applicability, so this view is not really asking us to give up anything with respect to our use of models for prediction, manipulation, and control. Our models of the mechanics of falling objects do not offer good counsel on what, exactly, will happen even to fairly solid, relatively heavy, though oddly shaped objects dropped from the Golden Gate Bridge into the water below. It's possible for a person to jump or fall from the bridge and be retrieved just beneath it very much alive, as happened to a real estate agent in 1988. More often, one does not survive the fall, as happened to the same real estate agent in 2003. Neither models of mechanics nor of biology will tell us exactly which outcome will result — even for the very same 'object' — because there is no good model that includes all the relevant forces. In this case, mechanical and biological models apply only partially at best.

What Cartwright does ask us to give up is what she takes to be the unsupported assumption that there *could be* such a model — that mechanics can *in principle* be universalized to be useful in those cases where it is currently of limited applicability. In order for models of (for instance) falling objects to be universalized, it must be the case that all instances of falling are relevantly similar. Whether some real case is enough like the model case, Cartwright argues, will have to be worked out for each new application. On this view it is anything but clear that we can build a model to fit every real or imagined situation. This is not a claim about our cognitive limits — Cartwright is not claiming that we cannot build models of some systems because their dynamics are too complex for us to measure or describe. She is arguing, rather, that we ought to consider in such cases whether what we have is a system that is genuinely and relevantly different than the ones we know how to deal with. Where this is so, we should not expect there to be any one small set of theories or models that will come to include all others. The best we can hope for is a patchwork of theories and models that will sometimes be compatible and sometimes will not.

By contrast to Cartwright's focus on models and their applicability, John Dupré's opposition to unity of science arguments focuses on the concepts used in different disciplines of science and is motivated by his view that essentialism about kinds is indefensible and thus that kind-membership is a much messier affair than we usually allow. He argues that most things objectively belong to more than one kind. Moreover, he thinks that privileging one kind-membership claim over another for the same individual is always unprincipled. Take a chicken (or all chickens), for example. Chickens are noticed by both biological taxonomists and cooks, but are chickens more fundamentally members of the taxonomic class 'Aves', or of the kind 'gustatory objects'? Both kinds, Dupré says, are objective, and there is no principled way to prefer one taxonomy to the other or to take one to be more basic. It will do no good, of course, to retreat to the position that one of these kinds is scientific while the other is not: we have neither a principle of demarcation nor reasons to think that science is more basic than cuisine.

For Dupré, though kind-membership is objective, it is also context relative. Is the thing I now have before me a common and domesticated instance of the taxonomic class 'Aves', or the sort of thing that a lot of people like to eat when it has been sautéed with mushrooms and port wine? One answer to this question that Dupré will endorse is 'yes'. Another is that arriving at a 'correct' or unambiguous division of objects into kinds requires one to specify one's underlying intent or theoretical perspective in carrying out the classification.

The upshot of Dupré's ontology for the unity of science debate is that the kind of hierarchical ordering that some unity theses rely upon is essentialist or idealist by his lights, and is therefore not to be found in the world. Sometimes one will get nice orderings, but only for a particular purpose, and the very same objects will often belong to some non-hierarchical ordering as well. Dupré points out that the parts of an automobile are hierarchically ordered only so long as we are interested in them *qua* parts of a car. Old pistons with their rings and wrist pins removed very often end up on the desks of autoshop managers and serve as instances of the kind 'ashtray' and 'paperweight'. When they do, they seem not to be part of a hierarchical ordering of parts.

Those unity of science theses that rely on seeing in past and present science some progress toward identifying the most basic kinds — the few microkinds in terms of which many or all macrokinds can or will be described, derived, or explained — will be frustrated if Dupré's ontology is accepted. On Dupré's picture of the world, identifying some kind of thing as most basic for some pursuit will not make it the most basic for all pursuits or even for all scientific pursuits. Put simply, Dupré's anti-unity thesis is that the world itself is radically disordered. We should not, then, expect any science that accurately describes the world to be itself so ordered as to be unified.

5 INTEGRATION INSTEAD OF UNITY

The underlying idea of both the theory-reduction model and Kitcher's revisionist account is that science will be unified through deductive relations. But a variety of scientific enterprises involve constructing bridges between theories without either one being reduced to the other. Lindley Darden and Nancy Maull saw the importance of integration without reduction and incorporated this characteristic when they advanced the notion of an *interfield theory.* Foundational to their account is the notion of a *field*, which they characterized in terms of the following elements:

> a central problem, a domain consisting of items taken to be facts related to that problem, general explanatory facts and goals providing expectations as to how the problem is to be solved, techniques and methods, and sometimes, but not always, concepts, laws and theories which are related to the problem and which attempt to realize the explanatory goals. [1977, 144]

By downplaying concepts, laws, and theories while emphasizing expectations, techniques, and methods, Darden and Maull departed significantly from traditional philosophical accounts. Their starting point was a field (this notion was first developed by Dudley Shapere [1974]) and its diverse characteristics, not theories that may or may not be part of what the field has to offer.[9] In examining cases in which two different fields became integrated, they arrived at the further notion of an *interfield theory*, "a different type of theory ... which sets out and explains

[9]Darden and Maull's notion of a field focused primarily on cognitive features: "a central problem, a domain consisting of items taken to be facts related to that problem, general explanatory facts and goals providing expectations as to how the problem is to be solved, techniques and methods, and sometimes, but not always, concepts, laws and theories which are related to the problem and which attempt to realize the explanatory goals [1977, 144]. But, as sociologists of science have emphasized, fields are also characterized by social structures — laboratories, departments, funding agencies, journals, and professional societies. There are also various informal networks, such as Derek de Solla Price sought to characterize with the notion of *invisible colleges* [1961; see also Crane, 1972; Chubin, 1982]. Recently techniques such as analysis of citation and co-authorship have been used to identify such networks [Wasserman and Faust, 1994]. These aspects of fields are shaped in part by social considerations but often play an important role in determining, for example, what problems are taken to be serious or what methods are accepted for addressing them. As a result, interfield connections involve more than just interfield theories but interfield communities, which often end up transforming the fields from which they originated.

the relations between fields". They identified several types of interfield relations: (a) structure-function, e.g., physical chemistry targets the structure of molecules while biochemistry describes their function; (b) physical location of a postulated entity or process, e.g., the chromosomes identified in cells by cytologists provide the physical location of the genes postulated by geneticists (a case that also exemplifies structure-function and part-whole relations); (c) physical nature of a postulated entity or process, e.g., biochemistry specifies the physical realization of entities postulated by the operon theory in genetics; (d) cause-effect; e.g., biochemical interactions are a cause of heritable patterns of gene expression.[10]

Such relations between different fields are not always obvious or straightforward to develop, since fields may conceptualize the phenomena they investigate in very different terms. Consider the construction of the interfield theory of vitamins, which successfully integrated research on nutritional requirements with the biochemistry of metabolism. Most B vitamins are either coenzymes or precursors of coenzymes that serve to transport hydrogen or phosphate groups from one macromolecule to another. But prior to the 1930s, neither nutrition researchers nor biochemists could recognize this function. For nutrition researchers, vitamins were a puzzle because they were required in the diet, but only in minute quantities. The working conception of nutrition from the mid-19th century was that nutrients were either burned to liberate energy or recruited into the structure of the animal's body (this was especially true of proteins, but also of fats). The minute quantity of vitamins required in a diet, however, would not provide for generating much energy or building much structure. Moreover, the only known components involved in metabolic reactions were carbohydrates, fats, and proteins and the enzymes that broke them down (catabolized them) into a succession of smaller molecules including pyruvate and succinate. With the rise of biochemical laboratory methods early in the 20th century, researchers learned that such reactions could be maintained in extracts of cells in the laboratory, but only if the substances that became known as coenzymes were provided. No one knew why until it was discovered in the 1930s that the energy released in catabolic metabolic reactions was harvested and stored by reversible reactions in active chemical groups of the coenzymes. For example, carrying hydrogen involved a reduction reaction (picking up hydrogen from a donor) followed by oxidation (handing off the hydrogen to a recipient). Since each active chemical group could reduce and oxidize repeatedly, it made sense that a great deal of work could be done under conditions of minimal replenishment. With this reconceptualization of biochemistry, an interfield theory relating nutrition and metabolism could be developed which helped guide further research in each field. For example, vitamin B_2 was a major component of the flavin nucleotide coenzymes and, in particular, contributed the active group that played such an essential role in harvesting energy. (For further discussion of this case see [Bechtel, 1984].)

[10]See Darden [1986] for an extension of this account to the multidisciplinary integration achieved by the synthetic theory of evolution in the 1930s.

Interfield theories sometimes serve simply to bridge existing disciplines, allowing practitioners in each discipline to utilize techniques developed and knowledge procured in the other. In the most interesting cases, however, constructing a bridge between fields or disciplines results in the construction of a new discipline. For example (see [Bechtel, 2006]), cell biology emerged after World War II from what had been a *terra incognita* between biochemistry and classical cytology. Its visionary pioneers developed techniques for using new instruments to tackle new problems. For instance, the electron microscope was used to identify cell components at a much smaller scale than previously possible and the ultracentrifuge was used to localize particular biochemical reactions in the newly discovered components. The methodological and theoretical bridges constructed between cytology and biochemistry gave rise to cell biology as a new discipline. Not all cases of successful interfield interaction result in new disciplines, however. If the existing disciplines are well-established and there is no uncharted territory requiring new instruments, interdisciplinary clusters such as cognitive science are more likely to result [Bechtel, 1986].

6 REDUCTION VIA MECHANISMS

Although philosophers have generally construed reduction as theory reduction, this notion fits poorly with what is scientists typically call 'reduction'. As Wimsatt [1976b] put it: "At least in biology, most scientists see their work as explaining types of phenomena by discovering mechanisms, rather than explaining theories by deriving them or reducing them to other theories, and *this* is seen as reduction, or as integrally tied to it."[11] To appreciate Wimsatt's claim, it is necessary to understand what is meant by a mechanism and by mechanistic explanation. These notions have been pursued since the late 1980s by an emerging school of philosophers of science focusing on biology rather than physics [Bechtel and Richardson, 1993; Glennan, 1996; 2002; Machamer *et al.*, 2000]. The following provides a basic conception of mechanism:

> A mechanism is a structure performing a function in virtue of its components parts, component operations, and their organization. The orchestrated functioning of the mechanism is responsible for one or more phenomena. [Bechtel and Abrahamsen, 2005]

A central feature of mechanistic explanations, and the one that makes them reductive, is that they involve decomposing the system responsible for a phenomenon into component parts and component operations. Given that parts and their operations are at a lower level of organization than the mechanism as a whole, mechanistic explanations appeal to a lower level than the phenomenon being explained. For most scientists and non-philosophers, such appeals to lower levels

[11] For Wimsatt, the complexity of mappings between lower- and upper-level entities establishes both the failure of translation as required in bridge principles and of reduction as a relation between theories [Wimsatt, 1975, 221].

are the hallmark of reduction. As we will see, though, lower-level components of a mechanism do not work in isolation and do not individually account for the phenomenon. Rather, they must be properly organized in order to generate the phenomenon. The most important feature of mechanistic explanation to bear in mind is that it seeks to explain why a mechanism as a whole behaves in a particular fashion under specific conditions. This strategy in no way undermines the reality of the phenomenon being explained; rather, it begins by treating the phenomenon as something that really occurs when the mechanism operates in a particular set of environments.

It is most convenient to introduce the mechanistic perspective on reduction by considering an example. One of the major activities of cells is the manufacture and export of proteins. Beginning around the middle of the 20th century, cell biologists together with biochemists and molecular biologists set out to explain how cells carry out this activity. Philosophers examining this case have focused especially on how DNA is transcribed into RNA, which then *codes* for the sequence of amino acids that comprise a protein. Even this part of the mechanism is extremely complex. For example, three types of RNA are involved. The sequence information is transcribed (by a complicated set of operations) into the sequence of base pairs comprising messenger RNA (mRNA). But to synthesize proteins, these must be *read* by ribosomes, which are complex structures composed of ribosomal RNA (rRNA) and proteins. They temporarily attach to mRNA strands and move along them. A third kind of RNA, transfer RNA (tRNA) forms bonds with particular free amino acids and transports them to the ribosome. There the ribosome creates peptide bonds between the last added amino acid and this new one before moving down the mRNA and repeating the process. (For an account of the discovery of this mechanism, see [Darden and Craver, 2002].) But this is only part of the mechanism. When proteins are synthesized for export from the cell, the ribosomes are attached to the membrane of the endoplasmic reticulum. The emerging strands are pushed across the membrane into the inner space of the endoplasmic reticulum and then transported to the Golgi apparatus. There they are encapsulated in another membrane and transported across a series of sacs (the saccules of the Golgi stack). There carbohydrates are combined with the proteins to create secretory particles, which are then excreted from the cell through the process of exocytosis [Whaley, 1975; Bechtel, 2006].

One important point to note from this example is that the components of the mechanism do different things than what the mechanism as a whole does. Individual lower-level components do not explain the overall performance of the mechanism. Individual enzymes, for example, catalyze particular reactions. They do not perform whole physiological activities such as protein synthesis. Only the mechanism as a whole is capable of generating the phenomenon, and then only under appropriate conditions. Herein lies the explanation for the need for bridge principles in the theory-reduction account — different vocabulary is needed to describe what the parts of a mechanism do than is required to describe what the mechanism as a whole does. The appropriate bridge in this case, however, is not a set of translation rules, but an account of how the operations of the parts of the mechanism are organized so as to yield the behavior of the whole mechanism.

One consequence of taking apart a mechanism that depends on organization to generate the phenomenon is that the investigator destroys the phenomenon itself. A not uncommon situation in science is that after investigators decompose a system they find they cannot readily put it back together again. Sometimes this is because they have neglected some important component. But more frequently it is because they have failed to recognize the specific mode of organization that was involved in the functioning mechanism. The simplest mode of organization is to relate the operations of different parts in a linear series. Understanding more than this simplest mode of organization has presented a serious challenge to humans [Bechtel and Richardson, 1993].

A simple but extremely powerful organizational principle is a negative feedback loop in which the product of an operation feeds back into an earlier operation, allowing for its regulation. (Recall that negative feedback was the central principle advanced by the cyberneticists and general systems theorists in their proposals to unify science.) We are all familiar with this kind of organization from mechanical systems in the home. In the heating system, for example, a thermostat monitors the output of an operation (the heating of the air) and, when the desired temperature has been reached, sends back a signal that stops the furnace from generating more heat. As familiar as negative feedback is today, it was a very difficult concept for engineers and scientists to acquire. It was reinvented numerous times, each in a specific application (for a discussion of the history of re-discovery of negative feedback, see [Mayr, 1970]). Ancient water clocks, for example, required that the water-supply tank be maintained at a constant level; in approximately 270 BCE, Ktesibios invented a feedback control system for such clocks. Windmills need to be pointed into the wind, and British blacksmith Edmund Lee developed the fantail as a feedback system to keep the windmill properly oriented. A temperature regulator for furnaces was developed by Cornelis Drebbel around 1624. Finally, James Watts' invention of a governor for his steam engine helped establish the principle as a general one for use in engineering. This was in large part a result of the mathematical analysis of such control systems in terms of differential equations developed by James Clerk Maxwell.

Recognizing negative feedback control in biological systems was equally difficult. Vitalists in the 19th century objected to mechanist accounts in physiology on the grounds that they could not conceive how a mechanism could behave in the manner biological organisms were known to behave.[12] In particular, organisms maintain themselves in the face of various assaults of their environment. Claude Bernard [1865] developed a framework for answering such objections by distinguishing between an *inner environment* in which the organs of an organism

[12]Bichat [1805], provides some of the most compelling arguments of such a type for vitalism. He focused, for example, on the apparent indeterminism in the responses of organisms to external stimuli and the tendency of organisms to behave in ways that resisted external forces that would kill them.

function and the *outer environment* in which the organism lives. He proposed that each organ in the body was designed to respond to specific changes in the internal environment so as to help maintain the *constancy of the internal environment.* As a result of the actions of the various organs, the inner environment provided a buffer against conditions in the external environment. Bernard, however, was not able to characterize in any detail how the organs each helped to maintain the constancy of the internal environment. Walter Cannon [1929] picked up this thread from Bernard and introduced the term 'homeostasis' (from the Greek words for 'same' and 'state') for the capacity of living systems to maintain a relatively constant internal environment. He also sketched a taxonomy of strategies through which animals are capable of maintaining homeostasis. The simplest involve storing surplus supplies in time of plenty, either by simple accumulation in selected tissues (e.g., water in muscle or skin), or by conversion to a different form (e.g., glucose into glycogen) from which reconversion in time of need is possible. Cannon noted that in most cases such conversions are under neural control. A second means of maintaining homeostasis is through negative feedback — measuring the effects of a continuous process and using that to alter the rate of its performance (e.g., measuring internal temperature and when it is too high or too low increasing or decreased the rate of blood flow by modifying the size of peripheral blood vesicles).

Negative feedback is frequently realized in biological systems as a result of cyclic organization in which the products of several successive chemical operations ultimately combine with some new input to produce an earlier intermediate. The citric acid cycle, first advanced by Krebs and Johnson [1937], provides an illustrative example (see figure 4). The ultimate function of the citric acid cycle is to enable synthesis of ATP, the macromolecule in which energy is stored in animal cells for use in such activities as muscle contraction. Specifically, energy is stored in a high-energy bond created by adding a phosphate group to ADP. A small amount of ATP is generated within the citric acid cycle itself (substrate-level phosphorylation), and a larger amount using the energy that is released by oxidative reactions in the cycle and transported, in the form of NADH or FADH, to another mechanism (oxidative phosphorylation). There is no point in performing the oxidations in the citric acid cycle at a rate that exceeds the system's capacity to synthesize ATP from ADP. Hence, when this happens, NADH and FADH build up and there is no NAD or FAD available to support further oxidations in the citric acid cycle. Thus, the rate of the citric acid cycle is regulated by means of negative feedback. The less ADP available, the less NAD and FAD is available, and therefore the less oxaloacetic acid is available to react with acetyl-CoA, the substrate that typically enters the cycle from other metabolic processes.

Although once the citric acid cycle was discovered its functional significance became apparent, the work leading to its discovery had other motivations. The spur to develop this and other cycles was the realization that the initially conceptualized linear pathway of reactions resulted in a product that, lacking hydrogen, could not be further oxidized directly. Recombination with something else was

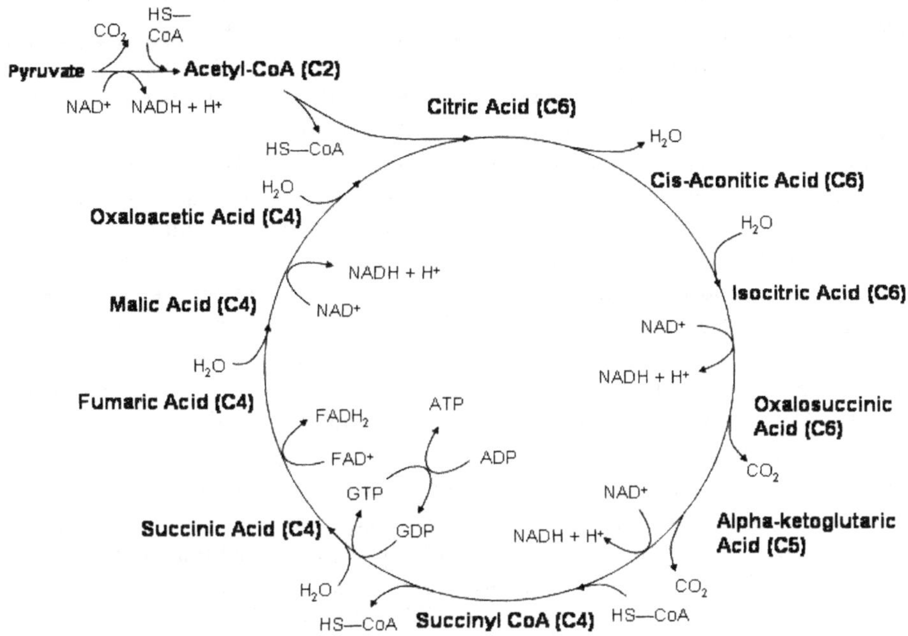

Figure 4. The citric acid cycle, a central biochemical reaction in cell metabolism. The crucial oxidation reactions are shown in the interior. When energy demands are low, there is no ADP available, which in turn means there is no NAD^+ or FAD^+ available (all supplied being taken up in NADH or $FADH_2$). This will result in no accumulation of oxaloacetic acid to react with acetyl-CoA, thereby bringing the reactions in the cycle to a halt. Trough such feedback, critical metabolites are conserved until they are needed to synthesize new ATP from ADP.

an expedient to overcome this obstacle. In short order biochemists discovered a number of cycles, such as the citric acid cycle, and began to appreciate cyclic organization as a common design principle in living organisms. But this was a hard-won battle since the focus remained on the overall production of the end product from the input, not the organization in between.

As difficult as it was to understand the significance of negative feedback, the importance of positive feedback was even more difficult to appreciate. At first positive feedback seemed not to be very functional since it appeared to lead to run-away mechanisms. That is, if the product of a mechanism spurred the mechanism to produce yet more of it, the process would continue until all supplies were exhausted. Yet, there are constrained contexts in which positive feedback is desirable. Particularly important are sets of reactions that function autocatalytically, with one reaction producing a catalyst for a second reaction, and it in turn producing a catalyst for the first reaction [Kaufmann, 1993; Maturana and Varela, 1980]. Theorists interested in the origins of life have been the leaders in exploring these ideas (see, for example, the intriguing models of Gánti [1975; 2003]), but they have yet to achieve major uptake in the broader scientific community.

It is easiest to recognize the role of organization in generating higher levels by considering the perspective of an engineer who has been asked to organize existing components in a new way to accomplish some task. When she has finished, she has built something new, perhaps something for which she could secure a patent. We would not expect the patent office to deny her a patent because all of the components were already known to her — they were also known to the others who failed to have the insight needed to develop the new mechanism. Thus, invention of a new organization alone is noteworthy. (In real life, an engineer would more often invent some of the components as well as their organization. However, at some level of decomposition the invented components would themselves be built from existing ones.)

Beyond organization, the environment is often key to understanding how a mechanism works. Mechanisms are not isolated systems, but depend on conditions in their environment. This is particularly the case for biological mechanisms as against physical machines that may be engineered to perform in an identical fashion over a wide range of conditions. With biological mechanisms evolved to operate in a specific range of environments, features of the environment may be co-opted into the mechanism's operation. Evolution is an opportunist, and if something can be relied upon in the mechanism's environment, then it does not have to be generated by the mechanism. Vitamins provide just one well-known example. Because our ancestors could generally count on the availability of vitamins in their foods, there was no evolutionary pressure for us to retain the ability to synthesize them. Nonetheless, insofar as such environmental factors are necessary for the functioning of the mechanism, mechanistic explanations need to focus on the mechanism's context, not just its internal configuration.

With this account of mechanisms and mechanistic explanation in place, we can consider further how they offer a fresh perspective. Unlike theory-reduction accounts, mechanistic reductionism neither denies the importance of context or of higher levels of organization nor appeals exclusively to the components of a mechanism in explaining what the mechanism does. The appeal to components, in fact, serves a very restricted purpose of explaining how, in a given context, the mechanism is able to produce a particular phenomenon. There are other differences as well. Whereas theory reduction is often treated as transitive, with higher-level theories ultimately being reduced to those at the lowest level, mechanistic reductions often proceed for only one or two iterations. Once investigators understand the operations performed by the parts and how the organization orchestrates their operation to produce the phenomenon, they generally have neither the desire nor the tools to pursue a further round of decomposition into subparts and suboperations. Moreover, it is not the case that detailed knowledge of how the component parts or subparts operate will already be available in lower-level disciplines, since, as we discussed above, these parts will be operating in specialized contexts not typically studied by practitioners of the lower-level science. While the study of mechanisms is reductionistic and can promote integration of knowledge from various disciplines, it does not promote a grand unificationist vision.

6.1 Rethinking Levels

The notion of levels plays a central role in all accounts of reduction, but it has not been fully explicated in any of them. In the early accounts of theory reduction, levels were associated with broad scientific disciplines, so that one sees reference to the physical level, the chemical level, etc. But just why the objects of physics, which range in size from the sub-atomic to the universe, comprise a level is left unspecified. Although still committed to the theory reduction framework, philosophers such as Causey approached levels from a more ontological perspective, emphasizing that lower levels deal with the parts of wholes studied at higher levels. Wimsatt develops this mereological perspective, making part-whole relations fundamental in distinguishing levels:

> By level of organization, I will mean here compositional levels — hierarchical divisions of stuff (paradigmatically but not necessarily material stuff) organized by part-whole relations, in which wholes at one level function as parts at the next (and at all higher) levels .... [Wimsatt, 1976a]

One limitation of compositional relations from Wimsatt's perspective is that they do not permit ordering of entities not part of the same part-whole hierarchy. Accordingly, Wimsatt also appeals to interactions between entities in identifying levels — entities interact principally with others at their own level and with entities at lower levels in terms of the complexes of which they are part. People, for example, interact primarily with other people, animals, plants, computers, furniture, etc., not the cells of other people or the chips of computer. Accordingly, Wimsatt comments: "*Levels of organization can be thought of as local maxima of regularity and predictability in the phase space of alternative modes of organization of matter*". [Wimsatt, 1994]

Wimsatt notes that the neat layering of levels breaks down at higher levels — individual humans do engage in relations with entities several times larger or smaller than themselves. Accordingly, he introduces the notions of *perspectives* and *causal thickets* for cases in which neat layering into levels breaks down. But the problems go deeper and calls into question the general project of conceiving of the natural world as layered in terms of levels. In biology it is routine for things of very different size-scales to interact. The transfer of energy released in basic metabolism to ATP, for example, is mediated by the transport of protons across the inner mitochondrial membrane, and its diffusion back. Yet the very membrane that is maintaining the proton gradient is also composed in part of protons. Protons are thus part of the very structure through which the protons are being transported. Thinking in terms of the operation of the mechanism, it is correct to say that the protons in the membrane are at lower level than those being transported across it.

Thinking in terms of mechanisms allows one to articulate a more limited but less problematic conception of levels. From the point of view of a given mechanism performing a particular function, the component parts into which a researcher decomposes it constitutes a lower level. If researchers decomposed these parts, they reach yet a lower level. This account allows for the denizens of a level to be of different sizes as long as they are working parts of the same mechanism. Moreover, it is compatible with viewing two structurally identical entities as at different levels if one performs its operations in a sub-mechanism of another — a proton that is being pumped across a membrane is at a higher level than one that is part of the membrane. But an important feature of this account of levels is that they are limited to the scope of the original mechanism.

One advantage of construing and limiting the notion of levels to levels of organization in mechanisms is that it permits a coherent account of the important idea that lies behind the problematic notion of downward causation [Campbell, 1974]. The important idea behind appeals to downward causation is that causal effects of interactions of higher-level entities have consequences for their component parts. Your DNA is a passenger on all your travels and some of your neurons are altered every time you learn something new. The notion of downward causation is problematic, though, since it seems to result in a problem of causal overdetermination — if we assume that there is a comprehensive account of causal interactions of entities at a lower level, then the effect is already determined regardless of any putative top-down effect [Kim, 1998]. One solution to this problem is to keep the notion of causation univocal by restricting it to intralevel cases and provide a different, constitutive account of interlevel relations within a mechanism [Craver and Bechtel, in press]. The intuition behind top-down causation can be maintained, but expressed in terms other than causation: the causal interactions of a mechanism with its environment (including other mechanisms) alters the mechanism itself. The changed condition of the parts and operations within the mechanism then propagate causal effects within the mechanism.[13]

A consequence of the mechanistic approach is surrendering the view that a complete causal story can be told at the lower level — all one can account for is changes in the mechanism as the parts operate and interact with each other under the conditions in which the mechanism is operating (some of these being set by the interaction of the environment with its environment). Since it does not have the resources to describe the way in which the mechanism engages its environment, the lower-level account of goings-on inside the mechanism cannot provide a complete account of all that is happening. Our discussion of the problems with global unity theses, though, suggests that the aspirations for a complete theory should

[13]On this view, so-called *bottom-up* causation works in the same manner — the operation of parts within the mechanism alters the condition of the mechanism itself, thereby altering the manner in which it engages its environment.

be surrendered anyway. What a mechanist requires is only that the causal effects at a given level within a mechanism can be explained — for example, that one can explain how, given the impingements on the brain from the environment, neural changes within it occur. This is precisely what molecular accounts of learning and memory strive to do [Craver and Darden, 2001]. The level of neural processes inside the brain is locally constituted — it is not part of a broad level that crosses mechanisms.

6.2 Within Level identities: Heuristic Identity Theory

In characterizing mechanisms we identified both parts and their operations. The research tools for decomposing mechanisms into their parts and operations are often different. As a result, the decompositions are often developed in different disciplines. For example, cytologists using various microscopes, identified various organelles in the cell, whereas biochemists, preparing homogenates and using various assays, identified chemical reactions. One of the accomplishments of modern cell biology was to establish that different cell functions were performed by specific cell structures, thereby localizing the function [Bechtel, 2006].[14] Since localization claims maintain that it is the same entity that constitutes a particular structure and has performs a specific operation, they are identity claims in the sense advanced by the mind-brain identity theory [Place, 1956; Feigl, 1958/1967; Smart, 1959] noted above. The identity theory is often construed as advancing a reduction of psychology to neuroscience, since neuroscience is at a lower level than psychology. From the point of view of mechanistic explanation, however, we can recognize that accounts of the part of the system and the operation it is performing are at the same level. For example, initial encoding of information to be stored as long-term episodic memories (an operation described by psychology) is an operation of the hippocampus (a structure identified by neuroscientists).

Although not themselves vehicles of reduction, since they are intralevel claims, identity claims play an important role in mechanistic research and ultimately help advance mechanistic reductions. One way to see this is to consider one of the major objections that critics raised to the mind-brain identity claim. They charged that at best empirical investigation could establish a correlation between the psychologically characterized phenomenon and a brain process, an objection that has been pressed anew in recent discussions of consciousness [Chalmers, 1996]. Despite the prevalence of the language "neural correlates" in recent presentations of empirical research concerning consciousness [Crick and Koch, 1998], most empirical researchers do not make a distinction between establishing a neural correlate and identifying the neural substrate. It is philosophers who insist in emphasizing that the empirical evidence cannot decide between correlation and causation. One import of making such a distinction is that a dualist can maintain that conscious

[14]Linking structural and functional accounts developed in different fields was one of Darden and Maull's major examples of an interfield theory. In general, interfield theorizing often culminates in accounts of mechanisms.

states are not material phenomena at all, but are simply correlated with brain processes.

When considered in the context of how identity claims typically figure in empirical research, however, the attempt to reconstrue them as correlation claims appears radically misguided. The reason is that they typically are not the conclusions of scientific investigations but heuristics for guiding further scientific discovery [McCauley, 1981]. Once an identity claim is made between a structural and a functional characterization of an entity, researchers use each characterization as a guide to elaborating the other. Discovery of an operation that cannot be linked to a part of the structure poses the question of whether that operation is indeed being performed and if so, by what component. Discovery of a component of a structure that does not seem to be performing any operation raises the question of whether it really is a working part and if so, what operation has been missed in extant functional decompositions. Such research invokes the converse of Leibniz's law of the identity of indiscernables, focusing instead on the indiscernability of identicals: what is learned about a structure or a function under one description must apply to it under the other, or one must revise the identity claim. Correlational claims, by contrast, impose no such burden. To indicate its constructive role in guiding further research, Bechtel and McCauley [Bechtel and McCauley, 1999; McCauley and Bechtel, 2001] speak of *heuristic identity theory*. Once an identity claim has fulfilled its heuristic function of guiding discoveries both on the structural and functional sides, the identity has been woven into the science and investigators who had taken advantage of the heuristic would not be tempted to consider it a mere correlation.

As noted above, identity claims are not themselves reductive since they relate different accounts of the same entity. They do, however, directly contribute to integration between different accounts of the phenomenon, often ones developed in different disciplines with different research techniques.

7 CASE STUDIES IN REDUCTION AND UNIFICATION ACROSS THE DISCIPLINES

Although we noted examples from various sciences to illustrate points in the previous sections, the focus was on the conceptual account and its continuity. Looking at actual cases of reduction and unification/integration reveals that they are quite diverse. In this final section we examine four cases that have been important in the discussion of reduction and unity. In each case we ask how the foregoing discussions applies and, in the last cases, identify foci that have not been sufficiently developed in accounts to date and should serve as topics for further philosophical investigation.

7.1 Temperature: Thermodynamics and Statistical Mechanics

At the end of Section 2.3 above, we pointed to the importance of the role played by boundary conditions and bridge principles in carrying out theory reductions of higher-level laws to lower-level ones. In this first case study we revisit this feature of reductions by a deeper look at the relationship between thermodynamics and statistical mechanics, the standard example of successful theory reduction since Nagel [1961]. As we saw in section 2.1, temperature in particular has long been regarded by many in the scientific and philosophical communities as completely explained in terms of the mean kinetic energy of lower-level particles (molecules): $2E/3 = kT$. Indeed, we now learn from some standard high school and university textbooks and from renowned physicists that temperature *just is* mean kinetic energy of the molecules that constitute the gas [Feynman, 1963, 39].

Several problems with this identity claim have been noted by philosophers and physicists, many of them having to do with boundary conditions. Philosopher Mark Wilson reminds us, for instance, that while the simple equality claim holds in the case of classical gases — the case Nagel emphasized — it is not anything like universal: "in point of fact, this temperature equation is generally false; the proportionality between temperature and kinetic energy is substance specific" [Wilson, 1985, 228].[15] As Nagel pointed out in developing his example, the kinetic theory of matter includes both the general postulates of statistical mechanics and more specific postulates appropriate to classical gases — those that are thermodynamically isolated, dilute, and in which the particles influence each other only by perfectly elastic collisions. The kinetic theory, of course, gives excellent predictive results for substances that are relevantly like those described by its postulates. But what about other kinds of substances or even non-dilute gases? Because of the way solids are constituted, for instance, the molecules cannot collide as they do in gases, but can only vibrate. Similar problems arise for other states of matter. It turns out that the observable macrophenomenon we call temperature is multiply realizable at the microlevel.

What this means for the quality of the reduction generally is not quite clear — except that there is good reason to think, as Lawrence Sklar puts it, that we "do not expect to 'deduce' or 'derive' thermodynamics from statistical mechanics in any simple minded way ..." [Sklar, 1974, 16]. In the case of temperature, there will not be just one reduction, but several, as boundary conditions for several states of matter, types of gases and for fluctuating energy situations will have to be specified. Some have argued that this situation causes no real problem for the reduction — we just need to be careful about specifying the boundaries of the reduction.

In addition, as we pointed out above, such specification relies importantly on empirical, rather than deductive, evidence. The descriptions of various states of matter and how they behave has been achieved experimentally, not deduced from

[15]As Lord Kelvin pointed out, it is possible, of course, to construct an absolute temperature scale — a scale on which what is being measured is not relative to what is being used to measure it. This is a separate issue from the one we are raising here.

the relevant lower-level theory. While statistical mechanics has thrown light on the knowledge gained from experiment, it is not the case that the relevant boundary conditions for temperature can be read off the axioms of statistical mechanics. Neither is it immediately clear how far from the 'ideal' boundary conditions a system can be before the lower-level laws cease to offer acceptably good predictions of the behaviour of that system at the higher level. This, too, must be investigated empirically, at least until standards are articulated.[16]

Given all this, even a 'successful' reduction in this seemingly simple case will turn out not to be as unificatory as many proponents of the theory-reduction model would have hoped. The reduction will be complicated, disjunctive, and empirically informed, rather than simple, general, and purely deductive. Indeed, the more general and unifying principles are actually those of classical thermodynamics, not the reductive bases.

It is worth noting that mechanistic reduction may provide a superior way to understand this case. The main problems noted above can be side-stepped: mechanistic reduction does not deny the importance of specifying the relevant context, neither does it demand that relations be deductive. Instead of an attempt at reduction that issues in a simple and powerful proportionality that fails to achieve full generality, a mechanistic explanation will be sensitive to boundary conditions in addition to the relations between higher- and lower-level phenomena and entities. This argues against unity, not for it, because we should not expect the physicist who works with concentrated gases to consult the physicist who works with dilute gases when she defines temperature for the systems on which she works. The simpler, better understood case has no obvious claim to epistemic superiority. On the contrary, *each* mechanistic explanation will be relatively substance specific and it is anything but clear that one is the best or more appropriate model for all the others.

The prospects for Darden and Maull-style integration also seem more promising than those for unity by theory reduction. Indeed, a great amount of integration has already taken place. Structure-function and cause-effect accounts on which relations between micro and macroproperties are specified are at the heart of thermal physics. So too are accounts from the perspective of the microlevel of the nature of features and processes at the macrolevel. These descriptions and accounts often represent the integration of different fields, of which thermodynamics and statistical mechanics are just one example.

7.2 Genes: Molecular Biology and Developmental Systems Theory

From what has been the primary exemplar case in philosophical accounts of reduction, we turn to one that we have also alluded to above and is currently capturing

[16]We have focused on temperature because of its familiarity and centrality in the reductionism literature, but problems with entropy have also been widely discussed as a possible confounder for the reduction of thermodynamics to statistical mechanics. For discussion see Sklar [1993] and Callender [1999].

both scientific and popular attention in the life sciences. Very near the end of the famous paper in which the outcome of their work on the structure of DNA is announced, Watson and Crick offer the following single-sentence paragraph: "It has not escaped our notice that the specific pairing [of bases] we have postulated immediately suggests a possible copying mechanism for the genetic material" [Watson and Crick, 1953]. With this was born a new emphasis on DNA as the ultimate source for knowledge about the macrofeatures of organisms. Biology soon had a new "central dogma" — DNA makes RNA makes protein — and with it an explicitly reductionist (gene-based) approach to accounting for all sorts of biological phenomena, including phenotypes [Dawkins, 1976], the evolution of morality [Ruse and Wilson, 1986], and even human belief in God [Hamer, 2004]. This approach quickly led to widespread accounts of macroproperties of organisms or groups of organisms in terms of genes. Some property P could be explained by or deduced from the presence (or absence) of the gene for P. Dean Hamer's recent claims about the gene for belief in god, or "self-transcendence," are a good example. Hamer argues that whether or not one believes in god is best predicted by whether or not one inherits the VMAT2 gene, the 'gene for' belief.

The gene-based approach, however, has important problems. As Oyama, Griffiths, and Gray [2001] have pointed out, privileging DNA's role in biological processes makes inheritance, evolution, and development, for instance, the mere passing on of DNA. On this view, DNA becomes the only relevant causal factor in these and other biological processes, and the locus of explanation for them. Richard Lewontin has pointed out on several occasions and at some length, however, that the central-dogma view cannot be the whole picture, because DNA can have no such causal efficacy. DNA, he contends, "is not self-reproducing", "makes nothing", and does not determine much, if anything, about organisms [Lewontin, 2000]. Without the rest of the cellular machinery of proteins and enzymes, DNA produces nothing at all. To extend a well-used metaphor, if DNA *codes for* this or that protein, there must be something that *reads* the code, something that *builds* what the code specifies, and perhaps most importantly, something that *writes* the code for the next iteration. DNA cannot do all this.

Another significant problem with the gene-based approach to accounting for macrofeatures is that being in possession of the full genome sequence does not by itself tell researchers much about the properties of the organism. Far from having a gene-for map that offers one-to-one correspondence of molecules to macrofeatures, we have learned that a great many genes have regulatory functions — they 'switch' other genes on and off rather than code for the manufacture of particular proteins. It is worth quoting the following passage from Karola Stotz and Adam Bostanci [2005]:

> *Gene regulation* means that there is always more involved in the production of the product than the coding sequence. In the case of *alternative* cis-*splicing* of exons and introns, one structure contains several modules that can be alternatively spliced together. One stretch of DNA may therefore give rise to several proteins. *Overlapping genes* and *al-*

> *ternative reading frames* entail that the "same" DNA sequence can yield different products. *Cotranscription* of adjacent DNA sequences blurs the boundaries between structural "genes". In the case of *trans-splicing*, one might say that two "genes" (if a gene is defined as a unit of transcription), are involved in coding for a single protein (or more than one products [sic] as in the case of *alternative* trans-*splicing*). Mechanisms such as *exon scrambling*, *exon repetition*, or *antisense*-trans-*splicing* further increase the divergence of DNA sequence and protein product. *mRNA editing* exchanges single nucleotides in the linear sequence. Last but not least, *protein splicing* changes the final product once more, but in this case by splicing so-called 'inteins' in and out of the final polypeptides of which proteins are composed.

The phenomenon of gene regulation clearly shows that in order to have good explanations of what genes are doing, we need to know what is being regulated and how. These explanations ask for more context than is available at the level of the molecular gene alone, and often come from physical chemistry rather than from genetics. This further suggests that privileging the gene as the locus of explanation is premature in at least some cases. There are also higher levels to consider: How did the genotype-phenotype map get to be the way it is? Why and how is it stable across generations?

Recently, developmental systems theory has emerged as a competitor for gene-based thinking about developmental biology. Proponents of developmental systems theory argue that development cannot be understood outside the framework of its neighbor disciplines and processes, and thus that the causal contexts of heredity and evolution cannot safely be ignored if developmental processes are to be explained. On this view, molecular genetics is just one part of a long and complex story — a story in which genetic goings-on do not make up the only plot.

The developmental systems approach rejects simple reduction of macrofeatures to molecular genetics and urges that there are very often several causal factors in a given developmental process. This viewpoint makes room for the kinds of alternatives to reduction discussed above. Mechanistic reduction, in particular, seems useful for explaining developmental processes in ways that do not neglect epigenetic influences. Mechanistic explanations, by their nature, account for phenomena in context and across levels or organization, rather than privileging a particular level.

This approach is exemplified by recent work on heterochrony — changes in the timing of events or processes during organismal development — as it applies to evolution. Researchers who have investigated differences in organisms that arise as a result of heterochrony have recognized that heterochrony is often not driven by the mere presence of some gene or other. Rather, there may be differences in the timing of gene expression or of the rates of expression. These processes are very often described in mechanistic terms (see, for instance, [Wray and Love, 2000; Tautz, 2000] and the review article by [Smith, 2003]), and researchers have not generally assumed or argued that in those cases where heterochrony can be mechanistically related to particular genes, gene products, or differences in the timing of gene expression, the observed differences can be explained at the molecular level. Even with the molecular part of the story in hand, if we are to apply what we know to evolutionary development, we will still want to know whether and how heterochrony leads to major evolutionary transitions, how the developmental process is regulated for embryos, and at what level(s) of organization this regulation is orchestrated. It is interesting to note that at present the best-known candidate for a developmental regulator in at least some organisms is the so-called somite clock. It is a kind of feedback mechanism responsible for the timing of segmentation in the vertebrate embryo that is usually described as operating at the cellular, rather than molecular, level [Pourquié, 1998; Dale and Pourquié, 2000].

There is also a strong case to be made that the proponents of developmental systems theory are calling for an explanatory strategy like the one advocated by Darden and Maull. We can see molecular genetics, embryology, cell biology, and other disciplines as fields that all have some relation to development, and the search for a better understanding of developmental systems as an attempt to specify interfield relations for particular developmental processes. There is no reason, however, to assume beforehand that the field concerned with the lowest level of organization is epistemically prior or more basic. Take, as a simple example, the well-known case of inheritance among diploid organisms. Studied from a molecular level, we only learn about gene variation at certain loci. Couple this knowledge, though, with the study of cellular mechanisms and we can begin to see why Mendel's second law holds: the process of meiosis regularly distributes each allele such that the assortment is independent of every other allele. Population genetics tells us still more of the story, informing us as to what the distributions of alleles will be when no outside forces are operating.

Choosing any one of these levels as primary artificially limits the inquiry in ways that may not be heuristically justifiable. At the cellular level, we can ask structure-function questions of the molecular level, as well as cause and effect questions. From the molecular and cellular perspectives we can ask about the physical processes that underlie the regularities captured by population genetics. We can also hope, as developmental systems theorists do, that not limiting ourselves to a single perspective will result in interfield theories that parlay knowledge at these various levels into a more thoroughgoing account of evolutionary development.

It is important to note that in the case of heterochrony and in the case of diploid inheritance, molecular genetics does not provide a sufficient account on its own. Rather, it requires interfield connections with developmental and evolutionary biology or explanations that pay attention to the important connections between the molecular, cellular, phenotypic, and population levels.

7.3 Historical Archaeology: Physical and Social Sciences

So far we have focused on the explanatory gain that results from integration of fields — interfield theories and accounts of mechanisms enable investigators to answer a multitude of questions that they could not otherwise address. But there is an additional virtue, one that has been clearly brought out by Alison Wylie [1999] in her account of historical archaeology. Drawing upon the insights of Ian Hacking [1983] on how scientists triangulate independent research techniques to secure reliable evidence even when they cannot directly establish the reliability of any one technique, Wylie shows how historical archaeologists are affecting such triangulation. The approaches of traditional history, which relies primarily on the analysis of documents, and archaeology, which has relied on the analysis of material remains of societies, are radically different. In many cases there is no potential for integrating them. Prehistoric civilizations have left no written documents and they have been the province of archaeologists. The material remains of more recent societies are often destroyed and historians have relied primarily on the analysis of documents to describe their history. But there are a range of early human societies for which both documents and material remains can be recovered. While practitioners of traditional history and traditional archaeology have tended to insist on the primacy of their own tools of investigation, starting after World War II a number of investigators attempted to integrate the two and have adopted the name *historical archaeology* for this integrated investigation.[17] In the U.S., for example, historical archaeologists tended to focus on early European settlement and the effects of these on native American peoples as well as subsequent expansion of the frontier and urbanization of the continent. Its institutional structure did not materialize until the late 1960s. They have attempted to weave together results from analysis of documents and archaeological remains.

As Wylie notes in describing the sometimes tempestuous relations between historical archaeologists and their home disciplines,

> A recurrent theme [sounded by advocates of historical archaeology] ... is an insistence that when events and conditions of life or historic periods are at issue, vastly more can be achieved by making conjoint use of the evidential, methodological, and theoretical resources of archaeology and documentary history than can be achieved by either field working in isolation from the other. [Wylie, 1999, 305]

What is significant is that the attempts to integrate sources often forced revisions in the accounts compiled from one source alone. By drawing upon archaeological methods to study the artifacts of a society, one is not just a filling in the historical record but procuring "substantially different, potentially transformative insights about the recent past" (p. 305). This stems from the fact that archaeology can provide evidence of people who do not show up in documentary records, illustrating the ways they lived their lives, which then provides a different perspective on the documents left by the cultures in question.

[17]The Society for Historical Archaeology was established in 1967 and began publishing the journal *Historical Archaeology* that year (see [Schuyler, 1978], for a discussion of these events in the U.S. and related developments in other countries during the same period).

Wylie's particular interest in historical archaeology is its potential to provide an illuminating example of how integrating the modes of investigation from multiple disciplines can both provide epistemic warrant beyond what each alone can produce and serve as a heuristic to encourage new inquiry. The key idea behind increased epistemic warrant is Whewell's [1840] notion of consilience of induction according to which results secured through independent lines of inquiry are more likely to be true than those relying on just one line of investigation. Wylie notes, however, that one cannot just assume that because evidence is advanced in two different disciplines that it represents independent evidence and emphasizes the need to tease apart difference in causal processes, independence of background knowledge and theories invoked, and disciplinary independence. These must be evaluated case by case. But she argues that historical archaeology does offer cases of such independent convergence of evidence and offers the convergence in dating by reliance on tree ring counts, radio-carbon decay, magnetic orientation, and evolution of stylistic traditions in documents:

> The disciplines that supply the relevant technologies of detection are certainly institutionally autonomous, and the content of their theories is substantially independent; it is unlikely that the assumptions that might produce error in the reconstruction of a date using principles from physics will be the same as those that might bias a date based on background knowledge from botany or socio-cultural studies of stylistic change. Finally, this independence in the content of the auxiliaries and in their disciplinary origins is especially compelling because it is assumed to reflect a genuine causal independence between the chemical, biological, and social processes that generated and transmitted the distinct kinds of material trace exploited by different dating techniques. [p. 310]

Securing different forms of evidence that can be used to evaluate and revise claims made by any one form of evidence is clearly an important aspect of integrating sciences that applies broadly. In entering the *terra incognita* [de Duve, 1984, 11] that then existed between classical cytology and biochemistry, pioneers in cell biology drew upon two new tools recently developed in physics and chemistry — the electron microscope and the ultracentrifuge. Each presented its own risk of artifact but their combined use, including the use of one to calibrate results from the other, provided investigators with the opportunity to develop an integrated structural and functional account of many basic cell mechanisms [Bechtel, 2006]. Integration thus can serve both an explanatory and an evidential role.

7.4 Language: Linguistics and Psycholinguistics

So far our examples have stemmed predominately from the physical and biological sciences, but we end with one that bridges into traditional areas of the humanities. This case also provides us a glimpse into the dynamics of integrating research efforts across disciplines. Many disciplines in the humanities, social sciences, and engineering focus their attention on products created, intentionally or unintentionally, by human beings. Literary, artistic, philosophical, and technical products typically are constructed intentionally by their authors. Languages and other symbol systems are typically not constructed intentionally, but are nonetheless the products of human activity. How do the disciplines that study these products relate to other disciplines in the physical, biological, and behavioral sciences? We will follow the analysis of Abrahamsen [1987] to discuss one such case: the relationship between linguistics (concerned with the formal structure of human languages) and psychology, especially cognitive psychology (concerned with the mental processes that enable cognitive systems, including humans, to perform their activities). Note that these are different enterprises and typically try to account for different phenomena using different theoretical constructs and appealing to different sources of evidence. Linguists are principally concerned with the structure of language, advance grammars to account for such structure, and test their grammars by their capacity to generate all and only the sentences of a particular language. Psychologists, on the other hand, attempt to explain the mental processes that enable individual language users to comprehend or produce sentences of their language.

Abrahamsen [1987] identifies three patterns in the relationship between linguistics and psychology in the 20th century: (1) boundary maintaining, in which the two disciplines pursued their in quiries independently, (2) boundary breaking, in which one discipline tried to usurp the territory of the other, and (3) boundary bridging, in which practitioners of the disciplines collaborated rather than competing for the same territory. Boundary-breaking episodes often attract the greatest attention. At the turn of the 20th century, psychology was a new and rapidly advancing discipline that attracted a number of young linguists seeking to move beyond the older traditions in their own discipline. What they encountered in psychology, however, was not a single view they could take back to linguistics but competing conceptual frameworks — notably the mechanistic cognitive framework of Johann Herbart and the antimechanistic idealist perspective of Wilhelm Wundt. Wundt [1900] himself addressed a host of issues in both linguistics proper (grammatical structure, phonological systems) and psycholinguistics (language acquisition, speech errors) whereas Herbart influenced linguistics through the applications of his work by the linguist Hermann Paul [1880]. As Blumenthal [1987] describes, these two approaches conflicted — Hobart's approach proceeded bottom-up from sentence elements invoking association techniques whereas Wundt's started with unified, often creative, mental representations and proceeded top-down. The conflict within psychology, according to Blumenthal, soon left linguists disillusioned and many opted to divorce linguistics from psychology [McCauley, 1987].

The second round of boundary-breaking interactions followed Chomsky's introduction of transformational grammar [Chomsky, 1957]. Chomsky viewed his approach to grammar not only as a revolution against structuralism in linguistics proper but also as a revolution against behaviorism in psychology [Chomsky, 1959]. Many psychologists, themselves striving to break free of the behaviorist tradition, eagerly followed Chomsky's lead. Notably, Miller [1962] sought to provide evidence for the psychological reality of transformations. This time it was psychologists who were to be disillusioned, as Chomsky repeatedly revised his grammars regardless of the evidence psychologists offered for their psychological reality ([Reber, 1987]; see also [McCauley, 1987]). Chomsky continued to break boundaries by characterizing many of his ideas as contributions to psychology, including his nativism, competence-performance distinction, and construal of linguistic grammars as accounts of human linguistic competence ([Chomsky, 1965; 1966; 1986], see discussion in [Abrahamsen, 1987]).

Abrahamsen contrasts such instances of boundary-breaking relations with ongoing boundary-bridging interaction between linguistics and psychology. She proposes that a boundary-bridging relation often holds between psycholinguistics, as a subdiscipline of psychology, and linguistics. In this boundary-bridging research, psycholinguists rely on linguists to provide specialized descriptions of, for example, phonemes, distinctive features, and phonological rules, while psycholinguists provide linguists with explanations (e.g., of universal characteristics of phonological systems) and evidence (e.g., for the psychological reality of certain linguistic accounts).[18] Abrahamsen observes, however, that the psycholinguist must often reformat the account provided by the linguist in order to make use of it. Some linguistic theories (e.g., augmented transition network grammars; lexical-functional grammars) require less adjustment than others (e.g., Chomsky's Standard Theory). Abrahamsen comments:

> The psychological studies benefit from ongoing involvement of linguists who are willing to consider psychological goals in addition to their own native goals as linguists. When these linguists carry out their work of linguistic description, they must satisfy two sets of constraints simultaneously, producing descriptions that can be easily applied in behavioral research as well as satisfy criteria of linguistic adequacy. [p. 373]

While boundary-breaking research as characterized by Abrahamsen would promote a unificationist conception of science, boundary-bridging research has far more limited aspirations. In some cases a cultural product discipline such as linguistics may simply provide a description of the phenomena for which psychologists then offer a mechanistic explanation. In other cases the understanding of the mechanism may explain certain linguistic phenomenon (e.g., multiply center embedded sentences such as *the dog the cat the mouse squeaked ate chased* are uncommon because they exceed the working memory capacity of humans). The results are interfield theories, not theory reductions.

[18]Abrahamsen generalizes this framework to many interdisciplinary relations. Subdisciplines of the physical sciences obtain specialized descriptions from the biological sciences, while biological sciences in turn appeal to these subdisciplines for explanation and evidence. The same, she proposes, is true of subdisciplines of the biological sciences with respect to the behavioral sciences, and of the subdisciplines of the behavioral sciences with respect to the cultural product disciplines (mathematics and engineering, humanities, and social sciences).

8 CONCLUSIONS

Visions of unifying all the sciences have been popular ever since the work of the ancient Greek philosophers. Such aspirations were prevalent in many of the historical proposals for unity with which we began this chapter. But the quest for unity can take make forms, often achieving integration rather than true unification. Perhaps the strongest vision of unity appeared in the theory-reduction model of the logical empiricists. This model was attractive because it suggested that logic might provide a powerful way to unite the results all scientific inquiries by showing higher-level theories to be derivable from lower-level ones. Not only were serious objections raised against this model, but as we have seen, much of the unity that appears to result is illusory. Even in the exemplar case of temperature, the bridge principles and boundary conditions have to be established empirically for each type of material in which heat is realized. For many years worries about multiple realizability provided the principal objections to the applicability of the theory-reduction account. A more troubling concern is that any lower-level theory that will provide a foundation from which to derive all higher-level theories will look very unlike contemporary lower-level theories, since it will have to incorporate all knowledge acquired at the higher levels. Altogether, the various objections to the theory-reduction have succeeded in moving it off center-stage in discussions about unity of science.

The problems confronting the theory-reduction model have led some philosophers to abandon the ideal of unity altogether. Cartwright emphasizes the plurality of models that investigators need to deal with the actual world, while Dupré focuses on the need for multiple different ways of categorizing phenomena, each of which is useful for different purposes. Kitcher remains a strong defender of the objective of theoretical unity, but even he has reduced it to the status of a regulative ideal. Still other philosophers, as we have shown, have adopted a reversionary perspective of advocating integration rather than advocating unity. This was the point of Darden and Maull's notion of an interfield theory — it integrates by bridging fields rather than establishing one complete unified theory. It is also exemplified in the notion of reduction which we have identified in the new mechanistic accounts of scientific explanation.

On mechanistic accounts, explanation consists in demonstrating how the orchestrated operation of the components of a mechanism enable the whole mechanism to perform a function in its environment. The conditions imposed on the mechanism from its environment remain a critical part of the explanation, so the higher-level account remains an autonomous component of any explanation. Further, there is no promise that the knowledge of how components behave in a mechanism will be unified with knowledge about how those components behave in other conditions. Lastly, organization turns out to be crucial in getting mechanisms to perform their function, and despite some key theoretical advances in understanding how negative and positive feedback systems enable dynamically organized mechanisms to maintain themselves, this inquiry is still in an early stage. Nonetheless, as the developments in the life sciences in the 20th century illustrate, there is great explanatory gain to developing models of mechanisms that integrate knowledge over several levels of organization. In discussing the more restrictive type of reduction that is achieved through understanding a mechanism, we also noted the need to rethink levels from the rather global perspective embraced in the theory-reduction account to a far more restricted sense in which the constituents of a given level are only determined as one takes a mechanism apart and establishes its working parts. Further, we noted that not all integration in mechanistic explanations is reductive — sometimes claims linking two characterizations of the same entity (e.g., a functional and a structural account) play an important heuristic role in fostering the development of science.

The kind of knowledge that results when investigators focus on mechanism is illustrated in the developmental systems account of how genetic information is linked to knowledge of biological traits — it is linked via an understanding of genetic regulation that relies on knowledge of the cellular machinery (especially the machinery of protein synthesis) which makes development possible. Our last two brief case studies bring out yet other important aspects of integration: the use of integration to overcome epistemic limitations and advance the epistemic warrant of research techniques and theories in each discipline and the dynamics of the process of interdisciplinary exchange (including boundary breaking as well as boundary bridging endeavors). Although we cannot follow up on these threads here, they point to very promising directions for further philosophical investigations of scientific integration.

ACKNOWLEDGEMENTS

We thank Adele Abrahamsen, Elihu Gerson, and Theo Kuipers for their very helpful comments and suggestions on earlier drafts of this paper.

BIBLIOGRAPHY

[Abrahamsen, 1987] A. A. Abrahamsen. Bridging boundaries versus breaking boundaries: Psycholinguistics in perspective. *Synthese*, 72(3): 355–388, 1987.

[Albert and Barabási, 2002] R. Albert, and A.-L. Barabási. Statistical mechanics of complex networks. *Review of Modern Physics*, 74: 47–97, 2002.

[Barabási and Albert, 1999] A.-L. Barabási, and R. Albert. Emergence of scaling in random networks. *Science*, 286: 509–512, 1999.

[Bechtel, 1984] W. Bechtel. Reconceptualization and interfield connections: The discovery of the link between vitamins and coenzymes. *Philosophy of Science*, 51: 265–292, 1984.

[Bechtel, 1986] W. Bechtel. The nature of scientific integration. In W. Bechtel (ed.), *Integrating Scientific Disciplines.* Dordrecht: Martinus Nijhoff, pages 3–52, 1986.

[Bechtel, 2001] W. Bechtel. Decomposing and localizing vision: An exemplar for cognitive neuroscience. In R. S. Stufflebeam (ed.), *Philosophy and the Neurosciences: A Reader.* Oxford: Basil Blackwell, pages 225–249, 2001.

[Bechtel, 2006] W. Bechtel. *Discovering Cell Mechanisms: The Creation of Modern Cell Biology.* Cambridge: Cambridge University Press, 2006.

[Bechtel and Abrahamsen, 2005] W. Bechtel, and A. Abrahamsen. Explanation: A mechanist alternative. *Studies in History and Philosophy of Biological and Biomedical Sciences*, 36: 421–441, 2005.

[Bechtel and McCauley, 1999] W. Bechtel, and R. N. McCauley. Heuristic identity theory (or back to the future): The mind-body problem against the background of research strategies in cognitive neuroscience. In S. C. Stoness (ed.), *Proceedings of the 21st Annual Meeting of the Cognitive Science Society*. Mahwah, NJ: Lawrence Erlbaum Associates, pages 67–72, 1999.

[Bechtel and Mundale, 1999] W. Bechtel, and J. Mundale. Multiple realizability revisited: Linking cognitive and neural states. *Philosophy of Science*, 66: 175–207, 1999.

[Bechtel and Richardson, 1993] W. Bechtel, and R. C. Richardson. *Discovering Complexity: Decomposition and Localization as Strategies in Scientific Research*. Princeton, NJ: Princeton University Press, 1993.

[Bernard, 1865] C. Bernard. *An Introduction to the Study of Experimental Medicine*. New York: Dover, 1865.

[Bichat, 1805] X. Bichat. *Recherches Physiologiques sur la Vie et la Mort* (3rd ed.). Paris: Machant, 1805.

[Bickle, 2003] J. Bickle. *Philosophy and Neuroscience: A Ruthlessly Reductive Account*. Dordrecht: Kluwer, 2003.

[Blumenthal, 1987] A. L. Blumenthal. The emergence of psycholinguistics. *Synthese*, 72(3), 313–323, 1987.

[Callender, 1999] C. A. Callender. Reducing statistical mechanics to thermodynamics: The case of entropy. *The Journal of Philosophy*, 96: 348–373, 1999.

[Campbell, 1974] D. T. Campbell. 'Downward causation' in hierarchically organised biological systems. In Dobzhansky (ed.), *Studies in the Philosophy of Biology*, Macmillan Press Ltd., 1974.

[Cannon, 1929] W. B. Cannon. Organization of physiological homeostasis. *Physiological Reviews*, 9: 399–431, 1929.

[Carnap, 1928] R. Carnap. *Der logische Aufbau der Welt*. Berlin: Weltkreis, 1928.

[Cartwright, 1980] N. Cartwright. Do the laws of physics state the facts? *Pacific Philosophical Quarterly*, 61: 64–75, 1980.

[Cartwright, 1983] N. Cartwright. *How the Laws of Physics Lie*. Oxford: Oxford University Press, 1983.

[Cartwright, 1999] N. Cartwright. *The Dappled World: A Study of the Boundaries of Science*. Cambridge: Cambridge University Press, 1999.

[Causey, 1977] R. L. Causey. *Unity of Science*. Dordrecht: Reidel, 1977.

[Chalmers, 1996] D. Chalmers. *The Conscious Mind*. Oxford: Oxford University Press, 1996.

[Chomsky, 1957] N. Chomsky. *Syntactic Structures*. The Hague: Mouton, 1957.

[Chomsky, 1959] N. Chomsky. Review of *Verbal Behavior*. *Language*, 35: 26–58, 1959.

[Chomsky, 1965] N. Chomsky. *Aspects of a Theory of Syntax*. Cambridge, MA: MIT Press, 1965.

[Chomsky, 1966] N. Chomsky. *Cartesian Linguistics: A Chapter in the History of Rationalist Thought*. Cambridge, MA: MIT Press, 1966.

[Chomsky, 1986] N. Chomsky. *Knowledge of Language: Its Nature, Origin, and Use*. New York: Praeger, 1986.

[Chubin, 1982] D. E. Chubin. *Sociology of Sciences: An Annotated Bibliography on Invisible Colleges, 1972-1981*. New York: Garland, 1982.

[Churchland, 1981] P. M. Churchland. Eliminative materialism and propositional attitudes. *The Journal of Philosophy*, 78: 67–90, 1981.

[Churchland, 1986] P. S. Churchland. *Neurophilosophy: Toward a Unified Science of the Mind-Brain*. Cambridge, MA: MIT Press/Bradford Books, 1986.

[Crane, 1972] D. Crane. *Invisible Colleges*. Chicago: University of Chicago Press, 1972.

[Craver and Bechtel, in press] C. Craver and W. Bechtel. Top-down causation without top-down causes. *Biology and Philosophy*, in press.

[Craver and Darden, 2001] C. Craver and L. Darden. Discovering mechanisms in neurobiology: The case of spatial memory. In P. McLaughlin (ed.), *Theory and Method in Neuroscience*. Pittsburgh, PA: University of Pittsburgh Press, pages 112–137, 2001.

[Crick and Koch, 1998] F. Crick and C. Koch. Consciousness and neuroscience. *Cerebral Cortex*, 8: 97–107, 1998.

[Dale and Pourquié, 2000] J. K. Dale and O. Pourquié. A clock-work somite. *Bioassays*, 22: 72–83, 2000.

[Darden, 1986] L. Darden. Relations amongst fields in the evolutionary synthesis. In W. Bechtel (ed.), *Integrating Scientific Disciplines.* Dordrecht: Martinus Nijhoff, pages 113–123, 1986.

[Darden and Craver, 2002] L. Darden and C. Craver. Strategies in the interfield discovery of the mechanism of protein synthesis. *Studies in the History and Philosophy of the Biological and Biomedical Sciences*, 33: 1–28, 2002.

[Darden and Maull, 1977] L. Darden and N. Maull. Interfield theories. *Philosophy of Science*, 43: 44–64, 1977.

[Dawkins, 1976] R. Dawkins. *The Selfish Gene.* Oxford: Oxford University Press, 1976.

[de Duve, 1984] C. de Duve. *A Guided Tour of the Living Cell.* New York: Scientific American Library, 1984.

[Dupré, 1983] J. Dupré. The disunity of science. *Mind*, 92: 321–346, 1983.

[Dupré, 1993] J. Dupré. *The Disorder of Things: Metaphysical Foundations of the Disunity of Science.* Cambridge, MA: Harvard University Press, 1993.

[Feigl, 1958/1967] H. Feigl. *The 'Mental' and the 'Physical': The Essay and a Postscript.* Minneapolis: University of Minnesota Press, 1958/1967.

[Feyerabend, 1962] P. K. Feyerabend. Explanation, reduction, and empiricism. In G. Maxwell (ed.), *Minnesota Studies in the Philosophy of Science.* Minneapolis, MN: University of Minnesota Press, Vol. III, pages 28–97, 1962.

[Feyerabend, 1963] P. K. Feyerabend. Mental events and the brain. *The Journal of Philosophy*, 60: 295–296, 1963.

[Feyerabend, 1970] P. K. Feyerabend. Against method: Outline of an anarchistic theory of knowledge. In M. R. a. S. Winokur (ed.), *Minnesota Studies in the Philosophy of Science.* Minneapolis, MN: University of Minnesota Press, Volume IV, pages 17–130, 1970.

[Feyerabend, 1975] P. K. Feyerabend. *Against method.* London: New Left Books, 1975.

[Feynman, 1963] R. P. Feynman. *The Feynman Lectures on Physics.* Reading, MA: Addison-Wesley Publishing Compan, 1963.

[Fodor, 1974] J. A. Fodor. Special sciences (or: the disunity of science as a working hypothesis). *Synthese, 28*, 97-115, 1974.

[Friedman, 1974] M. Friedman. Explanation and scientific understanding. *Journal of Philosophy*, 71: 5–19, 1974.

[Gánti, 1975] T. Gánti. Organization of chemical reactions into dividing and metabolizing units: The chemotons. *BioSystems*, 7: 15–21, 1975.

[Gánti, 2003] T. Gánti. *The Principles of Life.* New York: Oxford, 2003.

[Ghiselin, 2004] M. T. Ghiselin. Lorenz Oken and In T. Bach and O. Breidbach (eds.), *Naturphilosophie nach Schelling.* Stuttgart: Frommann-Holzboog, 2004, pages 433–457.

[Ghiselin and Breidbach, 2002] M. T. Ghiselin and O. Breidbach. Lorenz Oken and *Naturphilosophie* in Jena, Paris, and London. *History and Philosophy of the Life Sciences*, 24: 219–247, 2002.

[Glennan, 1996] S. Glennan. Mechanisms and the nature of causation. *Erkenntnis*, 44: 50–71, 1996.

[Glennan, 2002] S. Glennan. Rethinking mechanistic explanation. *Philosophy of Science*, 69: S342–S353, 2002.

[Gong and van Leeuwen, 2003] P. Gong and C. van Leeuwen. Emergence of a scale-free network with chaotic units. *Physical A: Statistical Mechanics and its Applications*, 321: 679–688, 2003.

[Goodman, 1955] N. Goodman. *Fact, Fiction, and Forecast.* Cambridge, MA: Harvard University Press, 1955.

[Hacking, 1983] I. Hacking. *Representing and Intervening.* Cambridge: Cambridge University Press, 1983.

[Hamer, 2004] D. Hamer. *The God Gene.* New York: Doubleday, 2004.

[Hempel, 1965] C. G. Hempel. Aspects of scientific explanation. In C. G. Hempel (ed.), *Aspects of Scientific Explanation and other Essays in the Philosophy of Science.* New York: Macmillan, pages 331–496, 1965.

[Hempel and Oppenheim, 1948] C. G. Hempel and P. Oppenheim. Studies in the logic of explanation. *Philosophy of Science*, 15: 137–175, 1948.

[Hennig, 1966] W. Hennig. *Phylogenetic Systematics*, (R. Zangerl, Trans.). Urbana: University of Illinois Press, 1966.

[Hooker, 1981] C. A. Hooker. Towards a general theory of reduction. *Dialogue*, 20: 38–59; 201–236; 496–529, 1981.

[Hull, 1972] D. L. Hull. Reduction in genetics — Biology or philosophy? *Philosophy of Science*, 39: 491–499, 1972.

[Hull, 1974] D. L. Hull. *The Philosophy of Biological Science*. Englewood Cliffs, NJ: Prentice-Hall, 1974.

[Kaufmann, 1993] S. A. Kaufmann. *The Origins of Order*. Oxford: Oxford University Press, 1993.

[Keijzer, 2001] F. Keijzer. *Representation and Behavior*. Cambridge, MA: MIT Press, 2001.

[Kelso, 1995] J. A. S. Kelso. *Dynamic Patterns: The Self Organization of Brain and Behavior*. Cambridge, MA: MIT Press, 1995.

[Kemeny and Oppenheim, 1956] J. G. Kemeny and P. Oppenheim. On reduction. *Philosophical Studies*, 7: 6–19, 1956.

[Kim, 1998] J. Kim. *Mind in a Physical World*. Cambridge, MA: MIT Press, 1998.

[Kitcher, 1981] P. Kitcher. Explanatory unification. *Philosophy of Science*, 48: 507–531, 1981.

[Kitcher, 1989] P. Kitcher. Explanatory unification and the causal structure of the world. In W. C. Salmon (ed.), *Scientific Explanation*. Minneapolis, MN: University of Minnesota Press, Vol. XIII, pages 410–505, 1989.

[Kitcher, 1999] P. Kitcher. Unification as a regulative ideal. *Perspectives on Science*, 7: 337–348, 1999.

[Krebs and Johnson, 1937] H. A. Krebs and W. A. Johnson. The role of citric acid in intermediate metabolism in animal tissues. *Enzymologia*, 4: 148–156, 1937.

[Kuhn, 1962/1970] T. S. Kuhn. *The Structure of Scientific Revolutions*, (Second ed.), Chicago: University of Chicago Press, 1962/1970.

[Kuipers, 2001] T. A. F. Kuipers. *Structures in Science*. Dordrecht: Kluwer, 2001.

[Landau, 1944] L. Landau. On the problem of turbulence. *Comptes Rendus d'Academie des Sciences, URSS*, 44: 311–314, 1944.

[Lewontin, 2000] R. Lewontin. *It ain't necessarily so: The Dream of the Human Genome and other Illusions*. New York: Basic Books, 2000.

[Machamer *et al.*, 2000] P. Machamer, L. Darden, and C. Craver. Thinking about mechanisms. *Philosophy of Science*, 67: 1–25, 2000.

[Maturana and Varela, 1980] H. R. Maturana and F. J. Varela. Autopoiesis: The organization of the living. In F. J. Varela (ed.), *Autopoiesis and Cognition: The Realization of the Living*. Dordrecht: D. Reidel, pages 59–138, 1980.

[Mayr, 1970] O. Mayr. *The Origins of Feedback Control*. Cambridge, MA: MIT Press, 1970.

[McCauley, 1981] R. N. McCauley. Hypothetical identities and ontological economizing: Comments on Causey's program for the unity of science. *Philosophy of Science*, 48: 218–227, 1981.

[McCauley, 1986] R. N. McCauley. Intertheoretic relations and the future of psychology. *Philosophy of Science*, 53: 179–199, 1986.

[McCauley, 1987] R. N. McCauley. The not so happy story of the marriage of linguistics and psychology: or why linguistics has discouraged psychology's recent advances. *Synthese*, 72: 341–353, 1987.

[McCauley, 1996] R. N. McCauley. Explanatory pluralism and the coevolution of theories in science. In R. N. McCauley (ed.), *The Churchlands and their Critics*. Oxford: Blackwell, pages 17–47, 1996.

[McCauley and Bechtel, 2001] R. N. McCauley and W. Bechtel. Explanatory pluralism and heuristic identity theory. *Theory and Psychology*, 11(6): 736–760, 2001.

[Milgram, 1967] S. Milgram. The small world problem. *Psychology Today*, 2: 60–67, 1967.

[Miller, 1962] G. A. Miller. Some psychological studies of grammar. *American Psychologist*, 17: 748–762, 1962.

[Nagel, 1961] E. Nagel. *The Structure of Science*. New York: Harcourt, Brace, 1961.

[Neurath, 1938] O. Neurath. Unified science as encyclopedic integration. In C. Morris (ed.), *International Encyclopedia of Unified Science*, Vol. I, Chicago: University of Chicago Press, 1938.

[Nickles, 1973] T. Nickles. Two concepts of intertheoretic reduction. *The Journal of Philosophy*, 70: 181–201, 1973.

[Oken, 1809] L. Oken. *Lehrbuch der Naturphilosophie*. Jena: Friedrich Frommann, 1809.

[Oken, 1831] L. Oken. *Lehrbuch der Naturphilosophie* (2nd ed.). Jena: Friedrich Frommann, 1831.

[Oppenheim and Putnam, 1958] P. Oppenheim and H. Putnam. The unity of science as a working hypothesis. In G. Maxwell (ed.), *Concepts, Theories, and the Mind-Body Problem*. Minneapolis: University of Minnesota Press, pages 3–36, 1958.

[Oyama *et al.*, 2001] S. Oyama, P. E. Griffiths, and R. Gray. What is developmental systems theory? In R. Gray (ed.), *Cycles of Contingency*. Cambridge, MA: MIT Press, 2001.

[Paul, 1880] H. Paul. *Principien der Sprachgeschichte*. Halle: Niemeyer, 1880.

[Place, 1956] U. T. Place. Is consciousness a brain process. *British Journal of Psychology*, 47: 44–50, 1956.

[Polger, 2004] T. Polger. *Natural Minds*. Cambridge, MA: MIT Press, 2004.

[Port and van Gelder, 1995] R. Port and T. van Gelder. *It's about Time*. Cambridge, MA: MIT Press, 1995.

[Pourquié, 1998] O. Pourquié. Clocks regulating developmental processes. *Current Opinion in Neurobiology*, 8: 665–670, 1998.

[Price, 1961] D. J. D. S. Price. *Science since Babylon*. New Haven: Yale University Press, 1961.

[Putnam, 1978] H. Putnam. *Meaning and the Moral Sciences*. London: Routledge and Kegan Paul, 1978.

[Pylyshyn, 1984] Z. W. Pylyshyn. *Computation and Cognition: Toward a Foundation for Cognitive Science*. Cambridge, MA: MIT Press, 1984.

[Quine, 1964] W. v. O. Quine. Ontological reduction and the world of numbers. *Journal of Philosophy*, 61: 209–216, 1964.

[Reber, 1987] A. S. Reber. The rise and (surprisingly rapid) fall of psycholinguistics. *Synthese*, 72(3): 325–339, 1987.

[Richardson, 1979] R. C. Richardson. Functionalism and reductionism. *Philosophy of Science*, 46: 533–558, 1979.

[Rorty, 1970] R. Rorty. In defense of eliminative materialism. *The Review of Metaphysics*, 24: 112–121, 1970.

[Rosenberg, 1994] A. Rosenberg. *Instrumental Biology and the Disunity of Science*. Chicago: University of Chicago Press, 1994.

[Rosenblueth *et al.*, 1943] A. Rosenblueth, N. Wiener, and J. Bigelow. Behavior, purpose, and teleology. *Philosophy of Science*, 10: 18–24, 1943.

[Ruse and Wilson, 1986] M. Ruse and E. O. Wilson. Moral philosophy as applied science. *Philosophy: The Journal of the Royal Institute of Philosophy*, 61: 173–192, 1986.

[Schaffner, 1967] K. Schaffner. Approaches to reduction. *Philosophy of Science*, 34: 137–147, 1967.

[Schaffner, 1969] K. F. Schaffner. The Watson-Crick model and reductionism. *British Journal for the Philosophy of Science*, 20: 325–348, 1969.

[Schuyler, 1978] R. L. Schuyler (ed.). *Historical Archaeology: A Guide to Substantive and Theoretical Contributions*. Farmingdale, NY: Baywood Publishing Company, 1978.

[Shapere, 1974] D. Shapere. Scientific theories and their domains. In F. Suppe (ed.), *The Structure of Scientific Theories*. Urbana: University of Illinois Press, 1974.

[Shapiro, 2004] L. Shapiro. *The Mind Incarnate*. Cambridge, MA: MIT Press, 2004.

[Sklar, 1967] L. Sklar. Types of inter-theoretic reduction. *British Journal for the Philosophy of Science*, 18: 109–124, 1967.

[Sklar, 1974] L. Sklar. Thermodynamics, statistical mechanics, and the complexity of reductions. In J. van Evra (ed.), *PSA 1974*. Dordrecht: Reidel, Vol. 32 of Boston Studies in the Philosophy of Science, pages 15–32, 1974.

[Sklar, 1993] L. Sklar. *Physics and Chance: Philosophical Issues in the Foundations of Statistical Mechanics*. New York: Cambridge, 1993.

[Smart, 1959] J. J. C. Smart. Sensations and brain processes. *Philosophical Review*, 68: 141–156, 1959.

[Smith, 2003] K. K. Smith. Time's arrow: Heterochrony and the evolution of development. *International Journal of Developmental Biology*, 47: 613–621, 2003.

[Stotz and Bostanci, 2005] K. C. Stotz and A. Bostanci. The representing genes project: Tracking the shift to "post-genomics". *New Genetics and Society*, 2005.

[Suppes, 1957] P. Suppes. *Introduction to Logic*. Princeton: van Nostrand, 1957.

[Suppes, 1981] P. Suppes. The plurality of science. In I. Hacking (ed.), *PSA 1978*. East Lansing, MI: Philosophy of Science Association, Vol. 2, pages 2–16, 1981.

[Tautz, 2000] D. Tautz. Evolution of transcriptional regulation. *Current Opinion in Genetic Development*, 10: 575–579, 2000.

[Taylor, 1967] C. Taylor. Mind-body identity, a side issue. *Philosophical Review*, 67: 201–213, 1967.

[Thelen and Smith, 1994] E. Thelen and L. Smith. *A Dynamical Systems Approach to the Development of Cognition and Action*. Cambridge, MA: MIT Press, 1994.

[von Bertalanfy, 1951] L. von Bertalanfy. *General Systems Theory: A new Approach to the Unity of Science*. Baltimore: Johns Hopkins Press, 1951.

[Wasserman and Faust, 1994] S. Wasserman and K. Faust. *Social Network Analysis: Methods and Applications*. New York: Cambridge, 1994.

[Watson and Crick, 1953] J. D. Watson and F. H. C. Crick. Molecular structure of nucleic acids. *Nature*, 171: 737–738, 1953.

[Watts and Strogratz, 1998] D. Watts and S. Strogratz. Collective dynamics of small worlds. *Nature*, 393: 440–442, 1998.

[Whaley, 1975] W. G. Whaley. *The Golgi Apparatus*, Vol. 2. New York: Springer-Verlag, 1975.

[Whewell, 1840] W. Whewell. *The Philosophy of the Inductive Sciences, founded upon their History*. London: J. W. Parker, 1840.

[Wiener, 1948] N. Wiener. *Cybernetics: Or, Control and Communication in the Animal Machine*. New York: Wiley, 1948.

[Wilson, 1985] M. Wilson. What is this thing called 'pain'? *Pacific Philosophical Quarterly*, 66: 227–267, 1985.

[Wimsatt, 1975] W. C. Wimsatt. Reductionism, levels of organization, and the mind-body problem. In I. Savodnik (ed.), *Brain and Consciousness*. New York: Plenum, pages 205–267, 1975.

[Wimsatt, 1976a] W. C. Wimsatt. Reductionism, levels of organization, and the mind-body problem. In I. Savodnik (ed.), *Consciousness and the Brain: A Scientific and Philosophical Inquiry*. New York: Plenum Press, pages 202–267, 1976.

[Wimsatt, 1976b] W. C. Wimsatt. Reductive explanation: A functional account. In J. van Evra (ed.), *PSA-1974*. Dordrecht: Reidel, pages 671–710, 1976.

[Wimsatt, 1994] W. C. Wimsatt. The ontology of complex systems: Levels, perspectives, and causal thickets. *Canadian Journal of Philosophy*, Supplemental Volume 20: 207–274, 1994.

[Woodger, 1952] J. H. Woodger. *Biology and Language*. Cambridge: Cambridge University Press, 1952.

[Wray and Love, 2000] G. Wray and C. Love. Developmental regulatory genes and echinoderm evolution. *Systematic Biology*, 49: 28–51, 2000.

[Wundt, 1900] W. Wundt. *Die Sprache*. Leipzig: Englemann, 1900.

[Wylie, 1999] A. Wylie. Rethinking unity as a "Working Hypothesis" for philosophy of science: How archaeologists exploit the disunities of science. *Perspectives on Science*, 7 (3): 293–317, 1999.

LOGICAL, HISTORICAL AND COMPUTATIONAL APPROACHES

Atocha Aliseda and Donald Gillies

1 INTRODUCTION

This chapter is concerned with logical, historical and computational approaches to the philosophy of science. We will deal with these various approaches in the historical order in which they were developed starting around 1920. In section 2 we will discuss how the logical approach to philosophy of science was introduced by the Vienna Circle, and developed by them and their followers and associates. The logical approach to philosophy of science remained dominant in the subject throughout the 1950s; but, from the early 1960's, it was challenged by a striking development of the historical approach. The historical approach was not introduced for the first time in the 1960s. On the contrary, it had been developed by Mach and Duhem much earlier, but, although Mach and Duhem are cited by the Vienna Circle as important influences on their philosophy, the Vienna Circle did not adopt the historical features of these two thinkers. In the excitement generated by the new logic of Frege and Russell, history of science seems to have been temporarily forgotten. But although the general idea of the historical approach is not new in the 1960s, that decade saw striking developments in this approach. After Kuhn, the analysis of scientific revolutions became a major problem for philosophy of science, while Lakatos applied the historical approach to mathematics for the first time. In section 3 we will give an account of the development of the historical approach from the early 1960s to the mid-1970s. After this time, a new factor enters the picture which seems to be changing society in many profound ways, and, in particular, is bringing about striking new developments in philosophy of science. This new factor is of course the development of computers. In section 4, we will show how various factors, including research in artificial intelligence (AI) led to the conception of science as problem solving in the 1970s. Then in section 5 we will trace the emergence in the 1980s and 1990s of logical and computational models for scientific inference and discovery. During this period, work in computer science brought about considerable developments in logic, leading to the introduction of new systems of logic, such as non-monotonic logic and abduction, which were unknown to the Vienna Circle. The new results have brought in to question earlier positions in the philosophy of science. For example some successes in the field of machine learning have provided a strong argument against Popper's claim

Handbook of the Philosophy of Science: General Philosophy of Science - Focal Issues, pp. 431–513.
Volume editor: Theo Kuipers. General editors: Dov M. Gabbay, Paul Thagard and John Woods.

that 'induction is a myth'. In general terms the new computer-based approaches allow the investigation in a formal way of the problem of discovery in science and so perhaps can be considered as closing the gap between the logical and historical approaches, which, up to the mid-1970s, tended to be seen as antagonistic.

2 THE LOGICAL APPROACH OF THE VIENNA CIRCLE AND THEIR FOLLOWERS FROM THE 1920S TO THE 1950S

In 1922 Moritz Schlick was appointed to the Mach-Boltzmann professorship of the inductive sciences at the University of Vienna. His arrival in Vienna that year marks the beginning of the Vienna Circle which was indeed known initially as the Schlick Circle.[1] Schlick organised a seminar which met once a fortnight on Thursday evenings in a room of the university building that housed the mathematics and physics institutes. Attendance at this seminar was by invitation only, and those who attended were the members of the Vienna Circle. They included Rudolf Carnap, Kurt Gödel, Hans Hahn, Otto Neurath, and Moritz Schlick himself. The building where they met has subsequently been restructured, but a room similar to and near their original meeting place has been turned into a kind of museum, and its walls are hung with photographs of the famous circle and their associates.

The views of the Vienna Circle in 1929 are set out in a pamphlet written mainly by Otto Neurath but with the help of some other members of the circle. In this work Neurath et al state very clearly the philosophical methodology which the circle employed. They write:

> The task of philosophical work lies in this clarification of problems and assertions, ... The method of this clarification is that of *logical analysis*; of it, Russell says (*Our Knowledge of the External World*, p. 4) that it "has gradually crept into philosophy through the critical scrutiny of mathematics ..."
>
> It is *the method of logical analysis* that essentially distinguishes recent empiricism and positivism from the earlier version that was more biological-psychological in its orientation. [1929, 8]

Here Neurath et al not only indicate their method (logical analysis), but also one of the main sources of their approach (Russell). Russell was certainly a major influence on the Vienna circle. In his memoir of Hahn, Menger, another member of the Vienna circle, writes: 'During the early 1920s he developed a great admiration for the works of Bertrand Russell. He reviewed some of them in the *Monatshefte für Mathematik und Physik*. In one of these reviews Hahn suggested that one day Russell might well be regarded as the most important philosopher of his time'

[1]There is now in Vienna a Vienna Circle Institute, directed by Friedrich Stadler, which publishes important works on the history of the Vienna Circle. Its website is `www.univie.ac.at/ivc`. Stadler [2001] gives an excellent scholarly and detailed account of the Vienna Circle. In this chapter we have also used the memoirs contained in Frank [1941] and Gadol [1982].

[Menger, 1980, xi]. Hahn also conducted a seminar on Russell and Whitehead's *Principia Mathematica* in the academic year 1924-5 during which the participants went through that work chapter by chapter.

It was perhaps mainly through Russell that the Vienna circle acquired its knowledge and love of logic, but there were other influences as well. Carnap records in his autobiography [1963, 5-6] that he studied under Frege. Carnap's involvement with Frege appears to have been rather by chance. His family lived in Jena and Carnap went to the University of Jena where Frege, although past 60, was only an associate professor of mathematics. Carnap writes:

> In the fall of 1910, I attended Frege's course "Begriffsschrift" (conceptual notation, ideography), out of curiosity, not knowing anything either of the man or the subject except for a friend's remark that somebody had found it interesting. [1963, 5]

Carnap himself was sufficiently interested to attend Frege's advanced course "Begriffsschrift II" in 1913 and his course Logik in der Mathematik in 1914. Carnap records that "Begriffsschrift II" was attended by 3 students: Carnap, a friend of Carnap's and a retired army major. One feels that in a modern 'efficiency'-minded university, Frege would have been forced into early retirement long before Carnap attended his courses. Who could have guessed at that time that Frege would later be lauded as the most important philosopher of his time?

Another important influence on the Vienna circle was Wittgenstein. The circle devoted itself to reading Wittgenstein's *Tractatus Logico-Philosophicus,* which had been published in 1921, 'paragraph by paragraph' during the academic year 1926-7 [Menger, 1980, xii]. Wittgenstein's *Tractatus* is of course a notable example of the application of the logic of Frege and Russell to philosophy.

Frege and Russell themselves had devoted their time much more to philosophy of mathematics than to philosophy of science. Frege had formulated the logicist thesis that mathematics could be reduced to logic, and had tried to demonstrate the correctness of this thesis — inventing a new system of logic in the process. Unfortunately Frege's system proved to be contradictory, but Russell tried to overcome this difficulty and to establish a consistent form of logicism. As Russell saw it, he had successfully applied the method of logical analysis and the new logic to the philosophy of mathematics, and he came to think that logic and logical analysis might be the essence of all philosophy. He expressed this point of view in a series of lectures which he gave at Harvard in March and April 1914, and which were published later that year with the title: *Our Knowledge of the External World.* In the preface, Russell wrote:

> The following lectures are an attempt to show, by means of examples, the nature, capacity, and limitations of the logical-analytic method in philosophy. This method, of which the first complete example is to be found in the writings of Frege, has gradually, in the course of actual research, increasingly forced itself upon me as something perfectly definite, capable of embodiment in maxims, and adequate, in all branches

> of philosophy, to yield whatever objective scientific knowledge it is possible to obtain. [1914, 7]

Lecture II was entitled: 'Logic as the Essence of Philosophy', and in it Russell said:

> The topics we discussed in our first lecture, and the topics we shall discuss later, all reduce themselves, in so far as they are genuinely philosophical, to problems of logic. This is not due to any accident, but to the fact that every philosophical problem, when it is subjected to the necessary analysis and purification, is found either to be not really philosophical at all, or else to be, in the sense in which we are using the word, logical. [1914, 42]

Russell's words had a profound impact on Carnap who records in his autobiography [1963, 13] that he read Russell's book, *Our Knowledge of the External World* in the winter of 1921. Carnap quotes a passage from this book which begins:

> The study of logic becomes the central study in philosophy: it gives the method of research in philosophy, just as mathematics gives the method in physics. [Russell, 1914, 243]

On this Carnap himself comments:

> I felt as if this appeal had been directed to me personally. To work in this spirit would be my task from now on! And indeed henceforth the application of the new logical instrument for the purposes of analyzing scientific concepts and of clarifying philosophical problems has been the essential aim of my philosophical activity. [1963, 13]

The Vienna Circle did continue the investigations of Frege and Russell into the philosophy of mathematics, but their main originality lay in the application of the method of logical analysis to the philosophy of science. This is the origin of the logical approach to the philosophy of science, which can be seen as an application to the philosophy of science of ideas originating in the philosophy of mathematics.[2] Neurath *et al.* formulate this aspect of the Vienna Circle's approach very clearly as follows:

> As we have specially considered with respect to physics and mathematics, every branch of science is led to recognise that, sooner or later in its development, it must conduct an epistemological examination of its foundations, a logical analysis of its concepts. [1929, 17]

[2]More details about this are to be found in Gillies and Zheng [2001], which also discusses the converse influence of the philosophy of science on the philosophy of mathematics. See particularly section 3, pp. 439-445.

Thus philosophers of science have the task of giving a logical analysis of the concepts of science. This essentially is what the logical approach to the philosophy of science amounts to.

The interest of the Vienna Circle in science is easily explained by the historical period in which they were living. The years 1900-30 were those of a great revolution in physics, which called into question the Newtonian mechanics which had been accepted for nearly two centuries, and gave birth to the new theories of relativity and quantum mechanics. There were personal contacts between members of the Vienna Circle and the leading physicists of the time. Schlick was a good friend of Einstein's, and the two of them engaged in a considerable correspondence concerning the philosophical interpretation of relativity. Neurath *et al.* list [1929, 20] three '*Leading representatives of the scientific world-conception*'. These are 'Albert Einstein, Bertrand Russell, Ludwig Wittgenstein.' This choice gives an excellent insight into the attitudes and interests of the Circle.

Let us now look at the logic which was used by the Vienna Circle for their logical analysis. Most of the Circle were inductivists and accepted inductive as well as deductive logic. Deductive logic was for nearly all of them the formal logic of Frege and Russell, together with the additions of later logicians such as Tarski. This was recently invented and had clearly superseded Aristotelian logic. It seemed at the time a powerful new tool for carrying out philosophical investigations. Fregean logic, or classical logic as it is now usually called, had at the time of the Vienna Circle only one opponent — the intuitionistic logic of Brouwer. Although several of the Vienna Circle did take an interest in Brouwer and intuitionism, it is fair to say that for them classical logic remained fundamental.

Turning now to inductive logic, a major problem for the circle was how evidence supported or confirmed scientific hypotheses. Thus investigations of confirmation and connected questions to do with probability were an important area of the Circle's activity. Confirmation theory was generally regarded as constituting an inductive logic which complemented deductive logic.

Nearly all the Vienna Circle accepted the distinction clearly formulated by Reichenbach between the *discovery* of scientific hypotheses and their *justification*. Moreover it was generally felt that discovery involved a creative act of a psychological and perhaps irrational nature. Thus the investigation of discovery fell outside the remit of philosophy of science, since philosophy of science consisted of logical analysis and no logical analysis could be given of discovery. Philosophers of science should therefore concentrate on giving a logical analysis of how scientific hypotheses are justified by evidence. Such a logical analysis constitutes confirmation theory or inductive logic. It is worth noting that this attitude to induction constitutes a considerable divergence from that of Bacon who regarded induction as principally a method of discovery.

The Vienna Circle's existence in Vienna was in fact quite short. Fascists seized power in Austria in 1934, and in 1936 Schlick was shot dead by a deranged student in the University of Vienna. After these events nearly all the other members of the Vienna Circle fled from Vienna — for the most part settling in the English-speaking world. This diaspora of the Vienna Circle had the effect of increasing rather than decreasing the influence of their ideas. In the period after the Second World War, the English-speaking world became the main centre for philosophy of science and the logical approach of the Vienna Circle came to dominate philosophy of science and perhaps the whole of philosophy as well. The high point of the influence of the logical approach may be the 1950s.

Two books which give an excellent illustration of the logical approach to the philosophy of science in this period are (1) Carnap's *Logical Foundations of Probability* which was published in 1950, but with a second edition in 1963, and (2) Hempel's *Aspects of Scientific Explanation and other Essays in the Philosophy of Science*, which was published in 1965, but is a collection of Hempel's essays written between 1942 and 1964. Carnap's book is a lengthy investigation of the central Vienna Circle topic of confirmation theory and inductive logic. Hempel was a younger follower of the Vienna Circle rather than an original member, but he became one of the leading advocates of their approach in the United States. His book contains an investigation of the logic of confirmation, but also of the logic of explanation, the logic of functional analysis, a logical appraisal of operationism, and so on.

Let us now consider two philosophers who might be considered as associates of the Vienna Circle because they had close links with the Circle without actually being members or followers. The first of these is Popper. Popper would certainly not have considered himself a supporter of the Vienna Circle since he criticized many of their ideas in very harsh terms. For example he was opposed to inductivism and even claimed that induction was a myth, while, conversely, he defended metaphysics against the Vienna Circle's claim that metaphysical statements were always meaningless. However, despite these differences, it remains true that Popper in his book of 1934 does adopt the Vienna Circle's logical approach to philosophy of science. He even, despite his dislike of inductive logic, develops a theory of corroboration in which corroboration, though not a probability function, can be defined in terms of probability. Moreover he supports the view of the Circle that discovery lies outside the sphere of philosophy of science. In fact Popper writes:

> ... the work of the scientist consists in putting forward and testing theories.
>
> The initial stage, the act of conceiving or inventing a theory, seems to me neither to call for logical analysis nor to be susceptible of it. The question how it happens that a new idea occurs to a man — whether it be a musical theme, a dramatic conflict, or a scientific theory — may be of great interest to empirical psychology; but it is irrelevant to the logical analysis of scientific knowledge. This latter is concerned not with *questions of fact* (Kant's *quid facti?*), but only with questions of *justification or validity* (Kant's *quid juris?*). Its questions are of the following kind. Can a statement be justified? And if so, how? Is it

> testable? Is it logically dependent on certain other statements? Or does it perhaps contradict them? [1934, 31]

There is something odd here since Popper in this passage explicitly denies that there can be a logic of scientific discovery. However, his book is actually entitled: *The Logic of Scientific Discovery.* Aliseda has thrown some light on this question in her [2004]. In fact the original German title of Popper's book was *Logik der Forschung*, whose literal translation into English would be ' Logic of Research'. But why did Popper, who was a master of the English language in his later period and who carefully and meticulously supervised the publication of his books in English, allow a mistranslation of his original title? The explanation we would like to propose is that Popper's approach to philosophy of science had changed between 1934 and 1959. As we will argue in more detail in section 3, he had moved, by 1959, towards a more historical approach to philosophy of science, and this made him more sympathetic to including the question of discovery within philosophy of science.

Another associate of the Vienna Circle was Quine who, as a young man, went from Harvard to Vienna for a period to study with the Circle. Like Popper, Quine criticized many of the key ideas of the Vienna Circle. In particular in his famous article: 'Two Dogmas of Empiricism', originally published in 1951 and reprinted in [Quine, 1953, 20-46], he criticizes the analytic-synthetic distinction and argues for a holistic view of science. However, like Popper in 1934, Quine, despite his criticisms of the Vienna Circle, accepts their general approach to philosophy. In fact this is shown very clearly in the title of his 1953 book: *From a Logical Point of View.* Quine's position as professor at Harvard and leading figure of the American philosophical establishment is indicative of the dominance in the United States at that time of the view of philosophy as logical analysis. Quine, however, unlike Popper, never moved towards the historical approach to philosophy, but remained faithful to the logical approach to the end of his life. Despite his admiration for logic, however, he did not take any notice of the new developments in logic which were brought about by the rise of computing and which we will discuss in section 5. Moreover Quine rarely if ever raised any questions about inductive logic and probability. Logic for him was always classical deductive logic with the occasional mention of intuitionistic logic.

3 THE CHALLENGE OF THE HISTORICAL APPROACH (C. 1960 TO THE MID 1970S)

The last hundred years have been characterised by very rapid change and this applies to philosophy of science as much as to anything else. We have seen that in the 1920s and 1930s, the logical approach to philosophy of science was a radical new idea and its proponents were liable to political persecution and even, in one case, assassination. By the 1950s the picture had completed changed. The logical approach was dominant in the dominant English-speaking world. Its advocates had become part of the establishment and many of them occupied prestigious chairs in the leading universities. The 1960s proved to be a turbulent decade in which establishment ideas were attacked throughout the world. Philosophy of science did not escape this trend, and in fact the dominant logical approach to the subject came to be challenged by the historical approach. The principal figures of this new movement in the subject which lasted roughly from the early 1960s to the mid-1970s were the later Popper, Kuhn, Lakatos and Feyerabend. Let us start by examining the case of Popper.

3.1 Popper's Historical Turn

We have already seen that Popper in his [1934] adopted the logical approach to the philosophy of science, but in his [1963] and [1972] there is a distinct shift to the historical approach. A study of the titles give an indication this change. The original title of Popper's [1934] was 'Logic of Research'. However the full title of his [1963] is 'Conjectures and Refutations. The Growth of Scientific Knowledge' and of his [1972] 'Objective Knowledge. An Evolutionary Approach.' Logic has been replaced by Growth and Evolution. It is not that Popper changed his basic circle of ideas, namely falsifiability, rejection of induction, partial rehabilitation of metaphysics, etc. However these ideas are presented in the later Popper in a context which is more historical than logical. Thus 1934, Section 21, pp. 84-6, is entitled: 'Logical Investigation of Falsifiability', while in Popper's 1963 falsification or refutation becomes part of the model of conjectures and refutations which is supposed to give an account of how scientific knowledge grows. Moreover Popper's later works contain many more allusions to episodes in the history of science, and sometime quite detailed analyses of historical case histories as well. An example of the latter is Popper's discussion of Galileo's erroneous theory of the tides in chapter 4 of 1972.

These rather qualitative observations can be supplemented by counting the number of references to history of science in Popper's works of various periods. We included in the count any scientist or mathematician whose work was carried out more than a third of a century before the publication of the volume in question. Using this criterion, it turns out that Popper's 1934 contains 0.33 references to history of science per page, his 1963 contains 0.71 such references per page, and his 1972 1.08 references per page. Thus the number of references which Popper makes to history of science more than doubles between 1934 and 1963 and increases by more than 50% between 1963 and 1972. The trend is unmistakeable. It is interesting to observe by way of comparison that Hempel's 1965 contains 0.075 references to history of science per page. This is not much more than a tenth of the figure for Popper's 1963. The distinction between the logical and historical approach to the subject shows up very clearly using this simple statistical criterion.[3]

[3]Hempel became a colleague of Kuhn's and a study of Kuhn's work influenced Hempel in the direction of the historical approach after 1965. An interesting account of this phase of Hempel's thought is to be found in Wolters [2003]. On p. 115 of this article, Wolters quotes the following

Let us now examine some of the differences between the logical and historical approaches to the philosophy of science. To begin with, the historical approach broadens the scope of philosophy of science to include questions about how scientific knowledge grows and develops and about how new scientific theories and entities are discovered. As we saw the logical approach to the subject, at least up to the 1950s, excluded questions of discovery as belonging to 'empirical psychology' rather than 'the logical analysis of scientific knowledge' [Popper, 1934, 31]. Some problems within philosophy of science are, however, common to the logical and historical approaches, but, generally speaking they are tackled in a different way in the two approaches. We can illustrate this by considering the problem of the way in which evidence confirms or disconfirms scientific hypotheses. As we have seen, this was a standard problem within the logical approach to the philosophy of science in the 1950s and early 1960s and is tackled by Carnap in his 1950 and Hempel in his 1965. Carnap begins by setting up a formal logical language. If a hypothesis h and evidence e are expressed in this language, then we can consider the degree of confirmation of h given e $[C(h, e)]$. Carnap proposes various c-functions as measures of $C(h, e)$ and tries to evaluate their merits. Hempel is not quite as formalistic as Carnap, but he uses a lot of logic. His most famous result in his 'Studies in the Logic of Confirmation' is of course the paradox of the ravens. It would generally be agreed that the hypothesis that all ravens are black is confirmed by observing a black raven. However, 'all ravens are black' is logically equivalent to 'all non-black things are non-ravens'. The latter should it seems, applying the same principle, be confirmed by observing any non-black thing which is a non-raven. So it looks as if 'all ravens are black' should also be confirmed by observing any non-black thing which is a non-raven. As Hempel says [1965, 15]: 'Consequently, any red pencil, any green leaf, any yellow cow etc., becomes confirming evidence for the hypothesis that all ravens are black.' I think it would be fair to say that literally hundreds of papers have been written trying to solve this paradox which remains to this day a favourite topic of philosophers of science.

We can contrast these logical approaches to the problem of confirmation with a historical approach to the same problem. Central to the historical approach to the philosophy of science is the use of case histories from the history of science. To tackle the question of confirmation such a case history would be selected. It could, for example, be Harvey's discovery of the circulation of the blood. Before Harvey's work, the Galenic theory was believed by the medical community. However there was shift in opinion and Harvey's new theory came to be accepted. The details of this change could be studied to see what bits of evidence Harvey brought forward in favour of his new theory and how convincing these various bits of evidence

passage from an interview with Hempel carried out in 1982. The following words of Hempel refer to Kuhn whom he first met in 1963: 'I was very much struck by his ideas. At first I found them strange and I had very great resistance to these ideas, his historicist, pragmatist approach to problems in the methodology of science, but I have changed my mind considerably about this since then.' Despite this change of mind, Hempel never became an important figure in the historical approach to the philosophy of science in the way that he certainly was in the logical approach to the subject.

were considered to be by his peers. Account could also be taken of what evidence was cited against Harvey's new theory by his opponents. In this way it could be seen what evidence was seen as confirming and to what extent, and what evidence was seen as disconfirming and to what extent. It would be hoped that, from such historical studies, general principles governing the confirmation of scientific hypotheses by evidence might be gleaned. Such a historical investigation obviously has a completely different character from the investigations of Carnap and Hempel, and this shows clearly some of the differences between the logical and the historical approaches to the philosophy of science.

Popper is interesting because of the shift in his ideas from a more logical approach in the 1930s to a more historical approach in the 1960s, and we will revisit his ideas in section 4.4 where his approach will be compared to that of Simon. However the person who did the most to promote the historical approach to philosophy of science in the 1960s was not Popper, but his younger contemporary Kuhn. We will now examine some of Kuhn's ideas and also how they were received by other philosophers of science — particularly by members of Popper's school.

3.2 Kuhn and his Critics

Kuhn begins his 1962 *The Structure of Scientific Revolutions* with the following rousing call for a historical approach to the philosophy of science:

> History ... could produce a decisive transformation in the image of science by which we are now possessed. That image has previously been drawn, even by scientists themselves, mainly from the study of finished scientific achievements as these are recorded in the classics and, more recently, in the textbooks from which each new scientific generation learns to practice its trade. ... This essay attempts to show that we have been misled by them in fundamental ways. Its aim is a sketch of the quite different concept of science that can emerge from the historical record of the research activity itself. [1962, 1]

We will now sketch the main ideas of Kuhn's essay. Kuhn's view is that science develops through periods of *normal science* which are characterised by the dominance of a *paradigm*, but which are interrupted by occasional revolutions during which the old paradigm is replaced by a new one. We will illustrate this theory by considering in turn the three scientific revolutions which constitute Kuhn's main examples. These are (i) the Copernican Revolution, (ii) the Chemical Revolution, and (iii) the Einsteinian Revolution.

(i) *The Copernican Revolution.* Kuhn's first book, published in 1957 was entitled: *The Copernican Revolution*, and it was probably this example more than any other which led him to his general model of scientific revolutions. From late Greek times until Copernicus, astronomy was dominated by the Aristotelian-Ptolemaic paradigm. The earth was considered to be stationary at the centre of the universe.

The different movements of sublunary and heavenly bodies were described by Aristotelian mechanics. The astronomer had to describe and predict the movements of the Sun, Moon and planets as accurately as possible, using the Ptolemaic scheme of epicycles. This was the normal science of the time.

Copernicus, however, challenged the dominant paradigm by suggesting that the Earth spun on its axis and moved round the Sun. He worked out this alternative theory in as detailed a mathematical fashion as Ptolemy's. His results were published in his book *De Revolutionibus Orbium Caelestium* in 1543, and this publication inaugurated a revolutionary period during which the old Aristotelian-Ptolemaic paradigm was overthrown and replaced by a new paradigm — first formulated in detail by Newton in *Philosophiae Naturalis Principia Mathematica* [1687].

(ii) *The Chemical Revolution.* The main theme of the chemical revolution was the replacement of the *phlogiston* theory by the *oxygen* theory, though there were many other important changes as well. According to the phlogiston theory, bodies are inflammable if they contain a substance called phlogiston, and this is released when the body burns. It was known that air was needed to support combustion, and the phlogiston theorists explained this by claiming that the air absorbed the phlogiston given off in combustion until it was saturated when combustion ceased. The phlogiston theory was also used to explain the calcination of metals. When a metal is heated in air, in many cases it turns into a powder known as the *calx*, e.g. iron -> rust. Conversely the calx is usually found in ores of the metal, and the metal itself could often be obtained by heating with charcoal. These transformations were explained by postulating that

calx + phlogiston = metal

When we heat a metal, phlogiston is given off, and the calx remains. Conversely when we heat the calx with charcoal, since charcoal is very rich in phlogiston because it burns easily, the phlogiston from the charcoal combines with the calx to give the metal. In the oxygen theory, burning is explained as the combination of the substance with oxygen. On this theory air is needed for combustion because it contains oxygen and combustion ceases when the oxygen is used up. The calx is identified with the oxide of the metal. So turning a metal into its calx by heating in air is explained by the equation

metal + oxygen = metal oxide

Similarly obtaining the metal by heating the calx with charcoal is explained by the equation

metal oxide + carbon = metal + carbon dioxide

The oxygen theory was developed by Lavoisier. At the beginning of his researches in 1772, he was already sceptical of the then dominant phlogiston theory. In the next decade or so, many experimental discoveries concerning gases were made. These discoveries were mainly owing to the English experimental chemists — particularly Priestley and Cavendish. However, these English chemists remained faithful to the phlogiston theory. For example Priestley prepared the gas which we now call: 'oxygen'. He observed that it supported combustion better than ordinary air, and concluded that it must be dephlogisticated air. On the phlogiston theory, air supports combustion by absorbing phlogiston. So air with the phlogiston removed (dephlogisticated air) will absorb more phlogiston and support combustion better. Lavoisier, on the other hand, reinterpreted the results of Priestley and the other English chemists in terms of his new and developing oxygen theory. Lavoisier's new paradigm for chemistry was set out in his *Traité élémentaire de chimie* of 1789, and within a few years it was adopted by the majority of chemists. Priestley, however, who lived until 1804, never gave up the phlogiston theory.

(iii) *The Einsteinian Revolution.* The triumph of the Newtonian paradigm initiated a new period of normal science for astronomy (c. 1700–c. 1900). The dominant paradigm consisted of Newtonian mechanics including the law of gravity, and the normal scientist had to use this tool to explain the motions of the heavenly bodies in detail — comets, perturbations of the planets and the moon, etc. In the Einsteinian revolution (c. 1900–c. 1920), however, the Newtonian paradigm was replaced by the special and general theories of relativity.

Kuhn introduced one further very important idea: *incommensurablity.* In many scientific revolutions, so Kuhn claimed, the new paradigm is incommensurable with the old paradigm. As he says:

> The normal-scientific tradition that emerges from a scientific revolution is not only incompatible but often actually incommensurable with that which has gone before. [1962, 102]

Incommensurablility was introduced at the same time by Feyerabend, and indeed the concept may have originated in conversations between Kuhn and Feyerabend in the years 1960 and 1961. Feyerabend's account of incommensurability, however, differs significantly from Kuhn's.

Kuhn's book on scientific revolutions was read with great interest by Popper and his school in London. Lakatos invited Kuhn to give a paper at an International Colloquium in the Philosophy of Science held at Bedford College, Regent's Park, London from 11 to 17 July 1965. A collection of papers which developed from the discussions at this conference was published 5 years later (Imre Lakatos and Alan Musgrave (eds.) *Criticism and the Growth of Knowledge*, Cambridge University Press, 1970). This volume contains an essay by Kuhn himself, a series of essays criticizing and developing some of Kuhn's ideas, and Kuhn's replies to his critics. Altogether it is a most important collection for understanding the development of the historical approach to the philosophy of science at that time. Many of Kuhn's critics in the volume came from Popper's school. There is an essay by Popper himself, and essays by Watkins and Lakatos. Feyerabend too could be considered as associated with the Popper school since he had studied with Popper and even spent a year as Popper's research assistant. By the mid-1960s, his thinking had diverged very considerably from Popper's, but he was still on very friendly terms with Lakatos. The volume, however, also includes critical essays by people who were definitely not members of the Popper school — notably Margaret Masterman. We will now turn to considering criticisms and developments of Kuhn's ideas which are to be found in this 1970 volume and also in subsequent discussions. Kuhn's views involve the three key concepts of (i) *paradigm*, (ii) *normal science*, and (iii) *incommensurability*, and we will consider these three concepts in turn.

Let us start then with 'paradigm', a term which was introduced by Kuhn into the philosophy of science. Many authors have criticized it for being too vague and ambiguous. Shapere, for example, in his review of Kuhn's *The Structure of Scientific Revolutions*, goes so far as to claim that Kuhn's relativism is 'a logical outgrowth of conceptual confusions ... owing primarily to the use of a blanket term [paradigm]' [1964, 393]. In her 1970 article 'The Nature of a Paradigm', Masterman is quite sympathetic to Kuhn, yet she says 'On my counting, he uses 'paradigm' in not less than twenty-one different senses in [Kuhn, 1962], possibly more, not less' [1970, 61]. She then proceeds to list the 21 senses. Kuhn took these criticisms somewhat to heart, and in his article 'Second Thoughts on Paradigms' [1974], he suggested replacing 'paradigm' by two new concepts, namely 'disciplinary matrix' and 'exemplar'. However, these terms have never proved as popular as the original term 'paradigm'.

There is a certain irony in Kuhn's doubts about the term 'paradigm' since no other philosophical term coined in the twentieth century has proved so popular among general writers. Typing 'Paradigm' into Google produced 15,700,000 pages in 0.24 seconds. The term is constantly recurring in newspapers and magazines in contexts ranging from politics to fashion. Just one example will serve to illustrate the ubiquity of the expression. In 1998 Levi Strauss's new brand developer proclaimed that 'loose jeans is not a fad, it's a paradigm shift.'[4]

Of course the fact that a term has become so popular with the general public by no means shows that it is suitable for use in philosophy of science which perhaps demands higher standards of rigour. However, despite Kuhn's own doubts, we will argue that the term is a useful one for philosophers of science. Our defence is based on the following famous passage from Aristotle:

> Our discussion will be adequate if it has as much clearness as the subject-matter admits of, for precision is not to be sought for alike in all discussions, ... it is the mark of an educated man to look for precision in each class of things just so far as the nature of the subject admits; ... [Nicomachean Ethics I iii 1094^b 12f]

Aristotle's point of view was supported in the twentieth century by Ramsey who wrote in his general essay on Philosophy:

[4]This pronouncement is quoted in [Klein, 2000, 70].

> The chief danger to our philosophy, apart from laziness and woolliness is *scholasticism*, the essence of which is treating what is vague as if it were precise and trying to fit it into an exact logical category. [1929, 269]

So although the notion of paradigm is indeed not very precise, this does not prove that it is unsuitable for use in philosophy of science. Indeed the search for more precise notions might lead, as Ramsey suggests, to scholasticism. Our claim is that the notion of paradigm has just the right degree of precision for the subject-matter in hand, that is to say for the analysis of how science develops. We will now try to substantiate this by looking in more detail at Kuhn's discussion of the notion.

In his 1962, Kuhn introduces the notion of paradigm as follows:

> ... achievements that some particular scientific community acknowledges for a time as supplying the foundation for its further practice ... today ... are recounted, though seldom in their original form, by science textbooks, elementary and advanced. These textbooks expound the body of accepted theory, illustrate many or all of its successful applications, and compare these applications with exemplary observations and experiments. Before such books became popular early in the nineteenth century (and until even more recently in the newly matured sciences), many of the famous classics of science fulfilled a similar function. Aristotle's *Physica*, Ptolemy's *Almagest*, Newton's *Principia* and *Opticks*, Franklin's *Electricity*, Lavoisier's *Chemistry*, and Lyell's *Geology* — these and many other works served for a time implicitly to define the legitimate problems and methods of a research field for succeeding generations of practitioners. They were able to do so because they shared two essential characteristics. Their achievement was sufficiently unprecedented to attract an enduring group of adherents away from competing modes of scientific activity. Simultaneously, it was sufficiently open-ended to leave all sorts of problems for the redefined group of practitioners to resolve.
>
> Achievements that share these two characteristics I shall henceforth refer to as "paradigms," ... [1962, 10]

We would like to draw particular attention to the connection which Kuhn makes in this passage between paradigms and textbooks. Since the early nineteenth century, paradigms have, according to Kuhn, been generally taught by means of textbooks. Before the nineteenth century, he thinks that many of the famous classics of science fulfilled a similar function. However, of the classics he mentions, some were not in fact used to teach a paradigm to students, while others were so used, and can to all intents and purposes be regarded as textbooks. Thus Newton's *Principia* was not the canonical text of Newtonian mechanics for the mainstream mathematicians of the 18th century, since these mathematicians preferred an approach more analytical and less geometrical than Newton's. Ptolemy's *Almagest* was certainly a classic of science, but it was also a textbook expounding the fruits of earlier work, though doubtless with many interesting additions by Ptolemy himself. Aristotle's *Physica* was actually used as a textbook in medieval universities.

If, therefore, we include under the term 'textbook' those classics of science which actually were used as textbooks, we can introduce what could be called the *textbook criterion for paradigms*. The suggestion is that, if historians wish to identify the paradigm of a group of scientists at a certain time and place, they should examine the textbooks which were used to teach the novices the knowledge needed to become fully recognised members of the group. The contents of these textbooks will then (more-or-less) define the paradigm accepted by the group.[5] This textbook criterion constitutes, in our view, a sufficient answer to those who complain that the notion of paradigm is too vague. The criterion in fact enables a historian of science to use the term 'paradigm' in quite a concrete and definite fashion.

Let us now turn from the concept of paradigm to the related concept of normal science. Here are a couple of quotations in which Kuhn picks out important features of normal science.

> When examining normal science ... we shall want finally to describe that research as a strenuous and devoted attempt to force nature into the conceptual boxes supplied by professional education. [Kuhn, 1962, 5]

> Normal science does not aim at novelties of fact or theory and, when successful, finds none. [Kuhn, 1962, 52]

Kuhn also describes the activity of the normal scientist as 'puzzle-solving' [1962, 36].

All this makes normal science sound a rather dreary and dogmatic affair. The Popperians and Feyerabend in Lakatos and Musgrave [1970] express strong hostility to the idea. Kuhn in his reply remarks quite wittily [1970, 233] that 'normal science ... calls forth some of the oddest rhetoric: normal science does not exist *and* is uninteresting.' This comment is really quite fair. The feature of normal science which is disliked by the Popperians and Feyerabend is the alleged consensus in commitment to a *single* paradigm. They think that science progresses better if there is competition between different theories, paradigms, or research

[5]The full grasp of a paradigm may also involve the ability to carry out experiments and observations. Thus we should perhaps take textbooks to include laboratory manuals and the like. Moreover, there lies beyond the reach of textbooks a certain amount of knowledge which can only be learnt by a kind of apprenticeship, e.g. practical training in the laboratory or in the field. Kuhn alludes to these matters when he speaks [1962, 41] of 'instrumental commitments that, as much as laws and theory, provide scientists with rules of the game.' In the light of this, the textbook criterion should be regarded as only approximate.

programmes. Moreover they think that competition rather than consensus is in fact the more usual state of science. These criticisms are well expressed by Feyerabend in his contribution to Lakatos and Musgrave [1970]. This has the interesting title 'Consolations for the Specialist' which suggests a way in which Kuhn's philosophy of science can be viewed. In the article, Feyerabend writes:

> More than one social scientist has pointed out to me that now at last he had learned how to turn his field into a 'science' ... . The recipe, according to these people, is to restrict criticism, to reduce the number of comprehensive theories to one, and to create a normal science that has this one theory as its paradigm. Students must be prevented from speculating along different lines and the more restless colleagues must be made to conform and 'to do serious work'. *Is this what Kuhn wants to achieve*? [1970, 198]

Feyerabend suggests that to prevent normal science getting off the ground [1970, 205]: 'we must be prepared to accept a *principle of proliferation*', and expresses [1970, 207]: 'the suspicion that normal or "mature" science, as described by Kuhn, *is not even a historical fact.*' After all, as Feyerabend goes on to say:

> ... why should we not start proliferating *at once* and *never* allow a purely normal science to come into existence? And is it too much to be hoped that scientists thought likewise, and that normal periods, if they ever existed, cannot have lasted very long and cannot have extended over large fields either? [1970, 207]

This is certainly a strong attack on both the existence and desirability of normal science, but Kuhn had already published some interesting remarks in its defence. Thus he says:

> ... history strongly suggests that, though one can practice science — as one does philosophy or art or political science — without a firm consensus, this more flexible practice will not produce the pattern of rapid consequential scientific advance to which recent centuries have accustomed us. [1959, 232]

Kuhn stresses that commitment to a paradigm may force scientists to investigate the natural world in a detail and depth which would not otherwise be achieved. This is one of the secrets of the success of normal science:

> By focusing attention upon a small range of relatively esoteric problems, the paradigm forces scientists to investigate some part of nature in a detail and depth that would otherwise be unimaginable. ... during the period when the paradigm is successful, the profession will have solved problems that its members could scarcely have imagined and would never have undertaken without commitment to the paradigm. And at least part of that achievement always proves to be permanent. [1962, 24–25]

This controversy between Feyerabend and Kuhn regarding the value of normal science is of very great interest and importance, and we will now try to assess the merits of the two sides in the debate and to suggest a possible compromise. Our discussion will hinge on the distinction between a paradigm becoming established in a community *for empirical reasons*, and a paradigm becoming established *by political methods*. Essentially we are going to argue that if a paradigm is established for empirical reasons, then normal science is likely to be fruitful and Kuhn is correct; whereas, if a paradigm is established by political methods when there are no good empirical reasons in its favour, then normal science is likely to be harmful and Feyerabend is right.

Let us start by considering the concept of empirical reasons. Suppose we have two competing theories A and B. Suppose that A explains a wide range of observations and known experimental results, while B explains much fewer of these, or does not explain them so well. Suppose that both A and B have been subjected to a number of severe experimental tests, and that A has passed them all with flying colours, while B has not done so well. We would then say that the evidence confirms (or corroborates)[6] A much more than B. If we write the degree of confirmation (or degree of corroboration) of A given the evidence (e) as $C(A, e)$, we could then say that $C(A, e)$ is much greater than $C(B, e)$. If under these circumstances A is preferred to B, we will say that A is preferred to B *for empirical reasons*. It is worth noting that this formulation assumes that the notions of confirmation or corroboration are coherent and that it is possible to develop a viable confirmation or corroboration theory. These assumptions have been questioned and we will consider some objections to confirmation theory later one. For the time being, however, we will assume that the notion of confirmation is coherent.

Perhaps the most standard example of normal science is constituted by the dominance of the Newtonian paradigm in astronomy and mechanics from the beginning of the 18th century to the beginning of the 20th century. With one qualification which will be made later, this can be regarded as a genuine example of normal science. So we can say that Feyerabend is wrong and that normal science has existed — at least on some occasions. Moreover it is also clear that the Newtonian paradigm was established for empirical reasons. When Newton's theory came to be accepted it had a very high degree of empirical confirmation and one that was much higher than any rival theory.[7] Now did this Newtonian normal science bring about progress in the period from c.1700 to c. 1900? It seems to us that it certainly did. To begin with there was a great deal of mathematical progress. The development of infinitesimal calculus and then of ε, δ analysis was mainly stimulated by the need to tackle ever more complicated problems in mechanics and astronomy, and this development brought a great advance to mathematics, enormously strength-

[6]We will use confirms and corroborates, confirmation and corroboration, etc. as synonyms.

[7]This is argued in detail in Gillies [1993] (see particularly pp. 218-20). However, we think it would be accepted by nearly all philosophers of science, excepting only the few who deny that the possibility of a viable concept of empirical confirmation.

ening the power of mathematical tools. Then in mechanics itself, we have a long list of developments: hydrodynamics, elasticity, Coriolis forces and the Foucault pendulum, the gryroscope and the wobbling and rotating of the Earth and other celestial bodies, etc. In astronomy, the paths of planets and comets were traced with ever more accuracy, and these calculations were extended to the stars. All these advances took place within the Newtonian paradigm and would have been impossible had Newton's theory not been generally accepted and taught. Kuhn's analysis of why normal science can succeed applies particularly well to one famous advance of this period — the discovery of Neptune. Kuhn emphasizes that normal science focuses [1962, 24]: 'attention upon a small range of relatively esoteric problems'. The esoteric problem which led to the discovery of Neptune arose because of small perturbations in the orbit of Uranus. Without the detailed development of the Newtonian mathematical apparatus, these perturbations would never have been detected. Nor would it have been possible to calculate that they could be caused by a hitherto unknown planet located in a specified position. The preceding developments of normal science were a precondition for the discovery of Neptune, and yet that discovery was a startling and dramatic one.

The example of the discovery of Neptune and others like it show that Kuhn was right to suggest that normal science could be fruitful and lead to advances. However, they also show that Kuhn was wrong in his depictions of normal science as inevitably a dreary and dogmatic affair. In particular, Kuhn gives a misleading picture in his well-known claim [1962, 52] that 'normal science does not aim at novelties of fact or theory and, when successful, finds none.' In reality many interesting novelties of fact and theory can appear within normal science. We will come back to this point in a different context in section 5.2, where the concept of abductive novelty is introduced.

The fact that normal science can be a more lively affair than Kuhn's account would lead us to believe, might help to promote a reconciliation between the Kuhnian and Popperian traditions. However this reconciliation should perhaps take a form somewhat different from the one that is usually suggested. In his contribution to the 1970 volume edited by Lakatos and Musgrave, Kuhn writes:

> I suggest then that Sir Karl has characterized the entire scientific enterprise in terms that apply only to its occasional revolutionary parts. [1970, 6]

As a matter of fact, however, it is questionable whether Popper's theory of conjectures and refutations gives a satisfactory account of scientific revolutions. The problem is that, although paradigms may be confirmed or disconfirmed by evidence, they cannot be directly refuted or falsified by evidence. Only low-level empirical generalisations, or specific, detailed models are subject to falsification by observation or experiment.[8] Thus a revolutionary change from an old to a new

[8]These claims would, we think, be accepted by most philosophers of science. Detailed arguments for them are to be found in [Gillies, 1993, 204-30].

paradigm needs a more complex characterisation than that of conjectures and refutations. On the other hand, patterns of conjectures and refutations do often occur in the context of normal science. So, in the example of the discovery of Neptune, Adams and Leverrier conjectured that the mysterious perturbations in the orbit of Uranus were caused by a hitherto unknown planet. They were able to develop this conjecture into a detailed form, which specified where this hypothetical planet should be, and so rendered the conjecture (in this specific form) refutable. In fact it was not refuted, as we know, but confirmed. However, if Neptune had not after all existed, the conjecture of Adams and Leverrier would have been refuted in due course.

Thus, contrary to Kuhn, we would like to suggest that Popper's methodology of conjectures and refutations applies not to revolutionary science, but to normal science. However, in order to adapt Popper's ideas in this way, we need to make a change to Popper's account of conjectures and refutations. Popper argues that scientists can put forward any arbitrary conjecture which is testable by experience. Indeed Popper urges scientists to put forward bold, sweeping conjectures. However, if we accept Kuhn's concept of normal science, then, during a period of normal science, a scientist cannot put forward any arbitrary testable conjecture (as Popper suggests), but only a conjecture which is compatible with the dominant paradigm. If a scientist puts forward a conjecture which contradicts the dominant paradigm, it is likely to be regarded as inadmissible by the rest of the scientific community. Admittedly in some cases, for example that of Copernicus, such a hypothesis may mark the beginning of a revolution, but, even if the hypothesis is vindicated in the long run, it is likely to be strongly opposed at first by the scientific community. So, to sum up, Popper's methodology of conjectures and refutations can be regarded as one of the principal patterns of development in normal science — provided the conjectures considered are limited to ones which are compatible with the dominant paradigm.

Naturally the formulation just given distinguishes rather too sharply between normal and revolutionary science, and this brings us to a consideration of a qualification which needs to be made to our example of Newtonian normal science in the period c. 1700 to c. 1900. This qualification affects not just this example, but the concept of normal science in general.

The qualification comes from Lakatos's paper: 'Newton's Effect on Scientific Standards' which was written in the years 1963-4, but not published until 1978 after Lakatos's death. This somewhat neglected but highly interesting paper, was written in the years immediately following the publication of *The Structure of Scientific Revolutions*, and contains an interesting criticism of Kuhn's notion of normal science. This criticism is concerned with developments in astronomy in the 18th century. Lakatos begins by saying that in 1746:

> ... Clairaut found that the progress of the Moon's apogee is in reality twice what would follow from Newton's theory, and he proposed an additional term to Newton's formula involving the inverse fourth power of the distance. [1978, 219]

In other words, in the face of an anomaly, Clairaut, one of the leading scientists of the time, suggested a modification of Newton's law of gravity. Now Newton's law of gravity was part of the dominant paradigm of the time, and so Clairaut was not acting as a normal scientist should have done. His suggestion did not prove successful, however, for, as Lakatos goes on to say:

> But as it turned out, Clairaut's mathematics was wrong, and in fact later a correct calculation was found among Newton's unpublished manuscripts. But even so, a small discrepancy remained: a "secular acceleration". In 1770 the Paris Academy put up a prize for the solution of this problem. Euler won this prize with an essay in which he first concluded that "it appears to be established, by indisputable evidence, that the secular inequality of the moon's motion cannot be produced by the [Newtonian] forces of gravitation", and he proposed a rival formula again involving an additional term, which, in a sequel published a year later, he tried to explain from the resistance of Cartesian ether. However, Laplace in 1787 showed that the problem can be solved *better* within the Newtonian research programme. [1978, 219]

This historical example does have some features which Kuhn attributes to normal science since it shows scientists [1962, 24]: 'focusing attention upon a small range of relatively esoteric problems'. However, it does not exhibit the respect which scientists are supposed to show to the dominant paradigm during a period of normal science. Once again a leading scientist (Euler) was prepared to modify Newton's theory of gravity in order to explain a small observational anomaly, although, once again, the suggestion proved to be unsuccessful. Lakatos comments as follows:

> Did Clairaut and Euler make a methodological blunder — as Kuhn would surely say — when they tried alternative research programmes to solve Newtonian puzzles and only wasted time, energy and talent? [1978, 219]

Of course the answer to Lakatos's rhetorical question is obvious. Clairaut and Euler acted very reasonably. As a matter of fact, their suggested modifications of Newtonian theory were not successful, but this could not have been known in advance. Moreover the challenge of Clairaut and Euler led the Newtonians to produce an explanation of the difficulty within their own framework.

But does an example of this sort show that we should condemn normal science and follow Feyerabend's strategy of trying always to proliferate alternative theories. This would, in our view, be too extreme a response. During the long period (c. 1700 to c. 1900) of Newtonian normal science, it would not, in our opinion, have helped scientific progress if scientists had devoted a great deal of time and energy to proliferating alternatives theories of mechanics, and then debating the value of these alternatives as compared to Newtonian mechanics. In fact it was only a long series of mathematical and empirical developments based upon

Newtonian mechanics which created the possibility of creating new systems of mechanics (relativity and quantum mechanics) in the twentieth century. So it does not seem correct to give up normal science in favour of a Feyerabendian 'anything goes' position. However, Lakatos's illuminating historical example does suggest that the dogmatism of normal science should not be too rigid. Scientists should consider the possibility of now and again introducing hypotheses which contradict the some features of the dominant paradigm. Such hypotheses may often prove unsuccessful, but occasionally they may be the beginning of some new and exciting revolutionary development. Moreover, by the same token, the scientific community should allow some dissidents who do not accept the general consensus. Some discipline may be required, but too much discipline can be counter-productive.

So far we have been arguing in favour of a (somewhat qualified) normal science on the assumption that the paradigm underlying that science has come to be accepted by the community for empirical reasons. Now, however, let us consider whether there could be cases of normal science where the paradigm becomes established by political means without their being any very strong empirical reasons in its favour. Feyerabend seems to suggest that this might be possible in the passage [1970, 198] which, because of its importance, we will quote again.

> More than one social scientist has pointed out to me that now at last he had learned how to turn his field into a 'science' ... . The recipe, according to these people, is to restrict criticism, to reduce the number of comprehensive theories to one, and to create a normal science that has this one theory as its paradigm. Students must be prevented from speculating along different lines and the more restless colleagues must be made to conform and 'to do serious work'. *Is this what Kuhn wants to achieve*? [1970, 198]

Now it is worth noting that in this passage Feyerabend speaks of 'more than one social scientist', and it could be claimed that this is unfair to Kuhn, since Kuhn explicitly states [1962, x] that his theory of normal science dominated by a single paradigm applies only to mature natural sciences and *not* to the social sciences. On the contrary, there are, according to Kuhn, competing schools of thought in every branch of the social sciences and we never find a single dominant paradigm. Feyerabend, however, does in effect recognize that Kuhn holds this position, and indeed takes it as a starting point for his criticism of Kuhn. Feyerabend's argument is that social scientists who had read Kuhn could be inspired to turn their field into a science by banning all schools of thought except one. But could such a strategy actually be carried out in the contemporary university system in order to produce a normal science based on a paradigm for which there is little or no empirical confirmation?

To make the question more specific, suppose that we have two general theories A and B which are potential paradigms in some area of research. Suppose further that there is little or no empirical confirmation of A, and that the degree of confirmation of A is less than that of B. Would it nonetheless be possible to establish a normal science based on A by political methods? The answer we suggest is that it would indeed be possible.

It is a characteristic of contemporary universities in the English-speaking world and many other countries that they are arranged in a fairly strict hierarchy with those at the top exercising a great deal of power and influence over the system as a whole. The first step in establishing A would therefore have to be that of the supporters of A gaining a majority of the positions in the top universities, and particularly of the senior positions such as full professorships. Once this is achieved, then establishing A in the system as a whole becomes relatively easy. Most academics aspire to a position in a top university which is not only more prestigious but offers better conditions such as a higher salary and more research time. Once the supporters of A have gained control of the top universities, it will become clear to any aspiring academic in the field (including graduate students just starting research) that they have a much better chance of getting to a good position if they adopt A rather than B. This will be a strong incentive for adopting A. Moreover the top universities exercise a great deal of control over appointments in other universities. For example, most of those who get lectureships will have done their graduate work in one of the top universities and so will have been trained in A rather than B. Of course there will inevitably be a few obstinate characters who adopt B rather than A. However their fate is likely to be an uncomfortable one. To begin with, they may fail to get a university job at all, and, if they do get a job in a university, it is likely to be one low down in the hierarchy. In such lowly universities, the staff have much worse conditions, usually having to teach for much longer hours and having much less research time. Thus the number of research hours available for research on B will be much less than those available for research on A, which makes progress in B less likely even if it is really the better theory.

However this does not end the methods available for ensuring the triumph of A. We have not yet spoken of the control of peer-reviewed journals. Contemporary journals, like contemporary universities, are arranged in a strict hierarchy. Once the supporters of A have established themselves in the top universities, they will find it easy to acquire the editorships of the most prestigious journals. Any papers submitted can then be sent for refereeing to the friends of the editor, i.e. supporters of A. So a paper which is based on theory B is very likely to be rejected. Thus the supporters of theory B will find that they are unable to publish in the most prestigious journals, but only in the less prestigious ones. Confined to low prestige universities and publishing in low prestige journals, it will naturally be concluded that their research is no good. If there is a research assessment exercise (as in Britain), their rating will be low, and therefore they will have their research time cut still further, and might even be sacked for incompetence. Given the grim fate which is likely to hang over the supporters of B, it is only to be expected that the vast majority of researchers will adopt A, and that a normal science based on A will be established. Our conclusion then is that it would be relatively easy in many contemporary universities, to establish a normal science by purely political means.

Our argument was that this is easy to do where there is a hierarchical ranking of the universities. If the universities were more equal with similar conditions of work and levels of prestige throughout, then it would be much harder to use the political methods just described.

So far we have spoken of contemporary universities, but, at other times, other less 'gentle' methods have been available for establishing a normal science by purely political methods. Dissident supporters of a rival approach to the dominant paradigm could have been handed over to the inquisition or sent to a labour camp.

It is, and has been, therefore eminently possible to establish, by purely political means, a dominant paradigm which has little or no empirical confirmation (or at least much less empirical confirmation than some rival), but which would nonetheless become the basis of a normal science research tradition. However, this is to speak hypothetically. We can still ask whether this possibility has ever actually occurred or is actually occurring. Here we enter a speculative and controversial area, about which there is likely to be disagreement. However, three possible examples of a normal science established by political methods do suggest themselves. First of all the Ptolemaic theory was the basic paradigm for astronomy among the Jesuits in the 17th century. Secondly Lysenkoism was the basic paradigm for research in biology in the Soviet Union under Stalin[9]. Thirdly neo-classical economics is the basic paradigm for economics in most contemporary universities. Fullbrook [2004] is a recent collection of 27 essays by different authors who criticize neo-classical economics from many points of view. These criticisms establish that there is indeed little or no empirical confirmation for this theory. Indeed because of the lack of realism of its basic assumptions, one could say that the degree of confirmation of the theory is negative. Yet neo-classical economics is unquestionably the dominant paradigm in most economics departments throughout the world.

These examples show that paradigms with little or no empirical confirmation, or at least with a degree of empirical confirmation much less than some rival, can indeed be established by political methods. We can agree with Feyerabend that normal science research founded on such a paradigm is unlikely to be fruitful. On the other hand if a paradigm comes to be accepted for empirical reasons, and has a degree of corroboration which is not only high but very much higher than any rival, then a normal science research tradition founded on such a paradigm may very well prove very fruitful, as was the case with research founded on the Newtonian paradigm in the period c. 1700 to c. 1900. In such a case Kuhn seems to be right and Feyerabend wrong.

At this point it might be objected that it is a little naïve to distinguish so sharply between paradigms which are accepted for empirical reasons, and those which are established by political methods. Surely, it could be said, that a mixture of the two processes occurs in most cases. Of course there is some truth in this, but it does not require a strong modification of the position here advocated. Strictly we should speak of paradigms which are accepted for predominantly empirical reasons as against those which are established principally by political methods. Our judge-

[9] A good account of Lysenkoism is to be found in Sheehan [1985].

ment is that most paradigms which have been established historically fall into one of these two categories, and there is not in reality a spectrum of intermediate cases. The reason for this is that the scientific community, if left to itself and not influenced by ideological/political factors originating from outside science, will accept paradigms predominantly for empirical reasons. Thus we can speak of science as a rational enterprise whose rationality is occasionally disturbed by powerful ideological/political currents coming from outside science. In the three examples we gave earlier of paradigms which were perhaps established purely by political means, it is clear that ideological/political factors were acting strongly. The Jesuits in the 17th century continued to hold to the Ptolemaic paradigm for astronomy for religious reasons because the Copernican paradigm was held to contradict the teachings of the Catholic Church. Stalin, who of course was quite ignorant of biology, took a liking to Lysenko's ideas, and this was sufficient for these ideas to become the dominant paradigm in the Soviet Union. As for neo-classical economics, its principal function is to justify contemporary neo-conservative economic policies. It is thus not surprising, in view of recent political trends, that it has come to be the dominant paradigm among economists, despite its empirical weaknesses.

Let us now turn to a key question which arises in this connection. We have drawn the distinction between paradigms accepted for empirical reasons, and those established by political methods. However this distinction is only valid if there can indeed be empirical reasons for accepting a paradigm, which, as we argued earlier, is equivalent to saying that a theory of confirmation or corroboration for scientific theories can be developed? But is it really possible to develop such a theory? Both the earlier logical tradition of the Vienna Circle and also Popper did accept the possibility of confirmation theory. Admittedly, as we have seen, Hempel produced some paradoxes of confirmation such as the famous paradox of the ravens. However, this certainly did not lead him, or others of the same way of thinking, to conclude that a confirmation theory could not exist. After all, in deductive logic, paradoxes such as Russell's paradox had arisen, but such paradoxes had not proved fatal to deductive logic. On the contrary, ways round these paradoxes, such as the theory of types or axiomatic set theory had been developed and seemed to function well. In the same way, it was held that the paradoxes of confirmation could also be overcome. Even Popper, despite his criticism of the Vienna Circle, accepted the existence of what he called: 'corroboration'. Popper used a different term from 'confirmation' to distinguish his theory from that the confirmation theory of Carnap. In this paper, we are using 'confirmation' and 'corroboration' as synonyms, and so prefer to speak of Popper having a different theory of confirmation (or corroboration) from Carnap's theory of confirmation (or corroboration). Certainly there were differences between the two theories. Carnap was a Bayesian and held that the confirmation function $C(h, e)$ satisfied the axioms of probability, while Popper held that $C(h, e)$ did not satisfy these axioms. There were other differences besides. Yet the two thinkers did both accept the possibility of developing a confirmation theory of some kind. If we turn to Kuhn, we find that in his [1962] he appears to reject the possibility of a confirmation theory.

Kuhn brings up the question in Chapter XII of his [1962] which is entitled: 'The Resolution of Revolutions'. Here Kuhn considers first Bayesian confirmation theories which he calls: 'probabilistic verification theories'. He has this to say about them:

> Few philosophers of science still seek absolute criteria for the verification of scientific theories. Noting that no theory can ever be exposed to all possible relevant tests, they ask not whether a theory has been verified but rather about its probability in the light of the evidence that actually exists. ... In their most usual forms, however, probabilistic verification theories all have recourse to one or another of the pure or neutral observation-languages ... If, as I have already urged, there can be no scientifically or empirically neutral system of language or concepts, then the proposed construction of alternate tests and theories must proceed from within one or another paradigm-based tradition. Thus restricted it would have no access to all possible experiences or to all possible theories. As a result, probabilistic theories disguise the verification situation as much as they illuminate it. [1962, 144–5]

Kuhn then goes on to consider Popper's views. He first makes the point that what he calls 'anomalies' have some points in common with what Popper calls 'falsifications'. However, Kuhn then continues:

> If any and every failure to fit were ground for theory rejection, all theories ought to be rejected at all times. On the other hand, if only severe failure to fit justifies theory rejection, then the Popperians will require some criterion of "improbability" or of "degree of falsification." In developing one they will almost certainly encounter the same network of difficulties that has haunted the advocates of the various probabilistic verification theories. [1962, 145–6]

These passages show that Kuhn in 1962 was very doubtful about the possibility of a confirmation theory either of the Bayesian or the Popperian kind.

Moreover if a confirmation theory is going to be useful in analysing scientific revolutions, it would be necessary to compare the degrees of confirmation given the evidence of the two competing paradigms. But can paradigms be compared in this way? Kuhn held that two competing paradigms are incommensurable, and this suggests that he would deny that they could be compared as to their respective degrees of confirmation (or corroboration) given the available evidence. In fact the Collins Dictionary of the English Language defines incommensurable as follows: 'incapable of being judged, measured or considered comparatively.' If this is what Kuhn meant by incommensurable, then it would follow that two paradigms could not be compared with regard to their degrees of empirical confirmation, and that consequently there could not be empirical reasons for preferring one to the other.

It would seem to follow from this that a new paradigm could only be established by political methods. Now, as we shall see when we come to discuss the incommensurability problem in section 3.4, Kuhn later denied that he had ever intended to use 'incommensurable' in such an extreme sense. However, established English usage certainly suggested that that was what he meant, and he was interpreted in this way by many of his contemporaries, including notably Lakatos who described Kuhn's position as follows:

> For Kuhn scientific change — from one 'paradigm' to another — is a mystical conversion which is not and cannot be governed by rules of reason and which falls totally within the realm of the *(social) psychology of discovery.* Scientific change is a kind of religious change. [1970, 9]

This was elaborated later in the same paper as follows:

> ... a new 'paradigm' emerges, incommensurable with its predecessor. There are no rational standards for their comparison. Each paradigm contains its own standards. The crisis sweeps away not only the old theories and rules but also the standards which made us respect them. The new paradigm brings a totally new rationality. There are no super-paradigmatic standards. The change is a bandwagon effect. Thus *in Kuhn's view scientific revolution is irrational, a matter for mob psychology.* [1970, 90–1]

Naturally Lakatos does not approve of Kuhn's position as thus interpreted. He (Lakatos) has this to say about it:

> If even in science there is no other way of judging a theory but by assessing the number, faith and vocal energy of its supporters, then this must be even more so in the social sciences: truth lies in power. Thus Kuhn's position vindicates, no doubt, unintentionally, the basic political *credo* of contemporary religious maniacs ('student revolutionaries').[10] [1970, 9–10]

Lakatos was determined to struggle against what he saw as Kuhn's irrationalism by developing:

> [a] position which, I think, may escape Kuhn's strictures and present scientific revolutions not as constituting religious conversions but rather as rational progress. [1970, 10]

In the next section, we will examine how Lakatos carried out this project, and how successful he was in achieving his goal of defending rationality.

---

[10]It is interesting that in this passage written in the late 1960s, Lakatos regards 'student revolutionaries' as the obvious contemporary examples of 'religious maniacs'. *Tempora mutantur.*

3.3 Lakatos and the Methodology of Scientific Research Programmes

Lakatos based his new approach to methodology on the concept of *research programme.* This had already been used for the purposes of analysing science within the Popperian school. In his [1934], Popper had defended the meaningfulness of metaphysics against the claim by the Vienna Circle that metaphysics is meaningless. One of Popper's most striking examples to support this thesis was that of *atomism.* Atomism was first introduced in the West by the pre-Socratic thinkers Leucippus and Democritus. It continued as a powerful trend in the ancient world with Epicurus in Greece and Lucretius in Rome. This ancient atomism was surely meaningful, but also definitely metaphysical.

Ancient atomism was revived in Western Europe in the seventeenth century, and discussed by the leading scientists of the day, but it was still at that time metaphysical rather than scientific. It was not till the nineteenth century with the work of Dalton, Maxwell and Boltzmann that atomism became scientific. These scientists were however influenced by the earlier metaphysical atomism, which shows that metaphysics can be, not only meaningful, but also helpful to science.

In his [1983] *Realism and the Aim of Science*, Popper develops his views on metaphysics by introducing the concept of a *metaphysical research programme* for science. Thus he says:

> ... atomism is an excellent example of a non-testable metaphysical theory whose influence upon science has exceeded that of many testable theories. [1983, 192]

And, after giving some further examples of metaphysical theories which have influenced science, he continues:

> Each of these metaphysical theories served, before it became testable, as a research programme for science. It indicated the direction of our search, and the kind of explanation that might satisfy us; and it made possible something like an appraisal of the depth of a theory. [1983, 192–3]

Although not published until 1983, this was written in 1956, and undoubtedly influenced Lakatos in the development of his new ideas on methodology. Lakatos, however, changed Popper's concept of metaphysical research programmes to that of *scientific research programmes.*

Lakatos uses two notions to characterise a scientific research programme. These are the *hard core* or *negative heuristic*; and the *positive heuristic.* We will deal with them in turn.

Lakatos explains the notion of *hard core* as follows:

> All scientific research programmes may be characterized by their '*hard core*'. The negative heuristic of the programme forbids us to direct the *modus tollens* at this 'hard core'. Instead, we must use our ingenuity to

> articulate or even invent 'auxiliary hypotheses', which form a *protective belt* around this core, and we must redirect the *modus tollens* to *these.* [1970, 48]

He then goes on to give the following example:

> In Newton's programme the negative heuristic bids us to divert the *modus tollens* from Newton's three laws of dynamics and his law of gravitation. This 'core' is 'irrefutable' by the methodological decision of its proponents: anomalies must lead to changes only in the 'protective' belt of auxiliary, 'observational' hypotheses and initial conditions. [1970, 48]

Here Lakatos is influenced by the Duhem thesis. Let T stand for the conjunction of Newton's three laws of motion and the law of gravitation, and A for some auxiliary assumptions. Duhem pointed out that we cannot derive observational results from T alone, but only from the conjunction of T and A. If O is derived, and not-O is observed, then we have the choice of changing T or A. Lakatos suggests that those working on the Newtonian programme should decide in advance to change A, and not T.

Let us now turn to the second of Lakatos' characterising notions — that of *positive heuristic.* Here are two passages in which he describes this concept.

> ... the positive heuristic consists of a partially articulated set of suggestions or hints on how to ... develop ... the research-programme ...' [1970, 50]

> Positive heuristic is thus in general more flexible than negative heuristic. ... It is better therefore to separate the 'hard core' from the more flexible metaphysical principles expressing the positive heuristic. [1970, 51]

We see here the influence on Lakatos of Popper's ideas about the possibility of metaphysical ideas helping science forward.

This then is a brief outline of Lakatos' concept of scientific research programme. Let us now turn to considering some criticisms of the notion. The first question which could be raised is whether the concept of scientific research programme really differs from that of paradigm. After all, Lakatos' Newtonian research programme with its hard core looks very like Kuhn's normal science based on the Newtonian paradigm. Has Lakatos done no more than express Kuhn's ideas in a different terminology? Our earlier discussion and attempted clarification of the concept of paradigm shows that this is not the case and that the two notions really are different. Moreover it makes clear how they differ.

A paradigm, we argued, consists of the assumptions shared by all those working in a given branch of science at a particular time. Historians can reconstruct the paradigm of a specific group at a particular time by studying the text-books used to instruct those wishing to become experts in the field in question. Thus a paradigm is what is common to a whole community of experts in a particular field at a particular time. By contrast only a few of these experts (or in the limit only one) may be working on a particular research programme. Characteristically only a handful of vanguard researchers are working on a specific research programme at a particular time. Historians who wish to reconstruct a research programme will look, not at textbooks in wide circulation, but at the writings of a few key figures. They will examine the notebooks, the correspondence, and the research publications of these leading figures, and, in this way, reconstruct the programme on which they were working. This shows clearly how research programmes differ from paradigms.

In a moment we would like to defend the concept of scientific research programme still further by arguing that it is not only different from the concept of paradigm, but is needed in addition to the concept of paradigm in order to give an adequate analysis of scientific revolutions. Before doing so, however, we would like to make a couple of further criticisms of the notion of scientific research programme which, we think, can result in some modifications and improvements of the concept.

The first of these criticisms is directed against the notion of hard core, and, oddly enough, is based on the example given earlier from Lakatos' paper: 'Newton's Effect on Scientific Standards.' It will be remembered that this paper, though first published in 1978, was largely written in the years 1963–4 when Lakatos' ideas were closer to Popper's than they became later on. In this paper Lakatos discusses the work of Clairaut and Euler in the 18th century. These two scientists would presumably have been working on what in Lakatos' terminology could be characterised as the Newtonian scientific research programme. Yet Clairaut in 1746 tried to explain the motion of the Moon's apogee by changing Newton's law of gravity, and a similar strategy was followed by Euler in 1770. So both Clairaut and Euler suggested changing the hard core of the programme rather than the auxiliary assumptions — contrary to the methodology of scientific research programmes. Admittedly these suggestions did not prove successful in the long run, but there is no *a priori* reason why they should not have succeeded.

Although the Clairaut and Euler example contradicts Lakatos' methodology of scientific research programmes, it does not perhaps constitute a very severe counter-example. Kvasz in his discussion of Lakatos' methodology [2002, 236] divides the hard core into a series of layers like an onion. The changes of Clairaut and Euler were, in Kavasz's terminology, *re-formulations* only affecting the outer layer of the onion. This suggests that some modification of Lakatos' approach is needed, but not too drastic a one.

Another general argument in the same direction is that the methodological decision which Lakatos recommends of rendering the hard core 'irrefutable' contradicts the open-mindedness, and lack of dogmatism, which should characterise the good researcher.

For these reasons, we suggest replacing the notion of the *hard core* of a programme by that of the *aim* or *goal* of the research programme. After all, scientific research is a conscious human activity, and so has a goal. Thus we can say that Clairaut and Euler were working on a research programme whose aim was to explain the motion of heavenly bodies using mechanical laws. The concept of the aim of a research programme is related to that of hard core, but is less dogmatic. It is possible to have an aim or goal without being certain that one can attain it, and it is moreover always possible to change the aim of an activity when the original aim is shown to be impossible.

This suggested change from 'hard core' to 'aim' is further supported by the fact that many notable scientific research programmes do not appear to have had anything resembling a hard core. A good example of this is the research programme which Lavoisier began around 1772. We are fortunate in having Lavoisier's own description of his research programme in a memorandum which he wrote probably on 20 February 1773.[11] Here he speaks of ' ... the long series of experiments that I intend to make on the elastic fluid that is set free from substances, either by fermentation or distillation or in every kind of chemical change, and also on the air absorbed in the combustion of a great many substances ... ...' There is nothing here at all like a 'hard core' for the research programme, but the programme certainly had an aim, because Lavoisier writes that these experimental investigations are 'in order to link our knowledge of the air that goes into combination or that is liberated from substances, with other acquired knowledge, and to form a theory.' Moreover he also says: 'The importance of the end in view prompted me to undertake all this work, which seemed to me destined to bring about a revolution in physics and chemistry.'

Another feature of Lavoisier's research programme is that he seems to have been largely uninfluenced by metaphysical considerations — unlike other scientists such as Kepler. Thus we could identify the positive heuristic of his programme as consisting of a range of experimental techniques and apparatus such as the pneumatic trough, burning glasses, furnaces, electric sparks, balances, etc. This example leads to our second criticism of Lakatos which is that, in his notion of positive heuristic, he is perhaps over-influenced by Popper's emphasis on metaphysics. As well as the 'metaphysical principles' mentioned by Lakatos, the positive heuristic could contain other things such as mathematical and experimental techniques.

Having suggested a few modifications in the concept of scientific research programme, we will now argue that this concept is needed *in addition to* that of paradigm in order to explain how paradigms come into existence. Kuhn has a rather romantic theory that a new paradigm is born in a flash of intuition. As he puts it:

> ... normal science ultimately leads only to the recognition of anoma-

[11] The quotations from Lavoisier's memorandum are taken from the English translation in [McKie, 1935, 120–3]. There has been some scholarly discussion about the date of this memorandum because it is clearly dated February 20, 1772, but is written on the opening pages of a laboratory note-book, dated from February 20 to August 28, 1773. We have adopted McKie's date of 1773 in view of the convincing arguments he presents for it in his [1935, 123–4].

> lies and to crises. And these are terminated, not by deliberation and interpretation, but by a relatively sudden and unstructured event like the gestalt switch. Scientists then often speak of the "scales falling from the eyes" or of the "lightning flash" that "inundates" a previously obscure puzzle, enabling its components to be seen in a new way that for the first time permits its solution. On other occasions the relevant illumination comes in sleep. No ordinary sense of the term "interpretation" fits these flashes of intuition through which a new paradigm is born. [1962, 121–2]

Now, there may indeed be a few cases in which paradigms are born in something like this fashion. The most convincing example is one suggested by Arthur Miller. If we regard the Bohr atom as a paradigm and quantum mechanics as the new paradigm which replaced it, then it does indeed seem that the basic ideas of the new quantum mechanics came to Heisenberg, if not in a 'lightning flash', then at least in a few months of feverish inspiration.[12] In general, however, a new paradigm is fashioned over a much longer period of time, and by a process which may involve flashes of inspiration, but which may also involve long periods of systematic and painstaking research. It is usually, in fact, work on research programmes by small groups, or, in the limit a single individual, which gives rise to a new paradigm.

Consider the case of Copernicus. He introduced a new research programme whose aim was to explain the motion of the heavenly bodies on the assumption that the Earth rotated on its axis, and moved once a year round a stationary Sun. Copernicus was indeed influenced by metaphysical ideas, more specifically by Pythagoreanism and Neo-Platonism which were both popular during the Renaissance period in which he lived. However, the positive heuristic of his programme contained some technical considerations. Copernicus used epicycles but deliberately eschewed the equants which had been used by Ptolemy. Copernicus' research programme was certainly not a paradigm, i.e. part of the preliminary 'text-book' instruction received by scientists training in the field. Indeed he was the only scientist working on his research programme. The theory which resulted from his long years of research did not become a paradigm either, though it was taken up by a few of the scientists working in the field. After Prolemy, the next paradigm to become generally accepted was the Newtonian, and this new paradigm, although it did contain Copernicus's heliocentric assumption, was in other ways, quite different from anything that Copernicus could have imagined. Moreover considerable work on further research programmes — those of Kepler, Descartes, Galileo, and Newton himself — were necessary before Copernicus' theory could be transformed into the new Newtonian paradigm. This example shows clearly that the concept of scientific research programme differs from that of paradigm, and that we need the concept of scientific research programme in order to explain how paradigms come into existence.

Thus far we have defended Lakatos' concept of scientific research programme —

[12]Some details about this example are to be found in [Miller, 1986, 127–43 & 248–54].

albeit in a somewhat modified form. However, we have defended this concept as filling a gap in Kuhn's account rather than as a replacement for Kuhn's account. It was of course the latter which Lakatos himself intended. As we explained earlier Lakatos held that [1970, 91] '... *in Kuhn's view scientific revolution is irrational, a matter for mob psychology.*', and that for Kuhn, scientific change occurs according to the 'political *credo*' that 'truth lies in power' [1970, 10]. Lakatos saw himself as defending the rationality of science against such doctrines. But did he succeed in this defence? This is the question which we must next consider.

In order to defend the rationality of science, Lakatos has to formulate some rational criteria for preferring one scientific research programme to another. This he does by distinguishing between *progressive* and *degenerating* research programmes. Progressiveness has both a theoretical and empirical character, and, as regards the empirical side, Lakatos stressed the exclusive importance of the production of *novel facts*. Thus he writes:

> The time-honoured empirical criterion for a satisfactory theory was agreement with the observed facts. Our empirical criterion for a series of theories is that is should produce new facts. *The idea of growth and the concept of empirical character are soldered into one.* [1970, 35]

and again [1970, 38]: 'the only relevant evidence is the evidence anticipated by a theory'.

Lakatos' views on novel facts were, however, criticized in a decisive fashion by his former pupil and then colleague at the London School of Economics — Zahar. Zahar wrote:

> Lakatos mentions the return of Halley's comet as a new fact anticipated by the Newtonian programme and, of course, I agree with him that the discovery of any new type of fact is the discovery of a novel fact. But, if we equate novelty simply with *temporal* novelty, we are driven into a paradoxical situation. We should, for example, have to give Einstein no credit for explaining the anomalous precession of Mercury's perihelion, because it had been recorded long before General Relativity was proposed. Similarly, we should have to say, contrary to informed opinion, that Michelson's experiment did not confirm Special Relativity and Galileo's experiments on free fall did not confirm Newton's theory of gravitation. Lakatos, who does not easily dismiss the judgements of physicists, is aware of this difficulty and tries to avert it by shifting his original view and saying that, in the light of a new theory, some known facts may 'turn into' novel ones. For example, whereas Balmer merely 'observed' that the hydrogen lines obey a certain formula, Bohr connected these lines with the energy levels of the electron in the hydrogen atom.
>
> However, Lakatos's modified notion of 'novel fact' is open to the following fatal objection. Any theory is a set of propositions connecting

> different terms and relations. We can always define the properties of a physical entity like mass through the relations which 'mass' bears to other concepts and notions within the theory. Consequently a new hypothesis will generally ascribe new meanings to old terms. For instance, any experimental consequence of relativity theory involving say mass, would trivially become the expression of a novel fact. Thus the fact that a steel ball rolling down a slope takes a certain time to reach the bottom, could become a novel fact when the steel ball is considered as having relativistic mass. This is obviously absurd. Therefore Lakatos's 1970 criterion for novelty is too liberal, while his 1968 criterion is too stringent. [1973, 101–2]

Zahar, however, is not just critical but makes a positive suggestion as to how the difficulty could be overcome. He proposes a redefinition of 'novel fact' which he states as follows:

> A fact will be considered novel with respect to a given hypothesis if it did not belong to the problem-situation which governed the construction of the hypothesis. ... Temporal novelty in a research programme is then a sufficient but not a necessary condition for novelty. [1973, 103]

The advantage of this definition of novel fact is that it allows us to say that the anomalous precession of Mercury's perihelion was a novel fact for Einstein's General Relativity because it was not part of the problem-situation which led Einstein to construct his new theory, or, in other words, it was not part of the heuristic of Einstein's programme. Lakatos accepted Zahar's modification of the concept of novel fact, but there do appear to be difficulties with Zahar's concept of novel fact as well. Zahar points out some of the consequence of his new concept as follows:

> This new criterion for novelty of facts also implies that the traditional methods of historical research are even more vital for evaluating experimental support than Lakatos had already suggested. The historian has to read the private correspondence of the scientist whose ideas he is studying; his purpose will not be to delve into the psyche of the scientist, but to disentangle the heuristic reasoning which the latter used in order to arrive at a new theory. Let us give an example. In Newton's time there was a well-known inverse square law for the intensity of light, Newton might have used some reasoning by analogy in order to propose that the gravitational 'intensity' is also distributed over the surface of a sphere and hence obeys an inverse square law; in this case Kepler's laws would support gravitational theory more strongly than if Newton had used them as his heuristic starting point. [1973, 103–4]

These points of Zahar's have rather counter-intuitive consequences. Suppose there had been another scientist (Dupont say) who was a contemporary of Newton and working like Newton on gravitational theory. We know that Newton used Kepler's laws as part of his heuristic, but let us suppose that Dupont was familiar with large parts of Descartes and Galileo, but, for some curious reason, quite ignorant of Kepler's work. Dupont used the analogy with light suggested by Zahar to develop the mechanics of Descartes and Galileo, and, in this way arrived at exactly the same theory as Newton. Only after formulating the theory did Dupont discover Kepler's work, and, being a man of great genius, he quickly showed that Kepler's laws in an approximate form followed from his new mechanics. According to Zahar, Dupont's theory would be much better supported by the evidence than Newton's theory — even though the two theories are identical. This seems to constitute a rather severe difficulty.

Moreover, in general terms, it seems rather questionable whether the scientific community needs to investigate the private correspondence of scientists in order to decide whether their theories are well-supported by experiment and observation. Suppose, for example, that a historian discovers a hitherto unknown notebook of Einstein's which reveals that Einstein spent several months considering possible explanations of the anomalous precession of Mercury's perihelion, and that this research was actually crucial for his later development of General Relativity. According to Zahar, this historical discovery should lead the scientific community to lower the empirical confirmation they have hitherto accorded to General Relativity. Surely, however, this would not be the case.

In the light of all this, there still seem to be unresolved problems concerning the notion of novel fact used in Lakatos' methodology of scientific research programmes. Let us leave these problems aside for the moment, however, as we have to consider some further difficulties. In Lakatos' framework, scientists have to choose on rational grounds between competing research programmes — R_1 and R_2 say. Now suppose that R_1 is degenerating and R_2 progressing, then it would seem rational for scientists to choose R_2 in preference to R_1. But now comes the difficulty. It could happen that R_1 having degenerated for a while, suddenly turns the corner and starts progressing again, while the opposite occurs with R_2. R_2 having been progressing, suddenly loses its momentum and begins to degenerate. Then the choice of R_2 rather than R_1 would turn out to have been the wrong one. Lakatos was aware of this difficulty and responded by proclaiming the end of instant rationality. He writes:

> It is very difficult to decide ... when a research programme has degenerated hopelessly; or when one of two rival programmes has achieved a decisive advantage over the other. There can be no "instant rationality". [1974, 149]

During the period 1968-74, Lakatos discussed the question of the rationality of science and other issues in the philosophy of science continually with Feyerabend with whom he was very friendly. Luckily their correspondence in these years has survived and has been published by Motterlini in his [1999], together with Lakatos' last lectures on scientific method. To simplify one could say that in the discussions between Lakatos and Feyerabend in this period, Lakatos attempted to defend the rationality of science whereas Feyerabend argued for its irrationality. In his 1975 book: *Against Method*, Feyerabend argued for the principle (p. 28): '*anything goes*', and the title of his 1987 book was: *Farewell to Reason*. This gives a good general idea of Feyerabend's position, while we have already seen that Lakatos was concerned to defend the rationality of scientific change against a threat thereto which he saw as coming from Kuhn. Those who are interested in details of the discussions between Feyerabend and Lakatos should consult the material in Motterlini's volume, including Motterlini's own admirable account of the controversy. Here we want to pick out just one point — that regarding the issue of the end of instant rationality. Feyerabend took up this issue in his criticism of Lakatos in 1975, where he writes:

> Considering a research programme in an advanced state of degeneration one will feel the urge to abandon it, and to replace it by a more progressive rival. This is an entirely legitimate move. *But it is also legitimate to do the opposite* and to retain the programme. ... it is ... unwise to reject research programmes on a downward trend because they might recover and might attain unforeseen splendour ... Hence, one cannot *rationally* criticize a scientist who sticks to a degenerating programme and there is no *rational* way of showing that his actions are unreasonable. ... the arguments that established the need for more liberal standards make it impossible to specify conditions in which a research programme *must* be abandoned, or when it becomes *irrational* to continue supporting it. *Any* choice of the scientist is rational, because it is compatible with the standards. 'Reason' no longer influences the actions of the scientist. [1975, 185–6]

Somewhat reluctantly perhaps one has to admit that Feyerabend gets the better of the argument here. His criticism, and the difficulties connected with the concept of novel fact which were described earlier, together show that Lakatos did not, with his methodology of scientific research programmes, succeed in what he had hoped to do, namely to construct a defence of the rationality of scientific change. Lakatos' failure in this respect is connected, in our view, with a feature of his thinking which we will now discuss.

One of the basic distinctions frequently made in the philosophy of science is between the *discovery* of scientific hypotheses and their *justification*. Most philosophers of science (including the present authors) accept this distinction, but a minority do challenge it, and Lakatos was among that minority. This is shown in the fact that Lakatos tried to reduce the problem of the *appraisal* of knowledge (a matter of justification) to that of the *growth* of knowledge (a matter of discovery). Lakatos puts forward this position in the following passage:

> But then two new problems arose. The *first* problem was the *appraisal of conjectural knowledge. ...* The *second* problem was *the growth of conjectural knowledge. ...*
>
> In this situation *two schools of thought emerged.* One school — *neoclassical empiricism* — started with the first problem and never arrived at the second. The other school — *critical empiricism* — started by solving the second problem and went on to show that this solution solves the most important aspects of the first too. [1968, 132–3]

Our own position in contrast to Lakatos' is that we must solve both problems — that of the *appraisal* of knowledge, and that of the *growth* of knowledge. Although the two problems are connected, they are distinct nonetheless, and a solution to the second problem does not solve the most important aspects of the first too. It was the erroneous assumption that it does, which, in our view, is the root cause of Lakatos' failure to defend successfully the rationality of scientific change. The same cause is also responsible for the difficulties in Zahar's development of Lakatos' position.

Our claim then is that philosophers of science have to develop not only a theory of the growth of science, but also a theory of the appraisal of scientific hypotheses. Lakatos' theory of scientific research programmes constitutes, in a modified form, an important contribution to the theory of the growth of science. However, it does not contribute to the theory of appraisal of scientific hypotheses. For that one needs to develop a theory of the confirmation or corroboration of scientific hypotheses by evidence — a theory which cannot be based on the methodology of scientific research programmes. In fact the appraisal of a scientific research programme as either progressing or degenerating may give little indication as to whether the theories of which it is composed are well-confirmed or strongly disconfirmed. To see why this is so, let us define two scientific research programmes:

$$\mathbf{R}_1 = (T_1, T_2, \ldots\ T_n)$$
$$\mathbf{R}_2 = (S_1, S_2, \ldots\ S_m)$$

The following situation is possible:

1. $\mathbf{R}_1$ makes very good progress, but the theory T_n is not very well confirmed. This case occurs (for n small) in the initial stages of many programmes, for example Bohr's programme.

2. $\mathbf{R}_2$ degenerates, but S_m has a very high degree of confirmation. An example of this case is given by the hidden variables programme in quantum mechanics. The attempts to replace standard quantum mechanics by a new and better theory have hitherto failed. So $S_1 = S_m$ = standard quantum mechanics, and this shows a total stagnation of the research programme. But S_m has a very high degree of confirmation, partly because of those experiments, for example Aspect's experiment, which were carried out in the context of work on $\mathbf{R}_2$.

Examples of this sort show that it is not possible to base a theory of the confirmation of scientific theories by evidence on the methodology of scientific research programmes. Lakatos' mistaken attempt to reduce the appraisal of scientific theories to the theory of the growth of science is, we think, the consequence of a notable aspect of his writings. Lakatos gives many examples of scientific and mathematical discoveries, but he never mentions the practical applications of science and mathematics. Reading only Lakatos, a Martian would have the impression that mathematics and science are intellectual amusements for humans similar to novels. He, she (or it!) would not be in the position to guess that mathematics and science are used in industry and commerce. Modern science, however, depends for its existence on the continual application of mathematics and science.

For the satisfactory application of science, we need a theory of the appraisal of scientific hypotheses which does not involve detailed considerations of how those hypotheses are discovered. To give just one obvious example, it is not permitted to sell a new medicine until the hypothesis that it is effective but has no harmful side effects has been very well confirmed. Indeed most governments specify the tests which must be carried out with satisfactory results before a company is allowed to put a new medicine on the market. Pharmaceutical companies are not, however, required to make public the heuristic research strategies which led them to their discoveries, and indeed they make every effort to keep these strategies secret.

Turning now to Zahar, we can agree that reading the private correspondence of Dupont and Newton may be highly relevant for uncovering the heuristic strategies which led them to their theories. However, given that they discovered the same theory (as in our hypothetical example), then these heuristic strategies would not be relevant to evaluating the experimental support for that theory. The scientific community would have to consider the degree to which this new theory is confirmed in the light of the observations and experimental results in the public domain, and to consider what further observations might be made and experiments carried out in order to test the new theory severely.

In fact the methodology of scientific research programmes, contrary to Lakatos' intention, would make science very vulnerable to manipulation by politics and ideology coming from outside the scientific community. Let us consider again two scientific research programmes $\mathbf{R}_1$ and $\mathbf{R}_2$ as defined above. This time, however, let us suppose that $\mathbf{R}_2$ has been making much better progress than $\mathbf{R}_1$, and that progress and confirmation are in agreement so that S_m is much better confirmed empirically than T_n. Suppose, however, that some powerful group favours the approach of $\mathbf{R}_1$ and dislikes that of $\mathbf{R}_2$ for political and ideological reasons. This group might use political methods, such as we described earlier, to ensure that almost all further research is done in $\mathbf{R}_1$ rather than $\mathbf{R}_2$. As a result the researchers in $\mathbf{R}_1$ produce a series of new theories $T_{n+1}, \ldots, T_{n+r}$. There is some slight improvement here so that by the criteria of the methodology of scientific research programmes, $\mathbf{R}_1$ can be said to be progressing slightly. Meanwhile, because almost no researchers are working on $\mathbf{R}_2$, no new theories are produced in that programme and it remains stuck at S_m. $\mathbf{R}_2$ would therefore be stagnant and $\mathbf{R}_1$ progressing.

So Lakatos and his followers would have to judge that $\mathbf{R}_1$was superior to $\mathbf{R}_2$ even though this result had clearly been brought about by political manipulation, and even though it might still be the case that S_m was much better corroborated by the evidence than T_{n+r}.

If however the scientific community has well established criteria for when and to what extent a theory is confirmed or disconfirmed by evidence, it could resist such political manipulation. Scientists could object that it is obviously unfair to give all the research funding to one research programme when another research programme has produced theories which are better confirmed empirically. This example shows once again that the defence of scientific rationality requires the development of a theory of empirical confirmation or corroboration, and that questions about confirmation are distinct from those concerning the progress, stagnation or degeneration or research programmes.

Thus our conclusion is that Lakatos did not solve the problem he set out to solve, that is the problem of whether scientific revolutions can be rational. However, he did create the tools for solving another problem namely the problem of how new paradigms are created, since new paradigms almost always come into existence as the result of work on one or more research programmes. This may seem a strange claim to make about Lakatos' ideas, but it is surprisingly in accordance with something Lakatos himself wrote, namely:

> After Columbus one should not be surprised if *one does not solve the problem one has set out to solve.* [1963-4, 90]

Our discussion has also led to the conclusion that the solution to the problem of the rationality of scientific revolutions (if there is one) must lie in the development of a theory of empirical confirmation or corroboration. This is a very major task and one which lies outside the scope of the present article. However, we can now consider one issue which arose in the 1960 and early 1970s and which we have ignored until now. That is the question of incommensurability. If 'incommensurable' is taken in the strong sense in which it is defined in the Collins English Dictionary, and if a new paradigm is incommensurable with an old one, then it would not be possible to compare the old and new paradigms as to their respective degrees of empirical confirmation. The whole project of defending the rationality of scientific revolutions in terms of confirmation theory would collapse. But is it really the case that in some or all scientific revolutions, the new paradigm is incommensurable with the old one in such a strong sense? This is something which we will investigate in the next sub-section.

3.4 The Incommensurability Problem

The concept of incommensurability (as applied to scientific theories) was introduced and developed by Feyerabend and Kuhn. This introduction was not independent, and indeed the concept emerged from discussions between them when they were both in the philosophy department in Berkeley in the years 1960 and

1961. Feyerabend in his 1970 has some interesting reminiscences of this period. He recalls that they had long discussions:

> Some of which were carried out in the now defunct *Café Old Europe* on Telegraph Avenue and greatly amused the other customers by their friendly vehemence. [1970, 198]

Later in the same article, Feyerabend says:

> I do not know who of us was the first to use the term 'incommensurable' in the sense that is at issue here. It occurs in Kuhn's '*Structure of Scientific Revolutions*' and in my essay 'Explanation, Reduction, and Empiricism' both of which appeared in 1962. [1970, 219]

Despite this joint origin of the concept of incommensurability, it was, as we shall see, developed in rather different ways by Feyerabend and Kuhn.

The issue of incommensurability has given rise to a great deal of often heated discussion among philosophers of science, and this discussion continues to the present day. So far in this section we have focussed on the historical approach to philosophy of science in the period form c. 1960 to the mid 1970s. This is a natural period to choose because interest in the historical approach to philosophy of science declined sharply after about 1975. There were a number of reasons for this, among which we can mention the sudden and unexpected death of Imre Lakatos from a heart attack on 2 February 1974, and the publication of Feyerabend's *Against Method* in 1975. Many philosophers of science saw this book of Feyerabend's as a *reductio ad absurdum* of the whole historical approach to philosophy of science. As a result of these and other factors, many of the discussions which we have described in this section petered out in the late 1970s. Discussions about incommensurability were an exception to this, because they continued unabated into the 1980s and 1990s. The reason for this is probably that the problem of incommensurability involved questions of language and meaning, and so fitted in with the linguistic philosophy which dominated the English speaking world. Ideas about language, meaning, translation, and reference which had been developed by Davidson, Kripke, Putnam and Quine were applied to the problem. As we shall see, neither Feyerabend nor Kuhn made much reference to language in their initial discussions of incommensurability, but Kuhn in his later period took a distinctly linguistic turn in accordance with the prevailing climate in philosophy at the time. Significantly in his 1983 Kuhn wrote:

> If I were now rewriting *The Structure of Scientific Revolutions*, I would emphasize language change more and the normal/revolutionary distinction less. [1983, 57]

The continuing interest in the problem of incommensurability is shown by a recent publication [Massimi, 2005] in which incommensurability is discussed in connection with the introduction of Pauli's Exclusion Principle.

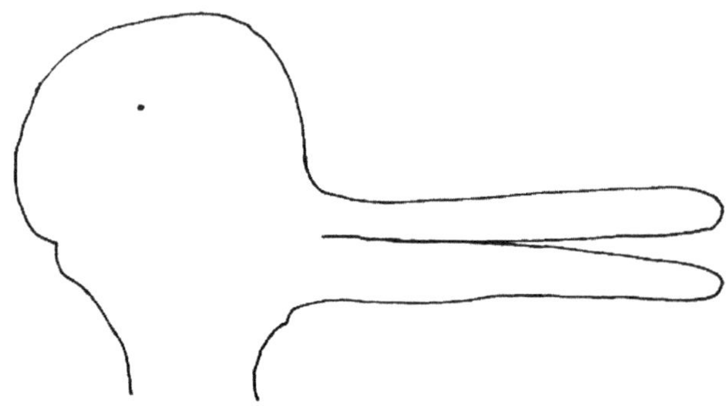

Figure 1. The Duck-Rabbit

Because of this situation, we will here not limit ourselves to discussions of incommensurability up to the mid-1970s, but take some account of how the debate has continued since then. The literature on this question is extensive and complicated, however, and we will only be able to discuss some of the arguments, choosing particularly those which relate to the question of the rationality of scientific change. The fundamental work for understanding the controversy as a whole is Sankey [1994] which gives a fine critical discussion of the most important works up to that date as well as Sankey's own contribution. Kuhn's contributions to the debate in the period from 1970 until his death in 1996 are conveniently collected in the 2000 volume: *The Road since Structure.*

Let us however begin by returning to the period c. 1960 to the mid 1970s and examine what Feyerabend and Kuhn said then about incommensurability. We will start with Kuhn. In his discussion of scientific revolutions in 1962, Kuhn claims that the new paradigm produced by such a revolution is incommensurable with the old paradigm. As he puts it:

> The normal-scientific tradition that emerges from a scientific revolution is not only incompatible but often actually incommensurable with that which has gone before. [1962, 102]

Kuhn's further discussions of incommensurability in his 1962 are not in terms of a change of language and meaning. He prefers to use the metaphor of the gestalt switch. Let us consider the duck-rabbit which is perhaps the most famous example of a gestalt switch because it was discussed by Wittgenstein in his [1953, 194]. Wittgenstein of course influenced Kuhn.

Figure 1 shows the duck-rabbit. The drawing can either be seen as a duck looking right or as a rabbit looking left. With a little practice, the viewer can make the 'gestalt switch' from seeing the drawing as a duck to seeing it as a rabbit at will. Kuhn compares the change of paradigm in a scientific revolution to a gestalt switch:

> ... at times of revolution, when the normal-scientific tradition changes, the scientist's perception of his environment must be re-educated — in some familiar situations he must learn to see a new gestalt. After he has done so the world of his research will seem, here and there, incommensurable with the one he had inhabited before. [1962, 111]

Kuhn does however make some qualifying remarks about the gestalt switch metaphor for he writes:

> That parallel can be misleading. Scientists do not see something *as* something else; instead they simply see it. ... In addition, the scientist does not preserve the gestalt subject's freedom to switch back and forth between ways of seeing. Nevertheless, the switch of gestalt, particularly because it is today so familiar, is a useful elementary prototype for what occurs in full-scale paradigm shift. [1962, 85]

Kuhn's reasons for regarding the parallel as misleading are not entirely convincing. The sentence S_1: 'The physicist sees the object as a tangent galvanometer' does not seem to differ greatly in meaning from the sentence S_2: 'The physicist sees a tangent galvanometer'. Moreover a scientist might preserve the gestalt subject's freedom to switch back and forth. Suppose a scientist first learns Newtonian mechanics and then Einsteinian mechanics. He or she might retain the capacity to switch back and forth between seeing the world as a Newtonian world and as an Einsteinian world.

We earlier criticized one use which Kuhn makes of the gestalt switch metaphor, namely his claim [1962, 121–2] that 'a new paradigm is born' through 'a relatively sudden and unstructured event like the gestalt switch.' We argued instead that new paradigms are born as the result of often long and painstaking work on a series of research programmes. However, once a new paradigm has been created, the process of switching from the old to the new paradigm can be compared quite accurately to a gestalt switch.

In his 1975, Feyerabend, like the early Kuhn, does not adopt a linguistic approach to the question of incommensurability. Indeed Feyerabend claims that it is not possible to give an explicit definition of incommensurability, and that the concept must be introduced by giving a number of different examples:

> As incommensurability depends on covert classifications and involves major conceptual changes it is hardly ever possible to give an explicit definition of it. ... The phenomenon must be shown, the reader must be led up to it by being confronted with a great variety of instances, and he must then judge for himself. [1975, 225]

Feyerabend follows the method by giving an interesting, and stimulating series of examples which are sometimes elaborated in considerable detail. These include the following: (1) the successive stages which, according to Piaget, children go through in their conception of the world [1975, 227–8], (2) the conceptual schemes of primitive tribes as compared with those of modern Westerners [1975, 249–51] — Feyerabend mentions particularly the studies of the Nuer by Evans-Pritchard, and (3) the change from the Homeric world-view to that of the Pre-Socratics [1975, 229–49 and 260–71]. In addition in his 1978, Feyerabend gives the striking example of the change brought about by waking up:

> ... waking up brings new principles of order into play and thereby causes us to perceive a waking world instead of a dream world ... [1978, 70]

We can see from this that incommensurability has for Feyerabend a general import, and is not restricted to changes brought about by scientific revolutions. However, he goes on to give examples of incommensurability in the scientific case as well. It is interesting to note, however, that Feyerabend is here somewhat more restrictive than Kuhn and denies that incommensurability was involved in the change from Ptolemy to Copernicus. He says:

> ... I never assumed that Ptolemy and Copernicus are incommensurable. They are not. [1975, 114]

Feyerabend's standard examples of incommensurability in science are the changes from Aristotelian mechanics to Newtonian mechanics, from Newtonian mechanics to relativistic mechanics, and from Newtonian mechanics to quantum mechanics [1975, 224–5 and 275–7].

Despite his reluctance to give a definition of incommensurability, Feyerabend comes close to doing so in his 1978. Significantly the passage occurs in a footnote. It is the following:

> ... mere *difference* of concepts does not suffice to make theories incommensurable in my sense. The situation must be rigged in such a way that the conditions of concept formation in one theory forbid the formation of the basic concepts of the other ... [1978, 68]

We can illustrate Feyerabend's idea here by considering the example of the transition from Aristotelian to Newtonian mechanics.[13] In Aristotelian mechanics every body in motion requires a force to move it along. If, for example, someone throws a stone, the thrower imparts to the stone a special force called 'impetus'. Impetus is then the force which continues to move the stone. The quantity of impetus gradually declines and the stone correspondingly ceases to move forward. Now at first sight we might think we could identify Aristotelian 'impetus' with Newtonian

[13] There are good discussions of this example in [Sankey, 1994, 88–89 and 109].

'momentum', but this would be a mistake. In Newtonian mechanics a body does not require a force to continue moving with uniform motion in a straight line. Momentum is a property of such a body but it is not a force moving it along. In fact there is no force in the Newtonian system moving the body along. The conditions of concepts formation in Newtonian mechanics forbid the formation of the Aristotelian concept of impetus.

Kuhn makes the point (e.g. in [1979, 205]) that the meaning of planet changed from Ptolemy to Copernicus. In Ptolemy's theory, the Sun and Moon were planets, but the Earth was not a planet. In Copernicus' theory, the Earth became a planet, but the Sun and Moon ceased to be planets. This is a significant change of meaning, but, as we have seen, it does not in Feyerabend's view lead to incommensurability. This is actually supported by Feyerabend's definition of incommensurability. We can define planet in the sense of Copernicus (or P_C) in terms of planet in the sense of Ptolemy (or P_P) as follows:

$$P_C(x) =_{def} [P_P(x) \mathbin{\&} \neg \{(x = \text{Sun}) \vee (x = \text{Moon})\}] \vee (x = \text{Earth})$$

It follows therefore that the conditions of concept formation in Ptolemy's theory do not forbid the formation of Copernicus' concept of planet. As the converse also holds, this shows that according to Feyerabend's definition of incommensurabililty, the change in the concept of planet does not lead to an incommensurability between Ptolemy's theory and Copernicus'. It may have been generalising from examples such as this that led Feyerabend to the conclusion that Ptolemy's theory is not incommensurable with Copernicus'. This is a significant conclusion because it shows that dramatic revolutionary changes of theory can occur without incommensurability, and that consequently incommensurability is not an essential feature of scientific revolution.

As the whole idea of incommensurability has been severely attacked by many philosophers of science, let us now say something in its favour. The discussions of the notion by Kuhn and Feyerabend which we have just described do bring out the fact that scientific revolutions can cause profound conceptual changes, and alter very significantly the way in which scientists see the world. The nature of these changes are well-illustrated by the series of interesting analogies which Kuhn and Feyerabend provide: Gestalt switches, the change from sleeping to waking, Piaget's stages in child development, primitive tribes compared to modern Western societies, Homer's world-view compared to that of the Pre-Socratics. These analogies do, in our opinion, illuminate the nature of the profound changes in world-view which can be brought about by scientific revolutions. They are in danger of being forgotten if incommensurability is analysed purely in terms of language and meaning in the manner which we will describe in a moment.

So far we have talked of scientific revolutions bringing about a change in the way scientists see the world, but both Kuhn and Feyerabend sometimes make the stronger claim that in a scientific revolution the world itself changes. Thus Kuhn says:

> Nevertheless, paradigm changes do cause scientists to see the world of their research-engagement differently. In so far as their only recourse to that world is through what they see and do, we may want to say that after a revolution scientists are responding to a different world. [1962, 110]

This view is supported with greater zeal by Feyerabend, who writes:

> ... we certainly cannot assume that two incommensurable theories deal with one and the same objective state of affairs (to make the assumption we would have to assume that both at least *refer* to the same objective situation. But how can we assert that 'they both' refer to the same situation when 'they both' never make sense together? ... Hence, unless we want to assume that they deal with nothing at all we must admit that they deal with different worlds and that the change (from one world to another) has been brought about by a switch from one theory to another. [1978, 70]

This view has some points in common with Kant's. Kant thought that the intersubjective world of human beings is partly the result of the way things are in themselves, and partly the result of the conceptual schemes used to process sensory input. Kant, however, thought that these conceptual schemes are the same for all times and all human beings (Euclidean geometry and the twelve categories). Kuhn and Feyerabend allow the possibility that different communities, or the same community at different times, can have different fundamental conceptual schemes, and conclude that the members of different communities may inhabit different worlds.

Kuhn in his later period quite explicitly adopted such a modified Kantian position. He writes:

> By now it may be clear that the position I'm developing is a sort of post-Darwinian Kantianism. Like the Kantian categories, the lexicon supplies preconditions of possible experience. But lexical categories, unlike their Kantian forebears, can and do change, both with time and with the passage from one community to another. [1990, 104]

This could be called 'Kant on Wheels'[14]. Kuhn, however, was undecided as to whether he should include Kant's concept of the thing in itself in his post-Darwinian Kantianism. In an earlier formulation of his new version of Kantianism, he definitely repudiates things in themselves, writing:

> The view toward which I grope would also be Kantian, but without "things in themselves" and with categories of the mind which could change with time as the accommodation of language and experience proceeded. A view of that sort need not, I think, make the world less real. [1979, 207]

[14]This is the title of Lipton's 2003 article which discusses Kuhn's later views.

However, in 1990, the thing in itself is reinstated:

> Underlying all these processes of differentiation and change, there must, of course, be something permanent, fixed, and stable. But, like Kant's *Ding an sich*, it is ineffable, undescribable, undiscussible. Located outside of space and time, this Kantian source of stability is the whole from which have been fabricated both creatures and their niches, both the "internal" and the "external" worlds. [1990, 104]

The general question of Kantianism and realism is a fascinating one, but its study requires consideration of some complicated issues in the theory of reference, and, in particular, an evaluation of different versions of the causal theory of reference. We will therefore not discuss this question further here[15], but rather return to what is the central theme of this section namely the problem of the rationality of scientific revolutions.

As we have seen, Lakatos took Kuhn's view to be that the change of paradigm in a scientific revolution was irrational and analogous to a religious conversion. We have already given some quotations from Kuhn [1962] which support this interpretation, and here are a few more. Kuhn writes:

> The man who embraces a new paradigm at an early stage must often do so in defiance of the evidence provided by problem-solving. ... A decision of that kind can only be made on faith. [1962, 157]

Moreover Kuhn hints that a scientific revolution may only seem to be an advance, because the victorious revolutionaries assert that an advance has been made:

> Revolutions close with a total victory for one of the two opposing camps. Will that group ever say that the result of its victory has been something less than progress? That would be rather like admitting that they had been wrong and their opponents right. To them, at least, the outcome of revolution must be progress, and they are in an excellent position to make certain that future members of their community will see past history in the same way. [1962, 165]

This looks very like the kind of 'might is right' doctrine which Lakatos was concerned to oppose.

After about 1970, the views of Kuhn and Feyerabend on this question begin to diverge. Feyerabend, as we have seen, took great pleasure in defending the thesis that science is irrational, and in proclaiming a 'farewell to reason'. Kuhn, on the contrary, drew back from the seemingly irrationalist implications of his earlier views, and seemed genuinely upset that his philosophy should have been taken up and developed in this sense. This divergence is already noticeable in the 1970 collection edited by Lakatos and Musgrave. As we have seen, Feyerabend in his contribution to this volume attacks Kuhn's views on normal science, but he stresses his agreement with Kuhn's views on incommensurability, writing:

[15]For the interested reader, there is an excellent discussion of these issues in Sankey [1994].

> With the discussion of incommensurability, I come to a point of Kuhn's philosophy which I wholeheartedly accept. [1970, 219]

Kuhn, however, in his 'Reflections on my Critics' in the same volume seems distinctly less enthusiastic about incommensurability. Perhaps he had already become worried by the interpretation which Lakatos had given to his views. At all events he writes:

> Such communication breakdown is important and needs much study. Unlike Paul Feyerabend (at least as I and others are reading him), I do not believe that it is ever total or beyond recourse. Where he talks of incommensurability *tout court*, I have regularly spoken also of partial communication, and I believe it can be improved upon to whatever extent circumstances may demand and patience permit, ... [1970, 232]

In 1976, Kuhn explicitly states that his earlier views on incommensurability had been misunderstood.

> Most readers of my text have supposed that when I spoke of theories as incommensurable, I meant that they could not be compared. But 'incommensurability' is a term borrowed from mathematics, and it there has no such implication. The hypotenuse of an isosceles right triangle is incommensurable with its side, but the two can be compared to any required degree of precision. What is lacking is not comparability, but a unit of length in terms of which both can be measured directly and exactly. In applying the term 'incommensurability' to theories, I had intended only to insist that there was no common language within which both could be fully expressed and which could therefore be used in a point-by-point comparison between them. [1976, 189]

Note that here Kuhn relates incommensurability to questions of language — something which he did not do in his 1962, and which is indicative of his linguistic turn. This linguistic approach is elaborated in his 1983 where he defines a 'modest version of incommensurability' in terms of the impossibility of translating some of the terms of one theory into the language of the other theory. This is how he puts it:

> Only for a small subgroup of (usually interdefined) terms and for sentences containing them do problems of translatability arise. The claim that two theories are incommensurable is more modest than many of its critics have supposed.
>
> I shall call this modest version of incommensurability 'local incommensurability'. ... The terms that preserve their meanings across a theory change provide a sufficient basis for the discussion of differences and for comparisons relevant to theory choice. [1983, 36]

Kuhn illustrates this by arguing that some of the terms of Newtonian mechanics cannot be translated into Aristotelian or Einsteinian mechanics (p. 44).

> ... Newtonian 'force' and 'mass' are not translatable into the language of a physical theory (Aristotelian or Einsteinian, for example) in which Newton's version of the second law does not apply. To learn any one of these three ways of doing mechanics, the interrelated terms in some local part of the web of language must be learned or relearned together and then laid down on nature whole. They cannot simply be rendered individually by translation.

This approach to incommensurability is not dissimilar to Feyerabend's 1978 definition which we described earlier.

Davidson and Putnam objected to the 'untranslatability' criterion on the grounds that we would not be able to understand an older theory unless we could translate its terms into our current language. However, Kuhn replied very reasonably to this objection that [1983, 39]: 'acquiring a new language is not the same as translating it into one's own.' Feyerabend made a similar reply to this objection pointing that we can learn a new language in the same way that children learn their first language.[16]

After these further clarifications of the concept of incommensurability, we will return to the key question of whether rational change from an old paradigm to a new incommensurable one is possible. Before doing so, however, it will be useful to look at one of the most interesting and original contributions to the discussion about incommensurability — Jane English's paper of 1978.

Many would think that there is no greater contrast among philosophers of science than that between Carnap on the one hand and Kuhn and Feyerabend on the other. Carnap is the extreme representative of the logical approach to philosophy of science. Nearly everything he considers is formalised in first order logic and a set of elaborate logical techniques is brought to bear upon it. Kuhn and Feyerabend on the other hand adopt the historical approach, basing their analysis on the history of science and proceeding informally without any use of formal logic. Despite these enormous differences, however, English argues that Carnap's partial-interpretation account of the meaning of theoretical terms has a very great deal in common with the views on meaning of Kuhn and Feyerabend and, in particular, gives rise to the same difficulties and counter-intuitive consequences. This is how she puts her thesis:

> Among the current views of the meaning of theoretical terms, Carnap's partial interpretation account and the meaning-change account of Kuhn and Feyerabend are usually thought of as antithetical. On the contrary, I will argue that the two have much in common. In particular, I will show that some of the major objections brought against

[16]For a fuller discussion, see [Sankey, 1994, 102–37].

> the meaning-change position apply equally to partial interpretation. [English, 1978, 57]

Let us therefore examine Carnap's partial interpretation account. Carnap of course begins his analysis of a scientific theory by formalising it in first order logic. He then decomposes the theory (TC say) into two parts. The first part contains the factual content of the theory, while the second part contains the meaning postulates. Each sentence of this second part is analytic, or true in virtue of meaning, and, in effect, the sentences of the second part, taken together, give an implicit definition of the theoretical terms of the theory. Carnap tried various ways of dividing a theory into these two parts, but finally decided on a method which uses the technical device of the Ramsey sentence. Give our theory TC, we form its Ramsey sentence R, and this represents the factual content of TC. The meaning postulates of TC are then given by $R \rightarrow TC$.

One interesting thing to note here is that although Carnap is employing the standard syntax of first order logic, he does not use the standard Tarskian semantics of first order logic. Let us illustrate Tarskian semantics by considering a mathematical example. Suppose we are dealing with a first order formalisation of Peano arithmetic. To give the formal symbols meaning using Tarskian semantics, we would first select a domain, which in this case would be the set N of natural numbers $\{1, 2, \ldots, n, \ldots\}$. Then to each of the individual constants of the theory we would assign a member of N. For example, there might be just one individual constant in this formalisation (a say) and we would assign to a the number 1. To each function letter in the formal theory we would assign a function over N. For example if there is a formal symbol $s()$, we might assign to $s()$ the function $+1$, so that $ss(a)$ would then stand for the number 3, and so on. To the 1-place predicate letters of the formal theory we would assign subsets of N. So to a predicate letter $O()$, we might assign the set of odd numbers, to a predicate letter $P()$, we might assign the set of prime numbers, and so on. In this way all the expressions of the formal theory are given meaning.

It is clear that Carnap does not use Tarskian semantics of this kind. He does not, for example, give 1-place predicate letters meaning by assigning to them subsets of a given domain, but rather by setting meaning postulates which implicitly define these 1-place predicates. His approach to meaning is in fact closer to that of Wittgenstein than to that of Tarski, for Carnap is in effect saying that the meaning of a predicate is given by the rules governing its use. It must be used in accordance with the meaning postulates. The fact that Carnap uses a Wittgensteinian approach may partly explain why his approach exhibits some strong resemblances to those of Kuhn and Feyerabend.

But why does Carnap, who was such a strong advocate of standard logic and of Tarski's ideas not use the standard Tarskian semantics? The answer is not far to seek, because a little reflection shows that it would be very difficult to apply Tarskian semantics to give a convincing account of meaning for formalised scientific theories. Such an application would result in a very artificial construction. Let us suppose, for example, that we have formalised Newtonian mechanics and are considering how to give meaning to the term $m(x)$ which is the formal equivalent of the mass of x. On a Tarskian approach, we might identify $m(x)$ with a function whose domain is the set of bodies and whose range is the set of positive real numbers. However, this definition of '$m(x)$' diverges completely form the way the term is in practice given meaning by physicists. Physicists explain the meaning of $m(x)$ to beginners by giving the laws governing masses, and the experimental and observational procedures used for determining the mass of a body. Without knowledge of these laws and procedures, the meaning of mass could not be grasped. Moreover it is not clear that the formal Tarskian approach is even coherent. It involves considering the set of bodies, for example, but is the concept of body clearly defined? It certainly is not. Let us consider an electromagnetic field, for example. A section of such a field would not normally be considered a body, but yet it could have a mass associated with it.

Our conclusion is that, while Tarskian semantics does indeed appear quite natural when dealing with mathematical examples, it seems, on the contrary, strained, artificial and inappropriate for handling the semantics of scientific theories. It is likely that this is why Carnap took a different approach when considering the question of the meaning of theoretical terms in a scientific theory. However, the result is interesting in a more general way. As we have seen, modern formal logic was developed by its pioneers (Frege, Peano, Russell, etc.) to handle mathematics. It was only later applied by the Vienna Circle and their followers to science. Now it may well be that many of the techniques of formal logic, while quite reasonable in the mathematical context, are inappropriate in a scientific context. This could be the reason for the appearance of some of the paradoxes which arose when applying formal logic to science, such as, for example, Hempel's paradox of the ravens.

Let us now, however, return to English's treatment of Carnap. Having explained Carnap's method of partial-interpretation of theoretical terms, she goes on to point out that it leads to some 'Kuhnian' consequences.

> Carnap's account here nicely supports Kuhn's meaning-change view. For instance, Kuhn relates in detail the history of the term 'compound'.[17] He claims that Dalton's assertion, "Compounds can be formed only in fixed proportions," and the pre-Daltonians' assertion, "Compounds can be formed in any ratios," did not contradict, because they meant different things by 'compound'. Dalton's predecessors included some of what we now call alloys, solutions, and suspensions under that term, whereas Dalton reserved it for things that follow his law. If we apply Carnap's method, Kuhn's interpretation results. Dalton is construed as saying, "If there is anything that ... and obeys the law of fixed proportions and ... then let us call it 'compound' ..." and his rivals are taken to say, "If there is anything that ... and combines in any ratios and ... then let us call *that* 'compound' ..." But then their theoretical statements "Compounds are formed only in

[17][Kuhn, 1962, 130–5]. This reference is given by English.

> fixed proportions" and "Compounds can be formed in any ratios" are both true; so they fail to contradict. [English, 1978, 70–71]

Indeed it would appear that Carnap's theory is more radical than Kuhn's, for as English says:

> This holism leads Carnap to an account of meaning change more extreme than Kuhn's. Since every postulate of the theory is represented in TC, any theoretical disagreement — not only disagreements in the most central assumptions — indicates a difference in meaning conventions. Although Kuhn has failed to specify how large a change must be to constitute a scientific revolution, he does hold that meanings are fixed despite small changes within "normal science." For Carnap, small changes as well as large are reflected in a change in the theory's Ramsey sentence, and thus in its meaning conventions. [1978, 71]

So on Carnap's account if a scientific theory T is changed even in a very slight way to produce a new theory T', then the terms of T' have different meanings form those of T. So no sentence of T' can contradict one of T. Thus the change from T to T' would appear to be an irrational leap of faith since T cannot be compared with T' to see which one is better confirmed by the evidence available. The problems of incommensurability seem to arise in a more extreme form in Carnap's account. Let us now see if they can be resolved.

Let us suppose then that we have two scientific theories T and T' — say Newtonian theory and Einsteinian theory. Since the theories are scientific, they will each contain a set of observation statements $\{O\}$ and $\{O'\}$. An observation statement is one whose truth-value, whether true or false, can in practice be decided by the scientific community on the basis of observation and experiment. Some philosophers of science maintain that there is a neutral observation language, but we will not make this assumption, which is anyway challenged by Kuhn and Feyerabend. We will assume to the contrary that the observations statements of T are made in the language of T, and those of T' in the language of T'. Thus if a particular observation statement is 'The mass of this body is 2.5 grams', we will assume that, within T, mass will be understood in a Newtonian sense yielding the observation statement O, while within T', mass will be understood in an Einsteinian sense yielding the observation statement O'. Now O and O' have different meanings, but, nonetheless, if we are dealing with an ordinary medium sized body moving with a low velocity, then the adherents of T' would certainly agree to give the same truth-value to O' as the adherents of T give to O on the basis of making the same observations and experiments. Thus these two observation statements would be ascribed the same truth-value by the two camps, a situation which we could describe by writing $O \sim O'$. Generalising we could establish a sequence of observation statements of T, $O_1, O_2, \ldots, O_n, \ldots$ say, and a corresponding sequence of observation statement of T', $O'_1, O'_2, \ldots, O'_n, \ldots$ say, such that $O_n \sim O'_n$. It now becomes easy to compare T and T' empirically. We work out how well T is confirmed (or disconfirmed) by the sequence $O_1, O_2, \ldots, O_n, \ldots$, and then how well T' is confirmed (or disconfirmed) by the sequence $O'_1, O'_2, \ldots, O'_n, \ldots$. If one of the two theories has a very much higher degree of confirmation than the other it becomes rational to accept it in preference to the other. No religious conversion, leap of faith, or political manoeuvring is needed here!

The same technique enables us to establish logical relations in an informal sense between T and T'. Suppose for example that T logically entails O_n and T' logically entails $\neg O'_n$, we can then say, speaking informally, that T contradicts T'. This can apply even if T and T' are formalised in two different systems S and S' within which the predicates of the two theories have different meanings. Of course if we work exclusively within the formal system S or within the formal system S', we cannot say that T and T' contradict each other. However the example of Gödel's incompleteness theorems surely shows that it is perfectly reasonable to extend reasoning outside a given formal system or formal systems, and to apply logic informally in this extension. The present examples shows that this technique should be used when applying logic to scientific theories, and that an exclusive reliance on say formal first order classical logic is not adequate for the philosophy of science.

Our conclusion then is that incommensurability is not such a monster threatening the rationality of scientific change as it might at first have appeared. On the contrary, it is quite easy to compare incommensurable theories both logically and empirically — provided one uses logic in a judicious fashion. But does this show that the question of incommensurability is not, after all, such an important one? Very different opinions have been expressed on this issue. Kuhn continued in his later period to believe in the importance of incommensurability. He wrote in 1990:

> No other aspect of *Structure* has concerned me so deeply in the thirty years since the book was written, and I emerge from those years feeling more strongly than ever that incommensurability has to be an essential component of any historical, developmental, or evolutionary view of scientific knowledge. [1990, 91]

Sankey on the other hand writes in the last few pages of his 1994 book: *The Incommensurability Thesis*.

> The overall thrust of my argument in this book is deflationary. Incommensurability is less of a problem than has generally been thought. The conceptual and semantical variance which initially gave rise to the idea of incommensurability do not threaten an unmitigated relativism of radically incompatible conceptual schemes. Nor do they force any concession upon an essentially realist view of the relation between scientific theory and extra-theoretic reality. ... there seems little point in saying that theories are incommensurable. [1994, 219 and 221]

On the whole we here side more with Sankey than with Kuhn, but would nonetheless like to make some observations in favour of the importance of incommensurability. There does seem some point in saying that two theories are incommensurable. This indicates that there is a radical conceptual shift in moving from one to the other. Moreover the study of such conceptual shifts has brought to light some interesting features of scientific change. It has shown that the logical analysis of science requires something more than the use of standard first order logic. The meaning of the theoretical terms in a scientific theory cannot be plausibly given using Tarskian semantics. If an alternative approach is adopted, such as Carnap's partial interpretation involving implicit definitions, then the study of the logical relations between two different scientific theories may well require considering both the formalisation of the theories in two different formal systems and then an examination of the relations between these formal systems. However the question of incommensurability is not simply one of logic and semantics. One of the most interesting features of the treatment of the question by Kuhn and Feyerabend was their stress on how incommensurable theories lead to different world views, and their attempts to illustrate the nature of such a change by a whole series of striking metaphors — gestalt switches, the change from sleeping to waking, the change from the Homeric world view to that of the pre-Socratics, and so on. These passages in Kuhn and Feyerabend are very insightful and illuminating, but unfortunately the dominance of the linguistic approach to philosophy often means that they are lost sight of. It is significant, for example, that English in her 1978 paper speaks of Kuhn's meaning-change view, and, in a passage already quoted cites [Kuhn, 1962, 130–5] as given an instance of this view (see our footnote 17). In this passage, however, Kuhn nowhere speaks of meaning. Instead he says things like the following:

> As a result, chemists came to live in a world where reactions behaved quite differently from the way they had before. [1962, 133]

Of course later in his life Kuhn did begin to speak of language, meaning, translation, etc. but this was because he too had fallen under the influence of the dominant linguistic paradigm in philosophy.

Another point in favour of the importance of the incommensurability problem is that many of the issues to which it has given rise could fruitfully be investigated further. The work of Kvasz (see his [1998; 1999; 2000]) is important here. We have argued that it is still worth speaking of theories or paradigms being incommensurable, because this indicates that there is a considerable conceptual change in passing from one theory or paradigm to the other. However the phrase 'a considerable conceptual change' is rather vague. Might there be conceptual changes of different magnitudes, and could we give some kind of classification of the size of these magnitudes? Kvasz takes up this problem in his 1999, which is concerned with the classification of scientific revolutions. He ingeniously suggests using perturbation theory as a device to measure the magnitude of the epistemic ruptures [1999, 219], and, as a result, comes up with three different kinds of epistemic rupture. Looking at the problem from a more linguistic point of view, we can say that a new theory or paradigm may well have a different language from the old theory or paradigm, but then the question arises of how the new language differs from the old, and how starting from the old language, a new language can be created. Kvasz tackles these problems in his 1998 and 2000. His 1998 is concerned with the history of geometry, and he shows that successive geometrical theories were expressed in different languages. He uncovers a mechanism by which a new language to express a new geometrical theory could be created. Following Wittgenstein in the *Tractatus*, Kvasz regards any language (L say) as having a form which is not expressible in the language. We can however incorporate the form of the language L into L thereby creating a new language L' say. Kvasz shows that this is precisely the way in which new languages for new geometrical theories were created, and in his 2000, he extends his results to mathematics as a whole. These investigations of Kvasz are closely connected to the issues which arose from the incommensurability problem, and his work shows that these issues can fruitfully be investigated further.

4 THE 1970S: SCIENCE AS PROBLEM SOLVING

4.1 Cognitive Science: The Emergence of a New Discipline

The Study of the Mind

The invention of computers brought forward a new dimension for the study of the mind. The study of cognition called for an integrated approach of theoretical as well as empirical disciplines, notably philosophy, psychology, linguistics, anthropology, neurosciences, and computer science.

The study of the mind was for a long time an exclusive topic for philosophy, going back to the Greeks and continuing into the 19th century, when the beginnings of experimental psychology emerged. In the 20th century, the gradual decline of the influence of behaviourism resulted in the preference of psychological theories taking into account mental representations and memory aspects. During the fifties, empirical results showed a limited capacity of the human mind, creating a point of contact between psychology and philosophy, for it introduced similar challenges to them. On the one hand, it opened new avenues to applied research, much of it focused on short-memory problems, and on the other, it introduced new epistemological questions, notably those having to do with the modelling of the generation and development of scientific knowledge.

Also during the fifties, but in this case between philosophy and computer science, representational theories of the mind emerged from the analogy of the mind as a computer [Turing, 1950; Fodor, 1975]. This idea served as a bridge between philosophy of mind and artificial intelligence (AI), the former providing the conceptual basis, the latter the tools to represent and manipulate knowledge.

The challenge of knowledge representation was at the core of all these disciplines and had logic as its main tool from the 50's to the 70's and 80's[18], when new logics emerged and proliferated in artificial intelligence. Moreover, the task of creating computational models of human intelligence put forward proposals such as the GPS (General Problem Solver), a program aiming at the mimicking of human problem solving. A task of such dimensions involved philosophy, psychology, and computer science, the constituent disciplines of cognitive science. A society of the cognitive sciences and a new journal emerged in the 70's and this kind of interdisciplinary research began to evolve. Pioneers such as John McCarthy, Marvin Minsky, Allen Newell, and Herbert Simon are the founding fathers of artificial intelligence. In addition, Noam Chomsky rejected behaviourist assumptions about language as a learned habit and proposed instead to explain language comprehension in terms of mental grammars consisting of rules [Chomsky, 1972; 1976]. All these researchers are to be regarded as the key founders of the field of Cognitive Science.

As far as the impact of computers on mainstream philosophy of science, although they had been around for some time, they are hardly mentioned before the seventies, as judged by the writings of Popper, Kuhn, Lakatos and Feyerabend. However, scientific knowledge and discovery constituted a substantial object of research for computational models of intelligent behaviour. Cognitive Scientists imported some of the problems of philosophy of science. An example of an interaction between computer science, artificial intelligence, logic and philosophy of science, is the development of machine learning, which casts doubt on Popper's claim that "induction is a myth", since computer programs began to be able to carry out induction successfully in some cases. Abduction was also studied as a form of explanatory reasoning, and new forms of computational representations and processes were devised for such an inference. More generally, there were considerable developments in logic with new logical systems such as non-monotonic logics (several of these will be analyzed in the next section). This meant that the logical approach to philosophy of science could be revived with a more powerful set of tools. Moreover, the scope of the logical approach could be extended to include discovery, though perhaps even at present this should be confined to discovery within a normal science context.

4.2 Philosophy of Science: Background Issues

This section aims firstly to set the scene for a refreshed view of the task of philosophy of science, that of analyzing science as a problem solving process, the guiding idea for a renewed enterprise in the seventies. Under this view, the analysis of scientific knowledge is addressed by questions having to do with the growth and evolution of knowledge, with the progress of science and the discovery and de-

[18]In the 1980's, a new paradigm emerged, namely connectionism and its companion neural networks. There was a change in the logical approach as well about that time. This was the appearance of probability (especially with the development of Bayesian networks). Most advocates of the logical approach these days would include probability, and interestingly the opposing connectionist approach also uses probability. The decade of 1990's is marked as the so called "antirepresentationalism" age. However, logic remained extensively used.

velopment of new theories, rather than centred in the fundamental concepts by themselves. The guiding principle of science as problem solving is however by no means genuinely new, for it pertains to marginal views in the philosophy of science, which up to the seventies, had no privileged place in the received view.

This view constitutes the challenge to broaden the scope of philosophy of science, but it is not until the 90's that issues like scientific discovery are decidedly in its research agenda. As is well-known, great philosophers and mathematicians have been brilliant exceptions in the study of discovery and development in science, and that their non conventional contributions to this field, although great inspirations, have not set new paradigms in the methodology of science. Therefore, before describing the proposals which shape the move to problem solving in the 70's, let us review the work of some of the main predecessors.

As far as logic goes in the seventies, the formal advances which shape the task of philosophy of science up to the 50's, that of giving a logical analysis of the concepts of science, were for some pretty much logically exhausted, and were for all often obscured by the historical analysis in vogue from the sixties. The logic used up to that date was fundamentally classical logic, which cannot help to account for the infallibility and the dynamics of scientific knowledge. It was not until the developments of logics in artificial intelligence in the eighties, that these new tools were exploited. However, there are important antecedents of work in logic which are more in accord with the new task of science, which is definitively marked by a new conception of logic altogether. Two figures from the turn of the 19th century are worthy of mention here, namely Bernard Bolzano (1781-1848) and Charles S. Peirce (1839-1914). As for authors in the 20th century, we will mention Polya, Hanson and Popper. We will illustrate some aspects of their proposals which will be relevant for our later discussion.

Bernard Bolzano

In his "Wissenschaftslehre" [1837], Bolzano engaged (among other things) in the study of different varieties of inference. One of Bolzano's goals was to show why the claims of science form a theory as opposed to an arbitrary set of propositions. For this purpose, he defines his notion of deducibility as a logical relationship extracting conclusions from premises forming *compatible propositions*, those for which some set of ideas make all propositions true when uniformly substituted throughout. In addition, compatible propositions must share *common ideas*. Restated in model-theoretic terms, Bolzano's notion of deducibility reads as follows (cf. [van Benthem, 1984]):

$T, C \Rightarrow E$ if

(1) The conjunction of T and C is consistent.

(2) Every model for T plus C verifies E.

Therefore, Bolzano's notion may be seen (anachronistically) as Tarski's consequence plus the additional condition of consistency. Bolzano does not stop here. A finer grain to deducibility occurs in his notion of *exact deducibility* which imposes greater requirements of 'relevance'. A modern version, may be transcribed (again, with some historical injustice) as:

$T, C \models^{+} E$ if

(1) $T, C \models E$

(2) There is no proper subset of T, T', or of C, C', such that $T', C' \models E$.

That is, the premise set (composed by T, C) must be 'fully explanatory' in that no subpart of it would do the derivation[19]. Bolzano's agenda for logic is relevant to the study of general non-monotonic consequence relations for several reasons. It suggests the methodological point that what we need is not so much proliferation of different logics as a better grasp of different styles of consequence.

Charles S. Peirce

Now let's turn to Charles S. Peirce. He proposed abduction to be the logic for synthetic reasoning, that is, a method to acquire new ideas. He was indeed the first philosopher to give to abduction a logical form, on a pair with deduction and induction. His formulation is reproduced as follows [Peirce 1931-1935, 5.189]:

The surprising fact, C, is observed.

But if A were true, C would be a matter of course.

Hence, There is reason to suspect that A is true.

For Peirce, three aspects determine whether a hypothesis is promising: it must be *explanatory*, *testable*, and *economic*. A hypothesis is an explanation if it accounts for the facts. Its status is that of a suggestion until it is verified, which explains the need for the testability criterion. Finally, the motivation for the economic criterion is twofold: a response to the practical problem of having innumerable explanatory hypotheses to test, as well as the need for a criterion to select the best explanation amongst the testable ones. The above formulation accounts for the explanatory aspect.

George Polya

The next reference, already within the 20th century, is G. Polya [1962], regarded as the modern founder of heuristics [Hintikka and Remes, 1974; 1976]. He analyzed mathematical problems and their relation to discovery. In the context of number theory, for example, a general property may be guessed by observing some relation as in:

[19]Notice that this leads to non-monotonicity (cf. 5.1). A consequence $\models$ is non-monotonic whenever $T \models b$ does not ensure $T, a \models b$. That is, the addition of new premises is no warrant for validity reservation. Here is an example: $T, a \rightarrow b,\ a \models^{+} b$, but it is not the case that T, $a \rightarrow b,\ a,\ b \rightarrow c \models^{+} b$.

$$3 + 7 = 10 \qquad 3 + 17 = 20 \qquad 13 + 17 = 30$$

Notice that the numbers 3,7,13,17 are all odd primes and that the sum of any of two of them is an even number. An initial observation of this kind eventually led Goldbach (with the help of Euler) to formulate his famous conjecture: '*Every even number greater than two is the sum of two primes*'. Moreover, Polya contrasts two types of arguments. A demonstrative syllogism in which from $A \Rightarrow B$, and B false, $\neg A$ is concluded, and a heuristic syllogism in which from $A \Rightarrow B$, and B true, it follows that A is more credible. The latter, of course, recalls Peirce's abductive formulation.

Russell Hanson and Karl Popper

Already in the 60's, an author emphasizing explanation as a process of discovery is Hanson [1961], who gave an account of patterns of discovery, recognizing a central role for retroduction (another name for abduction). Another intellectual inheritance from the past decade is the work of Popper in Conjectures and Refutations in 1963. In this work, the growth of scientific knowledge is the most important of the traditional problems of epistemology [Popper, 1934, 22]. His fallibilist position provided him with the key to reformulate the traditional problem in epistemology, which was focused on the reflection on the sources of our knowledge. Rather, for him, the focus should be on the advancement of knowledge. This concern is intimately related to his view of science as a problem solving activity: '*Science should be visualized as progressing from problems to problems — to problems of increasing depth. For a scientific theory — an explanatory theory — is, if anything, an attempt to solve a scientific problem, that is to say, a problem concerned with the discovery of an explanation*' [Popper, 1960, 179]. As we shall see, this view is in accord with Simon's famous slogan that *scientific reasoning is problem solving* made in research in cognitive psychology and artificial intelligence (to be later introduced). However, in regard to giving a logical account for discovery processes, Popper's position is broadly recognized as neglecting this kind of scientific practice as part of the methodology of science agenda, and rather regarding its study a business of psychology. (But as we shall see, under a broad view of discovery[20], Simon's and Popper's positions are not so far apart.).

4.3 Philosophy of Science: Discovery as Problem Solving

The work in the sixties most relevant to our discussion is the work by Lakatos (1963-4), namely *Proofs and Refutations*, a critical response to Popper's logic of scientific discovery:

[20]Here is a useful distinction between a narrow and a broad view of discovery. While the former view regards issues of discovery as those dealing exclusively with the initial conception of an idea, the latter view is that which deals with the overall process going from the conception of a new idea to its settlement as an idea subject for ultimate justification [Laudan, 1980].

> *There is no* ***infallibilist*** *logic of scientific discovery leading infallibly to results, but there is a fallibilistic logic of discovery which is the logic of scientific progress. But Popper, who has laid the basis for this logic of discovery was not interested in the meta-question of what is the nature of this investigation, so he did not realize that it is neither psychology nor logic, but an independent field, the logic of discovery,* ***heuristics*** [Lakatos, 1963-4, 167, our emphasis].

It is interesting to note that Lakatos was greatly inspired by the history of mathematics, paying particular attention to processes that created new concepts — often referring to G. Polya, one of his predecesors. Another key reference for the view of science as problem solving is the work of Laudan [1977], namely "Progress and its Problems", in which scientific progress is analyzed as a case of naturalization of science and in its relation to history as well as with problems having to do within the rationalist view of science. Still another important reference is the work of Rescher [1978], which introduces a *direction of thought.* Interestingly, this establishes a temporal distinction between 'prediction' and 'retroduction', by marking the precedence of the explanandum over the hypothesis in the latter case.

Does Scientific Discovery Have a Logic?

In principle, the pioneering work of Herbert Simon and his team shares the ideal on which the whole enterprise of artificial intelligence was initially grounded, namely that of constructing intelligent computers behaving like rational beings. In his essay *Does scientific discovery have a logic?* Simon sets himself the challenge to refute Popper's general argument, reconstructed for his purposes as follows: *'If "There is no such thing as a logical method of having new ideas", then there is no such thing as a logical method of having* ***small*** *new ideas'* [Simon, 1973, 327, my emphasis], and his strategy is precisely to show that an antecedent in the affirmative does not commit to an assessment of the consequent, as Popper seems to suggest. Thus, Simon converts the ambitious aim of searching for a logic of discovery revealing the process of discovery at large, into an unpretentious goal: *'Their modesty [of the examples dealt with] as instances of discovery will be compensated by their transparency in revealing the underlying process'* [Simon, 1973, 327].

This humble and brilliant move allows Simon to further draw distinctions on the type of problems to be analysed and on methods to be used. For Simon and his followers, scientific discovery is a problem-solving activity. To this end, a characterization of problems into those that are well structured versus those that are ill structured is provided, and the claim for a logic of discovery focuses mainly on the well-structured ones[21].

[21] A well structured problem is one for which there is a definite criterion for testing, and for which there is at least one problem space in which the initial and the goal state can be represented and all other intermediate states may be reached with appropriate transitions between them. An

Although there is no precise methodology by which scientific discovery is achieved, as a form of problem solving, it can be pursued via several methodologies. The key concept in all this is that of *heuristics*, the guide in scientific discovery which is neither totally rational nor absolutely blind. Heuristic methods for discovery are characterized by the use of selective search with fallible results. That is to say, while they provide no complete guarantee to reach a solution, the search in the problem space is not blind, but it is selective according to a predefined strategy. The authors further distinguish between 'weak' and 'strong' methods of discovery[22].

4.4 Simon and Popper Revisited

When confronting the work of two philosophical giants, we compare their views on inquiry in science, in order to explore to what extent their stances are close together. On the one hand, although Popper [1934] was genuinely interested in an analysis of new ideas in science, he rendered the very first process of the conception of an idea to be outside the boundaries of the methodology of science, and centered his efforts in giving an account of an ensuing process, concerned with the methods of analyzing new ideas logically, and accordingly produced his method of conjectures and refutations. Simon's [1973] aim was to simulate scientific discovery at large, giving an account both for the generation and evaluation of scientific ideas, convinced that the way to go was to give both an empirical and a normative account of discovery. The former to describe and then represent computationally the intellectual development of human discoveries made in science. The latter to provide prescriptive rules, mainly in the form of heuristic strategies to perform scientific discoveries.

Both authors hold a fallibilist position, one in which there is no certainty of attaining results and where it is possible to refute already assessed knowledge, in favour of new one that better explains the world. However, while for Popper there is one single method for scientific inquiry, the method of conjectures and refutations, for Simon there are several methods for scientific inquiry, for the discovery and justification processes correspond to several heuristic strategies, largely based on pattern seeking, the logic of scientific discovery. A further difference between these approaches is found in the method itself for the advancement of science, in what they regard to be the 'logic' for discovery. While for Popper ideas are generated by the method of blind search, Simon and his team develop a full theory to support the view that ideas are generated by the method of 'selective search'. Clearly the latter account allows for a better understanding of how theories and

ill-structured problem lacks at least one of the former conditions.

[22]The former is the type of problem solving used in novel domains. It is characterized by its generality, since it does not require in-depth knowledge of its particular domain. In contrast, strong methods are used for cases in which our domain knowledge is rich, and are specially designed for one specific structure. Weak methods include generation, testing, heuristic methods, and means-ends analysis, to build explanations and solutions for given problems.

ideas may be generated. Whether any of these methods corresponds to natural phenomena or rather belongs to the province of the artificial, is another question.

Popper's and Simon's approaches are close together, at least in so far as the following basic ideas are concerned: they both hold a fallibilist stance in regard to the well-foundedness of knowledge and view science as a dynamic activity of problem solving, in which the growth of knowledge is the main aspect to characterize, as opposed to the view of science as an static enterprise in search of the assessment of theories as true. But Popper failed to appreciate the philosophical potential of a normative theory of discovery, for he was blinded to the possibility of devising a logic for the development of knowledge. His view of logic remained static: *'I am quite ready to admit that there is a need for a purely logical analysis of theories, for an analysis which takes no account of how they change and develop. But this kind of analysis does not elucidate those aspects of the empirical science which I, for one, so highly prize'* [Popper, 1934, 50].

One reason that allows for the convergence of these two accounts, perhaps obvious by now, is that neither the "Friends of discovery" really account for the epistemics of creativity at large nor Popper neglects its study entirely. Both accounts fall naturally under the study of discovery — when a broad view is endorsed (cf. footnote 20) — and neither of them rejects the context of justification, or any other context for that matter. Therefore, it seems that when the focus is on the processes of inquiry in science, rather than on the products themselves, any possible division of contexts of research is doomed to fail sooner or later.

4.5 Logic: Logic for Problem Solving

The introduction of computers had a profound impact on logical research, to the extent of providing a new paradigm, that of viewing logic in a goal directed way. This idea led Bob Kowalski to propose 'logic programming' (together with Alain Colmerauer) in the early seventies. In his own words: "*The fundamental thesis of LP is that appropriate forms of logic can serve as a high level programming language*" [Kowalski, 1994, 38].

Logic programming is inspired by first-order logic, and it consists of logic programs, queries, and an underlying inferential mechanism known as resolution. It is implemented in (amongst others) the programming language Prolog. Roughly speaking, a Prolog program P is an ordered set of rules and facts. Rules are restricted to clause form:

$$A \Leftarrow L_1, \ldots, L_n$$

which contains one atom (A) in its consequent and a set of literals in its antecedent[23]. A is called the head and $L_1, \ldots, L_n$ is called the body of the program clause. A query q (theorem) is posed to program P to be solved (proved). If the query follows from the program, a positive answer is produced, and so the query

[23] An atom is an atomic formula. A literal is an atom or the negation of an atom.

is said to be successful. Otherwise, a negative answer is produced, indicating that the query has failed. However, the interpretation of negation is 'by failure'. That is, 'no' means 'it is not derivable from the available information in P — without implying that the negation of the query $\neg q$ is derivable instead. Resolution is an inferential mechanism based on refutation working backwards: from the negation of the query to the data in the program. In the course of this process, valuable by-products appear: the so-called 'computed answer substitutions', which give more detailed information on the objects satisfying given queries.

Kowalski's 1979 book "Logic for Problem Solving", was a key source which made logicians and theoretically-oriented computer scientists start to talk to one another. While logicians were at first shocked by a formal language sensitive to rule order and in demand of strange things like the "occur check" to warrant metalogical properties, computer scientists were having a hard time to digest a declarative programming language, one in which there was no distinction between data and program, as well as with the highly demanding logical rigour of a program specification. But gradually the idea of logic programming set in the curricula and soon thereafter, new courses were given and new research was carried out.

5 THE 1980S AND 1990S: LOGICAL AND COMPUTATIONAL MODELS FOR SCIENTIFIC INFERENCE AND DISCOVERY

5.1 Artificial Intelligence and Logic: Non-monotonic Logics

The invention of computers naturally brought forward the challenge to represent knowledge in a systematic manner, and in turn, confronted logicians and computer scientists with the problem of the formalization of non-monotonic reasoning, broadly conceived as the reasoning to conclusions on the basis of incomplete information. Just as in most of scientific reasoning, given more information, one must be capable to retract previously drawn inferences.

An early attempt to formalize this type of reasoning as part of a computer's reasoning mechanism, was proposed by John McCarthy in the 1970's; but it was until the 1980's when a proliferation of non-monotonic logics was at the core of logical and computational research. A multitude of logical systems was proposed, varying both in their logical approach (syntactic, semantic) as well as in their computational particular application. These applications concern three types of problems, namely puzzles and deductive databases (DB), default reasoning, and explanation-based reasoning.

The first of these applications points to one major problem of knowledge representation in a database, that is, the way to treat the ontological status of existing information, something which led to the assumption that it contains all relevant and true information needed to reason about. A prominent proposal was the *closed-world assumption* (CWA) (cf. [Brewka *et al.*, 1997]), aiming to capture that all of the non-given information is taken to be false[24]. Still, implicit information

[24] CWA (DB) $=$ DB $\cup\{\neg P(t)|$ DB $\not\models P(t)\}$, where $P(t)$ is a ground predicate instance (a

found in almost all commonsense reasoning puzzles (such as "the only way across the river is by the boat" in the famous missionaries and cannibals puzzle), was in need of explication. In the framework of second order logic, McCarthy [1980] proposed a solution known as *circumscription*. These two solutions however, lay outside the realm of classical logic and are rather committed to one general class of non-monotonic formalisms, namely model preference logics. These systems have the property of giving a characterization of logical consequence based on a (frequently defined in advance) class of preferred models, in which positive facts are minimized.[25]

The second of these applications shaped a prominent proposal for non-monotonic logics, namely *default logic*, put forward by Reiter [1980]. This formalism is based on the notion of a default, a *prima facie* justification of a conclusion, meaning that the inference is drawn on the basis of available information, likely to be defeasible in the presence of later conflicting information. More precisely, there is an initial set of defaults, validating all new consequences consistently generated. This idea was commonly formalized as a fixed-point equation, and accordingly these logics were referred to as *fixed-point logics* or *consistently-based logics*. It is worth mentioning that a special issue of the journal *Artificial Intelligence* (vol. 13, numbers 1 and 2, 1980) was devoted to these and other new formalisms[26].

As for the third computer application, devoted to explanation-based reasoning in problems involving diagnosis, it directly points to abductive reasoning, roughly defined as the reasoning from an observation to its possible explanations. As stated in the previous section, this type of reasoning was first prompted as such by Charles Peirce (1839-1914). His logical formulation (cf. 4.2), has played a key role in Peirce scholarship, and it has been the point of departure of many classic studies on abductive reasoning in artificial intelligence [Reggia *et al.*, 1985], such as in logic programming [Kakas *et al.*, 1995], knowledge acquisition [Kakas and Mancarella, 1990], and natural language processing [Hobbs *et al.*, 1990]. Nevertheless, these approaches have paid little attention to the elements of this formulation and none to what Peirce said elsewhere in his writings. In this field, the formulation has been generally interpreted as the following logical argument-schema:

$$\frac{\begin{array}{c} C \\ A \rightarrow C \end{array}}{A}$$

where the status of A is tentative (it does not follow as a logical consequence from

ground term or predicate is that containing no variables). That is, if a ground term cannot be inferred from the database, its negation is added to the closure. Cf. [Reiter, 1987] and [Brewka *et al.*, 1997].

[25] A later proposal in this direction was Shoham's [1988] notion of causal and default reasoning, which introduces a preference order on models, requiring that only the most preferred models of the premises be included in the models of the conclusion. And this again contrasts with Tarskian classical consequence, in which it is required that all models of the premises are included in the models for the conclusion.

[26] Cf. This subsection was largely based upon [Brewka *et al.*, 1997], in which a much more in-depth analysis is to be found.

the premises). However intuitive, this interpretation certainly captures neither the fact that C is surprising nor the additional criteria Peirce proposed (cf. 4.2). The additional Peircean requirements of testability and economy are not recognized as such in AI, but are nevertheless to some extent incorporated. Economy is carried out as a further selection process to produce the best explanation, since there might be several formulae that satisfy the above formulation but are not appropriate as explanations. As for the testability requirement, when the second premise is interpreted as logical entailment this requirement is trivialized, since given that C is true, in the simplest sense of 'testable', A will always be testable.

Axiomatic Theory of Consequence Relations

The proliferation of non-monotonic systems brought forward still another challenge to logicians, this time pointing to a methodological as well as to a demarcation question. On the one hand, there was a need for a common framework in order to analyze and compare all these new systems; but at the same time, there was an urgency to put some order, and establish the limits to the kind of systems accepted as logical. Logic had gone out of its mathematical domain with apparently no rules and clear cut ends:

> *"In an attempt to put some order in what was then a chaotic field, Gabbay asked himself what minimal properties do we require of a consequence relation $A_1, \ldots, A_n \vdash B$ in order for it to be considered as a logic. In his seminal paper [1985] he proposed the following:*
>
> *Reflexivity*: $\Delta, A \vdash A$
>
> *Restricted Monotonicity:* $\dfrac{\Delta \vdash A \qquad \Delta \vdash B}{\Delta, A \vdash B}$
>
> *Cut:* $\dfrac{\Delta, A \vdash B \quad \Delta \vdash A}{\Delta \vdash B}$
>
> *The idea is to classify non-monotonic systems by properties of their consequence relations. Kraus–Lehman–Magidor [1990] developed preferential semantics corresponding to various additional conditions on $\vdash$ and this has started the area now known as the axiomatic approach to non-monotonic logics".* [Ohlbach and Reyle, 1999]

This type of analysis started with Dana Scott [1971], and was inspired in the early works of logical consequence by Tarski and those of natural deduction by Gerard Gentzen [1934]. It describes a style of inference at a very abstract structural level, giving its pure combinatorics. The basic idea of an structural analysis (as it is also known) is the following: *A notion of logical inference can be completely characterized by its basic combinatorial properties, expressed by structural rules.* Structural rules are instructions which tell us, e.g., that a valid inference remains valid when we insert additional premisses ('monotonicity'), or that we may safely chain valid inferences ('transitivity' or 'cut'). The general format is that of logical sequents, with a finite sequence of premisses to the left, and one conclusion to the right of the sequent arrow ($\Delta \Rightarrow B$).

As already mentioned, this type of analysis has proved very successful in artificial intelligence for studying different types of plausible reasoning [Kraus *et al.*, 1990], and indeed as a general framework for inference, including non-monotonic consequence relations [Gabbay, 1985]. Another area where it has proved itself is dynamic semantics, where not one but many new notions of dynamic consequences are to be analyzed [van Benthem, 1994; 1996]. This new framework served to analyze and compare many proposed logical systems. One important contribution is that it goes beyond the view of classifying a set of logical systems for what they fail to validate — not surprisingly were labelled *non-monotonic logic* — and rather looks in a positive way for the properties that they do observe. However, the claim that these three specific rules proposed by Gabbay are valid in every system was refuted.[27]

Theory Change

When talking about theory change, an obvious related territory is found in theories of belief change in AI, mostly inspired by the work of Gärdenfors [1988], a work whose roots lie in the philosophy of science. These theories describe how to incorporate a new piece of information into a database, a scientific theory, or a set of common sense beliefs.

Given a consistent theory θ, called the belief state, and a sentence φ, the incoming belief, there are three *epistemic attitudes* for θ with respect to φ: either φ is accepted ($\varphi \in \theta$), φ is rejected ($\neg\varphi \in \theta$), or φ is undetermined ($\varphi \notin \theta$, $\neg\varphi \notin \theta$). Given these attitudes, three main operations may incorporate φ into θ, thereby effecting an epistemic change in our currently held beliefs:

> Expansion ($\theta + \varphi$)
> An accepted or undetermined sentence φ is added to θ.

[27] Ten years or so later on, Gabbay himself [1994] acknowledged the following: "*although some classification was obtained and semantical results were proved, the approach does not seem to be strong enough. Many systems do not satisfy restricted monotonicty. Other systems such as relevance logic, do not satisfy even reflexivity. Others have richness of their own which is lost in a simple presentation as an axiomatic consequence relation. Obviously, a different approach is needed, one which would be more sensitive to the variety of features of the systems in the field*". As is well-known in this field, Gabbay then moved to propose his *Labelled Deductive Systems*, certainly a much more robust framework for logical systems.

Still, the question remains in regard to what extent we may use the axiomatic theory of consequence relations as an attempt to provide a logical criterion of demarcation, and this seems to be a fertile area to explore. In [Aliseda, 2005b], the suggestion is that rather than aiming at an specific set of minimal rules, we should be asking for a *minimal schema set of structural properties* a system should satisfy to be considered a logical one. The particular proposal is that this schema set must consist of some forms of monotonicity, transivity or cut, and of reflexivity. And these forms, of course, need not be the same ones as those for classical logic.

Contraction ($\theta - \sigma$):
Some sentence σ is retracted from θ, together with enough sentences implying it.

Revision ($\theta^* \varphi$):
In order to incorporate a rejected φ into θ and maintain consistency in the resulting belief system, enough sentences in conflict with φ are deleted from θ (in some suitable manner) and only then φ added.

Of these operations, revision is the most complex one. It may indeed be defined as a composition of the other two. First contract those beliefs of θ that are in conflict with φ, and then expand the modified theory with sentence φ. While expansion can be uniquely defined, this is not so with contraction or revision, as several formulas may be retracted to achieve the desired effect. These operations are intuitively non-deterministic. The contraction operation per se cannot state in purely logical or set-theoretical terms which of the available formulae should be chosen. Therefore, an additional criterion must be incorporated in order to fix which formula to retract. Here, the general intuition is that changes on the theory should be kept 'minimal', in some sense of informational economy. Various ways of dealing with the latter issue occur in the literature[28].

In practice, however, full-fledged AI systems of belief revision can be quite diverse. Here are some aspects that help to classify them:

Representation of Belief States
Operations for Belief Revision
Epistemological Stance

Regarding the first, we find there are essentially three ways in which the background knowledge θ is represented: (i) belief sets, (ii) belief bases, or (iii) possible world models[29]. As for the second aspect, operations of belief revision can be given either constructively or via 'postulates'[30].

[28]We mention only that in [Gärdenfors, 1988]. It is based on the notion of *entrenchment*, a preferential ordering which lines up the formulas in a belief state according to their importance. Thus, we can retract those formulas which are 'least entrenched' first.

[29]A belief set (i) is a set of sentences from a logical language L closed under logical consequence. In this classical approach, expanding or contracting a sentence in a theory is not just a matter of addition and deletion, as the logical consequences of the sentence in question should also be taken into account. The second approach (ii) emerged in reaction to the first. It represents the theory θ as a *base for a belief set* B_θ, where B_θ is a finite subset of θ satisfying $Cons(B_\theta) = \theta$. (That is, the set of logical consequences of B_θ is the classical belief state). The intuition behind this is that some of the agent's beliefs have no independent status, but arise only as inferences from more basic beliefs. Finally, the more semantic approach (iii) moves away from syntactic structure, and represents theories as sets W_θ of possible worlds (i.e., their models). Various equivalences between these approaches have been established in the literature (cf. [Gärdenfors and Rott, 1995]).

[30]The former approach is more appropriate for algorithmic models of belief revision, the latter serves as a logical description of the properties that any such operations should satisfy. The two can also be combined. An algorithmic contraction procedure may be checked for correctness according to given postulates. [Say, one which states that the result of contracting θ with φ should be included in the original state ($\theta - \varphi \subseteq \theta$.)].

Finally, each approach takes an 'epistemological stance' with respect to justification of the incoming beliefs. Here are two major paradigms. A 'foundationalist' approach argues one should keep track of the justification for one's beliefs, whereas a 'coherentist' perspective sees no need for this, as long as the changing theory stays consistent and keeps its overall coherence.

Therefore, each theory of epistemic change may be characterized by its representation of belief states, its description of belief revision operations, and its stand on the main properties of belief one should be looking for[31]. In particular, the theory proposed in Gärdenfors [1988], known as the AGM paradigm after its original authors [Alchourrón, Gärdenfors and Makinson, 1985], represents belief states as theories closed under logical consequence, while providing 'rationality postulates' to characterize the belief revision operations, and finally, it advocates a coherentist view. The latter is based on the empirical claim that people do not keep track of justifications for their beliefs, as some psychological experiments seem to indicate (cf. [Harman, 1986]).

5.2 Cognitive Science and Philosophy of Science: Computational Philosophy of Science

Concrete and quite articulated computer programs of scientific discovery are found in the late eighties in the work of Simon and his team [Langley *et al.*, 1987]. These are the BACON system, one which simulates the discovery of quantitative laws in Physics (such as Kepler's laws and Ohm's law) and the GLAUBER program, which simulates the discovery of qualitative laws in Chemistry. In the same spirit, Paul Thagard proposes a new field of research, namely *Computational Philosophy of Science* [Thagard, 1988], an integrated approach of psychology, history and philosophy of science, all of it directed to questions of scientific discovery, having to do with its cognitive patterns, its place and time in the history of science and with core notions in the philosophy of science (such as explanation, confirmation, falsification, evaluation, induction, abduction and theory revision). Thagard's proposal, for example, puts forward the computational program PI (Processes of Induction) to model some aspects of scientific practice, such as concept formation and theory building. The general idea consists of the solution of a problem as a "match" between an initial and a final state. And when there is no match, several kinds of induction may be performed (generalization, abduction, concept formation, etc...).

We may identify at least two principles (1 and 2 below) and three claims (given 3) which characterize research and the computer programs found in the area of

[31]These choices may be interdependent. Say, a constructive approach might favor a representation by belief bases, and hence define belief revision operations on some finite base, rather than the whole background theory. Moreover, the epistemological stance determines what constitutes *rational epistemic change*. The foundationalist accepts only those beliefs which are justified, thus having an additional challenge of computing the reasons for an incoming belief. On the other hand, the coherentist must maintain coherence, and hence make only those minimal changes which do not endanger (at least) consistency.

computational philosophy of science, namely:

1. Scientific discovery is problem solving

2. The study of discovery is part of the methodological agenda of philosophy of science.

3. The computer programs are to be historically, psychologically and philosophically adequate.

The first principle is in accordance with the paradigm already identified as emerging in the 70's, that of science as problem solving, but in this case applied to discovery alone. Thus, the second principle states that problem solving is a notion to be handled within the methodology of science. In turn, this conceptual move suggests that existing computational and logical tools devised for other disciplines, like those existing in cognitive science and in artificial intelligence, may be imported and thus help bring some order to represent and model the aspects and machinery of scientific knowledge, its birth and development as well. Heuristic strategies are immersed in BACON's computer discovery simulation, machine learning is performing Popper's neglected induction, and abductive inference is modelling the epistemics of explanation generation.

The three claims given in (3) (identified in [Kuipers, 2001]), point to adequacy conditions for discovery computer programs. Ideally, a computer program that simulates discoveries should capture some aspects of its history, at least as far as a credible description of its development is told. Moreover, the design of this kind of computer programs should not overlook that it is, to some extent, a simulation of the real way a human would proceed. This implies some kind of cognitive commitment with its machinery, which gives sense to the psychological adequacy requirement. Finally, the computer program must be philosophically adequate in that there must exist some philosophical theory as a base for its epistemics. However, as noted by Kuipers, these claims "*may come into conflict with one another*" [Kuipers, 2001, 290].

The resulting enterprise is impressive in the unification of the disciplines involved, all for a common goal, that of giving an account of discovery and development of a privileged type of human knowledge, scientific knowledge. The methodological point is that the methods and heuristic strategies existing in computer science, have proved useful in artificial intelligence and cognitive simulation, and are used by several computer programs. All these tools have therefore been imported to philosophy of science to give a computer modeling account of processes such as explanation, confirmation, falsification, evaluation and discovery, and in general, of the modeling of the dynamics of scientific theories.

A major criticism to this whole enterprise however, is reflected in the debate of whether these computer programs really make new discoveries, for they seem to produce theories new to the program but not new to the world, and its discoveries seem spoon-fed rather than created. However, more recent research reveals that indeed the computer has been able to help produce important new research. One prominent example is reported in [Gillies, 1996, 50–55], and concerns the discovery of new laws about protein secondary structure. Further cases concern taxonomic discoveries in astrophysics, as well as qualitative laws in biochemical cancer research [Langley, 2000].

The Case of Abduction

Here we will deal with a particular case of scientific inference, that of inference to explanatory hypotheses, namely abduction. Research on abduction in artificial intelligence dates back to the seventies [Pople, 1973], but it is only fairly recently that it has attracted great interest, in areas like logic programming [Kakas *et al.*, 1995], knowledge assimilation [Kakas and Mancarella, 1990], and diagnosis [Poole *et al.*, 1987], to name just a few. It has been a topic of several workshops in artificial intelligence conferences (ECAI96, IJCAI97, ECAI98, ECAIOO) and model-based reasoning ones (MBR'98, MBR'01). It has also been at the center of some computer applied publications [Josephson and Josephson, 1994], and also present when compared with induction [Flach and Kakas, 2000]. Moreover, explanation based systems for computer applications were at the core of research in non-monotonic logics (cf. section 5.1). In all these places, the discussion about the different aspects of abduction has been conceptually challenging but also shows a (terminological) confusion with its close neighbour, induction.

We will now present the standard logical format for this inference, followed by the implementation given by the logic programming community, to continue by a proposal for a general taxonomy of this kind of reasoning. We then propose a particular interpretation, which conceives abduction as a process of epistemic change, a conception which goes beyond the interpretation of abduction as logical inference. We will finish by a brief coverage of the place of abduction in cognitive science, broadly conceived.

Abduction as Inference

The general trend in logic based approaches to abduction in AI interprets abduction as *backwards deduction plus additional conditions.* This brings it very close to deductive-nomological explanation in the Hempel style, witness the following format. What follows is the standard version of abduction as deduction via some consistent additional assumption, satisfying certain extra conditions. It combines some common requirements from the literature (cf. [Konolige, 1990; Kakas *et al.*, 1995; Mayer and Pirri, 1993; Aliseda, 1997] for further motivation):

Given a theory θ (a set of formulae) and a formula φ (an atomic formula), α is an explanation if

$\theta, \alpha \models \varphi$
α is consistent with θ

α is minimal[32]
α has some restricted syntactical form (usually an atomic formula or a conjunction of them).

An additional condition not always made explicit is that φ is not a logical consequence of θ. This says that the fact to be explained should not already follow from the background theory alone. Sometimes, the latter condition figures as a precondition for an *abductive problem* (cf. [Kakas *et al.*, 1995]).

Abduction as Computation in Logic Programming

Abduction emerges naturally in logic programming (cf. section 4.5) as a 'repair mechanism', completing a program with the facts needed for a query to succeed. This may be illustrated by the famous abductive rain example in Prolog:

Program P:
lawn-wet $\leftarrow$ rain.
lawn-wet $\leftarrow$ sprinklers-on.

Query q: lawn-wet.

Given program P, query q does not succeed because it is not derivable from the program. For q to succeed, either one (or all) of the facts 'rain', 'sprinklers-on', 'lawn-wet' would have to be added to the program. Abduction is the process by which these additional facts are produced. This is done via an extension of the resolution mechanism that comes into play when the backtracking mechanism fails. In our example above, instead of declaring failure when either of the above facts is not found in the program, they are marked as 'hypothesis', and proposed as those formulas which, if added to the program, would make the query succeed.

In actual Prolog abduction, for these facts to be counted as abductions, they have to belong to a pre-defined set of 'abducibles', and to be verified by additional conditions (so-called 'integrity constraints'), in order to prevent a combinatorial explosion of possible explanations[33].

[32] There are several ways to characterize minimality, cf. [Aliseda, 2006].

[33] In logic programming, the procedure for constructing explanations is left entirely to the resolution mechanism, which affects not only the order in which the possible explanations are produced, but also restricts the form of explanations, for rules cannot occur as abducibles, since explanations are produced out of sub-goal literals that fail during the backtracking mechanism. The additional restrictions select the best hypothesis. Thus, processes of both construction and selection of explanations are clearly marked in logic programming. (Another relevant connection here is to research in 'inductive logic programming' [Michalski, 1994], which integrates abduction and induction.). Logic programming does not use blind deduction. Different control mechanisms for proof search determine how queries are processed. This additional degree of freedom is crucial to the efficiency of the enterprise. Hence, different control policies will vary in the abductions produced, their form and the order in which they appear. To us, this variety suggests that the procedural notion of abduction is intensional, and must be identified with different practices, rather than with one deterministic fixed procedure.

A Taxonomy for Abduction

What we have seen so far may be summarized as follows. Abduction is a general process for producing explanations, with a certain inferential structure. We consider these two aspects to be of equal importance. Moreover, on the process side, we may distinguish between constructing possible explanations and selecting the best one amongst these. As for the logical form of abduction, we have found that it may be viewed as a threefold relation:

$$T, C \Rightarrow E$$

between an observation E, an abduced item C, and a background theory T. (Other parameters are possible here, such as a preference ranking — but these would rather concern the further selection process.) Against this background, we propose three main parameters that determine types of abduction. (i) An 'inferential parameter' ($\Rightarrow$) sets some suitable logical relationship among explananda, background theory, and explanandum. (ii) Next, 'triggers' determine what kind of abduction is to be performed: E may be a novel phenomenon, or it may be in conflict with the theory T. (iii) Finally, 'outcomes' (C) are the various products of an abductive process: facts, rules, or even new theories.

In the above schema, the notion of explanatory inference $\Rightarrow$ is not fixed. It can be classical derivability $\vdash$ or semantic entailment $\models$, but it does not have to be. Instead, we regard it as a parameter which can be set independently. It ranges over such diverse values as probable inference ($T, C \Rightarrow_{probable} E$), in which the explanans renders the explanandum only highly probable, or as the inferential mechanism of logic programming ($T, C \Rightarrow_{prolog} E$). Further interpretations include dynamic inference ($T, C \Rightarrow_{dynamic} E$, cf. [van Benthem, 1996]), replacing truth by information change potential along the lines of belief update or revision. Our point here is that abduction is not one specific non-standard logical inference mechanism, but rather a way of using any one of these.

As previously stated, for Peirce, abductive reasoning is triggered by a *surprising phenomenon*. The notion of surprise, however, is a relative one, for a fact E is surprising only with respect to some background theory T providing 'expectations'. What is surprising to me (e.g. that the canal bridge floor goes up from time to time) might not be surprising to a Dutch person. We interpret a surprising fact as one which needs an explanation. From a logical point of view, this assumes that the fact is not already explained by the background theory $T : T \not\Rightarrow E$.

Moreover, our claim is that one also needs to consider the status of the negation of E. Does the theory explain the negation of observation instead ($T \Rightarrow \neg E$)? Thus, we identify at least two triggers for abduction: *novelty* and *anomaly*:

Abductive Novelty: $T \not\Rightarrow E$, $T \not\Rightarrow \neg E$

E is novel. It cannot be explained ($T \not\Rightarrow E$), but it is consistent with the theory ($T \not\Rightarrow \neg E$).

Abductive Anomaly: $T \not\Rightarrow E$, $T \Rightarrow \neg E$

E is anomalous. The theory explains rather its negation ($T \Rightarrow \neg E$).

As already stated, novelty is the condition for an abductive problem. The suggestion in [Aliseda, 1997; 2006] is to incorporate anomaly as a second basic type[34]. Abducibles themselves come in various forms: facts, rules, or even theories. Sometimes one simple fact suffices to explain a surprising phenomenon, such as rain explaining why the lawn is wet. In other cases, a rule establishing a causal connection might serve as an explanation, as in our case connecting cloud types with rainfall. And many cases of abduction in science provide new theories to explain surprising facts. These different options may sometimes exist for the same observation, depending on how seriously we want to take it[35].

Once the above parameters get set, several kinds of abductive processes arise. For example, abduction triggered by novelty with an underlying deductive inference, calls for a process by which the theory is expanded with an explanation. The fact to be explained is consistent with the theory, so an explanation added to the theory accounts deductively for the fact. However, when the underlying inference is statistical, in a case of novelty, theory expansion might not be enough. The added statement might lead to a 'marginally consistent' theory with low probability, which would not yield a strong explanation for the observed fact. In such a case, theory revision is needed (i.e. removing some data from the theory) to account for the observed fact with high probability.

Our aim is to point out that several kinds of abductive processes are used for different combinations of the above parameters. (In Aliseda [1997] some procedures for computing different types of outcomes in a deductive format are explored in detail).

This taxonomy gives us the big picture of abductive reasoning. We can now see the patterns in a clearer focus. Varying the inferential parameter, we cover not only cases of deduction (plus additional conditions) but also statistical inferences. Different forms of outcomes will play a role in different types of procedures for producing explanations. In computer science jargon, triggers and outcomes are, respectively, preconditions and outputs of abductive devices, whether these be computational procedures or inferential ones.

Abduction as Belief Revision in Theory Change

Abductive reasoning may be seen as an epistemic process for belief revision. In this context an incoming sentence φ is not necessarily an observation, but rather a belief for which an explanation is sought. Existing approaches to abduction

[34]Of course, non-surprising facts (where $T \Rightarrow E$) should not be candidates for explanation. Even so, one might speculate if facts which are merely probable on the basis of T might still need explanation of some sort to further cement their status.

[35]Moreover, we are aware of the fact that genuine explanations sometimes introduce new concepts, over and above the given vocabulary. (For instance, the eventual explanation of planetary motion was not Kepler's, but Newton's, who introduced a new notion of 'force' — and then derived elliptic motion via the Law of Gravity.) Abduction via new concepts is outside the scope of our analysis. (Cf. [Thagard, 1992]) for an account of new concepts via conceptual combination).

usually do not deal with the issue of incorporating φ into the set of beliefs. Their concern is just how to give an account for φ. If the underlying theory is closed under logical consequence, however, then φ should be automatically added once we have added its explanation (which a foundationalist would then keep tagged as such).

Practical connections of abduction to theories of belief revision have often been noted. Of many references in the literature, we mention Aravindan and Dung [1994] (which uses abductive procedures to realize contractions over theories with 'immutability conditions'), and Williams [1994] (which studies the relationship between explanations based on abduction and 'Spohnian reasons').

Our claim however, is stronger. Abduction can function in a model of theory revision as a means of producing explanations for incoming beliefs. But also more generally, as defined above, it provides a model for epistemic change. Let us discuss some reasons for this. First, what we call the two 'triggers' for abductive reasoning correspond to two of the three epistemic attitudes of a formula introduced by Gärdenfors's (cf. section 5.1), viz., being undetermined or rejected. We did not consider accepted beliefs, since these do not call for explanation.

> φ is a novelty iff neither φ nor $\neg\varphi$ is a logical consequence of θ
>
> (φ is undetermined)
>
> φ is an anomaly iff φ is not a logical consequence of θ and $\neg\varphi$ is indeed a logical consequence of θ
>
> (φ is rejected)
>
> φ is an accepted belief
>
> (φ is a logical consequence of θ)[36].

In our account of abduction, both a novel phenomenon and an anomalous one involve a change in the original theory. The former calls for expansion and the latter for revision, which in turn involves contraction and then expansion. So, the basic operations for abduction are expansion and contraction. Therefore, both epistemic attitudes and changes in them are reflected in the presented abductive model.

However, our main concern is not the incoming belief φ itself. We rather want to compute and add its explanation α. But since φ is a logical consequence of the revised theory, it could easily be added. Here, then, are our abductive operations for epistemic change:

> Abductive Expansion
>
> Given a novel formula φ for θ, a consistent explanation α for φ is computed and then added to θ.

[36]The epistemic attitudes are presented in Gärdenfors [1988] in terms of membership (e.g., a formula φ is accepted if $\varphi \in \theta$). We defined them in terms of entailment, since theories are not necessarily closed under logical consequence.

Abductive Revision

Given a novel or an anomalous formula φ for θ, a consistent explanation α for φ is computed, which will involve modification of the background theory θ into some suitably new θ'. Again, intuitively, this involves both 'contraction' and 'expansion' (cf. section 5.1).

In its emphasis on explanations, our abductive model for belief revision is richer than many theories of belief revision[37]. Admittedly, though, not all cases of belief revision involve explanation, so our greater richness also reflects our restriction to a special setting.

Abduction in Cognitive Science

In cognitive science, abduction is a crucial ingredient in processes like inference, learning, and discovery, performed by people and computers to build theories of the world surrounding them. There is a growing literature on computer programs modelling these processes, and on abduction in particular.

A noteworthy reference is found in the earlier mentioned field of computational philosophy of science, and in broader computational cognitive studies of inductive reasoning [Thagard, 1988; 1992]. These studies distinguish several relevant mechanisms for hypotheses generation, indeed four kinds of abduction are implemented in the program PI (Processes of Induction): "*simple, existential, rule-forming, and analogical. Simple abduction produces hypotheses about individual objects ... Existential abduction postulates the existence of previously unknown objects... Rule-forming abduction produces rules that explain other rules, and hence is important for generating theories that explain laws. Finally, analogical abduction uses past cases of hypothesis formation to generate hypotheses similar to existing ones.*" [Thagard, 1992, 54]

This particular approach shows, on the one hand, the multiplicity of contexts in which abduction may appear, something which explains the need of a further distinction into abductive kinds, and on the other hand, it shows that it is closely related to other inferential processes, such as induction. In fact, simple abduction[38] seems to be a case of enumerative induction, perhaps one in which the conclusion is not a general or universal one.

5.3 *Philosophy of Science: Logics of Discovery*

The renewed enterprise of logics of discovery is based on two fundamental assumptions. The first of them is that the context of discovery allows, to some extent,

[37]Cf. [Aliseda, 2000] for a discussion of the type of belief revision theory this model of abduction corresponds to.

[38]According to Thagard: "the simplest case of abduction is one in which you want to explain why an object has some characteristic and you know that all objects with a particular property have that characteristic. Hence you conjecture that the object has the property in order to explain why it has the characteristic." [Thagard, 1992, 9]

a precise formal treatment. The second one claims that there is no single logical method in scientific practice in general, and with respect to abduction in particular. By this assumption, however, it is not claimed that it is possible to provide a logical analysis for all and every part of scientific inquiry. In this respect, the enterprise is as modest as the one proposed by Simon (cf. 4.3) and has no pretensions that it can offer either a logical analysis of great scientific discoveries, or put forward a set of logical systems that would provide general norms to make new discoveries. The aim is rather to lay down logical foundations in order to explore some of the formal properties under which new ideas may be generated and evaluated. The compensation we gain from this very modest approach is that we can gain some insight into the logical features of some parts of the scientific discovery and explanation processes. This is in line with a well-known view in the philosophy of science, namely that phenomena take place within traditions, something which echoes Kuhn's distinction between normal and revolutionary science. Hence, a general assumption is that a logical analysis of scientific discovery is for normal science, not denying there may be a place for some other kind of logical analysis of revolutionary science, but clearly leaving it out of the scope of this enterprise, at least for the moment.

The potential for providing a logic for scientific discovery is found in a normative account in the methodology of science, such as the one proposed by Simon (cf. 4.3)[39]. However, in this kind of computational approach, logic is identified with *pattern seeking methods*, a notion which fits very well their algorithmic and empirical approach to the question of a logic of discovery, but has little to do with providing logical foundations for their programs, either as conceived in the logico-mathematical tradition or as in artificial intelligence logical research.

Our claim is that logic, as understood in modern non-standard formal systems, has a place in the study of logics of discovery. By putting forward a logic for scientific discovery we claim no lack of rigour. What is clear is that standard deductive logic cannot account for abductive or inductive types of reasoning, but the present situation in logical research has gone far beyond the formal developments that deductive logic reached last century, and new research includes the formalization of several other types of reasoning, like induction and abduction. And a general goal could be to study the wider field of human reasoning while hanging on to these standards of rigour and clarity.

---

[39]From antiquity to the mid 19th century, researchers had in mind a logic of discovery of a descriptive nature, one that would capture and describe the way humans reason in science. These 'logics', however had little success, for they failed to provide such an account of discovery. Thereafter, when the search for a logic of discovery was abandoned, a normative account prevailed in favour of proposals of logics of justification. Regarding the approaches of Popper and Simon in this respect, it is clear that while Popper overlooked the possibility of a normative account of logics of discovery, Simon centered his efforts in the development of heuristic procedures, to be implemented computationally, but not on logics — per se — of discovery. In fact, there are proposals strictly normative and formal in nature, such as that found in Kelly [1997], in which one argues for a computational theory as the foundation of a logic of discovery, one which studies algorithmic procedures for the advancement of science. (Cf. [Aliseda, 2006] for an in-depth analysis of the development of logics of discovery).

In connection to philosophy of science, there are already some logical proposals that lay bridges between non-monotonic logics and Hempel's models for explanation. In [Tan, 1992] an inductive statistical model is constructed based on Reiter's default logic and in [Aliseda, 1997; 2006] two models of scientific explanation (deductive and statistical) are presented as cases of abductive logic, thus claiming that these models do not follow the canons of classical logic. These proposals naturally bring up the question of the properties of such enriched notions of consequence, which are in turn studied within the axiomatic theory of consequence relations (Cf. [Aliseda, 1997; 2006] for a structural characterization of abduction).

There are still many challenges ahead for the formal study of reasoning in scientific discovery, such as giving an integrated account of deductive, inductive, abductive and analogical styles of inference, the use of diagrams by logical means, and in general the device of logical operations for theory building and change. Already new logical research is moving into these directions. In Burger and Heidema [2002] degrees of 'abductive boldness' are proposed as spectrum for inferential strength, ranging from cases with poor background information to those with (almost) complete information. Systems dealing with several notions of derivability all at once have also been proposed. A formula may be 'unconditionally derived' or 'conditionally derived', the latter case occurring when a line in a proof asserts a formula which depends on hypotheses which may be later falsified, thus pointing to a notion of proof which allows for addition of lines which are non-deductively derived as well as for deletion of them when falsifying instances occur. This account is found in the framework of (ampliative) adaptive logics, a natural home for abductive inference [Meheus *et al.*, 2002] and for (enumerative) induction inference [Batens, 2005] alike, in which it is possible to combine deduction with ampliative steps [Meheus, 1999].

The formalization of analogical reasoning is still a growing area of research, without a precise idea of what exactly an analogy amounts to. Perhaps research on mathematical analogy [Polya, 1954], work on analogy in cognitive science [Helman, 1988] or investigations into analogical argumentation theory recently proposed for abduction [Gabbay and Woods, 2005], may serve to guide research in this direction. Finally, the study of 'diagrammatic reasoning' is a research field on its own right [Barwise and Etchemendy, 1995], showing that the logical language is not restricted to the two-dimensional left to right syntactic representation, but its agenda still needs to be expanded on research for non-deductive logics.

As for formal approaches for theory building and change applied to philosophy of science, one line of contemporary research in this direction, adopted by Aliseda [2005a], concerns the extension of a classical logical method to model empirical progress in science, as conceived by Theo Kuipers [1999; 2001]. In particular, the goal is to operationalize the task of *instrumentalist abduction*, that is, theory revision aiming at empirical progress. This particular account shows that evaluation and improvement of a theory can be modelled by (an extension of) the framework of Semantic Tableaux [Aliseda, 2005a].

All the above suggests that the use of non-standard logics to model processes in scientific practice, such as confirmation, falsification, explanation building, theory improvement and discovery is, after all, a feasible project. All of these logics are either characterized via structural rules, by an axiomatic or semantic approach as well as within dynamic theory revision systems. Nevertheless, this claim requires a broad conception of what logic is about as well as a modest account of what is it to be found.

6 CONCLUSIONS

Philosophy of science in the twentieth century has been remarkably rich and varied in character, and has passed through a series of quite radical changes. This chapter began by analysing the Vienna Circle approach which dominated the discipline from the 1920s to the end of the 1950s. This approach was based on the idea that philosophy of science should consist of the logical analysis of scientific theories. The logic to be used in this task might include inductive logic with the associated concepts of probability and confirmation, but, as far as deductive logic was concerned, it was limited to classical logic with the occasional reference to intuitionistic logic. The distinction between the context of discovery and the context of justification was generally accepted, and this, together with the narrow notion of what constituted logic, led to the conclusion that philosophy of science should concentrate on the justification of scientific theories, leaving the question of discovery to other disciplines.

The dominance of the logical approach was challenged in the 1960s by the emergence (or perhaps better re-emergence) of the historical approach to the philosophy of science. This allowed questions concerned with scientific discovery to be tackled once again, though they were approached from a historical rather than logical point of view. Some of the proponents of the historical approach (notably Kuhn and Feyerabend) went as far as to reject the logical approach (particularly confirmation theory) even in the context of justification. Others, however, (particularly Popper and Lakatos) were more sympathetic to logic, and sought to combine logic with the new historical approach.

From the mid-1970s we drew attention to a new force at work on the development of philosophy of science, namely the increasing importance of the computer. Computer science led to cognitive psychology and both together gave rise to the concept of science as problem solving — a conception to be found in both Simon and in Popper's works of this period. Simon, however, had more involvement with the computer, and in particular raised the question of whether computers could be programmed to make scientific discoveries. This helped to bring scientific discovery to the fore as a central problem in the philosophy of science, but, in contrast to the earlier historical period, the analysis of discovery could now be carried out with logical and computational techniques. Further developments in computer science, cognitive science and logic itself, provided a new set of tools of a logical and computational nature. The rather limited logic used by the Vienna Circle was now augmented by the discovery of quite new systems of logic such as non-monotonic logics. These strengthened the trend towards including issues of discovery as part of the philosophy of science agenda in the eighties and nineties. This period has been characterised by results and concrete proposals, for the representation and modelling of common sense reasoning, scientific inference, knowledge growth and discovery. In part, the emergence of new research in logic in the field of artificial intelligence, has helped to analyse, in a fresh and more powerful manner, issues of scientific inference in a rigorous and a systematic way.

As for the connection between logic and history, with the emergence of computer science a new contact with history is presented, which gave rise to computational philosophy of science, a place in which history and computing meet on a par. This creates the possibility of a partial synthesis between the logical and the historical approaches with a new computational element added to the mix. This has been part of the research agenda of a number of contemporary philosophers of science since about the mid 80's, but is by no means a privileged topic.

To conclude, the present setting, in which we have all logical, historical and computational approaches to philosophy of science, fosters the view that what we need is a balanced philosophy of science, one in which we take advantage of a variety of methodologies, such as logical, computational and also historical, all together giving a broad view of science. This view was already presented by Suppes [1951–69] in the late sixties. We know at present that logical models (classical or otherwise) are insufficient to completely characterize notions like explanation, confirmation or falsification in philosophy of science, but this fact does not rule out that some problems in the history of science may be tackled from a formal point of view. For instance, "*some claims about scientific revolutions, seem to require statistical and quantitative data analysis, if there is some serious pretension to regard them with the same status as other claims about social or natural phenomena*" [Suppes, 1951–69, 97]. In fact, *Computational Philosophy of Science* may be regarded as a successful marriage between historical and formal approaches. It is argued that although several heuristic rules have been derived from historical reconstructions in Science, they are proposed to be used for future research" [Meheus and Nickles, 1999].

This is not to say however, that historical analysis of scientific practice could be done in a formal fashion, or that logical treatment should contain some kind of "historic parameter" in its methodology, but we claim instead that these two views should share their insights and findings in order to complement each other.

ACKNOWLEDGMENTS

We are grateful to Ladislav Kvasz who read an earlier draft of sections 1, 2 and 3 carefully and made many useful suggestions which have been incorporated into the final version.

BIBLIOGRAPHY

[Alchourrón *et al.*, 1985] C. Alchourrón, P. Gärdenfors, and D. Makinson. 'On the logic of theory change: Partial meet contraction and revision functions'. *Journal of Symbolic Logic*, 50: 510–530, 1985.

[Aliseda, 1997] A. Aliseda. *Seeking Explanations: Abduction in Logic, Philosophy of Science and Artificial Intelligence.* PhD Dissertation, Philosophy Department, Stanford University. Published by the Institute for Logic, Language and Computation (ILLC), University of Amsterdam (ILLC Dissertation Series 1997–4). 1997.

[Aliseda, 2000] A. Aliseda. Abduction as epistemic change: A Piercean model in Artificial Intelligence. In P. Flach and A. Kakas (eds.), *Abductive and Inductive Reasoning: Essays on their Relation and Integration*, pages 45–58. Kluwer Academic Press, 2000.

[Aliseda, 2004] A. Aliseda. Sobre la lógica del descubrimiento científico de Karl Popper. Suplemento 11 (Monográfico Karl Popper), *Signos Filosóficos*, pages 115–130. Universidad Autónoma Metropolitana. México, 2004. Translated as "On Karl Popper's Logic of Scientific Discovery", in L. Magnani (ed.), *Model Based Reasoning in Science and Engineering*, King's College Publications, 2006.

[Aliseda, 2005a] A. Aliseda. Lacunae, empirical progress and semantic tableaux. In R. Festa, A. Aliseda, and J. Peijnenburg (eds.), *Confirmation, Empirical Progress, and Truth Approximation: Essays in Debate with Theo Kuipers* (Volume 1), pages 169–189. Poznan Studies in the Philosophy of the Sciences and the Humanities, Vol. 83, 2005.

[Aliseda, 2005b] A. Aliseda. What is a logical system? A commentary. In Artemov *et al.* (eds.), *We Will Show Them! Essays in Honour of Dov Gabbay on his 60th Birthday*, Volume 2. College Publications. King's College, 2005.

[Aliseda, 2006] A. Aliseda. *Abductive Reasoning: Logical Investigations into Discovery and Explanation.* Synthese Library, Vol. 330. Springer-Kluwer Academic Publishers, 2006.

[Aravindan and Dung, 1994] C. Aravindan and P. M. Dung. Belief dynamics, abduction and databases. In C. MacNish, D. Pearce, and L. M. Pereira (eds.), *Logics in Artificial Intelligence. European Workshop JELIA'94*, pages 66–85. Lecture Notes in Artificial Intelligence 838. Springer-Verlag, 1994.

[Aristotle, 1941] Aristotle. *Nicomachean Ethics.* English Translation by W. D. Ross in R. McKeon (ed.), *The Basic Works of Aristotle*, pages 927–1112, Random House, 1941.

[Barwise and Etchemendy, 1995] J. Barwise and J. Etchemendy. *Hyperproof.* Center for the Study of Language and Information (CSLI). Lecture Notes Series 42. Stanford, CA, 1995.

[Batens, 2005] D. Batens. A logic of induction. In R. Festa, A. Aliseda, and J. Peijnenburg (eds.), *Cognitive Structures in Scientific Inquiry: Essays in Debate with Theo Kuipers*, Volume 1, pages 221–247. Poznan Studies in the Philosophy of the Sciences and the Humanities, vol. 83, 2005.

[van Benthem, 1984] J. van Benthem. Lessons from Bolzano. Center for the Study of Language and Information. Technical Report CSLI-84-6. Stanford University. 1984. Later published as 'The variety of consequence, according to Bolzano'. *Studia Logica*, 44: 389–403, 1985.

[van Benthem, 1994] J. van Benthem. General dynamic logic. In D. M. Gabbay (ed.), *What is a Logical System?*, pages 107–140, Clarendon Press. Oxford, 1994.

[van Benthem, 1996] J. van Benthem. *Exploring Logical Dynamics.* CSLI Publications, Stanford University, 1996.

[Bolzano, 1837] B. Bolzano. *Wissenschaftslehre*, Seidel Buchhandlung, Sulzbach, 1837. Translated as *Theory of Science* by B. Torrel, edited by J. Berg. D. Reidel Publishing Company. Dordrecht, The Netherlands. 1973.

[Brewka *et al.*, 1997] G. Brewka, J. Dix and K. Konolige. *Non-Monotonic Reasoning: An Overview.* Center for the study of Language and Information (CSLI), Lecture Notes 73, 1997.

[Burger and Heidema, 2002] I. C. Burger and J. Heidema. Degrees of abductive boldness. In L. Magnani, N. J. Nersessian, C. Pizzi (eds.), *Logical and Computational Aspects of Model-based Reasoning*, Kluwer Applied Logic Series, pages 181–198. Kluwer Academic Publishers, 2002.

[Carnap, 1950] R. Carnap. *Logical Foundations of Probability.* University of Chicago Press, 1950. 2nd Edition, 1963.

[Carnap, 1963] R. Carnap. Intellectual autobiography. In P. A.Schilpp (ed.), *The Philosophy of Rudolf Carnap*, Library of Living Philosophers, Open Court, pages 3–84, 1963.

[Chomsky, 1972] N. Chomsky. *Language and Mind* (Enlarged Edition). New York: Harcourt Brace & Jovanovich, 1972.

[Chomsky, 1976] N. Chomsky. *Reflections on Language.* Glasgow: Fontana/Collins, 1976.

[Chomsky, 1993] N. Chomsky. *Lectures on Government and binding: The Pisa Lectures.* Mouton de Gruyter, Berlin. 7nd Edition, 1993.

[English, 1978] J. English. Partial interpretation and meaning change. *The Journal of Philosophy,* 75: 57–76, 1978.

[Feyerabend, 1970] P. Feyerabend. Consolations for the specialist. In Lakatos and Musgrave (eds.), pages 197–230, 1970.

[Feyerabend, 1975] P. Feyerabend. *Against Method. Outline of an Anarchist Theory of Knowledge.* 1975. Verso, 1984.

[Feyerabend, 1978] P. Feyerabend. *Science in a Free Society.* 1978. Verso, 1985.

[Feyerabend, 1987] P. Feyerabend. *Farewell to Reason.* Verso, 1987.

[Flach and Kakas, 2000] P. Flach and A. Kakas. *Abduction and Induction. Essays on their Relation and Integration.* Applied Logic Series. volume 18. Kluwer Academic Publishers. Dordrecht, The Netherlands, 2000.

[Fodor, 1975] J. Fodor. *The Language of Thought.* New York, Crowell, 1975.

[Frank, 1941] P. Frank. *Modern Science and its Philosophy.* 1941. Paperback edition. Collier Books, 1961.

[Fullbrook, 2004] E. Fullbrook (ed.) *A Guide to What's Wrong with Economics.* Anthem, 2004.

[Gabbay, 1985] D. M. Gabbay, Theoretical foundations for non-monotonic reasoning in expert systems. In K. Apt (ed.), *Logics and Models of Concurrent Systems,* pages 439–457. Springer-Verlag. Berlin, 1985.

[Gabbay, 1994] D. M. Gabbay (ed.). *What is a Logical System?* Clarendon Press. Oxford, 1994.

[Gabbay and Woods, 2005] D. M. Gabbay and J. Woods. *A Practical Logic of Cognitive Systems. The Reach of Abduction: Insight and Trial.* Amsterdam: North-Holland, volume 2, 2005.

[Gadol, 1982] E. Gadol (ed.). *Rationality and Science. A Memorial Volume for Moritz Schlick in Celebration of the Centennial of his Birth.* Springer Verlag, 1982.

[Galliers, 1992] J. L. Galliers. Autonomous belief revision and communication. In P. Gärdenfors (ed.), *Belief Revision,* pages 220–246. Cambridge Tracts in Theoretical Computer Science, Cambridge University Press, 1992.

[Gärdenfors, 1988] P. Gärdenfors. *Knowledge in Flux: Modeling the Dynamics of Epistemic States.* MIT Press, 1988.

[Gärdenfors and Rott , 1995] P. Gärdenfors and H. Rott. Belief revision. In D. M. Gabbay, C. J. Hogger and J. A. Robinson (eds.), *Handbook of Logic in Artificial Intelligence and Logic Programming.* Volume 4, Clarendon Press, Oxford Science Publications, 1995.

[Gentzen, 1934] G. Gentzen. *Recherches sur la deduction loguique.* Translaction of *Untersuchungen über das logische schliessen 1934,* R. Feys and J. Ladriere P.U.F., Paris, 121955, 1934.

[Gillies, 1993] D. A. Gillies. *Philosophy of Science in the Twentieth Century. Four Central Themes.* Blackwell, 1993.

[Gillies, 1996] D. A. Gillies. *Artificial Intelligence and Scientific Method.* Oxford University Press, 1996.

[Gillies and Zheng, 2001] D. A. Gillies and Y. Zheng. Dynamic interactions with the philosophy of mathematics. *Theoria,* 16: 437–459, 2001.

[Ginsberg, 1988] A. Ginsberg. Theory revision via prior operationalization. In *Proceedings of the Seventh Conference of the AAAI,* 1988.

[Haack, 1993] S. Haack. *Evidence and Inquiry. Towards Reconstruction in Epistemology.* Blackwell, Oxford UK and Cambridge, Mass., 1993.

[Hanson, 1961] N. R. Hanson. *Patterns of Discovery.* Cambridge at The University Press, 1961.

[Harman, 1986] G. Harman. *Change in View: Principles of Reasoning.* Cambridge, Mass. MIT Press, 1986.

[Helman, 1988] D. H. Helman (ed.). *Analogical Reasoning : Perspectives of Artificial Intelligence, Cognitive Science and Philosophy.* Dordrecht, Netherlands: Reidel, 1988.

[Hempel, 1965] C. G. Hempel. *Aspects of Scientific Explanation and other Essays in the Philosophy of Science.* The Free Press, 1965.

[Hintikka and Remes, 1974] J. Hintikka and U. Remes. *The Method of Analysis: Its Geometrical Origin and Its General Significance.* D. Reidel Publishing Company. Dordrecht, Holland, 1974.

[Hintikka and Remes, 1976] J. Hintikka and U. Remes. Ancient geometrical analysis and modern logic. In R. S. Cohen (ed.), *Essays in Memory of Imre Lakatos*, pages 253–276. D. Reidel Publishing Company. Dordrecht Holland, 1976.

[Hobbs *et al.*, 1990] J. R. S. Hobbs, M. Stickel, D. Appelt, and P. Martin. Interpretation as abduction. *SRI International, Technical Note* 499. Artificial Intelligence Center, Computing and Engineering Sciences Division, Menlo Park, CA, 1990.

[Josephson and Josephson, 1994] J. R. Josephson and S. G. Josephson. *Abductive Inference. Computation, Philosophy, Technology.* Cambridge University Press, 1994.

[Kakas and Mancarella, 1990] A. Kakas and P. Mancarella. Knowledge assimilation and abduction. In *Proceedings of the European Conference on Artificial Intelligence, ECAI'90* International Workshop on Truth Maintenance. Springer-Verlag Lecture Notes in Computer Science. Stockholm, 1990.

[Kakas *et al.*, 1995] A. C. Kakas, R. A. Kowalski, F. Toni. Abductive logic programming. *Journal of Logic and Computation*, 2(6): 719–770, 1995.

[Kelly, 1997] K. Kelly. *The Logic of Reliable Inquiry (Logic and Computation in Philosophy).* Oxford University Press, 1997.

[Klein, 2000] N. Klein. *No Logo.* Flamingo, 2000.

[Konolige, 1990] K. Konolige. A general theory of abduction. In: *Automated Abduction, Working Notes*, pages 62–66. Spring Symposium Series of the AAA. Stanford University, 1990.

[Kowalski, 1979] R. Kowalski. *Logic for Problem Solving.* North-Holland, 1979.

[Kowalski, 1994] R. Kowalski. Logic without model theory. In D. M. Gabbay (ed.), *What is a Logical System?*, pages 35–72. Clarendon Press. Oxford, 1994.

[Kraus *et al.*, 1990] S. Kraus, D. Lehmann, M. Magidor. Non-monotonic reasoning, preferential models and cumulative logics. *Artificial Intelligence*, 44: 167–207, 1990.

[Kuhn, 1957] T. S. Kuhn. *The Copernican Revolution. Planetary Astronomy in the Development of Western Thought.* 1957. Vintage Books. 1959.

[Kuhn, 1959] T. S. Kuhn. The essential tension: Tradition and innovation in scientific research. 1959. In [Kuhn, 1977, 225–39].

[Kuhn, 1962] T. S. Kuhn. *The Structure of Scientific Revolutions.* The University of Chicago Press, 1962 , 7th Impression, 1969.

[Kuhn, 1970] T. S. Kuhn. Reflections on my critics. In Lakatos and Musgrave (eds.), pages 231–78, 1970.

[Kuhn, 1974] T. S. Kuhn. Second thoughts on paradigms. 1974. In [Kuhn, 1977, 293–319].

[Kuhn, 1976] T. S. Kuhn. Theory change as structure change: Comments on the sneed formalism. 1976. In [Kuhn, 2000, 176–95].

[Kuhn, 1977] T. S. Kuhn. *The Essential Tension.* The University of Chicago Press, 1977.

[Kuhn, 1979] T. S. Kuhn. Metaphor in science. 1979. In [Kuhn, 2000, 196–207].

[Kuhn, 1983] T. S. Kuhn. Commensurability, comparability, communicability. 1983. In [Kuhn, 2000, 33–57].

[Kuhn, 1990] T. S. Kuhn. The road since structure. 1990. In [Kuhn, 2000, 90-104].

[Kuhn, 2000] T. S. Kuhn. *The Road since Structure.* The University of Chicago Press, 2000.

[Kuipers, 1999] T. Kuipers. Abduction aiming at empirical progress or even truth approximation leading to a challenge for computational modelling. *Foundations of Science*, 4: 307–323. 1999.

[Kuipers, 2001] T. Kuipers. *Structures in Science. Heuristic Patterns Based on Cognitive Structures.* Synthese Library 301, Kluwer AP, Dordrecht. The Netherlands, 2001.

[Kvasz, 1998] L. Kvasz. History of geometry and the development of the form of its language. *Synthese*, 116: 141–68, 1998.

[Kvasz, 1999] L. Kvasz. On classification of scientific revolutions. *Journal for the General Philosophy of Science*, 30: 201–32, 1999.

[Kvasz, 2000] L. Kvasz. Changes of language in the development of mathematics. *Philosophia Mathematica*, 8: 47–83, 2000.

[Kvasz, 2002] L. Kvasz. Lakatos' methodology between logic and dialectic. In G. Kampis, L. Kvasz and M. Stöltzner (eds.), *Appraising Lakatos. Mathematics, Methodology and the Man*, Vienna Circle Institute Library: Kluwer, pages 211–41, 2002.

[Lakatos, 1963-4] I. Lakatos. *Proofs and Refutations. The Logic of Mathematical Discovery.* 1963-4. Cambridge University Press, 1984.

[Lakatos, 1968] I. Lakatos. Changes in the problem of inductive logic. 1968. In [Lakatos, 1978b, 128-200].

[Lakatos, 1970] I. Lakatos. Falsification and the methodology of scientific research programmes. 1970. In [Lakatos, 1978a, 8–101].

[Lakatos, 1974] I. Lakatos. Popper on Demarcation and Induction. 1974. In [Lakatos, 1978a, 139–67].

[Lakatos, 1978] I. Lakatos. Newton's Effect on Scientific Standards. 1978. In [Lakatos, 1978a, 193–222].

[Lakatos, 1978a] I. Lakatos. *Philosophical Papers Volume I The Methodology of Scientific Research Programmes.* 1978. Edited by J. Worrall and G. Currie, Cambridge University Press, 1984.

[Lakatos, 1978b] I. Lakatos. *Philosophical Papers Volume II Mathematics, Science and Epistemology.* 1978. Edited by J. Worrall and G. Currie, Cambridge University Press, 1984.

[Lakatos and Musgrave, 1970] I. Lakatos and A. Musgrave (eds.), *Criticism and the Growth of Knowledge.* Cambridge University Press 1970.

[Langley, 2000] P. Langley. The computational support of scientific discovery. *International Journal of Human-Computer Studies,* 53: 393–410, 2000. Web: www.isle.org/~langley/pubs.html.

[Langley *et al.*, 1987] P. Langley, H. Simon, G. Bradshaw, and J. Zytkow. *Scientific Discovery.* Cambridge, MA:MIT Press/Bradford Books, 1987.

[Laudan, 1977] L. Laudan. *Progress and its Problems.* Berkeley, University of California Press, 1977.

[Laudan, 1980] L. Laudan. Why was the logic of discovery abandoned?. In T. Nickles (ed.), *Scientific Discovery, Logic and Rationality,* pages 173–183. D. Reidel Publishing Company, 1980.

[Lipton, 2003] P. Lipton. Kant on wheels. *Social Epistemology,* 17: 215–19, 2003.

[McCarthy, 1980] J. McCarthy. Circumscription: A form of non-monotonic reasoning. *Artificial Intelligence,* 13: 27–39, 1980.

[McKie, 1935] D. McKie. *Antoine Lavoisier. The Father of Modern Chemistry.* Victor Gollancz, 1935.

[Massimi, 2005] M. Massimi. *Pauli's Exclusion Principle: The Origin and Validation of a Scientific Principle.* Cambridge University Press, 2005.

[Masterman, 1970] M. Masterman. The nature of a paradigm. In [Lakatos and Musgrave, 1970, 59–89].

[Mayer and Pirri, 1993] M. C. Mayer and F. Pirri. First order abduction via tableau and sequent calculi. In *Bulletin of the IGPL,* vol. 1: 99–117, 1993.

[Meheus, 1999] J. Meheus. Model-based reasoning in creative processes. In L. Magnani, N. J. Nersessian, and J. Thagrad (eds.), *Model-Based reasoning in Scientific Discovery.* Kluwer Academic/Plenum Publishers. 1999.

[Meheus, 2002] J. Meheus. Ampliative adaptative logics and the foundation of logic-based approaches to abduction. In L. Magnani, N. Nersessian, C. Pizzi (eds.), *Logical and Computational Aspects of Model-based Reasoning,* Kluwer Applied Logic Series. Kluwer Academic Publishers, 2002.

[Meheus and Nickles, 1999] J. Meheus and T. Nickles. The methodological study of discovery and creativity — Some background. *Foundations of Science,* 4(3): 231–235, 1999.

[Meheus *et al.*, 2002] J. Meheus, L. Verhoeven, M. van Dyck, and D. Provijn. Ampliative adaptative logics and the foundation of logic-based approaches to abduction. In L. Magnani, N. Nersessian, and C. Pizzi (eds.), *Logical and Computational Aspects of Model-based Reasoning,* Kluwer Applied Logic Series, pages 39-72. Kluwer Academic Publishers, 2002.

[Menger, 1980] K. Menger. Introduction to Hans Hahn. *Empiricism, Logic, and Mathematics: Philosophical Papers,* ed. Brian McGuinness, Reidel, *ix-xviii,* 1980.

[Michalski, 1994] R. Michalski. Inferential theory of learning: Developing foundations for multistrategy learning. In *Machine Learning: A Multistrategy Approach,* Morgan Kaufman Publishers, 1994.

[Miller, 1986] A. I. Miller. *Imagery in Scientific thought. Creating 20th-Century Physics.* MIT Press, 1986.

[Motterlini, 1999] M. Motterlini. *For and Against Method. Imre Lakatos and Paul Feyerabend.* The University of Chicago Press, 1999.

[Neurath *et al.*, 1929] O. Neurath *et al. The Scientific Conception of the World. The Vienna Circle.* 1929. English translation, Reidel, 1973.

[Ohlbach and Reyle, 1999] H. J. Ohlbach and U. Reyle (eds.). Research themes of Dov Gabbay. In *Logic, Language and Reasoning*, pages 13–30. Kluwer Academic Publishers. 1999.

[Peirce, 1931-35] C. S. Peirce. *Collected Papers of Charles Sanders Peirce.* Volumes 1–6 edited by C. Hartshorne, P. Weiss. Cambridge, Harvard University Press. 1931–1935; and volumes 7–8 edited by A. W. Burks. Cambridge, Harvard University Press. 1958.

[Polya, 1954] G. Polya. *Induction and Analogy in Mathematics.* Vol I. Princeton University Press, 1954.

[Polya, 1962] G. Polya. *Mathematical Discovery. On Understanding, learning, and teaching problem solving.* Vol I. John Wiley & Sons, Inc. New York and London, 1962.

[Poole *et al.*, 1987] D. Poole, R. G. Goebel, and Aleliunas. Theorist: a logical reasoning system for default and diagnosis. In Cercone and McCalla (eds.), *The Knowledge Fronteer: Essays in the Representation of Knowledge.* Springer Verlag Lecture Notes in Computer Science, pages 331–352, 1987.

[Pople, 1973] H. E. Pople. On the mechanization of abductive logic. In: *Proceedings of the Third International Joint Conference on Artificial Intelligence* (IJCAI-73). San Mateo: Morgan Kauffmann, Stanford, CA, pages 147–152, 1973.

[Popper, 1934] K. R. Popper. *The Logic of Scientific Discovery.* 1934, 6th (revised) impression of the 1959 English translation, Hutchinson, 1972.

[Popper, 1960] K. R. Popper. The growth of scientific knowledge. 1960. In D. Miller (ed.), *A Pocket Popper.* Fontana Paperbacks. University Press, Oxford, 1983. This consists of Chapter 10 of K. Popper, (1963) *Conjectures and Refutations. The Growth of Scientific Knowledge.* 5th ed. London and New York, Routledge, 1963.

[Popper, 1963] K. R. Popper. *Conjectures and Refutations. The Growth of Scientific Knowledge.* Routledge & Kegan Paul, 1963.

[Popper, 1972] K. R. Popper. *Objective Knowledge. An Evolutionary Approach.* Oxford University Press, 1972.

[Popper, 1983] K. R. Popper. *Realism and the Aim of Science.* Hutchinson, 1983.

[Quine, 1953] W. V. O. Quine. *From a Logical Point of View.* 1953. 2nd Revised Edition, Harper Torchbooks, 1963.

[Ramsey, 1929] F. P. Ramsey. Last papers F. philosophy. 1929. In R. B.Braithwaite (ed.), *The Foundations of Mathematics and other Logical Essays by F. P. Ramsey*, Routledge & Kegan Paul, 4th Impression, 1965.

[Reggia *et al.*, 1985] J. A. Reggia, D. S. Nau and Y. Wang. A formal model of diagnostic inference I. Problem formulation and descomposition. *Inf. Sci.*, 37, 1985.

[Reiter, 1980] R. Reiter. A logic for default reasoning. *Artificial Intelligence*, 13, 1980.

[Reiter, 1987] R. Reiter. A theory of diagnosis from first principles. *Artificial Intelligence*, 32, 1987.

[Rescher, 1978] N. Rescher. *Peirce's Philosophy of Science. Critical Studies in His Theory of Induction and Scientific Method.* University of Notre Dame, 1978.

[Russell, 1914] B. Russell. *Our Knowledge of the External World as a Field for Scientific Method in Philosophy.* 1914. George Allen & Unwin, 1961.

[Sankey, 1994] H. Sankey. *The Incommensurability Thesis.* Avebury, 1994.

[Scott, 1971] D. Scott. On engendering an illusion of understanding, *Journal of Philosophy*, 68: 787–808, 1971.

[Shapere, 1964] D. Shapere. The structure of scientific revolutions. *Philosophical Review*, 73: 383–94, 1964.

[Sheehan, 1985] H. Sheehan. *Marxism and the Philosophy of Science: A Critical History.* Humanities Press, 1985.

[Shoham, 1988] Y. Shoham. *Reasoning about Change. Time and Causation from the standpoint of Artificial Intelligence.* The MIT Press, Cambridge, Mass, 1988.

[Simon, 1973] H. Simon. Does scientific discovery have a logic? In H. Simon, *Models of Discovery*, pages 326–337, 1973. A Pallas Paperback. Reidel, Holland. 1977. (Originally published in *Philosophy of Science*, 40: 471–480).

[Stadler, 2001] F. Stadler. *The Vienna Circle. Studies in the Origins, Development, and Influence of Logical Empiricism.* Springer, 2001.

[Suppes, 1951-69] P. C. Suppes. *Studies in the methodology and foundations of science. selected papers from 1951 to 1969.* D. Reidel. Dordrecht. The Netherlands, 1951-69.

[Tan, 1992] Y. H. Tan. *Non-Monotonic Reasoning: Logical Architecture and Philosophical Applications.* PhD Thesis, University of Amsterdam, 1992.

[Thagard, 1988] P. R. Thagard. *Computational Philosophy of Science.* Cambridge, MIT Press. Bradford Books, 1988.

[Thagard, 1992] P. R. Thagard. *Conceptual Revolutions.* Princeton University Press. 1992.

[Turing, 1950] A. M. Turing. computing machinery and intelligence. *Mind*, 59: 433–60, 1950.

[Williams, 1994] M. A. Williams. Explanation and theory base transmutations. In *Proceedings of the European Conference on Artificial Intelligence*, ECAI'94, pages 341–246, 1994.

[Wittgenstein, 1953] L. Wittgenstein. *Philosophical Investigations.* 1953. Basil Blackwell. 2nd Edition. 1958.

[Wolters, 2003] G. Wolters. Carl Gustav Hempel. Pragmatic Empiricist. In P. Parrini, W. C. Salmon, M. H. Salmon (eds.), *Logical Empiricism: Historical and Contemporary Perspectives*, University of Pittsburgh Press, pages 109–22, 2003.

[Zahar, 1973] E. Zahar. Why did Einstein's programme supersede Lorentz's? *British Journal for the Philosophy of Science*, 24: 95–123 and 223–262, 1973.

DEMARCATING SCIENCE FROM NON-SCIENCE

Martin Mahner

1 INTRODUCTION

Every field of inquiry deals with some subject matter: it studies something rather than nothing or everything. Thus it should be able to tell, at least roughly, what sort of objects it is concerned with and how its objects of study differ from those studied by other disciplines. If a discipline were unable to offer a characterization of its subject matter, we would be entitled to suspect that its representatives do not really know what they are talking about. Evidently, what holds for all fields of inquiry also holds for a particular discipline such as the philosophy of science. Therefore, it belongs to the job description, so to speak, of the philosopher of science to tell us what that "thing" called science is.

Yet whereas everyone seems to know intuitively which fields of knowledge are scientific (such as physics and biology) and which are not (such as astrology and palmistry), it has proved difficult to come up with a satisfactory demarcation criterion. Indeed, many of the demarcation criteria proposed by philosophers of science have proved to be unsatisfactory, for being either too narrow or too wide. In addition, due to the historical and sociological studies of science, many contemporary authors believe that there simply is, or even can be, no single criterion or set of criteria allowing for a clear-cut characterization of scientific vis-à-vis non-scientific areas of human inquiry. In particular, most contemporary philosophers doubt that there is a set of necessary and sufficient conditions demarcating science from non-science. It comes as no surprise therefore that, in a survey conducted with 176 members of the Philosophy of Science Association in the US, about 89% of the respondents denied that any universal demarcation criteria have been found [Alters, 1997].

Does this vindicate relativist views like Feyerabend's [1975] well-known anything-goes epistemology? Must we give up the attempt to descriptively partition the landscape of human cognition into scientific and nonscientific areas, as well as to tell genuine science from bogus science (pseudo-science)? Is, then, the philosophy of science unable to address the normative problem of why some form of human inquiry arrives at (approximately) true knowledge, whereas some other, purporting to be equally scientific, must be judged to produce only illusory knowledge?

Handbook of the Philosophy of Science: General Philosophy of Science - Focal Issues, pp. 515–575.
Volume editor: Theo Kuipers. General editors: Dov M. Gabbay, Paul Thagard and John Woods.

This situation illustrates the problems that a reasonably comprehensive analysis of the demarcation problem should address. To this end, let us restate three questions formulated by Thagard [1988, p. 157], which will guide the following analysis:

1. Why is it important to demarcate science and from what should it be distinguished?
2. What is the logical form of a demarcation criterion?
3. What are the units that are marked as scientific or nonscientific, in particular as pseudoscientific?

2 WHY DEMARCATION?

Let us begin with the second conjunct of the first question. Evidently, demarcating science means to demarcate it from nonscience. Yet, in so doing, how widely do we have to conceive of nonscience? In the broad sense, simply anything that is not science is nonscience: driving, swimming, cooking, dancing, or having sex are nonscientific activities. Now, the philosopher of science is not particularly interested in demarcating science from nonscientific activities such as these, although they involve learning and hence some cognition, leading in particular to procedural knowledge. Naturally, he will be interested first of all in cognitive activities and practices leading to propositional knowledge, i.e., explicit and clear knowledge that can be either true or false to some extent. Thus, we are primarily interested in nonscience in the sense of *nonscientific cognitive fields* involving hypotheses and systems of such (i.e., theories) as well as the procedures by means of which these are proposed, tested, and evaluated. Consequently, distinguishing science from nonscience in this narrower sense is not restricted to the classical science/metaphysics demarcation attempted by the neopositivists and Popper, but extends to all nonscientific epistemic fields.

The first reason why we should strive for a demarcation of science is theoretical: it is the simple fact stated in the beginning that every field of knowledge should be able to tell roughly what it is about, what its objects of study are. Unless the philosophy of science simply is nothing but epistemology in general, it should be able to distinguish scientific from nonscientific forms of cognition. Note that such a basic distinction between science and nonscience is not pejorative: it does not imply that nonscientific forms of cognition and knowledge are necessarily bad or inferior. Nobody doubts the legitimacy and value of the arts and humanities, for example.

The second reason, or rather set of reasons, why we ought to demarcate science from nonscience concerns in particular the normative aspect of distinguishing science from pseudoscience. Moreover, it is practical rather than theoretical, comprising aspects of mental and physical health, as well as culture and politics. Should we entrust our own as well as other peoples' health and even lives to diagnostic or therapeutic methods which have no proven effect? Should public health insurance cover magical cures? Should we even consider the possibility that clairvoyants search for missing children? Should we have dowsers search for people buried by an avalanche? Should we make sure that tax payers' money be spent only on funding scientific rather than pseudoscientific research? Should we demand that people living in a modern democratic society base their political decisions on scientific knowledge rather than superstition? Examples such as these show that the distinction of science and pseudoscience is vital not just to our physical, but also to our cultural and political life.

This aspect leads us to a third reason: the need of science education to teach what science is and how it works. To this end, the science educator needs input from the philosopher of science as to the nature of science [Alters 1997; Eflin *et al.*, 1999]. The science educator cannot just tell her students that nobody knows what science is, but that they are nonetheless supposed to learn science rather than pseudoscience. For all these reasons, we ought not to give up too readily when facing difficulties with the demarcation of science from nonscience, and in particular from bogus science.

3 HOW DEMARCATION?

In the history of the philosophy of science various demarcation criteria have been proposed (see Laudan [1983]). Let us briefly recapitulate some of the classical attempts at demarcation, starting with logical positivism. In tune with their linguistic focus the neopositivists' foremost goal was to distinguish sense from nonsense. A sentence was deemed to be (semantically) meaningful if, and only if, it was verifiable; otherwise, it was nonsense. Whereas, according to neopositivism, the statements of science are verifiable and thus meaningful, those of metaphysics and all other kinds of bad philosophy were not; they were just nonsense (see, e.g., [Wittgenstein, 1921; Carnap, 1936/37; Ayer, 1946]). To verify a sentence means to find out whether it is true, which requires that it be tested empirically. The central tenet, then, was that testability is a necessary condition of (semantic) meaning: meaning $\rightarrow$ testability.

One problem with this view is that it has things the wrong way. Indeed, to test a statement empirically, must we not know what it means, i.e., what it says, in the first place? To devise a test for a statement such as "unemployment increases crime", we must already know what that sentence means. Only thus can we handle the variables involved. Hence, meaning is in fact a necessary condition of testability rather than the other way around: testability $\rightarrow$ meaning [Mahner and Bunge, 1997]. As a consequence, nonscientific discourse can be semantically meaningful, although it may not be testable empirically. If a Christian tells us "Jesus walked on water", we know quite well what that means, although we cannot test this statement, which we may moreover regard as purely mythical. We may reject it for many reasons, but not for being nonsensical.

A logical and methodological objection against the verifiability thesis is the fact that it is rarely possible to verify a statement in the strict sense, i.e., to show that it is true. For example, we may easily verify or falsify a spatiotemporally restricted existential statement, such as "There is a pink elephant in my office". But if we are faced with general statements such as "For all X: if A then B", observing B does confirm A, but only inductively, never conclusively. The most cherished scientific statements, then, namely law statements, are not strictly verifiable. Hence the strong concept of conclusive verification was soon replaced by the weaker notion of confirmation [Carnap, 1936/37]. Nonetheless, having always been a critic of induction, Popper [1934/1959] suggested giving up the verifiability condition in favor of a falsifiability principle. Indeed, according to the modus tollens rule, observing not-B entails not-A. Thus, logically, falsification is conclusive, whereas verification is not. This logical asymmetry is the basis for Popper's famous demarcation criterion of falsifiability [Popper, 1963].

Critics have soon pointed out that not all scientific statements are universal: there are also unrestricted existential statements, such as "There are positrons" [Kneale, 1974; Bunge, 1983b]. These can be verified, e.g., in this case by coming up with at least one specimen of a positron; but they cannot be falsified, because we cannot search the entire universe to conclusively show that it does not contain even a single positron. Other critics have shown that scientists do not give up a theory as being unscientific just because there are some falsifying data, unless there is a better theory at hand, concluding that Popper's criterion does not match scientific practice [Lakatos, 1973; 1974].

In the light of such critique, Popper has later clarified his position, emphasizing that it is not practical falsifiability which is his concern, but instead logical falsifiability [Popper, 1994]. That is, a statement is *logically falsifiable* if there is at least one *conceivable* observation statement contradicting it. In other words, a statement is scientific only if it is not consistent with every possible state of affairs. In proposing falsifiability as a demarcation criterion, Popper had in mind examples such as Freudian psychoanalysis. According to psychoanalysis, the Oedipus complex is either manifest or repressed, so no possible observable state of affairs can count against it: it is unfalsifiable as a matter of principle. Again, critics were quick to point out that this does not hold for all of psychoanalysis (e.g., [Grünbaum, 1984]): while some claims are indeed unfalsifiable, many others are falsifiable and others have actually been falsified. The same holds for many other pseudosciences, such as astrology and creationism. For example, the central tenet of creationism, that a supernatural being created the world, is indeed unfalsifiable: it is compatible with every possible observation statement, for any state of affairs can be seen as exactly what the creator chose to do. Other and more specific creationist claims, however, such as that the earth is only 6000 years old, are falsifiable and falsified. Thus, the falsifiability criterion may be useful to weed out some claims as pseudoscientific, but it accepts too many falsifiable and falsified statements as scientific, although there are good reasons to regard them as pseudoscientific. For all these reasons, falsifiability has been almost unanimously rejected as *the* demarcation criterion (e.g., [Kuhn, 1970; Kitcher, 1982; Bunge, 1983b; Laudan, 1983; Siitonen, 1984; Lugg, 1987; Thagard, 1978; 1988; Rothbart, 1990; Derksen, 1993; Resnik, 2000]).

Being first of all a logical condition, falsifiability is an ahistorical criterion. The historical turn in the philosophy of science has suggested taking into account both the development of theories and their relation to rival theories. In so doing, it shifted the focus of demarcation from individual statements or hypotheses to entire theories. The classic approach is certainly Lakatos's [1970] notion of a research program. A research program is a historical sequence of theories, where each subsequent theory results form a semantical reinterpretation of its predecessor, or from adding auxiliary assumptions or other modifications. A research program is called *theoretically progressive* if each new theory has a larger content, e.g., by having greater explanatory or predictive power, than its predecessor. It is also *empirically progressive* if it is confirmed, i.e., if it actually leads to the discovery of some new fact. A research program is then called *progressive* if it is both theoretically and empirically progressive; otherwise it is called *degenerative*. Finally, Lakatos [1970] takes a research program to be *scientific* if it is at least theoretically progressive; otherwise it is *pseudoscientific*.

Critics like Laudan [1983] have objected that progress might as well occur in nonscientific fields, such as philosophy, and that some branches of science did not progress much during some periods in their history. And what if some science actually had discovered and explained everything in its domain that is to be discovered and that needs explanation; in other words, what if there were such a thing as "the end of science" as envisioned as the ultimate goal of science by Einstein [Holton, 1993; Haack, 2003], and as was more recently speculated on by Horgan [1995]? Would such a theory or discipline be no longer scientific just because it does not or rather cannot progress any more? Similarly, there is the opposite problem of radically new theories: can they be scientific without being part of an existing research program? Consequently, however useful the criterion of growth and progressiveness will be in many cases, it too cannot provide *the* decisive demarcation criterion.

Kuhn [1970] has suggested that we focus not so much on the testability of theories as on their problem-solving capacity. He illustrates his point in the case of astrology. Many predictions of astrology are testable and have failed, but astrology is not therefore a science as Popper's falsifiability criterion would allow for. According to Kuhn, this is because astrology has no puzzles to solve: even its failed predictions did not entice the astrological community to engage in problem-solving activity. At most, astrology has rules to apply, for it is essentially a craft — or rather a pseudotechnology. But if applying rules is simply a characteristic of technology rather than science, what distinguishes a scientific technique from a nonscientific one? Finally, although earlier authors like David Hilbert have already dealt with problems and pointed out that a wealth of problems is an indicator of a good science, what about the end-of-science scenario mentioned above? Would even a true theory become nonscientific if all problems surrounding it were solved?

Scientists would hardly think so.

Nevertheless, other authors too suggested focusing on problems, in particular on the permissible rules of asking questions and stating problems [Siitonen, 1984]. Obviously, the solution of a problem should enrich our knowledge and contribute to stating and solving further problems. Moreover, we may learn something about the current state of a field of inquiry by asking questions such as "What are the problems?", "How are these problems formulated?", and "Which efforts have been and can be used to solve these problems?" [Siitonen l.c., p. 347]. But again, all these considerations are far from providing a new demarcation criterion.

In a book on creationism, Kitcher [1982] focuses on three characteristics of science. First, the auxiliary hypotheses involved in the testing of any scientific theory are *independently testable* themselves, i.e., independently of the theory it is supposed to protect or of the particular case for which they were introduced. Second, scientific practices are *unified* wholes, not patchworks of isolated and opportunistic methods: they apply a small number of problem-solving strategies (if preferred, exemplars) to a wide range of cases and problems. Third, good scientific theories are *fertile* in the sense that they open up new areas of research. Thereby, one of the sources of fecundity is the incompleteness of scientific theories, so that some problems remain unresolved. Incompleteness and some unresolved problems are therefore not shortcomings of scientific theories but instead sources of progress.

Thagard [1988] lists five features characterizing science. As a method of inference, scientists use "correlation thinking"; that is, by means of various statistical procedures they infer causation, if any, from correlation (rather than from mere resemblance). They seek empirical confirmation and disconfirmation, and evaluate theories in relation to alternative theories, whereby these theories are consilient and simple. Finally, science progresses over time, i.e., it develops new theories explaining new facts. Thagard does not regard these features as both necessary and sufficient, but only suggests that they belong to the conceptual profile of science.

Rothbart [1990] attempts to formulate a metacriterion (or adequacy condition) for any demarcation criterion. This condition is the testworthiness of a hypothesis or theory, i.e., its plausibility to be selected for experimentation in the first place. To this end a hypothesis must fulfill certain eligibility requirements prior to testing. If it does not fulfill even one of these requirements, a hypothesis is untestworthy and hence unscientific. Actual demarcation is then obtained by specifying such eligibility requirements. One such requirement is that the proposed theory must account for all the facts that its rival background theory explains; another is that it must yield test implications that are inconsistent with those of its rival theory.

Vollmer [1993] distinguishes necessary and merely desirable features of a good scientific theory. The necessary conditions are noncircularity, internal consistency (noncontradiction), external consistency (compatibility with the bulk of well-confirmed knowledge), explanatory power, testability, and test success (confirmation). Among the desirable features are predictability and reproducibility, as well as fecundity and simplicity (parsimony). Predictability and reproducibility are not among the necessary conditions, for otherwise historical sciences, such as evolutionary biology, geology, cosmology, and of course human history, would not count as scientific because both their predictability and reproducibility are limited. However, even in such overall historical fields, not all events are unique, but repeatable at least in the sense that events *of the same kind* may reoccur on a more or less regular basis. Consequently, if the very nature of some event is of the repeatable kind, irreproducibility may still indicate that something is wrong with the given field's claim of being a science.

Reisch [1998] attempts to resuscitate the unity of science ideal of logical positivism, though not in its reductionist form. He suggests identifying the various theoretical and methodological interconnections of the sciences, which should result in what he calls a *network unification* of science and hence a *network demarcation*. An epistemic field that cannot be incorporated into the existing network of the established sciences without destroying it should be rejected as pseudoscientific. Again, such network demarcation does not draw a fixed boundary around the sciences, but allows for changes in what belongs to that network and what not. Finally, the neopositivist aspect of Reisch's approach consists in the claim that the specification of the interconnections among scientific fields is essentially a *scientific* form of demarcation rather than a philosophical one.

The result of the preceding overview is clear: neither is there a single criterion such as falsifiability to demarcate science from nonscience, nor is there a generally accepted set of necessary and sufficient criteria to do this job. However, *pace* Laudan [1983], this does not imply that no demarcation is possible. To see why, it will be useful to make a brief foray into the philosophy of biology, which faces a similar problem.

In the philosophy of biological systematics there has been a long debate concerning the ontological status and definition of biological species (see, e.g., [Mahner and Bunge, 1997]). The classical, essentialist view regards species as natural kinds defined by a set of necessary and sufficient properties. Against this view the anti-essentialists have argued that, due to the high genetic and morphological variety of organisms, there simply is no set of necessary and sufficient characters possessed by all and only the organisms of a given species, let alone higher taxonomic units (see, e.g., [Dupré, 1993]). Nevertheless, the organisms of a given species usually are both similar among each other and distinct from organisms belonging to different species.

The radical answer to this problem says that species should therefore not be conceived of as kinds at all, but rather as concrete supraorganismic individuals. Now, science too can be viewed as a concrete system, namely as a research community. In this case it is relatively easy to determine who is part of this community and who is not. But of course, science is more than that: in contrast to the sociologist of science, the philosopher of science is more interested in science as a collection of reliable knowledge items produced by following certain methodological standards. To this end, science is better regarded as a special *kind* of knowledge production, which can be demarcated from other kinds of knowledge acquisition. Now, if traditional essentialism with respect to kinds cannot be upheld at least in biology, we might try what in the philosophy of systematics could be called *moderate* species essentialism. This is the idea that biological species *can* be viewed as natural kinds, if only in a weaker sense defined by a variable *cluster* of features instead of a strict set of necessary and sufficient properties (see, e.g., [Boyd, 1999; Wilson, 1999]). Thus, whereas no single property need be present in all the members of the given species, there are always "enough" properties making these organisms belong to the given kind. (Forerunners of moderate essentialism are Wittgenstein's family resemblance concept, which was suggested for demarcation purposes by Dupré [1993], and Beckner's [1959] polythetic species definitions.)

Despite the unsolved problems concerning the formalization of such disjunctive characterizations [Mahner and Bunge, 1997], applying this approach to the demarcation of science might allow us to define science through a variable cluster of properties too, rather than through a set of necessary and sufficient conditions. For example, if we came up with ten conditions of scientificity (all of equal weight), we might require that an epistemic field fulfill at a minimum seven out of these ten conditions in order to be regarded as scientific, but it would not matter which of these ten conditions are actually met. According to the formula $N!/n!(N-n)!$, where $N = 10$ and $n = 7$, and adding the permutations for $n = 8, n = 9$, and $n = 10$, there would in this case be a total of 176 possible ways of fulfilling the conditions of scientificity.

In a similar vein, many authors have argued that, for demarcation purposes, we must do with a reasonable *profile* of any given field rather than with a clear-cut distinction (e.g., [Thagard, 1988; Derksen, 1993; Eflin *et al.*, 1999]). In other words, it will be worthwhile to attempt to come up with a whole battery of science indicators. Such a cluster of criteria should be as comprehensible as possible, and enable us to examine every possible field of knowledge by a list of marks noting the presence or absence of the relevant features, or the compliance or noncompliance with some, e.g. methodological, rule. On this basis we should be able to come to a well-reasoned (and hence rational) conclusion concerning the scientific or nonscientific status of a cognitive field.

4 CHARACTERIZING FIELDS OF KNOWLEDGE

As is obvious from the preceding section, scientificity has been ascribed to many items: individual statements, problems, methods, systems of statements (theories in the strict sense), entire practices (theories in the broad sense), historical sequences of theories and/or practices (research programs), and fields of knowledge. Given the notorious problems with the traditional demarcation criteria, it seems promising to try the most comprehensive approach, for it allows us to consider the many facets of the scientific enterprise, namely the fact that science is at the same time a body of knowledge and a system of people including their activities or practices, and hence something that did not come into existence ex nihilo, but has developed over several centuries from a mixed bag of ordinary knowledge, metaphysics and non- or at most pre-scientific inquiry. This most comprehensive approach is the one focusing on fields of knowledge (see, e.g., [Thagard, 1988]). As we shall see at the end, this approach has the advantage that, by demarcating entire fields of knowledge as scientific or nonscientific, it allows us to also evaluate individual components of such a field, like characteristic principles and methods, as being scientific or not.

Before we begin to determine whether or not a field of knowledge is scientific, we must first define what a field of knowledge is. In a chapter on pseudoscience, Thagard [1988] just refers to fields of knowledge, without, however, offering much of a characterization. In their work on "interfield theories", Darden and Maull [1977] point out that fields are characterized, for example, by a certain domain of facts as well as a number of problems, methods and theories concerning that domain. However, they do not use their characterization to demarcate between scientific and nonscientific, let alone pseudoscientific, fields. The most comprehensive characterization of epistemic fields has been proposed by Bunge [1983a; b], who has moreover explicitly used it for demarcation purposes [Bunge, 1982; 1983b; 1984]. For this reason, I shall rely heavily on his analysis, but will readily modify it whenever necessary to make it better suited to the task at hand.

4.1 Epistemic Fields

Roughly speaking, an epistemic field is a group of people and their practices, aiming at gaining knowledge of some sort. Thus, physics and theology, astronomy and astrology, psychology and parapsychology, evolutionary biology and creationism, art history and mathematics, medicine and economics, philosophy in general and epistemology in particular, as well as biology in general and genetics in particular are examples of epistemic fields. These examples show that epistemic fields, or, if preferred, cognitive disciplines can be more or less inclusive; in other words, they may be structured hierarchically. (Note that in the following we shall not distinguish between "field" and "discipline", although one might argue that the term "discipline" be reserved for denoting generally acknowledged or institutionalized fields.) They also indicate that the knowledge acquired in an epistemic field need neither be factual nor true: we may acquire knowledge about purely fictional rather than factual entities, and our knowledge may be false or illusory. (Thus, we do not adopt the classic definition of "knowledge" as "justified true belief", but rather the Popperian view that all knowledge is hypothetical, so that it can turn out to be either true or false.) Finally, it is immaterial whether the aim of our cognitive activities is either epistemic or practical, or both.

These examples of fields of knowledge just serve as a starting point for a more detailed characterization. In his characterization of epistemic fields, Bunge [1983a] considers ten aspects:

1. the group or *community* C of knowers or knowledge seekers;

2. the *society* S hosting the activities of C;

3. the *domain* or universe of discourse D of the members of C, i.e., the collection of factual or fictional objects the members of C refer to in their discourse;
4. the *philosophical background* or general outlook G, which consists of
   (a) an *ontology* or general view on the nature of things,
   (b) an *epistemology* or general view on the nature of knowledge, and
   (c) a *methodology*, *axiology* and *morality* concerning the proper ways of acquiring and handling knowledge;
5. the *formal background* F, which is a collection of logical or mathematical assumptions or theories taken for granted in the process of inquiry;
6. the *specific background* B, which is a collection of knowledge items (statements, procedures, methods, etc.) borrowed from other epistemic fields;
7. the *problematics* P, which is the collection of problems concerning the nature, value or use of the members of D, as well as problems concerning other components listed here, such as G or F;
8. the *fund of knowledge* K, which is the collection of knowledge items (propositions, theories, procedures, etc.) obtained by the previous and current members of C in the course of their cognitive activities;
9. the *aims* A, which are of course the cognitive, practical or moral goals of the members of C in the pursuit of their specific activities;
10. the *methodics* M, which is the collection of general and specific methods (or techniques) used by the members of C in their inquiry of the members of D.

Note that these aspects come in a certain logical order. For example, the method used to find out something in a given field depends on the problem to be solved, on what we already know and on our aims. Thus, Bunge analyzes an epistemic field E, for any given time, as an ordered set or, more precisely, a ten-tuple

$$\mathcal{E} = \langle C, S, D, G, F, B, P, K, A, M \rangle .$$

Since our emphasis here is on the usefulness of these coordinates for demarcation purposes, we can disregard the question of whether their order is optimal or whether an alternative order would be more adequate (e.g., exchanging P and K). Bunge calls the first three components of this ten-tuple the *material framework* of the given epistemic field, although he admits that this is a misnomer in the case of fields like mathematics and the humanities whose domains consists mostly or even exclusively of nonmaterial objects. In any case, C and S do consist of concrete objects, namely persons and systems of persons. Consisting mostly of abstract objects, the last seven components make up the *conceptual framework* of the field, which may as well be equated with Kuhn's notion of a paradigm or disciplinary matrix. This name too is a misnomer in some cases, because the methodics M need not only consist of rules and procedures as conceptual entities, but may also comprise material objects (artifacts) such as measuring instruments.

Most of the members of E will be obvious, such as D, G and M, but some remarks may nonetheless be helpful. For example, the two coordinates C and S indicate that cognition and knowledge are not self-existing, but activities of real people in a particular social environment. Only by taking these aspects into account can we do justice to the history, psychology, and sociology of knowledge. But why distinguish C from S, since C is actually a subsystem of S? Because the community C may have interesting sociological features worth examining and because it may emerge or go extinct, without necessarily having a serious effect on the entire society in which it exists or had existed. Think of L.R. Hubbard's scientology movement.

The problematics P and the aims A of an epistemic field are important characteristics, because the same domain may be studied by asking different questions, and with different aims. For example, biochemistry and molecular biology study virtually the same objects, namely certain classes of molecules, but they concern different problems: whereas biochemistry studies these molecules under purely chemical auspices, molecular biology is interested in the biological function of these molecules in living organisms. Similarly, the same object may be studied to simply learn more about it, or to control it by technical means. For example, the phylogeneticist may just be interested in the evolution of mosquitoes, whereas the applied entomologist and especially the ecotechnologist may be interested in how to control their population and restrict their geographical distribution.

4.2 Scientific epistemic fields

When speaking of science we are first of all interested in the *factual* (often called empirical) sciences, such as physics and chemistry, biology and psychology, as well as the social sciences. (Note that we prefer the expression "factual science" over "empirical science" because the advanced sciences are not just empirical, but have well-developed theoretical branches.) An epistemic field $\mathcal{S}$ is a (factually) *scientific* field if the elements of any ten-tuple $\langle C, S, D, G, F, B, P, K, A, M\rangle$ approximately satisfy the following conditions [Bunge, 1983b].

1. The community C of the field is a *research* community: it is a system of persons who share a specialized training, hold strong information links amongst each other, and initiate or continue a certain tradition of inquiry. Thus, every researcher belongs to either a local, regional, national, or international community of colleagues.

2. The society S hosting C supports or at least tolerates the activities of the persons in C. In particular, it allows for research free from authority, in that it does not proscribe which of its results have to be accepted as true, or else be rejected as false.

3. The domain D of a factual science deals exclusively with *concrete* entities (past, present and future), their properties and changes. These entities may be elementary particles, living beings, human societies, or the universe as a whole. Some of the entities hypothesized in a factual discipline may turn out not to exist really, but if they were real, they would be concrete (as opposed to abstract) entities.

4. That science rests on certain philosophical assumptions is rather uncontroversial. There is less agreement, however, as to which particular assumptions are characteristic of science. Let us therefore discuss some of the philosophical principles that are good candidates for membership in the general philosophical outlook G of any scientific field. To this end, consider a simple physiological experiment, which can be done in biology class (Fig. 1).

   Where is the hidden philosophy in this experiment? Unlike the solipsist or the follower of George Berkeley, the normal scientist does not assume that, when she is actually carrying out this experiment, it is occurring only in her mind. Nor does she suppose that a supernatural entity is producing the entire situation in her mind. We cannot prove that this is not actually the case, but it simply does not belong to the scientists' presuppositions. By contrast, the scientist takes it for granted that this experiment is occurring in an outer world existing independently of her mind, but including her as a part.

   Imagine we repeat this experiment several times under the same conditions. The first time, the gas produced would be helium, the second time oxygen, the third time no gas at all would appear. The fourth time, the entire setup would explode before even adding hydrogen peroxide, and the fifth time four of the test tubes would turn into chewing gum, whereas the fifth would fly off to the ceiling. For some reason such weird things do not happen. Instead, things remain the same under the same conditions. Moreover, the outcome of the experiment is *ceteris paribus* the same: the gas always consists of oxygen. Furthermore, its amount depends on the pH in the test tube, whereby the highest amount is produced at a pH of 8. Obviously, the properties of the things involved are constantly (i.e., lawfully) related. Imagine further that for some reason we do not get any gas out of the test tubes at all. In this case the scientist would not believe that the gas has disappeared into nothingness, but that there must be something wrong with the setup.

   Excluding effects coming out of nothingness or from some supernatural realm, the scientist further assumes that it is her adding hydrogen peroxide which causes the production of oxygen. In other words, by manipulating some part of the setup a certain effect can be produced, whereby the steps in this process are ordered: the steps in the causal chain follow each other rather than occurring capriciously. Furthermore, the scientist takes it for granted not only that no supernatural entities, like friendly fairies or evil

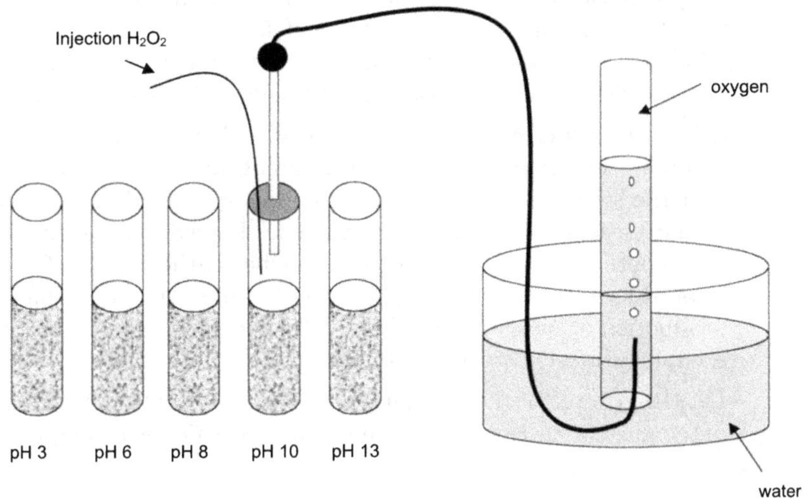

Figure 1. Take five test tubes filled with water and add a certain amount of yeast. Furthermore, by adding different amounts of hydrochloric acid (HCl) or caustic soda (NaOH) respectively, we arrange for a different acidity or alkalinity respectively in each tube, say, pH 3, pH 6, pH 8, pH 10, and pH 13. The yeast cells contain the enzyme catalase, which enables them to break down hydrogen peroxide into water and oxygen (i.e., $2H_2O_2 \rightarrow 2H_2O+O_2$). Upon adding a certain amount of hydrogen peroxide into one test tube after the other (by means of a syringe, for example), we each time close the tube and measure the amount of gas produced after 2 minutes by collecting it in a measuring tube, which is connected to the given test tube by a thin rubber hose. We do not need to specify the precise amounts and conditions here, because the basic setup of this experiment will be clear anyway (redrawn and modified from Knodel 1985, p. 39).

demons, meddle with the experiment either in the positive or in the negative, but also that they do not influence her own thinking, e.g., by making her hallucinate. And finally, she assumes that neither she herself nor anybody else can affect the setup by pure thinking or wishing alone, but only by acting; in other words, she takes it for granted that it is neither her own mind nor the mind of her colleague nor that of some little green alien on another planet which causes the outcome of the experiment.

In all this experimenting our scientist believes of course that she can get to know something about what is going on. Moreover, she also believes that the setup can be improved if necessary, and that thereby the precision of the measurement can be increased. Indeed, by varying and improving the experiment, she will find out that her earlier datum "The most oxygen is produced at a pH of 8.0" was not quite true, but that the maximum production occurs at a pH of 8.5. In other words, the initial finding was only an approximation to the real fact.

Let this sample of tacit assumptions suffice. It is time to extract some of the ontological, epistemological, and semantic *isms* or principles involved here.

a) Ontological assumptions

Despite the efforts of the positivists to denounce metaphysics as nonsense, it has long been acknowledged that science and metaphysics, though different, are related — and often even fruitfully so (see, e.g., [Agassi, 1964]). After all, one might argue that science is the emancipated daughter of metaphysics. As is indicated by the experiment described in the preceding, some minimal set of ontological tenets is presupposed even by modern science.

The first candidate is of course ontological realism, i.e., the thesis that there is a mind-independent world, whose inhabitants may become the subject matter of scientific investigation. Ontological realism is among the least controversial philosophical presuppositions of science, as is also indicated by Alters's [1997] survey mentioned earlier, in which about 90% of the interviewed philosophers agreed with the thesis that science presupposes realism. Note that ontological realism says nothing about whether this real world can be known and, if so, how and to which degree. This is a matter of epistemological realism. (It is, by the way, mostly the latter which is the target of antirealist criticism.)

The next assumption is ontological naturalism. This is the thesis that the inhabitants of the real world are exclusively natural as opposed to supernatural. Whether or not there is a transcendent world beyond our universe (if this very idea makes sense in the first place), our universe is causally closed, that is, there is no interaction with any possible other-worldly entities. Many philosophers of science would go even further and posit that there can be no interaction of concrete and spiritual as well as abstract entities either, even if the latter were natural ones — which reduces naturalism to materialism (e.g., [Armstrong, 1995; Mahner and Bunge, 1997]). Note that naturalism involves the parsimony principle (see Sect. 4.2, 4c). The least parsimonious view would be some sort of non-interventionism. This is the thesis that the universe is full of supernatural entities, but these have somehow agreed never to interfere with scientific measurements or experiments. Evidently, this view is quite arbitrary and nonparsimonious.

The third ontological ingredient of science is the principle of lawfulness. This is the hypothesis that the real world is not capricious, but behaves in a regular fashion. Indeed, if things behaved lawlessly, the world would resemble a cartoon movie in which everything can change into anything, forward and backward in time, in a completely arbitrary fashion. Presumably, there would be no living beings, no knowledge and no technology if the world were lawless. Note that the principle of lawfulness does not presuppose Laplacean determinism, because there are also stochastic processes — which follow probabilistic laws. Note further that, if laws as ontic regularities are distinguished from law statements purported to represent such laws, the various criticisms of the concept of natural law in science (e.g., by [Cartwright, 1983] and [Giere, 1999]) mostly concern the latter, i.e., the epistemological notion of a law. A too rigid traditional conception of natural law statements held by many philosophers of science, and our difficulties with idealization and approximation in representing real laws must not lead us to conclude that, as a consequence, there are no laws in the ontic sense, i.e., that the world behaves irregularly or even miraculously.

The fourth ontological presupposition is the principle of antecedence, which is often conflated with the causality principle. The antecedence principle maintains that causes precede their effects or, alternatively, that the presence is (causally or stochastically) determined by the past. By contrast, the principle of causality in the strict sense states that every event has an (external) cause producing the given event; more precisely, for every event e in some thing x, there is another event e' in some (other) thing $x' \neq x$, such that e' causes e. But since there are spontaneous (uncaused) events, such as exemplified by certain quantum events like radioactive decay, it is false as a universal principle. Nonetheless, in the case of our above experiment, we also need some version of the causality principle to account for the fact that our actions have some effect on the world.

The fifth ontological presupposition of science may be called the genetic or ex-nihilo-nihil-fit principle. Going back at least to Epicurus and Lucretius, this principle says that nothing comes out of nothing and nothing disappears into nothingness. Note that "nothing" here really means "nothing": even the curious vacuum field filling up empty space is *something* rather than nothing, for it can affect other things. (Note, incidentally, that this ontological assumption also affects physical cosmology: although one might be prepared to make exceptions for the universe as a whole, the genetic principle should encourage us to explore and prefer cosmological models, even big bang models, that do not assume a *creatio ex nihilo*, but presuppose some pre-existing state of the universe.)

Finally, there is the "no psi" principle [Broad, 1949; Bunge, 1983b], which is the postulate that minds or brain processes do not act directly on the things out there, but only through some motoric action of our body. Nobody could trust the readings of any measurement instrument or the results of any experiment if immediate mental forces and causes permeated the world.

These ontological principles must not be seen in isolation: they are a package deal. The idea that there are real and natural things, behaving lawfully and not popping out of, or into, nothing, is certainly the major metaphysical guide line of factual scientists. Note that these ontological and epistemological principles could all be false, which is why they are hypotheses or postulates, not ideological dogmas, as some critics of science tend to claim. However, both their eminent fertility and the extraordinary success of science justify that we accept them as true — for the time being. We might therefore call them the ontological default assumptions or, in some cases, metaphysical null hypotheses of factual science.

b) Epistemological assumptions

In order to do factual science, ontological realism must be combined with epistemological realism, i.e., the thesis that the real world can be known, if only approximately and imperfectly. Otherwise, scientists would just study the figments of their imagination, and technologists were unable to successfully alter real things, because this presupposes that at least some relevant properties of those things are known correctly.

Now, epistemological realism comes in different versions and strengths (see, e.g., the overview by Kuipers 2001, Ch. 2). We need not commit ourselves here to any position, although the most widely accepted version is likely to be what is often called scientific realism, which stipulates that we can know not just observables, but also unobservables. Elementary physics and evolutionary biology, for example, would make little sense without this assumption.

But what about instrumentalism, conventionalism, and other antirealist epistemological positions held on occasion by both scientists and philosophers of science? Are they not more parsimonious than realism? It is not just the claim that the majority of working scientists adopts realism in their daily work, but also the fact that, both in science and metascience, we should accept that position that has the greatest explanatory power and fecundity. In this regard realism beats instrumentalism, because the latter can explain neither the success nor, more importantly, the failure of scientific theories.

Moreover, whereas the instrumentalist cannot explain what the realist does and thinks, the realist is able to explain what the instrumentalist does. Thus realism subsumes instrumentalism [Vollmer, 1990; Kuipers, 2000]; see also [Kitcher, 1993] for an analysis of various antirealist arguments).

c) Methodological principles

A very general methodological maxim of any scientific approach is the principle of parsimony, also known as Ockham's Razor. It enjoins us not to multiply explanatory assumptions (entities, processes, causes, etc.) beyond necessity, in particular with respect to theoretical entities. It does not tell us, however, when such necessity obtains. Note that this principle is methodological, not ontological: it does not presuppose that nature is always and perhaps necessarily parsimonious, but that as inquirers we should begin with parsimonious assumptions. Note further that parsimony should not be readily equated with simplicity, such as the injunction to always prefer the simpler of two theories. After all, a theory can be simpler than another in many respects: it may be referentially simpler (having less qualitatively different referents), mathematically simpler, methodologically simpler (easier to test), or pragmatically simpler (easier to apply in a technological context). Simplicity in one such respect does not guarantee simplicity in another.

A second methodological principle is fallibilism or methodological skepticism. It is the acknowledgment of the fact that error is possible in all cognitive matters, so that our knowledge may be subject to criticism and, if possible, improvement and, if necessary, revision. We may highlight the latter by explicitly adding a "meliorist principle" [Bunge, 1983b] or a "principle of improvement of theories" [Kuipers, 2001].

d) Semantic assumptions

Most factual scientists maintain that their hypotheses, models and theories are true if they adequately represent the facts they refer to. That is, they subscribe to a correspondence theory of truth. Needless to say, the notion of truth is as tricky as many other concepts, so that there is no agreement among philosophers as to the appropriate truth concept in science [Weingartner, 2000]. Nevertheless, scientific realism is quite naturally associated with a correspondence concept of truth [Bunge, 1983b; Thagard, 1988; Devitt, 1996; Wilson, 2000]. Such a notion becomes easier to defend when we realize that the concept of correspondence truth provides just a semantic *definition* of "truth": it says nothing about how, and in particular how well, the truth of a hypothesis can be *known*. In other words, it does not provide a truth criterion. Truth criteria, such as evidential support, are not the business of semantics, but of methodology.

The concept of correspondence truth fits scientific practice even better when we realize that factual truth is in many cases not a dichotomy between true and false, but a matter of degrees. Models and theories often represent facts only in certain respects and moreover imperfectly so. Thus they correspond to facts only partially. Similarly, quantitative properties (represented by magnitudes) may be known only approximately, which is why scientists attempt to improve their measurement techniques. A realistic philosophy of science will therefore try to do justice to the idea of partial or approximate truth [Bunge, 1983b; Weston, 1992] and hence methods of truth approximation [Niiniluoto, 1987; Kuipers, 2000].

e) Axiological and moral assumptions

Most norms of science are built into its methodology. However, there are not only methodological values and norms, but also attitudinal and moral ones. Merton's [1973] expression "the ethos of science" captures this fact aptly, although his work is mostly concerned with attitudinal and moral norms that are not immediately relevant to the production of true knowledge (see below). To stress the fact that science has an internal system of values and corresponding norms, it may be useful to treat them all together. Thus, the researchers in a scientific field of knowledge are expected to accept the following values:

- *Logical values* such as the principle of noncontradiction and noncircularity. Together with the entire canon of valid reasoning, these are of course basic principles of rationality.
- *Semantical values* such as meaning definiteness, clarity, and maximal truth. Of course, a young or emerging scientific field may teem with vague and fuzzy concepts. But as it progresses and matures, in particular when it develops a theoretical branch, clarity and exactness are supposed to replace fuzziness. However heuristically fruitful vagueness may be in the beginning or in certain contexts, it may as well indicate that a field is degenerative rather than progressive.
- *Methodological values* such as testability (including the testability of the methods used in testing hypotheses, as well as the independent testability of auxiliary assumptions), explanatory power, predictability, reproducibility, and fecundity. Since these and other methodological categories are the main business of the philosophy of science, we shall not elaborate on them here.
- *Attitudinal-* and *moral values* such as critical thinking (or rationality in general), open-mindedness (but not blank-mindedness), universalism or objectivity (i.e., the requirement that ideas be evaluated independently of the personal, social or national characteristics of their proponents),

truthfulness, and acknowledgment of the work of others (e.g., by adequate citation).

As stated above, Merton's [1973] classic ethos of science concerns mostly attitudinal and moral values or norms, respectively. These are often abbreviated by the acronym CUDOS, which stands for four main norms: *c*ommunism (research results should be public property and accessible to everybody), *u*niversalism (see above), *d*isinterestedness (research should be uninfluenced by extra-scientific interests, and scientists should be emotionally detached from their subject matter), and finally *o*rganized *s*kepticism (scientists should be critical in particular towards their own work, and point out on their own weak spots or problematic parts). However, Merton's norms have been criticized for being too idealized and geared to an academic ivory tower situation (see [Ernø-Kjølhede, 2000] for an overview). Indeed, the history, psychology and sociology of science provide many examples that scientists have failed to follow one or more of these values. Like everyone else scientists are only human after all. Thus, individual scientists may be biased and jealous; they may intrigue against colleagues, or engage in nepotism; they often are emotionally attached to their subject matter in being passionate researchers, and they sometimes do not see the weak spots, if not flaws, in their own work; in particular, *pace* Popper, they are usually interested in having their hypotheses and theories confirmed, not refuted — after all, Nobel prizes are not awarded for the falsification of a theory. Moreover, the social and economic organization of scientific research has changed drastically during the past 50 years in that research institutions including universities are now run more like businesses, so that there is severe competition for funds and a strong pressure to focus on applied science and technology at the expense of basic science (see [Ziman, 1994]). For all these reasons Merton's classic ethos no longer describes realistically the behavior of scientists, however desirable his norms may still be from an ethical point of view (see also [Kuipers. 2001]). Finally, most of Merton's norms concern the professional social behavior of scientists in general, whereas the primary interest of the philosopher of science concerns those values and attitudes that are epistemologically relevant by contributing to gaining true knowledge, such as rationality, objectivity, and truthfulness.

In sum, the system of logical, semantical, methodological, and attitudinal ideals constitutes the *institutional rationality* of science [Settle, 1971], even though individual scientists may more or less often fail to behave rationally. (More on the problems of the rationality of science in [Kitcher, 1993].) And, however biased the individual scientist may be, the above values are also the basis for the *institutional objectivity* of science. As a consequence, basic science is value-free only in the sense that it does not make value judgments about its objects of study. In other words, basic science has no external value system.

This completes our extensive analysis of the philosophical outlook of a scientific field (condition 4), so that we proceed at last with our list of conditions characterizing an epistemic field as scientific.

5. The *formal background* F of a scientific field is a collection of up-to-date logical and mathematical theories used by the members of C in studying the items of D. This does not imply that scientificity is to be equated with formalization. All this criterion demands is that formal tools have to be handled correctly, and they must be adequate to tackle any given theoretical problem.

6. The *specific background knowledge* B is a collection of up-to-date and reasonably well-confirmed data, hypotheses, theories, or methods borrowed from adjacent fields. Every scientific field uses some knowledge from other scientific fields. For example, biology borrows knowledge from physics and chemistry. A science that borrows little from other fields is either very fundamental or very backward.

7. The *problematics* P is of course the collection of problems to be solved in the given field. It consists exclusively of epistemic questions on the nature and in particular on the lawful behavior of the objects in its domain D. It may also comprise problems concerning other components of its conceptual framework (e.g., the adequacy of methods, formalisms, and other background assumptions). If a discipline deals with practical problems, it is a technology, not a basic science.

8. The *fund of knowledge* K is a *growing* collection of up-to-date, testable and well-confirmed knowledge items (data, hypotheses, theories), gained by C and compatible with those in B. Even a young scientific field will possess some fund of knowledge, either taken over from ordinary knowledge or inherited from a parent science.

9. The *aims* A of the members of C of a field in basic science (as opposed to technology) are purely cognitive. They include, for example, the discovery and use of the laws of the members of D; the systematization of the knowledge in K (e.g., by constructing general theories); and the refinement of the methods in M.

10. The *methodics* M is a collection of empirical methods or techniques which may be used by the researchers in C in their study of the members of D, whereby "method" means a rule-directed procedure for collecting data or testing a theory. (Note that methods of reasoning, such as rules of inference or rules for evaluating theories, have been treated as belonging in G. Whence the distinction between methodics and methodology.) A scientific technique may be either concrete (i.e., involving instruments), such as electron microscopy, or conceptual (formal), such as the various statistical methods.

And they may be quite specific, such as Hennig's method of reconstructing phylogenies, or else more or less general, i.e., applicable in several fields, or for different purposes.

Among the methodological requirements for a technique to be scientific are the following. The functioning of these methods should be scrutable (e.g., by alternative procedures) and explainable by well-confirmed theories. (This may not be the case in a young field, but it should be achieved as the field matures. For example, when Galileo used his telescopes, optics was still too immature to fully explain their functioning.) And the techniques must be objective in the sense that every competent user is able to obtain roughly the same results.

It has been quite controversial whether there is such a thing as a scientific method in general (see, e.g., [Laudan, 1983; Haack, 2003]). If such a general method is expected to be a fool-proof procedure for delivering true and certain knowledge, then there is of course no such method. However, if we view the scientific method as an extremely general research *strategy*, then there may very well be a scientific method. For example, the sequence "problem–hypothesis–test–evaluation" reflects the general structure of any empirical scientific paper (introduction, methods, results, discussion), and may thus be seen as representing *the* scientific method [Bunge, 1983b]. However, if this definition is accepted, the scientific method is at best a necessary, but not a sufficient condition of scientificity: its application does not automatically turn one's inquiry into a scientific inquiry. Moreover, being extremely general, it is not an empirical method proper, so that it may as well be seen as belonging to the methodological rules in G.

In addition to the ten conditions of the ten-tuple $\langle C, S, D, G, F, B, P, K, A, M \rangle$ used in the preceding to characterize a scientific field, Bunge (1983b) requests that a scientific field satisfy two further conditions. These conditions take into account two aspects of science that have been emphasized by many philosophers of science: unity (consilience) and progressiveness.

11. The *systemicity condition.* There is at least one other field of research $\mathcal{S}'$ such that $\mathcal{S}$ and $\mathcal{S}'$ share some items in G, F, B, K, A and M; and either the domain D of one of the two fields $\mathcal{S}$ and $\mathcal{S}'$ is included in that of the other, or each member of the domain of one of the fields is a component of a system in the domain of the other [Bunge, 1983b, p. 198]. In simpler words, every scientific field has connections with other fields — a fact which allows for multi- and interdisciplinary research. This is due to the fact that nature is organized into several levels of complexity — levels that scientific disciplines may approach from various perspectives and with different aims and methods. Thus, despite all the differences in our cognitive interests, scientific disciplines form a network of approaches, striving for a unified — a consilient or convergent — view of nature, which need not be a reductionist

one [Kitcher, 1982; Bunge, 1983b; Bechtel, 1986; Thagard, 1988; Vollmer, 1993; Reisch, 1998]; for a dissenting view see [Dupré, 1993]. For this reason, new theories are evaluated not only on the basis of empirical tests, but also with regard to their overall compatibility with the well-confirmed background theories (external consistency). Although a new theory cannot by definition be compatible with every other theory, in particular its rivals, because it would otherwise not be a new theory, it must somehow allow to be accommodated within the totality of our knowledge. In Kuhnian terms: even if revolutionary, a new theory will cause only local or regional revolutions, never a total revolution turning upside down all existing fields at once.

12. The *changeability* or *progressiveness condition*. The membership of the conditions 5–10 changes, however slowly and meanderingly at times, *as a result of research* in the same field or as a result of research in neighboring disciplines. In Lakatosian words, the history of a scientific discipline must be progressive, at least on the whole. Even if science were to come to an end in the distant future, the history of a scientific discipline would have to show a certain amount of progress. (How the view that science is progressive can be defended against various antirealist objections has been shown by [Kitcher, 1993].)

This concludes the characterization of scientific epistemic fields. Note, firstly, that this characterization applies first of all to contemporary science, because many of its features have developed into their current state over the past 400 years. Consequently, it may not be fully applicable to 17th century science, for example. As for its future development, I doubt that the basic features and principles discussed above will evolve in a way that leads to their replacement by completely different principles, in particular their contraries. However, future development might consist in their improvement as well as in the discovery of some as yet unknown features and principles.

Note, secondly, that this characterization comprises both descriptive and normative aspects. Whereas the descriptive conditions provide diagnostic indicators, the normative ones will be the foundation for any judgment on the scientificity, or nonscientificity respectively, of an epistemic field.

What about science as a whole? Science as a whole is of course the totality of all individual scientific disciplines. If, as in the preceding, we represent each scientific field as a ten-tuple $\mathcal{S}_1 = \langle C_1, S_1, D_1, G_1, F_1, B_1, P_1, K_1, A_1, M_1 \rangle$, $\mathcal{S}_2 = \langle C_2, S_2, D_2, G_2, F_2, B_2, P_2, K_2, A_1, M_2 \rangle, \ldots, \mathcal{S}_n = \langle C_n, S_n, D_n, G_n, F_n, B_n, P_n, K_n, A_n, M_n \rangle$, science as a whole can be conceived of as the sum of these ordered sets: $\Sigma = \mathcal{S}_1 + \mathcal{S}_2 + \ldots + \mathcal{S}_n$. Similarly, we could characterize a *multidiscipline*, consisting of two or more scientific fields, as the sum of two or more ten-tuples representing them [Bunge, 1983b, p. 219]. In the case of a two-field multidiscipline this would be represented by: $\mathcal{S}_1 + \mathcal{S}_2 = \langle C_1 \cup C_2, S_1 \cup S_2, D_1 \cup D_2, G_1 \cup G_2, F_1 \cup F_2, B_1 \cup B_2, P_1 \cup P_2, K_1 \cup K_2, A_1 \cup A_2, M_1 \cup M_2 \rangle$. (Note that we represent the concrete systems C and S by their composition, i.e., the set of their components.

Otherwise we would need an operation of physical or mereological addition rather than simply the one of set-theoretical union.)

By contrast, an *interdiscipline* does not just consist of at least two fields retaining their identity, but it is a merger of fields attempting to approach a common domain from a unified point of view rather than from different angles. Therefore, an interdiscipline may be conceived of as the *intersection* of two or more fields.

The analysis of a scientific field as a ten-tuple also allows us to elucidate the notion of a scientific research project. In section 4.1 we have defined the conceptual framework of an epistemic field as a septuple $\mathcal{S}_c = \langle G, F, B, P, K, A, M \rangle$. A *research project* π within a scientific field $\mathcal{S}$ characterized by a conceptual framework $\mathcal{S}_c = \langle G, F, B, P, K, A, M \rangle$ is then the septuple $\pi = \langle g, f, b, p, k, a, m \rangle$, where every component is a subset of the corresponding component of S_c [Bunge, 1983b, p. 176].

How does Lakatos's notion of a research program fit into this conceptualization? According to Lakatos [1970], a research program is a historical sequence of theories. Now theories surely belong to the fund of knowledge K of a scientific discipline. But we must also include the reference class of the theory belonging in D, as well as the formalism used to built the theory, which belongs in F. Further, Lakatos also counts auxiliary and other relevant assumptions as belonging to a theory. These may belong either in B or in K. Thus, a theory ϑ at any given time t might be construed at least as a quadruple $\vartheta(t) = \langle d(t), f(t), b(t), k(t) \rangle$, and a research program ρ over a period τ, where $\tau = [t_1, t_n]$, as an ordered set of such quadruples, $\rho(\tau) = \langle \langle d(t_1), f(t_1), b(t_1), k(t_1) \rangle, \langle d(t_2), f(t_2), b(t_2), k(t_2) \rangle, \ldots, \langle d(t_n), f(t_n), b(t_n), k(t_n) \rangle \rangle$. Depending on what we take to belong to a theory, we might as well regard a research program as a sequence of research projects as defined in the previous paragraph. Or, disregarding the historical focus of Lakatos's concept, we might simply redefine "research program" in the broad sense of "research project" or even "conceptual framework" or "disciplinary matrix" as explicated above (see, e.g., [Kuipers, 2001] for an even broader conception of "research program"). I take this broader approach to be more useful for demarcation purposes than Lakatos's idea of a series of theories in themselves.

So much for a possible characterization of the notion of a scientific epistemic field, which views science in the sense of basic factual science. It is now time to take a look at other research fields which, though not factual sciences, are related to them: mathematics, technology, and the humanities.

4.3 Other Research Fields

4.3.1 Mathematics

In contrast to the factual sciences, mathematics as well as formal logics and semantics are often called *formal sciences.* Although they have much in common with the factual sciences, the question is whether these commonalities justify to regard them as sciences. In other words, the question is whether we should use the label "science" in the strict sense of factual science or in a broader sense including formal science and perhaps technology.

Let us quickly analyze the status of mathematics with regard to the twelve conditions listed in Section 4.2. In so doing, we shall mention only those conditions that show significant differences.

Clearly, the domain D of mathematics shows an important difference with factual science: all the referents of mathematics are abstract objects. Although we can apply mathematical concepts and theories to concrete things, their properties and processes, we do so only by interpreting them in factual terms. In this way we represent factual properties in formal terms. Pure mathematics does not deal with concrete objects.

The philosophical background G of mathematics is also quite different. To begin with, mathematics can do without ontological realism: it would work just as well if there were no mind-independent reality. Of course, most mathematicians are *de facto* also ontological realists, but this is not a necessary assumption for doing mathematics: mathematics can be done on the basis of a Platonist, nominalist, or constructivist ontology (see, e.g., [Agazzi and Darvas, 1997]). Being just as ontologically neutral as logics [Nagel, 1956], mathematics has no use for the other ontological assumptions of factual science either, except for the principle of lawfulness. Indeed, mathematicians also assume that the referents of their discourse "behave" lawfully, whether they be found in a Platonic realm of ideas or whether they be constructed by our minds. Depending on the philosophy of mathematics adopted, the mathematical Platonist will need a form of epistemological realism, whereas the constructivist can do without it.

A major difference lies in the semantic concept of truth in mathematics: dealing with abstract objects and thus purely formal properties, mathematics is in no need of a correspondence theory of truth and hence can do with a coherence theory of truth (recall Leibniz's *verités de raison*; see also [Bunge, 1983b]). Only the mathematical Platonists and empiricists may have use for a correspondence theory of mathematical truth. Still, mathematical truth is *de facto* established by formal coherence.

The methodological, attitudinal and moral values are by and large the same as in factual science. The major difference here lies in the notion of testability, which can only mean conceptual testability, not empirical testability. Moreover, testability in mathematics is stronger than empirical testability, because it allows for conclusive proof and disproof, whereas empirical testability only provides confirming or disconfirming instances.

As a consequence of the differences mentioned so far, there is another difference in the methodics M: mathematics uses no empirical, but only conceptual methods. (Even though some proofs obtained with the help of computers, such as that of the four color problem, may imitate empirical means in certain respects, they are still virtual and hence conceptual. Likewise, thought experiments, whether in mathematics or in the factual sciences, are conceptual means.) However, being extremely general, the scientific method, as defined in Sect. 4.2, seems to be used in mathematics as well.

As is obvious from the preceding, the main differences between mathematics and the factual sciences lie in the fact that it deals exclusively with abstract objects. On the other hand, mathematics too is a rigorous and progressive research field, consisting of a set of fruitfully interacting subfields.

4.3.2 Technology

In popular thinking, science and technology are often conflated. Worse, industrial production and marketing of technical goods is often equated with technology, which is in turn equated with science. So science gets often blamed for everything negative associated rightly or wrongly with the Western capitalist way of living. However closely these areas may be related *de facto*, the philosopher of science or of technology is of course interested in the question of whether science and technology can be distinguished *de jure*.

Borrowing again from Bunge [1983b], I shall propose the following distinctions. To begin with, the investigation of cognitive problems with *possible* practical relevance will be termed *applied science*. Thus, an applied science differs from its basic science partner mostly in its problematics (P) and aims (A). Further, its domain D will be narrower. For example, in contrast to human biology, medical research studies only those properties of humans that concern, directly or indirectly, matters of health. The same holds for clinical psychology as opposed to psychology in general.

If we now add the requirement that, on top of having discovered or studied some X which may be useful to produce (or else prevent) some Y, we actually *design* an artifact or a procedure to produce or else prevent Y, we arrive at technology. More precisely, *technology* may be defined as "the design of things or processes of possible practical value to some individuals or groups with the help of knowledge gained in basic or applied science" [Bunge, 1983b, p. 214].

Note first that, by making technology dependent on science, this definition distinguishes technology from the traditional crafts or *technics*, which are based solely on ordinary knowledge. Note further that this definition is so wide that it includes not only the classic fields of physical and chemical engineering, but also biological, psychological and social technologies. Thus, medicine, psychiatry, pedagogy, law, city planning, and management "science" are all technological fields.

Let us briefly review the coordinates of the ten-tuple $\langle C, S, D, G, F, B, P, K, A, M\rangle$ as to the differences between science and technology. As in the preceding section, only those showing significant differences will be mentioned. To begin with, although C is a research community, it is not as international and universalist as in the case of basic science, because patents and industrial secrets limit the circulation of technological knowledge. The domain D is both narrower and wider than in the case of applied science: it is narrower because it is concerned only with natural things which are useful for us, and it is wider because it includes not only natural things and processes but also artificial ones. The general outlook G

shares a realist and naturalist ontology and epistemology with basic science, as well as most of the other philosophical assumptions and values. The main difference lies in the fact that technology does not test so much for truth as for efficiency. Truth is relevant only as a means for design and planning. Finally, the ethos of technology differs from that of basic science: usually, it consists not in the free and disinterested search for knowledge, but in task-oriented work, often depending on the economic interests of some employer (see also [Ziman, 1994]). Obviously, the problematics P and the aims A are among the main differences: the problems and aims are practical rather than cognitive. Moreover, the aim of technology is not to discover new laws: it suffices to make use of known ones. Finally, technology is characterized by a coordinate of its own: in contrast to basic science, technology has not only an internal value system, but also an external one (V). That is, it attributes positive or negative values to natural or artificial things or processes, be it raw material or finished product. Thus, a technology is actually characterized by an eleven-tuple $\langle C, S, D, G, F, B, P, K, A, M, V \rangle$.

4.3.3 Humanities

In contrast to the social sciences, which study social systems (composed of human individuals) and their activities by empirical means, the humanities mostly abstract from these concrete individuals and groups as well as their activities and study their intellectual (including artistic) products, i.e., ideas or concrete artifacts. Inasmuch as the humanities study the activities of groups or individuals, these are usually of an artistic nature, such as a theatrical or musical performance. Accordingly, literature and literary criticism, languages (philology) and part of linguistics, art history and criticism, musicology, the history of ideas, religious studies, and philosophy belong to the humanities. On the other hand, some fields like history and archeology, as well as the history and sociology of religion belong — or should belong — to the social sciences. Similarly, part of linguistics is a social science too. And according to our classification, the law (jurisprudence) and pedagogy are not humanities but sociotechnologies (Sect. 5.2). These examples show that quite often there is an overlap between some social sciences and the humanities. In particular, some fields starting out as humanities may develop into sciences.

Again, a quick review of the ten coordinates of an epistemic field will be in order. To begin with, the humanities are clearly research fields with a specialized research community C. As just mentioned before, their domain D consists of ideas and artifacts rather than natural things and processes. Consequently, the humanities are consistent with either a naturalist-materialist or a Platonist outlook. As for epistemology, the natural approach is most likely a constructivist one, which can be either realist or antirealist. Furthermore, the humanities are open to the influence of subjectivist philosophies like phenomenology and hermeneutics. (And of course, in the field of philosophy, which has to provide its own metaphilosophy, just anything goes.) In sum, the philosophical outlook of the humanities is much more variegated than that of the sciences, and necessary connections, if any, with particular philosophical presuppositions are much less obvious. Presumably, the more aspects of the much straighter scientific outlook are adopted, the better the chances of bridging a humanistic field with a scientific one. Think of linguistics and comparative religion (*Religionswissenschaft*), which make contact with sociology, history, evolutionary biology, psychology and, more recently, even the neurosciences.

As for methodology and semantics, since the humanities deal with ideas and artifacts, which are not to be explained by natural laws and mechanisms but instead interpreted and comprehended, it is unclear which role the parsimony principle plays in the humanities. More complex views and interpretations may be preferred to simpler ones, just as conversely. Similarly, fallibilism may not be that important because there may be different reasonable perspectives and interpretations, without implying that therefore one of them is erroneous. Consequently, the notion of truth in the humanities is often contextual or relative rather than factual. The fact that Othello killed Desdemona is (fictionally) true only in the context of Shakespeare's story. Another author could easily write an alternative play in which Desdemona kills Othello, so that in this context the opposite would be true. On the other hand, inasmuch as the humanities are descriptive of certain (e.g., historical) facts, these descriptions can be correct or not in the correspondence sense.

What about the internal value system of the humanities? Rationalist humanities will certainly respect the standard logical values. But there are also irrationalist branches, in particular in philosophy and certain postmodernist cultural studies (see Sect. 5.2). Very often the semantical values of clarity and exactness cannot be heeded. This is due to the very nature of human thought and communication, which is far from unambiguous, whence the need for interpretation arises. However, if these semantical values are not accepted even as remote ideals, and fuzziness is instead turned into method, the line to obscurantism may easily be crossed.

Evidently, the methodological values of testability and explanatory power in the scientific sense are not part of the humanities. A certain view, reading, or interpretation may be open to criticism, but since it is neither true nor false, it cannot be tested for truth. At most, it is reasonable, plausible, sensible, or apposite. Explanatory power may be replaced by "comprehensive power" if we admit the hermeneutic goal of understanding in the humanities. On the other hand, fecundity is certainly also a value in the humanities, because humanistic understanding can be increased if some approach opens up new perspectives.

Whereas some attitudinal values are of course the same as in the sciences, others are different. For example, just as there may exist competing theories in the sciences, there may be competing interpretations in the humanities. Honesty requires at least mentioning the existence of such competing approaches, even though the researcher wants to focus on her own. The same holds for the adequate citation of sources, although the standards appear to be lower than in the natural sciences. For example, it seems to be much easier to survive peer-review when disregarding the work of disliked colleagues in a philosophical article than in a science paper. Furthermore, the value of universalism plays only a minor role, if any, in the humanities. For, naturally, the humanities are more inclined towards relativism, because many cultural items cannot be evaluated independently of the personal and cultural characteristics of their creators: they must be seen and understood in context. Finally, like technology, many humanities have an external value system: they attribute, for example, aesthetic values, meanings, and purposes to the objects in their domain D, because the latter are studied in their relation to humans.

The formal background of the humanities, if any, is of course small. Exceptions occurring for example in philosophy, such as mathematical logics and formal semantics, may be classified as formal sciences. On the other hand, other branches of analytical philosophy too are formal (like ontology), which indicates that they are science-oriented, though not full-fledged sciences.

The aims of the humanities can be either cognitive or practical, or both. In contrast to the sciences, however, they usually do not seek to find laws. Indeed, the "sciences of the mind" (*Geisteswissenschaften*) have been regarded as descriptive (idiographic) rather than law-finding (nomothetic). On the other hand, we have seen before that some humanities make contact with the sciences, so that such multi- and interdisciplinary ventures may be able to find some cultural or even aesthetic laws.

Obviously, a major difference with the sciences is found in the methodics M of the humanities. Naturally, except for some observation, their methods are mostly conceptual. Among these are some general methods unique to the humanities, such as the hermeneutic and dialectic "method" [Poser, 2001], although these are not methods in the strict sense of rule-guided procedures to attain a certain goal. (Here, "hermeneutics" does not mean philosophical hermeneutics, but only the traditional concept of text interpretation, or understanding of works of art, respectively. And the dialectic method concerns first of all the discoursive triad thesis-antithesis-synthesis, without presupposing the whole of dialectic philosophy.) If not objective in the sense that every competent user will get roughly the same results, these "methods" are at least intersubjective in that their results can be communicated to, and understood by, other people. The humanistic scholar may also borrow or apply certain techniques from the factual sciences, but this does not yet turn her field into a science. For example, the art historian may have some paint chemically analyzed, or some cloth radiocarbon-dated, without thereby changing the nature of her discipline.

In sum, compared to formal science and technology, the humanities show the greatest distance from factual science. But again, we emphasize that this is not a value judgment. When saying, for example, that the arts and humanities are not scientific, nobody claims that they are therefore objectionable or bad.

4.4 Conclusion

The factual and formal sciences, the technologies, and the humanities are all research fields producing genuine knowledge, which on the whole is either (approximately) true or else useful, and contributes to the understanding of the world and its inhabitants. For this reason, one might argue that they should all be included in a broad conception of science. This is for example done in the German intellectual tradition, where the name of almost any field of knowledge is dignified by the ending "-wissenschaft" (-science), including the humanities, which are called *Geisteswissenschaften* (sciences of the mind). So there is bioscience alongside "music science", just as there is computer science alongside "literature science". Consequently, if a practitioner of a *Geisteswissenschaft* is told that what he does is not science, he will most likely be offended. It comes as no surprise that such a broad, if not inflationary, construal of "science" aggravates the problem of demarcation (see, e.g., [Poser, 2001]).

By contrast, most other traditions and languages separate the arts and humanities from the sciences already terminologically, so that no offense is given by calling the humanities nonscientific. Yet even so, the question remains of what to do with mathematics and technology. While some authors include both of them in the sciences (e.g., [Kuipers, 2001] classifies them as explicative research programs and design programs, respectively, within a broad conception of a *scientific* research program), others assert that neither mathematics [Lugg, 1987] nor technology [Bunge, 1983b] are sciences. In any case, taking into account the preceding overview, the common post-positivist picture, which admits more categories than just sense (i.e., science) and nonsense (i.e., all the rest), may look like the one given in Fig. 2. One the one hand, there is science including mathematics and technology; on the other there is nonscience including the arts and humanities as good nonscience, so to speak, for it too is viewed as producing true, reliable, or at least valuable knowledge, respectively, and finally pseudoscience as bad nonscience, for its knowledge claims are unjustified.

We may refine this picture by adding protoscience and prototechnology, as well as ordinary knowledge. These straddle the lines between pseudoscience and science. A protoscience is expected to develop into a science proper by leaving behind its nonscientific (or even pseudoscientific) roots (see Sect. 6). And ordinary knowledge is mostly nonscientific and reliable, but contains illusory items on the one hand, and some knowledge adopted from the sciences on the other (Fig. 3). It is the task of the science educator to increase the share of the latter and to decrease that of pseudoscience and superstition.

We shall further refine this picture later on to reflect the distinctions made above between factual science, mathematics, technology, and the humanities. Before, however, we need to take a closer look at that kind of knowledge which is not just nonscientific but in fact unscientific or pseudoscientific.

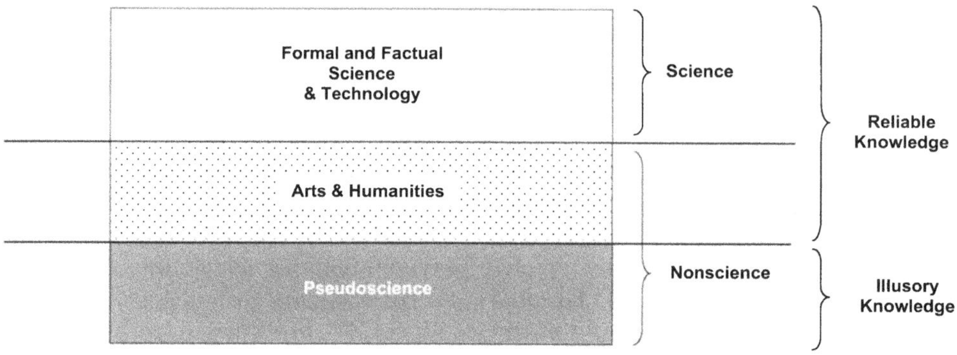

Figure 2. A common post-positivist picture of science and nonscience. As scientific research fields, mathematics, factual science (including psychology and social science), and technology are subsumed under the general label of "science". Nonscience divides into the arts and humanities (including philosophy) on the one hand, producing reliable or at least valuable knowledge, and pseudoscience on the other, offering nonreliable or illusory knowledge.

5 UNSCIENTIFIC FIELDS

As emphasized previously, calling an epistemic field *nonscientific* is not pejorative but descriptive. Calling it *unscientific*, however, is judgmental: it indicates that the given field cannot live up to its cognitive claims. Since there is no noun "unscience", an unscientific field is called a "pseudoscience". As usually defined, a pseudoscience is a particular form of nonscience, namely a nonscientific field whose practitioners, explicitly or implicitly, *pretend* to do science. Thus, to say that a field is pseudoscientific amounts to saying that it is a fake. In other words: While there is reliable or, if preferred, approximately true theoretical and practical nonscientific knowledge, the knowledge produced by pseudoscience is illusory. And since spreading bogus knowledge amounts to deception, pseudoscience has a moral dimension that other nonscientific fields lack. Therefore, a demarcation of science versus nonscience in general does not yet tell us how legitimate nonscientific fields are to be demarcated from pseudoscientific ones.

5.1 Characterizing Pseudoscience

For this reason, several authors have attempted to give not only a characterization of science as opposed to nonscience, but also of pseudoscience in particular [Thagard, 1978; 1988; Radner and Radner, 1982; Bunge, 1982; 1983b; 1984; Grove, 1985; Lugg, 1987; Derksen, 1993; 2001; Hansson, 1996; Wilson, 2000; Kuipers, 2001]. It will come as no surprise that the criticisms of such attempts parallel those leveled against any quick and clear-cut demarcation of science: though dealing with important aspects of pseudoscience, the proposed demarcation crite-

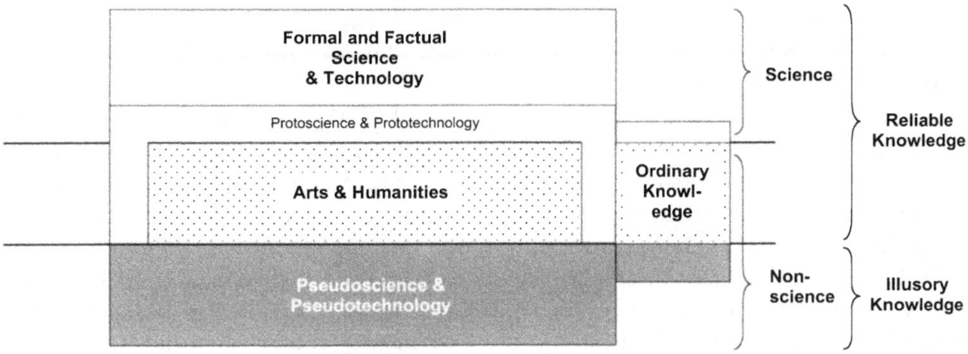

Figure 3. A refined post-positivist picture of science and nonscience, making room for ordinary knowledge as well as protoscience and prototechnology, which range from the pseudoscientific to the scientific.

ria do not combine to form a set of necessary and sufficient conditions, because they always leave some pseudosciences unscathed. Let us briefly review some such demarcation attempts.

Improving on his earlier demarcation proposal [Thagard, 1978], Thagard [1988, p. 170] contrasts his five characteristics of science mentioned in Section 3 with five features typical of pseudoscience. In pseudoscience, scientific correlation thinking is replaced by primitive resemblance thinking; empirical matters of confirmation and disconfirmation are neglected; practitioners of the field are oblivious to alternative theories; the theories are nonsimple and contain many ad hoc hypotheses; and there is no progress in doctrine and application. Thagard points out that these are indicators of pseudoscientificity, not necessary and sufficient criteria.

Grove [1985] gives four characteristics of pseudoscience. The first is the lack of an "independently testable framework of theory capable of supporting, connecting, and hence explaining their claims" (p. 237). The second is the lack of progress. Third, a pseudoscience is usually constructed in such a way that it is able to resist any possible counter-evidence; in other words, it is practically irrefutable (though it may be logically falsifiable). And fourth, according to Grove, not just irrefutability is a mark of pseudoscience but, more generally, their "total resistance to criticism".

Lugg [1987, p. 228] suggests regarding pseudosciences as "radically flawed practices, i.e., as radically flawed complexes of theories, methods and techniques". He maintains that, in the case of the pseudosciences, empirical matters are relatively unimportant, because their being conceptually flawed makes them unworthy of serious attention, whether or not their claims could actually be confirmed or disconfirmed. This is similar to Rothbart's [1990] claim that pseudoscientific theories are not testworthy. If we can already show by means of formal or informal logic that an argument or an approach is fallacious, there is no need to empirically test the hypotheses involved. Finally, according to Lugg, if pseudosciences are practices, they are social institutions, and realizing that they are such helps to explain their longevity and resilience.

Rationalistic approaches, such as Lugg's and Rothbart's, are likely to be rejected for smacking of dogmatism by those inclined towards empiricism. Can we really declare some theory untestworthy in an apriori manner? Is not empirical confirmation or disconfirmation the final arbiter of a theory? For example, Thagard [1988, p. 170] generously admits that, despite all the previous failures of astrology, future studies might find empirical support for astrology, although he takes that to be rather unlikely. By contrast, Kanitscheider [1991] maintains that there can be no such evidential support, because astrology is so defective theoretically that, even if there were strong empirical correlations between the star positions and human character and fortune, it could never explain these data by way of mechanisms that do not involve sheer magic. In other words, the empirical situation is irrelevant if the theory in question cannot even begin to explain the data at hand.

Derksen [1993] rejects the idea that it is theories, practices, or entire fields that are pseudoscientific. Instead he recommends examining the attitude or the pretensions of the individual pseudoscientist. After all, it is not a field that can have scientific pretensions, but only its practitioners, and only the latter can be blamed for not making good on these pretensions. Similarly, Kitcher [1993, p. 196] holds that "[t]he category of pseudoscientists is a psychological category. The derivative category of pseudosciences is derivatively psychological, not logical as philosophers have traditionally supposed. Pseudoscientists are those whose psychological lives are configured in a particular way. Pseudoscience is just what these people do." Whereas Kitcher has in mind the inflexible epistemic performance of American creationists, Derksen's analysis concerns the work of Freud. In his analysis Derksen [1993] lists seven attitudinal sins of the pseudoscientist. The first is the "dearth of decent evidence". Having scientific pretensions, the pseudoscientist will have to show respect for empirical evidence. But what he claims to be good evidence for his theory is in fact defective. For example, it is unclear how reliable Freud's clinical data are, because he did not ensure that they were not the result of his own suggestive questioning. (See also [Grünbaum, 1984].) The second sin consists in "unfounded immunizations", which result from selecting and tailoring the data until they fit the given theory; in other words, only particular interpretations of the data are accepted. This also happens in science but, there, immunization is based on well-confirmed theories rather than on unfounded ad hoc hypotheses.

Derksen calls the third sin the "ur-temptation of spectacular coincidence", which consists in ascribing a deeper significance to *prima facie* spectacular coincidences. The fourth sin is the application of a "magic method". That is to say, the pseudoscientist always has some magic method at hand by means of which he can generate all the data he needs. With regard to Freud, Derksen mentions the method of free association, the analysis of symbols and the interpretation of dreams, by which Freud was able to get any data he needed to support his ideas.

The fifth sin is the "insight of the initiate". This is not the claim that only the person with a specialized training can do proper research, since this holds also for science. Rather, it is the claim that the researcher has to overcome certain impediments and prejudices in order to be able to gain the knowledge and insight to be had in the given field. Thus, only the Freudian who underwent psychoanalysis himself is said to be able to practice psychoanalysis.

The sixth sin refers to the presence of an "all-explaining theory", i.e., "a theory that has ready answers to whatever happens". The seventh sin, finally, consists in "uncritical and excessive pretension". Here, "excessive" refers to the fact that, first, the pseudoscientist claims a much greater reliability of his knowledge than allowed for by the evidence (or rather the lack of evidence), and, second, that his pretensions concerning the importance of his theory are far too great. In a later paper, Derksen [2001] elaborates on these sins, offering seven further strategies typical of the "sophisticated pseudoscientist". In any case, although Derksen is right that, strictly speaking, only a person can have scientific pretensions, it seems rather unproblematic to abstract from these individual "sinful attitudes" and treat them as methodological rules, as is commonly done. The same holds in my view for Kitcher's [1993] psychologistic approach.

In a complex study of scientific research, which can be summarized only in a rather simplified way, Kuipers [2001, p. 247] defines pseudoscience as the combination of scientific pretensions and the neglect of the "principle of improvement of theories". The latter enjoins us to aim at more successful theories by eliminating the less successful ones. This improvement is supposed to occur within a research program (in the broad sense), i.e., we aim at better theories while keeping the hard core of the program intact. Only if this strategy fails should we try to adapt the hard core; and only if this strategy fails too, should we look out for a new research program. According to Kuipers, these rules may be seen as constituting *scientific* (or *methodological*) *dogmatism.* By contrast, *unscientific dogmatism* is characterized by the strict adherence to one or more central dogmas which are deemed to be in no need of improvement.

Although these authors do not quite agree on the characterization of pseudoscience, they provide important indicators of pseudoscientificity, useful for any analysis of any theory, practice, or field suspected of being pseudoscientific.

5.2 Pseudoscience or Parascience?

There is a fundamental problem, however, with the very definition of the term "pseudoscience". If it is an essential connotation of "pseudoscience" that it be a nonscientific field *with scientific pretensions*, what do we do with nonscientific fields that appear to be as defective as the classic pseudosciences, but do not claim to be scientific in the first place? As Hansson [1996] has rightly pointed out, many fields that are often subsumed under the label "pseudoscience" are not really such. Indeed, many areas in the vast realm of esoterics, occultism and New Age thinking do not pretend to be scientific at all. Some are even outright antiscientific: they reject the scientific approach to knowledge in favor of various "alternative ways of knowing". If not as completely wrong, the scientific world view is regarded at best as short-sighted and hence in dire need of "complementary" forms of cognition, such as "holistic", "spiritual" or "mystical" ones. Examples of such fields are various forms of "alternative healing" such as shamanism, or esoteric world views like anthroposophy (for further examples see [Carroll, 2003; Hines, 2003]; as well as the various articles in [Stalker and Glymour, 1989]). Obviously, the standard definition of a pseudoscience as a nonscientific field with scientific pretensions does not apply to such areas. Yet these esoteric fields do compete with science in claiming to produce, or have at their disposal, important factual knowledge that the "narrow-minded" scientific approach necessarily must overlook. Moreover, the alleged knowledge produced in these areas often collides head-on with well-confirmed scientific knowledge. For this reason, we must suspect that the "alternative knowledge" produced in such fields is just as illusory as that of the standard pseudosciences.

For these reasons it will be useful to have a different term which subsumes both the pseudosciences proper and all the other fields producing bogus knowledge. I suggest using the term *parascience* for this purpose. Note, though, that the term "parascience" is often used in a different sense, namely descriptively for a field of knowledge whose status as either a pseudoscience or a protoscience is still under debate. I shall disregard this descriptive usage here in favor of the normative one. Alternatively, we could as well give up the standard meaning of "pseudoscience" as a nonscientific field with scientific pretensions and conceive it in a broader sense to also cover all those areas dealing with bogus knowledge.

However, I shall stick here to the name "parascience", because it allows us to explore further distinctions, which are usually neglected in the demarcation literature. Thus, as a matter of principle, we can not only distinguish science from pseudoscience, but also pseudotechnology from paratechnics, and pseudohumanities from parahumanities. Recalling our earlier distinction between technology and technics, a pseudotechnology then would be a technological field based on some pseudoscience, whereas a paratechnic would just be a crackpot technic without any elaborate pseudoscientific background, or at most with a traditional magical background theory. A pseudohumanistic field would be one pretending to produce humanistic knowledge, although its business actually consists in sheer intellectual imposture or obscurantism. And a parahumanistic field, finally, would be the same, except for the fact that it does not pretend to be a field which should belong in the circle of the humanities. Finally, there is a category which contains all those fields that are neither pseudoscientific nor pseudo- or paratechnological nor pseudo- or parahumanistic. We have no choice but to call them parasciences in the narrow sense, in contradistinction to parascience in the broad sense as defined above (see Fig. 4). Having two notions of parascience is one of the disadvantages of the present analysis.

To see whether this extended typology is of any use, let us take a look at some examples. Considering these examples here does not imply that all of them

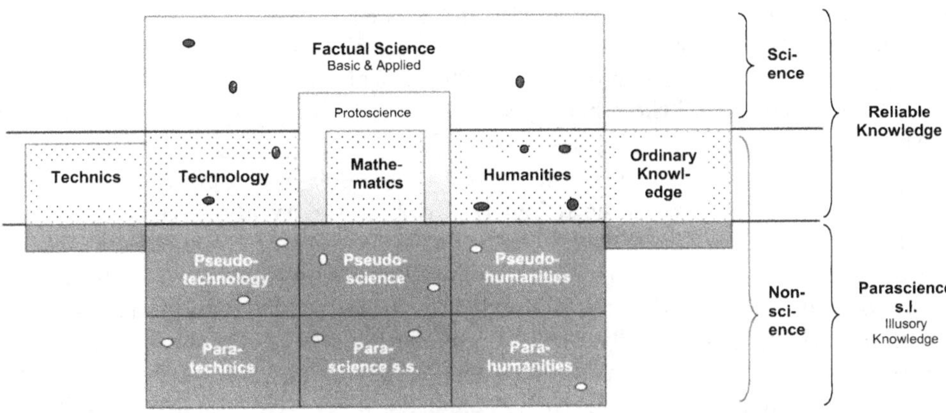

Figure 4. An extended typology of epistemic fields. In this typology only the basic and applied factual sciences are considered as strictly scientific, whereas technology, mathematics and the humanities are classed as nonscientific fields, though still close to the factual sciences. In any case they belong to the class of epistemic fields providing reliable knowledge. By contrast, the knowledge claims of the parasciences (*sensu lato*) are illusory: they do not enrich human knowledge, but pollute it. Protosciences are epistemic fields shading from the dubious into the scientific. The light gray shading indicates that by and large they are on the right track, although they are still burdened with nonscientific ideas or procedures. Ordinary (or everyday) knowledge and technics also lie in between the reliable and the mistaken. Note the gray spots on science's bright vest and the white spots on the dark attire of the parasciences. This indicates that the science/parascience distinction is not really a clear-cut black and white demarcation line, as suggested by this idealized diagram. There are pseudoscientific pockets within otherwise good sciences. These are sometimes labeled *pathological science*. And of course, some knowledge produced in science, technology, and the humanities has turned out to be false (without therefore being pseudoscientific), and not all knowledge in the parasciences need be false. Further explications in the text.

are correctly placed in the proposed category. Some of them certainly are, but the status of others is still under debate, so we may prefer to call them parascience *candidates*. Standard examples of pseudoscientific theories or fields are parapsychology, scientific creationism and intelligent design, psychoanalysis (as basic psychological theory), astrology (as a theory of human character), cryptozoology, Lyssenkoism, New Age physics, ufology, Däniken's archeology, Afrocentric history, and Sheldrake's morphogenetic fields theory (see [Shermer, 1997; 2002; Carroll, 2003; Hines, 2003]). A more recent suspect is the constructivist-relativist sociology of science [Gross and Levitt, 1994; Sokal and Bricmont, 1998; Bunge, 1999; Wilson, 2000]. All these fields pretend to be scientific, e.g., in using scientific methods.

By contrast, a parascience (in the narrow sense) does not claim to be scientific: it is just a field involving some (often traditional) theory about certain matters of facts. For example, traditional Chinese medicine involves a "biological" theory of the life energy *qi* flowing in meridians through the human body. The Indian theory of chakras asserts that the human body contains thousands of energy centers (chakras), which may be influenced by meditation (e.g., tantra). Similarly, the Western esoteric theory of reincarnation states that a personal soul really survives the body's death and can be reborn in some other body. (Note that the traditional Buddhist concept of reincarnation does not involve the survival of some spiritual substance.)

As for pseudotechnology, recall from section 4.3.2 that technology does not just consist of the classic physico-mechanical or engineering disciplines, but also of biological, psychological, and social technologies. All the fields attempting to come up with perpetua mobilia and other so-called free energy machines, with antigravitation devices and earth ray protection gizmos, count as pseudo-physicotechnologies. Likewise, sophisticated dowsing, which is based on pseudogeological assumptions, and water energizing on the basis of "quantum transformation" or other bogus concepts belong in pseudo-physicotechnology.

Examples of bio-medical pseudotechnologies are homeopathy, chiropractic, iridology, and biorhythmology. Candidates for psychological pseudotechnologies are psychoanalytical therapy, phrenological and graphological diagnosis, astrotherapy and horoscopes, neurolinguistic programming, and applied kinesiology. Finally, as pseudo-sociotechnologies have been regarded: Marxism as scientific socialism [Popper, 1959] as well as feminist technology and the so-called New Evidence Scholarship relying on subjective probabilities in jurisprudence [Bunge, 1999]. By contrast, mere paratechnics, i.e., procedures not based on some pseudoscience but at best on some parascience (in the narrow sense), are naive dowsing, faith healing, magic, voodoo, and prophetic techniques such as palmistry, Tarot, and I Ging.

What about pseudo- and parahumanities? Are there any examples at all? In section 4.3.3 we listed only some of the major differences between the humanities and the factual sciences. Since this does not constitute a positive and comprehensive characterization of the humanities, it does not enable us to demarcate genuine humanities from pseudo- and parahumanities. Thus, the following examples merely give some possible suspects, not the results of a detailed analysis. As pseudohumanities have been regarded: anthroposophy, theology, irrationalist philosophy (pseudophilosophy), and postmodernist cultural studies. Scientology may be another candidate. Parahumanities on the other hand might be hermetics, gnosticism, mysticism, and maybe traditional religions inasmuch as they make cognitive claims. These examples show the highly controversial nature of demarcating pseudo- and parahumanities. Even if this demarcation proves to be untenable or useless, it should at least provoke a detailed examination of the suspects involved before admitting them into the humanities or else refusing them entry.

Indeed, only few authors (e.g., [Kuipers, 2001]) have dared ask the question of whether, for example, theology is a pseudoscience, and whether there is such a thing as pseudophilosophy. Whereas Kuipers does not give an answer with respect to theology in his 2001 (see, however, [Kuipers, 2004]), he suggests that pseudophilosophy is the combination of philosophic pretensions with unscientific dogmatism. Philosophy reducing to nothing but exegesis, or the attempt to preserve the teachings of some master instead of developing and improving on them, would be examples of pseudophilosophy. Another example, not mentioned by Kuipers, could be irrationalist philosophy. For example, it is well known that Schopenhauer and many others accused Hegel of being a pseudophilosopher for writing utter nonsense, and the positivists, the critical rationalists and others have criticized some of the German philosophical tradition (e.g., Heidegger) for being obscurantist (see, e.g., [Albert, 1985; Edwards, 2004]). And recently, the French deconstructionists and others have been accused of being intellectual impostors [Sokal and Bricmont, 1998]. Be this as it may, if there is pseudophilosophy, it will be a pseudohumanistic field rather than a pseudoscientific one.

Theology is somewhat different, because the work of theologians ranges from the social sciences to the humanities. While working, for instance, in the field of comparative religion, text analysis, or sociology of religion, theologians do proper scientific and humanistic work — *de facto* and as individual researchers. Hence their individual work need not differ from religious studies or comparative religion (*Religionswissenschaft*), which can just as well be done by nontheologians. Presumably, the main problem with theology is institutional, because theology is by its very essence denominational: the theologian is the representative of some particular religion and is therefore expected to accept its creed as a given. The core of this belief system is not open to revision as a matter of principle, wherefore it must be regarded as a form of unscientific dogmatism. Thus, it is impossible that, as a result of internal progress in research, Christian theology will come to the conclusion that Christianity is actually false and Hinduism is true after all. For example, in the past 200 years the research of many theologians has contributed to demolishing the authority of the scriptures by putting them in a proper historical perspective, but this has not led them to abandon Christianity. Rather, it has spawned a hermeneutic industry of apologetics, attempting to save the Christian faith by reinterpreting and re-reinterpreting its tenets, often in unintelligible terms [Albert, 1985, Ch. 5]. Of course, the individual theologian may eventually change his mind and give up his belief, adopting another one or even becoming an atheist. But, unless he gets fired upon so doing, he has to leave his field if he wants to be consistent. Thus, it seems that, due to its fundamentally denominational and dogmatic nature, theology as an epistemic field is pseudoscientific or pseudohumanistic, respectively.

What about pathological science? In which category does it belong, or is it a category of its own? As mentioned in the legend of Fig. 4, pathological science concerns pockets or niches of pseudoscience still located within the sphere of science. In Fig. 4, this is indicated by the dark spots marring the field of science. Classic examples are the N-rays and polywater affairs. More recently cold fusion has been added to this list. But other theories and approaches within the sciences too have been regarded as pseudoscientific, such as steady state cosmology, the anthropic principle, the subjectivist interpretation of quantum theory, the quantum theory of measurement, evolutionary psychology, information processing psychology, and the research on race and IQ (see, e.g., [Bunge, 1982; 1983b; 1984; 1999; Shermer, 2002]). Some fields, like holocaust denial, have even somewhat branched off from academic historiography to form a specialized field of their own, which enforces the impression that they have turned into full pseudosciences [Shermer, 1997].

As for the corresponding white spots in the parascientific fields, they indicate that not every piece of knowledge in the parasciences need be false: we may find some true or useful items on occasion. An example is acupuncture. Although there is no hope for the magical theory of traditional Chinese medicine underlying the practice of acupuncture, there is some evidence that putting needles here and there has some effect on relieving certain forms of pain [Ernst *et al.*, 2001]. If this turns out to be true, acupuncture will become an area of biomedical research and explanation, which most likely will not have much in common with its parascientific origins. Finally, some parasciences, such as parapsychology, do use scientific methods for example, so that not everything occurring in an overall parascientific field need be unscientific.

So much for some possible examples illustrating the distinctions suggested in Fig. 4, and some qualifications concerning the idealizations involved. The purpose of this extensive typology is to show that in its standard definition the label "pseudoscience" fails to do justice to the wide variety of the parasciences. On the other hand, if we are only interested in distinguishing the genuine article from bunk, a simpler analysis will of course do, such as the one depicted in Fig. 3, in which, however, one might want to replace the terms "pseudoscience" and "pseudotechnology" by "parascience" and "paratechnology", respectively.

Having dealt with various parascience suspects, let us proceed at last with the characterization of parascience.

5.3 Characterizing Parascientific Fields

In the following analysis we shall try to develop a profile of parascience (in the broad sense) by applying the twelve criteria of scientificity listed in Sect. 4.2.

1. *Community C.* Faced with a parascience candidate, we need to examine whether there is in fact a real research community continuing a research tradition, or just a loose collection of individuals. If there really is a genuine system of persons, we need to check further whether this community engages in research, or whether it is just a group of believers.

   One of the few parasciences that does have a research community is parapsychology. Many others, by contrast, are belief communities: there is a single guru or a small number of authorities, surrounded by a more or less numerous crowd of followers, who do not engage in research, but at most in exegesis or application. Think of Immanuel Velikovsky's pseudocosmology, Erich von Däniken's pseudoarcheology and pseudohistory, Charles Berlitz's Bermuda triangle mystery, or Ron Hubbard's scientology.

2. *Society S.* The society hosting a community of researchers or else believers must at least tolerate its activities. However, political power can turn an epistemic field into a pseudoscience if it starts to proscribe what is to be accepted as true knowledge and what not, and if the people working in that field follow suit. Examples are *Deutsche Physik* (German physics) or, more generally, Aryan science in the Third Reich, and Lyssenkoism during the time of Stalinism and after. A contemporary example is creationism, which is adopted at the national level in official theocracies, or at least pushed at the regional or local level where conservative churches or fundamentalist religious groups of any color wield enough power (e.g., in Turkey, Iran, the US, and Russia). In the same vein, it is legitimate to ask whether the calls for a feminist science, based on the relativist-sociological "finding" that science is just an enterprise of white Western males, belong in the same category [Gross and Levitt, 1994; Bunge, 1999]. It may well be that women have somewhat different research interests, so that they focus on different problems. But as soon as we get to questions of method, testing, validity, and justification, there seems to be no leeway for "alternative" forms of science.

3. *Domain D.* The domains of parascientific fields often comprise dubious and ill-defined items, such as mysterious energies or vibrations, which have so far escaped detection. In other words, many parasciences still have to prove that the objects and processes they refer to in their discourse do exist really. Therefore, much of their domain is factually empty and consists mostly of speculative entities. An example is parapsychology, which has not been able to come up with a single unambiguous finding concerning the real existence of "psi" [Alcock, 2003; Hines, 2003].

   At first sight, hypothesizing unobserved or unobservable entities appears to be analogous to the theoretical entities posited in many scientific fields.

However, the difference is ontological, semantical and methodological: if not supernatural, the entities posited in many parascientific fields are by definition paranormal or, if preferred, paranatural, and they are often idle, arbitrary, or nonparsimonious, for not being embedded in some explanatory theory proper. Hence they are often ill-defined, i.e., they are so vague that it is unclear what is being tested — if there are serious tests at all. An example is the mysterious "psi" occurring in parapsychology, which is defined but negatively [Alcock, 2003]. For example, precognition is defined as seeing future events in a way that *cannot* be explained by contemporary science. Likewise, psychokinesis and telepathy involve interactions that *cannot* be accounted for by any mechanisms known to normal science. Moreover, parascientific entities are not hypothesized in a search for the best explanation (i.e., abductively, as it is often called), but they are often objects of prior beliefs, for which a justification is sought only if the belief is questioned by some skeptic. So whatever *prima facie* explanatory function they may have, the very same function could often be exerted by any other paranatural entity. In other words, paranatural entities are usually not specific enough for a satisfactory explanation (see, e.g., [Flew, 1990; Kanitscheider, 1991; Humphrey, 1999]).

4. *Philosophical background G.*

   (a) *Ontology.* The ontological aspects of parascience are often neglected in favor of its methodological problems. An early exception was the philosopher Charlie D. Broad, who was a firm believer in parapsychology. He pointed out that both science and our everyday practice presuppose various philosophical assumptions, which he called "basic limiting principles" [Broad, 1949]. He gave four main examples, three of which are ontological, one epistemological. His ontological principles were (i) the antecedence principle (effects cannot precede their causes); (ii) mind cannot directly act on matter without involving a brain event; and, conversely, (iii) the mind depends on the brain, i.e., a necessary condition of any mental event is an event in the brain of a living body. An epistemological consequence is (iv) that our ways of acquiring factual knowledge are limited to sensory experiences, i.e., a physical event does not directly act on our mind, but only through some intermediate events in our sensory organs and finally in our brain. (Note that (ii) and (iii) sound dualist — Broad was sympathetic to epiphenomenalism — but may be reformulated so as to be compatible with monistic mind-body theories.) Since he maintained that the existence of the various parapsychological phenomena like telepathy and precognition was established beyond doubt, Broad concluded that these basic limiting principles of science are refuted.

   The fact that some of the research Broad referred to was later shown to be fraudulent [Ludwig, 1978; Kurtz, 1985; Hines, 2003], and that sophisticated parapsychologists try to conceive of telepathy and precognition in a somewhat different manner, so as to retain at least *prima facie* a naturalist interpretation [Duran, 1990], does not invalidate this as a useful example of the ontological problems faced by most parasciences. Indeed, many of their claims can only be upheld by giving up basic ontological convictions, which have so far proven to be extraordinarily fruitful for scientific research.

The most radical departure from the ontological paradigm of factual science is the open supernaturalism espoused by creationism. Inasmuch as creationism stipulates a *creatio ex nihilo*, it also violates Lucretius's principle. It is unclear whether or not many other parascientific claims can be accommodated within ontological naturalism. In any case, they still violate much of what we know about the lawful behavior of things. Homeopaths, for example, claim that high dilutions that no longer contain even a single molecule of the given substance still have a potent pharmacological effect. If what we know about chemistry is roughly true, there can be no such effect. Homeopaths have learned to concede this objection, but now forward the protective hypothesis that, in the mandatory process of shaking the dilutions (called "dynamization"), somehow the relevant "information" of the given substance gets transferred to the solvent. So what produces the therapeutic effect is this "information". It goes without saying that this supposed information is ill-defined and perhaps even immaterial, because water chemistry tells us that any molecular structure formed by H_2O-clusters is too short-lived to do any informational work. Moreover, if water (or alcohol or whatever fluid) had a memory, why would it specifically remember only the information of the homeopathic substance rather than that of all the other chemicals it had contained previously?

Another example is Therapeutic Touch. By moving her hands about 10cm over the patient's body, the healer attempts to adjust the patient's "vital energy", whose "imbalance" is always among the causes of whatever disease is to be healed. Needless to say, biology has abandoned any idea of vital energies long ago.

These examples show that many of the ideas occurring in the parasciences and paratechnologies are not necessarily supernatural in the traditional sense of involving powerful personal entities like gods or demons, but nevertheless *paranatural* [Kurtz, 2000], in the sense that they are not compatible with the naturalist-materialist outlook of the factual sciences. If we enrich this standard naturalism with more and more paranatural elements, it remains unclear, when this results in destroying it altogether.

The only ontological principle that is rarely violated by the parasciences is ontological realism. Even the weirdest entities occurring in the domain of the parasciences are deemed to exist really after all. The same holds for epistemological realism, which is why we proceed with a look at the methodological principles in the following subsection.

These examples illustrate that the parasciences not only suffer from the methodological problem of lacking evidential support, but also from their incompatibility with the major metaphysical background assumptions, which belong to the general hard core — the hard hard core, so to speak — common to any scientific approach. (For an analysis of the ontological presuppositions of esoterics see [Runggaldier, 1996].)

(b) *Methodology.* It is rather obvious that both Ockham's razor and fallibilism are widely neglected in the parasciences. Indeed, many parasciences populate the universe with (often occult) entities that are not needed for a scientific explanation of the world around us. Examples are the many life or other energies and forces postulated by quack medicine and pseudophysics. Dowsers believe that there are not only earth rays, but that these also occur in certain grids, which can be measured and mapped. And occultism teems with ghosts and spirits. There is no indication that the nature or the number of such entities is restricted by considerations of parsimony in hypothetico-deductive reasoning: their only restriction seems to be due to the limits of their authors' imaginative powers. This is not to say that they serve no explanatory function: they certainly do. The point is, as mentioned earlier, that almost any other arbitrary alternative or additional entity would do just as well.

As for fallibilism, it too is evident that most parascientists are not willing to seriously consider the possibility that they may be in error. If we extend Settle's [1971, p. 185] diagnosis of magic to the parasciences in general, we might say that many parasciences are explanatorily complete and thus come with the air of certainty, whereas factual science is explanatorily incomplete and thus accompanied by corrigibility. This difference helps to explain why the former are so much more appealing to many than the latter. Obviously, an explanatorily complete field has no need for research and hence for improvement, let alone revision (see Kuiper's [2001] definition of pseudoscience mentioned above). As we shall see later on again, some parascientific fields do allow for some limited improvement, such as parapsychology and astrology. However, these changes are not due to an internal tradition of fallibilism, but they are the result of massive external criticism by mainstream scientists.

(c) *Semantics.* As a truth definition the correspondence notion of truth, being simply a companion of ontological realism, is adopted in most parascientific fields. The major difference between science and parascience lies in the question of what is acceptable as truth indicators. Now this belongs in methodology, not semantics, so it may suffice here to add that, beside the main question of what can be regarded as legitimate objective evidence, the parasciences often accept as indicators of truth also subjective "evidence", such as sheer belief or feeling, mystical vision, or other paranatural forms of experience.

(d) *Axiological and moral assumptions.* Different values manifest themselves in different behaviors of the individuals adopting these values. Thus, as mentioned in Section 5.1, Derksen [1993] has suggested analyzing the behavior and attitudes of the individual pseudoscientist, and Kitcher [1993] has recommended focusing on the psychology of the pseudoscientist. However enlightening this may be in some cases, in particular when taking a closer look at the founding father (or mother as the case may be) of some field, as Derksen did with Freud, it does not suffice to characterize the entire epistemic field. For example, it is possible for an individual to behave rationally within a magical belief system [Settle, 1971], whereas an individual scientist working in a rational tradition may on occasion behave irrationally. For this reason we better focus on the *institutional rationality*, or irrationality respectively, exhibited by the community C of some epistemic field, which is done best by examining the latter's general ethos or value system.

- *Logical values.* The canon of valid reasoning and thus the basic principles of rationality may be accepted officially, but they can be suspended whenever needed to save some claim. Lots of logical blunders occurring in the parasciences have been collected by various authors (see, e.g., [Schick and Vaughn, 1999; Wilson, 2000]). Since many of these occur in the context of justification, we shall give a sample in the subsection on methodological values.
- *Semantical values.* Meaning definiteness and clarity are rarely among the semantical values of the parasciences. Instead, vagueness and fuzziness are rampant, if not even seen as virtues by those cherishing the mysterious. We must also be prepared to encounter the meaningless, i.e., nonsense. (Note that scientists often are too quick in calling something nonsense, just because it is false. However, something that is false cannot be nonsense, because nonsense can be neither true nor false, for it has no semantic meaning in the first place.) Regrettably, since for most laymen many scientific theories are more or less incomprehensible, unintelligibility on the part of a parascientific theory may easily be mistaken for a sign of an authentic science.
- *Methodological values.* Many parasciences are characterized by methodological values and hence procedures of their own. These consist, for example, in certain rules of inference or rules of evaluating evidence which are quite often regarded as fallacious by philosophers of science. For this reason they either have been eliminated from science, or, if they occasionally reappear in some reasoning, are quickly detected and denounced as mistakes by the scientific community. Indeed, fallacious methods were described already by

19th century philosophers of science like Mill and Peirce, and many modern authors who attempted to demarcate pseudoscience by its peculiar inferential methods, have collected various fallacies as indicators of pseudoscientificity (e.g., [Radner and Radner, 1982; Giere, 1984; Thagard, 1988; Schick and Vaughn, 1999; Wilson, 2000]). Since these fallacies do constitute important parascience indicators, a quick sample will be in order.

The a priori method: Accept only those beliefs that are such that it is impossible to imagine that the contrary is true [Wilson, 2000]. In other words, a hypothesis is accepted and considered worthy of use for explanation not on the basis of empirical evidence, but because its proponents regard alternatives as inconceivable. Examples: von Däniken keeps repeating that he simply cannot imagine how some artifact could have been produced by ancient man without extraterrestrial help. The creationists (including the more recent branch of Intelligent Design) keep repeating that it is inconceivable how the natural process of evolution could have produced certain complex organs without divine design or even intervention.

The fallacy of competition: This is the claim that some parascientific theory should be admitted because it might become an alternative theory in the future. Yet, as Radner and Radner [1982] point out, competition is only among current alternatives: by referring to some unknown future science, one actually refuses to compete. Their very apt analogy is the attempt to participate in a marathon on roller skates, arguing that the marathon might be changed to a skating race in the future.

Simplistic elimination [Giere, 1984; Wilson, 2000]: Assuming there are two rival theories A and B, and they are the only possible alternatives, we may infer that A is true if B is false. Yet in reality there usually are many possible alternative theories that might explain the same fact. So if we are faced with two or more alternative theories, we must first make sure that they really are the only alternatives, and that they are not false all together. Thus, many supposed eliminations are fallacious, because they do not consider all possible alternatives. The creationists argue, for instance, that there are only two alternatives: evolutionary theory and the theory of divine creation. But if evolutionary theory, including all we know about the history of the universe, is false, then divine creation is not the only remaining alternative: it may well be then that life is coeval with an uncreated eternal universe. Ufologists argue that, since some strange sightings cannot be explained by the usual candidates such as satellites, balloons, aircraft, or bright planets, they must be due to extraterrestrial visitors. Yet there may also be unknown natural atmospheric processes causing a given UFO-sighting.

Anything-goes method [Wilson, 2000]: This is the argument that, since even a well-confirmed theory might possibly be false, we should not dismiss alternatives to it. So everything goes. If this were correct, the corollary would be that in fact nothing goes, because these supposed alternatives might likewise be false.
Method of authority [Wilson, 2000]: As pointed out earlier, many parasciences are belief systems rather than research fields. It comes as no surprise therefore that a rule "to accept as true what the relevant authority tells you" is wide-spread. Naturally, this holds in particular for religious or quasi-religious fields such as creationism, scientology, anthroposophy, or transcendental meditation.
Resemblance thinking [Thagard, 1988; Wilson, 2000]: This is the habit, already pointed out by John S. Mill, of inferring from the observation that *A resembles B*, that therefore *A causes B*. Prime examples of fields relying heavily on resemblance thinking are astrology and homeopathy. The latter's "law of similars", stating that like heals like, is even enshrined in the very name "homeo-pathy" (from the Greek *homoios*, similar).
The grab-bag approach to evidence [Radner and Radner, 1982]; see also the *blunderbuss argument* in [Wilson, 2000]): In evaluating the evidential support for some theory, we should not just look at the quantity of confirming instances, but first of all at their quality. Thus, we do not have to keep shooting canon balls in order to confirm the laws of motion. Of particular value, on the other hand, are data that were gathered after a theory had been proposed, and that were possibly even predicted by the theory; likewise with evidence that was produced under a variety of different conditions. Classical examples with regard to Newton's theory are the discovery of Uranus and Neptune, and the prediction of the return of Halley's comet. By contrast, it is typical of many parasciences that the sheer quantity of "evidence" makes up for the lack of quality of the individual data. For example, von Däniken pulls out artifact after artifact in favor of his "alien hypothesis"; the creationists keep listing complex biotic structures which impossibly could have come into existence naturally, i.e., by evolution; and the ufologists will report strange sighting after sighting. Moreover, as soon as one piece of such evidence has been rejected, either for being fallacious or forged, or for having been explained within a standard scientific context, the parascientist will simply continue to pull out data of the same kind and quality from his evidential grab bag, thereby keeping the skeptic busy for all times. Worse, the fact that scientists cannot always readily refute each and every item pulled out of the grab bag, is taken as a further reason for belief in the parascientific tenet in question.

- *Attitudinal values.* The attitudinal value system of the parasciences is as varied as the parasciences themselves. Thus, again, there are no universal features characterizing all the parasciences. Nonetheless, an attitudinal profile of parascience may include the following aspects. Parascientists pretend to be critical thinkers, but their canon of critical thinking is not the same as that of science and philosophy. In fact, many are just believers, not investigators. They also claim to be open-minded, but their open-mindedness does not extend to the possibility that the standard scientific view of nature is the correct one. Instead, it includes sympathy for the most outlandish claims, because to the parascientist open-mindedness often means "anything nonscientific goes", so that it amounts to blank-mindedness. Universalism and objectivity are not values in those fields dominated by authorities, or in which only the initiate has special access to the truth. Think of the various branches of occultism.

5. *Formal background F.* Concerning the formal background of any suspected parascience, we may ask questions such as the following: Are there any mathematical models? Is the mathematics in these models handled correctly? This is often not the case. In particular, in some pseudophysics such as the attempts at refuting the theory of relativity, the mathematics is defective, if not phoney. The same occurs in some social sciences, in particular sociology and economics, where pseudoquantitation may go unnoticed [Sorokin, 1956; Blatt, 1983; Bunge, 1999, Ch. 4]. The latter example illustrates once more that some research fields which on the whole are regarded as scientific may nonetheless exhibit some occasional pseudoscientific feature (Fig. 4).

6. *Specific background knowledge B.* In contrast to scientific fields, which borrow amply from adjacent disciplines, the parasciences are typically isolated enterprises. They presuppose some ordinary knowledge, and of course they borrow some science when needed. But note that the function of the scientific knowledge borrowed consists mostly in justifying the scientific pretensions of the given pseudoscience: it is easier to imitate science when you also use some well-accepted scientific knowledge. The scientific input is often not needed to advance the own field. Note also that the converse input does not obtain: scientific fields have hardly any use for knowledge produced in a parascientific field.

   Astrology, for example, accepts of course some basic astronomic facts, but disregards many others, in particular those that refute its own claims. Creationists rely heavily on biological knowledge, but only to prove the falsity of evolutionary theory. However, no scientific knowledge whatsoever can shed any light on the totally occult mechanism of divine creation. In other words, no scientific knowledge can advance creationist "theory".

The theory probably most often borrowed from the sciences is quantum theory, which has become an explanatory panacea for many parasciences, from New Age physics through parapsychology to holistic medicine [Grove, 1985; Stalker and Glymour, 1989]. For example, sophisticated parapsychologists have long abandoned stories of moving tables and telepathically communicating people. The naturalistically oriented part of current parapsychology claims that paranormal effects are microeffects rather than macroeffects, and that they can be accounted for by quantum theory. Telepathy, for instance, is no longer seen as a form of human communication, but at most as an instance of nonlocal correlations between some quantum events in two peoples' brains, or between a person's brain and some other object like a random number generator. It will come as no surprise then that the use of quantum theory in the parasciences often involves a serious distortion, in particular a return to long abandoned subjectivist interpretations. Moreover, one often uses the vocabulary of quantum theory but rarely its concepts [Stenger, 1995; Spector, 1999]. In sum, the motto is: if you don't know what it is and how it works, call in quantum theory to describe and explain it.

Note, incidentally, that in sophisticated parapsychology this move is due to the attempt to stay within the bounds of a naturalist ontology. At the same time, it presupposes a radically reductionist view, because it disregards the level structure of the world, i.e., the fact that macroobjects such as neural assemblages have systemic properties, so that their behavior is usually not influenced by microevents occurring at the quantum level. For example, neuroscientists know that mental processes, such as perception and thinking in general, involve millions, if not billions, of complexly interacting neurons and their coordinated activities at different organizational levels. The idea that quantum events occurring at the level of elementary particles or at most atoms should be able to influence these highly complex neuronal systems in a coordinated manner is extremely implausible [Beyerstein, 1987; Humphrey, 1999; Kirkland, 2000].

Parasciences sometimes also borrow ideas from other parasciences. A prime example is Carl G. Jung's concept of synchronicity, which is made use of both in sophisticated astrology and parapsychology. This is the idea that two events which have no causal connection are nonetheless "meaningfully" related (McGowan 1994; Carroll 2003; Hines 2003). Thus, if the quantum physical notion of nonlocal correlation cannot be called in as an ad hoc device to establish a connection between two (simultaneous) events, because what we have is just a coincidence, synchronicity will do the trick. For example, sophisticated astrologers have learned from the many scientific objections hurled at them during the past centuries: they nowadays admit frankly that the relation between humans and the various constellations of stars and planets is not a causal one. What saves the business though is the claim that the relation between the stars and humans is nonetheless a meaningful one, namely an instance of synchronicity. This neo-astrology then finds and interprets these meanings and explains them to its customers, turning the field into a form of astro-counseling. Note that this strategy is clearly ad hoc: it is not due to internal progress in astrology but a move to avert external criticism, making astrology immune against the standard astronomic objections without having to give up the "astro" in "astrology".

7. *Problematics P.* In the parasciences the collection of problems is usually small and mostly practical, for many parasciences are actually paratechnologies or paratechnics. Important questions about any parascience candidate are: Does it solve or help to solve problems other than its own? Do its problems arise from natural contexts, or are they artificial (fabricated)? Three examples might illustrate this problem concerning parascientific problems.

   Astrology mostly solves problems that would not exist without astrology in the first place. The only general and natural question that astrology tries to answer, namely the question why different people have different characters, is better answered by genetics, developmental psychology, and sociology. Moreover, the astrological answer is incompatible with the scientific one and thus does not enrich scientific knowledge. For the most part, however, astrology is a pseudotechnology, which has rules to apply, but no puzzles to solve [Kuhn, 1970]. In particular, the many failures of astrological predictions do not entice any problem-solving activity in the astrological community.

   The problems of von Däniken's pseudoarcheology too are fabricated rather than natural, because he preys on the natural problems of normal archeology and turns them into mysteries, which he claims can only be solved by his hypothesis about extraterrestrial visitors. Thus, von Däniken's hypothesis does not yield any new problems on its own: it is entirely parasitic on the pre-existing problems in other fields.

   Parapsychology started out with the natural problem of unusual human experiences, in particular at a time when spiritualism was *en vogue*. Some people sometimes do have anomalous (though nonpathological) experiences. The basic question therefore is whether all such anomalous experiences can be explained naturally (i.e., within the normal paradigm of scientific biopsychosociology), or whether we do need to enrich this paradigm with paranormal entities and processes to account for these unusual experiences. Yet, the more successful the normal sciences, including in more recent times the neurosciences, became in explaining anomalous experiences, the less needed were explanations referring to paranormal entities or processes. In this way, parapsychology practically lost its source of spontaneous or natural problems, although people keep experiencing unusual things. Not willing to give up the psi hypothesis in favor of the null hypothesis, parapsychologists started to fabricate new problems: they began studying arbitrary correlations between human subjects and virtually every possible other object, desperately looking for statistically significant deviations from chance expectation (i.e., anomalies), which can then be interpreted as evidence for psi. Since all the results from such — often quite sophisticated — studies are, if not negative, at best inconclusive, the consequence is the perpetual call for further research. Thus, parapsychology generates arbitrary problems of the sort "Could there be an anomalous correlation between x and y_1 or y_2 or $\ldots y_n$?" in order to keep itself alive. As Alcock [2003, p. 34] observes, the anomalies parapsychologists search for have never popped up in normal research. Thus, again, the contemporary problems of (sophisticated) parapsychology would not exist if it were not for the existence of parapsychology itself.

This may be the place to take a brief look at the role of anomalies in science and parascience. Normal scientists do not look for anomalies, they "hit them in the face" [Radner and Radner, 1982, p. 33; Alcock, 2003]. Indeed, every scientist who performs some measurement or experiment has certain expectations as to its outcome, in particular if the outcome is predicted by some theory. If the resulting data seriously deviate from these expectations, they constitute an anomaly. Although it takes more than just a few anomalies to initiate a scientific revolution, the importance of anomalies for theory change and hence scientific progress has been well known and discussed ever since the work of Kuhn [1962]. However, scientists are conservative in the sense that they will not give up an otherwise well-confirmed theory, let alone an entire research program, in favor of some alternative theory whose only merit is its ability to explain a certain anomaly. On top of explaining the given anomaly, the new rival theory must at least explain as much as does the standard theory.

By contrast, parascientists rejoice when they find anomalies. Their expectations are not those of an orderly and lawful world, but of a world teeming with mysteries. Therefore, they actively search for anomalies, which they can then turn into problems to be solved by their respective "alternative" theories. And these alternative theories are expected to revolutionize science. In so hoping, parascientists forget that no scientific revolution has ever been triggered from without. Nonetheless, there is even a field or rather a multi-field called *anomalistics*, which is exclusively devoted to the study of anomalies supposedly neglected by mainstream science. The main player in this field is the *Society of Scientific Exploration.*

8. *Fund of knowledge K.* The fund of knowledge of a parascience is not a growing collection of up-to-date and well-confirmed data and theories: it is usually small, it stagnates, it contains statements that are incompatible with well-confirmed scientific knowledge, and its hypotheses lack evidential support. For this reason, the knowledge in these fields is purely speculative and cannot be said to even approximate the truth, i.e., to roughly represent any real facts.

   A frequent feature of parascientific knowledge is its anachronistic character [Radner and Radner, 1982]. What many parascientists propagate as revolutionary new insights or at least as rival "scientific" theories is in fact very old news, so old indeed that they have long been discarded by science. For example, alternative medicine teems with mysterious vital energies that supposedly are out of balance when we are sick. Thus, the basic ideas of homeopathy only make sense when we go back 200 years when vitalism was still going strong in biology and medicine. Traditional Chinese medicine presupposes the existence of some vital energy (*qi* or *ch'i*), flowing in channels (meridians) unknown to biology. And the practitioners of therapeutic touch and reiki (*ki* is the Japanese equivalent of *qi*) claim that they treat the imbalances in the "human energy field", whereas the so-called prana healers refer to the Hinduist equivalent *prana*. The creationists still defend views that may have been legitimate 200 years ago. Then there are the pseudophysicists who still try to build perpetua mobilia or other so-called free energy machines as though thermodynamics were nonexistent, or who desperately strive to refute Einstein's two relativities in order to re-establish good old Newtonianism. Finally, astrology is another prime example of a world view that has been superseded for several hundred years.

9. *Aims A*. The aims of the parasciences are sometimes cognitive, but for the most part practical. That is, many parasciences are paratechnics or paratechnologies, such as astrology and alternative medicine. Yet even when the aims appear to be cognitive, the ultimate goal of many parasciences is often anthropocentric and quasi-religious (Alcock 1985), if not explicitly religious as in the case of creationism. *Prima facie* the goals of the creationists, such as the establishment of an alternative cosmology and history, appear to be cognitive rather than practical. But we may suspect that the ultimate goal is in fact personal salvation, which, in the fundamentalist world view, can only be achieved by a consistent way of life according to biblical literalism. Similarly, the spiritualist approach of esoterics wants to establish the multifarious spiritual connections of humans with the rest of the world. Often the ultimate goal is quite explicitly stated: the materialist world view of science is to be replaced with a spiritualist one. For example, one of the main figures in 20th century parapsychology, Joseph Banks Rhine, asserted that "little of the entire value system under which human society has developed would survive the establishment of a thoroughgoing philosophy of physicalism" (Rhine [1954/1978, p. 126]). This exemplifies how the aims of both science and parascience often depend on — conflicting — metaphysical outlooks.

10. *Methodics M*. The empirical methods used in the parasciences often are just as occult as the theoretical background assumptions. For example, an instrumental technique such as a pendulum used to diagnose some disease, presupposes some occult mechanism mediating between the healer and, say, the patient's "life energy". How can this method be checked? Interestingly, it can partly be checked scientifically, but it cannot be checked within the own theoretical system of the given field. In other words, in can partly be

tested externally, but not internally. For example, in a double-blind setup, someone claiming to be able to diagnose some specific disease by simply holding a pendulum over a photo of a patient, is given 25 photos of healthy persons and 25 photos of persons suffering from the given disease (i.e., neither the healer nor the experimenter knows which of the photos belongs in which group, and it is impossible to diagnose the given disease from merely looking at the peoples' faces on the photos). As yet, all experiments of such a kind have had negative results, i.e., the candidate's success rate has never been significantly above chance expectation.

Now this is of course a basic and objective scientific test which only checks whether or not the given technique works (not how it works if it did work in the first place). And it was imposed from the outside, because it does not belong to the methodics of the given parascience. So how can the functioning of the method be checked internally? Unsurprisingly, the healer herself might claim that she is able to check her diagnostic technique with alternative means. She may, for instance, use a dowsing rod, or perhaps just put her hand on the picture. In her normal environment all this will most likely combine with confirmation bias and subjective validation into the belief that her method is successful and reliable. However, as a matter of fact even within the own outlook of such a parascientific approach, the given method cannot be checked by other persons in the field, because her colleagues will not be able to reproduce her diagnosis. Indeed, every other person claiming the same ability will very likely come up with a different diagnosis, provided of course she does not know the earlier diagnosis of her colleague. There may be some overlap in the results due to chance, but by and large the success rate will not differ from mere guessing. In short, many techniques used in the parasciences are not objective in the sense that everyone applying the method will get the same results. This holds *a fortiori* for openly subjective methods like spiritual means of communication or mystical vision. The latter are not even methods in the sense of rule-guided procedures.

By contrast, in their attempt to imitate science, the pseudosciences often do use scientific methods. For example, the statistical methods used in sophisticated parapsychology are sometimes impeccable. Moreover, often even the general scientific method is followed as is obvious from the parapsychological journals. In so doing, many pseudosciences, in particular parapsychology and astrology, often exhibit a naive empiricist view of science: they believe that the application of scientific methods and techniques, including the scientific method as defined above, is sufficient to warrant the scientific status of their field. Indeed, in particular parapsychologists have learned a lot from their critics and have thus improved both their statistical sophistication and the precautions against fraud and self-deception. (Note again that these improvements are largely due to external pressure, not internal progress.) So they believe that what they do is proper science, and they reject the various methodological and other philosophical objections as sheer ideological dogmatism, failing to realize that conceptual criticism is part and parcel of science too.

11. *Systemicity.* The systemicity condition is one of the stronger indicators of parascientificity (recall Reisch's criterion of network demarcation mentioned in Sect. 3). Indeed, parasciences are isolated fields. They do not form a consilient system of knowledge; in particular, they make no contact with normal science. It is precisely because parascientific knowledge must be rejected as unfounded that it cannot enrich scientific knowledge. Moreover, parascientific knowledge often collides head-on with scientific knowledge: if parascientific theories were true, their scientific alternatives including those theories to which they are connected would be false. Thus, many parascientific theories would cause total or global revolutions: the entire edifice of scientific knowledge including the scientific paradigm as a whole would collapse. By contrast, contemporary scientific revolutions, if any, will only be local or regional revolutions, because too many things we have come to know during the past 400 years are reliable and must therefore be at least approximately true. Examples of fields calling for global revolutions are creationism and parapsychology. As for the latter, recall C. D. Broad's basic limiting principles, which underlie all modern science.

12. *Progressiveness.* According to the criterion of progressiveness, the membership of the conditions 5–10 changes, however slowly and meanderingly at times, *as a result of research* in the same field or as a result of research in neighboring disciplines. Obviously, many parascientific fields are plainly stagnant, which can be detected rather easily. This is due to the fact that many of them are not really research fields but instead belief systems.

    But of course, there are also some parasciences in which there is at least some minor change, and there are others which are actually research-oriented, such as parapsychology. Indeed, as mentioned before, research keeps parapsychology busy. However, despite its age of more than 120 years, it has not come up with a single conclusive finding [Kurtz, 1985; Hyman, 1989; Alcock, 2003]. Thus after 120 years it is still a field in search for its domain, and it desperately tries to gather hard data. Nonetheless, it has even produced some theories to explain certain supposedly paranormal events or experiences, respectively. It has also introduced plenty of ad hoc hypotheses to protect itself from criticism. An example is the idea of psi missing. If some experiment yields a score slightly above chance expectation, this is of course regarded as evidence for psi. Likewise, if some trial yields a below chance result, this too is seen as evidence for psi: in this case the subject's psi abilities somehow operate to avoid the target (psi missing). In this way any fluctuation around the exact chance expectation becomes evidence for psi. Given this situation, it seems that parapsychology is able to generate the appearance of progress, although a closer look reveals that this progress is just as illusory as the very domain of parapsychology. After all, can there be genuine progress when the given field does not even have a real domain?

5.4 Conclusion

We have now listed and examined a number of features characterizing parascientific fields. The features used in this characterization are of course of unequal weight: some are more decisive than others, so that their presence is a stronger indicator of a field's status. For example, a violation of some of the basic limiting principles in G carries more weight than some methodical flaw in M, which may be repairable more easily, provided the practitioners of the field care to. Since the above features are not jointly necessary and sufficient conditions, another open question is how many of these characteristics must at a minimum be present for a field to be parascientific. Insofar as such a condition is a necessary one, such as the logical requirement of noncontradiction, we may reject the given field as irrational on this one count. In most cases, however, a simple characterization of a parascience such as "it's all a matter of X", where X may stand for falsifiability, method, or attitude will not do. Indeed, we ought to be more careful and always attempt to prepare a comprehensive profile of the suspected field. Such a profile should allow us to come to a well-reasoned conclusion as to the scientific or parascientific status of the given field, although every such conclusion will differ in the reasons used as its premises.

The preceding analysis focused on epistemic fields as the central units of demarcation. However, a comprehensive profile of some parascientific epistemic field should also allow us to diagnose smaller units as parascientific, if they are the bearers of one or more characteristic features occurring in the profile. Such smaller units may be theories as systems of statements, which may be inconsistent or circular, or incompatible with the accepted background knowledge; individual hypotheses, which may be logically unfalsifiable; individual methods, which may have long been weeded out from the sciences for being defective; or some behavior or attitude of the representatives of the field, and so on. In this way we are justified in calling a theory, a hypothesis, a method, or a behavior unscientific. This is of particular importance when we are dealing with an epistemic field which we normally regard as scientific. For in such a case the philosopher of science may still detect some unscientific feature and denounce it as being pseudoscientific, calling for its repair or, if impossible, its elimination.

6 PROTOSCIENCE AND HETERODOXY

Calling some theory, approach or entire epistemic field parascientific is a strong and damning verdict. For this reason we must be quite careful in our judgment, which ought to be based on a diligent examination of the suspected theory or field. Now, whereas the philosopher of science may be more careful in such pursuit, scientists are sometimes less careful. Thus, many authors have warned us that the history of science should teach us sobering and humbling lessons concerning the science/pseudoscience demarcation (e.g., Toulmin 1984). First, it has always been too easy a temptation to reject a theory or approach as pseudoscientific just because it is heterodox, or maybe just because we do not like or understand it. Second, some theories that are declared pseudoscientific may actually turn out to be protoscientific, so that their possibly bright future could be endangered by an unfair judgment. Third, there is the historical problem of judging a certain field in retrospect: some field that may be clearly pseudoscientific today, may have been protoscientific at an earlier time and hence in a different scientific landscape.

A prime example is Alfred Wegener's hypothesis of contintental drift, which was initially rejected when proposed in 1915 and sometimes even derided, but eventually became the basis for the plate tectonics revolution in the 1960ies. Wegener's ideas were indeed protoscientific rather than pseudoscientific because he did not refer to untestable myths and mysteries like Velikovsky or von Däniken, but instead to geological and climatological data. And he did not behave like a pseudoscientist, for he admitted that his ideas were conjectural and that the main problem of his hypothesis was the unknown mechanism of continental drift. However, his geological colleagues also acted rationally in rejecting his hypothesis for being too implausible at that time (see [Kitcher, 1982; Radner and Radner, 1982]). Apart from the historical vindication of Wegener's protoscientific ideas, an assessment of Wegener's hypothesis in a pseudoscience profile would most likely have shown that even at their time his views were not pseudoscientific, but merely unorthodox [Edelman, 1988]. This indicates that it is not always true that we can determine the scientific status of a certain theory or field only retrospectively, e.g., by observing its historical progress or else degeneration.

A less favorable example is phrenology, which has been regarded as a protoscience leading to neuropsychology (Young 1970). Phrenology advanced the correct and fruitful idea that mental functions are localized in the brain, but was badly mistaken in the claim that these functions manifest themselves craniologically, i.e., as bulges on the skull. The latter made phrenological diagnosis a pseudotechnology, which, however, had some beneficial side-effects on the treatment of prisoners and the mentally ill [Hines, 2003]. In this case a retrospective analysis shows that a small part of phrenology led to progress, if only in a field that quickly emancipated itself from phrenology, whereas the larger part degenerated into a pseudoscience.

In the case of astrology opinions are divided. Apart from its defenders of course, even some philosophers of science are willing to grant astrology the status of a former protoscience (e.g., [Thagard, 1978]). Others maintain that astrology never was a protoscience, because even in antiquity educated people, like Strabo, Cicero and Ptolemy, clearly distinguished between astronomy and astrology, whether or not they believed in the latter [Culver and Ianna, 1988]. Moreover, it was obvious to many even back then that astrological predictions are unreliable for failing too frequently. And although some early scientists like Kepler practiced some astrology, they too kept it apart from science. Thus, it seems that despite various connections and flirtations between early astronomers and astrologers, astrology has long, if not always, been para- or even pseudoscientific, contributing nothing to astronomy or any other science.

These historical examples illustrate the need for a comprehensive analysis of any field or theory suspected of being a parascience. Even if we were wrong with our judgment at a given time, a genuine protoscience will sooner or later prove its fruitfulness and potential by developing into a full-fledged scientific field, propelled by successful research, or at least by giving rise to some scientific field.

But what exactly does "sooner or later" mean? We must ask this question because one of the most intriguing and sophisticated pseudosciences, namely parapsychology, has always claimed that it is actually a protoscience (or a pre-paradigmatic science, as some parapsychologists prefer to call it in Kuhnian terms), so that its classification as a pseudoscience would be unjustified. Now the birth of parapsychology as a field of research is usually taken to coincide with the establishment of the *Society for Psychical Research* in 1882, although earlier research in the area of spiritualism dates back to the 1850s [Kurtz, 1985]. Should a field still be regarded as a protoscience after more than 120–150 years? As mentioned several times in this chapter, parapsychology is a field still in search for a proper domain, because it has not succeeded in producing any findings that would convince its critics from mainstream psychology of the existence of some paranormal entities or processes [Hyman, 1989; Hines, 2003]. Worse, as Alcock [2003, p. 32] summarizes the situation: "...to the extent that parapsychology constitutes a 'field' of research, it is a field without a core knowledge base, a core set of constructs, a core set of methodologies, and a core set of accepted and demonstrable phenomena...". Does this not rather indicate that there is no such thing as psi (in other words, that the null hypothesis is true) and that the field is degenerative rather than protoscientific?

The same holds for astrology and creationism, which have also learned to exploit the "humbling lessons of history", claiming to be actually protosciences, which deserve to be granted their due chance of proving themselves full-fledged sciences. Yet if we are suspicious of a 120-150 years old protoscience, we are entitled to be even more skeptical of alleged protosciences that are thousands of years old.

A comprehensive profile of the epistemic field under consideration should also help to solve the problem of how to distinguish fruitful scientific heterodoxy from pseudoscientific deviation. In his foreword to the book "Scientists Confront Velikovsky" [Goldsmith, 1977], the famous science fiction author Isaac Asimov has coined the terms *endoheresy* and *exoheresy*. These terms capture nicely the gist of Section 5.3, namely the condition that a heresy must stay within the bounds of the scientific superparadigm, so to speak, if it is to be considered legitimate, even though the majority of the scientific community may reject it as mistaken or misguided. For example, in developmental biology there is a school called "developmental structuralism" [Webster and Goodwin, 1996], which takes genes to be relatively irrelevant for development, and hence seeks to explore the role of "universal laws of form" or "transformation laws" in development. Thus, it is attempted to describe the developing organism by field equations, reviving the earlier notion of a morphogenetic field. This structuralist approach is rejected or ignored by most developmental biologists, but it stays within the bounds of science, although some of the philosophical considerations of these authors seem to be in need of repair [Mahner and Bunge, 1997]. By contrast, the morphogenetic field hypothesis of the former biochemist Rupert Sheldrake is clearly an exoheresy, for it shows too many marks of pseudoscience and is irreparably esoteric [Carroll, 2003].

The preceding considerations result in the recommendation that both the sciences and the humanities ought to welcome endoheresies, because they form a valuable stock of alternative views, however implausible they may be at a given time. After all, it is too easy to be blinkered by orthodoxy which is reinforced by the routine of normal research. On the other hand, scientists must judge for themselves whether they wish to spend any time on investigating exoheresies. However, if not for scientific reasons, they should on occasion study exoheresies for educational purposes, explaining to the public why certain claims are parascientific and hence unworthy of serious attention. Although scientists may have very good reasons for rejecting exoheresies, they must keep explaining these reasons to the public in order to avoid the impression that their refusal to pay attention to parasciences is due to sheer dogmatism and arrogance. Thus, the advancement of the public understanding of science requires that we deal not only with science, but also with parascience.

7 CONCLUSION

Looking at the figures 2, 3 and 4, we notice that there are two main demarcation lines: the one between science and nonscience, and the other between reliable (approximately true) and illusory knowledge. Now some authors maintain that it is the latter which is the more important one (e.g., [Laudan, 1983; Haack, 2003]). After all, proper inquiry and proper standards of reasoning and evidence exist also outside science. For example, not only the philosopher arguing his case, but also the policeman investigating a crime knows (or at least should know) how to reason properly and how to distinguish good from bad evidence. As a consequence science would not differ in kind from other epistemic areas where common standards of rational and objective inquiry are practiced, but at most in the degree and thoroughness of their application [Haack, 2003]. Since determining when knowledge is gained in a proper way is the task of epistemology in general, it seems that the basic epistemological demarcation between knowledge and illusion is more important than that between science and nonscience.

This view usually rests on the idea that science is but an extended form of common sense, as both scientists like Thomas Huxley and Albert Einstein, and philosophers like John Dewey and Gustav Bergmann believed [Haack, 2003]. But unless the common sense of philosophers is totally different from everybody else's, this view is doubtful: there are good arguments for the contrary thesis that, in important respects, science transcends common sense and ordinary language, and therefore is quite "unnatural" [Wolpert, 1992]. The fact that so many people have serious difficulty in understanding scientific concepts, theories, and methods renders noncommonsensism more plausible than commonsensism. Yet even if scientific thinking were just extended common sense, it would still be the task of the philosopher of science to tell us how scientific cognition and knowledge differ from nonscientific cognition and knowledge.

In any case, wherever we eventually draw our lines, the important thing is to draw some line at all, so as not to surrender to relativism, arbitrariness, and irrationalism.

ACKNOWLEDGEMENTS

I am indebted to Mario Bunge, Michael Kary, Amardeo Sarma, Thomas Waschke, and Theo Kuipers for valuable suggestions and comments.

BIBLIOGRAPHY

[Agassi, 1964/1999] J. Agassi. The Nature of Scientific Problems and Their Roots in Metaphysics. In M. Bunge (ed.) *Critical Approaches to Science and Philosophy*, pp. 189–211. Transaction Publishers: New Brunswick, NJ 1999.

[Agazzi and Darvas, 1997] E. Agazzi and G. Darvas, eds. *Philosophy of Mathematics Today.* Kluwer: Dordrecht, 1997.

[Albert, 1985] H. Albert. *Treatise on Critical Reason.* Princeton University Press. Princeton, NJ, 1985.

[Alcock, 1985] J. Alcock. Parapsychology as a "Spiritual" Science. In: P. Kurtz (ed.), pp. 537–565, 1985.

[Alcock, 2003] J. Alcock. Give the Null Hypothesis a Chance. Reasons to Remain Doubtful about the Existence of Psi. In: J. Alcock, J. Burns, A. Freeman (eds.) *The Psi Wars: Getting to Grips with the Paranormal*, pp. 29–50. Imprint Academic: Exeter, 2003.

[Alters, 1997] B. J. Alters. Whose Nature of Science? *Journal of Research in Science Teaching* 34: 39–55, 1997.

[Armstrong, 1995] D. M. Armstrong. Naturalism, Materialism, and First Philosophy. In: P.K. Moser & J.D. Trout (eds.) *Contemporary Materialism*, pp. 35–50. Routledge: London, 1995.

[Ayer, 1946] A. J. Ayer. *Language, Truth and Logic.* Dover: New York, 1946.

[Bechtel, 1986] W. Bechtel, ed. *Integrating Scientific Disciplines.* Martinus Nijhoff: Dordrecht, 1986.

[Beckner, 1959] M. Beckner. *The Biological Way of Thought.* Columbia University Press: New York, 1959.

[Beyerstein, 1987] B. L. Beyerstein. Neuroscience and psi-ence. *Behavioral and Brain Sciences* 10: 571–572, 1987.

[Blatt, 1983] J. M. Blatt. How Economists Misuse Mathematics. In: A.S. Eichner (ed.) *Why Economics is not yet a Science*, pp. 166–186. M.E. Sharpe: Armonk, NY, 1983.

[Boyd, 1999] R. Boyd. Homeostasis, Species, and Higher Taxa. In: R.A. Wilson (ed.) *Species: New Interdisciplinary Essays*, pp. 141–185. MIT-Press: Cambridge, MA, 1999.

[Broad, 1949] C. D. Broad. The Relevance of Psychical Research to Philosophy. *Philosophy* 24: 291–309, 1949. [Also in J. Ludwig, ed. 1978]

[Bunge, 1982] M. Bunge. Demarcating Science from Pseudoscience. *Fundamenta scientiae* 3: 369–388, 1982.

[Bunge, 1983a] M. Bunge. *Treatise on Basic Philosophy, vol. 5: Epistemology and Methodology I: Exploring the World.* D. Reidel: Dordrecht, 1983.

[Bunge, 1983b] M. Bunge. *Treatise on Basic Philosophy, vol. 6: Epistemology and Methodology II: Understanding the World.* D. Reidel: Dordrecht, 1983.

[Bunge, 1984] M. Bunge. What Is Pseudoscience? *Skeptical Inquirer* 9(1): 36–46, 1984.
[Bunge, 1999] M. Bunge. *The Sociology-Philosophy Connection.* Transaction Publishers: New Brunswick, NJ, 1999.
[Carnap, 1936–37] R. Carnap. Testability and Meaning. *Philosophy of Science* 3: 419–471, 4: 1–40, 1936–37.
[Carroll, 2003] R. T. Carroll. *The Skeptic's Dictionary.* Wiley: New York, 2003. [Also online: www.skepdic.com]
[Cartwright, 1983] N. Cartwright. *How the Laws of Physics Lie.* Clarendon Press: Oxford, 1983.
[Culver and Ianna, 1988] R. B. Culver and P. A. Ianna. *Astrology: True or False? A Scientific Evaluation.* Prometheus Books: Buffalo, NY, 1988.
[Curd and Cover, 1998] M. Curd and J. A. Cover, eds. *Philosophy of Science. The Central Issues.* W.W. Norton: New York, 1998.
[Darden and Maull, 1977] L. Darden and N. Maull. Interfield Theories. *Philosophy of Science* 44: 43–64, 1977.
[Derksen, 1993] A. A. Derksen. The Seven Sins of Pseudo-Science. *Journal for General Philosophy of Science* 24: 17–42, 1993.
[Derksen, 2001] A. A. Derksen. The Seven Strategies of the Sophisticated Pseudo-Scientist: A Look into Freud's Rhetorical Toolbox. *Journal for General Philosophy of Science* 32: 329–350, 2001.
[Devitt, 1996] M. Devitt. *Realism and Truth.* Princeton University Press: Princeton, NJ.
[Dupré, 1993] J. Dupré. *The Disorder of Things. Metaphysical Foundations of the Disunity of Science.* Harvard University Press: Cambridge, MA, 1993.
[Duran, 1990] J. Duran. Philosophical Difficulties with Paranormal Knowledge Claims. In P. Grim (ed.), pp. 232–242, 1990.
[Edelman, 1988] N. Edelman. Wegener and Pseudoscience: Some Misconceptions. *Skeptical Inquirer* 12(4): 398–402, 1988.
[Edwards, 2004] P. Edwards. *Heidegger's Confusions.* Prometheus Books: Amherst, NY, 2004.
[Eflin *et al.*, 1999] J. T. Eflin, S. Glennan, and G. Reisch. The Nature of Science: A Perspective from the Philosophy of Science. *Journal of Research in Science Teaching* 36: 107–116, 1999.
[Ernø–Kjølhede, 2000] E. Ernø–Kjølhede. *Scientific Norms as (Dis)Integrators of Scientists?* MPP Working Paper 14, pp. 1–18. Copenhagen Business School: Copenhagen, 2000.
[Ernst *et al.*, 2001] E. Ernst, *et al.*, eds. *The Desktop Guide to Complementary and Alternative Medicine.* Mosby: Edinburgh, 2001.
[Feyerabend, 1975] P. Feyerabend. *Against Method.* New Left Books: London, 1975.
[Flew, 1990] A. Flew. Parapsychology: Science or Pseudoscience? In P. Grim (ed.), pp. 214–231, 1990.
[Giere, 1984] R. N. Giere. *Understanding Scientific Reasoning.* Holt, Rinehart & Winston. New York, 1984.
[Giere, 1999] R. N. Giere. *Science Without Laws.* University of Chicago Press: Chicago, IL, 1999.
[Goldsmith, 1977] D. Goldsmith, ed. *Scientists Confront Velikovsky.* Cornell University Press: Ithaca, NY, 1977.
[Grim, 1990] P. Grim, ed. *Philosophy of Science and the Occult.* State University of New York Press: Albany, NY, 1990.
[Gross and Levitt, 1994] P. Gross and N. Levitt. *Higher Superstition.* Johns Hopkins University Press: Baltimore, MD, 1994.
[Grove, 1985] J. W. Grove. Rationality at Risk: Science against Pseudoscience. *Minerva* 23: 216–240, 1985.
[Grünbaum, 1984] A. Grünbaum. *The Foundations of Psychoanalysis. A Philosophical Critique.* University of California Press: Berkeley, CA, 1984.
[Haack, 2003] S. Haack. *Defending Science — Within Reason.* Prometheus Books: Amherst, NY, 2003.
[Hansson, 1996] S. O. Hansson. Defining Pseudo-Science. *Philosophia naturalis* 33: 169–176, 1996.
[Hines, 2003] T. Hines. *Pseudoscience and the Paranormal.* Prometheus Books: Amherst, NY, 2003.
[Holton, 1993] G. Holton. *Science and Anti-Science.* Harvard University Press: Cambridge, MA, 1993.
[Horgan, 1995] J. Horgan. *The End of Science.* Addison-Wesley: Reading, MA, 1995.

[Humphrey, 1999] N. Humphrey. *Leaps of Faith. Science, Miracles and the Search for Supernatural Consolation.* Copernicus-Springer: New York, 1999.

[Kanitscheider, 1991] B. Kanitscheider. A Philosopher Looks at Astrology. *Interdisciplinary Science Reviews* 16: 258–266, 1991.

[Kirkland, 2000] K. Kirkland. Paraneuroscience? *Skeptical Inquirer* 24(3): 40–43, 2000.

[Kitcher, 1982] P. Kitcher. *Abusing Science. The Case Against Creationism.* MIT-Press: Cambridge, MA, 1982.

[Kitcher, 1993] P. Kitcher. *The Advancement of Science. Science without Legend, Objectivity without Illusion.* Oxford University Press: New York, 1993.

[Kneale, 1974] W. C. Kneale. The Demarcation of Science. In: P.A. Schilpp (ed.), pp. 205–217, 1974.

[Knodel, 1985] H. Knodel, ed. *Neues Biologiepraktikum Linder Biologie* (Lehrerband). [New Practical Instruction in Biology, Teachers' Edition] J.B. Metzler: Stuttgart, 1985.

[Kuhn, 1962] T. S. Kuhn. *The Structure of Scientific Revolutions.* University of Chicago Press: Chicago, IL, 1962.

[Kuhn, 1970] T. S. Kuhn. Logic of Discovery or Psychology of Research? In: I. Lakatos & A. Musgrave (eds.), pp. 1–24, 1970.

[Kuipers, 2000] T. A. F. Kuipers. *From Instrumentalism to Constructive Realism.* Kluwer: Dordrecht, 2000.

[Kuipers, 2001] T. A. F. Kuipers. *Structures in Science. Heuristic Patterns Based on Cognitive Structures.* Kluwer: Dordrecht, 2001.

[Kuipers, 2004] T. A. F. Kuipers. De logica van de G-hypothese. Hoe theologisch onderzoek wetenschappelijk kan zijn. In: K. Hilberdink (ed.) *Van God los? Theologie tussen godsdienst en wetenschap*, pp. 59–74. KNAW: Amsterdam, 2004.

[Kurtz, 1985] P. Kurtz, ed. *A Skeptic's Handbook of Parapsychology.* Prometheus Books: Buffalo, NY, 1985.

[Kurtz, 2000] P. Kurtz. The New Paranatural Paradigm. *Skeptical Inquirer* 24(6): 27–31, 2000.

[Lakatos, 1970] I. Lakatos. Falsification and the Methodology of Research Programmes. In: I. Lakatos & A. Musgrave (eds.), pp. 91–197, 1970.

[Lakatos, 1973] I. Lakatos. Science and Pseudoscience. In: M. Curd & J.A. Cover (eds. 1998), pp. 20–26, 1973. `http://www.lse.ac.uk/collections/lakatos/scienceAndPseudoscienceTranscript.htm;21.2.05`

[Lakatos, 1974] I. Lakatos. Popper on Demarcation and Induction. In: P.A. Schilpp (ed.), pp. 241–273, 1974.

[Lakatos and Musgrave, 1970] I. Lakatos and A. Musgrave, eds. *Criticism and the Growth of Knowledge.* Cambridge University Press: New York, 1970.

[Laudan, 1983] L. Laudan. The Demise of the Demarcation Problem. In: M. Ruse, ed. *But Is It Science? The Philosophical Question in the Creation/Evolution Controversy*, pp. 337–350. Prometheus Books: Buffalo, NY, 1988.

[Ludwig, 1978] J. Ludwig, ed. *Philosophy and Parapsychology.* Prometheus Books: Buffalo, NY, 1978.

[Lugg, 1987] A. Lugg. Bunkum, Flim-Flam and Quackery: Pseudoscience as a Philosophical Problem. *Dialectica* 41: 221–230, 1987.

[Mahner and Bunge, 1997] M. Mahner and M. Bunge. *Foundations of Biophilosophy.* Springer-Verlag: Berlin, Heidelberg, New York, 1997.

[McGowan, 1994] D. McGowan. *What is Wrong with Jung?* Prometheus Books, Buffaly, NY, 1994.

[Merton, 1973] R. Merton. *The Sociology of Knowledge.* University of Chicago Press: Chicago, 1973.

[Nagel, 1956] E. Nagel. *Logic Without Metaphysics.* Free Press: Glencoe, IL, 1956.

[Niiniluoto, 1987] I. Niiniluoto. *Truthlikeness.* D. Reidel: Dordrecht, 1987.

[Popper, 1959] K. R. Popper. *The Logic of Scientific Discovery.* Hutchinson: London, 1959.

[Popper, 1963] K. R. Popper. *Conjectures and Refutations.* Basic Books: New York, 1963.

[Popper, 1994] K. R. Popper. Zwei Bedeutungen von Falsifizierbarkeit [Two Senses of Falsifiability]. In: H. Seiffert & G. Radnitzky (eds.) *Handlexikon der Wissenschaftstheorie*, pp. 82–85. Deutscher Taschenbuch Verlag: München, 1994.

[Poser, 2001] H. Poser. *Wissenschaftstheorie. Eine philosophische Einführung* [Theory of Science: A Philosophical Introduction]. Reclam: Stuttgart, 2001.

[Radner and Radner, 1982] D. Radner and M. Radner. *Science and Unreason.* Wadsworth Publishing Company: Belmont, CA, 1982.

[Reisch, 1998] G. A. Reisch. Pluralism, Logical Empiricism, and the Problem of Pseudoscience. *Philosophy of Science* 65: 333–348, 1998.

[Resnik, 2000] D. B. Resnik. A Pragmatic Approach to the Demarcation Problem. *Studies in History and Philosophy of Science*, 31: 249–267, 2000.

[Rhine, 1954] J. B. Rhine. The Science of Nonphysical Nature. In: J. Ludwig (ed. 1978), pp. 117–127, 1954.

[Rothbart, 1990] D. Rothbart. Demarcating Genuine Science from Pseudoscience. In: P. Grim (ed.), pp. 111–122, 1990.

[Runggaldier, 1996] E. Runggaldier. *Philosophie der Esoterik* [Philosophy of Esoterics]. Kohlhammer: Stuttgart, 1996.

[Schick and Vaugn, 1999] T. Schick and L. Vaughn. *How to Think About Weird Things.* Mayfield Publishing Company: Mountain View, CA, 1999.

[Schlipp, 1974] P. A. Schilpp, ed. *The Philosophy of Karl Popper*, vol. 1. Open Court: La Salle, IL, 1974.

[Settle, 1971] T. Settle. The Rationality of Science versus the Rationality of Magic. *Philosophy of the Social Sciences* 1: 173–194, 1971.

[Shermer, 1997] M. Shermer. *Why People Believe Weird Things.* W. Freeman: New York, 1997.

[Shermer, 2002] M. Shermer, ed. *The Skeptic Encyclopedia of Pseudoscience*, 2 vols. ABC-Clio: Santa Barbara, CA, 2002.

[Siitonen, 1984] A. Siitonen. Demarcation of Science from the Point of View of Problems and Problem-Stating. *Philosophia naturalis* 21: 339–353, 1984.

[Sokal and Bricmont, 1998] A. Sokal and J. Bricmont. *Intellectual Impostures.* Profile Books: London, 1998.

[Sorokin, 1956] P. A. Sorokin. *Fads and Foibles in Modern Sociology and Related Sciences.* H. Regnery: Chicago, IL, 1956.

[Spector, 1990] M. Spector. Mind, Matter, and Quantum Mechanics. In: P. Grim (ed.), pp. 326–349, 1990.

[Stalker and Glymour, 1989] D. Stalker and C. Glymour, eds. *Examining Holistic Medicine.* Prometheus Books: Buffalo, NY, 1989.

[Stalker and Glymour, 1989b] D. Stalker and C. Glymour. Quantum Medicine. In: D. Stalker & C. Glymour (eds.), pp. 107–125, 1989.

[Stenger, 1995] V. Stenger. *The Unconscious Quantum.* Prometheus Books: Buffalo, NY, 1995.

[Thagard, 1978] P. Thagard. Why Astrology is a Pseudoscience. In: P. Asquith & I. Hacking (eds.) *PSA* 1978, vol. 1, pp. 223–234. East Lansing, MI: Philosophy of Science Association, 1978. [Also in: M. Curd & J.A. Cover (eds. 1998), pp. 27–37.]

[Thagard, 1988] P. Thagard. *Computational Philosophy of Science.* MIT-Press: Cambridge, MA, 1988.

[Toulmin, 1984] S. Toulmin. The New Philosophy of Science and the "Paranormal". *Skeptical Inquirer* 9(1): 48–55, 1984.

[Vollmer, 1990] G. Vollmer. Against Instrumentalism. In: P. Weingartner & G.J.W. Dorn (eds.) *Studies on Mario Bunge's Treatise*, pp. 245–259. Rodopi: Amsterdam, 1990.

[Vollmer, 1993] G. Vollmer. Wozu Pseudowissenschaften gut sind [What Pseudosciences Are Good For]. In: G. Vollmer, *Wissenschaftstheorie im Einsatz* [Philosophy of Science in Action]. Hirzel-Verlag: Stuttgart, 1993.

[Webster and Goodwin, 1996] G. Webster and B. C. Goodwin. *Form and Transformation.* Cambridge University Press: Cambridge, UK, 1996.

[Weingartner, 2000] P. Weingartner. *Basic Questions on Truth.* Kluwer: Dordrecht, 2000.

[Weston, 1992] T. Weston. Approximate Truth and Scientific Realism. *Philosophy of Science* 59: 53–74, 1992.

[Wilson, 2000] F. Wilson. *The Logic and Methodology of Science and Pseudoscience.* Canadian Scholars' Press: Toronto, 2000.

[Wilson, 1999] R. A. Wilson. Realism, Essence, and Kind: Resuscitating Species Essentialism? In: R.A. Wilson (ed.) *Species: New Interdisciplinary Essays*, pp. 187–207. MIT-Press: Cambridge, MA, 1999.

[Wittgenstein, 1921] L. Wittgenstein. *Tractatus Logico-Philosophicus.* Suhrkamp: Frankfurt 1921/1960.

[Wolpert, 1992] L. Wolpert. *The Unnatural Nature of Science.* Faber & faber: London, 1992.

[Young, 1970] R. M. Young. *Mind, Brain and Adaptation in the Nineteenth Century: Cerebral Localization and Its Biological Context from Gall to Ferrier*. Oxford University Press: Oxford, 1970.

[Ziman, 1994] J. M. Ziman. *Prometheus Bound*. Cambridge University Press. Cambridge, UK, 1994.

HISTORY OF THE PHILOSOPHY OF SCIENCE. FROM *WISSENSCHAFTSLOGIK* (LOGIC OF SCIENCE) TO PHILOSOPHY OF SCIENCE: EUROPE AND AMERICA, 1930–1960

Friedrich Stadler

"Philosophy of science without history of science is empty, history of science without philosophy of science is blind"
Imre Lakatos, 1974

PRELIMINARY REMARKS

There seems a general consensus in the scientific community that modern philosophy of science — as a subdiscipline of (scientific) philosophy — has emerged as a genuine academic research and teaching field as well as an institution only since the middle of the 20th Century.

Accordingly, we can reconstruct a process of differentiation and professionalization of philosophy of science from the ancient Greek philosophy (Pre-Socratics, Plato and Aristotle) via the rationalist and empiricist philosophers of the "Scientific Revolution" to the Enlightenment up to the (Neo-)Kantian versions of science-oriented philosophy. These developments lead to re-evaluations of a "Theory of Science" (Wissenschaftslehre) in the 19th century in close interaction with the rise of the empirical sciences between physics, physiology and psychology — as is typically illustrated with the philosopher-scientists Ernst Mach and Ludwig Boltzmann. In parallel, this dynamics of departure from, and interaction with traditional philosophy as a universal normative discipline was accompanied by a specific focus on the methods of the natural sciences in general, but also in the cultural and social sciences: historism as well as the "probabilist revolution" in the cultural sciences [Ringer, 1997]. There is a re-conceptualization of Empiricism and Rationalism [Santillana and Zilsel, 1941], which anticipated the formation of Logical Empiricism between the two World Wars in the 20th century. The context for this innovation was the so-called "Second Scientific Revolution" in science, with Einstein's Relativity Theory and Quantum Physics around Bohr and Schrödinger and the input of modern symbolic logic and set theory with Frege, Russell and

Handbook of the Philosophy of Science: General Philosophy of Science - Focal Issues, pp. 577–658.
Volume editor: Theo Kuipers. General editors: Dov M. Gabbay, Paul Thagard and John Woods.

Whitehead, Wittgenstein and Gödel. Generally, we find therein a permanent tension between a normative philosophy of science (methodology) on the one hand, and a descriptive history of science (theory dynamics) on the other, which indicates the later introduced distinction of the context of justification and the context of discovery as a main issue in this "Rise of Scientific Philosophy" [Reichenbach, 1938/1951].

It is not surprising, that most textbooks and the few handbooks or encyclopedias in the philosophy of science do not explicitly deal with the history of its own discipline, or are restricted regarding themes and time periods. The important comprehensive historical study on philosophy of science as a monograph from Greek philosophy up to contemporary issues is also limited to the natural sciences and its methodologies [Losee, 1972/2001]. Another restriction — with some exceptions [Serres, 1989; Collins, 1998] — is a strong European perspective disregarding the important contribution of the Chinese and Islamic world to the sciences and their philosophy. And this bias is re-inforced by the missing gender perspective, although in the meantime there are valuable contributions to the problem of women in philosophy and in the philosophy of science, especially feminist philosophy (of science) [Fricker and Hornsby, 2000].

Given this status quo of the fragmentary historiography and research and with reference to related contributions in this volume, the subsequent "History of Philosophy of Science" is restricted to a paradigmatic case study of transfer, transformation and institutionalization of philosophy of science from Europe to America, in the period from 1930 to 1960. This is done with reference to the standard volume on *The Intellectual Migration. Europe and America, 1930–1960* [Fleming and Bailyn, 1969], and to volumes on the origins and influence of Logical Empiricism in America and the history of the Vienna Circle leading up to 1938 [Stadler, 1997/2001; Giere and Richardson, 1996; Hardcastle and Richardson, 2004; Richardson and Uebel, 2006].

The point of departure of the following account is the rise of Logical Empiricism in Central Europe before the forced migration, and the end of this intellectual and institutional history is the placement of "The *Wiener Kreis* in America" [Feigl, 1969] in the Cold War period. The further development is characterized by the criticism of the so called "received view" and the re-transfer of a modified philosophy of science — as a mostly analytic and normative methodology — back to Europe, which was dealt with only in the last years. And these currents since Quine's "Two Dogmas" [1951] are a remarkable return of a hidden agenda in the described history of philosophy of science, namely the pragmatic and historical turn. These new insights allow us to reasonably speak of a re-union of the history *and* philosophy of science, or of a (cultural) history of philosophy of science, which does not privilege the context in comparison with the (meta-)theoretical dynamics.

1 THE EMERGENCE OF PHILOSOPHY OF SCIENCE: "WISSENSCHAFTSLOGIK" (LOGIC OF SCIENCE) BEFORE 1938

The emergence of the discipline known today as "philosophy of science" can be seen as converging with the process of the increasingly scientific status of philosophy, the so-called "rise of scientific philosophy" (Reichenbach 1951) in the inter-war years. Already in the programmatic text of the Vienna Circle (*Wissenschaftliche Weltauffassung. Der Wiener Kreis,* 1929), the autonomous regal discipline of philosophy had given way to an antimetaphysical, physicalist unified science. This idea was systematically elaborated in the thirties, most notably in Rudolf Carnap's writings. In the manifesto, reference had primarily been made to his *Logical Structure of the World* [Carnap, 1928] — as a constitutive system based on experience with logical analysis. A few years later the position he took in his *Logical Syntax* [1934a] found acceptance. The task of "Wissenschaftslogik" [1934b] is seen as lying in the study of science as a whole or in its disciplines:

> "The concepts, propositions, proofs, theories appearing in the various realms of science are analyzed — less from the perspective of the historical development of science or of the sociological and psychological conditions of its functioning, but more from a logical perspective. This field of work for which no generic term has been able to gain acceptance, could be called theory of science or to be more precise logic of science. Science is understood as referring to the totality of accepted propositions. This does not just include the statements made by scholars but also those of everyday life. There is no clear boundary line drawn between these two areas." [Carnap, 1934b, p. 5]

Here the distancing from traditional philosophy becomes highly salient, even if the role and function of a scientific *philosophy*, as linguistic analysis in Wittgenstein's sense, is not called into question. This new discipline is not so interested in propositions on the external world as the realm of the empirical disciplines (thing language), as in "science itself as an orderly structure of propositions", known as object language (ibid., p. 6) – accordingly, in the "sense" of the propositions and the "meaning" of concepts from a logical point of view. The realm of these concepts is limited either to the analytic propositions of logic/mathematics or to the empirical propositions of the sciences. This culminates in the view:

> "that the propositions of the logic of science are propositions of the logical syntax of language. Thus these propositions lie within the boundaries drawn by Hume, for logical syntax is ... nothing other than mathematics of language." (ibid.)

Before the logic of science as a "Wissenschaftslehre" (theory of knowledge or theory of science) was promulgated in the 19th century, for instance by Johann G. Fichte, Bernard Bolzano and Ernst Mach, the term "theory of science" was in circulation as an alternative to classical philosophy besides empirical disciplines

[Losee, 1980]. Nevertheless, it was the first time here that the so-called "overcoming of metaphysics through the logical analysis of language" [Carnap, 1931a] was propagated.

Carnap had combined the elaboration of this program of unified science in his *Logical Syntax of Language* [1934a] with its promulgation. As part of the internationalization of the Vienna Circle under way since 1929, two small books appeared almost at the same time in England, i.e., *The Unity of Science* [1934c] and *Philosophy and Logical Syntax* [1935] in the series "Psyche Miniatures" published by Kegan Paul. The former was an edition of the German article on physical language [Carnap, 1931b], reworked by the author and translated by Max Black. The latter united three lectures that Carnap had given at the University of London in October of 1934: "The Rejection of Metaphysics", "Logical Syntax of Language", "Syntax as the Method of Philosophy". These attempts to popularize "Logic of Science" in the Anglo-Saxon world were continued by the translation of *Logical Syntax* which appeared in 1937 in a expanded edition at the same English publisher [Carnap, 1937].

It is known that already in his *Logical Syntax* Carnap had been influenced by Polish and American logicians and philosophers of science (Tarski, Quine and Morris) to further develop the possible field of "Logic of Science". In addition to the syntactic dimension, he cited the semantic and pragmatic dimensions as future fields of work. Accordingly, he described the logic of science in his preface to the second edition as the "Analysis and Theory of the Language of Science":

> According to the present view, this theory comprises, in addition to logical syntax, mainly two further fields, i.e., semantics and pragmatics. Whereas syntax is purely formal, i.e., only studies the structure of linguistic expression, semantics studies the semantic relationship between expressions and objects or concepts; ... Pragmatics also studies the psychological and sociological relations between persons using the language and the expressions. [Carnap, 1968, VII]

With this new conceptualization of the logic of science, which already took place before the transfer of these ideas to the United States, we have also outlined the logical space for the philosophy of science as well as the terminological structure for the Unity of Science movement [1934ff]. Of course, Logical Empiricism before 1938 had no codified understanding of "logic of science" in relation to philosophy. Here, however, only the paradigmatic elements have been indicated which proved to be relevant later in the Anglo-American realm. In this context, I cannot dwell on the controversial protocol statement debate within the Vienna Circle in which various positions on the basic issue of knowledge were unearthed, e.g., in Edgar Zilsel's "Bemerkungen zur Wissenschaftslogik" (Notes on the Logic of Science) [1932/33]. This eventually led to a heated debate on fundamental questions in the epistemology of that time [Uebel, 1992].

The fact that there was, on both sides, a strong reception of European positivism centered around Ernst Mach and of American pragmatism focusing on William

James clearly shows that the trans-Atlantic process of communication did not suddenly begin in the 20th century. Rather, there had been a continuous process of international exchanges between related intellectual movements (positivism and pragmatism, operationalism and behaviorism) which became manifest in classical Logical Empiricism.

The direct contacts between Mach, William James and Paul Carus — the editor of the journals *Monist* and *Open Court* — paved the way for a strong convergence of Logical Empiricism and neo-pragmatism with Otto Neurath and Rudolf Carnap, on the one hand, and John Dewey and Charles Morris, on the other. The positive reception of Percy Bridgman's *The Logic of Modern Physics* [1927] has often been described, most notably by Philipp Frank [1949], as a milestone in this theoretical rapprochement. Even in psychology, there was direct cooperation, facilitated by Edward Tolman, Egon Brunswik and the Bühler-School, which led to a transfer of individual scholars and ideas [Fischer and Stadler, 1997; Smith, 1986]. (At that time, however, this did not mean that behaviorism predominated, since within the context of the *Encyclopedia of Unified Science*, psychoanalysis or cognitive psychology was seen as being at least equal, as Egon and Else Frenkel-Brunswik's attempts to integrate them show.)

The reception of philosophical ideas between the old continent and the United States is meanwhile well documented [Giere and Richardson, 1996; Hardcastle and Richardson, 2003].

The historian of science Gerald Holton, who played a seminal role in the forties in the Unity of Science Institute and as an assistant of Philipp Frank, has given a very apt reconstruction of these cognitive parallels and this transfer of knowledge in his "From the Vienna Circle to Harvard Square: The Americanization of a European World Conception" [1993]. This history of ideas, which also includes Quine, describes a growing internationalization best illustrated by the International Congresses of the *Encyclopedia of Unified Science* and the "Unity of Science Institute" founded by Frank. Holton characterizes the favourable conditions for Logical Empiricism in the United States from 1940 to 1969 metaphorically as an "ecological niche" in the New World and depicted these developments as an osmotic success story. Another (auto-)biographical study of Holton on this phenomenon already alluded in the title — "On the Vienna Circle in Exile" [1995[— to the possibility of the Viennese philosophy of science finding a dynamic (and transitory) context in American intellectual life. It should only be added here that these theoretical developments in the history of science called into question the dominant idea of Unified Science, both intensifying the inherent contradictions and overcoming them — which went almost unnoticed within this setting. If one takes into account the parallel reciprocal tie between positivism and pragmatism since the turn of the century leading up to the synthesis in today's analytic philosophy [Dahms, 1987/88; 1994], the non-linear and self-organizational theoretical dynamic becomes evident in all its historical and systematic complexity. If one wishes to make a qualitative assessment of the transfer of science through intellectual emigration against the background of the discourse of the scientific community, the actual contacts must be studied irrespective of the history of reception related to emigration history. This would include, for instance, Moritz Schlick's early lecture trips and visiting professorships. His two sojourns in the United States left traces as well as his presentation at the Seventh International Congress of Philosophy in Oxford in 1930 where he addressed the programmatic, linguistic-analytic turn in philosophy in his "The Future of Philosophy" lecture. Here he advocated the dissolution of the classical philosophical canon by drawing a functional distinction between scientific philosophy on the one hand and the related scientific theorizing on the other:

> "There will always be men who are especially fitted for analyzing the ultimate meaning of scientific theories, but who may not be skilful in handling the methods by which their truth or falsehood is ascertained. These will be the men to study and to teach philosophizing, but of course they would have to know the theories just as well as the scientist who invents them. Otherwise they would not be able to take a single step, they would have no object on which to work. A philosopher, therefore, who knew nothing except philosophy would be a knife without blade and handle. Nowadays a professor of philosophy very often is a man who is not able to make anything clearer, that means he does not really philosophize at all, he just talks about philosophy or writes a book about it. This will be impossible in the future. The result of philosophizing will be that no more book will be written about philosophy, but all books will be written in a philosophical manner." [Schlick, 1931, p. 116]

Here this necessary merging of philosophizing with the results of the so-called empirical sciences documents — in Wittgenstein's language — the transformation of philosophy as a regal discipline into the "maiden of sciences". This was the result of calling into question the existence of an autonomous area. It also comes as no surprise when one considers Schlick's allegiance to British empiricism and to the scientific orientation of philosophy since the turn of the century. His visiting professorsips in Stanford [1929] and Berkeley [1931/32] reinforced this anglophile leaning, but also paved the way for a gradual shift toward the United States which Herbert Feigl had already begun in 1930. Schlick reported on his impressions in his lecture "On the Scientific World View in the U.S.A." which he gave at the "Verein Ernst Mach" (Ernst Mach Society) in Vienna in 1930. Here he drew attention to the fact that there was a well-developed everyday rationalism, accompanied by favorable conditions for the scientific world view thanks to empirical psychology and John Dewey's pragmatism. [Schlick, 1930/31, p. 76]. It is thus no coincidence that Schlick already figured on the Advisory Board in the first issue of the quarterly *Philosophy of Science*, which had been founded in 1934, with his student Herbert Feigl in the Editorial Board — together with Rudolf Carnap. The institution behind this journal, edited by William M. Malisoff, was the, still existing, "Philosophy of Science Association" (PSA) as an

> "organized expression of the will of a fairly large body of intellectually competent individuals whose basic interest is in both science and philosophy, and particularly in their union... The Association humbly undertakes the task of uniting in one body the scattered elements available for this enterprise." (Announcement of the Publisher)

It should also be added that Schlick's assistant in Stanford was the young Paul Arthur Schilpp, who served as the editor of the important and influential book series "The Library of Living Philosophers" (including volumes on Carnap, Dewey, Lewis, Quine, Popper, Russell, Einstein, among others). With this series he contributed greatly to the continuation of the discussions in analytic as well as pragmatist philosophy of science.

Proceeding from the early forties as the beginning of the specific American philosophy of science, it is possible to reconstruct the intellectual conditions of the convergent development of Central European and US-American philosophy of science [Stadler, 2004, 227ff].

In the contemporary *Dictionary of Philosophy* [Runes, 1944] we find the relevant discussions of that time presented in various short entries. Here it becomes clear that the central contributions to the philosophy of science were written by Rudolf Carnap, Carl G. Hempel and Heinrich Gomperz. Carnap presents philosophy of science as

> "that philosophic discipline which is the systematic study of the nature of science, especially of its methods, its concepts and presuppositions, and its place in the general scheme of intellectual disciplines. No very precise definition of the term is possible since the discipline shades imperceptibly into science, on the one hand, and into philosophy in general, on the other. A working division of its subject-matter into three fields is helpful in specifying its problems, though the three fields should not be too sharply differentiated or separated." [Carnap, 1944, p. 284]

According to Carnap the three fields addressed here are the following:

> 1. A critical study of the method or methods of the sciences, of the nature of scientific symbols, and of the logical structure of scientific symbolic terms...
> 2. The attempted clarification of the basic concepts, presuppositions and postulates of the sciences, and the revelation of the empirical, rational, or pragmatic grounds upon which they are presumed to rest. ...
> 3. A highly composite and diverse study which attempts to ascertain the limits of the special sciences, to disclose their interrelations one with another, and to examine their implications so far as these contribute to a theory either of the universe as a whole or of some aspect of it. [Carnap, 1944, 284f.]

In a preceding section, Carnap had already subsumed today's science studies under "science of science" as "the analysis and description of science from various points of view, including logic, methodology, sociology, and history of science" [ibid]. In this connection he referred to his entry on "Scientific Empiricism" and the "Unity of Science" as "a wider movement, comprising besides Logical Empiricism other groups and individuals with related views in various countries" [ibid., p. 286].

The Unity of Science was also identified with internationalization. "Scientific Empiricism" was introduced as a transformation of Logical Empiricism. With this self-understanding, the institutionalization and further differentiation of philosophy of science took place — a development which had been anticipated by two decades of intellectual exchange between Europe and America. But in parallel to this transatlantic movement the European philosophy of science is to be addressed in more detail.

2 THE *WIENER KREIS* IN GREAT BRITAIN: EMIGRATION AND INTERACTION

2.1 Prologue

In 1968 the Austro-American philosopher of science Herbert Feigl (1902–1988) published a remarkable, largely autobiographical essay on "The *Wiener Kreis* in America". This historical and theoretical account of the Vienna Circle's emigration story was first anthologized in the second volume on *Perspectives in American History* [1968], edited by the "Charles Warren Center for Studies in American History" and then included in the standard volume on *The Intellectual Migration. Europe and America, 1930–1960*, published in 1969 by Harvard University Press — distributed in Great Britain by Oxford University Press. A last reprint can be found in Feigl's *Inquiries and Provocations. Selected Writings 1929–1974* [Feigl, 1981].

Together with his influential article on "Logical Positivism" (co-authored with Albert Blumberg in the *Journal of Philosophy)* that already appeared in 1931, the year of his definitive emigration to the USA — the above-mentioned publication marked a watershed in the historiography of Logical Empiricism as a paradigmatic intellectual history of forced migration — in addition to the autobiographical reports of Philipp Frank [1949] and Rudolf Carnap [1963] inter alia:

> It was this article, I believe, that affixed this internationally accepted label to our Viennese outlook ... Blumberg and I felt we had a mission in America, and the response to our efforts seemed to support us in this. We had, indeed, 'started the ball rolling', and for at least twenty years Logical Positivism was one of the major subjects of discussion, dispute, and controversy in United States philosophy. [Feigl, 1968, p. 646f.].

In his essay on one of the most influential philosophical movements in the field of philosophy of science coming from Central Europe to the USA Feigl reconstructed the intellectual and institutional trajectory of the Vienna Circle from a personal and professional perspective in what could best be described as a sort of philosophical "oral history". Starting with the origins and development in Vienna, Feigl describes the early contacts with American philosophers (Schlick 1929/1931 in Stanford and Berkeley) and the beginnings of the Vienna Circle's migration from the 1930s onwards. His own contacts with Dickinson Miller and Charles A. Strong (a son-in-law of John D. Rockefeller) enabled the Schlick student and gifted young philosopher of science to embark upon his brilliant academic career in Harvard with a Rockefeller Fellowship.

Because Carnap, Frank and most of the other members had to emigrate to the USA, we still lack a complementary account — a sort of "*Wiener Kreis* in Great Britain". Moreover, the most influential history of this transfer and intellectual transformation came from Alfred J. Ayer, who had attended the Vienna Circle in 1932–33, with his publications on the history and influence of the Viennese philosophy, especially with his booklet *Language, Truth and Logic* [1936a] — a publication that was influential into the postwar period. This was reinforced by Ayer's "The Vienna Circle" (in *The Revolution in Philosophy*, 1956) and his textbook volume on *Logical Positivism* [1959]. I do not want to deal with all the factors influencing the intellectual migration and cultural transformation of the Vienna Circle to Great Britain but only provide some significant material in order to criticize what I will call the *standard view* of "Logical Positivism" in England.

This widespread position has been challenged over the last decade in studies in the history and philosophy of science but we still lack a critical reconstruction analogous to the better researched topic of *Origins of Logical Empiricism* [Giere and Richardson, 1996] and *Logical Empiricism in North America* [Hardcastle and Richardson, 2003].

This *standard view* is determined on the one hand by Ayer's role as most important mediator and interpreter, and, on the other hand, by the extensive research on Ludwig Wittgenstein's impact on English analytic philosophy before and after World War II. The essence of this traditional historiographical account is that via Logical Positivism of the Vienna Circle (partly with Popper) postwar philosophy of science in Great Britain was directly influenced, whereas analytic philosophy, especially "ordinary language philosophy", was mostly motivated by Wittgenstein's late philosophy of the *Philosophische Untersuchungen*/*Philosophical Investigations* (published posthumously in 1953).

In the following account I want to show — in an admittedly cursory fashion — that

1. this traditional image of Logical Empiricism has shortcomings and is highly selective.

2. this distinction between two different currents (Vienna Circle vs. Wittgenstein) is rather artificial.

3. there was a flourishing communication among the dominant figures I have already mentioned, which seems to me at least an important correction of the usual history of reception.

From a *biographical* point of view this means that the players in this complex intellectual history have to be extended on both sides. Accordingly, we are dealing with a typical example of networking in the period from the 1930s to the 1960s with regard to "The *Wiener Kreis* in Great Britain". My intention is to show that, parallel to the American story there was another interconnected development and that it is futile to argue that "Continental", "British" and "American" branches all existed as separate movements.

2.2 Intersections and Interventions: Philosophy of Science and Analytic Philosophy between Central Europe and Great Britain

The process of interaction between the Continental and the British tradition in analytic/scientific philosophy did not suddenly begin in the 1930s: The key figure is without doubt Bertrand Russell, together with Einstein and Wittgenstein — "the leading representatives of the scientific world-conception" mentioned in the Vienna Circle's manifesto in 1929. Besides this context his early dispute with Alexius Meinong [1905] and his lifelong conflict-ridden involvement with the life and work of Ludwig Wittgenstein are well known, even if strongly contested in its interpretation.

It also seems plausible, that Russell was one of the most important partners for "Austrian philosophy" from the turn of the century until his American days, if we reconstruct this coincidence by overcoming the dominant stories on his relations to the Vienna Circle — and Wittgenstein, too (especially with the reception of his book *Our Knowledge of the External World*, [1914]).

I am not referring to the prehistory of the bilateral scientific-philosophical reflection as becomes manifest in the correspondence of Ernst Mach with Pearson, Whewell or the reception of John Stuart Mill with Theodor and Heinrich Gomperz ([Stadler, 2001, Ch. 3] and with reference to the US: [Holton, 1993]). This intellectual impact can be illustrated more precisely by the internationalization of Logical Empiricism in Europe and America, especially with a focus on the six "International Congresses for the Unity of Science" 1934–1941 organized in Paris (twice), Copenhagen, Cambridge (in England at Girton College), Harvard and Chicago:

At the latest in Paris in 1935 we encounter the first significant presentation of Logical Empiricism in an international context with an increasing overlap into the Anglo-Saxon world [Stadler, 2001, 363–371]. Russell gave a widely acclaimed opening address in which he presented "scientific philosophy" as a synthesis of logic and empiricism [Russell, 1936]. And for the American delegates Charles Morris had stressed the international cooperation of scientists, which indicated the growing Austro-American relations. But what about the British connection?

Already in January 1935 Neurath had organized an informal meeting on Logical Positivism in London (held at Belsize Park) with A. J. Ayer, G. E. Moore, Max Black, and Carl G. Hempel, which resulted in a still rather skeptical statement on common points in Vienna and Cambridge [Stebbing, 1935].

Afterwards in Paris, the young Ayer delivered a paper on "The Analytic Movement in Contemporary Philosophy" [1936b] referring to the analogous anti-metaphysical movements in Vienna and England since the turn of the century. He expressed his hope for a stronger interpenetration of science and philosophy — as opposed to the often deplored "Scientism" as "Infatuation of Philosophy with Science" [Sorell, 1991].

Otto Neurath gave a very positive account on the Paris congress, conveying his "impression that there was in fact something like a scholar's republic of logical empiricism" [Neurath, 1936, 377]. This, by the way, seems to be congruent with what was happening at the time: Robert Musil tried to get an invitation for the congress, the "Frankfurt School" sent Walter Benjamin, and Bert Brecht expressed his interest in cooperating with Neurath. The meeting of American pragmatists (Charles Morris), the English analytic philosophers (Susan Stebbing), the Polish logicians (Kasimir Ajdukiewicz) and of Italian scientific philosophers (Federigo Enriques) illustrated this tendency towards a unification of empiricism and rationalism, but also the international setting of this ambitious project.

A year later Russell [1936, 10f.], who had delivered his laudatio of Gottlob Frege in German (!), wrote, that "The congress of Scientific Philosophy in Paris in September 1935, was a remarkable occasion, and, for lovers of rationality a very encouraging one ...". Russell's review perfectly illustrates the international context of the already exiled "Vienna School", and particularly its rational-empiricist interpretation in the spirit of Galileo and Leibniz [ibid., 11]:

> Modern science arose from the marriage of mathematics and empiricism; three centuries later, the same union is giving birth to a second child, scientific philosophy, which is perhaps destined to as great a career. For it alone can provide the intellectual temper in which it is possible to find a cure for the diseases of the modern world.

Whether scientific philosophy was able to fulfill Russell's hopes is not for us to decide. Still his statement remains an impressive document of an atmosphere of optimism and awakening — one of the last ones before the war in Europe amidst a constantly growing tendency towards totalitarian "(final) solutions".

A large international committee for the International Congresses for the Unity of Science was formed, including the English members C. K. Ogden, Bertrand Russell, Susan Stebbing and Joseph H. Woodger. All of them will play a significant role in the convergence and divergence of Logical Empiricism and British philosophy (of science). The case of Woodger as biologist of the Unity of Science movement and the "Theoretical Biology Club" in Oxford is one of the issues which deserves further investigation.

The "Fourth International Congress for Unity of Science" on the main topic "Scientific Language" in Cambridge (UK, Girton College) was the last European meeting of the community in scientific philosophy, which took place in the framework of a larger Enyclopedia-oriented program some months after Austria's occupation by Nazi-Germany.

It also documents the high level of the dialogue between British and Central European proponents of scientific and analytic philosophy. At the same time it also provided a forum for constituting the international committee for the forthcoming congresses and the organizational committee for the *International Encyclopedia of Unified Science* [Carnap *et al.*, 1938ff].

In his inaugural address G.E. Moore focused on the historical reference point of Cambridge philosophy, i.e., Russell's and Whitehead's *Principia Mathematica,* but surprisingly without mentioning Wittgenstein, who was not present at the congress. Oxford Philosophy was represented by Gilbert Ryle, who discussed the practical and theoretical reasons for the "disunity of sciences".

Finally, one of the most important figures of this dialogue, namely Susan Stebbing (1885–1943), host and initiator of the congress, spoke about "Language and Misleading Questions", apparently in the spirit of Wittgenstein, but with a remarkable preference for Carnap's alternative:

> Since the conference is meeting in Cambridge and since its topic is 'Scientific Language', it seems to me not inappropriate to take for this inaugural address 'Language and Misleading Questions'. For it is, perhaps, to Wittgenstein more than to any other philosopher that the conception of philosophy as 'the critique of language' is due. His influence has, so I understand, now permeated Cambridge students of philosophy that to the outsider all their discussions appear to be concerned with investigation of language ... I have learnt even more from studying Carnap's writings. I have felt the attraction of the view that: 'an die Stelle des unentwirrbaren Problemgemenges, das man Philosophie nennt, tritt die Wissenschaftslogik.' [Stebbing, 1939–40, p. 1]

Despite this professed affiliation with the "Logic of Science" Stebbing concludes her paper with reference to Heinrich Hertz's *Principles of Mechanics*, which contains a linguistic critique of (metaphysical) questions and answers, with another (early) Wittgensteinian thought:

> "We want an answer to a question we have not asked. Our minds cease to be vexed when we find that the question is illegitimate; we no longer seek for an answer for there is no longer a question to be asked" [ibid., 6].

As can be seen from the congress report, however, the program focused on logical-analytical questions, with many special contributions on "scientific language". Inter alia, Otto Neurath who later should have to flee from the Netherlands (The Hague) to England in 1940, postulated many small scientific units as a logical starting point for the development of a future unified science, once again directing polemical attacks against one privileged "system" — as preference of 'Encyclopedism' vs. hierarchical , Pyramidism' [Neurath, 1983, esp. chapters 8–23].

Among the printed contributions of this congress the paper by the British-American philosopher Max Black on the "Relations between Logical Positivism and the Cambridge School of Analysis" is of special interest, because it offers a profound discussion from a British point of view of what Wittgenstein, the Vienna Circle and the Cambridge School have in common and what separates them (cf. [Skorupski, 1993]):

> ... the development of the analytical movement in England and of Logical Positivism are found to have much in common. They have had, roughly speaking, the same friends and the same enemies. The teachings of Wittgenstein, Russell, Moore and the earlier English empiricists have been among the most important formative influences of both. If Logical Positivists have proclaimed their attachment to the advance of science more loudly, the English movement, (...), has to some extent been permeated with the same values. There should be room for further fruitful interchange of opinions between the two movements. [Black, 1939/40, 33f.]

With this argument the translator of Frege and Carnap once more described the background for the relationships between the German-speaking and the Anglo-Saxon world in the philosophical field. But this convergence was interrupted for several years by the War, which became manifest in 1941, when the European participants of the "Sixth Congress for the Unity in Science" at the University of Chicago who had already registered met instead for a small conference on "Terminology" on October 3–5, 1941 at Linton Road in Oxford. This event was once again organized by Neurath who was living in exile, together with J. A. Lauwerys and Susan Stebbing, shortly after his release from the internment at the Isle of Man.

We do not possess proceedings, but Neurath's publications from this period on "Universal Jargon and Terminology" [1941] and "The Danger of Careless Terminology" (1941) seem to indicate the motivation and orientation of this joint activity – which ended significantly with another variation of his famous boat metaphor [Uebel, 1992; 2000]. This contribution was partly evoked by the critical remarks of Bertrand Russell in his William James Lectures *An Inquiry into Meaning and Truth* [1940], where the author writes in the preface: "I am, as regards method, more in sympathy with the logical positivists than with any other existing school". In order to avoid so-called 'ontological misinterpretations' as a consequence of a correspondence theory of truth Neurath accordingly directs his proposals towards the history and sociology of sciences (American neo-pragmatism), because "no judge is in the air who says which of us has the TRUTH" [Neurath, 1983, p. 229]. By the way, this nonfoundationalist attitude corresponds to his radical criticism of Schlick and Popper, thereby rejecting verificationism *and* falsificationism as absolute philosophies of science [Neurath, 1935].

In 1941 the Chicago congress united Americans and European emigrants as well as "Contributors from Europe" whose papers were presented in the absence of their authors — among them Friedrich Waismann and Martin Strauss.

From 1938 on we can note a renewed convergence of the *International Encyclopedia of Unified Science* (with Neurath, Morris and Carnap as main editors) and the transformation of the journal *Erkenntnis* [1930ff.], (ed. by Carnap and Reichenbach) into an English edition as *Journal of Unified Science* with the eighth and last volume. Since 1933 it had come under increasing pressure after the Nazis seized power. The first volume of the *International Encyclopedia*, with contributions by Neurath, Niels Bohr, John Dewey, Carnap and Russell again marked the beginning of the (uncompleted) project with the University of Chicago Press.

Here Russell wrote "On the Importance of Logical Form" [1938, p. 41] based on the instrument of mathematical logic for pure mathematics and the empirical sciences with the (somewhat Popperian) conclusion that

> "the unity of science ... is essentially a unity of method, and the method is one upon which modern logic throws much new light. It may be hoped that the Encyclopedia will do much to bring about an awareness of this unity".

When we look at the authors of the *Foundations of the Unity of Science* and also consider the members of the "Advisory Committee", we can detect a strong UK/US-Austrian dominance. A first analysis of the contents of the 8 volumes of the *Erkenntnis* shows a similar development as the six Congresses for the Unity of Science: among the English-language speaking authors of the journal from 1934 on — apart from the printed congress contributions — we find, for instance, Ayer, Black, Stebbing. The reviews on these publications reflect the increasing reception of the philosophy of science. From the first issue on, works by Ayer, Church, Lewis, Nagel, Quine and Woodger were presented within the context of what was still Central European "Wissenschaftslogik", the Logic of Science according to Carnap.

2.3 Transfer and Transformation: Circles and Networks Continued

Apart from these rather well known developments of the *Wiener Kreis* in the Anglo-Saxon world we can reconstruct another story of international relations in philosophy of science between the wars, which is also representative for the scientific communication within Europe and between Europe and America. Let me illustrate this phenomenon by describing these features with the intellectual networking before and after the forced migration:

It is worth noting that it was, above all, Moritz Schlick (1882–1936) (married to an American citizen Blanche Hardy) the founder and head of the Vienna Circle, who fostered early intellectual contacts with the English-speaking world: he visited England at least twice in the late 1920s as can be shown by his correspondence with Frank P. Ramsey [1927/28], who stood in close personal contact with both Wittgenstein and Schlick. Ramsey, who invited Schlick to the famous "Moral Sciences Club", discussed his controversy with Wittgenstein on the Philosophy of Mathematics:

> "I had a letter the other day from Mr Wittgenstein criticising my paper 'The Foundations of Mathematics' and suggesting that I should answer not to him but to you. I should perhaps explain what you may have gathered from him, that last time we didn't part on very friendly terms, at least I thought he was very annoyed with me (for reasons not connected with logic), so that I did not even venture to send him a copy of my paper. I now hope very much that I have exaggerated this, and that he may perhaps be willing to discuss various questions about which I should like to consult him. But from the tone of his letter and the fact that he gave no address I am inclined to doubt it." (Ramsey to Schlick, July 22, 1927).

And in one of his last letters before his premature death he reports to Schlick on Cambridge philosophy:

> "It is a great thing for us to have Wittgenstein here, he is such a great stimulus and has been doing most excellent work, quite destroying my notions on the Foundations of Mathematics. Apart from that I think the school of philosophy here is severing a little; there are more and better pupils, and a distinct improvement from the very low level we were at when you visited us." (Ramsey to Schlick, probably Dec.1929).

The *Journal of Philosophy* which has existed since the turn of the century dealt directly and indirectly, especially in the inter-war years, with the work of John Dewey and functioned in America as a moderate forum for the development from scientific philosophy to philosophy of science, as becomes clear in the contributions of Ernest Nagel, Willard Van Orman Quine, Carl G. Hempel or Nelson Goodman. This trend may be illustrated e.g., with Nagel's informative 'Impressions and Appraisals of Analytic Philosophy' (1936). His reports from Cambridge, Vienna, Prague, Warsaw and Lwow of the early thirties is a document of the advanced stage of internationalization in Europe and between Europe and America. Thus he correctly observes that in the 'Wiener Kreis' "significant shifts in positions taken have been made by some of its members..." [Nagel, 1936, 216ff.]. And the leftist American student of philosophy concludes,

> "in the first place, the men with whom I have talked are impatient with philosophic systems built in the traditionally grand manner. Their preoccupation is philosophy as analysis ... The intellectual temper cultivated by these men is that of ethical and political neutrality within the domain of philosophic analysis proper, however much they may be moved by the moral and social chaos which threatens to swallow the few extant intellectual oases they stand.

> In the second place, as a consequence of this conception of the task of philosophy, concern with formulating the method of philosophic analysis dominates all these places...without 'dogmatism and intellectual intolerance'... In the third place, students whose primary interest is the history of ideas will find that, with some important exceptions, they will profit little from talking of these men ... In the fourth place, what pertains to a common doctrine, the men to whom I refer subscribe to a common-sense naturalism".

Nagel who attended Schlick's lectures commented as follows on the sociological background of Vienna,

> "that although I was in a city foundering economically, at a time when social reaction was in the saddle, the views presented so persuasively from the Katheder were a potent intellectual explosive. I wondered how much longer such doctrines would be tolerated in Vienna...
>
> Analytic philosophy has thus a double function: it provides quiet green pastures for intellectual analysis, wherein its practitioners can find refuge from a troubled world...; and it is also a keen, shining sword helping to dispel irrational beliefs and to make evident the structure of ideas... it aims to make as clear as possible what it is we really know." (Ibid.).

And with special focus on the Cambridge philosophy around Moore and Wittgenstein, he admits the significance of the latter, "in spite of the esoteric atmosphere which surrounds" him.

Let me return to the transfer and transformation of Central European philosophy of science to the Anglo-Saxon world, where Great Britain has been featured rather unjustified primarily as a transition country. It is true that Carnap, through his contacts with Charles Morris from 1934 on, gradually found entry into American universities, but it is important to note that his books had already been translated and read in England before then.

On the invitation of Susan Stebbing, Carnap came from Prague to London where he delivered three lectures at the University of London in October 1934. Here he came into contact with Russell, Ogden, Woodger, Braithwaite – and, significantly, with the young philosophy student Max Black. The latter wrote his PhD. thesis on "The Theories of Logical Positivism" under the influence of Moore and Ramsey.

In his introduction Black dealt with the origins of the Viennese Circle, its relations to Wittgenstein, and the central semantic notion of meaning. This looks like an anticipation of Ayer's best-selling book in 1936 on Logical Positivism: *Language, Truth and Logic.* And in Carnap's introductory notes we read already in a clear and distinct diction that "we are not a philosophical school and that we put forward no philosophical theses whatsoever ... for we pursue Logical Analysis, but no Philosophy" [Carnap, 1934, pp. 21 and 29].

The second booklet in this series *Philosophy and Logical Syntax*, with the content of the three mentioned lectures was the first popularization of Carnap's *Logical Syntax* period since the beginning of the 1930s in Great Britain:

> "My endeavour in these pages is to explain the main features of the method of philosophising which we, the Vienna Circle, use, and, by using try to develop further. It is the method of logical analysis of science, or more precisely, of the syntactical analysis of scientific language. Only the method itself is here directly dealt with; our special views, resulting from its use, appear rather in the form of examples (for instance our empiricist and anti-metaphysical position in the first chapter, our physicalist position in the last)." [Carnap, 1935, p. 7]

Max Black, born 1909 in Baku (Russia), had studied mathematics and philosophy in Cambridge, Göttingen and London, and published the book *The Nature of Mathematics* in 1933 (advised by Moore and Ramsey). Later on he emigrated to the United States in 1940 where he began his career, first at the University of Illinois and as of 1946 as Professor of Philosophy at Cornell University. He became a leading figure in (British-American) analytic philosophy and was an important mediator between Logical Empiricism and the British tradition in philosophy of science – apart from Ayer's popularization and idiosyncratic interpretation of the Vienna Circle from 1936 on. (cf. the co-authored book with Ernst Gombrich and Julian Hochberg on *Art, Perception and Reality* in 1972). This function can be detected in his article "Relations between Logical Positivism and the Cambridge School of Analysis" [1939/40]. In describing on the common ground between analytical and common sense philosophy in England associated with Moore and Russell in the Vienna tradition, Black explicitly refers to Wittgenstein's influence in the 1930s:

> "During the last eight years Wittgenstein's influence upon younger English philosophers has been comparable with that exerted by Morris. In this the Tractatus has played less part than his lectures ... and oral discussions based upon Wittgenstein's later and more radical views." [Black, 1938/39, p. 32]

And he notes that this influence could be more closely linked with Schlick and Waismann than with Carnap and Neurath. This corroborates the thesis that in England analytic philosophy (with the subfield of ordinary language philosophy) were better able to gain acceptance than the philosophy of science related to the *International Encyclopedia* project. This development is confirmed by Friedrich Waismann's work after his immigration to the UK in 1937. His influence was, compared to Wittgenstein, unspectacular but continuous, if one takes into account the publication of his oeuvre. (cf. Waismann's Ayer-critical *Principles of Linguistic Philosophy*, 1965).

Notwithstanding all differences, Black underlined the convergence of both movements at that time, still hoping that there would be further productive cooperation:

"... There should be room for further fruitful interchange of opinions between the two movements." [Black, 1938/39, 34].

On the basis of this description it is not surprising that we can reconstruct a bilateral (and intercontinental) exchange of ideas also on the level of institutions and periodicals: The name of a philosophical journal published since 1933 in Oxford (with Basil Blackwell) indicates the program as such: *Analysis,* ed. by A. E. Duncan-Jones with the cooperation of Susan Stebbing and G. Ryle, issued six times a year, was founded under the influence of Moore, Russell, and Wittgenstein, followed by an "Analysis Society" in 1936. Besides Alfred Ayer and Max Black, also Carnap, Hempel and Schlick contributed to the early issues. After the war (the journal was suspended 1940–1947) we find amongst the authors, e.g., Friedrich Waismann (with 6 articles on 'Analytic-Synthetic'), and Karl Popper (on the Mind-Body-Problem).

Incidentally, together with another international project, the still existing journal *Synthese* 1936ff. published in the Netherlands and the already mentioned *Philosophy of Science,* in addition to the 1940 established *Philosophy and Phenomenological Research,* formed an extended international forum for the communication between the poles of (language-critical) analytic philosophy and philosophy of science.

In the context of these activities Black — although always maintaining a critical distance (a "friendly critic" in his own words) — paved the way for greater receptivity of the Vienna Circle — influenced and inspired mainly by the already mentioned ("Lizzy") Susan Stebbing: she contributed significantly to the intellectual acculturation of Logical Empiricism in Great Britain, but because of her early death in 1943 she regrettably fell into oblivion. She studied at Girton College, the University of London, before her teaching period at King's College in London (1913–1915), Bedford College (1915–1920), University of London (1920–1924), where she became the first woman professor of philosophy in Great Britain (1933–1943). While her colleagues remember her as being a passionate teacher, her philosophical writings document a highly profound knowledge of the empiricist and analytical tradition of Continental and English thinking: as president of the "Aristotelian Society" in 1933 she reinforced her presence on an institutional level, since she was also acquainted with Russell, Moore and Whitehead. As a supporter of Carnap, Neurath and Popper, and given her friendly relations with Wittgenstein she played the role of a go-between and mover and shaker in analytic philosophy of her time, as became clear in the philosophical lecture she gave to the British Academy on *Logical Positivism and Analysis* [1933]. In this lecture, she investigated the language-critical approach of the Vienna Circle and the *Tractatus* and compared it with the English empiricist tradition (from Russell, Moore to Ramsey). She argued that while all philosophy is concerned with language, she was rather skeptical that all philosophical problems are linguistic ones. Neurath welcomed one of her last lectures on "Men and the Moral Principles" [1944, 18f.] as follows in his notes in the complimentary copy, where Stebbing states:

> "Moral philosophy, I repeat, is not a science", but "Whatever maybe the case with politicians making weekend speeches in time of war, philosophers cannot afford to ignore the conditions of the problems set by the situations in which we live."

This appeal is consonant with the defense of democracy during World War II in her last book *Ideals and Illusions* [1941], which obviously impressed again the exiled Neurath in Oxford besides their philosophical affinity.

Only on the basis of this scientific communication and relationships can we fully appreciate the specific contribution of Alfred Jules Ayer as *the* chief interpreter and protagonist of the Vienna Circle and Wittgenstein I in England, esp. with his book *Language, Truth and Logic* (1936a). His primarily anti-metaphysical position already became manifest with his appearance at the Paris Congress of 1935, where he refers in his paper on "The Analytic Movement in Contemporary Philosophy" [1936b] to the analogous movement in Vienna and England since the turn of the century. The success of his book also influenced all other networks: still in 1955, the 11th imprinting of the second, enlarged (and critically revised) edition of this bestseller appeared. The 8 chapters addressed the following issues: I. The Elimination of Metaphysics, II. The (new) Function of Philosophy, III. The Nature of Philosophical Analysis, VI. The A Priori, V. Truth and Probability, VI. Critique of Ethics and Theology, VII. The Self and the Common World, and VIII. Solutions of Outstanding Philosophical Disputes.

In the Preface to the first edition we read that "*the views which are put forward in this treatise derive from doctrines of Bertrand Russell and Wittgenstein, which are themselves the logical outcome of the empiricism of Berkeley and Hume ...* " employing a modified verification principle for empirical hypotheses [Ayer, 1955, p. 31]. And he goes on to contextualize these assertions to the effect that, philosophizing is an activity of analysis which is associated in England with the work of G. E. Moore and his disciples, but, 'the philosophers whom I am in the closest agreement are those who compose the 'Vienna Circle' ... and of these I owe most to Rudolf Carnap.' [ibid., p. 32]. Additionally, Ayer not only expressed his indebtness to Gilbert Ryle and Isaiah Berlin, but also alluded to philosophical differences. Ultimately, he says: "*we must recognise that it is necessary for a philosopher to become a scientist, ..., if he is to make any substantial contribution towards the growth of human knowledge.*" [ibid., p. 153].

Was this really a part of *The Revolution in Philosophy* [1956], as Ayer later maintained in a collection on Bradley, Frege, Moore, Russell, Wittgenstein and the Analysis of the Ordinary Language Philosophy with reference to the Vienna Circle? Although his judgment is apparently ambivalent (e.g., he saw the Vienna Circle as a movement, as a thing of the past, to a certain extent, but on the contrary, he declared many of its ideas to be living on). Ayer remained a critic of the late Wittgenstein — and, by the way, of Popper too.

In summary, we can say that there was a lively culture of scholarly dialogue of Central European and English philosophers — with a stronger focus on analysis, as compared to the turn from "Wissenschaftslogik" to philosophy of science in the USA. But there were also mutual contacts since about 1900, which cannot be separated from what has been referred to as the Anglo-Saxon "sea change" [Hughes, 1975] proper. What we have here is a dynamic network on different levels (like personal contacts, publications, societies, conferences and institutions) with distinct convergences and divergences of ideas and theories. Moreover, it is a network that reflected the intellectual preoccupation with several philosophical and methodological disputes between thinkers from different countries: from the Austro-German *Methodenstreit*, the Positivism disputes to the foundational debates in mathematics and logic since the 1920s. But the style and form of theorizing changed under different social conditions and a new intellectual setting in the immigration countries and triggered a self-organizing set of innovation and academic exchange.

This hypothesis could be exemplified by case studies on the Bloomsbury Group, Wittgenstein's Cambridge and Neurath's Oxford — and, last but not least, Hayek's and Popper's London, which I want to briefly describe in the subsequent passages.

2.4 On the Neurath Connection (Oxford)

A striking example for such overlapping networking is the Neurath connection in England: The main promoter and proponent of the Unity of Science movement from 1934 on since his exile in the Netherlands re-established in the few years 1940-45 in England good old contacts with a remarkable intellectual and practical manifestation — and one is inclined to ask in the sense of counterfactual history: What would have been, if Neurath had survived the World War II period?

First of all, in Oxford he initiated Central European disputes on plan vs. market or socialism vs. liberalism (with Hayek) and philosophical relativism vs. absolutism (with Popper) again in the new context of the envisioned liberated postwar society. Therefore their relationships, significantly emerged already in the Viennese years, can be described as more or less conflict-ridden communication between family resemblance and distance. Given the limited space, I will only allude to Neurath vs. Popper before turning to Hayek and Neurath (cf. [Stadler, 2001, chapters 10.3–10.5]):

Besides sharing a rejection of Platonic social philosophy (*Republic*), seen as a legitimation for authoritarian and totalitarian ideas, including the *Führerkult* (which according to Hayek's Plato interpretation was also part of a specific English controversy) there was the controversial encounter of both personalities that began in the early twenties. Neurath immediately criticized Popper after the publication of his *Logik der Forschung* (1934) accusing him of being an advocate of an absolutist "pseudorationalism" [Neurath, 1935] — as he, incidentally, also rejected the verification endorsed by the 'Wittgenstein camp'. The options of 'unity of science' or 'unity of method' (which was also rejected by Hayek) appeared as main alternative. But there are also uncontested familiarities: it is Popper's appeal to planning for institutions in his exile publications on the *Open Society* (1945) and

Poverty of Historicism (1944/45), which highlight the differences with Hayek — marginalized by Popper given Hayek's total opposition towards any form of planning theory and practice.

As regards the relation of Neurath and Hayek, we can detect a truly unbridgeable gap in political and economical matters, but a remarkable convergence in methodology and even in epistemology.

The adamant Scottish liberal Hayek, although in earlier times fond of Mach's epistemology and Schlick's *General Theory of Knowledge* (1918/1925), distanced himself – together with Ludwig von Mises and Karl Popper — from the so called "positivist economics", esp. from all variations of planned economy (in kind) from the 1920s on. All of this opposition that had originated in Vienna culminated in a short dispute in the 1940s in the common English exile (although I admit Hayek coming to London in 1931 — as Wittgenstein returning 1929 to Cambridge – was not a typical emigrant). Hayek's articles on "Scientism and the Study of Society" (1942–1944) in *Economica* and the subsequent publication of his *Road to Serfdom* (1944) was the starting point for a renewed *Methodenstreit*: two cultures (natural science vs. the humanities) were the main options in scientific enterprise: Neurath published a remarkably moderate review of *Road to Serfdom*, by showing that Logical Empiricism was providing a "through and through" pluralist view towards "Planning for Freedom" [Neurath, 1942]. But Neurath's personal annotations in his own copy of *Road to Serfdom* are much more critical: "*His technique: Overstate a case, create car(r)ricature of it, then fight it and then kill it is either German or immoral etc.*"

The Central European social reformer is fighting against the extreme liberal economist. Nevertheless it is remarkable how these two different former Austrian intellectuals entered into a controversy abroad — with the common background of the tension between liberalism (as laissez-faire capitalism) and socialism (social democratic position). In short: it is the option between a liberalist international like the "Mont Pelerin Society" or the so-called "third way" between communism and capitalism which Popper preferred because of his Viennese progressive social liberalism [Hacohen, 2000]. Although Neurath tried to initiate a continuous discussion with Hayek between Oxford and Cambridge (where the London School of Economics had its wartime address), the latter refused to enter into details, even if agreeing "entirely with what you say on Plato. He is certainly the arch-totalitarian." (Hayek to Neurath, Febr. 2, 1945). Hayek was busy lecturing abroad and only moved back to London later. He was convinced that he had already dealt exhaustively with the issue of "Scientism". Two weeks after Hayek's last hesitant letter, Neurath died unexpectedly of a heart attack on Dec. 22, 1945: the dialogue between the adherents of plan and market shimmered through in the ensuing Hayek-Popper communication.

Mainstream historiography obscures the difference of these "ambivalent brothers in mind": first, Poppers's insistence on the unity of method for natural and social sciences, second, his preference for a limited planning for institutions, third, his adherence to a socially oriented welfare economy in the tradition of Austrian social reform.

Probably because of his personal indebtedness to, and acquaintance with, Hayek who essentially mediated Popper's engagement at LSE, Popper himself played down his differences with Hayek's social philosophy in New Zealand. This conclusion can be drawn on the basis of the published comments on Hayek's "Scientism". And although both directly/indirectly argue against Neurath under the (Cold War) labels of "objectivism", "collectivism" and "historicism", there can be no doubt about the shortcomings of the equation "scientism = historicism". In this field we lack further studies on the renewed interaction and relation of the Austrian School and Vienna Circle after their emigration, taking into account the controversial issues in "Red Vienna" between the wars.

But there is another hidden story to be revealed: it is the life and work of Otto Neurath and Marie Reidemeister (Neurath), who in their second exile systematically continued scientific relations during the 1930s. The unpublished memories of Marie illustrate their untiring efforts for the cause of the Encyclopedic movement and enlightened adult education via his *Isotype*-movement (*I*nternational *S*ystem *o*f *Ty*pographic *P*icture *E*ducation), and also his initiatives to continue the housing and settlement movement in his Vienna days. Neurath renewed all contacts (with Friedrich Waismann, Rose Rand, Friedrich Hayek, Susan Stebbing) from the continent and the US, and for a short time pursued his academic ambitions with significant publications and research projects that remained uncompleted. (Nemeth/Stadler 1996). A further look at his exile research library (which is now deposited at the Vienna Circle Institute in Vienna), once again established in England 1940ff., signals the continuity to develop the interwar plans including his experience with the fascist decade: books on 'International Planning for Freedom', 'Visual Education' and on 'Persecution and Toleration' were again on the agenda. And Hitler-Germany (and postwar German) education focusing on Plato and Kant — a topic which continued to preoccupy all participants. We realize Neurath's cooperation in a newly founded Fabian Society, production of Isotype-films with Paul Rotha for the anti-Nazi education and *Isotype* for visualization of public affairs as a contribution to the fight against totalitarianism.

Neurath's lectures at Oxford University (where also Ernst Cassirer and Friedrich Waismann taught in 1941) are forgotten, but his work on visual education and on modern social and economic museums has been further developed and integrated in the standards of today's work: at the University of Reading (Department of Graphic Communication and Typography) and partly at the British Natural History Museum.

2.5 The Unended Poker Story: Wittgenstein vs. Popper in Cambridge

Hayek serves as a link to this much more investigated and contested research topic because it was mainly "Wittgensteinians" who provided us with an alternative account of the 'genius' and his impact in Cambridge philosophy [Monk, 1990;

Hacker, 1996].

Hayek, a distant relative (cousin) to Wittgenstein, wrote a to date unpublished "Biographical Sketch" based on Russell's letters to Wittgenstein. (According to Hayek Wittgenstein's heirs refused any publication). Here I only want to indicate the essentials of my interpretation of 'Wittgenstein's Cambridge':

First, I do not share the inclusion of Wittgenstein in the traditional forced migration movement — as presented by H. Stuart Hughes (1975) focusing on the philosophical prologue with the transformation of Wittgenstein in Vienna into Wittgenstein in Cambridge on the basis of British idiosyncracies. Wittgenstein was not a typical emigrant of the interwar period, although he never would have attained any adequate academic position in Austria for philosophical and so-called "racial" reasons.

Second, even in England Wittgenstein appeared more as an apolitical and ahistorical philosopher (cf. [Sluga, 1999]) than an immigrated intellectual. He refused to become involved in Austrian exile organizations (as is reported by Engelbert Broda and Joseph Peter Stern).

Third, the philosophical importance of Friedrich Waismann independent from the old Viennese connection (Schlick, Wittgenstein, Waismann) should be made clear. This may be confirmed by the fact that Wittgenstein did not continue in England his former fruitful communication with Waismann, and refused to reestablish contact with his former partner and interpreter to the Vienna Circle. While Karl Popper's remembrances of this tragic relation are probably somewhat exaggerated, they indicate a typical feature of Wittgenstein's life and work.

Fourth, Wittgenstein was not a solitary thinker, even in England; as in his Vienna days he was once again active in intellectual circles and networks.

This brings me to another postwar Austro-English success story together with a philosophical hoax which also has its roots in the mid-thirties: The young Karl Popper was already welcomed around the Vienna Circle as an enrichment of the theoretical discussion — even if criticized by Neurath as 'official opposition' to Logical Empiricism. Like so many other figures of the movement, and for the same reasons, Popper had no chance of getting an adequate position at an Austrian university.

Therefore, he looked for a position abroad, preferably in the Anglo-Saxon world. Fortunately, once again, Susan Stebbing came to the rescue: on her invitation Popper delivered two lectures at Bedford College in 1935 on Alfred Tarski, thereby — in his own words — triggering Joseph Henry Woodger's interest in the Polish logician. In 1935/36 papers on probability followed this first presentation at Imperial College and further ones in Cambridge (with Moore) and Oxford (with Ayer, Ryle and Berlin). The decisive appearance was no doubt his talk on "The Poverty of Historicism" in Hayek's seminar at the LSE (inter alia with Ernst Gombrich who became his life-long friend and supporter). (By the way, in Oxford he also met the Austrian physicist and Nobel Laureate Erwin Schrödinger). During one session of the "Aristotelian Society" in 1936, to which Popper had been invited by Ayer, Russell spoke on "The Limits of Empiricism" on the basis of an inductivist approach. He appealed to a principle of induction, followed, as could be expected, by a controversial discussion with the adamant English neo-empiricists. (This is expressed again in Ayer's critique of Popper's falsificationism, to be found in his 1982 *Philosophy in the Twentieth Century*). After these contacts Woodger proposed that Popper apply for a lectureship in New Zealand, whereas Hayek via the "Academic Assistance Council", envisioned a position at LSE. As is well known, Popper decided for New Zealand, and recommended Friedrich Waismann for England. Regrettably, in his autobiography, Popper forgot the decisive positive role of Felix Kaufmann in establishing his English connections [Popper, 1974].

After Popper's second return to London when he accepted a position at LSE in 1946, with the help, in particular, of Hayek and Gombrich, he delivered a lecture on a "philosophical puzzle", organized by the "Moral Sciences Club" in Cambridge. Wittgenstein was present at this lecture. It was true that Popper had criticized Wittgenstein's concept of philosophy as the action of clearing up statements, claiming that all philosophical problems are essentially linguistic ones in his writings before 1945. Thus, it cannot be a surprise that his lecture with the rhetorical title "Are there philosophical problems?" provoked Wittgenstein to storm out of the room angrily, after having threatened Popper with a poker. This is only one side of the unended "poker story", reported primarily by the *agent provocateur* himself [Edmonds and Eidinow, 2001].

Let me close the thematic circle: what is striking here is, first, the fact that this topic is an old, classic Viennese one, and second, this stage two of the dispute was contextualized with the English philosophical peculiarities and thus transformed, which is confirmed in Popper's own triumphant words in relation to another lecture in Oxford — with a striking reference to World War II, and the Cold War discourse:

> One of the things which in those days I found difficult to understand was the tendency of English philosophers to flirt with nonrealistic epistemologies: phenomenalism, positivism, Berkeleyan or Machian idealism ('neutral monism'), sensationalism, pragmatism — these playthings of philosophers were in those days still more popular than realism. After a cruel war lasting for six years this attitude was surprising, and I admit that I felt that it was a bit 'out of date' (to use a historicist phrase). Thus, being invited in 1946–47 to read a paper in Oxford, I read one under the title 'A Refutation of Phenomenalism, Positivism, Idealism, and Subjectivism'. In the discussion, the defence of the views which I had attacked was so feeble that it made little impression. However, the fruits of this victory (if any) were gathered by the philosophers of ordinary language, since language philosophy soon came to support common sense. [Popper, 1974, 99f]

As is well known, Popper attacked Wittgenstein for his alleged immoral behaviour towards his former adherent and close partner Waismann — a story which has to be investigated in full length elsewhere. Notwithstanding all these oral histories of the communication in exile and with Popper finally settling in England, these controversies between Neurath, Hayek, Popper and Wittgenstein with a Viennese background of the classical Vienna Circle seems to me highly significant for the dynamic (in content and form) of intellectual migration — which was only partly determined by the unique external events of fascism, Nazism and the Holocaust.

If this is true, we can obtain a broader, subtler understanding of this period; which could also shed some light on the present state of intellectual life in Europe.

2.6 Frank P. Ramsey: Between Wittgenstein and the Vienna Circle

Frank Plumpton Ramsey (born February 22, 1903 in Cambridge, England, died in London on the 19th of January 1930) was certainly one of the most important and promising philosophers of the 20th century. Only his early and unexpected death at the age of 27 probably prevented him from becoming one of the leading figures in the philosophy of science and analytic philosophy — perhaps at par with Ludwig Wittgenstein, his lifelong close friend and also intellectual adversary.

It is well known that in his short life Ramsey immensely enriched philosophy and science with some profound and highly topical findings: the gifted student at Trinity College, Fellow at King's College and Lecturer at Cambridge University at least influenced Wittgenstein, Russell and Keynes as well as the Vienna Circle with his contributions on the foundations of mathematics, logic, and economics. Especially his significance for philosophy with its focus on the notions of truth, decision theory, belief and probability is worth mentioning. The intellectual context of Ramsey's thinking can also be illustrated with the famous Bloomsbury Group [Hintikka and Puhl, 1995].

Especially the period Ramsey spent in Vienna in 1924 and his contacts with the mathematician Hans Hahn, the physicist Felix Ehrenhaft, inter alia, draws attention to Ramsey's connection with the early Vienna Circle [Stadler, 2001]. Already in 1929, Ramsey was listed in the manifesto of the Vienna Circle and given credit for attempting to further develop Russell's logicism and cited as an author related to the Vienna Circle. There are references to his articles on "Universals" (1925), "Foundations of Mathematics" (1926), and "Facts and Propositions" (1927). And the proceedings of the "First Meeting on the Epistemology of the Exact Sciences in Prague" (September 15–17, 1929), mention Ramsey as one of the "authors closely associated with the speakers and discussions", together with Albert Einstein, Kurt Gödel, Eino Kaila, Viktor Kraft, Karl Menger, Kurt Reidemeister, Bertrand Russell, Moritz Schlick and Ludwig Wittgenstein. (*Erkenntnis* 1/1930–31, pp. 311, 329). But looking at the earlier communication of Ramsey with Wittgenstein and the Vienna Circle these references are not really surprising: whereas it is rather well known that Ramsey visited Wittgenstein in 1923 and 1924, his communication with Schlick and his probable participation in the Schlick Circle have not been fully appreciated.

Carnap's notes on the discussion in the Schlick Circle include Ramsey's definition of identity, the foundations of mathematics and probability. With reference to

July 7, 1927 we can read: "Discussion by Carnap and Hahn about Carnap's arithmetic and Wittgenstein's objection to Ramsey's definition of identity" [Stadler, 2001, pp. 238f]. Accordingly, Carnap reported on an earlier discussion (June 20, 1927) in the Wittgenstein group with Schlick and Waismann, in which the great "genius" (= Wittgenstein) also objected to Ramsey's notion of identity. Precisely this issue was on the agenda again 4 years later when Wittgenstein met Schlick and Waismann alone. (December 9, 1931, Ibid., p.441). His lifelong dealings with Ramsey is documented later on in Carnap's *Philosophical Foundation of Physics* (1966) with its special focus on the Ramsey sentence.

Another reference is worth mentioning here: commenting retrospectively on his article "The Role of Uncertainty in Economics" (1934) the mathematician Karl Menger, member of the Vienna Circle and the founder of the famous "Mathematical Colloquium", recognised the relevance of Ramsey's paper "Truth and Probability" (1931) — unknown to him at the time — for his own research, although distancing his own contribution from this study [Menger, 1979, p. 260]:

> "But the von Neumann-Morgenstern axioms as well as Ramsey's were based on the traditional concept of mathematical expectation and on the assumption that a chance which offers a higher mathematical expectation is always preferred to one for which the mathematical expectation is smaller. My study was not".

In connection with his stay in Vienna, there is another fact of Ramsey's life that merits attention: he underwent a (supposedly successful) psychoanalytic therapy with the lay psychoanalyst and historian of literature *Theodor Reik* (1888–1969), who, by the way, also gave him a book by the theoretical physicist Hans Thirring.

Ramsey, who invited Schlick to the "Moral Sciences Club" in Cambridge, discussed his personal controversy with Wittgenstein, which was triggered by his article "The Foundation of Mathematics" (1925). His description is also confirmed by Wittgenstein's critical and ambivalent comments on Ramsey in his *Diaries* (April 26, 1930) [Wittgenstein, 1999, S.20f].

These contacts continued, and in one of his last letters before his death, Ramsey reported to Schlick on Wittgensteins's impact on his own philosophy (namely in the sense that it "quite destroyed my notions on the Foundations of Mathematics") as well as on Cambridge philosophy in general. (Ramsey to Schlick, December 10, year not indicated).

It is no coincidence that Black many years later described Ramsey in the *Encyclopedia of Philosophy* (ed. by Paul Edwards), in the following terms, as

> "one of the most brilliant men of his generation; his highly original papers on the foundation of mathematics, the nature of scientific theory, probability, and epistemology are still widely studied. He also wrote two studies in economics, the second of which was described by J.M. Keynes as 'one of the most remarkable contributions to mathematical economics ever made'. Ramsey's earlier work led to radical

criticisms of A.N. Whitehead and Bertrand Russell's Principia Mathematica, some of which were incorporated in the second edition of the Principia. Ramsey was one of the first to expound the early teachings of Wittgenstein, by whom he was greatly influenced. In his last papers he was moving toward a modified and sophisticated pragmatism." [Black, 1939/40, Vol.7/8, p.65].

2.7 Between Unity and Disunity of Science: Family Resemblance and Distance with Otto Neurath, Friedrich A. von Hayek, and Karl Popper

Between the two World Wars, Austria was the center for two major scientific movements with international influence and recognition: on the one hand the Vienna Circle of Logical Empiricism in philosophy and methodology of science, and the Austrian School of Economics in social sciences, on the other. Although these two renowned traditions have of course been studied both historically and systematically, esp. over the last decade, we still lack research on both together, on their similarities and differences, mutual influences and interaction in the course of the development of science.

Thus the question arises (even it is only rhetorical) if we can view *The Philosophy of the Austrian School* [Cubeddu, 1993] or (the Philosophy) of the Vienna Circle [Stadler, 2001] as homogenous fields of research. Given the complexity and dynamic character of these so called "schools" it is necessary to examine their common ground, their similarities and differences with regard to socio-cultural background, theoretical development and methodological orientation.

At first sight they both are a manifestation of a typical "delayed" enlightenment. They share the fate of marginalization as a result of clerico-conservative, later on fascist and national-socialist repression. Conceived as an intellectual network it seems legitimate to describe both developments as interrelated scientific phenomena. Furthermore to unearth new aspects on the mutual interaction and influences between the two groups, which can be mirrored as overlapping intellectual circles.

Even if the opposition of Ludwig v. Mises and Friedrich August v. Hayek to Otto Neurath regarding the dualism of planned vs. market economy is well known, there remain many aspects of subtle common and differing features, whereas, on the other hand, the Hayek-Popper exchange of ideas cannot be characterized as an example of theoretical agreement. To show this, I will focus on the central methodological notion of "scientism". There was no consensus on this notion, as opposed to the critical view of Plato shared by these three thinkers.

The positions of mediating figures such as Felix Kaufmann and Richard von Mises show that this conflict was an immensely more complex debate. Furthermore this is documented by the strong influences on mathematical economy, esp. game and decision theory by the young Karl Menger's "Mathematisches Kolloquium" — even if we admit that this branch of the Austrian School only figured rather peripherally at that time.

The Vienna Circle undoubtedly contributed to the foundations of mathematics and logic and played a prominent role in the rise of John von Neumann's and Oskar Morgenstern's expanding economies and game theory, Abraham Wald's equilibrium theory, culminating in the formation of econometrics with Carnap's adherent Gerhard Tintner, who later applied Carnap's probability and logic.

A list of the publications relevant to (methodology of) economics and social science and the implicit/explicit treatment of the value problem in connection with the is/ought relation would already reveal the most neglected issues of research — apart from some recent studies (by Cubeddu [1993] and Robert Leonard [1998] on the "Menger-connection". In the research on the Vienna Circle we can detect only in the last few years guiding publications [Köhler and Leinfellner, 1997]). It is here that we can find the key concepts for a comparative study: normativity vs. descriptivity, theory vs. experience, reason and action, explanation vs. intuition (*Erklären* vs. *Verstehen*), foundations of natural and social sciences with psychology vs. *Logic of Scientific Discovery* (K. Popper). All these topics are based on an evolutionary and/or probabilistic approach.

In Karl Popper's sense we can state that the two main problems of epistemology, namely induction/deduction and the delineation of science from nonscience or metaphysics, form the heuristic and theoretical background for the theories of rationality and action addressed — as are also addressed in Schlick's "On the Foundation of Knowledge" [1934]. It's not a coincidence that still two decades later Fritz Machlup — as well as Morgenstern — was still reflecting on "The Problem of Verification in Economics" [1955] with reference to Felix Kaufmann's decision making "rules of procedure" in his *Methodology of Social Sciences* [1944].

Much less surprising is the revival of Hayekian cognitive science following the publication of *The Sensory Order* [1952], which has its roots in his early strong reception of Mach and Schlick, and which was (not officially) criticized by Popper because of its causal theory of mind [Birner, 1998]. This topic can also be seen as a sort of variation of the theme that is the all-embracing "Methodenstreit" since the beginning of the 20th century. And there remains the main question if there is a general "Theory of Valuation" [Dewey, 1939] that bridges the gap between theoretical concepts and empirical science or normative and descriptive aspects of human action. This would amount to the validity of the crucial meta-theoretical position called "methodological individualism" and part of the "Duhem-Neurath-Quine-thesis". The methodological tension of holism and individualism (with its inherent deontic logic) has overshadowed all substantial discussions in philosophy of science in general and specifically in the social sciences.

Since we can relate scientific world conceptions to all these methodological positions it also seems legitimate to reconstruct the background of *Weltanschaungen* for the (manifest) *Methodenstreit* in order to explore all the ideas underlying the so-called "progressive liberalism" (as attributed to Popper by [Hacohen, 1997]) and (Austro-Marxist) socialism very much inspired by the social reform concepts of Popper-Lynkeus combined with Mach's epistemology.

One misconception of Logical Empiricism to be destroyed is that of the disregard of ethics and value statements in the Vienna Circle. Thus Wolfgang Stegmüller, advancing Quine's influential "Two Dogmas of Empiricism" [1951] — namely the absolute distinction of analytic and synthetic statements and the empirical reduction of theoretical concepts — was critical of the replacement of ethics by meta-ethics (meta-ethical noncognitivism) as a consequence of the meaning criterion. On the contrary, the late Carnap himself (together with Richard Jeffrey) furthered the development "From Logical Empiricism to Radical Probabilism" [Jeffrey, 1993] through his probabilistic foundation of decision theory. But even before World War II, notwithstanding the verification or falsification criteria, different conceptions of ethical discourses were discussed — even if admittedly not at the center of Logical Empiricism. One may refer to the relevant publications of Felix Kaufmann — which also included purely economical writings — and in addition Viktor Kraft, Karl Menger, Richard von Mises, Otto Neurath, Josef Schächter, Moritz Schlick and Friedrich Waismann. Schlick for instance (*Problems of Ethics*, 1930) did consider ethics/aesthetics as philosophical and scientific sub-disciplines, presenting a naturalist ethics of imperative values in combination with an empirical psychological description of moral behavior and the meta-ethical analysis of concepts and statements. Viktor Kraft's *Foundations for a Scientific Analysis of Value* (first published in German 1937) also presented ethics as a scientific discipline, e.g. by analyzing value concepts normatively and factually, allowing value judgments to enter into relations of logical inference among each other and together with factual statements. Schlick's book — by the way — was an incentive for Karl Menger's *Moral, Wille, Weltgestaltung/ Morality, Decision and Social Organization* (1934). The central idea was the view that moral attitudes are based simply on decisions. Menger applied logico-mathematical thinking to human relations and associations resulting from diverse and even incompatible attitudes by completely avoiding evaluations. In this sense Menger deployed formal decision theory and a game-theoretic logic of groups — a kind of "socio-logic" — against the then predominant (Neo-Kantian) ethics of value and duty. This led to an empiristically "externalised ethics of decision". Menger's meta-theoretical "Principle of Tolerance" regarding the use of logics and scientific languages was one more cornerstone in the application of Logical Empiricism in the modern social sciences — far away from the dogmatic de-historization of the *Received View* of scientific theories between *Positivismusstreit* and analytic philosophy of science. This fits very well with Neurath's theory of social science best characterized by his boat metaphor directed against absolutist and dualist epistemologies — popularized in the Anglo-Saxon world by W. V. O. Quine:

> "Imagine sailors who, far out at sea, transform the shape of their clumsy vessel from a more circular to a more fishlike one. They make use of some drifting timber, besides the timber of the old structure, to modify the skeleton and the hull of their vessel. But they cannot put the ship in dock in order to start from scratch. During their work they stay on the old structure and deal with heavy gales and thundering

> waves. In transforming their ship they take care that dangerous leakages do not occur. A new ship grows out of the old one, step by step — and while they are building, the sailors may already be thinking of a new structure, and they will not always agree with one another. The whole business will go on in a way we cannot even anticipate today. That is our fate." [Neurath, 1939, 47].

2.8 Neurath and Popper: Relativism vs. Absolutism

Popper was aware of Neurath's life and work after World War I, as he himself reported in his autobiographical remarks [1974]. He remembered Neurath's involvement in the Bavarian revolution [1919/20] in connection with a planned economy based on full socialization and with reference to the semi-socialization program of Josef Popper-Lynkeus. Popper was inclined to sympathize with the latter's (utopian) project. Apart from differences in personality and mentality, on the one hand, the Marxist dissenter and politically oriented encyclopedist, and the critical rationalist philosopher on the other, Popper accused Neurath of having succumbed to utopianism, historicism and scientism as represented by the Vienna Circle and the *Ernst Mach Society.*

In his own words, Popper was very flattered that Neurath published his criticism as "Pseudorationalism of Falsification" [1935] and was not unpleased ("nicht unzufrieden") with this honorable attack, but surprisingly he never replied in a systematic way. Maybe because of Neurath's critique (in light of his methodological holism): "the absolutism of falsification ...is in many ways a counterpart against the absolutism of verification which Popper attacks". Popper's attempt to characterize Neurath's *Empirische Soziologie* (1931) in the context of historical prophecy fails to assess the author's foundation of social science. From the outset Neurath remained very skeptical of explanations on the basis of one method and one image of science without pragmatically relativizing the field of "Prediction and Induction" (1946): "Unity of Science" as represented in the ambitious project of the *International Encyclopedia of Unified Science* or "Unity of Method" — by the way contrary to Hayek — as explicated from Popper's *Logic of Scientific Discovery* to the *Open Society* and the *Poverty of Historicism* seemed to be alternative approaches in the history and philosophy of science. In Popper's own words:

> "Neurath and I had disagreed deeply on many and important matters, historical, political, and philosophical; in fact on almost all matters which interested us both except one — the view that the theory of knowledge was important for an understanding of history and political problems." [Popper, 1974, p. 56]

The direct confrontation of both opponents in their still unpublished correspondence shows a high level discussion in philosophy of science. At the same time one might wonder whether Popper did exaggerate the real differences between him and the so-called "positivists" [ter Hark, 2004] — a designation which Neurath so strongly opposed as a cliché — and underestimate any form of scientific cooperation between the new "encyclopedists". In this connection it is, indeed, surprising that also Critical Rationalism can be — counter-intuitively — a suitable tool for a planning methodology as Andreas Faludi tried to show in his book [1986]. And it is this aspect — namely Popper's appeal to planning for freedom or institutions (in his *Open Society* and *Poverty of Historicism*) — which highlights the differences marginalized by Popper due to Hayek's total opposition towards each form of planning theory and practice.

2.9 Neurath and Hayek: The Unbridgeable Gap (in Economics)

Hayek felt himself from the beginning of the 1920s on to be Neurath's opponent regarding economics: inspired by the Carl Menger's "conception of the spontaneous generation of institutions", by Ludwig von Mises' *Gemeinwirtschaft* and by Popper's anti-inductivist *Logik der Forschung* [1935] he lobbied against the so called "positivist economics" — with Neurath (1919: *Through War Economy to Economy in Kind = Durch die Kriegswirtschaft zur Naturalwirtschaft*) as the most suitable target.

Formerly, Hayek was impressed by Mach and Moritz Schlick, before he then began to oppose the Vienna Circle because its social science was dominated by Neurath. The intellectual divorce centered around planning theory, economy in kind, and generally speaking on the concept of value, which Hayek — erroneously — missed in Neurath's social science. All these indirect oppositions culminated in a short dispute in the 1940s in English exile. This conflict sheds more light on the alternative conceptions of social science and its methodology:

Neurath took the initiative, as of 1945, in communicating by correspondence:

> "Enclosing I am sending you a review of your book. I tried to discover, what we have in common - unfortunately you are rather "absolute" in your EITHER-OR attitude. On Plato you may find some remarks in the article enclosed." (11. Jan., 1945).

He defended Logical Empiricism as "through and through" PLURALIST — whereas accusing Plato of a totalitarian practice DIRECTLY — but did not really impress the reserved Hayek. Although we can reread Neurath's review of *Road to Serfdom* in *The London Quaterly of World Affairs* [1945, 121f] as a surprisingly moderate assessment of this anti-totalitarian pamphlet, it did at the same time offer a sophisticated justification of a special sort of "Planning for Freedom" [Neurath, 1942]. The latter proposed several possible solutions with a skeptical empiricist approach to find a third way between market economics and fascism — with happiness and prosperity as guiding notions.

Much more instructive seem to be Neurath's notations in Hayek's book:

> "There is some danger that planning as a fashion [may] be used by totalitarian groups for weakening the democratic behavior, which implies — muddle. Democracy — muddle — and victory. But that is not

> the muddle of slums, distressed wars, depressions etc. but multiplicity of decisions, freedom of societies, local authorities. The fascists try to discredit muddle and to praise order, unification, subordination as such, otherwise they cannot run the show!!
>
> Therefore we need an analysis of planning with reckoning in kind plus muddle!
>
> That lacks — therefore danger."

Scientifically more relevant is the controversy over the concept of "scientism" between Hayek and Neurath, which later also included Popper. There, Hayek dealt extensively with the applicability of the methods of natural science and "social engineering" to the problems of man and society (as directed against Mannheim, Neurath and maybe also in some sense against Popper).

In his "Scientism and the Study of Society" [1942–1944] — re-published in *The Counter-revolution in Science* [1952] — Hayek condemned the appraisal of natural science methodology as the only "scientific method". This does not hold, because "facts" in the natural and social sciences are totally different: on the one hand causally explicable, on the other they are mere unobservable "opinions" of the actors producing their "objects". Common sense via analogy is the central key for understanding in social science. These essays distanced Hayek from all forms of "objectivism", "behaviorism", directly referring to Neurath's "physicalism", accusing him of supporting *in natura* calculation (instead of calculation in terms of price and value) and taking "naively for granted that what appears alike to us will also appear alike to other people" (p.35). What surprises us at first sight is the omission of a critique of language and the merging of the theoretical and meta-theoretical levels of speaking about the external world. Despite of all these misunderstandings regarding "methods of science", Neurath was ultimately willing to agree with Hayek's conclusion, quoting Morris R. Cohen that *"the great lesson of humility which science teaches us, that we can never be omnipotent or omniscient, is the same as that of all great religions: man is not and never will be the god before whom he must bow down."* (p.39).

Although Neurath tried to start a discussion in which he referred to theoretical contributions on social science, Hayek refused to enter into a detailed discussion. From Cambridge (where LSE had its wartime address) to Oxford, Hayek wrote that he was *"by no means so much opposed to 'Logical Positivism' as you appear to think and with some members of your former group, particularly with Karl Popper, I find myself in complete agreement"* and — alluding to physicalism and *in natura* calculation — he continued to articulate his skeptical position towards Neurath, at the same time agreeing *"entirely with what you say on Plato. He certainly was the arch-totalitarian"*. (Hayek to Neurath, February 2, 1945).

2.10 Popper and Hayek: The Ambivalent Brothers in Mind

The deductive-hypothetical methodology was essentially directed against inductivism and/or apriorism, which were characteristic for the *Philosophy of the Austrian School* [Cubbeddu, 1993]. This move was promoted by Neurath, although there was no evidence of holism, collectivism or inductivism with any kind of prophecy in social science method. By the same token, we can reconstruct in the Vienna Circle's social philosophy with Kaufmann, Menger, Neurath and Richard von Mises (cautious) conceptions of empiricist methodologies with conventionalist tools. But these issues have not been investigated sufficiently so that we lack studies on the Hayek-Popper interaction. Already the critical review of Hayek's *Counter-Revolution* by Ernest Nagel (*Journal of Philosophy* 1952) should have been given greater recognition and should have provoked further discussions. (By the way, it is noteworthy that a structurally similar criticism has been raised by the Frankfurt School in the context of the "Positivismusstreit" from the thirties to the sixties — but this is a completely different story (cf. [Dahms, 1994; Uebel, 2000]).

It can be assumed that the motivational background of the scientism-controversy lies in the topicality of socialist planning theory between the wars (cf. Hayek's edition of *Collectivist Planning* in 1935). For insiders therefore the account of Richard von Mises seems adequate and more representative when he remarked in his book *Positivism* [1951] already finished in 1939 — a well-balanced re-evaluation of the Vienna Circle story from Mach to the high tide of Logical Empiricism — in his chapter on social sciences that

> "... neither the practical impossibility of experiments in the narrower sense nor the comparatively limited application of meta-mathematical methods is a specific feature of this field." (p.246)

Relating to classical economics he alludes to the shortcomings of terminology implying "eternal laws". In agreeing with Felix Kaufmann's *Methodenlehre der Sozialwissenschaften* [1936], Richard v. Mises acknowledged promising starting points for a rational treatment of economic problems in the theory of marginal utility, specifically in von Neumann's "economic games". Neurath's *Empirische Soziologie* [1931] — an alternative to the "polemic of neoliberalism against collectivist theories of economics" of Ludwig v. Mises' *Human Action* [1949] and Hayek's *Collectivist Economic Planning* [1935] — is presented by R. v. Mises as a sociology, which can not be separated from history. This account is in accordance with the prominent role of the Karl Menger jr. and his "Mathematical Colloquium" for the social sciences in interwar Vienna.

Already Morgenstern's article "Logistik und die Sozialwissenschaften" (*Zeitschrift für Nationalökonomie*, 1936) directed attention to the rich potential of the "new logic" (Karl Menger and Kurt Gödel) or "Logistics" (Russell and Whitehead) for economic research. There, Morgenstern explicitly endorsed the theory of types (Russell), axiomatics (Hilbert) as well as the use of an exact scientific language, of a so-called *Wissenschaftslogik* (logic of science) in Carnap's sense. He concludes his article by referring to the relevance of these methods for the social sciences as well, esp. for theoretical economics and political economy. To this end, he summed up the main ideas of Karl Menger's book *Morality, Decision and Social Organization* [1934]. Thus it is not surprising that we find Morgenstern's later collaborator John von Neumann participating at the Vienna Circle congress in Königsberg [1930] and in Karl Menger's "Mathematical Colloquium" in the twenties and thirties. Here we find some intellectual roots (issues such as experience and rationality, chance and determinism) for today's decision and game theories — extending to John Harsanyi's work on social theory as well as ethics.

The Vienna Circle's contribution to the foundation of probability calculation and theory (with Richard von Mises, but also Carnap's later inductive logic) and subsequent controversies between Karl Popper and Hans Reichenbach also inspired Abraham Wald's mathematical support of Richard von Mises' concept of probability. The latter also furthered mathematical economics by improving the equations of Walras and Cassel. Together with Morgenstern's first input-output model [1937] and von Neumann's equilibrium theory for expanding economies, the relevance of the "Mathematical Colloquium" is documented [Dierker and Sigmund, 1998]. This is also due to the fact that Menger himself was concerned with methodology of social science and "The Role of Uncertainty in Economics" as one of his articles in 1934 is entitled. And we fully agree with a contextualized analysis of Menger's significance for interwar social science [Leonard, 1998/99].

We should also add that Popper's participation in this context of mathematical economics (Hahn, Menger, Morgenstern, Tarski) is much stronger than in the paradigm of social science after 1945 as endorsed by Hayek, who still in 1937 ("Economics and Knowledge") was strongly interested in the mathematical foundations of social science. By the way, the evolutionary view of science can be traced back to the work of Mach and Boltzmann, but also to Popper's later position of *Objective Knowledge. An Evolutionary Approach* [1972].

All these influences are indirectly related to the above-mentioned dispute on "scientism", in which the central question was raised as to whether philosophy or philosophical foundation is needed for the methodology of the humanities and social sciences. Or alluding to the subtitle of Sorell's book *Scientism* [1991], one might ask whether scientism is the regrettable "infatuation of philosophy with science." At least it conveys the options of one, two or more scientific cultures with the inherent utopia of an unified methodology and theory or a science of man and nature.

For a better understanding of the lasting *Methodenstreit* in the 1930s and 1940s represented by the triangle Hayek-Popper-Neurath we first have to reconstruct the discussions in their socio-historical context, second to examine unpublished sources, third to distance ourselves from clichés about schools of thought and, finally to confront these results with today's research. In doing so we could fully appreciate the historical background together with internal theory dynamics without producing myths of partisanship. This would provide a rational option, namely a pluralist way for positioning this unsolved debate in an evolutionary context of theoretical fields.

3 VIENNA, PARIS AND THE "FRENCH CONNECTION": CONVENTIONALISM

In the last decade, a number of new publications on the transatlantic exchange of ideas have been presented by international scholars studying the Vienna Circle in the English-speaking countries.[1] Now it is high time to focus on the neglected "French Connection" in the philosophy of science.[2] That this connection has been somewhat neglected until now is all the more remarkable since we have known about the existence of close ties between Viennese and the Parisian intellectuals since the Fin de Siècle. Their exchange of ideas — studied by Ernst Mach — is most evident in the strong reception of Henri Poincaré and Pierre Duhem in the so-called "First Vienna Circle". This bilateral development in the Enlightenment discourse of the modern theory of science was already described by Philipp Frank in his book *Modern Science and its Philosophy* [1949], in which he underlined the threefold roots of logical empiricism in a more modern guise, making reference to English empiricism, French rationalism and American (neo-)pragmatism.[3]

What then evolved was, more specifically, a synthesis of empiricism and symbolic logic, with the Machian theory of science being refined by French conventionalism, along with an attempt to counter Lenin's critique of "empirico-criticism". Here Abel Rey's book *La théorie physique chez les physiciens contemporains* [1907] also sought to overcome mechanistic physics. It was ultimately Poincaré who tried to mediate between empirical description and analytic axiomatics of scientific terminology[4]:

According to Mach, the general principles of science are abbreviated economic descriptions of observed facts. For Poincaré, they are free creations of the human mind which say absolutely nothing about observed facts. The attempt to integrate both concepts in a coherent system was the origin of what was later to become known as logical empiricism.

This goal was attained with the help of Hilbert's axiomatics of geometry as a conventionalist system of "implicit definitions". This way Mach's philosophy could be integrated in the "new positivism" espoused by Henri Poincaré, Abel Rey and Pierre Duhem. The link between the new positivism and the old teachings of Kant and Comte consists in the demand that all abstract expressions of science — such as power, energy, mass — be interpreted as sense observations.[5]

As early as 1907, Pierre Duhem wrote the following in *La théorie physique, son objet et sa structure* (Aim and Structure of Physics), striking a similar chord as Mach :

---

[1]Cf. the most recent publications [Giere and Richardson, 1996; Hardcastle and Richardson, 2004.

[2]Anastasios Brenner, "The French Connection Conventionalism and the Vienna Circle", in: Michael Heidelberger / Friedrich Stadler (eds.), *History of Philosophy of Science. New Trends and Perspectives.* Dordrecht-Boston-London: Kluwer 2002, pp. 277–286.

[3]Philipp Frank, [1949].

[4]Frank, "Der historische Hintergrund", in: *ibid.,* p. 256.

[5]Frank, Ibid., p. 258f.

> "A physical theory is not an explanation. It is a system of mathematical propositions which can be derived from a small number of principles that serve to precisely depict a coherent group of experimental laws in a both simple and complete way."

This is followed by a crucial insight for the encyclopedia project: "The *experimentum crucis* is impossible in physics".[6] In spite of Duhem's metaphysical leanings his teachings became a framework of reference for further discussions between science and religion and, more generally, between science and ideologies. Reflecting further reciprocal influences, Frank compared Louis Rougier's publications with those of Moritz Schlick:

> "He proceeded from Poincaré, trying to incorporate Einstein in the 'new positivism' and wrote the best comprehensive critique of school philosophy ... the 'paralogisms of rationalism'." [Paris: Alcan 1920][7]

The physicist Marcel Boll translated writings by Rudolf Carnap, Hans Reichenbach, Moritz Schlick and Philipp Frank into French. The original influence exerted by Duhem was now to be reversed:

> "The French general Vouillemin (cf. C.E. Vouillemin, La logique de la science et l'Ecole de Vienne (Paris: Hermann 1935) recommended our group since we replaced the spelling "Science" by the more modest "science". ... The French neo-Thomists ... saw in logical positivism the destroyer of idealist and materialist metaphysics which for them were the most dangerous enemies of Thomism. To organize this international cooperation, a preliminary conference was held in Prague in 1934, in which Charles Morris and L. Rougier participated. The cornerstone was thus laid for the annual international congresses on the "Unity of Science."[8]

social science

With the heyday of the Vienna Circle in the interwar period, these European and transatlantic exchanges were increasingly consolidated while at the same time intellectual life in Germany and Austria had been disintegrating since 1930. Most significantly, however, direct recourse was taken both theoretically and practically to the French Encyclopédie of the 18th century in connection with the *International Encyclopedia of Unified Science* of the Logical Empiricists. Here it was mainly Otto Neurath who untiringly drew attention to the French intellectual precursors of his Unity of Science movement and was able to effectively implement this intellectual exchange before the outbreak of World War 2 at two international congresses in Paris (1935 and 1937) as late-enlightenment collective projects.[9] This comes as no

---

[6]Ibid., p. 259.

[7]Ibid., p. 291.

[8]Ibid., p. 291f.

[9]Stadler, *The Vienna Circle. Studies in the Origins, Development, and Influence of Logical Empiricism.* Vienna–New York: Springer 2001, pp. 363ff. and 377ff.

real surprise in view of the references to Comte, Poincaré and Duhem as precursors of the "scientific world view" in the programmatic manifesto of the Vienna Circle [1929].[10]

The "1st Congress for the Unity of Science in Paris in 1935", the "Congrès International de Philosophie Scientifique" held at the Sorbonne September 16-21, marked the first highpoint of the new philosophy of science of the Vienna Circle in exile. Already in late 1933 Neurath had conducted preliminary negotiations in Paris with Marcel Boll and Louis Rougier, still as representatives of the "Verein Ernst Mach" which disbanded in Vienna in 1934. These talks were then continued at a preliminary conference in Prague in 1934. Neurath's summary of the Paris conference reads like an extremely optimistic prognosis of the future of the "Republic of Scholars of the Logical Empiricism" and the "philosophique scientifique".

The first of the international congresses for the unity of science... was a success for logical empiricism vis-à-vis a larger public. The title "philosophie scientifique", which is so popular in France, aroused interest. The press constantly reported on the congress. Newspapers and journals dealt with it in sketches and interviews. This was all the more remarkable in view of the fact that, as Rougier and Russell had underlined in their introductory words, it was a conference whose task was to focus on a science without emotions. Some 170 persons from more than twenty countries had appeared and had shown a high degree of willingness to commit themselves to continuous cooperation. With their addresses at the opening of the congress in the rooms of the Institute for Intellectual Cooperation Rougier, Russell, Enriques, Frank, Reichenbach, Ajdukiewicz, Morris generated a living impression that there was such a thing as a republic of scholars of logical empiricism.[11]

The French institutions which co-organized the congress included the "L'institut International de Coopération Intellectuelle", the "Comité d'Organisation de L'Encyclopédie Francaise", the "Cité des Sciences", the "Institut d'Histoire des Sciences et des Techniques" as well as the "Centre International de Synthèse". The congress was documented in eight journals in the series "Actualités scientifiques et industrielles" published by the Parisian publisher Hermann & co (1936) with a number of French contributions. Bertrand Russell, who gave an appraisal of Frege in German in his opening address, remembered in retrospect a manifestation of rational-empirical thought in the tradition of Leibniz.[12] "The Congress of Scientific Philosophy in Paris in September 1935, was a remarkable occasion, and, for lovers of rationality a very encouraging one..." Neurath seconded this, arguing that "the individual sciences (should be) arranged next to each other by directly showing concrete relations and not indirectly by referring all of them to a common blurred conceptual system."[13]

The Congress unanimously committed itself to supporting the project of the

[10] *Wissenschaftliche Weltauffassung. Der Wiener Kreis.* Ed. Verein Ernst Mach. Vienna: Artur Wolf Verlag 1929. Reprint in: Fischer (ed.), loc.cit., p. 125 – 171.

[11] *Erkenntnis* 5, 1935, p. 377.

[12] Bertrand Russell, in: *Actes du Congrès International de Philosophie Scientifique.* Sorbonne, Paris 1935. Paris : Hermmann & co. 1936, p. 10.

[13] *Erkenntnis* 5, 1935, p. 381.

Encyclopedia of Unified Science, which had been organized by the Mundaneum Institute run by Neurath in The Hague. The committee of 37 included the French scholars Marcel Boll, H. Bonnet, E. Cartan, Maurice Frechet, J. Hadamard, P. Janet, A. Lalande, P. Langevin, C. Nicolle, Perrin, A. Rey and L. Rougier.

This event, which was also viewed as a testimony of the anti-fascist intellectuals, as can be seen in the interest shown by Robert Musil, Walter Benjamin and Bert Brecht, ultimately formed the basis for a stronger transcontinental development towards the cooperation of the German-, English- and French-speaking community of scholars who were primarily promoted by Neurath from his Dutch exile.[14]

The second round of the encyclopaedic renaissance was planned at the end of July 1937 to also take place in Paris as the "Third International Congress for the Unity of Science", after the organisational committee (Carnap, Frank, Joergensen, Morris, Neurath, Rougier) succeeded in obtaining a publisher's contract with the University of Chicago Press for the first two volumes of the "International Encyclopedia of Unified Science" (IEUS).

Moreover, a separate section on the "Unity of Science" (L'Unité de la Science: la Méthode et les méthodes) was organised in connection with the contemporaneous "Ninth International Congress of Philosophy" (Neuvième Congrès International de Philosophie — Congrès Descartes). Notwithstanding the theoretical differences on the conception of the "New Encyclopedia" between Carnap and Neurath (in particular on the notion of truth and probability), Neurath presented modern empiricism there as a type of heuristic puzzle aiming at a "mosaic of the sciences" in the following way[15]:

> "We can start out from the 'encyclopedia' as our model, and now observe how much we can achieve by a way of interconnection and logical construction and elimination of contradictions and unclarities. The synopsis of logical empiricism will be then the order of the day."

The main goal was thus to show "in addition to the existing great encyclopedias the logical framework of modern science"[16] with the construction of a sort of onion around a core consisting of 2 volumes with 20 introductory monographs, bringing forth a further 260 monographs, of which only 19 monographs were to appear in total due to the war."[17]

The complete realization of this project would have yielded 26 volumes with 260 monographs in English and French, supplemented by a ten-volume picture statistical "visual thesaurus" with global overviews in the spirit of Diderot and d'Alembert. The influence of history and sociology on the philosophy of science

[14]Cf. also Antonia Soulez. "The Vienna Circle in France", in: Friedrich Stadler (ed.), *Scientific Philosophy: Origins and Developments.* Dordrecht-Boston-London: Kluwer 1993, pp. 95 -112.

[15]Otto Neurath, "The New encyclopedia', in *Unified Science*, ed by Brian McGuinness. Dordrecht: Reidel, 1987, p. 136f.

[16]Ibid.

[17]Otto Neurath / Rudolf Carnap / Charles Morris (eds.), *Foundations of the Unity of Science. Toward an International Encyclopedia of Unified Science.* 2 vols. Chicago and London: The University of Chicago Press, 1971.

was also a sort of "science in context" meant to circumvent a strongly formalistic "scientism".

This encyclopedism was not intended to provide an absolute foundation of epistemology or "system" of sciences (neither with verification nor falsification as methodical instruments) but was to be more based on a broad everyday experience as the point of departure against the backdrop of uncertainty and indeterminacy, namely with the

> "Basic idea that one does not have any solid foundation, any system, that one has to keep trying on the basis of research and can experience the most unexpected surprises on later verification of many basic views that are used, is characteristic of the outlook that might be described as "encyclopedism"... As empiricists we will always proceed from our everyday expressions, and as empiricists we will use them again and again to verify our theories and hypotheses. These broad propositions with their many indeterminacies are the point of departure and the final point of our science."[18]

Now the question arises as to why it came to the rupture of this productive Austro-French cooperation and why this exchange was forgotten after this relatively successful history. Here I can only touch upon some of the reasons:

1. World War 2 destroyed a Central European late-Enlightenment culture of science, in particular in "Red Vienna".[19]
2. The ideologization in the wake of the second positivism debate (Horkheimer vs. Neurath) and the central focus of the project after 1945 being Louis Rougier who was a controversial figure in France prevented a reintegration in the community of scholars.[20]
3. Emigration, exile and the transfer of science to the Anglo-American world of scholars and the prevention of the return of ideas, reinforced by the third positivism debate within the context of the Cold War and the dominant *Dialectic of Enlightenment* (Horkheimer/Adorno) as well as the Marxist and the structuralist philosophers reinforced this rupture after 1938.
4. The preference for "German philosophy" of idealism and existentialism of the post-war years with a cliché of "positivism" and the belated research of the buried tradition of the philosophy of science since the turn of the century in France contributed to the rupture of a flourishing bilateral communication.
5. The integration of the 2nd Republic of Austria in the intellectual life of the West with a focus on the Anglo-American intellectual world additionally marginalized the "French Connection" after World War 2.

---

[18] Neurath, loc. cit., p. 213.

[19] *Wien und der Wiener Kreis. Orte einer unvollendeten Moderne. Ein Begleitbuch.* Ed. Volker Thurm-Nemeth and Elisabeth Nemeth, Vienna: WUV-Verlag 2003.

[20] Cf. Hans-Joachim Dahms, *Positivismusstreit.* Frankfurt/M.: Suhrkamp 1994.

An attempt has been made to help reduce the lacunae in the research and to study the common intellectual past with a view on its innovative potential for today's research.[21]

4 THE *WIENER KREIS* IN AMERICA: LOGICAL EMPIRICISM AND (NEO-)PRAGMATISM

4.1 Herbert Feigl and the Minnesota Center for the Philosophy of Science (MCPS) in Minneapolis

As the earliest immigrant of the Vienna Circle and a participant of the Unity of Science Movement at Harvard associated with Philipp Frank and, later, at the Boston Colloquium organized by Robert S. Cohen, *Herbert Feigl* (1902–1988) played a pivotal role in the transfer and development of "logical positivism" in the United States. With his research and teaching activities at the University of Iowa (1931–1940) and at the University of Minnesota in Minneapolis (from 1940), as well as his many visiting professorships on the East and West Coast, his functions as president of the "American Philosophical Association" and vice-president of the "American Academy of Arts and Science", Feigl became one of the most influential figures of the second generation of the Vienna Circle in the United States. The *Minnesota Center for the Philosophy of Science (MCPS)*, was founded in 1953 by Feigl, has published 18 volumes of the *Minnesota Studies in the Philosophy of Science* (MSPS) since 1956 and became a sort of training center for the history and philosophy of science. It is therefore all the more surprising that there has been hardly any research on the life, work and reception of this original thinker who has been associated with the Vienna, Harvard and Minneapolis circles. We only have the scanty autobiographical fragments and indirect references to publications to go by, which give us some idea of Feigl's great transatlantic impact [Feigl, 1981; Haller, 2003].

In his own reminiscences, Feigl first speaks of his revered teacher and founder of the Vienna Circle, Moritz Schlick:

> "Several members of the Circle had a reading knowledge of English, but Schlick, whose wife was American, spoke English perfectly. Some of the conversations at Schlick's house were in English, notably with such visitors as Roger Money-Kyrle but occasionally with Wittgenstein who was also fluent in English. Schlick was the first of our group to

[21] Two examples: The Moritz Schlick edition project at the Institute Vienna Circle: `http://www.univie.ac.at/Schlick-Projekt/` as well as the foundation of the "Austrian-French Society for Cultural and Scientific Cooperation // Societé franco-autrichienne pour la cooperation culturelle et scientifique". These activities follow the tradition of the Austrian (late) Enlightenment which has been marginalized in research. Cf. on this: Kurt Blaukopf, "Kunstforschung als exakte Wissenschaft. Von Diderot zur Enzyklopädie des Wiener Kreises", in: Friedrich Stadler (ed.), *Elemente moderner Wissenschaftstheorie. Zur Interaktion von Philosophie, Geschichte und Theorie der Wissenschaften.* Vienna-New York: Springer 2000, pp. 177-211.

> be invited to the United States. ... Schlick enjoyed his sojourn at Stanford, made many friends, and was promptly invited to another visiting professorship, this time (in 1931) to the University of California at Berkeley. Thus it came about that Schlick was the first to spread the Vienna 'gospel' (with a strong emphasis on Wittgenstein's ideas) in America. My own first journey to the United States occurred in September 1930, when I was fortunate to obtain an International Rockefeller Research Fellowship. This allowed me to work at Harvard University for about nine months." [Feigl, 1968, p. 643]

Feigl's own transit was finally made possible through his acquaintance with two American philosophers — Dickinson S. Miller and Charles A. Strong — as well as through his contact with the American student Albert Blumberg, who had written his dissertation under Schlick. All these contacts led Feigl to the conclusion that:

> "'Over there', I felt was a Zeitgeist thoroughly congenial to our Viennese position. It was also in 1929 that, I think through Blumberg's suggestion we became acquainted with Percy W. Bridgman's *Logic of Modern Physics* (1927). Bridgman's operational analysis of the meaning of physical concepts was especially close to the positivistic view of Carnap, Frank, and von Mises, and even to certain strands of Wittgenstein's thought." (Ibid., 645)

Feigl in his autobiographical account [1968] reported just as enthusiastically on his first impressions of his New World contacts in Harvard (including C. I. Lewis, Henry Sheffer, A. N. Whitehead, Susanne K. Langer, Paul Weiss, W. V. O. Quine — and also, once again, Karl Menger) as he had on his Bauhaus experience in Dessau. After his article "Logical Positivism", written together with Blumberg in 1931, Feigl saw a debate unfold which was going to last twenty years — a sort of "succés de scandale" in philosophy. (ibid., 647). As one of the first authors in the *Philosophy of Science,* "a periodical, for whose initiation I was in small part responsible" (ibid.), Feigl was already a part of the relevant discussions at a very early stage. Through Felix Kaufmann's intervention, Feigl was also able to teach at the "New School for Social Research", which strengthened the New York philosophy of science scene to which Carl G. Hempel also belonged. Communication also flourished on the West Coast. Feigl's contacts with the philosophers there — Hans Reichenbach, W. R. Dennes, Paul Marhenke, David Rynin and even Else Frenkel-Brunswik and Egon Brunswik (with a Unity of Science meeting in Berkeley in 1953) — made the *Rise of Scientific Philosophy* (Reichenbach 1951) a country-wide issue.

In Iowa, Feigl offered a course on philosophy of science for the first time. In some of the leading anthologies, e.g., *Philosophical Analysis* (ed. 1949) and *Readings in the Philosophy of Science* (ed. 1953), he prepared the ground for the reception of these ideas. The journal *Philosophical Studies,* founded as a counterpart to the British *Analysis,* was a successful parallel initiative, which is still being continued today. A private foundation enabled the MCPS to be established:

> "For a few years in the late forties and early fifties, Sellars and I, together with May Brodbeck, John Hospers, Paul Meehl, and D.B. Terrell, made up a discussion group in which occasionally visitors from other universities would participate. Gradually we came to think about organizing a more official center for research in the philosophy of science. Encouraged by the generous financial support of Louis W. and Maud Hill Family Foundation in St. Paul, the Minnesota Center for the Philosophy of Science was established in 1953. During the first few years the local staff members were Paul E. Meehl ..., Wilfrid Sellars ... and Michael Scriven ... In the fourteen years of its activities, the Center has enjoyed visits of various durations by many outstanding American, European, and Australian and New Zealand scholars. Our major publications (Minnesota Studies in the Philosophy of Science and Current Issues in the Philosophy of Science) have aroused considerable interest. Several younger generation philosophers have been our visitors, among whom have been Scriven, Adolf Grünbaum (Pittsburgh), Hilary Putnam (Harvard), N.R. Hanson (Yale), Wesley Salmon (Indiana), Karl R. Popper (London), Paul Feyerabend (Berkeley), Bruce Aunne (University of Massachusetts), Henryk Mehlberg (Chicago), George Schlesinger (Australia, now North Carolina) and Arthur Pap (Yale)." [Feigl, 1968, 664f]

This account underplays the important role of the MCPS, founded as Research Department at the College of Liberal Arts of the University of Minnesota, in the propagation of the philosophy of science, which Feigl only alludes to when referring to the influence of similar centers at Indiana University or the University of Pittsburgh. This, combined with the teaching activities at these universities exerted an influence on a number of student generations. With regard to the influence on Austrian thinkers, we should mention the stimulating activities of Arthur Pap and Paul Feyerabend who, together with Grover Maxwell, co-edited the only Feigl Festschrift to date, which appeared in 1966: *Mind, Matter and Method. Essays in Philosophy of Science in Honor of Herbert Feigl.*

There, in his "Biographical Sketch", we find one of the few tributes to Feigl and his center. Feyerabend had met Feigl for the first time in a Vienna coffee house in 1954 (at that time he was an assistant of Arthur Pap). This encounter was seen by Viktor Kraft's group as being a particularly enriching one. Feyerabend continued to praise Feigl's style of philosophizing, which, at the MCPS, reflected the high level of discussion. Referring to the internal life of the center, Feyerabend wrote the following:

> "The atmosphere at the Center, and especially Feigl's own attitude, his humor, his eagerness to advance philosophy and to get at least a glimpse at the truth, and his quite incredible modesty, made impossible from the very beginning that subjective tension that occasionally accompanies debate and that is liable to turn individual contributions

> into proclamations of faith rather than into answers to the question chosen. The critical attitude was not absent; on the contrary, one now felt free to voice basic disagreement in clear, sharp, straightforward fashion. The discussions were, and still are, in many respects similar to the earlier discussions in the Vienna Circle. The differences are that things are seen now to be much more complex than was originally thought and that there is much less confidence that a single, comprehensive empirical philosophy might one day emerge." [Feyerabend, 1966, p. 9]

The discussions that took place in this atmosphere were thus mainly devoted to the analysis of scientific theories with a strong focus on current research. In an epistemological sense, Feigl's personal preference for a critical realism found acceptance, resulting in a sort of "anti-Copenhagen mood" which prevented metaphysics from appearing obsolete a priori. (A similar development could be detected in the Harvard group associated with Frank). The list of participants at the MCSP is certainly impressive, since it practically covered the entire horizon of philosophy of science. In the publications named this broad spectrum is largely documented in the individual contributions. Feigl describes his acquaintance with the philosophical rebel in the following words:

> "I met Feyerabend on my first visit to Vienna after the war (my last previous visit was in 1935). This was in the summer of 1954 when Arthur Pap was a visiting professor at the University of Vienna. Feyerabend had been working as an assistant to Pap. Immediately, during my first conversation with Feyerabend, I recognized his competence and brilliance. He is, perhaps, the most unorthodox philosopher of science I have ever known. We have often discussed our differences publicly. Although the audiences usually sided with my more conservative views, it may well be that Feyerabend is right, and I am wrong." [Feigl, 1968, p. 668]

Against this background it is not surprising that Feyerabend published one of his first critical studies in the fourth volume in 1970 of the MCPS "Against Method: Outline of an Anarchistic Theory of Knowledge" which was a highly controversial attack on of the standard version of the philosophy of science. As opposed to the institutions in Harvard and Boston, the MCPS presented an almost complementary trend to psychology and the social sciences along with a fundamental analysis to be found, for instance, in Feigl's influential article "The 'Mental' and the 'Physical'" (1958). But the focus on the general status of scientific theories and on the "cognitive turn" was also characteristic of the field of study. The original self-description in the *Minnesota Studies*, which appeared in the volume *The Foundations of Science and the Concepts of Psychology and Psychoanalysis* [Feigl and Scriven, 1956] confirm this assessment:

> "The first volume of Minnesota Studies in the Philosophy of Science

> presents some of the relatively more consolidated research of the Minnesota Center for the Philosophy of Science and its collaborators. Established in the autumn of 1953 by a generous grant from the Hill Family Foundation, the Center has so far been devoted largely, but not exclusively, to the philosophical, logical, and methodological problems of psychology." [Feigl, 1956, V]

Fourty years later a new assessment provides a broader image of the MCPS's fields of study. In its 16th volume, *Origins of Logical Empiricism* [Giere and Richardson, 1996] a striking survey is given:

> "Minnesota Studies in Philosophy of Science is the world's longest running and best known series devoted exclusively to the philosophy of science. Edited by members of the Minnesota Center for the Philosophy of Science ... (MCPS) since 1956, the series brings together original articles by leading workers in the philosophy of science. The ... existing volumes cover topics ranging from philosophy of psychology and the structure of space and time to the nature of scientific theories and scientific explanation." (MCPS in Internet)

The various thematic volumes of the MSPS largely confirm this. The first volume includes studies on Logical Empiricism (Feigl), the methodological status of theoretical concepts (Carnap), a critique of psychoanalysis (Skinner), an account on radical behaviorism (Scriven), the principles of psychoanalysis (Ellis), motives and the unconscious (Flew), psychological tests (Crobach/Meehl), ego-psychology (Meehl), the logic of general behavioral system theory (Buck), the concept of emergence (Meehl/Sellars), empiricism and cognitive philosophy (Sellars) and the human sciences within the canon of science (Scriven). The critical reevaluation of philosophy of science leads Feigl to a modified image of its development since the Vienna Circle and at the same time indicated a qualitative transformation in terms of pluralization and relativism:

> "I have tried to convey my impression that the philosophy of science of logical empiricism, after twenty-five years of development, compares favorably with the earlier logical positivism, in that it is, firstly, more logical ... Secondly, it is more positive, i.e., less negativistic ... Thirdly, logical empiricism today is more empirical, in that it refrains from ruling out by decree ontologies or cosmologies which do not harmonize with the preconceptions of classical positivism. Alternative and mutually supplementary logical reconstructions of the meaning of cognitive terms, statements, and theories have come increasingly to replace the dogmatic attempts at unique reconstructions. Logical empiricism has grown beyond its adolescent phase. It is rapidly maturing, it is coming of age..." [Feigl, 1956, p. 34]

This statement on the task of a differentiated approach to philosophical research in the theory pool of the original Logical Empiricism gives us further evidence of an interdependent evolution of theories since the beginning of the thirties. This, together with an exchange of ideas and cooperation between Austria and America, transcend the classical boundaries of a traditional history of reception. A further area of study will emerge, if the as yet unpublished archival materials and the numerous activities of the MCPS are analyzed and interpreted in this context. Such a project has to focus on the Austro-American transfer, transformation and retransfer of the philosophy of science in the period from 1930 to 1960.

4.2 Rudolf Carnap, Karl Menger and the "Chicago Circle"

With his trips to America in 1929 and 1931/32, Schlick paved the way for his student Herbert Feigl, who decided very early in 1931 to emigrate to the US because of the increasing anti-Semitism in Austria. With his networking a door to America was opened for Rudolf Carnap, the most influential thinker of the Vienna Circle. On an organizational level, the Unity of Science movement became significantly international at the first congress in Paris 1935. Neo-pragmatism and Logical Empiricism found a platform thanks to Otto Neurath's untiring efforts together with Morris (who had become the driving force behind the semantically oriented synthesis of Logical Empiricism and pragmatism) and Carnap to launch the joint publication project of the "International Encyclopedia of Unified Science" (from 1938). The three philosophers were involved in the planning of the six "International Congresses for the Unity of Science" in Paris (1935 and 1937), Copenhagen (1936), Cambridge (1938), and already during World War II in Harvard (1939) and Chicago (1941).

The contact with Morris enabled Carnap to gradually emigrate to the United States. After a stay in London in 1934, Carnap traveled to the United States for the first time in December 1935 — this move was also motivated by the increasingly unbearable political atmosphere in Prague. Already the year before he had met Willard Van Orman Quine (Harvard) in Vienna and Prague, which was followed by an intense dialogue and continuous contact following his emigration to Chicago in 1936.

At the University of Chicago, Carnap and Morris held a regular colloquium, known as the "Chicago Circle", on methodological and interdisciplinary issues, even if the knowledge of modern logic was somewhat limited there.

With this development, Carnap broke with the original conception of the logic of science ("Wissenschaftslogik") understood as a logical syntax of language. Influenced by the work of Alfred Tarski, who immigrated from Warsaw in 1939, Carnap had undergone a "semantic turn" in the US by the time his *Introduction to Semantics* (1942) appeared. And the discussion of Quine's "Two Dogmas of Empiricism" (1951) drew from the early moment on Carnap's sensitivity to the question of the analytic/synthetic or theoretical/empirical dualism.

Carnap writes about this new circle in exile:

> "In Chicago Charles Morris was closest to my philosophical position. He tried to combine ideas of pragmatism and logical empiricism.

> Through him I gained a better understanding of the Pragmatic philosophy, especially of Mead and Dewey.
>
> For several years in Chicago we had a colloquium, founded by Morris, in which we discussed questions of methodology from scientists from various fields of science and tried to achieve a better understanding among representatives of different disciplines and greater clarity on the essential characteristics of the scientific method. We had many stimulating lectures; but, on the whole, the productivity of the discussions was somewhat limited by the fact that most of the participants ... were not sufficiently acquainted with logical and methodological techniques. It seems to me an important task for the future to see to it that young scientists, during their graduate education, learn to think about these problems both from a systematic and from an historical point of view." [Carnap, 1963, p 34f]

Referring to these meetings the editors of Karl Menger's *Reminiscences* add the following remarks:

> "Carnap and Morris had organized a discussion group, inevitably called the 'Chicago Circle', which met irregularly on Saturday mornings at the University of Chicago. As often as he could Menger came from South Bend and participated in the sessions of the Circle.
>
> The one tangible accomplishment of the Chicago Circle was to get some of its participants to write, and the University of Chicago Press to publish, the first monographs in the series called the International Encyclopedia of Unified Science. Apart from this the Circle suffered from an early series of blows from which, although it continued to meet in an desultory fashion until the 70's, it never fully recovered. The first of these was the departure of the noted linguist Leonard Bloomfield from the University of Chicago to become Sterling professor at Yale ... The next major and practically fatal blow ... was the war, which in the United States began in 1941, and which disrupted academic life in general." [Menger, 1994, pp. XIII f]

With respect to the early transatlantic communication on logical and mathematical issues, the role played by Karl Menger cannot be overemphasized. Through his journeys abroad and his publications in the thirties, both the results of his Viennese "Mathematical Colloqium" (1928–1936) as well his own works on scientific logic became known internationally.

Already in 1930 he lectured at Harvard, where he came into contact with philosophers like Percy Bridgman and Paul Weiss, who later founded the *Journal of Symbolic Logic* (from 1936 on), in which he and his famous disciple Kurt Gödel as well as Carnap published their work in the following years. Just as Bridgman was influenced by Mach's ideas, this was also true, surprisingly enough, of the Austrian economist Joseph Schumpeter, who joined Philipp Frank discussion group in Harvard.

It was Menger who lectured on Kurt Gödel's revolutionary work on completeness and consistency in the US before he decided to emigrate in 1937 from Vienna to Indiana (University of Notre Dame) because of the depressing political situation in Austria, especially after the murder of Moritz Schlick in June 22, 1936. From 1946 until his retirement in 1971, he taught at the Illinois Institute of Technology in Chicago, where he continued his Viennese project as the organizer and editor of the (less successful) *Reports of a Mathematical Colloquium.*

The last two "Congresses for the Unity of Science" in Harvard and Chicago functioned as a forum for the transfer of knowledge and the transformation of philosophy of science into the international Unity of Science Movement:

> "Quine wrote simply: 'Basically this was the Vienna Circle, with accretions, in international exile'. One might say that Mach's spirit had found a resting place in the New World at long last, and that the advance of the Vienna Circle had arrived at Harvard Square." [Holton, 1993, p. 62]

In summary, it is clear that a basis for dialogue between Vienna and Chicago in philosophy of science had been created on various levels already prior to the outbreak of World War 2 and the preceding cultural exodus from Austria. A path had been paved for the actual transfer of knowledge in the context of (direct and indirect) contacts, congresses and journals: the International Congresses for the Unity of Science (1935–1941) and the *International Encyclopedia of Unified Science* (1938–1962) provided a framework and forum for this scientific communication.

The final congress at the University of Chicago (September 2–6, 1941) ushered in the phase of internationally established philosophy of science — in spite of the fact that the program had been reduced as a consequence of the war. The Chicago congress united Americans and emigrants including "Contributions from Europeans" whose papers where presented in the absence of their authors. The discussions focused on "The Task of the Unification of Science", "Logic and Mathematics', "Psychology and Scientific Method', and brought together scholars from the encyclopedia project (like Morris, Neurath, Feigl, Carnap, Brunswik, Reichenbach, and Hans Kelsen) as well as new scientists such as Alfred Korzybski, Lewis Feuer, and Charles Stevenson.

From 1938 on, publications on these activities were edited by Neurath, Carnap and Morris as part of the *International Encyclopedia of Unified Science* (IEUS), a modernist project that extended into the 1960s but was to remain uncompleted.

At the same time, the Journal *Erkenntnis* edited by Carnap and Reichenbach, became international with the eighth (and last) volume as *Journal of Unified Science*, after it had come under pressure by the Nazi regime in 1933 [Spohn, 1991]. In 1938, the first volume of the IEUS, with contributions by Neurath, Niels Bohr, John Dewey, Bertrand Russell and Carnap, marked the beginning of the uncompleted project with 19 instead of 260 projected monographs published with the University of Chicago Press (Reprint of all 19 monographs: Neurath/Carnap/Morris 1970/71).

Even though the editors had very different ideas about the unification of the sciences, the project was continued also after the war although the death of Neurath (1945) and the Cold War resulted in the deterioriation of the whole enterprise of "late Enlightenment". The last path breaking contribution by Thomas Kuhn on *The Structure of Scientific Revolutions* (1962) can be seen as reflecting a change in the philosophy of science, characterized as a pragmatic or sociological turn with the philosophy of science embedded in a historical context.

It is really remarkable that Carnap expressed his appreciation of Kuhn's article in two letters (April 12, 1960 and April 28, 1962) as part of the *Encyclopedia* — a fact, which was obscured by subsequent historiography for many reasons to be discussed in a different context. (cf. [Reisch, 2004]).

Given the prehistory it is no surprise that the Advisory Committee of the IEUS documents a strong UK/US-Austrian bias. Accordingly, it becomes difficult to speak of an input-output or loss-gain transfer caused by the forced *Cultural Exodus from Austria* from 1938 [Stadler and Weibel, 1995]. We are rather dealing with a multilateral dynamic of science as transfer, transformation from Central Europe to Great Britain and America, which can be described as a parallel process of disintegration and internationalization.

We now come to a third important person, who was also an emigrant from Austria, the social scientist Hans Zeisel (1905–1992):

Zeisel, well known as a co-author of the study *Die Arbeitslosen von Marienthal* in 1933 (English edition: *Marienthal: The Sociography of an Unemployed Community,* 1971) studied law and political science at the University of Vienna before he was forced for political and "racial" reasons to leave Vienna for the U.S. after the *Anschluss* in 1938. In "Red Vienna" he had cooperated with Paul Lazarsfeld, Marie Jahoda as a member of the "Wirtschaftspsychologische Forschungsstelle" at the Institute of Psychology run by Charlotte and Karl Bühler of the University of Vienna. In New York he met Lazarsfeld again and worked as manager of media research at the McCann Erickson advertising agency (1943–1950) and was also an executive at the Tea Council (1950–1953). During this time he published the textbook *Say it with Figures* (1947) which became a standard reference book in social research with six editions and countless translations.

In 1953 Zeisel was appointed professor at the University of Chicago Law School, where he applied empirical social research to the field of (sociology of) law.

Together with Harry Kalven, Jr. he worked on the American jury system, esp. provided by the book *The American Jury* (1966). His *Prove it with Figures: Empirical Methods in Law and Litigation* (1997) is a sort of summary of his lifelong focus on social research as a tool for tackling legal problems. Apart from this focus, he, like Carnap, opposed the Cold War legislation and capital punishment for a number of reasons.

Zeisel also had studied philosophy under the influence of the Vienna Circle (both methodologically and scholarly), especially as a participant of Carnap's lectures in

Vienna. One year before he died in Chicago, March 7, 1992, he participated in the founding conference of the Institut Wiener Kreis/Vienna Circle Institute in 1991 ("Wien–Berlin–Prag: Der Aufstieg der wissenschaftlichen Philosophie" on the occasion of the centenaries of Rudolf Carnap, Hans Reichenbach and Edgar Zilsel). In his posthumously published contribution to the proceedings (Haller/Stadler 1993) entitled "Erinnerungen an Rudolf Carnap" (Remembrances of Rudolf Carnap) he presented personal recollections of his beloved teacher and his admiration of Vienna Circle's philosophy and methodology in general. (By the way, Herbert Feigl was a distant relative of Zeisel's family; he also took issue with the break with this tradition at the University of Vienna after 1945). Especially the dualism of facts and values seemed to him most important for his own work, inspired by the analysis of language and the anti-metaphysical orientation.

And Zeisel confirmed Carnap's autobiographical allusion (1963) to the fact that he was not really happy in Chicago because of the administration and the situation of the so-called "continental" philosophy there, dominated by Richard McKeon and Mortimer Adler. And he reports that Carnap only had few students and was somewhat isolated from the academic life – despite the "Chicago Circle" with Charles Morris.

Although Carnap left Chicago for Los Angeles when Zeisel was appointed there, we can see the strong impact of the Vienna Circle and the appreciation felt by the younger sociologist for the philosophy of Logical Empiricism in exile, which was only a continuation of the Vienna connection from the 1920's and 1930's:

> " ... alles zusammen hat er einen bedeutsamen Einfluß auf mein Leben (ausgeübt), auf meine Arbeit, und ich glaube auch auf viele andere Menschen. Sein steigender Weltruf ist nicht nur berechtigt, sondern entwickelt sich mit großer Klarheit, denn er war einer der großen Philosophen dieses Jahrhunderts, und als den haben wir ihn empfunden und geschätzt." [Zeisel, 1993, p. 223]

Summarizing this short account one might ask if there wasn't a sort of re-transfer of the "Chicago Circle" to Europe after 1945? One could first answer: there was a remarkable hidden impact, which could be the subject of future research focusing on following aspects:

1. Through the Austrian philosopher Wolfgang Stegmüller, who came into contact with Carnap after World War 2 the modern philosophy of science was introduced in the German speaking academia as (modified) analytic "Wissenschaftstheorie". `http://www.univie.ac.at/ivc/stegmueller`.

2. Hans Zeisel was an important figure, serving as both initiator and *spiritus rector* of the still existing Vienna-based "Institut für Kriminalsoziologie" (Institute of Legal and Criminal Sociology), founded in 1973 as one manifestation of the legal reform under the former Austrian Minister of Justice Christian Broda.

4.3 Philipp Frank and the "Institute for the Unity of Science"

One of the most important figures for the transfer, transformation and the further development of the Central European philosophy of science was the physicist/philosopher Philipp Frank (1884–1966) whose work has continued to have an indirect influence through his students up to this day.

Frank was Einstein's successor in Prague where he worked as full professor and director of the Institute for Theoretical Physics from 1912 until he had to emigrate in 1938. As a leading member of the Vienna Circle he significantly contributed, together with Carnap and Neurath, to the dissemination of Logical Empiricism. (Frank 1949) After giving a series of lectures on modern physics at various American universities, he taught at Harvard University until 1953 — first as lecturer and later as Hooper Fellow. Given his age he was no longer able to obtain an adequate position, yet his charisma and skills as an intellectual and as an organizer allowed the younger Herbert Feigl to become a central figure of intercontinental philosophy of science. Up until his death he played a seminal role in the Unity of Science movement and in organizing an innovative interdisciplinary forum for discussion at Harvard. Through these activities and functions and through his students (including, among others, Robert S. Cohen, Gerald Holton, Ernest Nagel) he also influenced an entire generation of young philosophers of science. The latter, in turn, contributed significantly to a critical further development of the philosophy of science, which had changed significantly as a result of the immigration of intellectuals. They have continued to shape the scientific scene through their academic positions and publications. It was, above all, Philipp Frank's achievement that science became a field of discourse, next to philosophy and religion, in the war-ravaged intellectual scene - in the heyday of the McCarthy era (Reisch 2004). Together with the American Academy of Arts and Sciences, he succeeded in positioning philosophy of science in a number of events. Frank wrote a very enlightening account of the development from Vienna, Berlin and Prague to "Harvard Square" (Introduction to 1949). This intellectual history was illustrated by Gerald Holton (1993 and 1995) in connection with a European-American reception. On his first teaching job in the USA, Frank wrote the following:

> "Since the fall of 1939 I have had the privilege of teaching at Harvard University not only mathematical physics but also the philosophy of science. This teaching has been a great experience for me and has been of great influence on my philosophical writing. I started with an audience of about fifteen students. Since this was an unusual subject I did not quite know what to tell them. I began by presenting to them the logical structure of physical theories as envisaged by logical empiricism. But very soon I noticed this was not the right thing to do. The frequent discussion that I had with the students showed me what they really wanted to know. By a process of interaction, a program was finally worked out that was a compromise between what I wanted to tell the students and what they wanted to know." [Frank, 1949, p.

50]

Based on this interaction, Frank also developed the curricula for science courses at Harvard, giving philosophy of science a more contextual orientation and including the historical and sociological perspectives. His efforts were backed by Harvard president J. B. Conant, who had strongly advocated a new approach to teaching science which he presented in his book *On Understanding Science* (1947). In his book *Science and Common Sense* (1951), he further refined these considerations with reference to the interdisciplinary discussion circles at Harvard, which Thomas Kuhn and Bernard I. Cohen were part of. In this connection, it is interesting that Frank saw all his writings after 1940 as influenced by his involvement with teaching at Harvard:

> "My point was now that the philosophy of science should, on the one hand, give to the science a more profound understanding in his own field, and on the other hand, be for all students a link between the sciences and the humanities, thus filling a real gap in our educational system." (Ibid., 51)

Having become increasingly interested in the issue of metaphysical interpretations of modern science as a result of his US acculturation, Frank began to systematically analyse the various metaphysical understandings of science in a logico-empirical and socio-psychological way. Since 1940 he was invited by Harlow Shapley, the director of the Harvard College Observatory, to the annual "Conferences of Science, Philosophy and Religion" which viewed science from a critical perspective:

> "I addressed this group several times between 1940 and 1947. My contributions were mostly around the question of whether 'relativism' of modern science is actually harmful to the establishment of the objective values in human life. I made an argument to prove that 'the relativism of science' has also penetrated every argument about human behavior. 'Relativism' is not responsible for any deterioration of human conduct. What one calls 'relativism' is rather the attempt to get rid of empty slogans and to formulate the goals of human life sincerely and unambiguously.' (Ibid., 52)

With this proclamation of an anti-absolutist understanding of science, the Einstein biographer Frank described "Science in Context" along similar lines as the *Encyclopedia* project and documented it in his book *Relativity — A Richer Truth* (1950). In the German edition of this 1952 publication, where he distinguished clearly between arbitrariness and convention in science and ethics on empirical grounds, there is a programmatic preface written by Albert Einstein.

In Harvard a so-called "Science of Science Discussion Group" had been formed as early as 1940/41 in which Frank also participated. (Hardcastle 2003, 170-196). It was obviously to serve as a model for the "Interscientific Discussion Group" that began meeting in 1944:

> "In the Fall of 1940 around Harvard University, an invitation was distributed to thirty-eight scientists from various fields, together with some logicians and methodologists present this year in Harvard. 'As an effort in the direction of debabelization' it began, 'the undersigned committee is organizing a supper-and-discussion-group to consider topics in the Science of Science' ". [Hardcastle, 1996, p. 24]

The participants of this monthly transdisciplinary discussion forum on theory and study of science included, in addition to Frank, Rudolf Carnap, Herbert Feigl, Willard Van Orman Quine, Richard von Mises, Alfred Tarski, Nelson Goodman, George Polya, Percy Bridgman, as well as the psychologists E. G. Boring, and S. S. Stevens and economist Joseph A. Schumpeter. Stevens had organized this forum after talks with Carnap and Frank.

This platform, a clear sign of the cooperation of the Vienna Circle in exile and American philosophy of science, paved the way for the scientific communication that was to follow and represented a sort of proto-circle of the expanded "Interscientific Discussion Group" that lead up to the Institute for the Unity of Science.

> "Between roughly 1940 and the end of the 1950's, the movement for a scientific philosophy in the USA flourished, pushed forward especially by the influx of arrivals from Europe. The main direction of the movement brought over from Europe was now identified most often by the slogans 'Unity of Science' and 'Unified Science', versions of the old terms Einheitswissenschaft and Gesamtauffassung which had been prominent in the manifestos of 1911–12 and 1929 as well as Carnap's Aufbau — a concept that had roots in the phenomenalistic monism of Mach." [Holton, 1993, 62]

This movement was seen as based on the *Encyclopedia of Unified Science* and the *Institute for the Unity of Science (IUS)* under the auspices of the *American Academy of Arts and Sciences* (AAAS) — one of the manifestations of the intellectual symbiosis between related European and American movements.

What was this new institution and what role did it play in the transfer of science already sketched ? The published statutes of the IUS read as follows:

> "The purposes for which the corporation is formed are to encourage the integration of knowledge by scientific methods, to conduct research in the psychological and sociological backgrounds of science, to compile bibliographies and publish abstracts and other forms of literature with respect to the integration of scientific knowledge, to support the International Movement for the Unity of Science, and to serve as a center for the continuation of the publications of the Unity of Science Movement." (*Synthese* 1947, VI, 158f., cited after [Holton, 1993, p. 72]

The AAAS Proceedings were used as a basis for publications as well as its infrastructure for diverse conferences and symposia. The participants of these considerably expanded discussion rounds show how the forum increasingly became more open, providing a setting in which the sciences could be discussed within a cultural and social context.

Even if Quine documented these regular meetings as a "Vienna Circle in exile" [Holton, ibid., p. 63], there had been a considerable leap in terms of quality and pluralization as becomes clear in the wide range of topics and in the composition of the participants. The "academization" of *Wissenschaftslogik*, [Dahms, 1987], applied to the proponents, but not to the organizational form outside of the universities. Holton is right in asking how an "ecological niche" was created for this hybrid scientific movement within two decades. Apart from the anti-metaphysical and empiricist orientation of pragmatism, a number of related factors can also be mentioned here, e.g.,the preceding personal contacts on an university level, the private organizations promoting science such as the Rockefeller Foundation (e.g., for Feigl in Harvard) and the anti-Nazi outlook of the Scientific Community — all factors that should be evaluated more closely. The most important factor was ultimately the recognition of the high quality of émigré scholars, which could be exemplified by the integration of Philipp Frank as a consequence of Bridgman's and Conant's efforts.

Whether this intellectual osmosis also provides an adequate explanation of the qualitative transformation and paradigm shift that has taken place since the 'received view' remains to be seen. In any case, it is certain that this interaction significantly enriched philosophy both theoretically and methodologically.

The personal composition of the "Interscientific Discussion Group" (IDG), which met in Harvard since 1944, was an interesting mixture of older and younger scientists of American and European origin. Most of them spoke at both of the Unity of Science Congresses in Harvard and Chicago [Holton, 1995]. Personal invitation letters from the committee (Percy W. Bridgman, Walter Cannon, Philipp Frank, Philippe LeCorbeiller, Wassily W. Leontief, Harlow Shapley and George Uhlenbeck) were also sent to Karl W. Deutsch, Roman Jakobson, Willard Van Orman Quine, Charles Morris, Richard von Mises, Ernest Nagel, Giorgo de Santillana, Victor F. Weisskopf and Norbert Wiener, who were requested to participate regularly. The group described itself as follows:

> "Our group consists of persons in different fields who feel that the extreme specialization within science demands as its corrective an interest in the entire scientific edifice. We plan to hold meetings from time to time in which discussions of different topics will be led by competent scholars." (Inter-Scientific Discussion Group, December 30, 1944, cited after [Holton, 1995, p. 284]

In the program planned for 1945, three large areas, "Logic of Science", "Psychology" and "Sociology of Science", were announced along with further subtopics. This reflected both an internal and external perspective on philosophy of science and confirmed psychology and sociology of science anticipating Kuhn's work. Holton, at the time IDG secretary, saw this program in retrospect as continuing the tradition of the Vienna Circle, the Verein Ernst Mach and their programmatic manifesto of 1929. The speakers and the themes of these early meetings comprised the history of philosophy (Santillana), psychoanalysis and social science (Talcott Parsons), mathematics/statistics (Richard von Mises), cybernetics (Norbert Wiener), biology and social science (Georg Wald) and science in general (C. J. Ducasse) — in the context of Morris's conception of semiotics (syntax, semantics, pragmatics).

The osmotic process of discussion between the immigrant philosophers of the first wave since 1930, of the second wave from 1938 on and of the American scientific community became an experimental setting for a future history *and* philosophy of science. Sociology of science (Talcott Parsons) was represented along with philosophy of economics (Paul Samuelson, Gottfried Haberler, Joseph Schumpeter), cybernetics (Norbert Wiener), mathematics (Gustav Kuerti), and political science (Karl W. Deutsch), psychology (Gordon Allport), and history of science (I. B. Cohen, G. de Santillana). The fact that philosophy of science in a more narrow sense also had strong presence in American philosophy (e.g., C. I. Lewis, W. V. O. Quine) was decisive for a theory dynamics that involved both diffusion and confrontation of different movements. Lecture themes such as Simplicity of Science, What is Science?, Psychoanalysis and Social Sciences, Sense and Nonsense in Modern Statistics, Biology and Social Behavior, Living Organisms and the Second Principle of Thermodynamics, Stability and Flux in the Living Organism, Relation of Hypothesis and Experiment, served to stimulate the interdisciplinary dialogue. A highlight was certainly Oskar Morgenstern's presentation of his and John von Neumann's *Theory of Games and Economic Behavior* (1944) on February 28, 1944. This book had an immense influence on modern social sciences. Here a line of reception became manifest for which the ground had been laid in the Vienna years.

To complete this theoretical development the following could be added: After emigrating back to Austria, Morgenstern tried, after World War 2, to bring modern social scientific research back to Austria; together with Paul Lazarsfeld, he founded the Viennese "Institute for Advanced Study" in 1963.

This informal IDG circle gave rise to the need for a continuous, expanded forum, which resulted in the initiative for founding the "Institute for the Unity of Science" (IUS). Philipp Frank's efforts resonated particularly well with Harvard president James Conant, since they related to his General Education Program — a sort of survey course on the scientific disciplines. With the help of the Rockefeller Foundation, the IUS — a sort of international variant of the former Vienna Ernst Mach Society — was established in 1947 — in cooperation with the AAAS, officially based in Boston. The self-description reads as follows:

> "This Institute is a non-profit corporation which has offices in Ithaca, New York and Boston, Massachusetts. The Charter says, 'The purpose for which the corporation is formed, are to encourage the integration of knowledge by scientific methods, to conduct research in the psychological and sociological backgrounds of science, to compile bibliographies

> and publish abstracts and other forms of literature with respect to the integration of scientific knowledge, to support the international movement for the unity of science, and to serve as a center for the continuation of the publications of the unity of science movement.' The Institute attempts to stimulate interest in these issues among college students, college faculties, and among the public at large. The Institute has arranged an essay contest for college students and young college graduates. It is editing the Encyclopedia of Unified Science, published by the University of Chicago Press. It is starting research projects in the fields of semantics, logic of science, and sociology of science. It arranges discussion groups and meetings at several places in the United States. It is a part of the International Union for the Philosophy of Science. It cooperates with the International Society for Significs (psycholinguistic studies) in Amsterdam and is organizing, together with this society, an international meeting in Amsterdam. In cooperation with the European societies for the philosophy of science..., this Institute publishes communications in the international journal 'Synthese' which is published in Amsterdam and is the central organ of these groups...
>
> The Institute cooperates also with the movement for general education, which attempts to integrate the college curriculum and to break down the barriers between the departments. The Institute arranges lectures and courses at different places in the United States." (Announcement of the IUS, cited after [Holton, 1995, p. 288]

In this self-description there are at least three important factors. First, the international nature of the activities in the U.S. and of those between the American-European institutions, second, the interdisciplinary orientation and third, the strongly public-oriented and educational political motivation resembling the popularization efforts of the Vienna Circle.

In a theoretical sense, it is interesting to note the presence of *Wissenschaftslogik* ("logic of science"), together with the sociology of science, as well as the reference to the Dutch representatives of the Significs movement (Gerrit Mannoury), which Neurath worked together with in exile until 1940. This connection, which has hardly been taken into account until now was mentioned explicitly in the journal *Synthese* which was published from 1936 to 1939 and after World War, from 1946 on. The contributors to this organ included Gustav Bergmann, E. W. Beth, P. W. Bridgman, L. E. J. Brouwer, Rudolf Carnap, J. Clay, R.S. Cohen, Karl W. Deutsch, C.G. Hempel, G. Holton, W. McCulloch, Karl Menger, Charles Morris, Ernest Nagel, Otto Neurath, Karl Popper, W. V. O. Quine, N. Rashevsky, Moritz Schlick, Herbert A. Simon, Friedrich Waismann as well as Philipp Frank, who contributed to *Synthese* (6, 1947/48). And in the book series "Synthese Library", also launched in 1959, one finds an early *Festschrift* for Rudolf Carnap [Kazemier and Vuysje, 1962] and the first volume of the "Boston Studies in the Philosophy of Science" [Wartofsky, 1963].

The international orientation of the early Unity of Science movement was already reflected in a separate *Synthese* supplement which appeared as *Unity of Science Forum* in the period 1936–1939 under the auspices of the International Institute for the Unity of Science (headed by Frank, Morris and Neurath) at the Hague. For Neurath, this institution served as a platform for the *Encyclopedia* project, while at the same time it was an important organizational bridge in the difficult years in Dutch exile. In its embryonic stage, it was also the model for the IUS in the U.S.A. after European science had been ravaged by National Socialism. The small brochures also included a report by Neurath on the "Fourth International Congress for Unity of Science" in Cambridge (August 1938) and relevant articles or reviews on the Unity of Science Movement. The editors' board of *Synthese* had welcomed the *Unity of Science Forum* already in 1936.

This was only a brief digression to the Dutch background of the IUS, which ten years later was able to work under considerably more favorable conditions.

Already the composition of the Board of Trustees assured a more successful point of departure: Philipp Frank as president, Charles Morris and Ernst Nagel as vice-presidents, Milton R. Konvitz as treasurer and the further members Percy W. Bridgman, Egon Brunswik, Rudolf Carnap, Herbert Feigl, Carl G. Hempel, Hudson Hoagland, Roman Jakobson, Willard Van Orman Quine, Hans Reichenbach, Harlow Shapley and Stanley S. Stevens. This was a renowned group of older and younger generation scientists, both emigrants and American scholars from philosophy, natural and social sciences, who negotiated and coordinated lectures, meetings, conferences and publications from 1948 to 1966.

If we now analyse the related archival material, we note that a more consistent line was taken towards the themes addressed in both natural science and social science. This also meant that issues were dealt with also from a historical and sociological perspective: symposia titles such as "Science and Value", "Logic and the Sociology of Science", "Social Physics" or "Current Issues in the Philosophy of the Sciences" reflected this more open approach and self-reflection within the sciences. Of the individual lectures, the following merit mention: the one given by Roman Jakobson (linguistics), a series on the problem of meaning in the individual disciplines (W. V. O. Quine, P.W. Deutsch), information theory (D. Gabor). A central event was the Boston "Conference on the Validation of Scientific Theories" from December 27 to 30, 1953. The proceedings were edited by Philipp Frank in a book with the same title, at Beacon Press in Boston. The sections included — Acceptance of Scientific Theories, The Present State of Operationalism, Freud's Psychoanalytic Theory, Organism and Machine as well as Science as a Social and Historical Phenomenon — illustrate the cognitive process of transformation in the Philosophy of Science beyond a syntactic/semantic *"Wissenschaftslogik"*. In his introduction, Frank described the situation determined by an increasingly critical view of science as follows:

> "In order to obtain a basis from which one can pronounce sound judgment about this situation one should put the question: In what sense

> does science search for the 'truth' about the universe? ... What are the criteria under which we accept a hypothesis or a theory? If we put this or a similar question, we shall see soon that these criteria will contain, to a certain extent, the psychological and sociological characteristics of the scientist, because they are relevant for the acceptance of any doctrine. In other words, the validation of 'Theories' cannot be separated neatly from the values which the scientist accepts. This is true in all fields of science, over the whole spectrum ranging from geometry and mechanics to psychoanalysis." [Frank, 1956, VII f.]

Here the psycho-sociological turn can easily be recognized as a programmatic requirement for every future history and philosophy of science. The conference was sponsored by the Institute for the Unity of Science as an event of the American Association for the Advancement of Science (AAAS). The individual contributions reflect the pragmatic and operational dimension in the validation and corroboration of theories. Comparing the natural and human sciences from a cross-disciplinary perspective, Warren McCulloch noted:

> "Cybernetics has helped to pull down the wall between the great world of physics and the ghetto of the mind." [Ibid., X]

To elaborate these general principles in more concrete terms, historians of science and scholars were invited; among them: Henry Guerlac, Alexandre Koyré, Robert S. Cohen and E.G. Boring. The conference was headed by R. J. Seeger, H. Margenau, H. Feigl, G. Wald and G. Holton who together had directly or indirectly worked on a radical reform of the philosophy of science, which had already been in the making for three decades.

Against this background, Gerald Holton's argument must be reconsidered. Was it really only Alexandre Koyré's philosophy of science (epistemology) that brought about a loss of meaning in empiricist unified science? The cited publications from the fifties written by Frank and others around him — in connection with the AAAS publications (*Proceedings* and *Daedalus*) indicate that there was actually a trend towards a sort of "Science of Science". The thematic issue "Science and the Modern World View" of *Daedalus* (winter 1958) clearly reflects the complementary contribution made to the symposium volume mentioned above. This volume was presented by Holton himself in his role as editor-in-chief of the Academy on the occasion of the retirement of Bridgman and Frank. The *Weltanschauung* analyses demonstrating an internalist philosophy of science that had become relativized and pluralized are impressing documents of this evolutionary and multi-facetted transition from the received or standard view to the non standard view of scientific theories, or to put it differently, the transition from text to context that was to be "proclaimed " ten years later by a younger generation of scholars as a decisive event [Suppe, 1977].

The conferences organized by AAAS together with IUS from 1951 to 1954 and the proceedings published in the four volumes of *Contributions to the Analysis and Synthesis of Knowledge* show the broad spectrum of approaches in the recent discussion on the unity/diversity of sciences [Galison and Stump, 1997].

Alluding to the *Old Testament* Frank described the dangers of Babylonian linguistic confusion in the 20th century in his contribution on "Contemporary Science and the Contemporary World View":

> "As a matter of fact, the view that science is the product of abstraction from our rich and full experience is rather misleading. It has become more clear by the evolution of science in our century that the principles of science are not dehydrated abstractions but a system of symbols that is produced by the creative imagination of the scientist." [Frank, 1958, p. 59]

Referring to Richard von Mises' *Positivism. An Essay in Human Understanding* (1938/1951), Frank once again gave — in analogy to poetry — a non-foundationalist and relativistic account of science reflecting a certain continuity with the thirties:

> "We have seen that the main activity of science does not exist in producing abstractions from experience. It consists in the invention of symbols and in the building of a symbolic system from which our experience can be logically derived. This system is the work of creative imagination which acts on the basis of our experience."
>
> Subsequently he concludes with the late Wittgenstein: 'One might give the name 'philosophy' to what is possible before all new discoveries and inventions'." [Ibid., p. 65]

This elegant attempt to rehabilitate philosophy, science and philosophy of science in a critical phase governed by public skepticism vis-à-vis science and the controversy over *The Two Cultures* [Snow, 1959] provided crucial impulses for the numerous accounts of *Science and Antiscience* [Holton, 1993] and for today's historiography situated between modernism and postmodernism (cf. [Galison, 1996]).

It would thus be clearly problem-oriented if the eternal question as to the unity or disunity of science had already then been dealt with in various ways — parallel to and correlating with the *Encyclopedia of Unified Science* — be it in the guise of "social physics" (J. Q. Stewart), the unification of systems theory (N. Wiener) or sign theory (Ch. Morris). Holton's more recent publications, including *The Scientific Imagination* [1978], *The Advancement of Science, and its Burdens* [1986], *Science and Anti-Science* [1993b], are also manifestations of this problem located in the specific constellation of science, society and *Weltanschauung.*

It seems as if the external factors of this evolutionary, cognitive process slowly led to a convergence of the "Vienna Circle in America" [Feigl, 1968] with American philosophy of science. It would thus be inaccurate to say that the emergence of a younger generation (Quine, Kuhn, Hanson, Feyerabend, et al.) had resulted in the demise of a philosophical school

The fact that in recent years a new generation has joined ranks in the *History of Philosophy of Science* as well as the *History and Philosophy of Science* (cf.

HOPOS or IVC) documents this historico-pragmatic turn in the philosophy of science.

After Frank and Bridgman, Robert S. Cohen has been a seminal figure in the transmission and further development of this Austro-American philosophy of science. Together with Marx W. Wartofsky, he was active already in the fifties as scholar and secretary of the IDG and the IUS. He finally founded his own forum, the Boston Colloquium for the Philosophy of Science, where also the pioneers of this movement, Frank and others, appeared. Since 1959 this institution has contributed to the continuity and criticism of the philosophy of science. After the IUS was dissolved, the remaining funds were directed to the journal *Philosophy of Science* and the newly founded *Philosophy of Science Association* [Holton, 1995, p. 279]. Holton's own recollections as an immediate participant of this movement in the forties and fifties illustrate in an exemplary way what is generally described abstractly as history of reception:

> "As I fully appreciated at the time, for a young person, participating in these activities was immensely stimulating. Perhaps precisely because of the high density of superb intellectuals, the various leading members of the group, brought together by remnants of the Vienna Circle, made no effort to accept any uncomfortable agreements, but relished in the most wide-ranging debates. I never felt that I had to follow, or to struggle against, any doctrinaire master. When my own first historical studies convinced me of the need to add Thematic Analysis to the older tool-kit of the historian and the philosopher of science, I sensed only encouragement, instead of the kind of opposition one might have expected from rock-hard logical empiricists. If I had to characterize the members of that group in one sentence, I would focus on their unlimited curiosity and their generosity of spirit, a generosity which seemed founded on their ever-youthful self-confidence. When future historians study the philosophy of science during the middle part of this century, I hope they, too, will remember this." [Holton, 1995, p. 279]

4.4 Robert S. Cohen and the "Boston Center for the Philosophy and History of Science"

The physicist Robert S. Cohen was active at Columbia University, in the Division of War Research and on the Communications Board of the U.S. Joint Chiefs of Staff during World War 2. He met most of the members of the Vienna Circle personally after 1938, and, as already mentioned above, he was involved in the organizational work of IUS in its final phase. (On his life and work, see [Gavroglu *et al.*, 1995]).

As early as 1953 he participated in the conference "The Validation of Scientific Theories" where he contributed a paper on "Alternative Interpretations of the

History of Science" [1956]. Here his later, untiring work as organizer, editor and scholar is already alluded to:

> "I am neither a historian nor a sociologist, and at a symposium of the unity of science movement I can only join with those, who are deploring the lack of detailed studies in this history of the social relations of science. I can only regret that the sociology of knowledge, especially of science, has remained so long outside that movement's sweep, and so largely in the hands of metaphysically oriented phenomenologists and other speculative thinkers. The early death of Edgar Zilsel, a pioneer in the sociological treatment of science, left his work tragically incomplete." [Cohen, 1956, p. 219]

His references to Zilsel and also to H. Guerlac, E.G. Boring and A. Koyré stake off the area for Cohen's idea of a history and philosophy of science and for his *Boston Studies in the Philosophy of Science* [1963 ff.] book series which grew out of the Colloquium and is still being continued today. This series has presented a volume on Edgar Zilsel's various works on *The Social Origins of Modern Science* [Raven *et al.*, 2000]. It thus follows a line leading back to the early phase of the Boston Center (BCPS) and the Boston Studies (BSPS).

The first BSPS volume edited by Marx Wartofsky — published in the *Synthese Library* — presented the contributions of the initial phase of 1961/62. It was a very pluralist program ranging from the mind-body problem, scientific language and concept formation, logical foundations of physics (Philipp Frank, among others), modal logic (W. V. O. Quine, Saul Kripke, among others), quantum theory, falsificationism and holism in the philosophy of science (Adolf Grünbaum) to experience and language (Noam Chomsky). An interdisciplinary survey was presented along with comments reflecting the principles of the colloquium:

> "Initiated in 1960 as an inter-university interdisciplinary faculty group, the Colloquium is intended to foster creative and regular exchange of research and opinion, to provide a forum for professional discussion in the philosophy of science, and to stimulate the development of academic programs in philosophy of science in the colleges and universities of metropolitan Boston. The base of the Colloquium is our philosophic and scientific community, as broad and heterodox as the academic, cultural and technological complex in and about this city." [Cohen and Wartofsky, 1963, VII]

The second voluminous volume of the *Boston Studies*, that was edited as *Proceedings of the BCPS* of 1962–64 by Robert S. Cohen and Marx Wartofsky, and printed by Humanities Press in New York, was dedicated to Philipp Frank and appeared in 1965, one year before his death. In the preface, the editors paid tribute to Frank as an individual and describe the orientation of the colloquium in keeping with Frank's scientific life work:

> "Our Colloquium construes the philosophy of science broadly, as he has advised us to do. We try to discuss open problems in the foundations of science, and, wherever relevant, to bring material from the history and cultural relations of science to bear upon such problems. We try also to talk with each other across all boundaries of discipline, to include scholars from philosophy, logic and mathematics, the physical and biological sciences, history and the social sciences, and the humanities as well." [Cohen and Wartofsky, 1965, VII]

The subsequent contributions of Frank's students, including Peter G. Bergmann, Rudolf Carnap, R. Fürth, Gerald Holton, Edwin C. Kemble, Henry Margenau, Hilda von Mises, Ernest Nagel, Raymond Seeger (on behalf of the National Science Foundation) and Kurt Sitte, illustrate the range of his intellectual charisma, his life-long efforts to convey science, the merging of science and philosophy, as well as the contextualization of science – as he demonstrated convincingly in his Einstein biography [Frank, 1947]. Gerald Holton once again underscored the importance of Frank's anthology *Between Physics and Philosophy* [1941] as a link to European philosophy of science after Ernst Mach and Henri Poincaré. Seeger's comment is interesting in the sense that it refers to science politics. In connection with the "National Science Foundation Program on the History and Philosophy of Science" he took account of Frank's suggestions and advice.

The contributions to the Festschrift ran the whole gamut of the history and philosophy of science in keeping with this general orientation, and also included articles and comments. Among the advocates of this approach we already find "young wild" thinkers (Norwood R. Hanson, Paul Feyerabend, Hilary Putnam) who criticized the proposition-oriented theory of science. The volume also included a contribution of Herbert Marcuse "On Science and Phenomenology" on Husserl with a comment by Aron Gurwitsch.

When we examine the subsequent BSPS volumes — as a reflection of the philosophy of science in the USA since the beginning of the sixties — we see the programmatic text on the themes, authors and editors of the series confirmed:

> "The series Boston Studies in the Philosophy of Science was conceived in the broadest framework of interdisciplinary and international concerns. Natural scientists, mathematicians, social scientists and philosophers have contributed to the series as have historians and sociologists of science, linguists, psychologists, physicians, and literary critics. Along with the principal collaboration of Americans, the series has been able to include works by authors from many other countries around the world. As European science has become world science, philosophical, historical, and critical studies of that science have become of universal interest as well.
>
> The Editors believe that philosophy of science should itself be scientific, hypothetical as well as self-consciously critical, human as well as rational, skeptical and undogmatic while also receptive to discussion

> of first principles. One of the aims of the Boston Studies, therefore, is to develop collaboration among scientists and philosophers. However, because of this merging, not only has the neat structure of classical physics changed, but, also, a variety of wide-ranging questions have been encountered. As a result, philosophy of science has become epistemological and historical: once the identification of scientific method with that of physics had been queried, not only did biology and psychology come under scrutiny, but so did history and the social sciences, particularly economics, sociology, and anthropology.
>
> Boston Studies in the Philosophy of Science look into and reflect on all these interactions in an effort to understand the scientific enterprise from every viewpoint."

This text, taken from the cover of the 1985 anniversary volume (Cohen/Wartofsky, eds.) can be read as the intellectual legacy of the Harvard group around Frank. It also reformulates the demand for a view of the natural and social sciences encompassing in principle all the sciences, with a relativisation of the methodological and meta-theoretical dualism of the "two cultures". Seen in this light, the very direct, dialectical argument of Marx Wartofsky is plausible. In a 1994 lecture he spoke on the influence of the exiled Vienna Circle in the Boston region, on "Invariance through Transformation: The Boston Adventures of the 'Wiener Kreis', 1960–1994". [Gould and Cohen, 1994]

In our connection, the volumes of the BSPS, which to date number more than 200, are particularly informative. They were published as *Proceedings of the Boston Colloquium for the Philosophy of Science* (volumes 1-5, 13-14, 31) or dealt with the theories of Logical Empiricism and their reception (cf. volumes 6, 8, 1, 9, 3, 7, 39, 53, 76, 87, 118, 132, 133, 168). In addition, the current BCPS program figures significantly – next to the *Minnesota and Pittsburgh Centers for the Philosophy of Science* – as the only institution following the tradition of the *Encyclopedia of Unified Science* and the IUS.

In the cited volume marking the 25th anniversary of the interdisciplinary work of the international BCPS 160-1985 (Cohen/Wartofsky, eds. 1985), this quarter of a century was summed up as follows:

> "The Boston Colloquium for the Philosophy of Science began 25 years ago as an interdisciplinary, interuniversity collaboration of friends and colleagues in philosophy, logic, the natural sciences and the social sciences, psychology, religious studies, arts and literature, and the often celebrated man-in-the-street. Boston University came to be the home base. Within a few years, proceedings were seen to be candidates for the journal Synthese within the Synthese Library, both from the D. Reidel Publishing Company of Dordrecht, then and now in Boston and Lancaster too. Our Colloquium was inheritor of the Institute for the Unity of Science, itself the American transplant of the Vienna Circle, and we were repeatedly honored by encouragement and participation

of the Institute's central figure, Philipp Frank." [Cohen and Wartofsky, 1985, VII]

In addition to the *Proceedings* which were selected and reworked following the discussions at the Boston Center, the series also includes outside volumes (monographs and anthologies — a series that was first published by Reidel Publishing Company (later: Kluwer Academic Publishers, today: Springer) in Holland. It is a collective undertaking that dates back to the pre-war period. The selection made for the anniversary volume ("Invariance through Transformation") sheds interesting light on the self-image of the series: Adolf Grünbaum wrote on holism in the philosophy of science, Hilary Putnam on explanatory models in linguistics, Nelson Goodman on the epistemological argument, Stephen Toulmin on conceptual revolutions in natural science, Herbert Feigl on empiricism, Robert S. Cohen and Marx Wartofsky on the limitations of science and historical epistemology, respectively, Carl Hempel on values and objectivity in science, Abner Shimony on the philosophy of Bohr, Heisenberg and Schrödinger. The fact that texts by Herbert Marcuse and Noam Chomsky were also included once again reflects the openness of the project which did not adhere to an orthodox logico-empiricist line of research.

Here, we can only focus on the above-mentioned BSPS proceedings and a few volumes which more or less directly relate to the history of the reception or the transformation of the Vienna Circle in exile. The first five volumes document the intellectual spectrum of the Boston Colloquium in the years 1961–1968. (The third volume ("In Memory of Norwood Russell Hanson") focused on a scholar who, along with Kuhn, Toulmin and Feyerabend, was a staunch critic of the internalist philosophy of science. Dedicated to the re-evaluation of the work of Ernst Mach the physicist and philosopher, the sixth volume [Cohen and Seeger, 1970] shows the productive reception of Mach's work leading all the way up to Feyerabend. It also confirmed Holton's reconstruction (1993) of the intercontinental Mach/Boltzmann reception since the turn of the century and in so doing prepared the way for further more in-depth analyses of Mach research. Related studies can be found in volume 143 of the BSPC, in *Ernst Mach — A Deeper Look* [Blackmore, 1992] and in studies on Boltzmann [Blackmore, 1995].

This critical reassessment of the Central European "*Wissenschaftslogik*" and its reception can be found in the commemorative Carnap volume published in 1970 [Buck and Cohen, 1970]. Similarly, the volume of Helmholtz's epistemological writings originally published by Paul Hertz and Moritz Schlick [Cohen and Elkana, 1977], the volume in memory of Imre Lakatos [Cohen *et al.*, 1976], the two volumes by Herbert A. Simon (*Models of Discovery and Other Topics in the Methods of Science*, 1977) or Nelson Goodman's *The Structure of Appearance* [1977] that followed the line of reception of Carnap's work. Reference should also be made to those editions that document, as translations, the historical and sociological expansion of philosophy of science in the sense alluded to above: *Cognition and Fact. Materials on Ludwik Fleck* [Cohen and Schnelle, 1986], *Philosophy, History and Social Action. In Honor of Lewis Feuer* [Hook *et al.*, 1988], *Beyond Rea-*

son. Essays on the Philosophy of Paul Feyerabend [Munévar, 1991], *The Natural Sciences and the Social Sciences* [Cohen, 1994].

This history of reception has come full circle in the documentation of the rediscovery and further study of Central European philosophy of science over the past twenty years after it was interrupted by the forced emigration of the leading protagonists. Here *Rediscovering the Forgotten Vienna Circle. Austrian Studies on Otto Neurath and the Vienna Circle* [Uebel, 1991] deserves mention. The international orientation of the BSPS is also underscored by the many publications on the history and philosophy of science outside of the Anglo-American world. Such studies have helped to overcome the mental barriers between East and West and the North–South hierarchy and to cultivate a dialogue of the scientific community without political and socio-economic restrictions.

The *Festschrifts* for both of the two leading figures of the BCPS, Robert S. Cohen and Marx Wartofsky (1994 and 1995) reflect the personal aspects involved in this particular history of knowledge transfer and reception.

4.5 Felix Kaufmann, John Dewey and Edgar Zilsel: Between Phenomenology, Pragmatism and Sociology of Science

The *New School for Social Research* in New York, founded by Alvin Johnson in 1919, became a classical university of emigrants. In 1933, the University in Exile with its "Graduate Faculty of the Political and Social Science" offered a platform for German-speaking social scientists. In the following decades it also played an important role in the further development of the philosophy of science in New York. (On the history of the New School see [Rutkoff and Scott, 1986; Krohn, 1987]). A number of generations of American and Central European scientists had taught and studied on Fifth Avenue since the thirties, contributing to a project of modern social science with philosophical underpinnings [Srubar, 1988]. The Austrian contribution to this project, while limited to a few scholars, is significant for the transfer of scientific and philosophical ideas. Within the Unity of Science movement the history of science and social science were dealt with in an academic setting. The convergence of logical empiricism, phenomenology and neo-pragmatism became manifest in research, teaching and publications, most notably in the journals *Social Research* and *Philosophy and Phenomenological Research.* The Viennese mathematician, philosopher of law and social scientist Felix Kaufmann (1895–1949), the "phenomenologist of the Vienna Circle" played a seminal role in this history of reception which has received little attention to date. (On the life and work of Kaufmann [Zilian, 1990; Stadler, 1997a]).

Kaufmann had studied law and philosophy in Vienna. From 1922 to 1938 he was a lecturer of legal philosophy at the University of Vienna's School of Law, while at the same time he worked as a manager. He frequented a number of Viennese intellectual circles — from the Vienna Circle, the Kelsen School to the (Ludwig von) Mises Circle and the "Geist-Kreis" associated with F. A. von Hayek. Even before he emigrated, Kaufmann practiced interdisciplinary thought, mediating between various positions (e.g., between Husserl and the Vienna Circle or between understanding and explication). Because of his Jewish background and liberalism, he became part of the *Cultural Exodus* from Austria [Stadler and Weibel, 1995]. At the age of 43 he succeeded in securing an academic position at the New School in New York, together with his old Viennese friend Alfred Schütz: first as "associate" and from 1944 on as "full professor" for philosophy at the Graduate Faculty until his untimely death in 1949. He made an effort to come into contact with John Dewey and to discuss his ideas with him, but the latter did not cooperate as Kaufmann had hoped. He was also co-editor of the quarterly *Philosophy and Phenomenological Research*, an organ for interdisciplinary discussion, published from 1940 on, after the Dutch *Synthese* was discontinued at the outbreak of the war. The following members of the Editorial Board deserve mention since they were to a greater or lesser extent involved in this Unity of Science movement: C.J. Ducasse, Aron Gurwitsch, Charles Hartshorne, Wolfgang Köhler and Alexandre Koyré. This pluralism is also reflected in the names of the contributors to volume 6 of the journal (June 1946) whose articles were also related in a certain way: Gustav Bergmann, Rudolf Carnap, Horace M. Kallen, Felix Kaufmann, Alexandre Koyré, Richard von Mises, Ernest Nagel, Alfred Schütz and Donald Williams. The discussions ranged from induction and probability to the Unity of Science. Horace Kallen, dean and representative of "cultural pluralism" at the *New School*, contributed an impressive obituary on Otto Neurath [Kallen, 1946]. The volume also included shorter discussions and book reviews (e.g., on Kaufmann's *Methodology of the Social Science* by V. J. McGill or Reichenbach's *Philosophic Foundation of Quantum Mechanics* by Victor Lenzen) which rounded off this controversial discourse on philosophy in general and philosophy of science in particular. It comes as no surprise that the same volume includes Rudolf Carnap, Fritz Machlup, Ludwig von Mises, Günther Stern (= Günther Anders) among members of the "International Phenomenological Society" such as Felix Kaufmann and Alfred Schütz.

Austrian philosophers of science and social scientists were also represented in the official organ of the Graduate Faculty, *Social Research. An International Quarterly of Political and Social Science*, where they figured as authors and members of the Editorial Board (Felix Kaufmann and Ernst Karl Winter). There we find the *Graduate Faculty* lecture programs which have unfortunately not been considered up until now. The 1940/41 curriculum, for instance, includes a joint seminar on "Methodology of the Social Sciences" by Max Wertheimer, Gerhard Kolm, Kurt Riezler and Felix Kaufmann. The title of the seminar was later to become the title of a book published in 1944. In sociology, Kaufmann contributed "Forecast and Prediction in the Natural and Social Sciences", "Modern Philosophy and Value" and in philosophy, "The Logic of Pragmatism", "Analysis of Dewey's Logic". Later, Charles Morris and Otto Neurath's son Paul taught at the New School as visiting professors. Up until his death, Kaufmann published in the two journals named (as well as in the *Journal of Philosophy*), presenting above all his methodology of the social sciences within the context of a unified science. He made use of a methodology based on linguistic critique and phenomenological and analytical elements. In the year of his death, Kaufmann's article "The Issue of Ethical Neutrality in Political Science" appeared. This was followed, posthumously, by a long survey with the title "Basic Issues in Logical Positivism" [1950], which provided a sort of overview of the develepent of the philosophy of science from Vienna to New York.

A short analysis of Kaufmann's lecture program in the forties — New School and Graduate Faculty — show that he tried to cover the areas of "Science and Philosophy", "History and Modern Theory of Knowledge", "Philosophical Introduction to Scientific Method" and even value theory. At the same time, the official philosophy figured centrally with Morris Cohen's skeptical contribution "Scientific Method" in the *Encyclopedia of the Social Sciences* for the *New School*, together with the contributions by Horace Kallen and Sidney Hook [Rutkoff and Scott 1986, 75ff].

> "Between them, Cohen, Kallen, and Hook made versions of pragmatism the unofficial philosophy of the New School, and the school's unofficial sponsorship of the Encyclopedia of the Social Sciences reinforced its advocacy. Together with Dewey, these three comprised the core of a distinguished group of New York philosophers who dominated American philosophy between the world wars." [Ibid., 78]

Here the (neo)pragmatic background of New York which was to play a significant role in the contact with Logical Empiricism is once again addressed. The subsequent turn to phenomenological social theory (lifeworld-oriented, "understanding" social science), represented most notably by Alfred Schütz, was not counted out here.

If one reads both accounts of the *New School*, one sees that Kaufmann's own contribution to the philosophy of science was marginal. At the same time, as a continental liberal in the tradition of German enlightenment (also as a student of Hans Kelsen), he was a central figure in the interdisciplinary discussion group on liberalism and democracy against the background of the Nazi catastrophe.

> "Felix Kaufmann, ... addressed these issues from a different perspective, taking issue with Horace Kallen's assertion that democracy could thrive only where there was a tradition of liberal politics. In his reexamination of Dewey's German Philosophy and Politics, Kaufmann discussed the ways in which he thought American philosophers, particularly Dewey and Santayana, had unfairly treated the tradition of German idealism. Kaufmann defended Kant's philosophical and ethical positions against Dewey, ... Kaufmann instead argued that Kant was the embodiment of the German Enlightenment, the philosopher of reason." [Rutkoff and Scott 1986, 137f.]

In spite of his early death, Kaufmann seems to have exerted a lasting influence on American academic life in the wake of the Schütz reception [Zilian, 1990].

A similar history of reception, but with a much more tragic turn, can be found in the case of another "scholar in exile", namely the Viennese mathematician, sociologist of science, and educator *Edgar Zilsel* (1891–1944).

Zilsel, one of the pioneers of an externalist history and philosophy of science was not able to find an adequate academic position after emigrating to the United States. He had to make ends meet under the most difficult circumstances at various colleges and with the help of insufficient grants (e.g., 1939–41, through Horkheimer's "Institute for Social Research"), before finally committing suicide out of desperation and exhaustion resulting from his work on his large project on the "Social Origins of Modern Science". (On his life and work cf. Haller/Stadler, eds. 1993, in particular the articles by Dahms and Fleck). Already in Vienna, Zilsel had started working on his ambitious project on *The Social Origins of Modern Science* [Krohn, 1976; Zilsel, 1990; Dahms, 1993]. The comparison of the original plan with its actual realization clearly shows how this innovative project was carried out. The existing parts of the study in German were expanded in English, the language in which they were then published [Raven *et al.*, 2000]. These fragments also include the study "Problems of Empiricism" [1941] which was integrated into the *Encyclopedia of Unified Science.* In view of his marginalized position, already evident in Vienna, but even more pronounced in the United States, Zilsel can certainly be described as a "case of failed transfer of knowledge" [Fleck, 1993], even if the reasons for this have yet to be analysed. American sociology of knowledge was not so much influenced by Zilsel's fundamental, fragmentary studies as by Robert K. Merton's work from the year of Zilsel's emigration (1938) on. This is true in spite of the fact that both Zilsel and Merton appeared at the "Fifth International Congress for the Unity of Science" at Harvard in 1939 and Zilsel published three years later his study on the problems of empiricism in the *Encyclopedia of Unified Science.* At the Harvard congress he summarized his studies as follows:

> "In the period from the end of the Middle Ages until 1600 the university scholars and the humanistic literati are rationally trained but they do not experiment as they despise manual labor. Many more or less plebian craftsmen experiment and invent but lack methodical rational training. About 1600, with the progress of technology, the experimental method is adopted by rationally trained scholars of the educated upper class. So the two components of scientific search are united at last: modern science is born. The whole process is embedded in the advance of early capitalist economy which weakens collective-mindedness, magical thinking, traditions, and the belief in authority, which furthers mundane, rational, and causal thinking, individualism and rational organization." [Zilsel, 1939]

It is only today that studies in the history of science have, in retrospect, shown the relevance of Zilsel's oeuvre in a larger context [Raven, 2000]:

> "In the early forties Edgar Zilsel published a number of important and well-known essays on the emergence of science. These essays have

given rise to the so-called Zilsel thesis. But Zilsel published a couple of smaller and far less well-known essays. These essays are directed particularly against the efforts of South-West-German Neo-Kantianism..., Dilthey's philosophy of life, and interpretative sociology. ... His main argument is that philosophers from cultural science and the humanities proceed on the basis of a false understanding of natural science. It looks as though these two sets of essays do not seem to have much in common. Closer investigation of Zilsel's life and work reveals, however, that for Zilsel at least, there is an inner connection. The essays on the emergence of modern science are, in fact, a case study aimed at showing that law-like explanations in history are, indeed, possible; something that the other sets of essays argued in the abstract... We show how these two projects are not only complementary but in fact form part of an overarching motive of Zilsel: to argue the modernity of the socio-historical sciences."

Edgar Zilsel can be seen as a case of a slow, highly delayed transfer of knowledge through emigration. His findings have only been unearthed in contemporary history of science and science studies and now belong to the main stream together with Merton, Fleck, Kuhn and Feyerabend — even if one is often not aware of the background from which they emerged.

4.6 Epilogue: Continuity and Break in the Philosopy of Science — Boston, Pittsburgh, and Vienna

In North America the most important institutions continue to be the Philosopy of Science Association (PSA), the Boston Center and especially the Center for Philosophy of Science at the University of Pittsburgh, which was founded in 1960 by the philosopher of science Adolf Grünbaum, inspired by Feigl's Minnesota Center. (`http://www.pitt.edu/~pittcntr`).

Adolf Grünbaum (born 1923 in Köln, Germany) — currently the Andrew Mellon Professor of Philosopy of Science, Research Professor of Psychiatry, and Chairman of the Center for Philosophy of Science at the University of Pittsburgh — is one of the most distinguished and influential scholars working on philosophy of physics (space and time), the theory of scientific rationality, the philosophy of psychiatry, and the critique of theism in the tradition of Logical Empiricism (Carnap, Feigl, Hempel, and above all Hans Reichenbach). As the President of the International Union for the History and Philosophy of Science (2006/07) he continues his outstanding academic and scholarly career in this field [Cohen and Laudan 1983/1992; Earman *et al.*, 1993]. Because of its extraordinary importance the Pittsburgh Center will be desribed separately by the author on the occasion of the transfer of Adolf Günbaum's private papers to the Vienna Circle Institute, where the Robert S. Cohen collection is already located.

Regarding the long neglected return of philosopy of science back to Europe after World War II, research has been launched only in the recent years.

This is in contrast to the recent decades, where analytic philosophy and philosophy of science has become a paradigm for research and teaching in philosophy, also in the German speaking world. With the forced emigration (principally to the USA and UK) of the Vienna, Berlin and Prague Circles, representatives of Logical Empiricism disappeared almost entirely from Germany and Austria. None of them returned after the war, for there were no official invitations to do so. Nevertheless, it is possible to reconstruct some aspects of the transformation and belated return of the philosophy of science to the places of its Central European origins. The focal point of investigation is directed at some philosophers of science, who were mainly responsible for its transfer, transformation and retroactive development: Rudolf Carnap, Herbert Feigl, Wolfgang Stegmüller and the members of a Viennese post-war discussion circle around Viktor Kraft, with Paul Feyerabend and the US-Visiting Professor Arthur Pap. Feigl was the first member of the Vienna Circle to emigrate (in 1931) to the USA and to introduce Logical Empiricism into American academia. Through his contacts to European scholars after the war, he — together with Carnap — were of most importance for introducing philosophy of science in Austria and Germany. In parallel, the Viennese group around Viktor Kraft and Bela Juhos (the "Third Vienna Circle") was another attempt to revive the banished philosophy of science in the country of its origin. In the context of a hostile atmosphere Kraft tried to re-establish lost contacts and to take up international developments. Another proponent of the Kraft Circle, Wolfgang Stegmüller, succeeded because of his philosophical commitment not in Austria, but in Munich, where he founded a school of philosophy of science in close contact with Carnap and Feigl, a school which continues to be influential to this day.

We can speak of the forgotten "Third Vienna Circle", as a so far hidden story of the survival and return of philosophy of science in the Cold War phase. (Fischer/Stadler 2006).

This process was initiated by Viktor Kraft (1880–1975), who — after being dismissed by the Nazis in 1938 and working in inner emigration during the war — founded and led the so called "Kraft-Kreis" (Kraft-Circle) 1949-1953, and who contacted again some former members of the Vienna Circle (Herbert Feigl, Philipp Frank, Rudolf Carnap), and Karl Popper.

This discussion group at the Vienna based "Institut für Wissenschaft und Kunst" as well as the "Austrian College/Forum Alpbach", both still existing, was a remarkably short renaissance of the Viennese heritage in the Philosophy of Science. It exerted influence on the second wave of emigré philosophers of science after World War II — like Paul Feyerabend, who wrote his dissertation "Zur Theorie der Basissätze" under Kraft; Ernst Topitsch, who took over a chair at the University of Heidelberg; and Wolfgang Stegmüller, who succeeded at the University in Munich after being rejected by the Universities of Innsbruck and Vienna. The main results of this Circle are documented in the *Festschrift* of Kraft [1960]. Ten years later another re-transfer of philosophy of science took place at the "Institute for Advanced Study" in Vienna with Carnap, Feigl and Popper as visiting lecturers.

A decisive event in this context was Kraft's invitation of Arthur Pap as a visiting professor to Vienna in 1953/54, where he published the book *Analytische Erkenntnistheorie* [1955] with the assistance of Feyerabend, and dedicated to the Vienna Circle. Together with Feigl's article "Existential Hypotheses" [1950] this book formed the philosophical background to be debated controversially in the "Kraft Circle" (*inter alia* with Elisabeth Anscombe, Walter Hollitscher, Bela Juhos, and, by the way, with one appearance of Ludwig Wittgenstein). The main issue on the agenda was realism, especially the existence of an external world.

From a philosophical point of view the central debate was on realism vs. phenomenalism in the philosophy of science, to be continued at Feigl's "Minnesota Center for Philosophy of Science", and which obviously influenced the participating Feyerabend. From a broader perspective of the *Methodenstreit* we can identify the dualism of the hypothetico-deductive (critical or constructive) realism and inductive phenomenalism. On the one hand, the transfer of this controversy to England and America obscured its origins in the "Third Vienna Circle" and was later on overshadowed by Karl Popper's dominant preference for realism and objectivism. On the other hand, the return and modified transformation of Analytic Philosophy and Philosophy of Science back to the German speaking countries was realized by Wolfgang Stegmüller as a late consequence of the Forum Alpbach, and the Kraft Circle.

In the long run, the founding of the Vienna based "Institut Wiener Kreis" (Vienna Circle Institute) in 1991 is a late outcome of the emigration and return of the philosophy of science from the city of its origin. (`http://www.univie.ac.at/ivc`). And by the end of 2006 a "European Philosophy of Science Association" (EPSA) was founded in Vienna — as a sort of counterpart and partner of the Philosophy of Science Association (PSA), which has been the successful American institution for the philosophy of science since 1934.

5 CONCLUDING REMARKS

1. The transfer, transformation and impact of Central European, in particular Austrian, German and Polish philosophy of science in the period 1930–1960 did not take place abruptly. Rather, it involved a continuous brain drain which was reinforced by the mass exodus that set in around 1938. The early ties to Anglo-American philosophy of science prepared the ground for a pronounced convergence between "Wissenschafslogik" and the history and philosophy of science in the United States. Already in the twenties, there was a trend towards internationalization. With the dominance of neo-pragmatism/behaviorism in the context of American philosophy, the "*Wiener Kreis* in America" [Feigl, 1968] became relatively successful.

2. In spite of the fact that there was no significant intellectual remigration after 1945, there was a considerable re-transfer of knowledge primarily influenced by the contribution of former emigrants. The belated (re)discovery of the

philosophy of science in this context took place in Austria's Second Republic in connection with the paradigm of "Austrian philosophy". However, there has also been an increasing disciplinary specialization and autonomization of the *Wissenschaftstheorie* which no longer could be seen as reflecting the model of a comprehensive history and philosophy of science. The fact that the emigrated philosophers of science did not return to Central Europe was also related to the phenomenon that they had been completely uprooted from the German-speaking world.

3. From an intellectual perspective, it is striking that the transformation of the philosophy of science as a break with the so called *Received View* was already anticipated in the development of the history and philosophy of science in the thirties — long before Kuhn's path breaking book on *The Structure of Scientific Revolutions* [1962/1970]. Here one already finds a number of themes and methodological principles more or less anchored in the program of the *Encyclopedia of Unified Science*, from 1930–1960, as a result of the pragmatic, historical and naturalistic turns in the philosophy of science. This was accompanied by a development towards pluralism and relativism.

4. For historiography, these findings suggest a need to bring together exile and emigration studies with history of science research (including psychology and sociology of science) within the framework of contemporary history [Stadler, 1998/2001]. One of the most significant insights of this new perspective is the futility of linear cause-effect models given the fact that it is impossible to detect qualitative "units of impact" in the cognitive sphere. Science must be regarded as a largely complex, self-organizing project within a socio-political context. The above mentioned persons and institutions, along with a comparative account of the international history of the disciplines, provide the basic elements for a historical study of science and its philosophy. This research, however, cannot dispense with the theoretical core, i.e., representative scientific texts. The complementarity of text and context is thus postulated for historical studies.

5. In this sense, history of science can be regarded as a constitutive part of an interdisciplinary historiography. It must address from a common perspective disparate fields of human, social and natural sciences as cultural phenomena. If one draws conclusions from the *Cultural Exodus*, it becomes clear how positively factors such as migration, mobility and internationality influenced the development of philosophy and science. To do justice to history, the development from *Wissenschaftslogik*, via *Philosophy of Science*, to today's *Wissenschaftstheorie*, must be contextualized.

ACKNOWLEDGEMENT

This article is part of a research project on "The Banishment and Return of the Philosophy of Science", funded by the Austrian Science Fund (FWF), P18066-G04: `http://www.univie.ac.at/ivc/stegmueller`.

BIBLIOGRAPHY

[Achinsten and Barker, 1969] P. Achinstein and S. F. Barker, eds. *The Legacy of Logical Positivism.* Studies in the Philosophy of Science, Baltimore: The Johns Hopkins Press, 1969.

[Albert, 1994] H. Albert. Wissenschaft in Alpbach, in *Forum Alpbach*, pp. 17-22, 1994.

[Ash and Sellner, 1996] M. Ash and A. Sellner, eds. *Forced Migration and Scientific Change. Emigre German-Speaking Scientists and Scholars after 1933.* Cambridge: Cambridge University Press, 1996.

[Ash and Sellner, 1996b] M. Ash and A. Sellner. Forced Migration and Scientific Change after 1933, in: [Ash and Sellner, 1996, pp. 1–22].

[Auer *et al.*, 1994] A. Auer, Behrendt, Flora, and Knapp, eds. *Das Forum Alpbach 1945-1994. Die Darstellung einer Europäischen Zusammenarbeit.* Hrsg. von Alexander Auer. Wien: Ibera Verlag European University Press, 1994.

[Ayer, 1936a] A. J. Ayer. *Language, Truth and Logic*, London: Victor Gollancz Ltd, 1936. (15th impression 1955).

[Ayer, 1936b] A. J. Ayer. The Analytic Movement in Contemporary Philosophy, in: *Actes du Congres International de Philosophie Scientifique* VIII, Paris: Hermann & Cie, 1936.

[Ayer, 1959] A. J. Ayer.*Logical Positivism.*Glencoe, Ill.: The Free Press, 1959.

[Ayer, 1970] A. J. Ayer. *Sprache, Wahrheit und Logik.* Hrsg. van H. Herring. Stuttgart: Reclam, 1970.

[Ayer, 1982] A. J. Ayer. *Philosophy in the Twentieth Century*, London: Weidenfeld and Nicolson, 1982.

[Ayer *et al.*, 1956] A. Ayer, W. C. Kneale, G. A. Paul, D. F. Pears, P. F. Strawson, G. J. Warnock and R. A. Wollheim. . *The Revolution in Philosophy.* With an Introduction by Gilbert Ryle. London: Macmillan & Co. Ltd, 1956.

[Bernard and Stadler, 1997] J. Bernard and F. Stadler, (Hrsg). *Neurath: Semiotische Projekte und Diskurse.* Wien: OGS, 1997.

[Black, 1939/40] M. Black. Relations between Logical Positivism and the Cambridge School of Analysis, *Erkenntnis/Journal of Unified Science* VIII, 24-35, 1939/40.

[Black *et al.*, 1972] M. Black, E. H. Gombrich, and J. Hochberg. *Art, Perception, and Reality*, Baltimore-London: The Johns Hopkins Press, 1972.

[Blackmore, 1992] J. Blackmore, ed. *Ernst Mach - A Deeper Look. Documents and New Perspectives.* Dordrecht-Boston-London: Kluwer, 1992.

[Blumberg and Feigl, 1931] A. E. Blumberg and H. Feigl. Logical Positivism, *Journal of Philosophy* 28, 281-296, 1931.

[Brunswik, 1952] E. Brunswik. *The Conceptual Framework of Psychology*, in: Neurath/Carnap/Morris (Eds.) 1970, 655-760, 1952.

[Buck and Cohen, 1970] R. C. Buck and R. S. Cohen, eds. *PSA 1970. In Memory od Rudolf Carnap.* Dordrecht-Boston: Reidel, 1970.

[Carnap, 1932] R. Carnap. Oberwindung der Metaphysik durch logische Analyse der Sprache, in: *Erkenntnis* II, 91-105, 1932.

[Carnap, 1934] R. Carnap. *The Unity of Science.* Translated with an Introduction by M. Black, London: Kegan Paul, 1934.

[Carnap, 1934a/1968] R. Carnap. *Logische Syntax der Sprache.* Wien: Springer, 1934a/1968.

[Carnap, 1934b] R. Carnap. *Die Aufgabe der Wissenschaftslogik.* Wien: Gerold & Co, 1934.

[Carnap, 1934c] R. Carnap. *The Unity of Science.* London: Kegan Paul, 1934.

[Carnap, 1935] R. Carnap. *Philosophy and Logical Syntax*, London: Kegan Paul, 1935.

[Carnap, 1936] R. Carnap. Von der Erkenntnistheorie zur Wissenschaftslogik, in: *Actes de Congres International de Philosophie Scientifique* I. Paris: Hermann & Cie., 36-41, 1936.

[Carnap, 1937] R. Carnap. *The Logical Syntax of Language.* London: Kegan Paul, 1937.

[Carnap, 1942] R. Carnap. *Introduction to Semantics.* Cambridge, Mass.:Harvard University Press, 1942.

[Carnap, 1963] R. Carnap. Intellectual Autobiography, in *The Philosophy of Rudolf Carnap*, P. A. Schilpp, ed., pp. 1–84. La Salle, Ill.: Open Court, 1963.

[Carnap, 1993] R. Carnap. *Mein Weg in die Philosophie.* Übersetzt und mit einem Nachwort sowie einem Interview hrsg. von Willy Hochkeppel. Stuttgart: Reclam, 1993.

[Carnap *et al.*, 1938] R. Carnap, Ch. Morris, and O. Neurath, eds. *International Encyclopedia of Unified Science* 1-19, 1938. Reprint: *Foundations of the Unity of Science*, Chicago and London: University of Chicago Press, 2 Volumes, 1970-71.

[Cohen, 1994] I. B. Cohen. *Revolutionen in der Naturwissenschaft.* Frankfurt/M.: Suhrkamp, 1994.

[Cohen, 1956] R. S. Cohen. Alternative Interpretations of the History of Science, in: Frank (ed.), 218-132, 1956.

[Cohen, 1963a] R. S. Cohen, ed. *Boston Studies in the Philosophy of Science.* Dordrecht-Boston-London: Kluwer, 1963.

[Cohen and Laudan, 1983] R. S. Cohen and L. Laudan, eds. *Physics, Philosophy and Psychoanalysis. Essays in Honor of Adolf Grünbaum.* Dordrecht-Boston-Lancaster: D.Reidel, 1983.

[Cohen and Wartofsky, 1963b] R. S. Cohen and M. W. Wartofsky. Preface, in: Cohen (Ed.), Vllf, 1963.

[Cohen and Wartofsky, 1965a] R. S. Cohen and M. W. Wartofsky. Preface, in: Cohen and Wartofsky, VII, 1965b.

[Cohen and Wartofsky, 1965b] R. S. Cohen and M. W. Wartofsky, eds. *Boston Studies in the Philosophy of Science. Volume Two: In Honor of Philipp Frank.* New York: Humanities Press, 1965.

[Cohen and Wartofsky, 1985] R. S. Cohen and M. W. Wartofsky, eds. *A Portrait of Twenty-Five Years.* Boston Colloquium for the Philosophy of Science 1960-1985. Dordrecht-Boston-London: Reidel, 1985.

[Conant, 1951] J. B. Conant. *Science and Common Sense.* New Haven:Yale University Press, 1951.

[Coser, 1984] L. Coser. *Refugee Scholars in America. Their Impact and their Experiences.* New Haven-London: Yale University Press, 1984.

[Coser, 1988] L. Coser. Die 6sterreichische Emigration als Kulturtransfer Europa - Amerika, in: Stadler (Hrsg.), 93-101, 1988.

[Creath, 1990] R. Creath, ed. *Dear Carnap, Dear Van: The QuineCarnap Correspondence and Related Work.* Berkeley-Los AngelesLondon: University of California Press, 1990.

[Dahms, 1987] H.-J. Dahms. Die Emigration des Wiener Kreises, in: Stadler (Hrsg.), 66-122, 1987.

[Dahms, 1988] H.-J. Dahms. Die Bedeutung der Emigration des Wiener Kreises für die Entwicklung der Wissenschaftstheorie in: Stadler (Hrsg.), 155-168, 1988.

[Dahms, 1993] H.-J. Dahms. Edgar Zilsels Projekt 'The Social Roots of Science' und seine Beziehungen zur Frankfurter Schule, in: Haller and Stadler (Hrsg.), 474-500, 1993.

[Dahms, 1994] H.-J. Dahms. *Positivismusstreit. Die Auseinandersetzungen der Frankfurter Schule mit dem logischen Positivismus, dem amerikanischen Pragmatismus und dem kritischen Rationalismus.* Frankfurt/M.: Suhrkamp, 1994.

[Dahms, 1997] H.-J. Dahms. Positivismus, Pragmatismus, Enzyklopadieprojekt, Zeichentheorie, in: Bernard/Stadler (Hrsg.), 1997.

[Danneberg *et al.*, 1994] L. Danneberg, A. Kamiah and L. Schafer (Hrsg.). *Hans Reichenbach und die Berliner Gruppe.* Braunschweig: Vieweg, 1994.

[Dawson, 1997] J. W. Dawson. *Logical Dilemmas. The Life and Work of Kurt Gödel.* Wellesley, MA: A K Peters, 1997.

[De-Pauli-Schimanovich *et al.*, 1995] W. DePauli-Schimanovich, E. Kohler, and F. Stadler, eds. *The Foundational Debate. Complexity and Constructivity in Mathematics and Physics.* Dordrecht-Boston-London: Kluwer, 1995.

[Dewey, 1951] J. Dewey. *Theory of Valuation.* The University of Chicago Press. (= *International Encyclopedia of Unified Science* II/4), 1951.

[Earman *et al.*, 1993] J. Earman, A. I. Janis, G. J. Massey, and N. Rescher, eds. *Philosophical Problems of the Internal and External Worlds. Essays on the Philosophy of Adolf Grünbaum.* University of Pittsburgh Press and Universitätsverlag Konstanz, 1993.

[Edmonds and Eidinow, 2001] D. J. Edmonds and J. A. Eidinow. *Wittgenstein's Poker. The Story of a Ten-Minute Argument between two Great Philosophers*, London: Faber and Faber, 2001. German edition: *Wie Wittgenstein Karl Popper mit dem Feuerhaken drohte. Eine Ermittlung,* Stuttgart-München: DVA *Einheitswissenschaft* 1992. Hrsg. von Joachim Schulte und Brian McGuinness. Mit einer Einleitung von Rainer Hegselmann. Frankfurt/M.: Suhrkamp.

[Faludi, 1986] A. Faludi. *Critical Rationalism and Planning Methodology.* London: Pion Limited, 1986.

[Farber, 1950] M. Farber. *Philosophic Thought in France and the United states. Essays Representing Major Trends in Contemporary French and American Philosophy.* New York: University of Buffalo Publ, 1950.

[Feigl, 1936] H. Feigl. Sense and Nonsense in Scientific Realism, in: *Actes du Congres International de Philosophie Scientifique* III. Paris: Hermann & Cie., 50-56, 1936.

[Feigl, 1956a] H. Feigl. Preface, in: Feigl and Scriven (Eds.), V-VII, 1956.

[Feigl, 1956b] H. Feigl. Some Major Issues and Developments in the Philosophy of Science of Logical Empiricism, in: Feigl and Scriven (Eds.), 3-37, 1956.

[Feigl, 1968] H. Feigl. The *Wiener Kreis* in America, in *Charles Warren Center for Studies in American History*, ed., Perspectives in American History, Vol. II: Harvard University Press, 630-673, 1968.

[Feigl, 1969] H. Feigl. The *Wiener Kreis* in America, in Fleming, D. and Bailyn, B., eds., The *Intellectual Migration. Europe and America, 1930 – 1960.* Cambridge, Mass.: Harvard University Press, 630-673, 1969. Reprinted in Feigl 1981, 57-94.

[Feigl, 1981] H. Feigl. *Inquiries and Provocations. Selected Writings, 1929-1974.* Ed. by Robert S. Cohen. Dordrecht-Boston-London: Reidel, 1981.

[Feigl and Blumberg, 1931] H. Feigl and A. Blumberg. Logical Positivism. A New Movement in European Philosophy, in: *Journal of Philosophy* 28, 281-296, 1931.

[Feigl and Brodbeck, 1953] H. Feigl and M. Brodbeck, eds. *Readings in the Philosophy of Science.* New York: Appleton-Century-Crofts, 1953.

[Feigl and Scriven, 1956] H. Feigl and M. Scriven, eds. Minnesota Studies in the Philosophy of Science. Volume I: *The Foundations of Science and the Concepts of Psychology and Psychoanalysis.* Minneapolis: University of Minnesota Press, 1956.

[Feigl and Sellars, 1949] H. Feigl and W. Sellars, eds. *Readings in Philosophical Analysis.* New York: Appleton-Century-Crofts, 1949.

[Felderer, 1993] B. Felderer (Hrsg.) *Wirtschafts- und Sozialwissenschaften zwischen Theorie und Praxis. 30 Jahre Institut far Höhere Studien in Wien.* Heidelberg: Physica-Verlag, 1993.

[Felt *et al.*, 1995] U. Felt, H. Nowotny, and K. Taschwer. *Wissenschaftsforschung. Eine Einführung.* Frankfurt/M.-New York: Campus, 1995.

[Feyerabend, 1955] P. K. Feyerabend. Wittgenstein's 'Philosophical Investigations', in: The *Philosophical Review* 64, 449-483, 1955. Auch in: Feyerabend 1981, 293-325.

[Feyerabend, 1966] P. K. Feyerabend. Herbert Feigl: A Biographical Sketch, in: [Feyerabend and Maxwell, 1966, pp. 3–16].

[Feyerabend, 1978] P. K. Feyerabend. *Der wissenschaftstheoretische Realismus und die Autorität der Wissenschaften. Ausgewählte Schriften,* Band 1. Braunschweig-Wiesbaden, 1978.

[Feyerabend, 1981] P. K. Feyerabend. *Probleme des Empirismus. Schriften zur Theorie der Erklärung, der Quantentheorie und der Wissenschaftsgeschichte. Ausgewählte Schriften,* Band 2. Braunschweig-Wiesbaden: Vieweg, 1981.

[Feyerabend, 1995] P. K. Feyerabend. *Zeitverschwendung.* Frankfurt/M.: Suhrkamp, 1995.

[Feyerabend and Maxwell, 1966] P. K. Feyerabendand G. Maxwell, eds. *Mind, Matter, and Method. Essays in Philosophy and Science in Honor of Herbert Feigl.* Minneapolis: University of Minnesaota Press, 1966.

[Fischer, 1995] K. R. Fischer (Hrsg.). *Das goldene Zeitalter der Österreichischen Philosophie. Ein Lesebuch.* Wien: WUV-Verlag, 1995.

[Fisher and Stadler, 1997] K. R. Fischer and F. Stadler (Hrsg.). *'Wahrnehmung und Gegenstandswelt'. Zum Lebenswerk von Egon Brunswik (1903-1955).* Wien-New York: Springer, 1997.

[Fischer and Stadler, 2006] K. R. Fischer and F. Stadler (Hrsg.). *Paul Feyerabend – Ein Philosoph aus Wien.* Wien-New York: Springer, 2006.

[Fischer and Wimmer, 1993] K. R. Fischer and F. M. Wimmer (Hrsg.). *Der geistige Anschluß. Philosophie und Politik an der Universität Wien 19301950.* Wien: WUV-Verlag, 1993.

[Fleck, 1993] C. Fleck. Marxistische Kausalanalyse und funktionale Wissenschaftssoziologie. Ein Fall unterbliebenen Wissenstransfers, in: Haller/Stadler (Hrsg.), 501-524, 1993.

[Fleming and Bailyn, 1969] D. Fleming and B. Bailyn, eds. *The Intellectual Migration. Europe and America, 1930-1960.* Cambridge: Harvard University Press, 1969.

[Frank, 1947] P. Frank. *Einstein - His Life and Times.* New York: Knopf, 1947.

[Frank, 1949] P. Frank. *Modern Science and its Philosophy.* Harvard University Press, 1949.

[Frank, 1950] P. Frank. *Relativity – A Richer Truth.* Preface by Albert Einstein. Boston: Beacon Press, 1950. German edition: Zürich: Pan Verlag 1952.

[Frank, 1952] P. Frank. *Wahrheit - relativ oder absolut?* Mit einem Vorwort von Albert Einstein. ZUrich: Pan-Verlag, 1952.

[Frank, 1956] P. Frank, ed. *The Validation of Scientific Theories.* Boston: The Beacon Press, 1956.

[Frenkel-Brunswik, 1996] E. Frenkel-Brunswik. *Studien zur autoritaren Persönlichkeit.* Hrsg. und eingelitet von Dietmar Paier. Graz: Nausner & Nausner, 1996.

[Fuller, 2000] S. Fuller. Thomas Kuhn. *A Philosophical History for Our Time.* University of Chicago Press, 2000.

[Galison and Stump, 1996] P. Galison and D. Stump, eds. *The Disunity of Science.* Stanford University Press, 1996.

[Gavroglu *et al.*, 1995] K. Gavroglu, J. Stachel and M. W. Wartofsky, eds. *Physics, Philosophy and the Scientific Community; Science Politics and Social Practice; Science Mind and Art.* 3 Vlms. In Honor of Robert S. Cohen. Dordrecht-Boston-London: Kluwer, 1995.

[Giere, 1996] R. N. Giere. From Wissenschaftliche Philosophie to Philosophy of Science, in: [Giere and Richardson, 1996, pp. 335–354].

[Giere and Richardson, 1996] R. N. Giere and A. W. Richardson, eds. Minnesota Studies in the Philosophy of Science. Volume XVI: *Origins of Logical Empiricism.* Minneapolis-London: University of Minnesota Press, 1996.

[Gould and Cohen, 1994] C. C. Gould and R. S. Cohen, eds. *Artifacts, Representations and Social Practice. Essays for Marx Wartofsky.* Dordrecht-Boston-London: Kluwer, 1994.

[Gower, 1987] B. Gower, ed. *Logical Positivism in Perspective. Essays on Language, Truth and Logic.* London-Sydney: Croom Helm, 1987.

[Hacker, 1996] P. M. S. Hacker. *Wittgenstein's Place in Twentieth Century Analytic Philosophy,* Oxford: Basil Blackwell, 1996.

[Hacohen, 2000] M. Hacohen. *Karl Popper – The Formative Years, 1902 – 1945. Politics and Philosophy in Interwar Vienna,* Cambridge University Press, 2000.

[Haller, 1988] R. Haller. Die philosophische Entwicklung in Osterreich am Beginn der Zweiten Republik, in: [Stadler, 1987/88, pp. 157–180].

[Haller, 1993] R. Haller. *Neopositivismus. Eine historische Einführung in die Philosophie des Wiener Kreises.* Darmstadt: Wissenschaftliche Buchgemeinschaft, 1993.

[Haller and Stadler, 1988] R. Haller and F. Stadler (Hrsg.). *Ernst Mach - Werk und Wirkung.* Wien: Hölder-Pichler-Tempsky, 1988.

[Haller and Stadler, 1993] R. Haller and F. Stadler (Hrsg.). *Wien-Berlin-Prag. Der Aufstieg der wissenschaftlichen Philosophie.* Wien: Hölder-Pichler-Tempsky, 1993.

[Hardcastle and Richardson, 2004] G. Hardcastle and A. W. Richardson, eds. *Logical Empiricism in North America.* Minneapolis-London: University of Minnesota Press, 2004.

[Hark, 2004] M. ter Hark. *Popper, Otto Selz and the Rise of Evolutionary Epistemology.* Cambridge: Cambridge University Press, 2004.

[Hanisch, 1995] E. Hanisch. *Der lange Schatten des Staates. Österreichische Gesellschaftsgeschichte im 20. Jahrhundert.* Wien: Ueberreuter, 1995.

[Hayek, 1942] F. A. Hayek. Scientism and the Study of Society, *Economica,* IX-XI, 1942.

[Hayek, 1944] F. A. Hayek. *The Road to Serfdom,* London: Routledge, 1944. (Fiftieth Anniversary Edition: University of Chicago Press 1994).

[Hayek, 1945] F. A. Hayek and O. Neurath. Correspondence 1945. Vienna Circle Archives, Haarlem (NL).

[Hardcastle, 1996] G. Hardcastle. The Science Of Science Discussion Group at Harvard, 1940-41. Abstract. First International HOPOS Conference, Roanoke, Virgina, 19-21 April 1996, 24f.

[Hegselmann, 1988] R. Hegselmann. Alles nur Mißverständnisse? Zur Vertreibung des Logischen Empirismus aus Osterreich und Deutschland, in: Stadler (Hrsg.), 188-203, 1988.

[Heidelberger and Stadler, 2002] M. Heidelberger and F. Stadler, eds. *History of Philosophy of Science. New Trends and Perspectives.* Dordrecht-Boston-London: Kluwer 2002.

[Helbling and Wagnleitner, 1992] W. Helbling and R. Wagnleitner, eds. *The European Emigrant Experience in the U.S.A.* Tübingen: Gunter Narr, 1992.

[Hintikka and Puhl, 1995] J. Hintikka and K. Puhl, eds. *The British Tradition in 20th Century Philosophy.* Vienna: Holow-Pichler-Tempsky, 1995.

[Holton, 1993] G. Holton. From the Vienna Circle to Harvard Square: The Americanization of a European World Conception, in: [Stadler, 1993, pp. 47–74].

[Holton, 1993b] G. Holton. *Science ant Anti-Science.* Cambridge, Mass.: Harvard University Press, 1993.

[Holton, 1994] G. Holton. *Thematic Origins of Scientific Thought. Kepler to Einstein.* Revised Edition. Harvard University Press, 1994.

[Holton, 1995] G. Holton. On the Vienna Circle in Exile: An Eyewitness Report, in: [DePauli-Schimanovich *et al.*, 1995, pp. 269]-292].

[Hughes, 1975] H. S. Hughes. *The Sea Change: The Migration of Social Thought, 1930-1965.* New York, 1975

[Hughes, 1961] H. S. Hughes. *Consciousness and Society. The Reorientation of European Social Thought 1890 - 1930,* New York: Vintage Books, 1961.

[Hughes, 1983] H. S. Hughes. Social Theory in a New Context, in: [Jackman and Borden, 1983, pp. 111-122].

[Jackman and Borden, 1983] J. C. Jackman and C. M. Borden, eds. *The Muses Flee Hitler. Cultural Transfer and Adaption 1930-1945.* Washington, D.C.: Smithsonian Institution Press, 1983.

[Juhos, 1965] B. Juhos. Gibt es in Osterreich eine wissenschaftliche Philosophie?, in: *Österreich - Geistige Provinz?,* pp. 232-244, 1965.

[Kallen, 1946] H. M. Kallen. Postscript: Otto Neurath, 1882-1945, in: *Philosophy and Phenomenological Research.* VI/4, 529-533, 1946.

[Kamlah, 1983] A. Kamlah. Die philosophiegeschichtliche Bedeutung des Exils (nichtmarxistischer) Philosophen zur Zeit des Dritten Reiches, in: *Dialektik* 7, 29-43, 1983.

[Katz, 1991] B. Katz. The Acculturation of Thought: Transformations of the Refugee Scholar in America, in: *Journal of Modern History* 63, 740-752, 1991.

[Kaufmann, 1936] F. Kaufmann. *Methodenlehre der Sozialwissenschaften.* Wien: Springer, 1936.

[Kaufmann, 1944] F. Kaufmann. *Methodology of the Social Sciences.* New York-London: Oxford University Press, 1944.

[Kaufmann, 1949] F. Kaufmann. The Issue of Ethical Neutrality in Political Science, in: *Social Research* 16, 344-352, 1949.

[Kaufmann, 1950] F. Kaufmann. Basic Issues in Logical Positivism, in: [Farber, 1950, pp. 565-588].

[Kaufmann, 1978] F. Kaufmann. *The Infinite in Mathematics. Logicomathematical Writings.* Ed. by Brian McGuinness. With an Introduction by Ernest Nagel. Dordrecht-Boston-London: Reidel, 1978.

[Kazemier and Vuysje, 1962] B. H. Kazemier and D. Vuysje, eds. *Logic and Language. Studies Dedicated to Professor Carnap on the Occasion of his Seventeenth Birthday.* Dordrecht: Reidel, 1962.

[Kendler, 1989] H. Kendler. The Iowa Tradition, in: *American Psychologist* 44/8, 1124-1132, 1989.

[Kitcher, 2001] P. Kitcher. *Science, Truth, and Democracy.* Oxford University Press, 2001.

[Koertge, 1998] N. Koertge, ed. *A House Built on Sand. Exposing Postmodernist Myths about Science.* Oxford University Press, 1998.

[Koppelberg, 1987] D. Koppelberg. *Die Aufhebung der analytischen Philosophie. Quine als Synthese von Carnap und Neurath.* Frankfurt/M.: Suhrkamp, 1987.

[Kraft, 1950] V. Kraft. *Der Wiener Kreis. Der Ursprung des Neopositivismus. Ein Kapitel der jüngsten Philosophiegeschichte.*Wien: Springer, 1950.

[Kroner, 1988] H.-P. Kroner. Überlegungen zur Wirkungsgeschichte der deutschsprachigen wissenschaftlichen Emigration, in: [Stadler, 1988, pp. 82-92].

[Krohn, 1987] C.-D. Krohn. *Wissenschaft im Exil. Deutsche Sozial und Wirtschaftswissenschaftler in den USA und die New School for Social Research.* Frankfurt/M.: Campus, 1987.

[Langer, 1988] J. Langer, (Hrsg.). *Geschichte der österreichischen Soziologie. Konstituierung, Entwicklung und europäische Bezüge.* Wien: Verlag für Gesellschaftskritik, 1988.

[Lazarsfeld, 1993] P. F. Lazarsfeld. The Pre-History of the Vienna Institute for Advanced Studies (1973), in: [Felderer, 1993, pp. 9-50].

[Lehrer and Marek, 1997] K. Lehrer and J. C. Marek, eds. *Austrian Philosophy. Past and Present. Essays in Honor of Rudolf Haller.* Dordrecht-Boston-London: Kluwer, 1997.

[Leinfellner, 1988] W. Leinfellner. Oskar Morgenstern, in: [Stadler, 1988, pp. 416-424].

[Leinfellner, 1993] W. Leinfellner. Der Wiener Kreis und sein Einfluß auf die Sozialwissenschaften, in: [Haller and Stadler, 1993, pp. 593-618].

[Leinfellner *et al.*, 1997] W. Leinfellner, *et al*, eds. *Game Theory, Experience, Rationality. In Honor of John C. Harsanyi.* Dordrecht-BostonLondon: Kluwer, 1997.

[Losee, 1980] J. Losee. *A Historical Introduction to the Philosophy of Science.* Oxford-New York-Toronto-Melbourne: Oxford University Press, 1980.

[Mach, 1905] E. Mach. *Erkenntnis und Irrtum. Skizzen zur Psychologie der Forschung.* Leipzig: J.A.Barth, 1905. English: *Knowledge and Error. Sketches on the Psychology of Enquiry.* With an Introduction by Erwin N. Hiebert. Dordrecht-Boston 1976.

[Misak, 1995] C. J. Misak. *Verificationism. Its History and Prospects.* London-New York: Routledge, 1995.

[Marin, 1978] B. Marin. *Politische Organisation sozialwissenschaftlicher Forschungsarbeit. Fallstudie zum Institut für Höhere Studien.* Wien: Braumüller, 1978.

[McCulloch, 1956] W. S. McCulloch. Mysterium Iniquitatis - Of Sinful Man Aspiring into the Place of God, in: [Frank, 1956, pp. 159–170].

[Menger, 1934] K. Menger. *Moral, Wille und Weltgestaltung. Grundlegung zur Logik der Sitten.* Wien: Springer, 1934. Reprint: Hrsg. von Uwe Czaniera, Frankfurt/M: Suhrkamp 1997.

[Menger, 1994] K. Menger. *Reminiscences of the Vienna Circle and the Mathematical Colloquium.* Ed. by L. Golland/B. McGuinness/A. Sklar. Dordrecht-Boston-London: Kluwer, 1994.

[Misak, 1995] C. J. Misak. *Verificationism. Its History and Prospects.* London-New York: Routledge, 1995.

[Mises, 1951] R. von Mises. *Positivism. An Essay in Human Understanding.* New York: Dover. Ursprünglich 1939 erschienen unter dem Titel: *Kleines Lehrbuch des Positivismus. Einführung in die empiristische Wissenschaftsauffassung.* Den Haag: Van Stockum & Zoon, 1951. Reprint: 1990 Frankfurt/M: Suhrkamp, hrsg. von Friedrich Stadler.

[Molden, 1981] O. Molden. *Der andere Zauberberg. Das Phänomen Alpbach.* Wien-München-Zürich-New York: F. Molden, 1981.

[Monk, 1990] R. Monk. *Ludwig Wittgenstein. The Duty of Genius*, London: Jonathan Cape, 1990.

[Morgenstern, 1936] O. Morgenstern. Logistik und Sozialwissenschaften, in: *Zeitschrift far Nationalökonomie* 7, 1-24, 1936.

[Morgenstern and von Neumann, 1944] O. Morgenstern and J. von Neumann. *Theory of Games and Economic Behavior.* Princeton: Princeton University Press, 1944.

[Morris, 1935] C. Morris. Some Aspects of American Scientific Philosophy, in: *Erkenntnis* 5, 142-150, 1935.

[Morris, 1936] C. Morris. Opening Speech (For the American Delegates), "Semiotic and Scientific Empiricism", in: *Actes du Congres International de Philosophie Scientifique* I. Paris: Hermann & Cie., 22 bzw. 42-56, 1936.

[Morris, 1938] C. Morris. *Foundations of the Theory of Signs.* VIm. 1/2, 1938. Auch in Neurath/Carnap/Morris 1970, 77-138.

[Nagel, 1936] E. Nagel. Impressions and Appraisals of Analytic Philosophy in Europe, in: *Journal of Philosophy* 33, 1936. Auch in Nagel 1956, 191-246.

[Nagel, 1936] E. Nagel. Impressions and Appraisals of Analytic Philosophy, Journal of Philosophy XXXIII. Reprint: Nagel, E. 1956. *Logic without Metaphysics and other Essays in the Philosophy of Science*, Glencoe, Ill.: The Free Press, 191-246, 1936.

[Nagel, 1956] E. Nagel. *Logic without Metaphysics and other Essays in the Philosophy of Science.* Glencoe: The Free Press, 1956.

[Neurath, 1935] O. Neurath. Pseudorationalismus der Falsifikation *Erkenntnis* V, 353-365, 1935. Reprint: Neurath 1983, 121-131.

[Neurath, 1936] O. Neurath. Erster Internationaler Kongress für Einheit der Wissenschaft in Paris 1935, *Erkenntnis* 5, 377-428, 1936.

[Neurath, 1941] O. Neurath. Universal Jargon and Terminology, in: *Proceedings of the Aristotelian Society* 41, 127-148, 1941. In deutscher Übersetzung in Neurath 1981, 901-924.

[Neurath, 1942] O. Neurath. International Planning for Freedom, *The New Commonwealth Quaterly*, April/July 1942, 23-28, 1942. Reprint: Neurath 1973, 422-440. (Neurath was member of the Editorial Board of the NCQ and its successor, *The London Quaterly of World Affairs*).

[Neurath, 1945] O. Neurath. The Road to Serfdom, *The London Quaterly of World Affairs*, Jan. 1945, 121f.

[Neurath, 1973] O. Neurath. *Empiricism and Sociology*. Ed. by M. Neurath and R.S. Cohen, Dordrecht-Boston: Reidel, 1973.

[Neurath, 1981] O. Neurath. *Gesammelte philosophische und methodologische Schriften.* 2 Bande. Hrsg. von Rudolf Haller und Heiner Rutte. Wien: Hölder-Pichler-Tempsky, 1981.

[Neurath, 1983] O. Neurath. *Philosophical Papers 1913-1946*. Ed. by R.S. Cohen and M. Neurath. Dordrecht-Boston-Lancaster: Reidel, 1983.

[Neurath, 1987] O. Neurath. The New Encyclopedia, in: *Unified Science*, p. 136f. Ed. by Brian McGuiness. Dordrecht: Reidel, 1987.

[Neurath *et al.*, 1938] O. Neurath, E. Brunswik, C. L. Hull, G. Mannoury, and J. H. Woodger. *Zur Enzyklopadie der Einheitswissenschaft. Vorträge.* Den Haag: VanStockum & Zoon, 1938. Wiederabgedruckt in: Einheitswissenschaft 1992.

[Neurath *et al.*, 1970] O. Neurath, R. Carnap,and C. Morris, eds. *Foundations of the Unity of Science. Toward an International Encyclopedia of Unified Science.* 2 Vlms. Chicago-London: The University of Chicago Press, 1970.

[Österreich, 1965] *Österreich- - Geistige Provinz?* 1965. Wien-Hannover-Bern: Forum Verlag.

[Österreicher, 1995] *Österreicher im Exil. USA 1938-1945.* DÖW (Hrsg.) 1995. Eine Dokumentation. 2 Bande. Einleitung, Auswahl und Bearbeitung: Peter Eppel. Wien: Osterreichischer Bundesverlag.

[Pap, 1955] A. Pap. *Analytische Erkenntnistheorie. Kritische Übersicht aber die neueste Entwicklung in den USA und England.* Wien: Springer, 1955.

[Popper, 1934] K. R. Popper. *Logik der Forschung. Zur Erkenntnistheorie der modernen Naturwissenschaft*, Wien: Verlag Julius Springer, 1934. Published: 1935 (= Schriften zur wissenschaftlichen Weltauffassung, ed. by Ph. Frank and M. Schlick, Vol. 9). First English edition: *The Logic of Scientific Discovery*, London: Hutchinson & Co. New York: Basic Books 1959.Sixth (revised) impression 1972.

[Popper, 1974] K. R. Popper. Intellectual Autobiography, in *The Philosophy of Karl Popper.* Ed. by P.A. Schilpp, La Salle, Ill.: OpenCourt, 3-181, 1974.

[Pabisch, 1989] P. Pabisch, ed. *From Wilson to Waldheim.* Riverside: Ariadne Press, 1989.

[Platt and Hoch, 1996] J. Platt and P. K. Hoch. The Vienna Circle in the United States and Empirical Research Methods in Sociology, in: [Ash and Sollner, 1996, pp. 224-245].

[Quine, 1951] W. V. O. Quine. Two Dogmas of Empiricism, in: *Philosophical Review* 60, 20-43, 1951. Auch in Quine 1953.

[Quine, 1953] W. V. O. Quine. *From a Logical Point of View. 9 Logico-Philosophical Essays.* Cambridge, Mass.: Harvard University Press, 1953.

[Ramsey, 1923] F. P. Ramsey. Correspondence with Moritz Schlick. Schlick Papers, Vienna Circle Archives, Haarlem, NL, 1923.

[Raven and Krohn, 1996/2000] D. Raven and W. Krohn. Edgar Zilsel. His Life and Work, in: [Raven *et al.*, 2000, pp. xix-lxi].

[Raven *et al.*, 2000] D. Raven, W. Krohn, and R. S. Cohen, eds. Edgar Zilsel, *The Social Origins of Modern Sciences.* Dordrecht-Boston-London: Kluwer, 2000.

[Regis, 1989] E. Regis. *Einstein, Godel & Co. Genialitat und Exzentrik Die Princeton-Geschichte.* Basel-Boston-Berlin: Birkhauser, 1989

[Reichenbach, 1938] H. Reichenbach. *Experience and Prediction. An Analysis of the Foundations and the Structure of Knowledge.* The University of Chicago Press 1938. Deutsch: *Erfahrung und Prognose. Eine Analyse der Grundlagen und der Struktur der Erkenntnis.* Mit Erläuterungen von Alberto Coffa. Braunschweig-Wiesbaden: Vieweg 1983.

[Reichenbach, 1951] H. Reichenbach. *The Rise of Scientific Philosophy.* University of California Press 1951. Deutsch: *Der Aufstieg der wissenschaftlichen Philosophie.* Berlin:Grunewald: Herbig 1951

[Reisch, 1995] G. Reisch. *A History of the International Encyclopedia of Unified Science.* Ph.D. Thesis. Chicago 1995.

[Reisch, 2005] G. Reisch. *How the Cold War Transformed Philosopy of Science. To the Icy Slopes of Logic.* Cambridge University Press 2005.

[Richardson and Uebel, 2006] A. Richardson and T. Uebel, eds. *The Cambridge Companion of Logical Empiricism.* Cambridge University Press, 2006.

[Ringer, 1997] F. Ringer. *Max Weber's Methodology. The Unification of the Cultural and Social Sciences.* Harvard University Press, 1997.

[Runes, 1944] D. Runes, ed. *The Dictionary of Philosophy.* London: Routledge, 1944.

[Russell, 1905] B. Russell. On Denoting, in *Mind* 14, 479-473, 1905.

[Russell, 1914] B. Russell. *Our Knowledge of the External World as a Field for Scientific Method* in Philosophy, London: Open Court, 1914.

[Russell, 1936] B. Russell. The Congress of Scientific Philosophy, in *Actes du Congrés de Philosophie scientifique*, Sorbonne 1935, Paris: Hermann & Cie, 10-12, 1936.

[Russell, 1938] B. Russell. On the Importance of Logical Form, 1938. Reprint: Carnap, Morris, Neurath 1971, 39ff.

[Russell, 1940] B. Russell. *An Inquiry into Meaning and Truth*, London: Allen & Unwin, 1940.

[Rutkoff and Scott, 1986] P. M. Rutkoff and W. B. Scott. *New School. A History of The New School for Social Research.* New York-London: The Free Press-Collier Macmillan Pub, 1986.

[Schlick, 1930/31] M. Schlick. Ü ber wissenschaftliche Weltauffassung in den Vereinigten Saaten von Amerika, in: *Erkennntis* 1, 75f, 1930/31.

[Schlick, 1930] M. Schlick. Die Wende der Philosophie, *Erkenntnis* I, 4-11, 1930.

[Schlick, 1931] M. Schlick. The Future of Philosophy, *Proceedings of the Seventh International Congress of Philosophy*, held at Oxford, 1930, London, 112-116, 1931.

[Schurz and Dorn, 1993] G. Schurz and G. J. W. Dorn. Report: After Twenty Years. Die Entwicklung der Wissenschaftstheorie in Osterreich 1971-1990, in: *Journal for General Philosophy of Science* 24/2, 315-347, 1993.

[Skorupski, 1993] J. Skorupski. *English-Language Philosophy 1750 to 1945.* Oxford-New York: Oxford University Press, 1993.

[Sluga, 1999] H. Sluga. What Has History to Do with Me? Wittgenstein and Analytic Philosophy, *Inquiry,* 41, 99-121, 1999.

[Smith, 1986] L. D. Smith. *Behaviorism and Logical Positivism. A Reassessment of the Alliance.* Stanford: Stanford University Press, 1986.

[Somerville, 1936] J. Somerville. Logical Empiricism and the Problem of Causality in Social, in: *Erkenntnis* 6, 405-411, 1936.

[Sorell, 1991] T. Sorell. *Scientism. Philosophy and the Infatuation with Science*, London-New York: Routledge, 1991.

[Snow, 1959] C. P. Snow. *The Two Cultures: and a Second Look.* An Expanded Version of the two Cultures and the Scientific Revolution. Cambridge University Press 1959/1964.

[Sokal and Bricmont, 1998] A. Sokal and J. Bricmont. *Fashionable Nonsense. Postmodern Intellectuals' Abuse of Science.* New York: Picador, 1998. French edition: *Impostures Intellectuelles.* Paris: Editions Odile Jacob. German edition: *Eleganter Unsinn. Wie Denker der Postmoderne die Wissenschaften mißbrauchen.* München: Beck 1999.

[Spohn, 1991] W. Spohn, ed. *Erkenntnis Orientated: A Centennial Volume for Rudolf Carnap and Hans Reichenbach.* Dordrecht-Boston-London: Kluwer, 1991.

[Srubar, 1988] I. Srubar, (Hrsg.). *Exil, Wissenschaft, Identität. Die Emigration deutscher Sozialwissenschaftler 1933-1945.* Frankfurt/M.: Suhrkamp, 1988.

[Stadler, 1987/88] F. Stadler (Hrsg.). *Vertriebene Vernunft. Emigration und Exil österreichischer Wissenschaft*, 1987/88. 2 Bande. Wien-München: Jugend und Volk. 2. Auflage Münster: LIT Verlag 2004.

[Stadler, 1988] F. Stadler (Hrsg.). *Kontinuität und Bruch 1938-1945/1955. Beitrage zur österreichischen Kultur- und Wissenschaftsgeschichte*, 1988. Wien-München: Jugend und Volk. 2. Auflage Münster: LIT Verlag 2004.

[Stadler, 1990] F. Stadler, (Hrsg.). Richard von Mises. *Kleines Lehrbuch des Positivismus. Einführung in die empiristische Wissenschaftsauffassung.* Frankfurt/M.: Suhrkamp, 1990.

[Stadler, 1993] F. Stadler, ed. *Scientific Philosophy: Origins and Developments.* Dordrecht-Boston-London: Kluwer, 1993.

[Stadler, 1997] F. Stadler. *Studien zum Wiener Kreis. Ursprung, Entwicklung und Wirkung des Logischen Empirismus im Kontext.* Frankfurt/M.: Suhrkamp, 1997.

[Stadler, 1997a] F. Stadler (Hrsg.). *Phänomenologie und Logischer Empirismus. Zentenarium Felix Kaufmann.* Wien-New York: Springer, 1997.

[Stadler, 1997b] F. Stadler, (Hrsg.). *Wissenschaft als Kultur. Osterreichs Beitrag zur Moderne.* Wien-New York: Springer, 1997.

[Stadler, 1997c] F. Stadler. Die andere Kulturgeschichte. Am Beispiel von Emigration und Exil der österreichischen Intellektuellen 19301945", in: Steininger/Gehler (Hrsg.) II, 499-558, 1997.

[Stadler, 2001] F. Stadler. *Studien zum Wiener Kreis. Ursprung, Entwicklung und Wirkung des Logischen Empirismus im Kontext*, Frankfurt/M.: Suhrkamp. (Sonderausgabe), 2001.

[Stadler, 2001b] F. Stadler. *The Vienna Circle. Studies in the Origins, Development, and Influence of Logical Empiricism,* Wien-New York: Springer, 2001.

[Stadler, 2001c] F. Stadler. *The Vienna Circle. Studies in the Origins, Development, and Influence of Logical Empiricism.* Wien-New York: Springer, 2001.

[Stadler, 2004] F. Stadler. Transfer and Transformation of Logical Empiricism: Quantitative and Qualitative Aspects, in [Hardcastle and Richardson, 2004].

[Stadler and Weibel, 1995] F. Stadler and P. Weibel, eds. *The Cultural Exodus from Austria*, Wien-New York: Springer, 1995.

[Stebbing, 1933] S. Stebbing. *Logical Positivism and Analysis, Annual Philosophical Lecture, British Academy, London:* Oxford University Press, 1933.

[Stebbing, 1935] S. Stebbing. Notes on an Informal Conference on Logical Positivism, held at Belsize Park, London, 5-6th January, 1935 in Vienna Circle Archives, Haarlem, NL, Neurath papers, 1935.

[Stebbing, 1939-40] S. Stebbing. Language and Misleading Questions, *Erkennntis/The Journal of Unified Science*, VIII, 1-6, 1939-40.

[Stebbing, 1944] S. Stebbing. *Ideals and Illusions*, London: Watts and Co, 1944.

[Stebbing, 1944a] S. Stebbing. *Men and Moral Principles.* L.T. Hobhouse memorial Trust Lectures, No. 13, London: Oxford University Press, 1944.

[Stegmüller, 1979] W. Stegmüller. *Rationale Rekonstruktion von Wissenschaft und ihrem Wandel. Mit einer autobiographischen Einleitung.* Stuttgart: Reclam, 1979.

[Steininger and Gehler, 1997] R. Steininger and M. Gehler, eds. *Osterreich im 20. Jahrhundert. Ein Studienbuch in zwei Banden.* Wien-Köln-Weimar: Böhlau, 1997.

[Strauss, 1991] H. A. Strauss. Wissenschaftsemigration als Forschungsproblem, in: Strauss u.a. (Hrsg.), 9-23, 1991.

[Strauss *et al.*, 1991] H. A. Strauss, K. Fischer, Ch. Hoffmann, and A. Söllner, eds. *Die Emigration der Wissenschaften nach 1933. Disziplingeschichtliche Studien.* München-London-New York-Paris: K.G. Saur, 1991.

[Suppe, 1977] F. Suppe, ed. *TheSstructure of Scientific Theories.* Urbana-Chicago: University of Illinois Press, 1977.

[Thiel, 1984] C. Thiel. Folgen der Emigration deutscher und österreichischer Wissenschaftstheoretiker und Logiker zwischen 1933 und 1945, in: *Berichte zur Wissenschaftsgeschichte* 7, 227–256, 1984.

[Topitsch, 1960] E. Topitsch (Hrsg.). *Probleme der Wissenschaftstheorie. Festschrift für Viktor Kraft.* Wien: Springer, 1960.

[Uebel, 1991] T. E. Uebel, ed. *Rediscovering the Forgotten Vienna Circle. Austrian Studies on Otto Neurath and the Vienna Circle.* Dordrecht-Boston-London: Kluwer, 1991.

[Uebel, 1992] T. E. Uebel. *Overcoming Logical Positivism from Within. The Emergence of Neurath's Naturalism in the Vienna Circle's Protocol Sentence Debate.* Amsterdam-Atlanta, 1992.

[Uebel, 2000] T. E. Uebel. *Vernunftkritik und Wissenschaft. Otto Neurath und der Erste Wiener Kreis.* Wien-New York: Springer, 2000.

[Veröffentlichungen, 1991] *Veröffentlichungen* 1991ff. Veröffentlichungen des Instituts Wiener Kreis. Hrsg. von Friedrich Stadler. Wien-New York: Springer, 1991.

[Vienna Circle, 1993] *Vienna Circle Institute Yearbooks.* Ed. by Friedrich Stadler. Dordrecht-Boston-London: Kluwer, 1933.

[Waismann, 1965] F. Waismann. *The Principles of Linguistic Philosophy*, ed. by R. Harré, London-Melbourne-Toronto: Macmillan, 1965.

[Waismann, 1976] F. Waismann. *Logik, Sprache, Philosophie.* Hrsg. von G.P. Baker/B. McGuinness/J. Schulte. Stuttgart: Reclam, 1976.

[Warnock, 1971] G. J. Warnock. *Englische Philosophie im 20. Jahrhundert.* Stuttgart: Reclam, 1971.

[Wartofsky, 1963] M. W. Wartofsky, ed. *Boston Studies in the Philosophy of Science. Proceedings of the Boston Colloquium for the Philosophy of Science 1961/1962.* Dordrecht: Reidel, 1963.

[Wissenschaftliche, 1919] *Wissenschaftliche Weltauffassung: Der Wiener Kreis* 1929. Translation: The Scientific Conception of the World: The Vienna Circle, Dordrecht-Boston-London: Reidel 1973. Reprinted in Neurath, O. 1973, 299-318.

[Zecha, 1970] G. Zecha. Die gegenwärtige Situation der Wissenschaftstheorie in Osterreich, in: *Zeitschrift far allgemeine Wissenschaftstheorie* 1/2, 284-321, 1970.

[Zilian, 1990] H. G. Zilian. *K1arheit und Methode. Felix Kaufmanns Wissenschaftstheorie.* Amsterdam-Atlanta 1990.

[Zilian, 1997] H. G. Zilian. Felix Kaufmann - Leben und Werk, in: [Stadler, 1997a, pp. 9-22].

[Zilsel, 1932/33] E. Zilsel. Bemerkungen zur Wissenschaftslogik, in: *Erkenntnis* 3, 143-161, 1932/33.

[Zilsel, 1939] E. Zilsel. Preprint: The Social Roots of Science, Harvard, 1939.

[Zilsel, 1942] E. Zilsel. Problems of Empiricism, in: [Neurath *et al.*, 1970, pp. 803-844].

[Zilsel, 1972] E. Zilsel. *Die Entstehung des Geniebegriffs. Ein Beitrag zur Ideengeschichte des Frühkapitalismus.* Tübingen: Mohr 1926. Reprint: Hildesheim-New York: Georg Olms, 1972.

[Zilsel, 1976] E. Zilsel. *Die sozialen Ursprünge der neuzeitlichen Wissenschaft.* Hrsg von Wolfgang Krohn. Frankfurt/M.: Suhrkamp, 1976.

[Zilsel, 1990] E. Zilsel. *Die Geniereligion. Ein kritischer Versuch über das moderne Persönlichkeitsideal, mit einer historischen Begründung.* Hrsg und eingeleitet von Johann Dvorak. Frankfurt/M.: Suhrkamp., 1990

INDEX

CPSIA information can be obtained
at www.ICGtesting.com
Printed in the USA
LVHW100710310520
656209LV00026B/361

9 780444 515483